## Ramification Groups of Local Fields

Ramification groups of local fields are essential tools for studying boundary behavior in geometric objects and the degeneration of Galois representations. This book presents a comprehensive development of the recently established theory of upper ramification groups of local fields with imperfect residue fields, starting from the foundations. It also revisits classical theory, including the Hasse–Arf theorem, and offers an optimal generalization via log monogenic extensions. The conductor of Galois representations, defined through ramification groups, has numerous geometric applications, notably the celebrated Grothendieck–Ogg–Shafarevich formula. A new proof of the Deligne–Kato formula is also provided; this result plays a pivotal role in the theory of characteristic cycles.

With a foundational understanding of commutative rings and Galois theory, graduate students and researchers will be well equipped to engage with this rich area of arithmetic geometry.

TAKESHI SAITO is a Professor of Mathematics at the School of Mathematical Sciences, University of Tokyo, specializing in arithmetic geometry. He is a recipient of the Algebra Prize of the Mathematical Society of Japan (1998) and the Spring Prize of the Mathematical Society of Japan (2001), and is the Israel Gelfand Chair in Mathematics at the Institut des Hautes Études Scientifiques, Université Paris-Saclay (2024–2026).

NEW MATHEMATICAL MONOGRAPHS



All the titles listed below can be obtained from good booksellers or from Cambridge University Press. For a complete series listing visit www.cambridge.org/mathematics.

18. T. Downarowicz *Entropy in Dynamical Systems*
19. C. Simpson *Homotopy Theory of Higher Categories*
20. E. Fricain and J. Mashreghi *The Theory of H(b) Spaces I*
21. E. Fricain and J. Mashreghi *The Theory of H(b) Spaces II*
22. J. Goubault-Larrecq *Non-Hausdorff Topology and Domain Theory*
23. J. Śniatycki *Differential Geometry of Singular Spaces and Reduction of Symmetry*
24. E. Riehl *Categorical Homotopy Theory*
25. B. A. Munson and I. Volić *Cubical Homotopy Theory*
26. B. Conrad, O. Gabber and G. Prasad *Pseudo-reductive Groups (2nd Edition)*
27. J. Heinonen, P. Koskela, N. Shanmugalingam and J. T. Tyson *Sobolev Spaces on Metric Measure Spaces*
28. Y.-G. Oh *Symplectic Topology and Floer Homology I*
29. Y.-G. Oh *Symplectic Topology and Floer Homology II*
30. A. Bobrowski *Convergence of One-Parameter Operator Semigroups*
31. K. Costello and O. Gwilliam *Factorization Algebras in Quantum Field Theory I*
32. J.-H. Evertse and K. Győry *Discriminant Equations in Diophantine Number Theory*
33. G. Friedman *Singular Intersection Homology*
34. S. Schwede *Global Homotopy Theory*
35. M. Dickmann, N. Schwartz and M. Tressl *Spectral Spaces*
36. A. Baernstein II *Symmetrization in Analysis*
37. A. Defant, D. García, M. Maestre and P. Sevilla-Peris *Dirichlet Series and Holomorphic Functions in High Dimensions*
38. N. Th. Varopoulos *Potential Theory and Geometry on Lie Groups*
39. D. Arnal and B. Currey *Representations of Solvable Lie Groups*
40. M. A. Hill, M. J. Hopkins and D. C. Ravenel *Equivariant Stable Homotopy Theory and the Kervaire Invariant Problem*
41. K. Costello and O. Gwilliam *Factorization Algebras in Quantum Field Theory II*
42. S. Kumar *Conformal Blocks, Generalized Theta Functions and the Verlinde Formula*
43. P. F. X. Müller *Hardy Martingales*
44. T. Kaletha and G. Prasad *Bruhat–Tits Theory*
45. J. Schwermer *Reduction Theory and Arithmetic Groups*
46. J. Kotus and M. Urbański *Meromorphic Dynamics I*
47. J. Kotus and M. Urbański *Meromorphic Dynamics II*
48. S. Garibaldi, H. P. Petersson and M. L. Racine *Albert Algebras over Commutative Rings*
49. D. Pellicer *Abstract Chiral Polytopes*
50. C. Xu *K-stability of Fano Varieties*
51. D. Denisov, D. Korshunov and V. Wachtel *Markov Chains with Asymptotically Zero Drift*
52. B. Simon *Phase Transitions in the Theory of Lattice Gases*
53. P. Lefèvre and H. Queffélec *A Primer, and Beyond, on Composition Operators on the Unit Disk*

# Ramification Groups of Local Fields with Geometric Applications

TAKESHI SAITO
*University of Tokyo*

CAMBRIDGE
UNIVERSITY PRESS

CAMBRIDGE
UNIVERSITY PRESS

Shaftesbury Road, Cambridge CB2 8EA, United Kingdom

One Liberty Plaza, 20th Floor, New York, NY 10006, USA

477 Williamstown Road, Port Melbourne, VIC 3207, Australia

314–321, 3rd Floor, Plot 3, Splendor Forum, Jasola District Centre,
New Delhi – 110025, India

Cambridge University Press is part of Cambridge University Press & Assessment,
a department of the University of Cambridge.

We share the University's mission to contribute to society through the pursuit of education, learning and research at the highest international levels of excellence.

www.cambridge.org
Information on this title: www.cambridge.org/9781009617536
DOI: 10.1017/9781009617567



When citing this work, please include a reference to the DOI 10.1017/9781009617567

First published 2026

*A catalogue record for this publication is available from the British Library*

*A Cataloging-in-Publication data record for this book is available from the Library of Congress*

ISBN 978-1-009-61753-6 Hardback

# Contents

# Preface

Finite extensions of algebraic number fields ramify at finitely many primes. The primes that ramify in an extension carry crucial information about the extension. For example, the primes that ramify in the cyclotomic extension $\mathbf{Q}(\zeta_n)$ generated by a primitive $n$th root $\zeta_n$ of unity for an integer $n \not\equiv 2 \bmod 4$ are precisely the prime divisors of $n$. The inertia groups at primes dividing $n$ generate the Galois group and this fact implies the surjectivity of the canonical injection $\mathrm{Gal}(\mathbf{Q}(\zeta_n)/\mathbf{Q}) \to (\mathbf{Z}/n\mathbf{Z})^\times$, or equivalently the irreducibility of cyclotomic polynomials. Similarly, for a finite morphism $X \to Y$ of algebraic curves, if the extension of the function fields is separable, there are finitely many points that ramify in the covering.

We distinguish two types of ramifications: tame ramification and wild ramification. Tame ramification is easier to understand and allows geometric intuition similar to the ramification of finite coverings of Riemann surfaces. Tame ramification is determined essentially by the ramification index.

Wild ramification means that the ramification index is divisible by the characteristic of the residue field in the classical setting where the extension of residue fields is separable, and is much more complicated. In the case of cyclotomic extensions, the ramification at $p$ is wild if and only if $n$ is divisible by $p^2$. Understanding wild ramification inevitably involves additive groups of some vector spaces such as the space of differential forms over the residue fields.

To understand wild ramification, the key tool we use is filtration by ramification groups. In the case of cyclotomic fields, if we identify the Galois group with $(\mathbf{Z}/n\mathbf{Z})^\times$ by the canonical isomorphism, the congruence modulo powers of prime divisors $p$ of $n$ defines a filtration of the inertia subgroup at each ramified prime $p$. The fact that this filtration actually gives the ramification groups is the first case of the Hasse–Arf theorem that is at the core of classical ramification theory.

A main application of the filtration is to the conductor of Galois representations. The integrality of the conductor is proved as a consequence of the Hasse–Arf theorem. We briefly introduce two examples of application of the theory of the conductor, one in number theory and one in algebraic geometry.

In the Langlands correspondence between Galois and automorphic representations, the conductor of a Galois representation equals the level of the corresponding automorphic representation. For example, the proof of Fermat's last theorem is completed by proving the modularity of elliptic curves over $\mathbf{Q}$: For an elliptic curve $E$ over $\mathbf{Q}$, there exists a modular form $f$ such that the $L$-function $L(E, s)$ defined by the Galois representation on the Tate module $T_\ell E$ equals $L(f, s)$. In this correspondence, the conductor of $E$ defined by the ramification of $T_\ell E$ equals the level of $f$.

In a geometric setting, the conductor appears as the local term in the Grothendieck–Ogg–Shafarevich formula computing the Euler–Poincaré characteristic $\chi_c(U, \mathcal{F})$ of a locally constant étale sheaf $\mathcal{F}$ on a smooth algebraic curve $U$ over an algebraically closed field of characteristic $p > 0$. The local terms are the contributions of points at infinity $X - U$ of the smooth compactification $X$ of $U$ where the sheaf $\mathcal{F}$ ramifies.

In arithmetic geometry, beyond extensions of algebraic number fields or coverings of algebraic curves, we are interested in coverings of schemes of finite type over $\mathbf{Z}$ or those of varieties of higher dimension. Then, contrary to the local fields in the classical setting, the residue fields of local fields are no longer perfect and the extension of residue fields may be inseparable. This implies that the theory of wild ramification is required to cover extensions such that the extension of residue fields are not separable.

As a geometric application in this general setting, we will give a proof of the Deligne–Kato formula [48] computing the dimension of the space of nearby cycles on a curve over a local field. A special case of the formula plays a crucial role in the proof of the existence of the characteristic cycle of a constructible sheaf on a smooth scheme over a perfect field [74]. As another example of application, the characteristic cycle is known to be computed in codimension 1 by using ramification groups of local fields with imperfect residue fields [74].

Thus, the main subject of this book is the filtrations by ramification groups of the Galois group of an extension of local fields with no restriction on the residue fields. There are two kinds of ramification groups,lower numbering and upper numbering. They have properties that show sharp contrasts:

(1) The lower numbering ramification groups are defined in a fairly elementary way.

Let $L$ be a finite Galois extension of a local field $K$ and $G$ be the Galois group. Then, the inertia subgroup $I \subset G$ is defined as the kernel of the action of $G$ on the residue field $E = O_L/\mathfrak{m}_L$. As a generalization, a decreasing filtration $G'_i \subset G$ is defined as the kernel of the action of $G$ on $O_L/\mathfrak{m}_L^i$ indexed by integers $i \geqslant 1$. The lower ramification group $G_i \subset G$ is defined as its slight modification obtained by replacing the quotients of the additive group by those of the multiplicative group $L^\times/(1 + \mathfrak{m}_L^i)$. From this definition, we can easily deduce the following properties:

(2) The graded quotients $\mathrm{Gr}_i G = G_i/G_{i+1}$ are $\mathbf{F}_p$-vector spaces for $i \geqslant 1$ and $G_i = 1$ for sufficiently large $i$.
(3) For the subextension $M \subset L$ corresponding to a subgroup $H \subset G$, we have $H_i = H \cap G_i$.

However, in the case where $M$ is a Galois extension of $K$ and hence $H \subset G$ is a normal subgroup:

(4) There is no general way to express the ramification groups $\overline{G}_i \subset \overline{G}$ of the quotient $\overline{G} = G/H = \mathrm{Gal}(M/K)$ out of $G_i \subset G$.

In the classical case where the extension of residue fields is separable, we exceptionally have such a relation, known as a theorem of Herbrand. We generalize the relation by introducing a notion of log monogenic extensions. Many classical theorems in ramification theory, including the theorem of Herbrand, the conductor-discriminant formula, and the Hasse–Arf theorem are stated under the assumption that the extension of residue fields is separable. We will reformulate them as certain inequalities and prove that the equalities are equivalent to the extension being log monogenic.

Properties (3) and (4) indicate that:

(5) The lower numbering filtration is an invariant of $L$ rather than $K$.

We will see in Proposition 5.3.2 that the graded quotients $\mathrm{Gr}_i G$ for $i \geqslant 1$ are related to vector spaces over the residue field $E = O_L/\mathfrak{m}_L$ not over $F = O_K/\mathfrak{m}_K$.

On the other hand:

(1′) The definition of the upper numbering filtration requires a heavy geometric machinery.

We present a heuristic observation in the introduction to Part V using the terminology of rigid geometry leading to the definition of upper numbering groups in [1], inspired by [22, Proposition A.6.2, Remarques A.6.3, A.6.4]. The actual definition in this book is given entirely in the language of schemes and no knowledge of rigid geometry is required. Still, it requires dilatations, stability of integral models, the going down theorem in commutative algebra, and so on. The upper ramification subgroups $G^r \subset G$ are indexed by positive rational numbers $r > 0$ and the symbols $r+$ for rational numbers $r \geqslant 0$. It is a non-trivial fact that $G^1$ equals the inertia subgroup $I$ and $G^{1+}$ equals its unique $p$-Sylow subgroup $P$ called the wild inertia subgroup.

(1″) In the classical case where the extension of residue fields is separable, the upper numbering filtration and the lower numbering filtration are the same up to renumbering controlled by a rescaling function denoted $\varphi_{L/K}$.

In general, there is no such relation. An elementary counterexample is given in Exercise 3.1 as compositions of Artin–Schreier extensions.

The following properties of upper ramification groups are highly non-trivial:

(2′) There exist finitely many rational numbers $1 = r_0 < r_1 < \cdots < r_n$ such that $G^r$ are constant on the intervals $[r_i+, r_{i+1}]$ for $i = 0, \ldots, n-1$ and $G^r = 1$ on $[r_n+, \infty)$.

(2″) The graded quotients $\mathrm{Gr}^r G = G^r/G^{r+}$ are $\mathbf{F}_p$-vector spaces for $r > 1$.

The proof of (2′) requires the reduced fiber theorem and the stable reduction theorem of curves. The proof of (2″) is by reduction to the classical case where the residue field is perfect and relies on (2) and (1″) above. The reduction requires a study of the first homology group $\mathrm{H}_1(L_{E/O_K})$ of the cotangent complex which plays the role of the cotangent space at the closed point of the spectrum of the discrete valuation ring. The most difficult case is where $p$ is a uniformizer and $dp = 0$ in the usual sense. The introduction of the vector space $\mathrm{H}_1(L_{E/O_K})$ allows us to avoid this difficulty.

Statements (3) and (4) are switched as follows.

(3′) For the subextension $M \subset L$ corresponding to a subgroup $H \subset G$, there is no general way to express the ramification groups $H^r \subset H$ out of $G^s \subset G$.

In the case where $M$ is a Galois extension of $K$ and hence $H \subset G$ is a normal subgroup, the following property is built in the definition of upper numbering subgroups:

(4′) The ramification groups $\overline{G}^r \subset \overline{G}$ of the quotient $\overline{G} = G/H = \mathrm{Gal}(M/K)$ equal the images of $G^r \subset G$.

The exceptional classical case of (4) is a consequence of (4′) and (1″). The property (4′) indicates:

(5′) The upper numbering filtration is an invariant of $K$ and is defined on the absolute Galois group $G_K = \mathrm{Gal}(K_{\mathrm{sep}}/K)$.

In Chapter 14, we will construct an injection called the *characteristic form* from the dual group of the graded quotients $\mathrm{Gr}^r G$ for $r > 1$ to a twist of the vector space $H_1(L_{E/O_K})$ that behaves as if it is a scalar extension of that over the residue field $F = O_K/\mathfrak{m}_K$.

For abelian extensions, Kato defined another filtration [50], which we call the cohomological filtration, using the cup product of abelian characters of the Galois group with the multiplicative group valued in the Brauer group. In the classical case where the residue field is finite, this cohomological filtration is given as the images of $1 + \mathfrak{m}_K^n \subset K^\times$ by the reciprocity morphism $K^\times \to G$ of local class field theory. A feature of the cohomological filtration is the property that the filtration is indexed by integers. Consequently, a comparison of the lower numbering filtration with cohomological filtration implies integrality of the conductor.

For cyclic extensions, we prove in Section 5.3 a relation between the cohomological filtration and the lower numbering filtration called the Hasse–Arf theorem. We will state and prove the Hasse–Arf theorem as equivalent conditions to an equality expressing the invariant of cohomological filtration using the lower numbering filtration. We first prove the theorem in the case where the degree equals the characteristic of the residue field by explicit computation for Artin–Schreier extensions in the equal characteristic case and for Kummer extensions in the mixed characteristic case. The general case is proved by introducing a stronger version due to Kato formulated in terms of a certain twisted differential form called the *refined Swan conductor* defined in [50].

The cohomological filtration shares property (5′) above with the upper numbering filtration. The refined Swan conductor is a twisted differential form of the residue field $F = O_K/\mathfrak{m}_K$ with logarithmic pole. We prove in the last chapter of the book a fact proved in [52] that the cohomological filtration and the refined Swan conductor are in fact equal to logarithmic variants of the upper numbering filtration and of the characteristic form.

In the equal positive characteristic case, by Artin–Schreier–Witt theory describing cyclic extensions with degrees powers of $p$, the cohomological filtration and the logarithmic variant of the upper numbering filtration admit a concrete description in terms of a filtration on Witt vectors defined by Brylinski [16]. In the mixed characteristic case, we present Kato's theory of the cohomological filtration without giving the details of the proof that relies heavily on computations of $p$-adic cohomology.

The book consists of sixteen chapters grouped into six parts. Each chapter usually starts with sections preparing some generalities on more basic subjects. In later sections, we apply them to the study of ramification.

In the first two parts, we study the lower numbering filtration and the cohomological filtration. We introduce the two filtrations in Part I and we compute them explicitly for cyclic extensions of degree $p$ in Part II. We prove the Hasse–Arf theorem asserting that an equality linking the two filtrations for cyclic extensions holds if and only if the cyclic extension is log monogenic, introduced in Part I.

In the next two parts, we study applications to Galois representations and to geometric invariants. In Part III, we introduce the conductor of Galois representations using the lower numbering filtration and the rescaling function $\varphi$, and deduce the integrality of the conductor from the Hasse–Arf theorem. In Part IV, we study geometric applications and prove the Grothendieck–Ogg–Shafarevich formula and the Deligne–Kato formula.

In the last two parts, we study the upper numbering filtration. In Part V, after a short introduction on heuristic observation on the construction of the upper numbering filtration using rigid geometry, we construct the filtration in the language of schemes and study its basic properties together with a logarithmic variant. In Part VI, we study graded quotients using the group structure in the geometric construction and introduce characteristic forms defining an injection from the dual group of the graded quotients to twisted cotangent spaces.

The last two parts studying the upper numbering filtration are almost independent of the first four parts. The geometric results in Chapters 8 and 10 are not used in the rest of the book.

At the end of each section, an exercise is given. Most of them are explicit computations of basic examples. Solutions are given at the end of the book. At the end of each chapter, historical notes are included. They are rather personal recollections and the author apologizes for not mentioning many important works. The majority of Chapters 2, 5, 6, 7, 10, and 16 are due to Kazuya Kato and the majority of Chapters 11–14 are the result of collaboration with Ahmed Abbes.

Let us now describe the contents of each chapter in more detail. In many books, including [78], finite extensions of complete discrete valuation fields are studied. In this book, we replace complete discrete valuation fields by a generalization, henselian discrete valuation fields defined algebraically. Similarly as complete discrete valuation fields, finite extensions of henselian discrete valuation fields are again henselian discrete valuation fields. In Chapter 1, we prepare some generalities on henselian local rings, characterized by Hensel's lemma.

We introduce lower ramification groups in Chapter 1. We prove two formulas under the assumption that the extension is log monogenic, which covers the classical setting where the residue field extension is separable. The first is the relation with the ramification groups of a quotient group, which implies the theorem of Herbrand. The second is the relation with the different, called the conductor-discriminant formula. The latter is derived from computation of the trace mapping of the quotient ring of the polynomial ring by a monic polynomial.

In the classical case where the residue field is finite, the reciprocity morphism of local class field theory is defined by the cup-product with values in the Brauer group, which is canonically identified with $\mathbf{Q}/\mathbf{Z}$ in this case. Chapter 2 begins with an introduction to Brauer groups as Galois cohomology and a brief summary of local class field theory, except for the theorem of existence. A classical theorem of Hasse gives a relation between the image of the filtration $1 + \mathfrak{m}_L^n$ by the reciprocity morphism and the ramification subgroups.

In Chapter 2, we give the definition of the cohomological filtration by Kato generalizing this classical construction using Galois cohomology. A key construction due to Kato is that of the refined Swan conductor giving an injection from the graded quotients to twisted spaces of differential forms with log poles. We prepare the basic setup in Chapter 2 but postpone the proof of the existence of the refined Swan conductor in the general case to Chapter 16.

We compute explicitly the lower numbering filtration, the cohomological filtration, and the refined Swan conductor for cyclic extensions of degree $p$ using Artin–Schreier theory in the equal characteristic case and Kummer theory in the mixed characteristic case in Chapter 3. This computation makes the first step in the proof of the Hasse–Arf theorem in Chapter 5 and is also used in the proof of Epp's theorem in Chapter 9.

To prove the Hasse–Arf theorem, we formulate a stronger form involving the refined Swan conductor. To reduce the general case to the basic case proved in Chapter 3, we prepare generalities on the trace of differential forms in Chapter 4. After establishing that it is non-vanishing for log monogenic extensions and its compatibility with the corestriction, we prove the Hasse–Arf theorem as equivalent conditions to an equality for the Swan conductor in Chapter 5. An inequality is proved in general in Chapter 2 using a property of the norm mapping for cyclic extensions of degree $p$.

In Chapter 6, under the classical assumption that the extension of residue fields is separable, we introduce the conductor of Galois representations and study its fundamental properties. We define the conductor by using the break decomposition and the rescaling function. We prove the integrality of the conductor. In the rank one case, this is a consequence of the Hasse–Arf theorem. The general case is deduced from the rank one case by the induction formula. The induction formula is proved by another description of the conductor in terms of the Swan character. This is used as the definition in many books, including [79].

In Chapter 7, we introduce a generalization for log monogenic extensions of the Swan conductor originally defined by Kato in a slightly less general case. In this setting, the refined Swan conductor is defined as a certain twisted differential form of the residue field. This generalization shares the same properties as the conductor studied in Chapter 6. The refined Swan conductor will be used in Chapter 10 to state and prove the Deligne–Kato formula. The refined Swan conductor turns out to be a special case of the characteristic form constructed in general in Chapter 15.

In Chapter 8, we prove the Grothendieck–Ogg–Shafarevich formula computing the Euler–Poincaré characteristic of an étale sheaf on a curve over an algebraically closed field. The conductor defined in Chapter 6 appears as the local term in the formula. The Grothendieck–Ogg–Shafarevich formula is deduced from the Lefschetz trace formula for an endomorphism of a curve. We prove the Lefschetz trace formula in Section 8.1 admitting some basic properties of étale cohomology.

Using the study of cyclic extensions in Chapter 3, we give a proof in Chapter 9 of Epp's theorem: For a transcendental extension of discrete valuation fields, there exists a finite extension of the base field such that the composition field has ramification index 1 under a certain mild assumption. In a geometric situation, Epp's theorem implies the reduced fiber theorem. This theorem implies stability of the integral model, which is a crucial ingredient in the definition in Chapter 12 of the upper numbering ramification groups. The reduced fiber theorem also plays an important role in the statement of the Deligne–Kato formula.

In Chapter 10, after recalling the definition and some basic properties of the nearby cycles functor, we state and give a new local proof of the Deligne–Kato formula computing the dimension of the space of nearby cycles of an étale sheaf on a curve over a local field. The statement uses the refinement of the conductor introduced in Chapter 7. Similarly to the proof of the Grothendieck–Ogg–Shafarevich formula, the Deligne–Kato formula is derived from a trace formula and the dimension formula in the constant coefficient case. These two formulas are deduced directly from the Picard–Lefschetz formula in the case where the curve over the discrete valuation ring is nodal, that is, with at most nodes as singularities. The general case is reduced to this case by a local version due to Temkin of the stable reduction theorem of curves. The new point in the proof in this book is that we take a blow-up by avoiding taking compactification.

From Chapter 11 on, we study upper ramification groups. We define the upper ramification groups in Chapters 11 and 12. To define subgroups $G^r$ for $r > 0$ and $G^{r+}$ for $r \geqslant 0$ of the Galois group $G = \mathrm{Gal}(L/K)$, it suffices to define quotient functors $F^r$ and $F^{r+}$ of the fiber functor $F$ sending subextensions

$M \subset L$ to the finite set $F(M) = \mathrm{Mor}_K(M, L)$ and to verify that $F^r$ and $F^{r+}$ satisfy certain properties. For $r > 0$, to define the set $F^r(M)$, first we take an immersion $T = \mathrm{Spec}\, O_M \to Q$ into a smooth scheme over $S = \mathrm{Spec}\, O_K$ and construct a dilatation $Q^{[r]}_{\overline{S}}$, with respect to the multiplicity $r$, of the base change $Q_{\overline{S}} = Q \times_S \overline{S}$ to $\overline{S} = \mathrm{Spec}\, O_{K_s}$ for the ring of integers in a separable closure. Then, $F^r(M)$ is defined as the set $\pi_0(Q^{(r)}_{\overline{F}})$ of connected components of the geometric closed fiber of the normalization $Q^{(r)}_{\overline{S}}$ of $Q^{[r]}_{\overline{S}}$. For $r \geqslant 0$, the immersion $T \to Q$ induces a morphism $T^{\natural}_{\overline{S}} \to Q^{(r)}_{\overline{S}}$ of normalizations and the set $F^{r+}(M)$ is defined as the image of the restriction of the morphism on the closed fibers.

In Chapter 11, we study this geometric construction. The fact that the functor $F^r(M)$ is defined independently of the choice of the immersion $T \to Q$ is a consequence of a homotopy property of the construction of dilatations. The existence of the stable integral model is a consequence of the reduced fiber theorem proved in Chapter 9. The property of the functors $F^r$ and $F^{r+}$ that ensures the definition of the subgroups $G^r$ and $G^{r+}$ is a consequence of the going down theorem in commutative algebra.

We compute the geometric construction explicitly in the case where the ring $O_L$ of integers is generated by a single element and observe the appearance of the rescaling function $\varphi_{L/K}$. This implies property (1″). We also prove the equalities $G^1 = I$ and $G^{1+} = P$. After recalling the general form of the reduced fiber theorem, we prove property (2′) to close Chapter 12.

In Chapter 13, we study a logarithmic variant of the construction in Chapter 12. First, we introduce some generalities on log smooth schemes over $S = \mathrm{Spec}\, O_K$ and on log smooth extensions of $K$ without using the general theory of log geometry. Then, by replacing an immersion $T \to Q$ into a smooth scheme by an exact immersion into a log smooth scheme, we define a logarithmic variant of the upper ramification groups. In the second half of Chapter 13 we focus on Artin–Schreier–Witt extensions in characteristic $p > 0$. We give a geometric proof of the fact that Brylinski's filtration induces the logarithmic upper numbering filtration.

In Chapters 14 and 15, we study the graded quotients $\mathrm{Gr}^r G = G^r / G^{r+}$ for $r > 1$. We prove property (2″) in Chapter 14 and we study a construction of an injection of the dual group to a twisted "cotangent" space in Chapter 15. Chapter 14 begins with the study of $\mathrm{H}_1(L_{E/S})$, which plays the role of the cotangent space at the closed point of $S = \mathrm{Spec}\, O_K$.

The tool we use to study the graded quotients $\mathrm{Gr}^r G$ is the reduced geometric closed fiber $Q^{(r)}_{\overline{F}} \to Q^{[r]}_{\overline{F},\mathrm{red}}$ of the morphism to the dilatation from the normalization. The target $Q^{[r]}_{\overline{F},\mathrm{red}}$ is a twist of the fiber $N_{T/Q} \otimes_{O_L} \overline{F}$ of the conormal sheaf of an immersion $T = \operatorname{Spec} O_L \to Q$ into a smooth scheme over $S = \operatorname{Spec} O_K$. If the immersion is minimal in the sense that $\dim Q$ is minimal, the canonical injection $\operatorname{Tor}_1^{O_L}(\Omega^1_{O_L/O_K}, \overline{F}) \to N_{T/Q} \otimes_{O_L} \overline{F}$ is an isomorphism and a connected component $Q^{[r]}_{\overline{F},\mathrm{red}}$ is canonically identified with a vector space denoted $\Theta^{(r)\circ}_{L/K,\overline{F}}$. This space is linked to $\mathrm{H}_1(L_{\overline{F}/S})$ by the injection $\operatorname{Tor}_1^{O_L}(\Omega^1_{O_L/O_K}, \overline{F}) \to \mathrm{H}_1(L_{\overline{F}/S})$ defined by the functoriality of cotangent complexes.

By the compatibility (4′) with the quotient, the study of the graded quotients $\mathrm{Gr}^r G$ is reduced to the study of the last piece $G^r$ of the filtration satisfying $G^{r+} = 1$. Under this condition, the crucial property is that the finite morphism $Q^{(r)}_{\overline{F}} \to Q^{[r]}_{\overline{F},\mathrm{red}}$ is étale. This, for a minimal immersion $T \to Q$, defines a $G^r$-torsor $\Phi^{(r)\circ}_{L/K,\overline{F}} \to \Theta^{(r)\circ}_{L/K,\overline{F}}$. By using the functoriality of this $G^r$-torsor, we first prove that $G^r$ is abelian.

In the classical case where the residue field is perfect, the relation (1″) of the upper and lower ramification groups together with (2) implies property (2″). We prove the general case of (2″) by reduction to this case. By the functoriality of the construction in the previous paragraph, this is proved by constructing an extension of the henselian discrete valuation field with perfect residue field such that the morphism on the "cotangent" space $\mathrm{H}_1(L_{E/S})$ is an injection.

By the description of the extension group of a vector space by a finite $\mathbf{F}_p$-vector space prepared in the first sections in Chapter 14, the definition of the characteristic form is reduced to showing that the $G^r$-torsor $\Phi^{(r)\circ}_{L/K,\overline{F}} \to \Theta^{(r)\circ}_{L/K,\overline{F}}$ is in fact an isogeny of group schemes. In this book, we prove this result only in the case where $K$ is of characteristic $p > 0$ and $r$ is an integer, and refer to [76] in the general case. The case above is reduced to the case where $K$ is a local field at the generic point of a smooth divisor in a smooth scheme over a perfect field. In this geometric case, the group structure arises at the boundary of the dilatation, from the groupoid structure of self-products.

In Chapter 16, we study a logarithmic variant. We also study the cohomological filtration for abelian extensions and its relation with the logarithmic variant. In the equal characteristic $p > 0$ case, the cohomological filtration is also induced by Brylinski's filtration on Witt vectors. This and the result in Chapter 13 imply the equality of the logarithmic upper numbering filtration and the cohomological filtration. We give a proof with Kato of the equality in the mixed characteristic case as well. By the functoriality of the logarithmic characteristic form and the refined Swan conductor, the proof is reduced to the case where the ramification index is one and the ring of integers is generated by a single element. In this case, the equality follows from the Hasse–Arf theorem and a variant of property $(1'')$. To assure the existence of a suitable extension in the reduction, we apply a variant of Epp's theorem proved in Chapter 9. In the final section, we sketch an outline of the proof by Kato of the existence of refined Swan conductor.

Throughout the book, we assume basic knowledge of commutative algebra, as found in [10]. We recall basic properties of étale morphisms in Section 1.1. Here, we also refer to some results in [67] and [31]. A logarithmic variant of quasi-finite flat morphisms of complete intersection is studied in Section 4.1 to define the trace of differential forms with log poles following [59]. Although we also use the language of schemes, in the most part, it is limited to local theory on affine schemes. Exceptions are Chapters 8 and 10 where we use proper curves and blow-ups respectively. In Chapter 8, we also use étale cohomology, admitting its basic properties. We briefly recall the definition and basic properties of nearby cycles in Section 10.2.

In Chapter 2, we freely use Galois cohomology as in [78, Chapitre X] including cup-product and the inflation-restriction sequence. We also use some fundamental facts on differential forms in characteristic $p > 0$. In Part III, we assume basic knowledge of representations of finite groups as in [79, Partie III]. We briefly recall the theory of Brauer trace of modular representations in Section 6.1.

In Chapter 12, it is not necessary but will be useful to be familiar with the terminology of Galois categories and fiber functors. In Chapters 13 and 16, we introduce the terminology of log geometry but knowledge of its generality is not required. In the second half of Chapter 13, we also use Artin–Schreier–Witt theory and Witt vectors. In Chapter 14, we use some elementary homological algebra including Tor and cotangent complexes. We only use basic properties and it is not necessary to go back to the construction. We also use interpretation of $\mathrm{H}^1$ in terms of torsors in Chapter 14 and Artin–Schreier coverings in characteristic $p > 0$. In Chapter 15, we use terminology of group schemes and groupoids.

The author thanks Hiroki Kato for helping to prepare the manuscript of Chapter 10. The author thanks Ahmed Abbes, Teiji Okami, and Yuri Yatagawa for careful reading and numerous and helpful comments. The author thanks Okami for a suggestion to improve statements in Sections 4.4 and 5.1. The author thanks Luc Illusie and Kazuya Kato for encouragement.

# PART I

# RAMIFICATION OF HENSELIAN DISCRETE VALUATION FIELDS

We study elementary properties of ramification of finite extensions of henselian discrete valuation fields. Complete discrete valuation fields are henselian discrete valuation fields and, similarly to the former, rings of integers of finite extensions of the latter are again discrete valuation rings.

We define the lower ramification groups of the Galois groups of finite Galois extensions of henselian discrete valuation fields. We prove an equality giving a relation with the filtration on quotient groups for log monogenic extensions. The classical case where the extension of the residue fields is separable is a special case of log monogenic extensions. We also define the different of finite separable extensions and show its relation with lower ramification groups of log monogenic extensions. These relations will imply the compatibility with quotient groups of the Swan conductor in Part III.

We introduce a filtration on the character group of the abelian quotient of the absolute Galois group. This is defined as the dual of the filtration on the multiplicative group, using the pairing with values in the Brauer group defined by the cup-product of Galois cohomology. In the classical case where the residue field is finite, the pairing induces the reciprocity mapping of local class field theory.

In the same way as the $p$-torsion part of the unramified part of the character group receives a surjection from the additive group of the residue field by the Artin–Schreier theory, the $p$-torsion part of the unramified part of the Brauer group receives a surjection from the group of differential forms of the residue field with logarithmic poles. Using this fact, we study the graded pieces of the cohomological filtration and introduce the refined Swan conductor.

# 1

# Finite Extensions

The ramification theory for complete discrete valuation fields works equally for henselian discrete valuation fields. The main reason to consider henselian discrete valuation rings is the equivalence (1) $\Leftrightarrow$ (3) in Proposition 1.4.5 on the integral closures in finite extensions.

Before introducing henselian local rings in Section 1.3, we prepare fundamental properties of étale morphisms of rings in Section 1.1. We study basic properties of finite extensions of henselian discrete valuation fields in Section 1.4.

We define the lower ramification groups in Section 1.6. The inertia group $G_0$ and the wild inertia group $G_1$ correspond to the maximal unramified extension and the maximal tamely ramified extension introduced in Section 1.5. The lower ramification groups provide a tool to study the wild inertia group. We introduce the notion of log monogenic extensions in Section 1.8 as a generalization of the classical setting where the extension of residue fields is separable. We extend a classical theorem of Herbrand on the relation between the lower numbering filtration and that on quotient groups to log monogenic extensions.

We define the different in Section 1.9 and give a criterion in terms of the different for an extension to be unramified or tamely ramified. We will compute the different explicitly using the derivative of the minimal polynomial of a generator of the ring of integers and the computation of the trace mapping prepared in Section 1.2. We also study in Section 1.10 some basic properties of the norm mapping and the filtration on the multiplicative groups.

## 1.1 Étale Algebras

A ring means a commutative ring with 1. A morphism of rings sends 1 to 1. For finiteness conditions on morphisms of rings, we refer to [33, Chapitre 0, Section 6.3].

**Definition 1.1.1** Let $A \to B$ be a morphism of rings.

(1) The $B$-module $\Omega^1_{B/A}$ of differential forms is defined as $J \otimes_{B\otimes_A B} B$ where $J$ is the kernel of the surjection $B \otimes_A B \to B$ sending $x \otimes y$ to $xy$.
(2) We say that $B$ is of finite presentation (resp. of finite type) over $A$ if there exists an isomorphism $A[X_1, \ldots, X_n]/(f_1, \ldots, f_m) \to B$ (resp. a surjection $A[X_1, \ldots, X_n] \to B$) of rings over $A$ [33, (6.3.4), (6.3.5)].

**Proposition 1.1.2** *Let $A \to B$ be a morphism of rings.*

(1) *The $B$-module $\Omega^1_{B/A}$ represents the functor* $\mathrm{Der}_A(B, -)\colon (B\text{-modules}) \to (\text{Sets})$ *sending a $B$-module $M$ to the module*

$$\mathrm{Der}_A(B, M) = \{d \in \mathrm{Hom}_A(B, M) \mid d(xy) = xdy+ydx \text{ for every } x, y \in B\}$$

*of derivations.*
(2) *Let $B = A[X_1, \ldots, X_n]/(f_1, \ldots, f_m)$ and define a $B$-linear morphism $D\colon B^n \to B^m$ by the Jacobian matrix $(\partial f_i/\partial X_j) \in \mathrm{M}(m, n; B)$. Then, the elements $dX_1, \ldots, dX_n \in \Omega^1_{B/A}$ define an isomorphism* $\mathrm{Coker}(D^\vee\colon B^m \to B^n) \to \Omega^1_{B/A}$ *of $B$-modules.*

*Proof* (1) For a $B$-module $M$, regard $B \oplus M$ as a ring over $B$ by $xy = 0$ for $x, y \in M$. Then, the mapping $\mathrm{Mor}_A(B, B \oplus M) \to \mathrm{Hom}_A(B, M)$ from the set of morphisms of rings over $A$ sending $f$ to $\mathrm{pr}_2 \circ f$ induces a bijection $P = \{f \in \mathrm{Mor}_A(B, B \oplus M) \mid \mathrm{pr}_1 \circ f = 1_B\} \to \mathrm{Der}_A(B, M)$. Let $i_1, i_2\colon B \to B \otimes_A B$ be the morphisms defined by $i_1(b) = b \otimes 1$ and $i_2(b) = 1 \otimes b$, and regard $B \otimes_A B$ as a ring over $B$ by $i_2$. Then, by the universality of tensor product, the set $P$ is identified with $Q = \{g \in \mathrm{Mor}_B(B \otimes_A B, B \oplus M) \mid \mathrm{pr}_1 \circ g \circ i_1 = 1_B\}$ by sending $g$ to $g \circ i_1$. Since $(B \otimes_A B)/J^2 = B \oplus (J/J^2)$, the set $Q$ is further identified with $\mathrm{Hom}_B(J/J^2, M)$. Hence the assertion follows.

(2) Let $x_i \in B$ be the images of $X_i$. It suffices to show that, for a $B$-module $M$, the mapping $\mathrm{Der}_A(B, M) \to M^n$ sending a derivation $d$ to $(dx_i)$ induces a bijection $\mathrm{Der}_A(B, M) \to \mathrm{Ker}(D\colon M^n \to M^m)$. In the case $m = 0$, $s_1, \ldots, s_n \in M$ uniquely defines a derivation $A[X_1, \ldots, X_n] \to M$ sending $g$ to $\sum_j (\partial g/\partial X_j)s_j$. In the general case, the derivation $d\colon A[X_1, \ldots, X_n] \to M$ defined by $s_1, \ldots, s_n \in M$ factors though the surjection $A[X_1, \ldots, X_n] \to B$ if and only if $df_i = \sum_j (\partial f_i/\partial X_j)s_j$ is 0 for $i = 1, \ldots, m$. This means that $(s_1, \ldots, s_n) \in M^n$ lies in the kernel of $D$. Hence, the assertion follows. □

We recall a functorial characterization of rings of finite presentation.

**Lemma 1.1.3** ([33, Proposition (6.3.11)]) *Let $A \to B$ be a morphism of rings. The following conditions are equivalent:*

(1) *$B$ is of finite presentation over $A$.*
(2) *For every filtered inductive system $(C_\lambda)_{\lambda\in\Lambda}$ of rings over $A$, the canonical mapping $\varinjlim_{\lambda\in\Lambda} \mathrm{Mor}_A(B, C_\lambda) \to \mathrm{Mor}_A(B, \varinjlim_{\lambda\in\Lambda} C_\lambda)$ is a bijection.*

**Definition 1.1.4** Let $A \to B$ be a morphism of rings.

(1) We say that $B$ is étale over $A$ if $B$ is of finite presentation over $A$, if $B$ is flat over $A$, and if $\Omega^1_{B/A} = 0$.
(2) Let $\mathfrak{q}$ be a prime ideal of $B$. We say that $B$ is étale over $A$ at $\mathfrak{q}$ if there exists an element $b \in B - \mathfrak{q}$ such that $B[1/b]$ is étale over $A$.

A ring $B$ over $A$ is étale over $A$ if and only if it is étale over $A$ at every prime ideal. If $B$ is étale over $A$, for another ring $A'$ over $A$, the tensor product $B' = B \otimes_A A'$ is étale over $A'$. Conversely, if $A'$ is faithfully flat over $A$ and if $B'$ is étale over $A'$, then $B$ is étale over $A$.

We give examples of rings étale over $A$. First, we recall a flatness criterion.

**Proposition 1.1.5** ([31, Chapitre $0_{\mathrm{IV}}$ (15.1.16) *b*) $\Leftrightarrow$ *c*)]) *Let $f: A \to B$ be a local morphism of noetherian local rings and let $k$ be the residue field of $A$. Let $f_1, \ldots, f_n \in \mathfrak{n}$ be elements of the maximal ideal of $B$ and set $\overline{B}_i = B/(f_1, \ldots, f_i)$ for $i = 0, \ldots, n$. Then, the following conditions are equivalent:*

(1) *For every $i = 1, \ldots, n$, the image $\overline{f}_i \in \overline{B}_{i-1}$ is a non-zero divisor and $\overline{B}_n$ is flat over $A$.*
(2) *For every $i = 1, \ldots, n$, the image $\overline{f}_i \in \overline{B}_{i-1} \otimes_A k$ is a non-zero divisor and $B$ is flat over $A$.*

**Lemma 1.1.6** *Let $A$ be a ring, $P \in A[X]$ be a polynomial, and $P'$ be the derivative. Then the ring $B = A[X]/(P)[P'^{-1}]$ is étale over $A$.*

*Proof* Since $B$ is isomorphic to $A[X,Y]/(P, P'Y - 1)$ over $A$, it is of finite presentation over $A$. By replacing $A$ by the subring of $A$ generated by the coefficients of $P$ over $\mathbf{Z}$, we may assume that $A$ is noetherian. Further, $B$ is flat over $A$ by Proposition 1.1.5 (2) $\Rightarrow$ (1). Since the $B$-module $\Omega^1_{B/A}$ is generated by $dX$ and its annihilator is $(P') = B$ by Lemma 1.1.2(2), the assertion follows. □

To give a local description of étale morphisms, we show some properties of idempotents.

**Lemma 1.1.7** *Let $A$ be a ring.*

(1) *For an ideal $I \subset A$, the following conditions are equivalent:*

(a) *$I$ is finitely generated and $I = I^2$.*

(b) *There exists an idempotent $e \in A$ such that $I = (1 - e)$.*

(2) *Let $B$ be a ring of finite type over $A$ and $B \to A$ be a morphism over $A$ such that $\Omega^1_{B/A} \otimes_B A = 0$. Then there exists an idempotent $e$ of $B$ such that* $\mathrm{Ker}(B \to A) = (1 - e)B$.

*Proof* (1) (a) $\Rightarrow$ (b): Let $x_1, \dots, x_n$ be a system of generators of $I$. Since $I = I^2$, there exists $a_{ij} \in I$ such that $x_i = \sum_{j=1}^n a_{ij}x_j$ for $i = 1, \dots, n$. Namely, the matrix $\boldsymbol{A} \in M(n, A)$ with entries $a_{ij} \in I$ and the column vector $\boldsymbol{x} \in A^n$ with entries $x_i \in I$ satisfy $\boldsymbol{x} = \boldsymbol{A}\boldsymbol{x}$. Hence, we have $e = \det(1_n - \boldsymbol{A}) \equiv 1 \bmod I$ and $e \cdot \boldsymbol{x} = 0$. Thus, the ideal $I' = (e)$ satisfies $A = I' + I$ and $II' = 0$. Hence we obtain $A = A/I \times A/I'$ by the Chinese remainder theorem, and $e \in A$ corresponds to $(1, 0)$.

(b) $\Rightarrow$ (a): The ideal $I$ is generated by an idempotent $1 - e$.

(2) Let $I$ be the kernel of the surjection $B \to A$ and $J$ be the kernel of the surjection $B \otimes_A B \to B$. Since $B$ is of finite type over $A$, the ideal $I \subset B$ is finitely generated. Since $I = JB$ with respect to the morphism $B \otimes_A B \to B$ induced by $B \to A$ on the second factor, and $\Omega^1_{B/A} = J/J^2$, we have $I/I^2 = J/J^2 \otimes_B A = 0$. Since $I = I^2$, there exists an idempotent $e \in B$ such that $\mathrm{Ker}(B \to A) = (1-e)B$ by (1). □

Étale algebras over a field are products of finite separable extensions.

**Proposition 1.1.8** *Let $k$ be a field and $B$ be a ring of finite type over $k$.*

(1) *For a prime ideal $\mathfrak{q} \subset B$, the following conditions are equivalent:*

(a) *$B$ is étale over $k$ at $\mathfrak{q}$.*

(b) *The quotient ring $B/\mathfrak{q}$ is a finite separable extension of $k$ and $B$ is isomorphic to a product ring $B/\mathfrak{q} \times B'$.*

(2) *The following conditions are equivalent:*

(a) *$B$ is étale over $k$.*

(b) *$B$ is isomorphic over $k$ to a product of finitely many finite separable extensions of $k$.*

*Proof* (1) (b) $\Rightarrow$ (a): We may assume that $B$ is a finite separable extension of $k$. Then we have a separable irreducible polynomial $P \in k[X]$ and an isomorphism $k[X]/(P) \to B$ over $k$. Hence, the assertion follows from Lemma 1.1.6.

(a) $\Rightarrow$ (b): First, we assume that $k$ is an algebraically closed field. Let

$\mathfrak{n} \supset \mathfrak{q}$ be a maximal ideal. Then, by Lemma 1.1.7(2), we have an isomorphism $B \to B/\mathfrak{n} \times B'$. Hence, we have $\mathfrak{n} = \mathfrak{q}$ and the assertion follows in this case.

Let $\overline{k}$ be an algebraic closure of $k$. By the case where $k$ is an algebraically closed field, every point of Spec $(B_{\mathfrak{q}} \otimes_k \overline{k})$ is an isolated point and $B_{\mathfrak{q}} \otimes_k \overline{k}$ is reduced. Hence its underlying set is the same as Spec $(B/\mathfrak{q}) \otimes_k \overline{k}$. Since this is quasi-compact, we have $B_{\mathfrak{q}} \otimes_k \overline{k} = (B/\mathfrak{q}) \otimes_k \overline{k}$ and this is a product of finitely many copies of $\overline{k}$. Further, $B \otimes_k \overline{k}$ is isomorphic to a product $(B/\mathfrak{q}) \otimes_k \overline{k} \times B_1$. Hence $B/\mathfrak{q}$ is a finite separable extension of $k$ and $B$ is isomorphic to a product ring $B/\mathfrak{q} \times B'$.

(2) (a) $\Rightarrow$ (b): By (1) (a) $\Rightarrow$ (b), every point of Spec $B$ is an isolated point. Since Spec $B$ is quasi-compact, the assertion follows.

(b) $\Rightarrow$ (a): Clear from (1) (b) $\Rightarrow$ (a). □

We prove the local description below using Zariski's main theorem.

**Theorem 1.1.9** (Zariski's main theorem, [67, Chapitre IV, Théorème 1, Corollaire 1]). *Let $f: A \to B$ be a morphism of rings such that $B$ is of finite type over $A$. Let $\mathfrak{q} \subset B$ be a prime ideal and $\mathfrak{p} = f^{-1}(\mathfrak{q}) \subset A$. Assume that $\mathfrak{q}$ is an isolated point of the fiber* Spec $B \otimes_A k(\mathfrak{p})$.

*Then, there exist a subring $B' \subset B$, finite over $A$, and an element $b \in B' - (B' \cap \mathfrak{q})$ such that the inclusion $B'[1/b] \to B[1/b]$ is an isomorphism.*

**Theorem 1.1.10** ([31, Théorème (18.4.6)], [67, Chapitre V, §1, Théorème 1.1]). *Let $f: A \to B$ be a morphism of rings such that $B$ is of finite presentation over $A$. Let $\mathfrak{q} \subset B$ be a prime ideal and $\mathfrak{p} = f^{-1}(\mathfrak{q}) \subset A$. Then, the following conditions are equivalent:*

(1) *$B$ is étale over $A$ at $\mathfrak{q}$.*
(2) *There exist a monic polynomial $P \in A_{\mathfrak{p}}[T]$, a maximal ideal $\mathfrak{n}$ of $C = A_{\mathfrak{p}}[T]/(P)$ not containing $P'$, and an isomorphism $C_{\mathfrak{n}} \to B_{\mathfrak{q}}$ over $A_{\mathfrak{p}}$.*

*Proof* (1) $\Rightarrow$ (2): By Proposition 1.1.8(1) and Theorem 1.1.9, we may assume that $B$ is finite over $A$. By replacing $A$ by the localization $A_{\mathfrak{p}}$, we may assume that $A$ is local and that $\mathfrak{p}$ is the maximal ideal.

By Proposition 1.1.8(1), $B/\mathfrak{q}$ is a finite separable extension of $F = A/\mathfrak{p}$ and $B/\mathfrak{p}B$ is a product ring $B/\mathfrak{q} \times B'$. Hence, there exists $x \in B - \mathfrak{q}$ such that $B/\mathfrak{q}$ is generated by $x$ over $A/\mathfrak{p}$ and that the image of $x$ in $B'$ is 0. Since $B$ is finite over $A$, the element $x$ is integral over $A$ and the subring $A[x] \subset B$ generated by $x$ is also finite. Let $n = \dim_F A[x] \otimes_A F$. Then $1, x, \ldots, x^{n-1}$ is a basis of $A[x] \otimes_A F$ and the $A$-module $A[x]$ is generated by $1, x, \ldots, x^{n-1}$ by Nakayama's lemma. Hence, there exist a monic polynomial $P \in A[X]$ of degree $n$ and a surjection $C = A[X]/(P) \to A[x] \subset B$.

Let $\mathfrak{n} \subset C$ be the inverse image of $\mathfrak{q} \subset B$. Since $\mathfrak{q} \in \operatorname{Spec} B$ is a unique point above $\mathfrak{n} \in \operatorname{Spec} C$ and since the morphism $C/\mathfrak{n} \to B/\mathfrak{n}B = B/\mathfrak{q}$ is an isomorphism, the finite morphism $C_\mathfrak{n} \to B \otimes_C C_\mathfrak{n}$ is a surjection by Nakayama's lemma. The quotient ring $B \otimes_C C_\mathfrak{n}$ of $C_\mathfrak{n}$ is a local ring and hence we have $B \otimes_C C_\mathfrak{n} = B_\mathfrak{q}$. Since $C$ and an open neighborhood of $\operatorname{Spec} B$ are flat of finite presentation over $A$, and since $C/\mathfrak{n} = C_\mathfrak{n}/\mathfrak{p}C_\mathfrak{n} \to B/\mathfrak{q} = B_\mathfrak{q}/\mathfrak{p}B_\mathfrak{q}$ is an isomorphism, the local morphism $C_\mathfrak{n} \to B_\mathfrak{q}$ is flat by [31, Théorème (11.3.10)] and is injective. Hence, the morphism $C_\mathfrak{n} \to B_\mathfrak{q}$ is an isomorphism. Since $\Omega^1_{C/A} \otimes_C k(\mathfrak{n}) = 0$, we have $P' \notin \mathfrak{n}$.

(2) $\Rightarrow$ (1): The isomorphism $C_\mathfrak{n} \to B_\mathfrak{q}$ of local rings is extended to an isomorphism $C[1/c] \to B[1/b]$ over $A$ for $b \in B - \mathfrak{q}$, $c \in C - \mathfrak{n}$ by Lemma 1.1.3. Hence, the assertion follows from Lemma 1.1.6. □

We omit the proof of the following functorial characterization of étale morphisms.

**Theorem 1.1.11** ([31, Théorème (17.6.1)], [67, Chapitre V, §1, Théorème 2, Corollaire 1]). *Let $f: A \to B$ be a morphism of rings such that $B$ is of finite presentation over $A$. The following conditions are equivalent:*

(1) *$B$ is étale over $A$.*
(2) *For every ring $C$ over $A$ and every ideal $I \subset C$ such that $I^2 = 0$, the surjection $C \to C/I$ induces a bijection $\operatorname{Mor}_A(B, C) \to \operatorname{Mor}_A(B, C/I)$ of the sets of morphisms of rings over $A$.*

**Lemma 1.1.12** *Let $A \to B$ be a morphism of rings and assume that $B$ is free of finite rank as an $A$-module. Then the functor $F$ sending a ring $C$ over $A$ to the set $\operatorname{Idem}(B \otimes_A C)$ of idempotents is representable by a ring $E$ étale over $A$.*

*Proof* Let $b_1, \ldots, b_n$ be a basis of the $A$-module $B$. Let $A[x_1, \ldots, x_n]$ be a polynomial ring and let $x = \sum_{i=1}^n x_i \otimes b_i \in A[x_1, \ldots, x_n] \otimes_A B$ be the universal section. Then the functor $F$ is represented by the quotient $E$ of $A[x_1, \ldots, x_n]$ by the ideal generated by the coefficients of $b_1, \ldots, b_n$ in $x^2 - x \in A[x_1, \ldots, x_n] \otimes_A B$. The ring $E$ is of finite presentation over $A$.

We show that $E$ is étale over $A$. In the case $B = A$ and $b_1 = 1$, we have $E = A[X]/(X^2 - X)$. Since $E$ is isomorphic to $A \times A$ by the Chinese remainder theorem, $E$ is étale over $A$.

We show the general case. Since the case $B = A$ is already proved, for every ring $C$ over $A$ and every ideal $I \subset C$ satisfying $I^2 = 0$, the surjection $C \to C/I$ induces a bijection $\operatorname{Idem} C \to \operatorname{Idem} C/I$ by Theorem 1.1.11 (1) $\Rightarrow$ (2). Applying this to the surjection $B \otimes_A C \to B \otimes_A C/I$, we see that $F(C) \to F(C/I)$ is also a bijection. Hence, by Theorem 1.1.11 (2) $\Rightarrow$ (1), $E$ is étale over $A$. □

**Exercise 1.1** Let $p > 2$ be a prime number and $\zeta_p$ be a primitive $p$th root of 1. Find the prime ideals of $\mathbf{Z}[\zeta_p]$ where $\mathbf{Z}[\zeta_p]$ is not étale over $\mathbf{Z}$.

## 1.2 Trace

**Lemma 1.2.1** *Let $A$ be a ring and $B$ be a ring over $A$. Assume that $B$ is a free $A$-module of finite rank $n$ and let $b_1, \ldots, b_n$ be a basis of the free $A$-module $B$. Let $f_1, \ldots, f_n \in \mathrm{Hom}_A(B, A)$ be the dual basis of $b_1, \ldots, b_n$. Then, for the trace mapping $\mathrm{Tr}_{B/A} \in \mathrm{Hom}_A(B, A)$, we have*

$$\mathrm{Tr}_{B/A} = \sum_{i=1}^{n} b_i \cdot f_i. \tag{1.2.1}$$

*Proof* For $b \in B$, if we set $bb_j = \sum_{i=1}^{n} a_{ij} b_i$, we have $\sum_{i=1}^{n} (b_i f_i)(b) = \sum_{i=1}^{n} f_i(bb_i) = \sum_{i=1}^{n} a_{ii} = \mathrm{Tr}_{B/A} b$. □

We compute the trace mapping for the quotient ring of the polynomial ring by a monic polynomial.

**Proposition 1.2.2** *Let $A$ be a ring and let $P = X^n + a_1 X^{n-1} + \cdots + a_n \in A[X]$ be a monic polynomial of degree $n$. Let $b \in B = A[X]/(P)$ be the image of $X \in A[X]$ and define $Q = c_1 X^{n-1} + \cdots + c_n \in B[X]$ by $P = (X - b) \cdot Q$.*

(1) *The sequence $c_1 = 1, c_2, \ldots, c_n$ is a basis of the free $A$-module $B$.*

(2) *Let $f_1, \ldots, f_n \in \mathrm{Hom}_A(B, A)$ be the dual basis of the basis $c_1 = 1, c_2, \ldots, c_n$ as a free $A$-module. Then $f_n \in \mathrm{Hom}_A(B, A)$ defines an isomorphism $B \to \mathrm{Hom}_A(B, A)$ of free $B$-modules of rank 1 and we have $f_i = b^{n-i} f_n$ for $i = 1, \ldots, n$.*

(3) *We have $\mathrm{Tr}_{B/A} = P'(b) \cdot f_n$ in $\mathrm{Hom}_A(B, A)$. In particular, if $B$ is étale over $A$, we have*

$$\mathrm{Tr}_{B/A} \frac{b^i}{P'(b)} = \begin{cases} 1 & \text{if } i = n-1, \\ 0 & \text{if } i = 0, \ldots, n-2. \end{cases} \tag{1.2.2}$$

*Proof* (1) Define an invertible matrix

$$\boldsymbol{A} = \begin{pmatrix} 1 & 0 & \cdots & 0 \\ a_1 & \ddots & \ddots & \vdots \\ \vdots & \ddots & \ddots & 0 \\ a_{n-1} & \cdots & a_1 & 1 \end{pmatrix}.$$

The equality $QX = P + Q \cdot b$ means $c_k = a_{k-1} + c_{k-1}b$ for $k = 1, \ldots, n$. By induction on $k$, this implies that

$$\begin{pmatrix} c_1 \\ c_2 \\ \vdots \\ c_n \end{pmatrix} = \boldsymbol{A} \begin{pmatrix} 1 \\ b \\ \vdots \\ b^{n-1} \end{pmatrix}.$$

Since $1, \ldots, b^{n-1}$ is a basis of the $A$-module $B$, the assertion follows.

(2) The equality $b \cdot Q = QX - P$ means that we have

$$b \begin{pmatrix} c_1 \\ c_2 \\ \vdots \\ c_n \end{pmatrix} = \begin{pmatrix} -a_1 & 1 & \cdots & 0 \\ -a_2 & 0 & \ddots & \vdots \\ \vdots & \vdots & \ddots & 1 \\ -a_n & 0 & \cdots & 0 \end{pmatrix} \begin{pmatrix} c_1 \\ c_2 \\ \vdots \\ c_n \end{pmatrix}. \tag{1.2.3}$$

Its transpose gives

$$b \begin{pmatrix} f_1 \\ f_2 \\ \vdots \\ f_n \end{pmatrix} = \begin{pmatrix} -a_1 & -a_2 & \cdots & -a_n \\ 1 & 0 & \cdots & 0 \\ \vdots & \ddots & \ddots & \vdots \\ 0 & \cdots & 1 & 0 \end{pmatrix} \begin{pmatrix} f_1 \\ f_2 \\ \vdots \\ f_n \end{pmatrix}.$$

This means that $f_i = b^{n-i} f_n$ for $i = 1, \ldots, n$ and that $f_n$ defines an isomorphism $B = A[X]/(P) \to \mathrm{Hom}_A(B, A)$ of $B$-modules.

(3) By (2) and Lemma 1.2.1, we have $\mathrm{Tr}_{B/A} = \sum_{i=1}^{n} c_i b^{n-i} f_n = Q(b) f_n$. Since $P' = Q + (X - b)Q'$, we have $Q(b) = P'(b)$.

If $B$ is étale, then $P'(b)$ is invertible and we have $f_n = 1/P'(b) \cdot \mathrm{Tr}_{B/A}$. This means (1.2.2) holds. □

As an application of Proposition 1.2.2, we show that rings étale over normal rings are again normal.

**Proposition 1.2.3** *Let $A$ be a ring and $P \in A[X]$ be a monic polynomial. Set $B = A[X]/(P)$ and let $b \in B$ be the image of $X$. If $A$ is normal, then $B' = B[P'(b)^{-1}]$ is also normal.*

*Proof* By replacing $A$ by the local ring at any prime ideal, we may assume that $A$ is an integrally closed integral domain. Let $K$ be the fraction field of $A$ and set $L = B \otimes_A K$ and $L' = B' \otimes_A K$. Since $B'$ is flat over $A$, the morphism $B' \to L'$ is an injection. First, we reduce it to the case where the surjection $L \to L'$ is an isomorphism.

Since $A$ is integrally closed, the minimal polynomial $Q \in K[X]$ of the image $b_1 \in L'$ of $b$ is contained in $A[X]$. The image $B_1 \subset L'$ of $B$ is identified with the quotient ring $A[X]/(Q)$ of $B = A[X]/(P)$ since $Q$ divides $P$. Since the surjection $B \to B_1$ induces a surjection $B' \to B_1[P'(b_1)^{-1}]$ of subrings of $L'$, we may identify $B' = B_1[P'(b_1)^{-1}]$. If we define $R \in A[X]$ by $P = QR$, then we have $P'(b_1) = Q'(b_1)R(b_1)$. Hence, $B' = B_1[P'(b_1)^{-1}]$ is a localization of $B_1[Q'(b_1)^{-1}]$. Thus, by replacing $P$ by $Q$, we may assume $L = L'$.

Since $L = B' \otimes_A K$ is étale over $K$ by Lemma 1.1.6, the ring $L$ is a product of finitely many finite separable extensions of $K$ by Proposition 1.1.8(2) (a) $\Rightarrow$ (b). Let $C \subset L$ be the normalization of $B$. Since $A$ is normal, we have $\mathrm{Tr}_{L/K} C \subset A$. Hence, the injection $\mathrm{Tr}_{B/A} \colon B \to \mathrm{Hom}_A(B, A)$ is extended to an injection $C \to \mathrm{Hom}_A(B, A)$. Thus, we obtain injections $B' = B[P'(b)^{-1}] \to C[P'(b)^{-1}] \to \mathrm{Hom}_A(B, A) \otimes_B B[P'(b)^{-1}]$. Since the composition is an isomorphism by Proposition 1.2.2, the first morphism $B' \to C[P'(b)^{-1}]$ is an isomorphism and $B'$ is normal. □

**Corollary 1.2.4** *Let $B$ be a ring étale over $A$.*

(1) *If $A$ is normal, then $B$ is also normal.*

(2) *If $A$ is a discrete valuation ring with the maximal ideal $\mathfrak{m}$ and if $\mathfrak{n} \subset B$ is a prime ideal such that the inverse image is $\mathfrak{m}$, then the local ring $B_{\mathfrak{n}}$ is a discrete valuation ring.*

*Proof* (1) By Theorem 1.1.10, we may assume that $B = A[X]/(P)[P'^{-1}]$ for a monic polynomial $P \in A[X]$. Hence, the assertion follows from Proposition 1.2.3.

(2) By (1), $B_{\mathfrak{n}}$ is a normal local noetherian ring. Since $B_{\mathfrak{n}}$ has exactly two prime ideals, the maximal ideal $\mathfrak{n}B_{\mathfrak{n}}$ and 0, it follows that $B_{\mathfrak{n}}$ is a discrete valuation ring. □

**Exercise 1.2** Let the notation be as in Proposition 1.2.2 and its proof. Compute $\mathrm{Tr}_{B/A} c_i$ for $i = 1, \ldots, n$ and show that

$$\begin{pmatrix} \mathrm{Tr}_{B/A} 1 \\ \mathrm{Tr}_{B/A} b \\ \vdots \\ \mathrm{Tr}_{B/A} b^{n-1} \end{pmatrix} = \mathbf{A}^{-1} \begin{pmatrix} n \\ (n-1)a_1 \\ \vdots \\ a_{n-1} \end{pmatrix}.$$

## 1.3 Henselian Local Rings

If you are interested only in complete discrete valuation rings, you may skip this section and read henselian as complete in the later sections, admitting the characterization of henselian discrete valuation rings in Proposition 1.4.5 and Hensel's lemma, Lemma 1.4.7. Henselian local rings are defined as follows.

**Definition 1.3.1** Let $A$ be a local ring and $\mathfrak{m}$ be the maximal ideal.

(1) We call a ring $B$ étale over $A$ equipped with a morphism $B \to A/\mathfrak{m}$ over $A$ an étale neighborhood of $\mathfrak{m}$.
(2) We say that $A$ is henselian if, for every étale neighborhood $B$ of $\mathfrak{m}$ and $\mathfrak{n} = \operatorname{Ker}(B \to A/\mathfrak{m})$, the morphism $A \to B_{\mathfrak{n}}$ to the localization is an isomorphism.

**Lemma 1.3.2** *Let $A$ be a local ring and let $B$ and $B \to B/\mathfrak{n}$ be an étale neighborhood of the maximal ideal $\mathfrak{m} \subset A$. Let $A \to A'$ be a local morphism of local rings and let $\mathfrak{m}' \subset A'$ be the maximal ideal. Then the following conditions are equivalent:*

(1) *There exists a morphism $B \to A'$ over $A$ such that the inverse image of $\mathfrak{m}'$ equals $\mathfrak{n}$.*
(2) *The morphism $A' \to B'_{\mathfrak{n}'}$ to the local ring at the maximal ideal $\mathfrak{n}' \subset B' = B \otimes_A A'$ above $\mathfrak{m}'$ and $\mathfrak{n}$ is an isomorphism.*

*Further, the morphism in (1) is unique.*

*Proof* By replacing $A \to B$ by $A' \to B \otimes_A A'$, we may assume $A' = A$. Then, since a section $\operatorname{Spec} A \to \operatorname{Spec} B$ of an étale morphism $A \to B$ is an open immersion by Lemma 1.1.7(2), the conditions are equivalent.

The morphism in (1) is induced by the inverse of the isomorphism in (2) and is unique. □

A field is a henselian local ring, as we will see in Corollary 1.3.4.

**Proposition 1.3.3** ([67, Chapitre 7, §3, Proposition 3]) *Let $A$ be a local ring and $F = A/\mathfrak{m}$ be the residue field. Then, the following conditions are equivalent:*

(1) *$A$ is henselian.*
(2) *Let $P \in A[X]$ be a polynomial and let $\overline{a} \in F$ be a root of $\overline{P} \in F[X]$ satisfying $\overline{P}'(\overline{a}) \neq 0$. Then there exists a root $a \in A$ of $P$ satisfying $a \equiv \overline{a} \bmod \mathfrak{m}$.*
(2′) *Let $P \in A[X]$ be a monic polynomial and let $\overline{a} \in F$ be a simple root of $\overline{P} \in F[X]$. Then there exists a root $a \in A$ of $P$ satisfying $a \equiv \overline{a} \bmod \mathfrak{m}$.*

*In conditions* (2) *and* (2′), *the solution a congruent to* $\overline{a}$ *is unique.*

*Proof* (1) ⇒ (2): By Lemma 1.1.6, the ring $B = A[X]/(P)[P'^{-1}]$ is étale over $A$. Hence, $B$ is an étale neighborhood of $\mathfrak{m}$ with respect to the surjection $B \to F$ defined by the solution $\overline{a}$ of $\overline{P}$. Since $A$ is henselian, the morphism $A \to B_{\mathfrak{n}}$ to the localization at $\mathfrak{n} = \mathrm{Ker}(B \to F)$ is an isomorphism. Hence, the composition $B \to B_{\mathfrak{n}} \to A$ with its inverse defines a unique root $a$ of $P$ lifting $\overline{a}$ by Lemma 1.3.2 (2) ⇒ (1).

(2) ⇒ (2′): If $\overline{a}$ is a simple root of $\overline{P}$, we have $\overline{P}'(\overline{a}) \neq 0$.

(2′) ⇒ (1): Let $B$ be an étale neighborhood of $\mathfrak{m}$ and $\mathfrak{n} = \mathrm{Ker}(B \to F)$. By Theorem 1.1.10 (1) ⇒ (2), we may assume that $B = A[X]/(P)[P'^{-1}]$ for a monic polynomial $P \in A[X]$. Let $\overline{a} \in F$ be the image of $X$ by $B \to F$. Then $\overline{P}'(\overline{a}) \neq 0$ and the morphism $B/\mathfrak{n} \to F$ defined by $\overline{a}$ is lifted to a section $B \to A$ defined by a lifting $a$ of $\overline{a}$ by (2′). This induces an isomorphism $B_{\mathfrak{n}} \to A$ by Lemma 1.3.2 (1) ⇒ (2). □

A complete noetherian local ring is a henselian local ring.

**Corollary 1.3.4** *Let* $A = \varprojlim_n A/\mathfrak{m}^n$ *be a complete noetherian local ring. Then A is henselian. In particular, a field is a henselian local ring. In other words, a complete noetherian local ring A satisfies* (2) *in Proposition* 1.3.3.

The last property in Corollary 1.3.4 is called Hensel's lemma.

*Proof* We show that $A$ satisfies condition (2) in Proposition 1.3.3. The case where $A$ is a field is clear. Let $\mathfrak{m} \subset A$ be the maximal ideal. Since, by Lemma 1.1.6, $B = A[X]/(P)[P'^{-1}]$ is étale over $A$, the natural mapping $\mathrm{Mor}_A(B, A/\mathfrak{m}^{n+1}) \to \mathrm{Mor}_A(B, A/\mathfrak{m}^n)$ is a bijection for every $n \geqslant 1$ by Theorem 1.1.11. Hence, by induction on $n$, there exists a unique solution $a_n \in A/\mathfrak{m}^n$ of $P = 0$ satisfying $a_n \equiv \overline{a} \bmod \mathfrak{m}$. Then $a = (a_n)_n \in A = \varprojlim A/\mathfrak{m}^n$ is a solution of $P = 0$. □

We give another criterion for a local ring to be henselian.

**Proposition 1.3.5** ([67, Chapitre 7, §3, Proposition 3]) *Let A be a local ring and* $F = A/\mathfrak{m}$ *be the residue field. Then the following conditions are equivalent:*

(1) *A is henselian.*
(2) *Let B be a ring finite over A. Then B is isomorphic to a finite product* $\prod_{i \in I} B_i$ *of local rings.*

*Proof* (1) ⇒ (2): First, we show that a field $A = K$ satisfies condition (2). Let $B$ be a ring over $K$ of finite dimension as a $K$-vector space. We prove that condition (2) is satisfied by induction on the dimension $\dim_K B$. If $\dim = 0$, then $B = 0$. Assume $B \neq 0$ and let $\mathfrak{m} \subset B$ be a maximal ideal. Then there exists an integer $n \geqslant 1$ such that $\mathfrak{m}^{n+1} = \mathfrak{m}^n$. By Lemma 1.1.7(1), $B$ is decomposed as a product $B = B/\mathfrak{m}^n \times B'$. Since $B/\mathfrak{m}^n$ is a local ring and $\dim_K B' < \dim_K B$, the assertion follows.

Assume that $A$ is henselian and let $\mathfrak{m}$ be the maximal ideal. Let $B$ be a ring finite over $A$. First assume that $B$ is free of finite rank as an $A$-module. Let $F$ be the functor in Lemma 1.1.12 for $B$ over $A$ and let $E$ be the ring étale over $A$ representing $F$. Since $B/\mathfrak{m}B$ is of finite dimension over $A/\mathfrak{m}$, it is isomorphic to a product $\prod_i (B/\mathfrak{m}B)_i$ of finitely many local rings. Let $e_i \in B/\mathfrak{m}B$ be the corresponding primitive idempotents. Then each $e_i \in \operatorname{Idem} B/\mathfrak{m}B = F(A/\mathfrak{m})$ defines a morphism $E \to A/\mathfrak{m}$ over $A$.

Since $E$ is étale over $A$, the ring $E$ is an étale neighborhood of $\mathfrak{m}$ with respect to the morphism $E \to A/\mathfrak{m}$ corresponding to $e_i$. Since $A$ is henselian, for the kernel $\mathfrak{n}_i = \operatorname{Ker}(E \to A/\mathfrak{m})$, the morphism $A \to E_{\mathfrak{n}_i}$ is an isomorphism and the corresponding morphism $E \to E_{\mathfrak{n}_i} \to A$ defines an idempotent $\widetilde{e_i} \in F(A) = \operatorname{Idem} B$ lifting $e_i$. Then the product $B \to \prod_i B_i$ of the projections $B \to B_i = B/(1 - \widetilde{e_i})$ is an isomorphism since it is a morphism of projective $A$-modules of finite type whose $\otimes_A F$ is an isomorphism. Since $B_i/\mathfrak{m}B_i = (B/\mathfrak{m}B)_i$ is a local ring, $B_i$ is also a local ring.

We show the general case. Let $b_1, \ldots, b_n$ be a system of generators of $B$ over $A$ and let $P_1, \ldots, P_n \in A[X]$ be monic polynomials satisfying $P_i(b_i) = 0$. Then $B' = A[X_1, \ldots, X_n]/(P_1(X_1), \ldots, P_n(X_n))$ is free of finite rank as an $A$-module and the morphism $B' \to B$ over $A$ sending $X_i$ to $b_i$ is a surjection. Since $B'$ is already shown to be a product of finitely many local rings, its quotient $B$ is also a product of finitely many local rings.

(2) $\Rightarrow$ (1): We show that condition (2′) in Proposition 1.3.3 is satisfied. If $P \in A[X]$ is a monic polynomial, then $B = A[X]/(P)$ is a free $A$-module. If $\overline{a} \in F$ is a simple root of $\overline{P}$, then $\overline{B} = F[X]/(\overline{P})$ is isomorphic to a product $F \times \overline{B}_1$. If $B$ is a product of local rings, we have $B = B_0 \times B_1$ such that $B_0 \otimes_A F = F$. Since $B_0$ is a free $A$-module, the canonical morphism $A \to B_0$ is an isomorphism by Nakayama's lemma. □

**Proposition 1.3.6** ([67, Chapitre 8, Théorème 1]) *The inclusion*

$$(\text{henselian local rings}) \to (\text{local rings}) \tag{1.3.1}$$

*of a full subcategory has a left adjoint. If* $A^{\mathrm{h}}$ *denotes the image of a local ring* $A$ *by the adjoint functor, the canonical morphism* $A \to A^{\mathrm{h}}$ *is faithfully flat and induces an isomorphism on the residue fields.*

For a local ring $A$, we call its image by the adjoint functor of (1.3.1) the henselization of $A$ and write it as $A^{\mathrm{h}}$.

*Proof* Let $A$ be a local ring and $F$ be the residue field. There exists a cofinal filtered system $(A \to A_i \to F)_{i\in I}$ of étale neighborhoods of the maximal ideal $\mathfrak{m}$ of $A$ by Theorem 1.1.10 (1) $\Rightarrow$ (2) and Lemma 1.3.2. Set $\widetilde{A} = \varinjlim_{i\in I} A_i$. Then the kernel $\widetilde{\mathfrak{m}}$ of the induced morphism $\widetilde{A} \to F$ is a maximal ideal and $\widetilde{A}$ is a local ring. Since $A_i$ are flat over $A$, the local morphism $A \to \widetilde{A}$ is faithfully flat.

We show that the local ring $\widetilde{A}$ is henselian. Let $\widetilde{B}$ be an étale neighborhood of the maximal ideal $\widetilde{\mathfrak{m}}$ of $\widetilde{A}$ and let $\widetilde{\mathfrak{n}} \subset \widetilde{B}$ be the kernel of $\widetilde{B} \to \widetilde{A}/\widetilde{\mathfrak{m}} = F$. Then, since $\widetilde{B}$ is of finite presentation over $\widetilde{A}$, by [31, Théorème (11.2.6) (ii)], there exist $i \in I$, a ring $B_i$ étale over $A_i$, and an isomorphism $B_i \otimes_{A_i} \widetilde{A} \to \widetilde{B}$. Since $B_i$ is an étale neighborhood of $\mathfrak{m}$ and since $(A_i)_{i\in I}$ is cofinal, there exist $j \geqslant i$ and a morphism $B_i \to A_j$. This induces a morphism $\widetilde{B} \to \widetilde{A}$ and an isomorphism $\widetilde{A} \to \widetilde{B}_{\widetilde{\mathfrak{n}}}$ by Lemma 1.3.2 (1) $\Rightarrow$ (2).

Let $B$ be a henselian local ring with maximal ideal $\mathfrak{n}$ and let $A \to B$ be a local morphism. We show that there exists a unique local morphism $\widetilde{A} \to B$ compatible with $A \to B$. Let $i \in I$. Then $B_i = A_i \otimes_A B$ is étale over $B$ and the kernel of $A_i \to F$ defines a maximal ideal $\mathfrak{n}_i$ of $B_i$ such that $B/\mathfrak{n} \to B_i/\mathfrak{n}_i$ is an isomorphism. Since $B$ is henselian, the morphism $B \to B_{i,\mathfrak{n}_i}$ to the localization is an isomorphism. Hence, there exists a unique morphism $A_i \to B$ over $A$ by Lemma 1.3.2 (2) $\Rightarrow$ (1). They form an inductive system of morphisms and define $\widetilde{A} \to B$. The uniqueness follows from the construction. □

**Corollary 1.3.7** *Let $A \to B$ be a finite morphism of local rings.*

(1) *If $A$ is a henselian local ring, then $B$ is also a henselian local ring.*

(2) *Let $A^{\mathrm{h}}$ and $B^{\mathrm{h}}$ be the henselizations. Then the morphism $B \otimes_A A^{\mathrm{h}} \to B^{\mathrm{h}}$ is an isomorphism.*

*Proof* (1) Let $C$ be a ring finite over $B$. Then, since $C$ is finite over $A$, the ring $C$ is a product of local rings by Proposition 1.3.5 (1) $\Rightarrow$ (2). Hence $B$ is henselian by Proposition 1.3.5 (2) $\Rightarrow$ (1).

(2) Since the ring $B \otimes_A A^{\mathrm{h}}$ finite over $A^{\mathrm{h}}$ has a unique prime ideal above the maximal ideal of $A^{\mathrm{h}}$, it is a local ring and hence is henselian by (1). By the universality of $B^{\mathrm{h}}$, we obtain a morphism $B^{\mathrm{h}} \to B \otimes_A A^{\mathrm{h}}$ going the other way. By the universality of the henselizations and the tensor product, they are inverse to each other. □

**Exercise 1.3** Show that the field $\mathbf{Q}_p$ of $p$-adic numbers contains all the $(p-1)$th roots of 1.

## 1.4 Finite Extensions

We study elementary properties of finite extensions of henselian discrete valuation fields.

**Definition 1.4.1** We call the fraction field $K$ of a discrete valuation ring $O_K$ a discrete valuation field. A generator of the maximal ideal $\mathfrak{m}_K$ is called a uniformizer.

Let $K$ and $L$ be discrete valuation fields and $K \to L$ be a morphism of fields. If $K \to L$ induces a local morphism $O_K \to O_L$ of the discrete valuation rings, we say that $L$ is an extension of discrete valuation field $K$.

If $L$ is an extension of discrete valuation field $K$, the ramification index $e_{L/K}$ is the unique integer $e \geqslant 1$ satisfying $\mathfrak{m}_L^e = \mathfrak{m}_K O_L$ for the maximal ideals $\mathfrak{m}_K \subset O_K$ and $\mathfrak{m}_L \subset O_L$. The degree $[E : F]$ of extensions of residue fields is denoted $f_{L/K}$.

We say that a finite extension $L$ is a totally ramified extension of $K$ if the normalization $O_L$ is a discrete valuation ring, if the morphism $O_K/\mathfrak{m}_K \to O_L/\mathfrak{m}_L$ of residue fields is an isomorphism and if $e_{L/K} = [L : K]$.

If $O_K$ is henselian (resp. complete), we call $K$ a henselian (resp. complete) discrete valuation field.

The normalized valuation $K^\times \to \mathbf{Z}$ is denoted by $\mathrm{ord}_K$. The residue field $O_K/\mathfrak{m}_K$ of a discrete valuation ring $O_K$ will be denoted by $F$. By Corollary 1.3.4, a complete discrete valuation field is a henselian discrete valuation field.

**Proposition 1.4.2** *Let $K$ be a discrete valuation field with residue field $F$, and $L$ be a finite extension of $K$ of degree $n$. Assume that the integral closure $O_L$ of $O_K$ is a discrete valuation ring and let $E$ be the residue field of $L$.*

(1) *The residue field $E$ of $L$ is a finite extension of $F$ and the dimension of the $F$-vector space $O_L/\mathfrak{m}_K O_L$ is $e_{L/K} \cdot [E : F] \leqslant n = [L : K]$.*
(2) *Assume that $O_L$ is finite over $O_K$ and set $f_{L/K} = [E : F]$. Then the equality $e_{L/K} \cdot f_{L/K} = n = [L : K]$ holds and the norm mapping $\mathrm{N}_{L/K} : L^\times \to K^\times$ induces a commutative diagram of exact sequences:*

$$\begin{array}{ccccccccc} 0 & \longrightarrow & E^\times & \longrightarrow & L^\times/(1+\mathfrak{m}_L) & \longrightarrow & \mathbf{Z} & \longrightarrow & 0 \\ & & \downarrow{\scriptstyle e_{L:K}\cdot \mathrm{N}_{E/F}} & & \downarrow{\scriptstyle \mathrm{N}_{L/K}} & & \downarrow{\scriptstyle f_{L/K}} & & \\ 0 & \longrightarrow & F^\times & \longrightarrow & K^\times/(1+\mathfrak{m}_K) & \longrightarrow & \mathbf{Z} & \longrightarrow & 0 \end{array} \qquad (1.4.1)$$

(3) *Further assume that the residue field $E$ of $L$ is a purely inseparable exten-*

*sion of the residue field $F$ of $K$. Then, for $b \in L^\times$, we have*

$$N_{L/K} b \equiv (-1)^{(n-f)\cdot \mathrm{ord}_L b} \cdot b^n \quad \bmod 1 + \mathfrak{m}_L. \tag{1.4.2}$$

*Proof* (1) Let $E$ be the residue field of $L$. The $F$-vector space $O_L/\mathfrak{m}_K O_L$ is a successive extension of $\mathfrak{m}_L^i/\mathfrak{m}_L^{i+1}$ for $i = 0, \ldots, e-1$. If $x_1, \ldots, x_n \in O_L$ are liftings of elements of $E$ linearly independent over $F$, and if $s$ is a uniformizer of $L$, then the images of $x_i s^j$, $i = 1, \ldots, n$, $j = 0, \ldots, e-1$, are linearly independent over $F$ in an $F$-vector space $O_L/\mathfrak{m}_K O_L$ and the submodule $\langle x_i s^j; i = 1, \ldots, n, j = 0, \ldots, e-1\rangle \subset O_L$ is a free $O_K$-module of rank $n \cdot e \leqslant [L : K]$. Hence, $E$ is a finite extension of $F$ and we have $[E : F] \cdot e_{L/K} = \dim_F O_L/\mathfrak{m}_K \leqslant [L : K]$.

(2) If the torsion-free $O_K$-module $O_L$ is of finite type, it is a free module and its rank $[L : K]$ equals the dimension $[E : F] \cdot e_{L/K}$ of the $F$-vector space $O_L/\mathfrak{m}_K O_L$.

For (1.4.1), it suffices to show that the diagram

$$\begin{array}{ccccccc}
E^\times & \longleftarrow & O_L^\times & \xrightarrow{\subset} & L^\times & \xrightarrow{\mathrm{ord}_L} & \mathbf{Z} \\
{\scriptstyle e_{L:K}\cdot N_{E/F}}\downarrow & & {\scriptstyle N_{O_L/O_K}}\downarrow & & \downarrow{\scriptstyle N_{L/K}} & & \downarrow{\scriptstyle f_{L/K}} \\
F^\times & \longleftarrow & O_K^\times & \xrightarrow{\subset} & K^\times & \xrightarrow{\mathrm{ord}_K} & \mathbf{Z}
\end{array}$$

is commutative. Since $O_L \otimes_{O_K} F$ is a successive extension of $E$-vector spaces $\mathfrak{m}_L^i/\mathfrak{m}_L^{i+1}$, $i = 0, \ldots, e_{L/K} - 1$, of dimension 1, the left square is commutative. Since $L = O_L \otimes_{O_K} K$, the middle square is commutative. Since the $O_K$-module $\mathrm{Coker}(b\colon O_L \to O_L)$ is of length $[E : F] \cdot \mathrm{ord}_L b$ for $b \in O_L - \{0\}$, the right square is commutative.

(3) Since $L^\times$ is generated by uniformizers, we may assume that $b = s$ is a uniformizer of $L$. Let $t \in K$ be a uniformizer and define a unit $u \in O_L^\times$ by $s^e = ut$. Let $a_1, \ldots, a_f \in O_L$ be liftings of a basis of $E$ over $F$. Then $a_i s^j$, $1 \leqslant i \leqslant f$, $0 \leqslant j \leqslant e-1$, is a basis of $O_L$ over $O_K$. By computing the determinant of the matrix representing the multiplication by $s$ with respect to this basis, we obtain $N_{L/K} s \equiv (-1)^{f(e-1)\cdot f} N_{E/F}(u) t^f \bmod 1 + \mathfrak{m}_K$.

By the assumption that $E$ is a purely inseparable extension of $F$, we have $N_{E/F}(u) = u^f \equiv (s^e/t)^f \bmod 1 + \mathfrak{m}_L$. Thus, we obtain (1.4.2). □

Totally ramified extensions are generated by roots of Eisenstein polynomials. We say that a monic polynomial $P \in O_K[X]$ of degree $n$ is an Eisenstein polynomial if $P \equiv X^n \bmod \mathfrak{m}_K$ and if the constant term $t = P(0)$ is a uniformizer of $K$.

**Lemma 1.4.3** *Let $K$ be a discrete valuation field.*

(1) *Let $P \in O_K[X]$ be an Eisenstein polynomial. Then $L = K[X]/(P)$ is a totally ramified extension of $K$, the integral closure $O_L$ is $O_K[X]/(P)$, and the class $x$ of $X$ is a uniformizer.*
(2) *Conversely, let $L$ be a totally ramified extension of $K$ and $x$ be a uniformizer of $O_L$. Then the minimal polynomial $P \in O_K[X]$ of $x$ is an Eisenstein polynomial and $O_L = O_K[x]$.*

A generalization of Lemma 1.4.3(2) to extensions with separable extension of residue fields is given in Lemma 1.8.3(1).

*Proof* (1) Let $B = O_K[X]/(P)$. Then, $v(\sum_{i=0}^{n-1} a_i x^i) = \min_i(i + n \cdot \mathrm{ord}_K a_i)$ defines a discrete valuation and $B$ is a discrete valuation ring with uniformizer $x$. Hence, $L$ is a totally ramified extension and $B$ is the integral closure.

(2) We have $O_L/\mathfrak{m}_K O_L = F[x]/(x^n)$. Hence, we have $\deg P = [L : K] = n$ and $P \equiv X^n \bmod \mathfrak{m}_K$. The valuation of the constant term $P(0) = \mathrm{N}_{L/K}(-x)$ is 1 by Proposition 1.4.2(1). Hence, $P$ is an Eisenstein polynomial and the subring $O_K[x] \subset O_L$ is isomorphic to $O_K[X]/(P)$ and equals $O_L$ by 1. □

**Example 1.4.4** Let $K$ be a henselian discrete valuation field of characteristic $p > 0$, $F$ be the residue field, and $t$ be a uniformizer.

(1) Let $L$ be a cyclic extension defined by the Artin–Schreier equation $x^p - x = 1/t^n$ for an integer $n = pm - 1 > 0$ congruent to $-1$ modulo $p$. Then $s = t^m x$ is a root of the Eisenstein polynomial $S^p - t^{(p-1)m} S - t$. Hence, $L$ is a totally ramified extension of $K$ and $s$ is a uniformizer.
(2) Let $v \in O_K$ be an element such that $\bar{v} \in F - F^p$. Let $L$ be a cyclic extension defined by the Artin–Schreier equation $x^p - x = v/t^n$ for a multiple $n = pm > 0$ of $p$. Then $u = t^m x$ is a root of the polynomial $U^p - t^{(p-1)m} U - v$. Hence, the residue field $E = F(u)$ is a purely inseparable extension of the residue field $F$ of $K$.

**Proposition 1.4.5** *Let $K$ be a discrete valuation field and $O_K$ be the valuation ring. Then the following conditions are equivalent:*

(1) *$O_K$ is henselian.*
(2) *Let $L$ be a finite separable extension of discrete valuation field $K$. If the ramification index $e_{L/K}$ is 1 and if the morphism $O_K/\mathfrak{m}_K \to O_L/\mathfrak{m}_L$ of residue fields is an isomorphism, then $K \to L$ is an isomorphism.*
(3) *Let $L$ be a finite extension of $K$. Then the integral closure $O_L$ of $O_K$ in $L$ is a discrete valuation ring.*

*Proof* (1) ⇒ (3): The integral closure $O_L$ is the inductive limit $\varinjlim_\lambda B_\lambda$ of subrings finite over $O_K$. Since $O_K$ is henselian and since the extension $L$ has a unique idempotent, each $B_\lambda$ is local and the morphisms $B_\lambda \to B_\mu$ are local. Hence, their limit $O_L$ is also local.

We show that $O_L$ is a discrete valuation ring. Assume first that $L$ is a separable extension of $K$. Since $O_K$ is noetherian, the integral closure $O_L \subset L$ of $O_K$ is finite over $O_K$. Hence, $O_L$ is a normal noetherian local ring and has exactly two prime ideals, 0 and the maximal ideal. Therefore, $O_L$ is a discrete valuation ring in this case. By replacing $K$ by the separable closure in $L$, we may assume that $L$ is a purely inseparable extension of $K$. Let $q = [L : K]$ be the degree of the extension. Then, since $(1/q)\mathrm{ord}_K x^q$ is the unique valuation of $L$ extending $\mathrm{ord}_K$, an element $x \in L$ is integral over $O_K$ if and only if $x^q \in O_K$. Hence, the integral closure $O_L \subset L$ of $O_K$ is a discrete valuation ring.

(3) $\Rightarrow$ (2): By (3), the normalization $B$ of $O_K$ in $L$ is a discrete valuation ring. Hence, the subring $B \subset O_L \subsetneq L$ equals $O_L$. Since $L$ is a finite separable extension, $B = O_L$ is finite over $O_K$. Hence, the morphism $O_K \to O_L$ is an isomorphism by Nakayama's lemma.

(2) $\Rightarrow$ (1): Let $B$ be a ring étale over $O_K$ and $\mathfrak{n} \subset B$ be the maximal ideal such that $F = O_K/\mathfrak{m}_K \to B/\mathfrak{n}$ is an isomorphism. Then the local ring $B_\mathfrak{n}$ is a discrete valuation ring by Corollary 1.2.4(2) and its fraction field $L$ is a finite separable extension of $K$ by Proposition 1.1.8(1) (a) $\Rightarrow$ (b). Since $e_{L/K} = 1$ and $F \to B/\mathfrak{n}$ is an isomorphism, the morphisms $K \to L$ and $O_K \to B_\mathfrak{n}$ are isomorphisms by (2). □

**Corollary 1.4.6** *Let $K$ be a discrete valuation field with residue field $F$. Let $L$ be a finite extension of $K$, and $O_L$ be the integral closure of $O_K$.*

(1) *If $K$ is henselian, then $O_L$ is a henselian discrete valuation ring.*
(2) *If $K$ is complete, then $O_L$ is finite over $O_K$ and $L$ is also complete.*

*Proof* (1) By Proposition 1.4.5 (1) $\Rightarrow$ (3), the integral closure $O_L \subset L$ of $O_K$ is a discrete valuation ring. Let $L'$ be a finite extension of $L$. Then, since $L'$ is a finite extension of $K$, the integral closure $O_{L'}$ of $O_K$ is a discrete valuation ring by Proposition 1.4.5 (1) $\Rightarrow$ (3). Since $O_{L'}$ is the integral closure of $O_L$, it follows that $O_L$ is a henselian discrete valuation ring by Proposition 1.4.5 (3) $\Rightarrow$ (1).

(2) By Corollary 1.3.4, $K$ is henselian and by (1), $O_L$ is a discrete valuation ring. Since $O_L/\mathfrak{m}_K O_L$ is of finite dimension by Proposition 1.4.2, there exist finitely many elements in $O_L$ generating the completion $O_{\hat{L}} = \varprojlim_n O_L/\mathfrak{m}_K^n O_L$ as an $O_K$-module by Nakayama's lemma. Hence, the injection $O_L \to O_{\hat{L}}$ is an isomorphism of $O_K$-modules of finite type. □

For henselian discrete valuation rings, we will use Proposition 1.3.3 (2) in the following generalized form. We will refer to this as Hensel's lemma.

**Lemma 1.4.7** *Let $K$ be a henselian discrete valuation field and $P \in O_K[X]$ be a polynomial. Let $a \in O_K$ and assume $P'(a) \neq 0$ and $\mathrm{ord}_K P(a) > 2 \cdot \mathrm{ord}_K P'(a)$. Let $n = \mathrm{ord}_K P'(a) < m$ be integers satisfying $n + m \leqslant \mathrm{ord}_K P(a)$. Then there exists a unique solution $x \in O_K$ of $P(x) = 0$ satisfying $x \equiv a \bmod \mathfrak{m}_K^m$.*

*Proof* Let $t$ be a generator of $\mathfrak{m}_K^{m-1}$. Since $P(a+tX) \equiv P(a)+P'(a)tX \bmod t^2$ and $P'(a)t$ divides $P(a)$ and $t^2$, we have $Q(X) = P(a+tX)/P'(a)t \in O_K[X]$. Since $Q(0) \equiv P(a)/P'(a)t \equiv 0 \bmod \mathfrak{m}_K$ and $Q'(0) \equiv 1 \bmod \mathfrak{m}_K$, there exists a unique solution $x \in \mathfrak{m}_K$ of $Q(x) = 0$ by Proposition 1.3.3 (1) $\Rightarrow$ (2). This means that there exists a unique solution $x \equiv a \bmod \mathfrak{m}_K^m$ of $P(x) = 0$. □

The henselization and the completion have the same absolute Galois groups.

**Proposition 1.4.8** ([9, Lemme 2.2.1]) *Let $K$ be a henselian discrete valuation field and let $\widehat{K} = (\varprojlim_n O_K/\mathfrak{m}_K^n) \otimes_{O_K} K$ be the completion of $K$. Then the completion defines an equivalence of categories*

$$(\text{finite separable extensions of } K) \to (\text{finite separable extensions of } \widehat{K}). \tag{1.4.3}$$

*For every finite separable extension $L$ over $K$, the canonical morphism*

$$O_L \otimes_{O_K} O_{\widehat{K}} \to O_{\widehat{L}} \tag{1.4.4}$$

*is an isomorphism.*

*Proof* Let $L$ be a finite separable extension of $K$. By Proposition 1.4.5 and Corollary 1.4.6(2), the normalization $O_L$ is a discrete valuation ring and the tensor product $O_L \otimes_{O_K} O_{\widehat{K}}$ is the completion of $O_L$ and hence is a discrete valuation ring. Consequently, $L \otimes_K \widehat{K}$ is the completion $\widehat{L}$ of $L$ and is a finite separable extension of $\widehat{K}$. Thus, the morphism $G_{\widehat{K}} \to G_K$ of absolute Galois groups is a surjection.

To prove the injectivity of $G_{\widehat{K}} \to G_K$, it suffices to show the essential surjectivity of the functor (1.4.3). Namely, we show that for every finite separable extension $M$ of $\widehat{K}$, there exist a finite separable extension $L$ of $K$ and an isomorphism $L \otimes_K \widehat{K} \to M$. Take an element $a \in O_M$ such that $M = \widehat{K}(a)$ and let $P \in O_{\widehat{K}}[X]$ be the minimal polynomial. Since $M$ is a separable extension, we have $P'(a) \neq 0$. Let $m > n = \mathrm{ord}_M P'(a)$ be an integer such that $\mathfrak{m}_M^m \subset \mathfrak{m}_K \cdot O_{\widehat{K}}[a] \subset O_M$.

Take a monic polynomial $Q \in O_K[X]$ such that $Q \equiv P \bmod \mathfrak{m}_K^{n+m}$. We have $Q(a) \equiv P(a) = 0 \bmod \mathfrak{m}_M^{n+m}$. Since $Q'(a) \equiv P'(a) \bmod \mathfrak{m}_M^{n+m}$, we have $\mathrm{ord}_M Q'(a) = \mathrm{ord}_M P'(a)$. Hence, by Hensel's lemma, there exists a unique solution $b \equiv a \bmod \mathfrak{m}_M^m$ of $Q(b) = 0$. Since $b \in a+\mathfrak{m}_M^m \subset O_{\widehat{K}}[a]$, the solution

$b$ of $Q$ defines a morphism $O_{\widehat{K}}[X]/(Q) \to O_{\widehat{K}}[a]$ over $O_{\widehat{K}}$. Since this is a morphism of $O_{\widehat{K}}$-modules of the same rank and $b \equiv a \bmod \mathfrak{m}_M^m \subset \mathfrak{m}_K O_{\widehat{K}}[a]$, it is an isomorphism by Nakayama's lemma. If we set $L = K[X]/(Q)$, we obtain an isomorphism $L \otimes_K \widehat{K} \to M$. Hence, $L$ is a finite separable extension of $K$ satisfying the required condition. □

The henselization and the completion are related as follows.

**Proposition 1.4.9** *Let $K$ be a discrete valuation field and let $O_K^{\mathrm{h}}$ be the henselization of $O_K$.*

(1) *The henselization $O_K^{\mathrm{h}}$ is a discrete valuation ring and the canonical morphism $O_K \to O_K^{\mathrm{h}}$ induces an isomorphism on the completions.*
(2) *We identify the fraction field $K^{\mathrm{h}}$ of $O_K^{\mathrm{h}}$ as a subfield of the completion $\widehat{K}$ by the isomorphism in* (1). *Then we have $K^{\mathrm{h}} = \{x \in \widehat{K} \mid x$ is separably algebraic over $K\}$.*

*Proof* (1) Write $O_K^{\mathrm{h}} = \varinjlim_{i \in I} B_i$ as an inductive limit of étale neighborhoods of the maximal ideal $\mathfrak{m} \subset O_K$ as in the proof of Proposition 1.3.6. Let $\mathfrak{n}_i \subset B_i$ be the kernel of the surjection $B_i \to F = O_K/\mathfrak{m}$. The local ring $B_{i,\mathfrak{n}_i}$ is a discrete valuation ring by Corollary 1.2.4(2) and the ramification index over $O_K$ is 1 by Proposition 1.1.8(1) (a) ⇒ (b). Hence, the limit $O_K^{\mathrm{h}} = \varinjlim_{i \in I} B_{i,\mathfrak{n}_i}$ is also a discrete valuation ring of ramification index 1 over $O_K$. Since the residue field of $O_K^{\mathrm{h}}$ is $F$ by Proposition 1.3.6, the morphism $O_K/\mathfrak{m}_K^n \to O_K^{\mathrm{h}}/\mathfrak{m}_{K^{\mathrm{h}}}^n$ is an isomorphism for every integer $n \geqslant 1$ and the morphism $\widehat{O}_K = \varprojlim_n O_K/\mathfrak{m}_K^n \to \varprojlim_n O_K^{\mathrm{h}}/\mathfrak{m}_{K^{\mathrm{h}}}^n = \widehat{O}_K^{\mathrm{h}}$ of completions is an isomorphism.

(2) Since $K^{\mathrm{h}}$ is an inductive limit of finite separable extensions of $K$, every element of $K^{\mathrm{h}}$ is separably algebraic over $K$. To show the other inclusion, it suffices to show that if $L$ is a finite separable extension of $K^{\mathrm{h}}$ inside $\widehat{K}$, then we have $L = K^{\mathrm{h}}$. This follows from the equivalence of categories (1.4.3) for $K^{\mathrm{h}} \subset \widehat{K}$. □

**Exercise 1.4** Let $p$ be a prime number and $n \geqslant 1$ be an integer. Let $\zeta_{p^n} \in \mathbf{Q}_p(\zeta_{p^n})$ be a primitive $p^n$th root of 1. Find the minimal polynomial of $\zeta_{p^n} - 1$ over $\mathbf{Q}_p$ and show that $\zeta_{p^n} - 1$ is a uniformizer of $\mathbf{Q}_p(\zeta_{p^n})$. Compute the degree of the extension $\mathbf{Q}_p(\zeta_{p^n})$ over $\mathbf{Q}_p$.

## 1.5 Tamely Ramified Extensions

We study finite extensions of henselian discrete valuation fields either without or with little ramification.

**Definition 1.5.1** Let $K$ be a henselian discrete valuation field and let $L$ be a finite extension of $K$. Let $F$ and $E$ be the residue fields of $K$ and $L$ respectively.

(1) Assume that $L$ is a separable extension. We say that $L$ is an unramified extension of $K$ if $E$ is a separable extension of $F$ and if $e_{L/K} = 1$. We say that $L$ is a tamely ramified extension of $K$ if $E$ is a separable extension of $F$ and if the ramification index $e_{L/K}$ is invertible in $F$.
(2) If $L$ is not a tamely ramified extension, we say that $L$ is wildly ramified. Assume that the residue field is of characteristic $p > 0$. We say that $L$ is a purely wildly ramified extension of $K$ if $[L : K]$ is a power of $p$ and if $E$ is a purely inseparable extension of $F$.

**Lemma 1.5.2** *Let $K$ be a henselian discrete valuation and let $F$ be the residue field of $K$.*

(1) *For every finite separable extension $L$ of $K$ and the integral closure $O_L$ of $O_K$, the following conditions are equivalent:*
    (a) *$O_L$ is étale over $O_K$.*
    (b) *$L$ is an unramified extension of $K$.*
(2) *Let $L$ be an unramified extension of $K$. Then, for every extension $K'$ of henselian discrete valuation field $K$, a composition field $L' = LK'$ is an unramified extension of $K'$ and $O_{L'}$ is generated by $O_L$ over $O_{K'}$.*

*Proof* (1) (a) $\Rightarrow$ (b): By Proposition 1.1.8(2) (a) $\Rightarrow$ (b), $O_L \otimes_{O_K} F$ is a finite separable extension of $F$. Hence, $L$ is an unramified extension of $K$.

(b) $\Rightarrow$ (a): The normalization $O_L$ is a free $O_K$-module of finite rank. Hence, $O_L$ is of finite presentation and flat over $O_K$. Since $\Omega^1_{O_L/O_K} \otimes_{O_K} F = \Omega^1_{O_L \otimes_{O_K} F/F} = 0$, we have $\Omega^1_{O_L/O_K} = 0$ by Nakayama's lemma. Hence, $O_L$ is étale over $O_K$.

(2) Since $O_{K'}$ is henselian, the finite ring $O_L \otimes_{O_K} O_{K'}$ over $O_{K'}$ is decomposed as a product $\prod_i B_i$ of local rings. Since $O_L$ is étale over $O_K$ by (1) (b) $\Rightarrow$ (a), each $B_i$ is étale over $O_{K'}$. Hence, by (1) (a) $\Rightarrow$ (b), $L'_i = B_i \otimes_{O_{K'}} K'$ is an unramified extension of $K'$. □

**Proposition 1.5.3** *Let $K$ be a henselian discrete valuation field and let $F'$ be a finite separable extension of the residue field $F$.*

(1) *There exists an unramified extension $K'$ of $K$ with residue field $F'$.*
(2) *Let $K'$ be an unramified extension of $K$ with residue field $F'$. Then the functor* (extensions of henselian discrete valuation field $K$) $\to$ (sets) *sending $L$ to* $\mathrm{Mor}_F(F', E)$*, where $E$ denotes the residue field of $L$, is representable by $K'$.*

*Proof* (1) Let $a \in F'$ be a generator over $F$ and let $P \in O_K[X]$ be a lifting as a monic polynomial of the minimal polynomial of $a$. Then $O_{K'} = O_K[X]/(P)$ is étale over $O_K$ and hence is normal by Corollary 1.2.4(1). Hence, $O_{K'}$ is the ring of integers in a finite separable extension $K'$ of $K$ and is a discrete valuation ring by Proposition 1.4.5. Therefore, by Lemma 1.5.2(1) (a) $\Rightarrow$ (b), $K' = O_{K'} \otimes_{O_K} K$ is a finite unramified extension of $K$ and the residue field is $F'$.

(2) We show that the canonical mapping $\mathrm{Mor}_K(K', L) \to \mathrm{Mor}_F(F', E)$ is a bijection. Let $a \in O_{K'}$ be a lifting of a generator of $F'$ over $F$, and let $P \in O_K[X]$ be the minimal polynomial of $a$. We may identify $\mathrm{Mor}_K(K', L) = \{x \in O_L \mid P(x) = 0\}$ and $\mathrm{Mor}_F(F', E) = \{x \in E \mid \overline{P}(x) = 0\}$. Since $\overline{P} \in F[X]$ is a separable polynomial, the reduction defines a bijection $\{x \in O_L \mid P(x) = 0\} \to \{x \in E \mid \overline{P}(x) = 0\}$ by Hensel's lemma. □

**Corollary 1.5.4** *Let $K$ be a henselian discrete valuation field and $L$ be a finite extension of $K$.*

(1) *There exists a largest subextension $M_u \subset L$ unramified over $K$. The residue field of $M_u$ is the separable closure of $F$ in $E$.*
(2) *Let $K' \subset L$ be a subextension. Then $K'M_u$ is the largest unramified extension of $K'$ in $L$.*

We call $M_u$ the maximal unramified extension of $K$ in $L$.

*Proof* (1) Let $F'$ be the separable closure of $F$ in $E$. By Proposition 1.5.3, there exists an unramified extension $K' \subset L$ with residue field $F'$. If $K'' \subset L$ is another unramified extension with residue field $F'' \subset E$, then $F''$ is a subfield of $F'$ and $K''$ is a subfield of $K'$ by Proposition 1.5.3.

(2) Since the residue field $E$ of $L$ is a purely inseparable extension of the separable closure $F'$ of $F$ in $E$, it is a purely inseparable extension of the residue field of $K'M_u$. Since $K'M_u$ is an unramified extension of $K'$ by Lemma 1.5.2(2), the assertion follows. □

If the residue field $F$ is of characteristic 0, then $L$ is a totally and tamely ramified extension of the maximal unramified extension $M_u$.

**Proposition 1.5.5** *Let $K$ be a henselian discrete valuation field and let $L$ be a finite extension of $K$. Assume that the residue fields $F$ and $E$ of $K$ and $L$ are of characteristic $p > 0$.*

(1) *Assume that $L$ is a tamely ramified extension of $K$ and let $K'$ be an extension of henselian discrete valuation $K$. Then, a composition field $L' = LK'$ is a tamely ramified extension of $K'$. The ramification index $e_{L'/K'}$ equals $e_{L/K}/\gcd(e_{L/K}, e_{K'/K})$.*

(2) *Assume that $E$ is a purely inseparable extension of $F$ and let $m \mid e_{L/K}$ be a divisor prime to $p$. Then there exists a unique totally ramified subextension $K' \subset L$ of degree $m$ over $K$. Further, there exists a uniformizer $s$ of $K'$ such that $s^m$ is a uniformizer of $K$.*

*Proof* Existence in (2). Let $t \in K$ be a uniformizer, let $s \in L$ be an element of valuation $\mathrm{ord}_K s = 1/m$, and define a unit $u \in O_L^\times$ by $t = us^m$. By the assumption that $E$ is a purely inseparable extension, the quotient group $E^\times/F^\times$ is $p$-power torsion and is an $\mathbf{Z}_{(p)}$-module. Since $m$ is prime to $p$, there exist units $v \in O_L^\times$ and $w \in O_K^\times$ such that $u \equiv v^m/w \bmod \mathfrak{m}_L$. Since $O_L$ is henselian and $m$ is prime to $p$, we may assume that $u = v^m/w$ by Hensel's lemma. Hence, replacing $s$ by $vs$ and $t$ by $wt$, we may assume that $u = 1$. Then $L$ contains an $m$th root $s$ of $t$ and a totally ramified subextension $K[X]/(X^m - t)$ of degree $m$ as a subfield.

(1) By Proposition 1.5.3 and Corollary 1.5.4(2), after replacing $K$ by the maximal unramified extension $M_u$ in $L$ and $K'$ by $M_u K' \subset L'$, we may assume that $L$ is a totally ramified extension of $K$ of degree $m$ prime to $p$.

First assume that $m = e_{L/K}$ divides $e_{K'/K}$. By the existence in (2) proved above, there exists a uniformizer $t \in K$ and an isomorphism $O_K[X]/(X^m - t) \to O_L$. By the assumption $m \mid e_{K'/K}$, there exist $y \in O_{K'}$ and $v \in O_{K'}^\times$ such that $t = vy^m$. Then, the normalization of $O_L \otimes_{O_K} O_{K'} = O_{K'}[X]/(X^m - vy^m)$ is a ring $O_{K'}[Z]/(Z^m - v)$ étale over $O_{K'}$. Since $O_L[Z]/(Z^m - v)$ is further finite over $O_{K'}$, the tensor product $O_{K'}[Z]/(Z^m - v) \otimes_{O_{K'}} K'$ is a product of finitely many unramified extensions of $K'$ by Proposition 1.1.8(2) (a) $\Rightarrow$ (b) and Lemma 1.5.2(1) in this case.

Thus, by the decomposition $m = dm'$ where $d = (m, e_{K'/K})$, it is reduced to showing the case where $m$ is prime to $e_{K'/K}$. Since $m = e_{L/K}$ divides $e_{L'/K} = e_{L'/K'} e_{K'/K}$ and is prime to $e_{K'/K}$, it follows that $m$ divides $e_{L'/K'} \leqslant [L' : K'] \leqslant [L : K] = m$. Hence, $L'$ is a totally ramified extension of degree $m$ of $K'$ in this case.

Uniqueness in (2). Assume that $K'' \subset L$ is another totally ramified subextension of degree $m$. Then the composition $K'K'' \subset L$ is an unramified extension of $K'$ by (1). Since $E$ is a purely inseparable extension of $F$, we have $K'K'' = K'$ and the assertion follows. □

**Corollary 1.5.6** *Let $K$ be a henselian discrete valuation field and $L$ be a finite extension of $K$.*

(1) *There exists a largest subextension $M_t \subset L$ tamely ramified over $K$. The extension $M_t$ is a totally ramified extension of the maximal unramified*

*extension $M_u$ and the ramification index $e_{M_t/M_u}$ is the prime-to-p part of $e_{L/K}$.*

*The extension of L is a purely wildly ramified extension of $M_t$.*

(2) *Let $K' \subset L$ be a subextension. Then $K'M_t$ is the largest tamely ramified extension of $K'$ in L.*

We call $M_t$ the maximal tamely ramified extension of $K$ in $L$ and $w_{L/K} = [L : M_t]$ the wild degree of $L$ over $K$.

*Proof* (1) Let $M_u \subset L$ be the maximal unramified extension and let $m$ be the prime-to-$p$ part of $e_{L/K}$. By Proposition 1.5.5(2), there exists a totally ramified separable extension $K' \subset L$ of $M_u$ of ramification index $m$. If $K'' \subset L$ is another tamely ramified extension, the composition $K'K'' \subset L$ is an unramified ramified extension of $K'$ by Proposition 1.5.5(1). Since $E$ is a purely inseparable extension of the residue field $F'$ of $K'$, we have $K'' \subset K'K'' = K'$ as in the proof of Corollary 1.5.4(1). Hence, $K'$ is the largest tamely ramified extension in $L$.

The ramification index of $L$ over $M_t$ is a power of $p$ and the residue field $E$ is a purely inseparable extension of the residue field of $M_t$. The separable closure $M_s \subset L$ of $K$ is a finite separable extension of $M_t$ of ramification index a power of $p$, and the residue field extension is purely inseparable. Hence, $[M_s : M_t]$ is a power of $p$ by Proposition 1.4.2(2). Since $L$ is a purely inseparable extension of $M_s$, the degree $[L : M_s]$ is also a power of $p$ and the assertion follows.

(2) By Proposition 1.5.5(1), $K'M_t$ is a tamely ramified extension of $K'$. Since $L$ is a purely widely ramified extension of $M_t \subset K'M_t$, the assertion follows. □

The following lemma means that to focus on wild ramification, we may replace the base field by a tamely ramified extension.

**Lemma 1.5.7** *Let L be a finite extension of a henselian discrete valuation field K and let $M \subset L$ be the maximal tamely ramified subextension of K. Let $K'$ be a tamely ramified extension and $L' = LK'$. Then $M' = MK' \subset L'$ is the maximal tamely ramified subextension of $K'$ and $L \otimes_M M' \to L'$ is an isomorphism.*

*Proof* By Proposition 1.5.5(1), $M'$ is a tamely ramified extension of $K'$. By replacing $K$ by $M$, we may assume that $M = K$ and $M' = K'$. We show that $L'$ is a purely widely ramified extension of $K'$. It suffices to consider the cases where $K'$ is an unramified extension and a totally ramified extension respectively. In the unramified case, we have an isomorphism $O_L \otimes_{O_K} O_{K'} \to O_{L'}$ by Lemma 1.5.2(2) and the assertion follows. Assume that $K'$ is a totally ramified extension of $K$ of degree $m$ invertible in $F$. Then $L'$ is a totally ramified extension of $L$ of the same degree $m$ and $L \otimes_M M' \to L'$ is an isomorphism. □

**Definition 1.5.8** Let $K$ be a henselian discrete valuation field and let $L$ be a finite Galois extension of $K$ of Galois group $G = \mathrm{Gal}(L/K)$. Let $M_u \subset M_t \subset L$ be the maximal unramified extension and the maximal tamely ramified extension respectively. We call the corresponding subgroups $I \supset P$ of $G$ the inertia subgroup and the wild inertia subgroup respectively.

Since the extensions $M_u$ and $M_t$ are stable under the action of $G$, the subgroups $I$ and $P$ are normal subgroups. If the residue field $F$ is of characteristic $p > 0$, the wild inertia group $P$ is a unique $p$-Sylow subgroup of $I$. If the residue field $F$ is of characteristic 0, the wild inertia group $P$ is trivial.

**Proposition 1.5.9** *Let $K$ be a henselian discrete valuation field and let $L$ be a finite Galois extension of $K$ of Galois group $G = \mathrm{Gal}(L/K)$. Let $I \supset P$ be the inertia subgroup and the wild inertia subgroup of $G$ corresponding to the maximal unramified extension and the maximal tamely ramified extension $M_u \subset M_t \subset L$ respectively.*

(1) *The residue field $F'$ of $M_u$ is a Galois extension of $F$ and the action of $G$ on $F'$ defines an isomorphism $G/I \to \mathrm{Gal}(F'/F)$.*
(2) *The action of $I$ on $\mathfrak{m}_{M_t}/\mathfrak{m}_{M_t}^2$ defines an injection*

$$I/P \to F'^{\times}. \tag{1.5.1}$$

*and the quotient $I/P$ is a cyclic group of order $m = [M_t : M_u]$.*

*If the residue field $F$ is of characteristic $p > 0$, then the index $m = [I : P]$ equals the prime-to-$p$ part of $e_{L/K}$ and the subgroup $P \subset I$ is a unique $p$-Sylow subgroup.*

The ramification groups $G_i \subset G$ defined in the next section will be essentially a filtration on the wild inertia subgroup $P = G_1$.

*Proof* (1) The canonical morphism $G/I = \mathrm{Gal}(M_u/K) \to \mathrm{Aut}_F F'$ is an isomorphism by Proposition 1.5.3. Hence, $\#\mathrm{Aut}_F F' = [F' : F]$ and $F'$ is a Galois extension of $F$. Further, the morphism $G/I \to \mathrm{Gal}(F'/F)$ is an isomorphism

(2) Since the action of $I$ on $F'$ is trivial, its action on an $F'$-vector space $\mathfrak{m}_{M_t}/\mathfrak{m}_{M_t}^2$ of dimension 1 defines a character $I \to F'^{\times}$. By Proposition 1.5.5(2), there exist uniformizers $t \in M_u$ and $x \in M_t$ satisfying $x^m = t$. Since $m$ is invertible in $F'$, the reduction $\{\zeta \in M_u^{\times} \mid \zeta^m = 1\} \to \{\zeta \in F'^{\times} \mid \zeta^m = 1\}$

defines an isomorphism by Hensel's lemma. Since $M_t = M_u(x)$ is generated by an $m$th root $x$ of $t$, the character $I \to F'^{\times}$ induces an isomorphism $I/P = \mathrm{Gal}(M_t/M_u) \to \mu_m = \{\zeta \in F'^{\times} \mid \zeta^m = 1\}$.

The index $m$ equals the prime-to-$p$ part $e_{M_t/M_u}$ of $e_{L/K}$. Since the normal subgroup $P = \mathrm{Gal}(L/M_t)$ of $I$ is a $p$-group by Corollary 1.5.6(1) and is of index $[I : P] = m$ prime to $p$, it is a unique $p$-Sylow subgroup of $I$. □

For a separable closure $K_s$ of a henselian discrete valuation field $K$, the maximal unramified extension and the maximal tamely ramified extension $K_u \subset K_t \subset K_s$ are defined to be the unions of finite unramified extensions and finite tamely ramified extensions.

**Corollary 1.5.10** *Let $K$ be a henselian discrete valuation field, and $K_u \subset K_t \subset K_s$ be the maximal unramified extension and the maximal tamely ramified extension in a separable closure of $K$. Let the inertia group $I \subset G = \mathrm{Gal}(K_s/K)$ and $P \subset I$ be the corresponding subgroups. Then the quotient $G/I$ is identified with the absolute Galois group $\mathrm{Gal}(F_s/F)$ of the residue field $F$ and $I/P$ is identified with the projective limit $\varprojlim_{p \nmid m} \mu_m$. The subgroup $P \subset I$ is a unique pro-$p$ Sylow subgroup and the quotient $G/P$ is isomorphic to the semi-direct product $G/I \ltimes I/P$.*

*Proof* By taking the limit of finite subextensions in $K_s$, we obtain $K_u \subset K_t \subset K_s$ and $P \subset I \subset G$. The residue field of $K_u$ is a separable closure $F_s$ of $F$ by Proposition 1.5.3 and we obtain isomorphisms $G/I \to \mathrm{Gal}(K_u/K) \to \mathrm{Gal}(F_s/F) = G_F$. Since $K_t$ is the composite extension $K_u \cdot K(t^{1/m}, p \nmid m)$ for a uniformizer $t \in K$, we obtain isomorphisms $I/P \to \mathrm{Gal}(K_t/K_u) \to \varprojlim_{p \nmid m} \mu_m$ and $G/P \to \mathrm{Gal}(K_t/K) \to \mathrm{Gal}(K_u/K) \ltimes \mathrm{Gal}(K_t/K_u) \to G_F \ltimes I/P$. Since $P$ is a pro-$p$ group and $I/P$ is prime to $p$, $P$ is the unique pro-$p$ Sylow subgroup of $I$. □

The following property will be used to reduce a proof for finite separable extensions to that for extensions of degree $p$.

**Corollary 1.5.11** *Let $L$ be a finite separable extension of a henselian discrete valuation field $K$ and let $M \subset L$ be the maximal tamely ramified extension. Let $N$ be a Galois closure of $L$ over $K$, $M' \subset N$ be the maximal tamely ramified extension, and $L' = LM'$ be the composition. Then $L'$ is a tamely ramified extension of $L$ and there exist subextensions $M' = L'_0 \subset L'_1 \subset \cdots \subset L'_e = L'$ such that $L'_i$ is a wildly ramified extension of $L'_{i-1}$ of degree $p$ for $i = 1, \ldots, e$.*

*Proof* Since the Galois group $\mathrm{Gal}(N/M')$ is a $p$-group, the assertion follows from Lemma 1.5.12(2). □

**Lemma 1.5.12** *Let $G$ be a $p$-group of order $p^n$.*

(1) *There exists a sequence $G = G_0 \supsetneqq G_1 \supsetneqq \cdots \supsetneqq G_n = 1$ of normal subgroups such that $[G, G_i] \subset G_{i+1}$ and $[G_i : G_{i+1}] = p$ for $i = 0, \ldots, n-1$.*

(2) *Let $H$ be a subgroup of $G$ of index $[G : H] = p^k$. Then there exists a sequence $G = H_0 \supsetneqq H_1 \supsetneqq \cdots \supsetneqq H_k = H$ of subgroups such that $[H_i : H_{i+1}] = p$ for $i = 0, \ldots, k-1$.*

*Proof* (1) We may assume that $G \neq 1$. Since the conjugate action of $G$ on $G$ itself has at least one fixed point and the cardinality of fixed points is conjugate to the order of $G$ modulo $p$, the center of $G$ is non-trivial. Hence, the assertion follows by induction on the order of $G$.

(2) It suffices to consider $H_i = G_i H$ and remove the repeating ones. □

**Exercise 1.5** Let $p$ be a prime number and $n \geqslant 1$ be an integer. Assume that $n = mp^e$ is the product of an integer $m \geqslant 1$ prime to $p$ and a power $p^e > 1$ of $p$.

(1) Find the maximal unramified extension and the maximal tamely ramified extension of $\mathbf{Q}_p$ in $\mathbf{Q}_p(\zeta_n)$.

(2) Determine the Galois group $G = \mathrm{Gal}(\mathbf{Q}_p(\zeta_n)/\mathbf{Q}_p)$, the inertia group, and the wild inertia group $P \subset I \subset G$ as subgroups of $(\mathbf{Z}/n\mathbf{Z})^\times = (\mathbf{Z}/m\mathbf{Z})^\times \times (\mathbf{Z}/p^e\mathbf{Z})^\times$.

## 1.6 Lower Ramification Subgroups

We define the filtration by lower ramification groups on the Galois group of a Galois extension of henselian discrete valuation fields.

**Definition 1.6.1** Let $K$ be a henselian discrete valuation field and let $L$ be a finite Galois extension of $K$ of Galois group $G = \mathrm{Gal}(L/K)$.

For an integer $i \geqslant 1$, we define a normal subgroup $G_i \subset G$ by

$$G_i = \mathrm{Ker}(G \to \mathrm{Aut}(L^\times/(1 + \mathfrak{m}_L^i))) \tag{1.6.1}$$

and call $G_i$ the $i$th ramification group. To distinguish them from the upper ramification groups defined in Section 12.3, we call them the lower ramification groups. We set $G_0$ to be the kernel $\mathrm{Ker}(G \to \mathrm{Aut}\, O_L/\mathfrak{m}_L)$.

Define a function $i_G : G \to \mathbf{N} \cup \{\infty\}$ by

$$i_G(\sigma) = \begin{cases} i & \text{if } \sigma \in G_i - G_{i+1} \text{ for } i \geqslant 1, \\ 0 & \text{if } \sigma \in G - G_1, \\ \infty & \text{if } \sigma = 1. \end{cases}$$

Note that in the classical case where $L$ is a totally ramified extension of $K$, the definition of $i_G(\sigma)$ here is shifted by 1 from that in [78, page 70]. This shift by 1 is explained by Corollary 1.6.6(1). For a subgroup $H \subset G$ corresponding to a subextension $M \subset L$, we have $H_i = G_i \cap H$ for every $i \geqslant 0$. There is another filtration called the non-logarithmic ramification groups defined by $G'_i = \text{Ker}(G \to \text{Aut}(O_L/\mathfrak{m}_L^i))$. We compare this with that in Definition 1.6.1 in Proposition 1.6.5.

**Proposition 1.6.2** *Let $L$ be a finite Galois extension of a henselian discrete valuation field $K$. Let $F$ and $F' \subset E$ be the residue fields of $K$ and of the maximal unramified extension $M_u \subset L$, and let $q = e_{L/M_t}$ be the ramification index for the maximal tamely ramified extension $M_t \subset L$.*

(1) *The subgroup $G_0 \subset G$ equals the inertia subgroup $I \subset G$ corresponding to $M_u$. The action of $G$ on the residue field $E$ of $L$ defines an isomorphism $G/G_0 \to \text{Aut}(E/F)$.*
(2) *The subgroup $G_1 \subset G$ equals the wild inertia subgroup $P \subset G$ corresponding to $M_t$. The character $I/P \to F'^{\times}$ (1.5.1) equals the $q$th power of the character*

$$G_0 \to E^{\times} \tag{1.6.2}$$

*defined by the action on an $E$-vector space $\mathfrak{m}_L/\mathfrak{m}_L^2$ of dimension 1.*

*Proof* (1) Since $E$ is a purely inseparable extension of the separable closure $F' \subset E$ of $F$, the restriction defines an injection $\text{Aut}(E/F) \to \text{Gal}(F'/F)$. Hence, by Proposition 1.5.9(1), we have $G_0 = I$ and the morphism $G/G_0 \to \text{Aut}(E/F)$ is an isomorphism.

(2) We have an exact sequence $0 \to E^{\times} \to L^{\times}/(1 + \mathfrak{m}_L) \to \mathbf{Z} \to 0$ of $G$-modules. Since $G_0 = I = \text{Ker}(G \to \text{Aut}\, E) = \text{Ker}(G \to \text{Aut}\, E^{\times})$ by (1), the exact sequence defines the character (1.6.2) and its kernel equals $G_1$.

To show the last assertion, we may assume that $F$ is of characteristic $p > 0$ and that $q | [L : M_t]$ is a power of $p$ by Corollary 1.5.6. By the canonical isomorphisms $\mathfrak{m}_{M_t}/\mathfrak{m}_{M_t}^2 \otimes_{F_s} E \to \mathfrak{m}_L^q/\mathfrak{m}_L^{q+1} \leftarrow (\mathfrak{m}_L/\mathfrak{m}_L^2)^{\otimes q}$, the character (1.5.1) equals the $q$th power of the character (1.6.2). Since the character (1.6.2) is of order prime to $p$, the kernel $G_1$ of (1.6.2) equals the kernel $P$ of the character (1.5.1). □

In the following, we assume that the residue field $F$ is of characteristic $p > 0$. We study the graded quotients $\mathrm{Gr}_i G = G_i/G_{i+1}$ for $i \geqslant 1$.

**Proposition 1.6.3** *Assume that the residue field $F$ is of characteristic $p > 0$. Let $i \geqslant 1$ be an integer.*

(1) *The action of $G_i$ on $(1+\mathfrak{m}_L)/(1+\mathfrak{m}_L^{i+1})$ is trivial.*
(2) *The action of $G_i$ on $L^\times/(1+\mathfrak{m}_L^{i+1})$ defines a morphism*

$$\delta\colon G_i \to \mathrm{Hom}(L^\times/(1+\mathfrak{m}_L), \mathfrak{m}_L^i/\mathfrak{m}_L^{i+1}) \tag{1.6.3}$$

*such that the kernel equals $G_{i+1}$. The image of $\sigma \in G_i$ is induced by the morphism $\delta_\sigma\colon L^\times \to \mathfrak{m}_L^i/\mathfrak{m}_L^{i+1}$ sending $b$ to $\sigma(b)/b - 1 \bmod \mathfrak{m}_L^{i+1}$.*

*Proof* (1) For $\sigma \in G_i$ and $b \in \mathfrak{m}_L - \{0\}$, we have

$$\frac{\sigma(1+b)}{1+b} = 1 + \frac{\sigma(b)-b}{1+b} = 1 + \frac{b}{1+b}\left(\frac{\sigma(b)}{b} - 1\right) \in 1 + \mathfrak{m}_K^{i+1}.$$

(2) The subgroups $(1+\mathfrak{m}_L^i)/(1+\mathfrak{m}_L^{i+1}) \subset (1+\mathfrak{m}_L)/(1+\mathfrak{m}_L^{i+1})$ of $L^\times/(1+\mathfrak{m}_L^{i+1})$ are stable by the action of $G$, and the actions of $G_i$ on $L^\times/(1+\mathfrak{m}_L^i)$ and on $(1+\mathfrak{m}_L)/(1+\mathfrak{m}_L^{i+1})$ are trivial by (1). Hence, we obtain an injection $G_i/G_{i+1} \to \mathrm{Hom}(L^\times/(1+\mathfrak{m}_L), (1+\mathfrak{m}_L^i)/(1+\mathfrak{m}_L^{i+1}))$. By identifying $(1+\mathfrak{m}_L^i)/(1+\mathfrak{m}_L^{i+1})$ with $\mathfrak{m}_L^i/\mathfrak{m}_L^{i+1}$, we obtain (1.6.3). □

**Corollary 1.6.4** *Let $i \geqslant 1$ be an integer.*

(1) *The morphism* (1.6.3) *induces an injection*

$$\mathrm{Gr}_i G \to \mathrm{Hom}(L^\times/(1+\mathfrak{m}_L), \mathfrak{m}_L^i/\mathfrak{m}_L^{i+1}) \tag{1.6.4}$$

*compatible with the actions of $G$. The graded quotient $\mathrm{Gr}_i G = G_i/G_{i+1}$ is an $\mathbf{F}_p$-vector space.*
(2) *We have $[P, G_i] \subset G_{i+1}$. The injection* (1.6.4) *is compatible with the conjugate action of $I/P$ on $\mathrm{Gr}_i G$ and the actions $I/P$ on $L^\times/(1+\mathfrak{m}_L)$ and on the $E$-vector space $\mathfrak{m}_L^i/\mathfrak{m}_L^{i+1}$ by the ith power of the character $I/P \to E^\times$* (1.6.2).

*Proof* (1) Since the kernel of (1.6.3) is $G_{i+1}$, we obtain an injection (1.6.4). For $\sigma \in G_i$, $\tau \in G$, and $b \in L^\times$, we have

$$\delta_{\tau\sigma\tau^{-1}}\bar{b} \equiv \frac{\tau\sigma\tau^{-1}(b)}{b} - 1 = \tau\left(\frac{\sigma(\tau^{-1}b)}{\tau^{-1}b} - 1\right) = \tau(\delta_\sigma\overline{\tau^{-1}b}) = \tau(\delta_\sigma(\tau^{-1}\bar{b})).$$

Hence, the injection (1.6.4) is compatible with the actions of $G$. Since $\mathfrak{m}_L^i/\mathfrak{m}_L^{i+1}$ is an $E$-vector space, the graded quotient $\mathrm{Gr}_i G$ is an $\mathbf{F}_p$-vector space.

(2) Since the action of $P$ on the target of the injection (1.6.4) is trivial, the

conjugate action of $P$ on $G_i/G_{i+1}$ is trivial. This means $[P, G_i] \subset G_{i+1}$. The compatibility on the injection (1.6.4) follows from (1). □

We compare the two definitions of filtrations.

**Proposition 1.6.5** *For integers $i \geqslant 1$, define normal subgroups $G'_i \subset G$ by*

$$G'_i = \mathrm{Ker}(G \to \mathrm{Aut}(O_L/\mathfrak{m}_L^i)).$$

(1) *We have* $G'_i = \mathrm{Ker}(G \to \mathrm{Aut}(O_L/\mathfrak{m}_L^i)^\times)$.
(2) *Let $s \in L$ be a uniformizer. Then*

$$G'_{i+1} \subset G_i = \{\sigma \in G'_i \mid \sigma(s)/s \equiv 1 \bmod \mathfrak{m}_L^i\} \subset G'_i. \tag{1.6.5}$$

*Proof* (1) The inclusion $G'_i \subset \mathrm{Ker}(G \to \mathrm{Aut}(O_L/\mathfrak{m}_L^i)^\times)$ is clear. Since the action of $\sigma$ on $O_L/\mathfrak{m}_L^i - (O_L/\mathfrak{m}_L^i)^\times = \mathfrak{m}_L/\mathfrak{m}_L^i$ is trivial if and only if that on $(1 + \mathfrak{m}_L)/(1 + \mathfrak{m}_L^i) \subset (O_L/\mathfrak{m}_L^i)^\times$ is trivial, we have the other inclusion.

(2) We have $G_i \subset \mathrm{Ker}(G \to \mathrm{Aut}(O_L/\mathfrak{m}_L^i)^\times) = G'_i$ since $(O_L/\mathfrak{m}_L^i)^\times \subset L^\times/(1 + \mathfrak{m}_L^i)$. By the exact sequence $0 \to (O_L/\mathfrak{m}_L^i)^\times \to L^\times/(1 + \mathfrak{m}_L^i) \to \mathbf{Z} \to 0$ and (1), we have the equality in (1.6.5). For $\sigma \in G'_{i+1}$, we have $\sigma(s) \equiv s \bmod \mathfrak{m}_L^{i+1}$. Hence, the equality in (1.6.5) implies $G'_{i+1} \subset G_i$. □

**Corollary 1.6.6** (1) *Assume that the residue field $E$ is a separable extension of $F$. Then, for $i \geqslant 1$, we have $G_i = G'_{i+1}$ and the morphism* (1.6.4) *induces an injection*

$$\delta\colon \mathrm{Gr}_i G \to \mathfrak{m}_L^i/\mathfrak{m}_L^{i+1}. \tag{1.6.6}$$

*For $\sigma \in G_0$ and a uniformizer $s$ of $L$, we have*

$$i_G(\sigma) = \mathrm{ord}_L\Big(\frac{\sigma(s)}{s} - 1\Big). \tag{1.6.7}$$

(2) *Assume $e_{L/K} = 1$. Then, for $i \geqslant 1$, we have $G_i = G'_i$ and the morphism* (1.6.4) *induces an injection*

$$\delta\colon \mathrm{Gr}_i G \to \mathrm{Hom}(E^\times, \mathfrak{m}_L^i/\mathfrak{m}_L^{i+1}). \tag{1.6.8}$$

*Further, if $O_L = O_K[u]$, for $\sigma \in G_0$ we have*

$$i_G(\sigma) = \mathrm{ord}_L(\sigma(u) - u). \tag{1.6.9}$$

*Proof* (1) Since $G_0 = G'_1$, by replacing $G$ by $I = G_0$ we may assume $G = I$. Then we have $E = F$ and $L$ is totally ramified over $K$. Since $O_L = O_K[s]$, we have $G'_{i+1} = \{\sigma \in G'_i \mid \sigma(s) \equiv s \bmod \mathfrak{m}_L^{i+1}\}$. This equals $G_i$ by the equality in (1.6.5).

For $\sigma \in G_i = G'_{i+1}$, its image $\delta_\sigma\colon L^\times/(1 + \mathfrak{m}_L) \to \mathfrak{m}_L^i/\mathfrak{m}_L^{i+1}$ by the injection (1.6.4) annihilates $E^\times \subset L^\times/(1+\mathfrak{m}_L)$ and defines an element $\mathrm{Hom}(\mathbf{Z}, \mathfrak{m}_L^i/\mathfrak{m}_L^{i+1}) = \mathfrak{m}_L^i/\mathfrak{m}_L^{i+1}$. Thus, we obtain an injection (1.6.6).

By $O_L = O_K[s]$, we have $G_i = G'_{i+1} = \{\sigma \in G \mid \mathrm{ord}_L(\sigma(s) - s) \geqslant i+1\}$ and (1.6.7).

(2) Since $e_{L/K} = 1$, a uniformizer $s \in O_K$ is a uniformizer of $L$ and then the equality in (1.6.5) shows $G_i = G'_i$. For $\sigma \in G_i = G'_i$, its image $\delta_\sigma\colon L^\times/(1+\mathfrak{m}_L) \to \mathfrak{m}_L^i/\mathfrak{m}_L^{i+1}$ by the injection (1.6.4) annihilates the subgroup generated by a uniformizer $s \in K$ and defines an element $\mathrm{Hom}(E^\times, \mathfrak{m}_L^i/\mathfrak{m}_L^{i+1})$.

If $O_L = O_K[u]$, we have $G_i = G'_i = \{\sigma \in G \mid \mathrm{ord}_L(\sigma(u) - u) \geqslant i\}$ and (1.6.9). □

**Example 1.6.7** Let $K$ be a henselian discrete valuation field of characteristic $p > 0$, $F$ be the residue field, and $t \in K$ be a uniformizer.

(1) Let $L$ be the Artin–Schreier extension defined by $x^p - x = 1/t^n$ for $n = pm - 1 > 0$, and $s = t^m x \in L$ be a uniformizer as in Example 1.4.4(1). Let $\sigma \in G = \mathrm{Gal}(L/K)$ be the generator defined by $\sigma(x) = x + 1$. Since $\sigma(s)/s - 1 = 1/x$, we have $i_G(\sigma) = -\mathrm{ord}_L x = n$. Hence, $i = n$ is the unique integer satisfying $G_i = G$ and $G_{i+1} = 1$.

(2) Let $L$ be the Artin–Schreier extension defined by $x^p - x = v/t^n$ for $n = pm > 0$ and for $v \in O_K$ such that $\bar{v} \in F - F^p$. Let $u = t^m x$ be the generator of the residue field $E = F(u)$ as in Example 1.4.4(2). Let $\sigma \in G = \mathrm{Gal}(L/K)$ be the generator defined by $\sigma(x) = x+1$. Since $\sigma(u)/u - 1 = 1/x$, we have $i_G(\sigma) = -\mathrm{ord}_L x = m$. Hence, $i = m$ is the unique integer satisfying $G_i = G$ and $G_{i+1} = 1$.

**Exercise 1.6** Let $p$ be a prime number and $n \geqslant 1$ be an integer. Identify the Galois group $G = \mathrm{Gal}(\mathbf{Q}_p(\zeta_{p^n})/\mathbf{Q}_p)$ with $(\mathbf{Z}/p^n\mathbf{Z})^\times$. Compute the lower numbering filtration $G_i$.

## 1.7 Ramification Groups and Subextensions

We study the ramification groups and those of quotient groups. Recall that for a finite Galois extension $L$ of a henselian discrete valuation field $K$, the subgroup $G_1 \subset G = \mathrm{Gal}(L/K)$ is the wild inertia subgroup $P$. Further, for a subextension $L' \subset L$ corresponding to a subgroup $H \subset G$, the intersection $G_1 \cap H$ equals $H_1 \subset H = \mathrm{Gal}(L/L')$.

**Definition 1.7.1** Let $L$ be a finite Galois extension of a henselian discrete valuation field $K$ and set $G = \mathrm{Gal}(L/K)$. Let $L' \subset L$ be a subextension corresponding to a subgroup $H \subset G$. Then, for $\sigma \in G_1 - H_1$, let

$$i_{L'}(\sigma) = \min\left\{\mathrm{ord}_L\left(\frac{\sigma(x)}{x} - 1\right) \mid x \in L'^{\times}\right\} \geqslant 1. \tag{1.7.1}$$

If $L' = L$, we have $i_G(\sigma) = i_L(\sigma)$ for $\sigma \in G_1 - \{1\}$. More generally, if $L' \subset L$ is a Galois extension of Galois group $\overline{G} = G/N$, then we have $e_{L/L'} \cdot i_{\overline{G}}(\overline{\sigma}) = i_{L'}(\sigma)$ for $\sigma \in G_1 - N_1$ and its image $\overline{\sigma} \in \overline{G}_1 - \{1\}$. In this section and the next we study the relation between the functions $i_{K'}$ and $i_{L'}$ for subextensions $K' \subset L'$ of a Galois extension $L$ of $K$.

**Proposition 1.7.2** *Let $L$ be a Galois extension of a henselian discrete valuation field $K$ and set $G = \mathrm{Gal}(L/K)$. Let $L' \subset L$ be a subextension corresponding to a subgroup $H \subset G$. Let $\sigma \in G_1 - H_1$.*

(1) *Assume that $s \in L'$ satisfies $p \nmid \mathrm{ord}_{L'} s$ and that the extension $E'$ of $F$ is generated by the classes of units $u_1, \dots, u_n \in O_{L'}^{\times}$. Then*

$$i_{L'}(\sigma) = \min\left(\mathrm{ord}_L\left(\frac{\sigma(s)}{s} - 1\right), \mathrm{ord}_L(\sigma(u_1) - u_1), \dots, \mathrm{ord}_L(\sigma(u_n) - u_n)\right). \tag{1.7.2}$$

(2) *Let $L'' \subset L$ be a tamely ramified extension of $L'$. Then $i_{L'}(\sigma) = i_{L''}(\sigma)$.*

*Proof* (1) Let $i$ denote the right-hand side of (1.7.2). By (1.7.1), we have $i_{L'}(\sigma) \leqslant i$. By replacing $K$ by the maximal tamely ramified extension $M \subset L'$, we may assume that the ramification index $e_{L'/K}$ is a power of $p$. Then, $\mathrm{ord}_{L'} s$ is prime to $e_{L'/K}$ and there exist integers $a, b$ such that, for a uniformizer $t$ of $K$, $z = s^a t^b$ is a uniformizer of $L'$. Since $\mathrm{ord}_L(\sigma(z) - z) - 1 = \mathrm{ord}_L(\sigma(z)/z - 1) = \mathrm{ord}_L(\sigma(s)/s - 1)$, and since $O_{L'} = O_K[z, u_1, \dots, u_n]$ by Nakayama's lemma, we have $i_{L'}(\sigma) \geqslant i$.

(2) Since $i_{L'}(\sigma) \geqslant i_{L''}(\sigma)$ is clear, we set $i = i_{L'}(\sigma)$ and prove $i \leqslant i_{L''}(\sigma)$. It suffices to prove the cases where $L''$ is an unramified extension of $L'$ and $L''$ is a totally ramified extension of $L'$ respectively.

Assume first that $L''$ is an unramified extension of $L'$ and let $u$ be a generator of $O_{L''}$ over $O_{L'}$. By (1), it suffices to show that $\sigma(u) - u \equiv 0 \bmod \mathfrak{m}_L^i$. Let $f \in O_{L'}[X]$ be the characteristic polynomial of $u$. Since $\sigma(f) \equiv f \bmod \mathfrak{m}_L^i$ and $\sigma(f)(\sigma(u)) = 0$, we have $f(\sigma(u)) \equiv 0 \bmod \mathfrak{m}_L^i$. Since $L''$ is an unramified extension of $L'$, we have $f(\sigma(u)) \equiv f'(u)(\sigma(u) - u) \bmod (\sigma(u) - u)^2$ and $f'(u)$ is a unit of $O_{L''}$. Hence, we have $\sigma(u) - u \equiv 0 \bmod \mathfrak{m}_L^i$ as required.

Next, assume that $L''$ is a totally ramified extension of $L'$ of degree $m$ prime to $p$. Then, by Proposition 1.5.5(2), there exist uniformizers $s, t$ of $L'', L'$ satisfying $s^m = t$. Then, since $\sigma(t)/t = (\sigma(s)/s)^m$, we have $\sigma(t)/t - 1 \equiv m(\sigma(s)/s - 1) \bmod (\sigma(t)/t - 1)\mathfrak{m}_L$, and we also have $i \leqslant i_{L''}(\sigma)$ in this case. □

We show that tamely ramified base extension only multiplies the ramification index to $i_{L'}$.

**Corollary 1.7.3** *Let $L$ be a Galois extension of a henselian discrete valuation field $K$ and let $K'$ be a tamely ramified extension. Let $L' = LK$ and let $e_{L'/L}$ be the ramification index. Let $G = \mathrm{Gal}(L/K)$ and $G' = \mathrm{Gal}(L'/K')$ be the Galois groups.*

(1) *The canonical injection $G' \to G$ induces an isomorphism $G'_1 = P' \to G_1 = P$ on the wild inertia subgroups.*
(2) *Let $L_1 \subset L$ be a subextension corresponding to $H \subset G$, and let $L'_1 = K'L_1 \subset L'$ corresponding to $H' = H \cap G' \subset G'$. For $\sigma \in P - (P \cap H)$ and the corresponding element $\sigma' \in P' - (P' \cap H')$, we have $i_{L'_1}(\sigma') = e_{L'/L} \cdot i_{L_1}(\sigma)$.*

*Proof* Let $M \subset L$ be the maximal tamely ramified extension and set $M' = MK' \subset L'$.

(1) By Lemma 1.5.7, the induced morphism $P' = \mathrm{Gal}(L'/M') \to P = \mathrm{Gal}(L/M)$ is an isomorphism.

(2) By considering the Galois closure of $K'$ over $K$, it suffices to prove the case where $K'$ and hence $L'$ is a Galois extension of $K$. Then the assertion follows by Proposition 1.7.2(2). □

**Lemma 1.7.4** *Let $L$ be a Galois extension of the henselian discrete valuation field $K$ of Galois group $G$. Let $K' \subset L'$ be subextensions of $L$ and let $G' \supset H$ be the corresponding subgroups of $G$. Let $\sigma \in G - G'$, and let $T \subset G'$ be a complete set of representatives of $G'/H$.*

(1) *Assume that $L'$ is a totally ramified extension of $K'$ and let $s$ be a uniformizer of $L'$. Let $t = \mathrm{N}_{L'/K'} s$ be a uniformizer of $K'$. Then*

$$\frac{\sigma(t)}{t} - 1 \equiv (-1)^{[L':K']-1} \prod_{\tau \in T} \left( \frac{\sigma\tau(s)}{s} - 1 \right) \mod \mathfrak{m}_L^{i_{K'}(\sigma)+e_{L/L'}}. \tag{1.7.3}$$

(2) *Assume that $e_{L'/K'} = 1$ and that the residue field $E'$ of $L'$ is a purely inseparable extension of the residue field $F'$ of $K'$ generated by a unit $u \in O_{L'}$. Let $v = \mathrm{N}_{L'/K'} u$. Then*

$$\frac{\sigma(v)}{v} - 1 \equiv \prod_{\tau \in T} \left( \frac{\sigma\tau(u)}{u} - 1 \right) \mod \mathfrak{m}_L^{i_{K'}(\sigma)+1}. \tag{1.7.4}$$

*Proof* (1) Let $n = [L' : K']$ and $P = X^n + a_1 X^{n-1} + \cdots + a_n = \prod_{\tau \in T}(X - \tau(s)) \in O_{K'}[X]$ be the minimal polynomial of $s$. We have $\sigma P = X^n + \sigma(a_1)X^{n-1} + \cdots + \sigma(a_n) = \prod_{\tau \in T}(X - \sigma\tau(s))$. Since $P(s) = 0$, we have

$(\sigma P - P)(s) = \sum_{i=1}^{n}(\sigma(a_i) - a_i)\cdot s^{n-i} = (\sigma P)(s) = \prod_{\tau\in T}(s - \sigma\tau(s))$. Since $t = (-1)^n a_n$, we obtain

$$(-1)^n\Big(\frac{\sigma(t)}{t} - 1\Big) + \sum_{i=1}^{n-1}\frac{a_i}{t}\Big(\frac{\sigma(a_i)}{a_i} - 1\Big)\cdot s^{n-i} = \frac{s^n}{t}\prod_{\tau\in T}\Big(1 - \frac{\sigma\tau(s)}{s}\Big). \quad (1.7.5)$$

In the summation on the left-hand side, we ignore the terms where $a_i = 0$. Since $P \in O_{K'}[X]$ is an Eisenstein polynomial, the coefficients of $s^{n-i} \in \mathfrak{m}_{L'}$ on the left-hand side are contained in $\mathfrak{m}_L^{i_{K'}(\sigma)}$ and we have $s^n/t = s^n/\mathrm{N}_{L/K'}s \equiv (-1)^{n-1} \bmod \mathfrak{m}_{K'}$ by (1.4.2).

(2) Let $q = [L' : K']$ and $Q = X^q + b_1X^{q-1} + \cdots + b_q \in O_{K'}[X]$ be the minimal polynomial of $u \in O_{L'}$. Similarly to the proof of (1), we have $Q = \prod_{\tau\in T}(X - \tau(u))$ and $(\sigma Q)(u) = \prod_{\tau\in T}(u - \sigma\tau(u)) = (\sigma Q - Q)(u) = \sum_{i=1}^{q}(\sigma(b_i) - b_i)\cdot u^{q-i}$. Since $v = (-1)^q b_q$, we obtain

$$(-1)^q\Big(\frac{\sigma(v)}{v} - 1\Big) + \sum_{i=1}^{q-1}\frac{b_i}{v}\Big(\frac{\sigma(b_i)}{b_i} - 1\Big)\cdot u^{q-i} = \frac{u^q}{v}\prod_{\tau\in T}\Big(1 - \frac{\sigma\tau(u)}{u}\Big). \quad (1.7.6)$$

We ignore the terms where $b_i = 0$ in the summation on the left-hand side. Since $v \in O_{K'}$ is a unit and $Q \equiv X^q + b_q \bmod \mathfrak{m}_{K'}$, the coefficients of $u^{q-i}$ on the left-hand side are contained in $\mathfrak{m}_L^{i_{K'}(\sigma)+1}$. □

**Proposition 1.7.5** *Let $L$ be a Galois extension of a henselian discrete valuation field $K$ and $G = \mathrm{Gal}(L/K)$ be the Galois group. Let $K' \subset L'$ be subextensions of $L$ and let $G' \supset H$ be the corresponding subgroups of $G$. For $\sigma \in G_1 - G'_1$ and for a complete set $T \subset G'$ of representatives of $G'_1/H_1$, we have*

$$i_{K'}(\sigma) \leqslant \sum_{\tau\in T} i_{L'}(\sigma\tau). \quad (1.7.7)$$

In the next section, we give a sufficient condition for the equality to hold. First, we prove a basic case.

**Lemma 1.7.6** *Let the notation be as in Proposition 1.7.5 and assume that $L$ is purely wildly ramified over $K$. Assume further that $[L' : K'] = p$.*

(1) *We have an inequality in* (1.7.7).
(2) *Let $s \in O_{L'}$ be a generater over $O_{K'}$ and let $z = \mathrm{N}_{L'/K'}s$. Assume that $s$ is a uniformizer if $L'$ is totally ramified over $K'$. If the equality in* (1.7.7) *holds, then we have $i_{K'}(\sigma) = \mathrm{ord}_L((\sigma(z)/z) - 1)$ and $i_{L'}(\sigma\tau) = \mathrm{ord}_L((\sigma\tau(s)/s) - 1)$ for every $\tau \in G'$.*

*Proof* Let $s \in O_{L'}$ be as in (2). Since $O_{L'} = O_{K'}[s]$, for $\tau \in G'$, we have $i_{L'}(\sigma\tau) = \min\{i_{K'}(\sigma), \mathrm{ord}_L((\sigma\tau(s)/s) - 1)\}$ by Proposition 1.7.2(1). If

$i_{L'}(\sigma\tau) = i_{K'}(\sigma)$ for some $\tau \in G'$, we have $i_{K'}(\sigma) = i_{L'}(\sigma\tau) < \sum_{\tau\in T} i_{L'}(\sigma\tau)$ and a strict inequality in (1.7.7).

Assume $i_{L'}(\sigma\tau) = \mathrm{ord}_L((\sigma\tau(s)/s) - 1)$ for every $\tau \in G'$. Since $z = \mathrm{N}_{L'/K'}s \in K'$, we have $i_{K'}(\sigma) \leqslant \mathrm{ord}_L((\sigma(z)/z) - 1)$. By Lemma 1.7.4, we have either

$$i_{K'}(\sigma) = \mathrm{ord}_L\Big(\frac{\sigma(z)}{z} - 1\Big) = \sum_{\tau\in T} \mathrm{ord}_L\Big(\frac{\sigma\tau(s)}{s} - 1\Big) = \sum_{\tau\in T} i_{L'}(\sigma\tau)$$

or both $\mathrm{ord}_L(\sigma(z)/z - 1)$ and $\sum_{\tau\in T} \mathrm{ord}_L(\sigma\tau(s)/s - 1)$ are strictly larger than $i_{K'}(\sigma)$. In the latter case, we have a strict inequality,

$$i_{K'}(\sigma) < \sum_{\tau\in T} \mathrm{ord}_L\Big(\frac{\sigma\tau(s)}{s} - 1\Big) = \sum_{\tau\in T} i_{L'}(\sigma\tau),$$

in (1.7.7). □

*Proof of Proposition* 1.7.5 By Proposition 1.7.2(2), we may assume that $L$ is a purely wildly ramified extension of $K$ and hence $G$ is a $p$-group. We prove the assertion by induction on the index $[G' : H]$. By Lemma 1.5.12(2), we may assume that (1.7.7) holds for a subextension $L'' \subset L'$ over $K'$ such that $[L' : L''] = p$, and let $H' \supset H$ be the corresponding subgroup of $G'$. Let $S \subset H'$ be a complete set of representatives of $H'/H$. Then $TS \subset G'$ is a complete set of representatives of $G'/H$. Since (1.7.7) for the extension $L'$ over $L''$ is proved in Lemma 1.7.6, we have

$$i_{K'}(\sigma) \leqslant \sum_{\tau\in T} i_{L''}(\sigma\tau) \leqslant \sum_{\tau\in T}\sum_{\rho\in S} i_{L'}(\sigma\tau\rho) = \sum_{\tau\in TS} i_{L'}(\sigma\tau)$$

and (1.7.7) holds for $L'$ over $K'$. □

We study a relation of ramification subgroups with those of a quotient group.

**Definition 1.7.7** Let $L$ be a Galois extension of a henselian discrete valuation field $K$ and set $G = \mathrm{Gal}(L/K)$. Let $N \subset G$ be a normal subgroup and let $K'$ be the corresponding Galois extension of $K$. We define a continuous piecewise linear function $\varphi = \varphi_{L/K'} \colon [0, \infty) \to [0, \infty)$ by

$$e_{L/K'} \cdot \varphi(j) = \sum_{\tau\in N} \min(i_G(\tau), j). \tag{1.7.8}$$

**Proposition 1.7.8** *Let $L$ be a Galois extension of $K$ of Galois group $G$. Let $N \subset G$ be a normal subgroup and let $K'$ be the corresponding Galois extension of $K$. We identify the quotient group $\overline{G} = G/N$ with the Galois group* $\mathrm{Gal}(K'/K)$ *and define a continuous piecewise linear function $\varphi = \varphi_{L/K'} \colon [0, \infty) \to [0, \infty)$ by* (1.7.8).

(1) *Let* $\sigma \in \overline{G}$ *be a non-trivial element and let* $j(\sigma)$ *denote* $\max(i_G(\tau), \tau \in \sigma \subset G)$. *Then*

$$e_{L/K'} \cdot i_{\overline{G}}(\sigma) \leqslant \sum_{\tau \in \sigma} i_G(\tau) \tag{1.7.9}$$

*and*

$$i_{\overline{G}}(\sigma) \leqslant \varphi_{L/K'}(j(\sigma)). \tag{1.7.10}$$

(2) *Let* $i, j \in \mathbf{N}$ *be integers satisfying* $i \geqslant \varphi(j)$. *Then* $\overline{G}_i$ *is a subgroup of the image of* $G_j$.

In the next section we give a sufficient condition for the equalities to hold.

*Proof* (1) We may assume $\overline{\sigma} \in \overline{G}_1$. For (1.7.9), it suffices to apply Proposition 1.7.5 to $L' = L$.

Let $\tau_0 \in \sigma \subset G$ be an element satisfying $j(\sigma) = i_G(\tau_0)$. If $\tau \in N$ satisfies $i_G(\tau) \geqslant i_G(\tau_0)$, then $\tau\tau_0 \in \sigma$ satisfies $i_G(\tau\tau_0) \geqslant i_G(\tau_0)$. Since $j = i_G(\tau_0)$ is the largest, we have $i_G(\tau\tau_0) = i_G(\tau_0) = j$. If $\tau \in N$ satisfies $i_G(\tau) < i_G(\tau_0)$, then $\tau\tau_0 \in \sigma$ satisfies $i_G(\tau\tau_0) = i_G(\tau) < i_G(\tau_0) = j$. Hence, the right-hand side of (1.7.9) equals that of (1.7.8) for $j = j(\sigma)$.

(2) For $\sigma \in \overline{G}$, the condition $\sigma \in \overline{G}_i$ implies $i_{\overline{G}}(\sigma) \geqslant i \geqslant \varphi(j)$. Since $\varphi$ is strictly increasing, this implies $j(\sigma) \geqslant j$ by (1.7.10). This inequality means that $\sigma$ is in the image of $G_j$. □

The following inequality will be used in the proof of the Hasse–Arf theorem, Theorem 5.3.4.

**Corollary 1.7.9** *Let* $L$ *be a purely wildly ramified cyclic extension of degree* $p^e \geqslant p^2$ *and let* $M$ *be the unique subextension of* $[L : M] = p$. *Let* $G = \mathrm{Gal}(L/K)$ *and* $\overline{G} = \mathrm{Gal}(M/K)$ *be the Galois groups, and let* $i = g_{L/K}$ *and* $i' = g_{M/K}$ *be the greatest integers satisfying* $G_i \neq 1$ *or* $\overline{G}_{i'} \neq 1$ *respectively. Then*

$$g_{M/K} < f_{L/M} \cdot g_{L/K}. \tag{1.7.11}$$

*Proof* Let $\sigma \in G$ be an element of order $p^2$ and set $h = i_G(\sigma) \geqslant 1$. Then $h$ is the greatest integer such that $\sigma \in G_h$. Since $\mathrm{Gr}_h G$ is annihilated by $p$ by Corollary 1.6.4(1), we have $h < g_{L/K}$. Let $\sigma$ also denote the image in $\overline{G} = G/\langle\sigma^p\rangle$ by abuse of notation. Then we have $i_{\overline{G}}(\sigma) = g_{M/K}$. By Proposition 1.7.8, we have $e_{L/M} \cdot g_{M/K} \leqslant \sum_{\tau \in \langle\sigma^p\rangle} i_G(\sigma\tau) = p \cdot h < p \cdot g_{L/K}$. Dividing both sides by $e_{L/M}$, we obtain (1.7.11). □

**Lemma 1.7.10** *We keep the assumption in Proposition* 1.7.2. *Let* $s \in L'^{\times}$ *be* *an element satisfying*

$$\operatorname{ord}_L\left(\frac{\sigma(s)}{s}-1\right)=i_{L'}(\sigma)\geqslant 1. \tag{1.7.12}$$

*Then either of the following conditions is satisfied:*

(1) $\gcd(p, e_{L'/K}) \nmid \operatorname{ord}_{L'} s$.

(2) *Assume that* $a \in K^\times$, $b \in L'^\times$ *satisfy* $\operatorname{ord}_{L'} s = e_{L'/K} \cdot \operatorname{ord}_K a + p \cdot \operatorname{ord}_{L'} b$ *and define a unit by* $u = s/ab^p \in O_{L'}^\times$. *Then the class of* $u$ *is not contained in the subfield* $F(E'^p) \subset E'$ *of the residue field* $E'$ *of* $L'$.

*Proof* Let $i = i_{L'}(\sigma) \geqslant 1$. We may assume that $\gcd(p, e_{L'/K})$ divides $\operatorname{ord}_{L'} s$. Then, since $\sigma(s)/s = \sigma(u)/u \cdot (\sigma(b)/b)^p \equiv \sigma(u)/u \bmod (1+\mathfrak{m}_L^i)^p \subset 1+\mathfrak{m}_L^{i+1}$, we may further assume that $s = u$ is a unit. If $u \equiv \sum_j a_j v_j^p$ for $a_j \in O_K$ and $v_j \in O_{L'}^\times$, we have $\sigma(u)/u - 1 = u^{-1} \cdot \sum_j (a_j v_j^p \cdot ((\sigma(v_j)/v_j)^p - 1)) \in O_L((1+\mathfrak{m}_L^i)^p - 1) \subset \mathfrak{m}_L^{i+1}$. □

**Exercise 1.7** Let $K = \mathbf{F}_p(v)((t))$ and define a cyclic extension $L$ of degree $p^2$ by the Artin–Schreier–Witt equation $(x^p, y^p) - (x, y) = (a, b)$ for $a = v/t^p$ and $b = 1/t^{p^3-1}$. Let $G = \operatorname{Gal}(L/K)$ and let $K' \subset L$ be the subextension corresponding to the subgroup $N \subset G$ of order $p$. Compute the function $\varphi$ defined by (1.7.8) and show that we have strict inequalities in (1.7.9) and (1.7.10).

## 1.8 Log Monogenic Extensions

Classically, the ramification groups are studied under the assumption that the extension of the residue field is separable. We introduce a generalization that implies an equality in (1.7.7).

**Lemma 1.8.1** *Let* $L$ *be a finite extension of a henselian discrete valuation field* $K$. *Assume that* $L$ *is a purely wildly ramified separable extension, namely, the maximal tamely ramified extension* $M \subset L$ *equals* $K$. *Then the following conditions are equivalent:*

(1) *The morphism* $L^\times/(1+\mathfrak{m}_L)L^{\times p} \to K^\times/(1+\mathfrak{m}_K)K^{\times p}$ *induced by the norm mapping* $\mathrm{N}_{L/K}\colon L^\times \to K^\times$ *is non-trivial.*

(2) *There exist a uniformizer* $t$ *of* $K$ *and a uniformizer* $s$ *of* $L$ *such that the residue field* $E$ *of* $L$ *is generated by* $u = s^{e_{L/K}}/t$ *over* $F$.

*Proof* By Proposition 1.4.2(2), we have a commutative diagram

$$\begin{array}{ccccccccc} 0 & \longrightarrow & E^\times/E^{\times p} & \longrightarrow & L^\times/(1+\mathfrak{m}_L)L^{\times p} & \longrightarrow & \mathbf{Z}/p\mathbf{Z} & \longrightarrow & 0 \\ & & \big\downarrow{\scriptstyle [L:K]} & & \big\downarrow{\scriptstyle \mathrm{N}_{L/K}} & & \big\downarrow{\scriptstyle f_{L/K}} & & \\ 0 & \longrightarrow & F^\times/F^{\times p} & \longrightarrow & K^\times/(1+\mathfrak{m}_K)K^{\times p} & \longrightarrow & \mathbf{Z}/p\mathbf{Z} & \longrightarrow & 0 \end{array} \tag{1.8.1}$$

of exact sequences. The right vertical arrow is non-zero if and only if $L$ is a totally ramified extension. In this case (1) and (2) are satisfied. The left vertical arrow is non-zero if and only if $e_{L/K} = 1$ and the purely inseparable extension $E$ is generated by a single element $u$ over $F$. In this case (1) and (2) are also satisfied.

Assume that the left and right vertical arrows are zero. Then, condition (1) is equivalent to the condition that the boundary mapping $\mathbf{Z}/p\mathbf{Z} \to F^\times/F^{\times p}$ is non-trivial. This means that, for uniformizers $t, s$ of $K, L$ respectively, $v = \mathrm{N}_{L/K}s/t^{f_{L/K}}$ is not an element of $F^{\times p}$. Since $\mathrm{N}_{L/K}s \equiv s^{[L:K]} \bmod 1 + \mathfrak{m}_L$ by Proposition 1.4.2(3), this is further equivalent to the condition that the purely inseparable extension $E$ is generated by $u = s^{e_{L/K}}/t$. □

**Definition 1.8.2** Let $L$ be a finite separable extension of a henselian discrete valuation field $K$ and let $M \subset L$ be the maximal tamely ramified extension of $K$. We say that $L$ is a log monogenic extension of $K$ if the purely wildly ramified extension $L$ over $M$ satisfies the equivalent conditions in Lemma 1.8.1.

The terminology will be explained by Corollary 13.1.7. This notion covers two extreme cases.

**Lemma 1.8.3** *Let $L$ be a finite separable extension of a henselian discrete valuation field $K$. Let $M \subset L$ be the maximal tamely ramified extension of $K$.*

(1) *Assume that the residue field $E$ of $L$ is a separable extension of the residue field $F$ of $K$. Then $L$ is a log monogenic extension of $K$, and $O_L$ is generated by a single element over $O_K$.*
(2) *Assume that the ramification index $e_{L/K}$ is prime to $p$. Then the following conditions are equivalent:*
    (a) *The residue field $E$ of $L$ is generated by a single element over the residue field $F$ of $K$.*
    (b) *$L$ is a log monogenic extension of $K$.*
(3) *Assume that $L$ is a log monogenic extension and $p \mid e_{L/K}$. Then, for every uniformizer $t$ of $M$ and every uniformizer $s$ of $L$, the unit $u = s^{e_{L/M}}/t$ is a generator of the residue field $E$.*
(4) *If $[L : K] = p$, then $L$ is a log monogenic extension of $K$.*

*Proof* (1) The assumption implies that $L$ is a totally ramified extension of $M$. Hence $L$ is a log monogenic extension of $K$.

Assume that the separable extension $E$ of $F$ is generated by the image $\bar{x}$ of $x \in O_L$ and let $P \in O_K[X]$ be a monic lifting of the minimal polynomial of $\bar{x}$ over $F$. Since $P'(x) \in O_L$ is invertible, for a uniformizer $s$ of $O_L$ there exists $y \in O_L$ such that $s = P(y)$ and $y \equiv x \bmod \mathfrak{m}_L$ by Hensel's lemma. Then $y$ is a generator of $O_L$ over $O_K$ by Nakayama's lemma.

(2) (a) $\Rightarrow$ (b): Let $t$ be a uniformizer of $M$ and let $u \in O_L^\times$ be a unit generating $E$ over $F$. Since $e_{L/K}$ is prime to $p$, $e_{L/M} = 1$ and $s = ut$ is a uniformizer of $L$. The class of the unit $u = s/t$ is a generator of the purely inseparable extension $E$ of $F_M \supset F$.

(b) $\Rightarrow$ (a): There exists a unit $u \in O_L^\times$ generating the purely inseparable extension $E$ over the separable closure $F_M$ of $F$. Hence the assertion follows from Lemma 1.8.4 (2) $\Rightarrow$ (1).

(3) In this case the left vertical arrow in (1.8.1) for the extension $L$ over $M$ is 0. Hence the proof of Lemma 1.8.1 shows the assertion.

(4) It suffices to consider the case $M = K$. If $L$ is a totally ramified extension, then it is log monogenic by (1). If $e_{L/K} = 1$, then it is log monogenic by (2). □

**Lemma 1.8.4** *Let $E$ be a finite inseparable extension of a field $F$ of characteristic $p > 0$. Let $F'$ be the separable closure of $F$ in $E$. Then the following conditions are equivalent:*

(1) *The extension $E$ of $F$ is generated by a single element.*
(2) *The extension $E$ is generated by a single element over $F'$.*
(3) *The $E$-vector space $\Omega^1_{E/F}$ is of dimension 1.*

*Proof* (2) $\Rightarrow$ (1): We may assume that $F$ is an infinite field. It suffices to show that $E$ contains at most finitely many subextensions. Let $E_1$ be a subextension. Then $E_1$ is a purely inseparable extension of $F_1 = F' \cap E_1$ and $q = [E_1 : F_1]$ is a power of $p$. We have that $E_1 \subset F_{1,q} = \{x \in E \mid x^q \in F_1\}$ and $E_1 F' \subset F_{1,q} F'$ are subfields of the unique purely inseparable extension of $F'$ of degree $q$. Hence, $E_1 = F_{1,q}$ is uniquely determined by $F_1$ and $q$.

(1) $\Rightarrow$ (3): If $E = F(u)$, the $E$-vector space $\Omega^1_{E/F}$ is generated by $du$.

(3) $\Rightarrow$ (2): Let $u \in E$ be an element such that $\Omega^1_{E/F} = \Omega^1_{E/F'}$ is generated by $du$. Then $E$ is a separable and purely inseparable extension of $F'(u)$ since $\Omega^1_{E/F'(u)} = 0$. Hence $E$ is generated by $u$ over $F'$. □

**Proposition 1.8.5** *Let $L$ be a finite separable extension of a henselian discrete valuation field $K$. Let $M \subset L$ be the maximal tamely ramified extension of $K$, and let $F_M \subset E$ be the residue fields.*

(1) *Let $K'$ be a tamely ramified extension. Then $L$ is a log monogenic extension of $K$ if and only if $L' = LK'$ is a log monogenic extension of $K'$.*

(2) *Assume that $L$ is a log monogenic extension. Let $L' \subset L$ be a subextension and let $M' = M \cap L' \subset L'$ be the maximal tamely ramified extension of $K$. Let $F_{M'} \subset E'$ be the residue fields. If $[L' : M'] \leqslant [E : F_M]$, we have $p \nmid e_{L'/K}$ and $[L' : M'] = [E' : F_{M'}]$. If $[L' : M'] \geqslant [E : F_M]$, then $E$ is a separable extension of $E'$.*

*Proof* (1) Let $F' \subset E'$ be the residue fields of $K' \subset L'$. The composition field $M' = MK' \subset L'$ is the maximal tamely ramified extension of $K'$. By replacing $K$ and $K'$ by $M$ and $M'$, we may assume $K = M$ and $K' = M'$. The inclusions induce an isomorphism $L \otimes_K K' \to L'$. Hence, $e_{L/K} = e_{L'/K'}$ and the morphisms on the residue fields induce an isomorphism $E \otimes_F F' \to E'$.

The extension $L$ is totally ramified over $K$ if and only if $L'$ is totally ramified. Hence we have the equivalence in this case. In the case $e_{L/K} = e_{L'/K'} = 1$, the purely inseparable extension $E$ of $F$ is generated by a single element if and only if $E'$ over $F'$ is also, by Lemma 1.8.4 (1) $\Leftrightarrow$ (3). Hence, we also have the equivalence in this case. Assume $e_{L/K} = e_{L'/K'} \geqslant p$ and $[E : F] = [E' : F'] \geqq p$. Then the left and right vertical arrows in the diagram (1.8.1) are 0. Since $F'$ is a separable extension of $F$, the morphism $F^\times/F^{\times p} \to F'^\times/F'^{\times p}$ is an injection. Hence, the boundary mapping $\mathbf{Z}/p\mathbf{Z} \to F^\times/F^{\times p}$ of the diagram (1.8.1) is non-zero if and only if $\mathbf{Z}/p\mathbf{Z} \to F'^\times/F'^{\times p}$ is non-zero.

(2) First, we show the case $M \subset L'$. Let $t$ and $s$ be uniformizers of $M$ and $L$ such that the residue field $E$ is generated by $u = s^{e_{L/M}}/t$ over $F_M$. Let $s' = \mathrm{N}_{L/L'}s \in L'$. We have $s'/s^{[L:L']} \equiv 1$.

Assume $[L' : M] \leqslant [E : F_M] = q \cdot [L' : M]$. Then $u^q \equiv s'/t^q \in L'$ is a generator of a unique subextension $E' \subset E$ such that $[E' : F_M] = [E : F_M]/q = [L' : M]$. Hence, the residue field of $L'$ is $E'$ and we have $e_{L'/M} = 1$.

Assume $[L' : M] \geqslant [E : F_M]$ and let $e' = e_{L/M}/[L : L']$. Then $u \equiv s'^{e'}/t \in L'$ is a generator of $E$ over $F_M$. Hence, the residue field of $L'$ is $E$ and $L$ is a totally ramified extension of $L'$.

We show the general case. Since $L' \otimes_{M'} M \to L'M$ is an isomorphism, the assertion for $L'$ follows from that for $L'M$. □

**Corollary 1.8.6** *Let $L$ be a finite separable extension of $K$.*

(1) *Assume that $L$ is a log monogenic extension of $K$. If $K \subset K' \subset L' \subset L$ are subextensions, then $L'$ is a log monogenic extension of $K'$.*

(2) *Assume that $L$ is purely wildly ramified over $K$. Let $K = K_0 \subset K_1 \subset \cdots \subset K_n = L$ be a sequence of subextensions such that $[K_i : K_{i-1}] = p$. If $K_i$ is*

*a log monogenic extension of $K_{i-2}$ for every $i = 2, \dots, n$, then $L$ is a log monogenic extension of $K$.*

*Proof* (1) We may assume that $K$ equals the maximal tamely ramified extension $M \subset L$ by Proposition 1.8.5(1). Since $\mathrm{N}_{L/K}$ is the composition $\mathrm{N}_{K'/K} \circ \mathrm{N}_{L'/K'} \circ \mathrm{N}_{L/L'}$, the assertion follows.

(2) Let $F_i$ be the residue field of $K_i$. By Lemma 1.8.3(1), we may assume that there exists an integer $1 \leqslant i \leqslant n$ such that $e_{K_i/K_{i-1}} = 1$. Let $1 \leqslant i \leqslant n$ be the largest integer such that $e_{K_i/K_{i-1}} = 1$. Then $L$ is a totally ramified extension of $K_i$ and we have $e_{K_i/K} = 1$ by Proposition 1.8.5(2). Let $u \in F_i$ be a generator over $F_{i-1}$. For $2 \leqslant j \leqslant i$, we have $F_j^p \subset F_{j-1}$ but $F_j^p \not\subset F_{j-2}$. Hence, by induction, we have $F_j = F_{j-1}(u^{p^{i-j}})$ and $F_i = F(u)$. Thus, if $i = n$ then $L$ is a log monogenic extension of $K$.

Assume $i < n$. Then, for uniformizers $s_{i+1}, s_{i-1}$ of $K_{i+1}, K_{i-1}$ and for $u = \mathrm{N}_{K_{i+1}/K_i} s_{i+1}/s_{i-1}$, we have $F_i = F_{i-1}(u)$. Let $s, t$ be uniformizers of $L, K$ and $v = \mathrm{N}_{L/K_i} s/t$. Then, since $\mathrm{N}_{L/K_{i+1}} s$ and $t$ are uniformizers of $K_{i+1}, K_{i-1}$, we have $F_i = F_{i-1}(v) = F(v)$ and $L$ is a log monogenic extension of $K$. □

For log monogenic extensions, we have an equality in (1.7.7).

**Proposition 1.8.7** *Let $L$ be a log monogenic Galois extension of $K$ of Galois group $G$.*

1. *Let $s$ be a uniformizer of $L$. If $p \nmid e_{L/K}$, assume that $s = ut$ for a uniformizer $t$ of the maximal tamely ramified extension $M \subset L$ and a unit $u$ lifting a generator of $E$ over the residue field $F_M$. Then, for $\sigma \in P - \{1\}$, we have $i_G(\sigma) = \mathrm{ord}_L(\sigma(s)/s - 1)$.*
2. *Let $N \subset G$ be a normal subgroup and let $K'$ be the corresponding Galois extension of $K$. We identify the quotient group $\overline{G} = G/N$ with the Galois group $\mathrm{Gal}(K'/K)$. Then, for $\sigma \in \overline{G} \neq 1$, we have*

$$e_{L/K'} \cdot i_{\overline{G}}(\sigma) = \sum_{\tau \in \sigma} i_G(\tau). \tag{1.8.2}$$

*Proof* (1) Let $M \subset L$ be the maximal tamely ramified extension and let $t$ be a uniformizer of $M$. If $p \mid e_{L/K}$, a unit $u \in O_L^\times$ satisfying $u \equiv s^{e_{L/M}}/t \bmod \mathfrak{m}_L$ is a generator of the residue field $E$ of $L$ over the residue field $F_M$ of $M$ by Lemma 1.8.3(3). Hence, we have $O_L = O_M[s, u]$ by Nakayama's lemma. For $\sigma \in P - \{1\}$, we have $i_G(\sigma) = \min(\mathrm{ord}_L(\sigma(s)/s - 1), \mathrm{ord}_L(\sigma(u)/u - 1))$ by Proposition 1.7.2(1). Since $\sigma(u)/u \equiv (\sigma(s)/s)^{e_{L/M}} \bmod \mathfrak{m}_L^{i_G(\sigma)+1}$ by Proposition 1.6.3(2), we have $\mathrm{ord}_L(\sigma(u)/u - 1) \geqslant \mathrm{ord}_L(\sigma(s)/s - 1) = i_G(\sigma)$.

(2) Since $\overline{G}_1 = \overline{P}$ is the image of $G_1 = P$, we may replace $K$ by the maximal tamely ramified extension $M \subset L$ by Proposition 1.7.2(2). Hence, we may

assume that $L$ is purely wildly ramified over $K = M$. Let $t, s$ be uniformizers of $M, L$ and $u \in O_L^\times$ be a unit generating the residue field $E$ over $F_M$ such that $u \equiv s^{e_{L/M}}/t \bmod \mathfrak{m}_L$ as in (1). Then we have $i_G(\tau) = \mathrm{ord}_L(\tau(s)/s - 1)$ by (1).

First, we consider the case $[K' : K] \geqslant [E : F]$. Then $L$ is a totally ramified extension of $K'$ by Proposition 1.8.5(2) and $z = \mathrm{N}_{L/K'}s$ is a uniformizer of $K'$. Further, we have $i_{\overline{G}}(\sigma) = \mathrm{ord}_{K'}(\sigma(z)/z - 1)$ also by (1) applied to $K'$ over $K$. Hence, we have the equality (1.8.2) by Lemma 1.7.4(1).

Next, assume $[K' : K] \leqslant [E : F]$. Then we have $e_{K'/K} = 1$ by Proposition 1.8.5(2). Since $G$ is a $p$-group, we may take a subextension $K' \subset L' \subset L$ such that $[L' : K] = [E : F]$ by Lemma 1.5.12(1). Let $H \subset N$ be the subgroup corresponding to $L'$ and let $z = \mathrm{N}_{L/L'}s$ be a uniformizer of $L'$. Then we may assume that $u = z/t \in O_{L'}^\times$ is a generator of $O_{L'}$ over $O_K$ and $v = \mathrm{N}_{L'/K'}u$ is a generator of $O_{K'}$ over $O_K$. Thus, we have $i_{\overline{G}}(\sigma) = \mathrm{ord}_{K'}(\sigma(v)/v - 1)$. Let $S \subset N$ be a complete set of representatives of $N/H$ and take a lifting of $\sigma$ to $G$ also denoted $\sigma$ by abuse of notation. Since $\sigma\rho(u)/u = \sigma\rho(z)/z$, by Lemma 1.7.4(2) and 1.7.4(1), we have

$$e_{L/K'} \cdot i_{\overline{G}}(\sigma) = \mathrm{ord}_L\left(\frac{\sigma(v)}{v} - 1\right) = \sum_{\rho \in S} \mathrm{ord}_L\left(\frac{\sigma\rho(z)}{z} - 1\right)$$
$$= \sum_{\tau \in H} \mathrm{ord}_L\left(\frac{\sigma\tau(s)}{s} - 1\right).$$

The right-hand side is equal to that of (1.8.2) by (1). □

For a log monogenic Galois extension, we have equalities in Proposition 1.7.8.

**Corollary 1.8.8** *Let $L$ be a log monogenic Galois extension of $K$ of Galois group $G$. Let $N \subset G$ be a normal subgroup and let $K'$ be the corresponding Galois extension of $K$. We identify the quotient group $\overline{G} = G/N$ with the Galois group* $\mathrm{Gal}(K'/K)$ *and define a continuous piecewise linear function $\varphi = \varphi_{L/K'} \colon [0, \infty) \to [0, \infty)$ by (1.7.8).*

(1) *Let $\sigma \in \overline{G}$ be a non-trivial element and let $j(\sigma)$ denote $\max(i_G(\tau), \tau \in \sigma \subset G)$. Then $i_{\overline{G}}(\sigma) = \varphi_{L/K'}(j(\sigma))$.*
(2) *Let $i, j \in \mathbf{N}$ be integers satisfying $i = \varphi(j)$. Then $\overline{G}_i$ equals the image of $G_j$.*

The case of Corollary 1.8.8(2) where the extension of residue fields is separable is called the theorem of Herbrand.

*Proof* (1) By the same argument as in the proof of Proposition 1.7.8(1), the right-hand side of (1.8.2) equals that of (1.7.8) for $j = j(\sigma)$.

(2) For $\sigma \in \overline{G}$, the condition $\sigma \in \overline{G}_i$ means $i_{\overline{G}}(\sigma) \geqslant i = \varphi(j)$. Since $\varphi$ is

strictly increasing, this is equivalent to $j(\sigma) \geqslant j$ by (1). This inequality means that $\sigma$ is in the image of $G_j$. □

We have a converse of Proposition 1.8.7(2) as follows. A similar converse of Corollary 1.8.8(2) does not hold as in Exercise 1.8 (2).

**Proposition 1.8.9** *Let $L$ be a purely wildly ramified Galois extension of a henselian discrete valuation field $K$. Let $G = \mathrm{Gal}(L/K) = G_0 \supset G_1 \supset \cdots \supset G_n = 1$ be a sequence of subgroups and let $K = K_0 \subset K_1 \subset \cdots \subset K_n = L$ be the corresponding subextensions. Assume that $[G_m : G_{m+1}] = p$ for $m = 0, \ldots, n-1$.*

*For $m = 0, \ldots, n-2$, let $\sigma_m \in G_m - G_{m+1}$ and let $T_m \subset G_{m+1}$ be a complete set of representatives of $G_{m+1}/G_{m+2}$. Assume that we have*

$$i_{K_{m+1}}(\sigma_m) = \sum_{\tau \in T_m} i_{K_{m+2}}(\sigma_m \tau) \tag{1.8.3}$$

*for $m = 0, \ldots, n-2$. Then $L$ is a log monogenic extension of $K$.*

*Proof* By Corollary 1.8.6(2), we may assume that $[L : K] = p^2$. Let $K_1 = K'$, $\sigma_0 = \sigma$, and $G_1 = N \subset G$. As in Lemma 1.7.6(2), let $s \in O_L$ be a uniformizer if $L$ is totally ramified over $K'$ and a generator of $O_L$ over $O_{K'}$ otherwise. Then, as in the proof of Proposition 1.7.8(1), by Lemma 1.7.6 the assumption implies $i_{K'}(\sigma) = \mathrm{ord}_L(\sigma(z)/z - 1)$ for $z = \mathrm{N}_{L/K'}(s)$.

Let $F$, $F'$, and $E$ be the residue fields of $K$, $K'$, and $L$. First consider the case $e_{L/K'} = 1$. Then, since $z$ is a unit of $O_{K'}$, by Lemma 1.7.10 the equality $i_{K'}(\sigma) = \mathrm{ord}_L(\sigma(z)/z - 1)$ implies that $e_{K'/K} = 1$ and that $z \equiv s^p$ is a generator of the purely inseparable extension $F' \supsetneqq F$. Hence, $E = F'(s) = F(s, s^p)$ is generated by $s$ over $F$ and $L$ is a log monogenic extension of $K$ in this case.

Assume that $L$ is a totally ramified extension of $K'$. It suffices to consider the case $e_{K'/K} = 1$. Let $t$ be a uniformizer of $K$ and set $u = z/t$. Since $i_{K'}(\sigma) = \mathrm{ord}_L(\sigma(z)/z - 1) = \mathrm{ord}_L(\sigma(u)/u - 1)$, by Lemma 1.7.10 the unit $u$ is a generator of the purely inseparable extension $F \supsetneqq F$ and $L$ is a log monogenic extension of $K$. □

**Exercise 1.8** Let $p$ be a prime number and $K = \mathbf{F}_p(v)((t))$. Let $a = v/t^p$ and define a cyclic extension $K'$ over $K$ by the Artin–Schreier equation $x^p - x = a$.

(1) Find cyclic extensions $L_1$ and $L_2$ of degree $p^2$ over $K$ containing $K'$ satisfying the following properties: $L_1$ is a log monogenic extension of $K$ and $L_2$ is not a log monogenic extension of $K$; $L_1$ and $L_2$ are totally ramified extensions of $K'$.

(2) Let $L'$ be the Artin–Schreier extension of $K$ defined by $y^p - y = 1/t^{p-1}$ and set $L = L'K'$. Let $G = \mathrm{Gal}(L/K) \supset N = \mathrm{Gal}(L/K')$, $\overline{G} = \mathrm{Gal}(K'/K) =$

$G/N$, and let $\sigma$ be the generator of $\mathrm{Gal}(L/L')$ satisfying $\sigma(x) = x + 1$. Compute the function $\varphi_{L/K'}$ in (1.7.8) and $j(\overline{\sigma})$, and show that $L$ is not a log monogenic extension of $K$. Show that, if $p = 2$, $\overline{G}_i$ is the image of $G_j$ for every pair of integers satisfying $i = \varphi(j)$.

## 1.9 Different

We study the morphism $\mathrm{Tr}_{L/K}$ for finite separable extensions $L$ of a henselian discrete valuation field $K$.

**Proposition 1.9.1** *Let L be a finite separable extension of a henselian discrete valuation field K. Let F and E denote the residue fields of K and L respectively.*

(1) *We have* $\mathrm{Tr}_{L/K}(O_L) \subset O_K$, $\mathrm{Tr}_{L/K}(\mathfrak{m}_L) \subset \mathfrak{m}_K$, *and the diagram*

$$\begin{array}{ccc} O_L & \longrightarrow & E \\ {\scriptstyle \mathrm{Tr}_{L/K}}\downarrow & & \downarrow{\scriptstyle e_{L/K}\cdot \mathrm{Tr}_{E/F}} \\ O_K & \longrightarrow & F \end{array} \tag{1.9.1}$$

*is commutative. The* $O_L$*-modules*

$$\mathrm{Hom}_{O_K}(\mathfrak{m}_L, \mathfrak{m}_K) = \mathfrak{m}_L^{-1}\mathfrak{m}_K \cdot \mathrm{Hom}_{O_K}(O_L, O_K) \subset \mathrm{Hom}_{O_K}(O_L, O_K)$$

*are free of rank* 1.

(2) *The following conditions are equivalent:*
   (a) *L is tamely ramified over K.*
   (b) $\mathrm{Tr}_{L/K}$ *is a basis of the free* $O_L$*-module* $\mathrm{Hom}_{O_K}(\mathfrak{m}_L, \mathfrak{m}_K)$.
   (c) *The morphism* $\mathrm{Tr}_{L/K}\colon O_L \to O_K$ *is a surjection.*

(3) *The following conditions are equivalent:*
   (a) *L is unramified over K.*
   (b) $\mathrm{Tr}_{L/K}$ *is a basis of the free* $O_L$*-module* $\mathrm{Hom}_{O_K}(O_L, O_K)$.

*Proof* (1) Since $\mathrm{Tr}_{L/K}$ is the base change of $\mathrm{Tr}_{O_L/O_K}\colon O_L \to O_K$, we have $\mathrm{Tr}_{L/K}(O_L) \subset O_K$. Since the $F$-vector space $O_L/\mathfrak{m}_K O_L$ is a successive extension of $\mathfrak{m}_L^i/\mathfrak{m}_L^{i+1}$ for $i = 0, \dots, e-1$, the diagram (1.9.1) is commutative. The inclusion $\mathrm{Tr}_{L/K}(\mathfrak{m}_L) \subset \mathfrak{m}_K$ follows from the commutative diagram (1.9.1).

Since $\mathrm{Hom}_{O_K}(O_L, O_K)$ is a free $O_K$-module of rank $[L : K]$, it is a free $O_L$-module of rank 1. Hence, $\mathrm{Hom}_{O_K}(\mathfrak{m}_L, \mathfrak{m}_K) = \mathfrak{m}_L^{-1}\mathfrak{m}_K \cdot \mathrm{Hom}_{O_K}(O_L, O_K)$ is also a free $O_L$-module of rank 1.

(2) (a) $\Leftrightarrow$ (c): By the commutative diagram (1.9.1) and by Nakayama's lemma, the surjectivity of $\mathrm{Tr}_{L/K}\colon O_L \to O_K$ is equivalent to the surjectivity of

$e_{L/K} \cdot \mathrm{Tr}_{E/F} : E \to F$. The latter condition is equivalent to $E$ being a separable extension of $F$ and $e_{L/K}$ being invertible in $F$.

(b) ⇔ (c): Since $\mathrm{Hom}_{O_K}(\mathfrak{m}_L, \mathfrak{m}_K) = \mathfrak{m}_L^{-1}\mathrm{Hom}_{O_K}(O_L, \mathfrak{m}_K)$, we obtain an exact sequence

$$0 \to \mathrm{Hom}_{O_K}(O_L, \mathfrak{m}_K) \to \mathrm{Hom}_{O_K}(\mathfrak{m}_L, \mathfrak{m}_K) \to \mathrm{Hom}_F(E, F) \to 0$$

where the second arrow maps $f \in \mathrm{Hom}_{O_K}(\mathfrak{m}_L, \mathfrak{m}_K) \subset \mathrm{Hom}_{O_K}(O_L, O_K)$ to the induced morphism $\overline{f} : E \to F$. Hence, condition (b) is also equivalent to $e_{L/K} \cdot \mathrm{Tr}_{E/F}$ being non-zero by Nakayama's lemma.

(3) Since

$$\mathrm{Hom}_{O_K}(\mathfrak{m}_L, \mathfrak{m}_K) = \mathfrak{m}_L^{-1}\mathfrak{m}_K \cdot \mathrm{Hom}_{O_K}(O_L, O_K) \subset \mathrm{Hom}_{O_K}(O_L, O_K)$$

and the equality holds if and only if $e_{L/K} = 1$, the assertion follows from (2) (a) ⇔ (b). □

We define the different and its logarithmic variant for a finite separable extension of discrete valuation fields.

**Definition 1.9.2** Let $L$ be a finite separable extension of a henselian discrete valuation field $K$. Define the different and the logarithmic different $D_{L/K} \subset D^{\log}_{L/K} \subset O_L$ to be the ideals satisfying

$$O_L \cdot \mathrm{Tr}_{L/K} = D_{L/K} \cdot \mathrm{Hom}_{O_K}(O_L, O_K) = D^{\log}_{L/K} \cdot \mathrm{Hom}_{O_K}(\mathfrak{m}_L, \mathfrak{m}_K).$$

Since $\mathrm{Hom}_{O_K}(\mathfrak{m}_L, \mathfrak{m}_K) = \mathfrak{m}_L^{-1}\mathfrak{m}_K \cdot \mathrm{Hom}_{O_K}(O_L, O_K)$, we have $D_{L/K} = \mathfrak{m}_L^{-1}\mathfrak{m}_K \cdot D^{\log}_{L/K}$.

**Lemma 1.9.3** *Let $L$ be a finite separable extension of a henselian discrete valuation field $K$.*

(1) *We have* $\mathrm{Tr}_{L/K}((D^{\log}_{L/K})^{-1}) = O_K$ *and* $\mathrm{Tr}_{L/K}(\mathfrak{m}_L \cdot (D^{\log}_{L/K})^{-1}) = \mathfrak{m}_K$. *There exists a generator $b$ of $(D^{\log}_{L/K})^{-1}$ satisfying* $\mathrm{Tr}_{L/K} b = 1$.

(2) *Let $M \subset L$ be a subextension over $K$. Then*

$$D_{L/K} = D_{M/K} \cdot D_{L/M}, \tag{1.9.2}$$

$$D^{\log}_{L/K} = D^{\log}_{M/K} \cdot D^{\log}_{L/M}. \tag{1.9.3}$$

The equalities (1.9.2) and (1.9.3) are called the chain rules.

*Proof* (1) The linear mapping $\mathrm{Tr}_{L/K}$ is a generator of the $O_L$-module $D^{\log}_{L/K} \cdot \mathrm{Hom}_{O_K}(\mathfrak{m}_L, \mathfrak{m}_K)$. This means that we have $\mathrm{Tr}_{L/K}(\mathfrak{m}_L \cdot (D^{\log}_{L/K})^{-1}) \subset \mathfrak{m}_K$ and $\mathrm{Tr}_{L/K}((D^{\log}_{L/K})^{-1}) \not\subset \mathfrak{m}_K$. Since $\mathrm{Tr}_{L/K}(\mathfrak{m}_K \cdot (D^{\log}_{L/K})^{-1}) \subset \mathrm{Tr}_{L/K}(\mathfrak{m}_L \cdot$

$(D^{\log}_{L/K})^{-1})$, the first inclusion implies $\mathrm{Tr}_{L/K}((D^{\log}_{L/K})^{-1}) \subset O_K$. Hence we obtain the equalities.

By $\mathrm{Tr}_{L/K}((D^{\log}_{L/K})^{-1}) = O_K$, there exists an element $b \in (D^{\log}_{L/K})^{-1}$ satisfying $\mathrm{Tr}_{L/K} b = 1$. Since $\mathrm{Tr}_{L/K}(\mathfrak{m}_L \cdot (D^{\log}_{L/K})^{-1}) = \mathfrak{m}_K$, the element $b$ is not contained in $\mathfrak{m}_L \cdot (D^{\log}_{L/K})^{-1}$ and is a generator of $(D^{\log}_{L/K})^{-1}$.

(2) The equality (1.9.2) follows from the canonical isomorphism

$$\mathrm{Hom}_{O_K}(O_M, O_K) \otimes_{O_M} \mathrm{Hom}_{O_M}(O_L, O_M) \to \mathrm{Hom}_{O_K}(O_L, O_K)$$

and $\mathrm{Tr}_{M/K} \circ \mathrm{Tr}_{L/M} = \mathrm{Tr}_{L/K}$. The second equality (1.9.3) follows from the first equality (1.9.2). □

We state a criterion in terms of the different for the extension to be unramified or tamely ramified.

**Lemma 1.9.4** *Let $L$ be a finite separable extension of a henselian discrete valuation field $K$.*

(1) *The following conditions are equivalent:*
   (a) *$L$ is tamely ramified over $K$.*
   (b) *We have $D^{\log}_{L/K} = O_L$.*
(2) *The following conditions are equivalent:*
   (a) *$L$ is unramified over $K$.*
   (b) *We have $D_{L/K} = O_L$.*

*Proof* They follow from Proposition 1.9.1(2) and (3) respectively. □

We compute the different assuming that the ring of integers is generated by a single element over $O_K$.

**Proposition 1.9.5** *Let $K$ be a henselian discrete valuation field and let $L$ be a finite separable extension of $K$.*

(1) *Assume that $O_L$ is generated by $b \in O_L$ over $O_K$ and let $P \in O_K[X]$ be the minimal polynomial of $b$. Then the different $D_{L/K}$ is generated by $P'(b)$.*
(2) *Assume that $L$ is a Galois extension of Galois group $G$. Then*

$$\mathrm{ord}_L D^{\log}_{L/K} \leqslant \sum_{\sigma \in G,\, \sigma \neq 1} i_G(\sigma). \tag{1.9.4}$$

(3) *Assume that $L$ is a Galois extension of Galois group $G$. Then we have an equality*

$$\mathrm{ord}_L D^{\log}_{L/K} = \sum_{\sigma \in G,\, \sigma \neq 1} i_G(\sigma) \tag{1.9.5}$$

*if and only if $L$ is a log monogenic extension of $K$.*

The relation (1.9.5) between the different and the lower ramification groups is called the conductor–discriminant formula.

*Proof* (1) By Proposition 1.2.2, we have $O_L \cdot \mathrm{Tr}_{L/K} = P'(b) \cdot \mathrm{Hom}_{O_K}(O_L, O_K)$.

(2) By replacing $K$ by the maximal tamely ramified extension $M \subset L$, we may assume that $M = K$. Then $G$ is a $p$-group. By Lemma 1.5.12(1), there exists a sequence $K = K_0 \subset K_1 \subset \cdots \subset K_n = L$ of Galois extensions such that $[K_i : K_{i-1}] = p$. By the chain rule (1.9.3), we have

$$\mathrm{ord}_L D^{\log}_{L/K} = \sum_{i=1}^{n} e_{L/K_i} \cdot \mathrm{ord}_{K_i} D^{\log}_{K_i/K_{i-1}}. \tag{1.9.6}$$

For each $i = 1, \ldots, n$, either $K_i$ is a totally ramified extension of $K_{i-1}$ of degree $p$ or $e_{K_i/K_{i-1}} = 1$ and the residue field $F_i$ is a purely inseparable extension of degree $p$ over $F_{i-1}$. In the first case, let $s_i$ be a uniformizer of $K_i$ and let $P_i \in O_{K_{i-1}}[X]$ be the minimal polynomial. In the other case, let $s_i$ be a generator of $O_{K_i}$ and let $P_i \in O_{K_{i-1}}[X]$ be the minimal polynomial. In either case, by (1), we have

$$\mathrm{ord}_{K_i} D^{\log}_{K_i/K_{i-1}} = \mathrm{ord}_{K_i} P_i'(s_i)/s_i^{p-1}. \tag{1.9.7}$$

Let $G_i \subset G$ be the subgroup corresponding to $K_i$. We identify the Galois group $\mathrm{Gal}(K_i/K_{i-1})$ with $G_{i-1}/G_i$. Since $P'(s) = \prod_{\sigma \in G_i - \{1\}}(s - \sigma(s))$, we have

$$\mathrm{ord}_{K_i} P_i'(s_i)/s_i^{p-1} = \sum_{\sigma \in G_i - \{1\}} i_{G_i}(\sigma). \tag{1.9.8}$$

Substituting (1.9.7) and (1.9.8) into (1.9.6), we obtain

$$\mathrm{ord}_L D^{\log}_{L/K} = \sum_{i=1}^{n} \sum_{\sigma \in G_i - \{1\}} e_{L/K_i} \cdot i_{G_i}(\sigma). \tag{1.9.9}$$

Since $G - \{1\} = \coprod_{i=1}^{n}(G_i - G_{i-1})$, we obtain (1.9.4) from (1.7.7) and (1.9.9).

(3) If $L$ is a log monogenic extension of $K$, using the equality (1.8.2) in place of (1.7.7), we obtain (1.9.5) from (1.9.9).

Assume conversely that we have (1.9.5). We may assume that $L$ is purely wildly ramified over $K$ and hence $G$ is a $p$-group. Since we have an inequality (1.9.4) in general, the equalities (1.9.9) and (1.9.5) imply that we have equalities (1.8.3) for every subextension $K' \subset L' \subset L$ Galois over $K$. Hence, by Lemma 1.5.12(1) and Proposition 1.8.9, $L$ is a log monogenic extension of $K$. □

**Corollary 1.9.6** *Let $L$ be a log monogenic Galois extension of $K$ of Galois* *group* $G = \mathrm{Gal}(L/K)$. *Let* $H \subset G$ *be a subgroup and let* $M \subset L$ *be the corresponding subextension. Then*

$$e_{L/M} \cdot \mathrm{ord}_M D^{\log}_{M/K} = \mathrm{ord}_L D^{\log}_{L/K} - \sum_{\tau \in H - \{1\}} i_G(\tau).$$

*Proof* By the chain rule, Lemma 1.9.3(2), we have $e_{L/M} \cdot \mathrm{ord}_M D^{\log}_{M/K} + \mathrm{ord}_L D^{\log}_{L/M} = \mathrm{ord}_L D^{\log}_{L/K}$. The equality follows from Proposition 1.9.5(3). □

**Example 1.9.7** Let $K$ be a henselian discrete valuation field of characteristic $p > 0$ and $t \in K$ be a uniformizer. Let $L$ be the Artin–Schreier extension defined by $x^p - x = 1/t^n$ for $n = pm - 1 > 0$, and $s = t^m x \in L$ be a uniformizer as in Example 1.4.4. Since the minimal polynomial $P$ of $s$ is $S^p - t^{(p-1)m} S - t$ and $P'(s) = -t^{(p-1)m}$, we have $\mathrm{ord}_L D_{L/K} = \mathrm{ord}_L P'(s) = (p-1)pm$ and $\mathrm{ord}_L D^{\log}_{L/K} = (p-1)n$. Since $i_G(\sigma) = n$ as in Example 1.6.7, the equality $\mathrm{ord}_L D_{L/K} = \sum_{i=1}^{p-1} i_G(\sigma^i)$ (1.9.5) holds.

**Exercise 1.9** Let $p$ be a prime number and $n \geqslant 1$ be an integer.

(1) Find a generator of the different $D_{\mathbf{Q}_p(\zeta_{p^n})/\mathbf{Q}_p}$ and verify the equality (1.9.5) for the extension $\mathbf{Q}_p(\zeta_{p^n})$ over $\mathbf{Q}_p$.
(2) Find a generator $b$ of the inverse $(D^{\log}_{\mathbf{Q}_p(\zeta_{p^n})/\mathbf{Q}_p})^{-1}$ of the logarithmic different such that $\mathrm{Tr}_{\mathbf{Q}_p(\zeta_{p^n})/\mathbf{Q}_p} b = 1$.

## 1.10 Norm

We study the norm mapping $\mathrm{N}_{L/K} \colon L^\times \to K^\times$ for a finite extension $L$ of a henselian discrete valuation field $K$. We consider the filtration of $L^\times$ defined by $1 + \mathfrak{m}_L^i$ for integers $i \geqslant 1$ and study the images $\mathrm{N}_{L/K}(1 + \mathfrak{m}_L^i) \subset K^\times$.

**Lemma 1.10.1** *Let $L$ be a finite extension of $K$. For every integer $i \geqslant 0$, we have*

$$\mathrm{N}_{L/K}(1 + \mathfrak{m}_K^{i+1} O_L) \subset \mathrm{N}_{L/K}(1 + \mathfrak{m}_K^i \mathfrak{m}_L) \subset 1 + \mathfrak{m}_K^{i+1}.$$

*For $i \geqslant 1$ and $x \in \mathfrak{m}_K^i O_L$, we have*

$$\mathrm{N}_{L/K}(1 + x) \equiv 1 + \mathrm{Tr}_{L/K} x \mod \mathfrak{m}_K^{i+1}. \tag{1.10.1}$$

*Proof* Let $b \in L$, and $\mathrm{N}_{L/K}(X + b) = X^n + a_1 X^{n-1} + \cdots + a_n \in K[X]$ be the characteristic polynomial of $-b$. Then we have $a_1 = \mathrm{Tr}_{L/K} b$, $a_n = \mathrm{N}_{L/K} b$ and

$$\mathrm{N}_{L/K}(1 + xb) = 1 + a_1 x + a_2 x^2 + \cdots + a_n x^n \tag{1.10.2}$$

for $x \in K$. If $b \in \mathfrak{m}_L$, the multiplication by $b$ on the $F$-vector space $O_L/\mathfrak{m}_K O_L$ is nilpotent and we have $a_1, \ldots, a_n \in \mathfrak{m}_K$. Thus, we have the second inclusion by (1.10.2). The first inclusion follows from $\mathfrak{m}_K O_L \subset \mathfrak{m}_L$

If $b \in O_L$, we have $a_1, \ldots, a_n \in O_K$. Hence the congruence (1.10.1) follows from (1.10.2) for $x \in \mathfrak{m}_K^i$ and $a_1 = \mathrm{Tr}_{L/K} b$. □

We show an inclusion in the opposite direction.

**Lemma 1.10.2** *Let $L$ be a finite separable extension of $K$ of degree $n$. Let $c \in \mathfrak{m}_L$ be a non-zero element, $j = \mathrm{ord}_L c > 0$, and let $\mathrm{N}_{L/K}(X + c) = X^n + a_1 X^{n-1} + \cdots + a_n \in O_K[X]$ be the characteristic polynomial of $-c$. Assume $\mathrm{Tr}_{L/K} c = a_1 \neq 0$ and let $i = \mathrm{ord}_K a_1 \geqslant 1$. Assume further that $\mathrm{ord}_K a_q > i$ for every $q = 2, \ldots, n = [L : K]$. Then*

$$1 + \mathfrak{m}_K^i \subset \mathrm{N}_{L/K}(1 + \mathfrak{m}_L^j). \tag{1.10.3}$$

*Proof* The polynomial

$$P = \frac{1}{a_1} \cdot (\mathrm{N}_{L/K}(1 + cX) - 1) = X + \sum_{q=2}^{n} \frac{a_q}{a_1} X^q$$

has coefficients in $\mathfrak{m}_K$ except for degree 1 by the assumption $\mathrm{ord}_K a_q > i$ for every $q = 2, \ldots, n$. Let $z = a_1 w$ be any element of $\mathfrak{m}_K^i$. Since $O_K$ is henselian and $P(w) \equiv w$, $P'(w) \equiv 1 \bmod \mathfrak{m}_K$, there exists $x \in O_K$ such that $P(x) = w$ by Hensel's lemma. Since $\mathrm{N}_{L/K}(1 + cx) = 1 + a_1 P(x) = 1 + z$, we obtain the inclusion (1.10.3). □

We have a characterization of tamely ramified extensions.

**Proposition 1.10.3** *Let $L$ be a finite separable extension of a henselian discrete valuation field $K$.*

(1) *Let $d_{L/K}$ be the valuation of the logarithmic different $D_{L/K}^{\log}$, $e_{L/K}$ the ramification index, and let $i \geqslant 1$ be an integer satisfying $e_{L/K} \cdot i > d_{L/K}$. Then*

$$1 + \mathfrak{m}_K^{2i} \subset \mathrm{N}_{L/K}(1 + \mathfrak{m}_K^i \mathfrak{m}_L) \subset \mathrm{N}_{L/K}(1 + \mathfrak{m}_L). \tag{1.10.4}$$

(2) *The following conditions are equivalent:*

   (a) *$L$ is a tamely ramified extension of $K$.*
   (b) *We have $\mathrm{N}_{L/K}(1 + \mathfrak{m}_K^i O_L) = 1 + \mathfrak{m}_K^i$ for every $i \geqslant 1$.*
   (c) *We have $\mathrm{N}_{L/K}(1 + \mathfrak{m}_K O_L) = 1 + \mathfrak{m}_K$.*

*If these equivalent conditions hold, we have $\mathrm{N}_{L/K}(1 + \mathfrak{m}_K^i \mathfrak{m}_L) = 1 + \mathfrak{m}_K^{i+1}$ for every $i \geqslant 0$.*

*Proof* (1) Let $a \in \mathfrak{m}_K^i$ be an element with valuation $\operatorname{ord}_K a = i$, and let $b \in (D_{L/K}^{\log})^{-1}$ be a generator satisfying $\operatorname{Tr}_{L/K} b = 1$ as in Lemma 1.9.3(1). Let $\operatorname{N}_{L/K}(X + b) = X^n + X^{n-1} + a_2 X^{n-2} + \cdots + a_n$ be the characteristic polynomial of $-b$. Then, by $e_{L/K} \cdot i > d_{L/K}$, we have $ab \in \mathfrak{m}_L$ and hence $a^q a_q \in \mathfrak{m}_K$ for $q \geqslant 2$. For $c = a^2 b \in \mathfrak{m}_K^i \mathfrak{m}_L$, the coefficients of $\operatorname{N}_{L/K}(X+c) = X^n + a^2 X^{n-1} + a^4 a_2 X^{n-2} + \cdots + a^{2n} a_n$ satisfy $2i = \operatorname{ord}_K a^2 < \operatorname{ord}_K a^{2q} a_q$ for $q = 2, \ldots, n$. Hence, we have (1.10.4) by Lemma 1.10.2.

(2) (a) $\Rightarrow$ (b): Assume that $L$ is a tamely ramified extension and $i \geqslant 1$. By Lemma 1.10.1(2), we have $\operatorname{N}_{L/K}(1 + \mathfrak{m}_K^i O_L) \subset 1 + \mathfrak{m}_K^i$. Since $D_{L/K}^{\log} = O_L$ by Lemma 1.9.4(1) (a) $\Rightarrow$ (b), there exists a unit $b \in O_L^\times$ satisfying $\operatorname{Tr}_{L/K} b = 1$ by Lemma 1.9.3(1). Let $a \in \mathfrak{m}_K^i$ be a generator. Then, for a generator $c = ab$ of $\mathfrak{m}_K^i O_L$, we have $\operatorname{Tr}_{L/K} c = a \operatorname{Tr}_{L/K} b = a$ and $\operatorname{N}_{L/K}(X + c) \equiv X^n + aX^{n-1} \bmod \mathfrak{m}_K^{2i}$. Hence, we have the other inclusion by Lemma 1.10.2.

(b) $\Rightarrow$ (c) is trivial.

(c) $\Rightarrow$ (a): Assume $\operatorname{N}_{L/K}(1 + \mathfrak{m}_K O_L) = 1 + \mathfrak{m}_K$. For $x \in \mathfrak{m}_K O_L$, we have $\operatorname{N}_{L/K}(1 + x) \equiv 1 + \operatorname{Tr}_{L/K} x \bmod \mathfrak{m}_K^2$ by (1.10.1). Hence, $\operatorname{Tr}_{L/K}$ induces a surjection $\mathfrak{m}_K O_L \to \mathfrak{m}_K / \mathfrak{m}_K^2$. By Nakayama's lemma, $\operatorname{Tr}_{L/K} \colon \mathfrak{m}_K O_L \to \mathfrak{m}_K$ is a surjection. Hence, the assertion follows from Proposition 1.9.1(2) (c) $\Rightarrow$ (a).

The equality $\operatorname{N}_{L/K}(1 + \mathfrak{m}_K^{i+1} O_L) = 1 + \mathfrak{m}_K^{i+1}$ implies $\operatorname{N}_{L/K}(1 + \mathfrak{m}_K^i \mathfrak{m}_L) = 1 + \mathfrak{m}_K^{i+1}$ by Lemma 1.10.1(2). □

We compute the norm mapping for cyclic extension of degree $p$. This will be used both in the first case and in the induction step of the proof of Theorem 5.3.4 through Proposition 1.10.5(3) and Corollary 1.10.6 respectively.

**Lemma 1.10.4** *Let $L$ be a cyclic extension of degree $p$ of $K$ of Galois group $G$. Let $d_{L/K}$ be the valuation of the logarithmic different $D_{L/K}^{\log}$, $e_{L/K}$ the ramification index, and $f_{L/K} = [E : F]$. Let $i$ and $j$ be integers satisfying $j - 1 \geqslant e_{L/K} \cdot (i - 1) - d_{L/K}$. Let $x \in \mathfrak{m}_L^j$ and let $X^p + a_1(x) X^{p-1} + \cdots + a_p(x) \in K[X]$ be the characteristic polynomial of $-x$.*

(1) *We have*

$$\operatorname{Tr}_{L/K} x = a_1(x) \in \mathfrak{m}_K^i, \quad \operatorname{N}_{L/K} x = a_p(x) \in \mathfrak{m}_K^{f_{L/K} \cdot j}. \tag{1.10.5}$$

(2) *For $2 \leqslant q \leqslant p-1$, we have $a_q(x) \in \mathfrak{m}_K^i$. Further, if $2j - 1 \geqslant e_{L/K} \cdot i - d_{L/K}$, we have $a_q(x) \in \mathfrak{m}_K^{i+1}$.*

*Proof* (1) Since $\operatorname{Tr}_{L/K}(\mathfrak{m}_L^{-(d_{L/K} + e_{L/K} - 1)}) = \operatorname{Tr}_{L/K}(D_{L/K}^{-1}) \subset O_K$ by Lemma 1.9.3(1), the inequality $j + d_{L/K} + e_{L/K} - 1 \geqslant e_{L/K} \cdot i$ implies $\operatorname{Tr}_{L/K}(\mathfrak{m}_L^j) \subset \mathfrak{m}_K^i$. By Proposition 1.4.2(2), the condition $x \in \mathfrak{m}_L^j$ implies $\operatorname{N}_{L/K} x \in \mathfrak{m}_K^{[E:F] \cdot j}$.

(2) Let $2 \leqslant q \leqslant p-1$. Let $S_q$ denote the set of subsets of $G = \mathrm{Gal}(L/K)$ with $q$ elements. Then we have $a_q(x) = \sum_{A \in S_q} \prod_{\sigma \in A} \sigma(x)$. The group $G$ acts on $S_q$ without a fixed point. Let $T_q \subset S_q$ be a complete set of representatives with respect to the action of $G$. Then the right-hand side equals $\sum_{A \in T_q} \mathrm{Tr}_{L/K}(\prod_{\sigma \in A} \sigma(x))$. Since $\#A = q \geqslant 2$, we have $\prod_{\sigma \in A} \sigma(x) \in \mathfrak{m}_L^{2j} \subset \mathfrak{m}_L^j$ and $\mathrm{Tr}_{L/K}(\prod_{\sigma \in A} \sigma(x)) \in \mathfrak{m}_K^i$ as in the proof of (1).

Further, if $2j + d_{L/K} + e_{L/K} - 1 \geqslant e_{L/K} \cdot (i+1)$, we have $\mathrm{Tr}_{L/K}(\prod_{\sigma \in A} \sigma(x)) \in \mathrm{Tr}_{L/K}(\mathfrak{m}_L^{2j}) \subset \mathfrak{m}_K^{i+1}$. □

**Proposition 1.10.5** *Let $K$ be a henselian discrete valuation field and let $L$ be a wildly ramified cyclic extension of $K$ of degree $p$ of Galois group $G$. Let $d_{L/K}$, $e_{L/K}$ denote the valuation of the logarithmic different and the ramification index, and let $g_{L/K} \geqslant 1$ denote the largest integer $i$ such that $G_i \neq 1$.*

(1) *We have $d_{L/K} = (p-1) \cdot g_{L/K}$. The function $\varphi = \varphi_{L/K}$ in (1.7.8) is given by*

$$e_{L/K} \cdot \varphi(j) = \begin{cases} pj & \text{if } j \leqslant g_{L/K}, \\ j + d_{L/K} & \text{if } j \geqslant g_{L/K}. \end{cases} \tag{1.10.6}$$

(2) *Let $j > g_{L/K}$ be an integer and let $i = \lceil \varphi(j) \rceil$ be the smallest integer satisfying $j \leqslant e_{L/K} \cdot i - d_{L/K}$. Then*

$$\mathrm{N}_{L/K}(1+\mathfrak{m}_L^j) = 1+\mathfrak{m}_K^i \quad \text{and} \quad \mathrm{Tr}_{L/K}(1+\mathfrak{m}_L^j) = 1+\mathfrak{m}_K^i. \tag{1.10.7}$$

*If $i = \varphi(j)$ is an integer, then the diagram*

$$\begin{array}{ccc} \mathfrak{m}_L^j/\mathfrak{m}_L^{j+1} & \longrightarrow & 1+\mathfrak{m}_L^j/1+\mathfrak{m}_L^{j+1} \\ {\scriptstyle \mathrm{Tr}_{L/K}} \downarrow & & \downarrow {\scriptstyle \mathrm{N}_{L/K}} \\ \mathfrak{m}_K^i/\mathfrak{m}_K^{i+1} & \longrightarrow & 1+\mathfrak{m}_K^i/1+\mathfrak{m}_K^{i+1} \end{array} \tag{1.10.8}$$

*is commutative.*

(3) *Let $1 \leqslant j \leqslant g_{L/K}$ be an integer and set $1 \leqslant i = \varphi(j) = f_{L/K} \cdot j \leqslant f_{L/K} \cdot g_{L/K}$. Then $\mathrm{N}_{L/K}(1+\mathfrak{m}_L^j) \subset 1+\mathfrak{m}_K^i$, $\mathrm{N}_{L/K}(1+\mathfrak{m}_L^{j+1}) \subset 1+\mathfrak{m}_K^{i+1}$ and the diagram*

$$\begin{array}{ccc} \mathfrak{m}_L^j/\mathfrak{m}_L^{j+1} & \longrightarrow & (1+\mathfrak{m}_L^j)/(1+\mathfrak{m}_L^{j+1}) \\ \downarrow & & \downarrow {\scriptstyle \mathrm{N}_{L/K}} \\ \mathfrak{m}_K^i/\mathfrak{m}_K^{i+1} & \longrightarrow & (1+\mathfrak{m}_K^i)/(1+\mathfrak{m}_K^{i+1}) \end{array} \tag{1.10.9}$$

*where the left vertical arrow is $\mathrm{Tr}_{L/K} + \mathrm{N}_{L/K}$ if $j = g_{L/K}$ and is $\mathrm{N}_{L/K}$ if $1 \leqslant j < g_{L/K}$ is commutative.*

*Proof* (1) Since $[L : K] = p$, $L$ is a log monogenic extension of $K$ and we have $d_{L/K} = (p-1) \cdot g_{L/K}$ by Proposition 1.9.5(3). The equality (1.10.6) follows from this.

(2) Let $j > g_{L/K}$ be an integer not necessarily satisfying $e_{L/K} \mid d_{L/K} + j$ and let $i > f_{L/K} \cdot g_{L/K}$ be the smallest integer satisfying $j \leqslant e_{L/K} \cdot i - d_{L/K}$. Since $i$ is the smallest, we have $j - 1 \geqslant e_{L/K} \cdot (i-1) - d_{L/K}$. In the notation of Lemma 1.10.4, for $x \in \mathfrak{m}_L^j$ we have

$$\mathrm{N}_{L/K}(1+x) = 1 + a_1(x) + a_2(x) + \cdots + a_p(x). \tag{1.10.10}$$

We have $a_q(x) \in \mathfrak{m}_K^i$ for $q = 1, \ldots, p$ by $j - 1 \geqslant e_{L/K} \cdot (i-1) - d_{L/K}$, $f_{L/K} \cdot j \geqslant i$, and Lemma 1.10.4. Hence, by (1.10.10), we have an inclusion $\mathrm{N}_{L/K}(1+\mathfrak{m}_L^j) \subset 1 + \mathfrak{m}_K^i$ in (1.10.7).

To show the other inclusion, by replacing $j$ by $j' = e_{L/K} \cdot i - d_{L/K} \geqslant j$ if necessary, we may assume $e_{L/K} \mid j + d_{L/K} = e_{L/K} \cdot i$. Let $b \in D_{L/K}^{\log -1}$ be a generator such that $\mathrm{Tr}_{L/K} b = 1$ as in Lemma 1.9.3(1). Since $2j - 1 \geqslant j = e_{L/K} \cdot i - d_{L/K}$, for $a \in \mathfrak{m}_K^i$, $c = ab \in \mathfrak{m}_L^j$, and for $q = 2, \ldots, p-1$, we have $a_q(c) \in \mathfrak{m}_K^{i+1}$ by Lemma 1.10.4(2). For $q = p$, we have $\mathrm{ord}_K a_p(c) = \mathrm{ord}_K \mathrm{N}_{L/K} c \geqslant f_{L/K} \cdot j > i$ since $j > g_{L/K}$. Thus, by Lemma 1.10.2, we obtain the other inclusion $\mathrm{N}_{L/K}(1+\mathfrak{m}_L^j) \supset 1 + \mathfrak{m}_K^i$ and the first equality of (1.10.7) is proved.

The second one is just for the record and follows from Lemma 1.9.3(1). If $i = \varphi(j)$ is an integer, we have $i + 1 = \lceil \varphi(j+1) \rceil$ and the commutative diagram (1.10.8) follows from (1.10.10) and the second assertion in Lemma 1.10.4(2).

(3) We have $d_{L/K} = (p-1) \cdot g_{L/K} \geqslant (p-1)j$ and hence $j - 1 \geqslant pj - e_{L/K} - d_{L/K} = e_{L/K} \cdot (i-1) - d_{L/K}$. Since $f_{L/K} \cdot j = i$, similarly to the proof of (2), we have $\mathrm{N}_{L/K}(1+\mathfrak{m}_L^j) \subset 1 + \mathfrak{m}_K^i$ by Lemma 1.10.4(1). From this and (2) in the case $j = g_{L/K}$, we have $\mathrm{N}_{L/K}(1+\mathfrak{m}_L^{j+1}) \subset 1 + \mathfrak{m}_K^{i+1}$.

We have $2j - 1 \geqslant j \geqslant p \cdot j - d_{L/K} = e_{L/K} \cdot i - d_{L/K}$ for $j \leqslant g_{L/K}$ and $j - 1 > e_{L/K} \cdot (i-1) - d_{L/K}$ for $j < g_{L/K}$. Hence, the commutative diagram (1.10.9) follows from Lemma 1.10.4(2) similarly to the proof of (2). □

**Corollary 1.10.6** *Let $K$ be a henselian discrete valuation field and let $L$ be a wildly ramified cyclic extension of $K$ of Galois group $G$. Let $d_{L/K}$, $e_{L/K}$ denote the valuation of the logarithmic different and the ramification index, and let $g_{L/K} \geqslant 1$ denote the largest integer such that $G_i \neq 1$. Then, for an integer $j > g_{L/K}$ and the smallest integer $i = \lceil (j + d_{L/K})/e_{L/K} \rceil$ satisfying $j \leqslant e_{L/K} \cdot i - d_{L/K}$, we have*

$$\mathrm{N}_{L/K}(1+\mathfrak{m}_L^j) = 1 + \mathfrak{m}_K^i \quad \text{and} \quad \mathrm{Tr}_{L/K}(1+\mathfrak{m}_L^j) = 1 + \mathfrak{m}_K^i. \tag{1.10.11}$$

*If $i = (j + d_{L/K})/e_{L/K}$ is an integer, then the diagram*

$$\begin{array}{ccc} \mathfrak{m}_L^j/\mathfrak{m}_L^{j+1} & \longrightarrow & 1+\mathfrak{m}_L^j/1+\mathfrak{m}_L^{j+1} \\ {\scriptstyle \mathrm{Tr}_{L/K}}\downarrow & & \downarrow{\scriptstyle \mathrm{N}_{L/K}} \\ \mathfrak{m}_K^i/\mathfrak{m}_K^{i+1} & \longrightarrow & 1+\mathfrak{m}_K^i/1+\mathfrak{m}_K^{i+1} \end{array} \tag{1.10.12}$$

*is commutative.*

*Proof* By Proposition 1.10.3(2), we may replace $K$ by the maximal tamely ramified extension $K'$ in $L$; we may assume that $L$ is purely wildly ramified and $[L : K]$ is a power of $p$. We show the case $[L : K] = p^e$ by induction on $e$. The case $[L : K] = p$ is Proposition 1.10.5(2). The second equality in (1.10.11) is just for the record and follows from Lemma 1.9.3(1).

Assume $e \geqslant 2$ and let $M$ be the subextension such that $[L : M] = p$. Let $j' = \lceil (j+d_{L/M})/e_{L/M} \rceil$ be the smallest integer satisfying $j \leqslant e_{L/M} \cdot j' - d_{L/M}$. Then, by Proposition 1.10.5(2), we have

$$\mathrm{N}_{L/M}(1+\mathfrak{m}_L^j) = 1+\mathfrak{m}_M^{j'}.$$

If $(j + d_{L/M})/e_{L/M}$ is an integer, then the diagram (1.10.12) for the extension $L$ over $M$ is commutative.

Since $d_{L/K} = d_{L/M} + e_{L/M} \cdot d_{M/K}$, the integer $i$ is the smallest one satisfying $j' \leqq e_{M/K} i - d_{M/K}$. Since $j > g_{L/M}$ and $d_{L/M} + g_{L/K} = p \cdot g_{L/K}$ by (1.9.5), the inequality $f_{L/M} \cdot g_{L/K} > g_{M/K}$ (1.7.11) implies $e_{L/M} \cdot j' \geqslant j + d_{L/M} > g_{L/K} + d_{L/M} = p \cdot g_{L/K} = e_{L/M} f_{L/M} \cdot g_{L/K} > e_{L/M} \cdot g_{M/K}$. Hence, by the induction hypothesis, we have

$$\mathrm{N}_{M/K}(1+\mathfrak{m}_M^{j'}) = 1+\mathfrak{m}_K^{i}.$$

If $(j + d_{L/K})/e_{L/K}$ is an integer then $(j + d_{L/M})/e_{L/M} = ((j + d_{L/K})/e_{L/K}) \cdot e_{M/K} - d_{M/K}$ is an integer and the diagram (1.10.12) for the extension $M$ over $K$ is also commutative. □

**Corollary 1.10.7** *Let $K$ be a henselian discrete valuation field and assume that the residue field $F$ is algebraically closed.*

(1) *Let $L$ be a cyclic extension of degree $p$ and fix an isomorphism $G = \mathrm{Gal}(L/K) \to \mathbf{F}_p$. Let $g_{L/K}$ be the largest integer $i$ such that $G_i = G$ as in Proposition* 1.10.5, *and define an injection $G \to (1+\mathfrak{m}_L^i)/(1+\mathfrak{m}_L^{i+1})$* (1.6.6) *for $i = g_{L/K}$ by sending $\sigma \in G$ to $\sigma(s)/s$ for a uniformizer $s$ of $L$.*

*Then, for $i = j = g_{L/K}$, there exists a unique isomorphism*

$$\begin{array}{ccccccc} 0 \to G & \xrightarrow{(1.6.6)} & (1+\mathfrak{m}_L^i)/(1+\mathfrak{m}_L^{i+1}) & \xrightarrow{\mathrm{N}_{L/K}} & (1+\mathfrak{m}_K^j)/(1+\mathfrak{m}_K^{j+1}) & \to 0 \\ \downarrow & & \downarrow a & & \downarrow b & \\ 0 \to \mathbf{F}_p & \longrightarrow & F & \xrightarrow{x\mapsto x^p - x} & F & \to 0 \end{array} \quad (1.10.13)$$

*of exact sequences such that the right and the middle vertical arrows $a, b$ are isomorphisms of $F$-vector spaces. The diagram*

$$\begin{array}{ccccc} \mathfrak{m}_L^i/\mathfrak{m}_L^{i+1} & \xrightarrow{\text{can}} & (1+\mathfrak{m}_L^i)/(1+\mathfrak{m}_L^{i+1}) & \xrightarrow{a} & F \\ {\scriptstyle -\mathrm{Tr}_{L/K}}\downarrow & & & & \| \\ \mathfrak{m}_K^j/\mathfrak{m}_K^{j+1} & \xrightarrow{\text{can}} & (1+\mathfrak{m}_K^j)/(1+\mathfrak{m}_K^{j+1}) & \xrightarrow{b} & F \end{array} \quad (1.10.14)$$

*is commutative.*

(2) *Let $L$ be a finite separable extension of $K$. Then the norm mapping $\mathrm{N}_{L/K}\colon L^\times \to K^\times$ is a surjection.*

*Proof* (1) We have a commutative diagram (1.10.9) with the left vertical arrow $\mathrm{Tr}_{L/K}+\mathrm{N}_{L/K}$. Hence, for every choice of isomorphisms $a, b$ for the right and the middle vertical arrows, the right square is commutative with the bottom arrow sending $x$ to $ux^p + vx$ for some $u, v \in F^\times$. Hence, the kernel of $\mathrm{N}_{L/K}\colon (1 + \mathfrak{m}_L^i)/(1+\mathfrak{m}_L^{i+1}) \to (1+\mathfrak{m}_K^j)/(1+\mathfrak{m}_K^{j+1})$ is of order $p$. Let $\sigma \in G$ be the generator corresponding to 1 and define an isomorphism $a\colon (1+\mathfrak{m}_L^i)/(1+\mathfrak{m}_L^{i+1}) \to F$ of $F$-vector spaces by sending $\sigma(s)/s$ to 1 for a uniformizer $s$ of $L$. Then, we obtain the commutative left square and this defines the right isomorphism $b$ uniquely. Since the linear part of $\mathrm{N}_{L/K}\colon (1 + \mathfrak{m}_L^i)/(1 + \mathfrak{m}_L^{i+1}) \to (1 + \mathfrak{m}_K^j)/(1 + \mathfrak{m}_K^{j+1})$ is identified with $\mathrm{Tr}_{L/K}\colon \mathfrak{m}_L^i/\mathfrak{m}_L^{i+1} \to \mathfrak{m}_K^j/\mathfrak{m}_K^{j+1}$, the diagram (1.10.14) is commutative.

(2) By replacing $L$ by its Galois closure over $K$, we may assume that $L$ is a Galois extension. Set $n = [L : K] = e_{L/K}$. Since the norm mapping induces the identity on $\mathbf{Z} = L^\times/O_L^\times \to \mathbf{Z} = K^\times/O_K^\times$ and the $n$th power mapping on $E^\times \to F^\times$ by Proposition 1.4.2(2), it suffices to show that $1+\mathfrak{m}_L \to 1+\mathfrak{m}_K$ is a surjection. If $L$ is tamely ramified, the assertion follows from the last assertion of Proposition 1.10.3(2) for $i = 0$. Hence, by replacing $K$ by the maximal tamely ramified extension in $L$, we may assume that the Galois group $G = \mathrm{Gal}(L/K)$ is a $p$-group. By Lemma 1.5.12(1), we may assume that $L$ is a totally ramified cyclic extension of degree $p = \operatorname{char} F > 0$.

Let $g_{L/K}$ be the largest integer $i$ such that $G_i = G$ as in (1). Then, by Proposition 1.10.5(3), we have $\mathrm{N}_{L/K}(1 + \mathfrak{m}_L^{g_{L/K}+1}) = 1 + \mathfrak{m}_K^{g_{L/K}+1}$. For $i = j =$

$g_{L/K}$, the morphism $\mathrm{N}_{L/K}\colon (1+\mathfrak{m}_L^i)/(1+\mathfrak{m}_L^{i+1}) \to (1+\mathfrak{m}_K^j)/(1+\mathfrak{m}_K^{j+1})$ is a surjection by (1) and hence we have $\mathrm{N}_{L/K}(1+\mathfrak{m}_L^{g_{L/K}}) = 1+\mathfrak{m}_K^{g_{L/K}}$. For $1 \leqslant i = j < g_{L/K}$, the diagram

$$\begin{array}{ccc} \mathfrak{m}_L^i/\mathfrak{m}_L^{i+1} & \longrightarrow & (1+\mathfrak{m}_L^i)/(1+\mathfrak{m}_L^{i+1}) \\ {\scriptstyle \mathrm{N}_{L/K}}\big\downarrow & & \big\downarrow{\scriptstyle \mathrm{N}_{L/K}} \\ \mathfrak{m}_K^j/\mathfrak{m}_K^{j+1} & \longrightarrow & (1+\mathfrak{m}_K^j)/(1+\mathfrak{m}_K^{j+1}) \end{array} \tag{1.10.15}$$

is commutative. Since $\mathrm{N}_{L/K}$ on the left vertical arrow is the $p$th power mapping, the left vertical arrows are surjections by the assumption that $F$ is algebraically closed. Hence we have $\mathrm{N}_{L/K}(1+\mathfrak{m}_L) = 1+\mathfrak{m}_K$ as required. □

**Exercise 1.10** Let $p$ be a prime number and $n \geqslant 1$ be an integer. Show the inclusion $\mathrm{N}_{\mathbf{Q}_p(\zeta_{p^n})/\mathbf{Q}_p}\mathbf{Q}_p(\zeta_{p^n}^{\times}) \supset \langle p\rangle\cdot(1+p^n\mathbf{Z}_p)$. (We will see in Exercise 2.2 that the inclusion is in fact an equality.)

## Historical Notes

The characterization by the different of tamely ramified extensions given in Lemma 1.9.4(1) is proved in [78, Chapitre III, §7, Proposition 13] under the assumption that $O_L$ is generated by a single element. The proof of the general case given here is taken from [1, Proposition A.3]. The equality (1.2.2) is attributed to Euler in [78, Chapitre III, §6, Lemme 2].

# 2

# Cohomological Filtration

The cohomological filtration on the character group of the absolute Galois group of a local field is defined in Section 2.3. The definition is based on the pairing defined as the cup-product with the multiplicative group with values in the Brauer group defined in Section 2.1. In the classical case where the residue field is finite, the pairing induces the reciprocity mapping from the multiplicative group to the abelianized absolute Galois group, as we study in Section 2.2. For log monogenic abelian extensions, the Hasse–Arf theorem connecting the lower numbering filtration and the cohomological filtration will be proved later in Section 6.3.

To study the graded quotients of the cohomological filtration, we introduce the refined Swan conductor of wildly ramified characters. This is defined as a morphism to the space of differential forms with log poles introduced in Section 2.4. The definition of the refined Swan conductor is based on a canonical morphism from the space of differential forms of the residue field with log poles to the Brauer group defined in Section 2.5 as an analogue of the Artin–Schreier theory.

## 2.1 Brauer Group

Let $K$ be a field. We fix a separable closure $K_s$ of $K$ and let $G_K = \mathrm{Gal}(K_s/K)$ denote the absolute Galois group. For an abelian group $A$ with discrete topology and a continuous action of $G_K$, let $\mathrm{H}^q(K, A)$ denote the Galois cohomology defined as $\mathrm{H}^q(G_K, A)$. For the multiplicative group $\mathbf{G}_m$, the first cohomology $\mathrm{H}^1(K, \mathbf{G}_m)$ defined as $\mathrm{H}^1(G_K, K_s^\times)$ vanishes by Hilbert 90 [78, Chapitre X, §1, Proposition 2]. For an abelian group $A$ and an integer $n \geqslant 1$, let $A[n]$ denote the $n$-torsion part $\{a \in A \mid na = 0\}$.

We define the Brauer group of $K$ by

$$\operatorname{Br} K = \mathrm{H}^2(K, \mathbf{G}_m).$$

Since the Brauer group is defined as a Galois cohomology, it is a torsion group. If $n \geqslant 1$ is an integer invertible in $K$, the Kummer sequence $0 \to \mu_n \to \mathbf{G}_m \to \mathbf{G}_m \to 0$ induces an isomorphism $\mathrm{H}^2(K, \mu_n) \to \operatorname{Br} K[n]$ to the $n$-torsion part since $\mathrm{H}^1(K, \mathbf{G}_m) = 0$.

Let $X_K$ denote the character group

$$X_K = \mathrm{H}^1(K, \mathbf{Q}/\mathbf{Z}) = \operatorname{Hom}_{\mathrm{cont}}(G_K^{\mathrm{ab}}, \mathbf{Q}/\mathbf{Z})$$

of the abelian quotient $G_K^{\mathrm{ab}}$. We identify $X_K$ with $\mathrm{H}^2(K, \mathbf{Z})$ by the boundary morphism $\mathrm{H}^1(K, \mathbf{Q}/\mathbf{Z}) \to \mathrm{H}^2(K, \mathbf{Z})$ defined by the exact sequence $0 \to \mathbf{Z} \to \mathbf{Q} \to \mathbf{Q}/\mathbf{Z} \to 0$. If $K$ is of characteristic $p > 0$, since

$$\mathrm{H}^q(K, \mathbf{G}_a) = 0 \tag{2.1.1}$$

for $q \geqslant 1$ and the additive group $\mathbf{G}_a = K_s$ [78, Chapitre X, §1, Proposition 1], the Artin–Schreier sequence $0 \to \mathbf{F}_p \to \mathbf{G}_a \overset{x \mapsto x^p - x}{\to} \mathbf{G}_a \to 0$ defines an isomorphism

$$K/\{x^p - x \mid x \in K\} \to \mathrm{H}^1(K, \mathbf{F}_p). \tag{2.1.2}$$

By identifying $\mathbf{F}_p$ with $\frac{1}{p}\mathbf{Z}/\mathbf{Z} \subset \mathbf{Q}/\mathbf{Z}$, we identify $\mathrm{H}^1(K, \mathbf{F}_p)$ with the $p$-torsion part $X_K[p] \subset X_K$. Then (2.1.2) defines a surjection $a \colon K \to X_K[p]$.

The cup-product $\mathrm{H}^2(K, \mathbf{Z}) \times \mathrm{H}^0(K, \mathbf{G}_m) \to \mathrm{H}^2(K, \mathbf{G}_m)$ [78, Chapitre VIII, §3] defines a pairing

$$\{\ ,\ \} \colon X_K \times K^\times \to \operatorname{Br} K. \tag{2.1.3}$$

For $\chi \in X_K$ and $x \in K^\times$, if we regard $x$ as a morphism $\mathbf{Z} \to \mathbf{G}_m$ of $G_K$-modules, then the cup-product $\{\chi, x\}$ is the image of $\chi \in \mathrm{H}^2(K, \mathbf{Z})$ by the morphism $x_* \colon \mathrm{H}^2(G_K, \mathbf{Z}) \to \mathrm{H}^2(G_K, \mathbf{G}_m)$, by the functoriality of cup-product [78, Chapitre XI, Annexe Lemme 1]. For an extension $L$ of $K$, let

$$\operatorname{Br}(L/K) = \operatorname{Ker}(\operatorname{Res}_{L/K} \colon \operatorname{Br} K \to \operatorname{Br} L)$$

denote the kernel of the restriction morphism. For $\chi \in X_K$ and $x \in L^\times$, let $\{\chi, x\} \in \operatorname{Br} L$ denote the cup-product with the image of $\chi$ by the restriction morphism $\operatorname{Res}_{L/K} \colon X_K \to X_L$.

**Proposition 2.1.1** *Let $L$ be a Galois extension of Galois group $G = G_K/G_L$.*

(1) *The inflation morphism $\mathrm{H}^2(G, L^\times) \to \mathrm{H}^2(G_K, K_s^\times)$ induces a canonical isomorphism*

$$\mathrm{H}^2(G, L^\times) \to \operatorname{Br}(L/K). \tag{2.1.4}$$

(2) *Let $\chi \in X_K$ and assume that $L$ is the cyclic extension of $K$ corresponding to the kernel of $\chi$. Then the sequence*

$$L^\times \xrightarrow{\mathrm{N}_{L/K}} K^\times \xrightarrow{\{\chi,\ \}} \mathrm{Br}\, K \xrightarrow{\mathrm{Res}_{L/K}} \mathrm{Br}\, L \qquad (2.1.5)$$

*is exact.*

*Proof* (1) By Hilbert 90, the inflation–restriction sequence [78, Chapitre VII, §6, Proposition 5] gives an exact sequence $0 \to \mathrm{H}^2(G, L^\times) \to \mathrm{H}^2(G_K, K_s^\times) \to \mathrm{H}^2(G_L, K_s^\times)$.

(2) Since the Galois group $G = \mathrm{Gal}(L/K)$ is cyclic, the cup-product

$$\{\chi,\ \}\colon \widehat{\mathrm{H}}^0(G, L^\times) = K^\times/\mathrm{N}_{L/K}L^\times \to \mathrm{H}^2(G, L^\times) = \mathrm{Br}(L/K)$$

(2.1.4) is an isomorphism by [78, Chapitre VIII, §4, Proposition 6, Corollaire et Remarque suivante]. □

If $K$ is a henselian discrete valuation field, we say that $\chi \in X_K$ is unramified (resp. tamely ramified) if $\chi$ is in the kernel of $X_K \to X_{K_u}$ (resp. of $X_K \to X_{K_t}$) where $K_u$ (resp. $K_t$) is the maximal unramified (resp. tamely ramified) extension. We identify the subgroup of unramified characters with the image of the injection $X_F \to X_K$ induced by the isomorphism $\mathrm{Gal}(K_u/K) \to \mathrm{Gal}(F_s/F)$. A henselian discrete valuation field with algebraically closed residue field has the trivial Brauer group.

**Corollary 2.1.2** (1) *Let $K$ be a henselian discrete valuation field, and assume that the residue field $F$ is algebraically closed. Then the Brauer group* $\mathrm{Br}\, K$ *is trivial.*

(2) *Let $F$ be a finite field. Then* $\mathrm{Br}\, F = 0$.

(3) *Let $K$ be a henselian discrete valuation field and let $\chi \in X_K$ be a tamely ramified character. Then $\{\chi, 1 + \mathfrak{m}_K\} = 0$ in* $\mathrm{Br}\, K$.

*Proof* (1) It suffices to show that, for every finite Galois extension $L$ of $K$, the restriction morphism $\mathrm{Br}\, K \to \mathrm{Br}\, L$ is an injection. Since $\mathrm{Gal}(L/K)$ is an extension of a cyclic group by a $p$-group, we may assume that $L$ is a cyclic extension of $K$ by Lemma 1.5.12(1). Then, by Proposition 2.1.1(2), we have an exact sequence $0 \to K^\times/\mathrm{N}_{L/K}L^\times \to \mathrm{Br}\, K \to \mathrm{Br}\, L$. Since $K^\times/\mathrm{N}_{L/K}L^\times = 0$ by Corollary 1.10.7, the assertion follows.

(2) Set $q = \#F$. For every finite extension $E$ of degree $n = [E : F]$, the norm mapping is the morphism sending $x \in E^\times$ to its $\sum_{i=0}^{n-1} q^i = (q^n - 1)/(q - 1)$th power. Hence, $\mathrm{N}_{E/F}\colon E^\times \to F^\times$ is a surjection. Thus, $\mathrm{Br}\, F \to \mathrm{Br}\, E$ is an injection as in the proof of (1). Hence, we obtain $\mathrm{Br}\, F = 0$.

(3) For the cyclic extension $L$ corresponding to the kernel of $\chi$, we have

$\{\chi, \mathrm{N}_{L/K}L^\times\} = 0$ in $\operatorname{Br} K$ by Proposition 2.1.1(2). Since $1 + \mathfrak{m}_K \subset \mathrm{N}_{L/K}L^\times$ by Proposition 1.10.3(2) (a) $\Rightarrow$ (c), the assertion follows. □

**Proposition 2.1.3** *Let $K$ be a henselian discrete valuation field with residue field $F$. Let $K_u \subset K_s$ be the maximal unramified extension in a separable closure $K_s$.*

(1) *The inflation morphism $\mathrm{H}^2(F, K_u^\times) \to \mathrm{H}^2(K, K_s^\times)$ defines an isomorphism*

$$\mathrm{H}^2(F, K_u^\times) \to \operatorname{Br}(K_u/K). \tag{2.1.6}$$

(2) *The surjection $K_u^\times \to K_u^\times/(1 + \mathfrak{m}_{K_u})$ induces an isomorphism*

$$\mathrm{H}^2(F, K_u^\times) \to \mathrm{H}^2(F, K_u^\times/(1 + \mathfrak{m}_{K_u})). \tag{2.1.7}$$

(3) *The exact sequence $0 \to O_{K_u}^\times \to K_u^\times \to \mathbf{Z} \to 0$ defines an exact sequence*

$$0 \to \operatorname{Br} F \to \operatorname{Br}(K_u/K) \to X_F \to 0. \tag{2.1.8}$$

*For a uniformizer $t \in K$, the morphism $X_F \to \operatorname{Br}(K_u/K)$ sending $\chi \in X_F \subset X_K$ to $\{\chi, t\}$ defines a splitting of the exact sequence.*

(4) *Assume that the residue field $F$ is perfect. Then (2.1.8) gives an exact sequence*

$$0 \to \operatorname{Br} F \to \operatorname{Br} K \to X_F \to 0. \tag{2.1.9}$$

*Proof* (1) We identify $\operatorname{Gal}(K_u/K)$ with the absolute Galois group $G_F = \operatorname{Gal}(F_s/F)$. Then, it follows from Proposition 2.1.1(1).

(2) A uniformizer of $K$ defines a splitting of the exact sequence $0 \to O_{K_u}^\times \to K_u^\times \to \mathbf{Z} \to 0$ of $G_F$-modules. Hence, the assertion follows from the isomorphism $\mathrm{H}^2(F, O_{K_u}^\times) \to \mathrm{H}^2(F, F_s^\times)$ [63, Remark III, 3.11(a)].

(3) The exact sequence $0 \to O_{K_u}^\times \to K_u^\times \to \mathbf{Z} \to 0$ induces an exact sequence $0 \to F^\times \to K_u^\times/(1 + \mathfrak{m}_{K_u}) \to \mathbf{Z} \to 0$ of $G_F$-modules, and a uniformizer of $K$ defines a splitting. Applying the functor $\mathrm{H}^2(F, -)$ and identifying $\mathrm{H}^2(F, K_u^\times/(1+\mathfrak{m}_{K_u}))$ with $\operatorname{Br}(K_u/K)$ by the isomorphisms (2.1.6) and (2.1.7), we obtain (2.1.8).

(4) By the assumption that $F$ is perfect, the residue field $F_s$ of $K_u$ is an algebraic closure of $F$. Hence, we have $\operatorname{Br} K_u = 0$ by Corollary 2.1.2(1) and the injection $\operatorname{Br}(K_u/K) \to \operatorname{Br} K$ is an isomorphism. □

**Corollary 2.1.4** *Let $K$ be a henselian discrete valuation field and let $K_1$ be an extension of a discrete valuation field. Assume that $e_{K_1/K} = 1$, and that the residue field of $K_1$ is a rational function field $F(T)$. Assume further that the residue field $F$ of $K$ is perfect and* $\operatorname{Br} F = 0$. *Then the canonical morphism* $\operatorname{Br} K \to \operatorname{Br} K_1$ *is an injection.*

The assumption that $F$ is perfect and $\operatorname{Br} F = 0$ is known to be redundant, but we will prove it only under the assumption.

*Proof* By Proposition 2.1.3(3) and (4), we have a commutative diagram

$$\begin{array}{ccc} \operatorname{Br} K_1 & \longrightarrow & X_{F(T)} \\ \uparrow & & \uparrow \\ \operatorname{Br} K & \longrightarrow & X_F \end{array}$$

The lower horizontal arrow is an isomorphism by the assumption that $\operatorname{Br} F = 0$. Since the right vertical arrow is an injection, the left vertical arrow is also an injection. □

For a finite extension $L$ of $K$, we have a morphism $\operatorname{Cor}_{L/K} \colon \operatorname{Br} L \to \operatorname{Br} K$ called the corestriction going the other way. This and $\operatorname{Cor}_{L/K} \colon X_L \to X_K$ satisfy the following properties, whose proof we omit.

**Lemma 2.1.5** *Let $L$ be a finite extension of $K$.*

(1) [78, Chapitre VII, §7, Proposition 6] *If $n = [L : K]$, the composition* $\operatorname{Cor}_{L/K} \circ \operatorname{Res}_{L/K} \colon \operatorname{Br} K \to \operatorname{Br} L \to \operatorname{Br} K$ *equals multiplication by $n$. If $L$ is a Galois extension of Galois group $G$, the composition* $\operatorname{Res}_{L/K} \circ \operatorname{Cor}_{L/K} \colon \operatorname{Br} L \to \operatorname{Br} K \to \operatorname{Br} L$ *equals the norm mapping* $\mathrm{N}_G = \sum_{\sigma \in G} \sigma$.
(2) *For $\chi \in X_K$ and $b \in L^\times$, we have* $\operatorname{Cor}_{L/K}\{\chi, b\} = \{\chi, \mathrm{N}_{L/K} b\}$. *For $\chi \in X_L$ and $a \in K^\times$, we have* $\operatorname{Cor}_{L/K}\{\chi, a\} = \{\operatorname{Cor}_{L/K} \chi, a\}$.
(3) *Assume that $K$ is of characteristic $p > 0$. Then the morphisms* (2.1.2) *make a commutative diagram*

$$\begin{array}{ccc} L & \xrightarrow{a_L} & \mathrm{H}^1(L, \mathbf{F}_p) \\ {\scriptstyle \operatorname{Tr}_{L/K}} \downarrow & & \downarrow {\scriptstyle \operatorname{Cor}_{L/K}} \\ K & \xrightarrow{a_K} & \mathrm{H}^1(K, \mathbf{F}_p) \end{array}$$

**Exercise 2.1** Compute the Brauer group of the real number field $\mathbf{R}$.

## 2.2 Local Class Field Theory

As preparation for the definition of the reciprocity mapping (2.2.3), we study the character group $X_F$ for a finite field $F$.

**Lemma 2.2.1** *Let $F$ be a finite field of characteristic $p$ and of order $q = \#F$.*

(1) *The evaluation at the qth-power Frobenius* $\varphi_F \in G_F = \mathrm{Gal}(F_s/F)$ *defines an isomorphism* $\mathrm{ev}_{\varphi_F} : X_F \to \mathbf{Q}/\mathbf{Z}$.

(2) *The diagram*

$$\begin{array}{ccc} F & \xrightarrow{a} & X_F \\ {\scriptstyle \mathrm{Tr}_{F/\mathbf{F}_p}}\downarrow & & \downarrow{\scriptstyle \mathrm{ev}_{\varphi_F}} \\ \mathbf{F}_p & \xrightarrow{1/p} & \mathbf{Q}/\mathbf{Z} \end{array} \tag{2.2.1}$$

*is commutative, where* $a\colon F \to X_F$ (2.1.2) *is defined by the Artin–Schreier sequence.*

*Proof* (1) Since $\varphi_F$ defines an isomorphism $\widehat{\mathbf{Z}} = \varprojlim \mathbf{Z}/n\mathbf{Z} \to G_F$, the morphism $\mathrm{ev}_{\varphi_F} : X_F \to \mathbf{Q}/\mathbf{Z}$ is an isomorphism.

(2) Identify $\mathbf{F}_p$ with $(1/p)\mathbf{Z}/\mathbf{Z} \subset \mathbf{Q}/\mathbf{Z}$ by sending 1 to $1/p$, and let $\#F = p^e$. Then, for $x \in F$ and a solution $t$ of $t^p - t = x$, we have $a(x)(\varphi_F) = t^q - t = x + x^p + \cdots + x^{p^{e-1}} = \mathrm{Tr}_{F/\mathbf{F}_p} x$. Hence, the diagram (2.2.1) is commutative. □

We define the reciprocity mapping (2.2.3).

**Definition 2.2.2** Let $K$ be a henselian discrete valuation field with finite residue field $F$.

(1) Define a morphism

$$\mathrm{inv}_K : \mathrm{Br}\, K \to \mathbf{Q}/\mathbf{Z} \tag{2.2.2}$$

to be the composition of the morphisms $\mathrm{Br}\, K \to X_F$ (2.1.9) and $\mathrm{ev}_{\varphi_F} : X_F \to \mathbf{Q}/\mathbf{Z}$.

(2) Define the reciprocity morphism

$$\mathrm{rec}_K : K^\times \to \mathrm{Hom}(X_K, \mathbf{Q}/\mathbf{Z}) = G_K^{\mathrm{ab}} \tag{2.2.3}$$

to be the morphism defined by the pairing $X_K \times K^\times \to \mathrm{Br}\, K \xrightarrow{\mathrm{inv}} \mathbf{Q}/\mathbf{Z}$.

We make explicit the definition of the morphisms inv and rec.

**Lemma 2.2.3** *Let $K$ be a henselian discrete valuation field with finite residue field $F$.*

(1) *The morphism* $\mathrm{inv}_K : \mathrm{Br}\, K \to \mathbf{Q}/\mathbf{Z}$ *is an isomorphism.*

(2) *Let* $t \in K$ *be a uniformizer. Then the morphism* $X_F = \mathbf{Q}/\mathbf{Z} \to \mathrm{Br}\, K$ *sending* $\chi$ *to* $\{\chi, t\}$ *is the inverse of the isomorphism* $\mathrm{inv}_K$. *For* $\chi \in X_F$, *we have*

$$\mathrm{inv}_K\{\chi, t\} = \chi(\varphi_F). \tag{2.2.4}$$

(3) *For* $\chi \in X_K$, *the diagram*

$$\begin{array}{ccc} K^\times & \xrightarrow{\mathrm{rec}_K} & \mathrm{G}_K^{\mathrm{ab}} \\ {\scriptstyle \{\chi,\ \}}\downarrow & & \downarrow{\scriptstyle \chi} \\ \mathrm{Br}\, K & \xrightarrow{\mathrm{inv}_K} & \mathbf{Q}/\mathbf{Z} \end{array} \tag{2.2.5}$$

*is commutative, and for* $x \in K^\times$ *we have*

$$\chi(\mathrm{rec}_K x) = \mathrm{inv}_K \{\chi, x\}. \tag{2.2.6}$$

*The diagram*

$$\begin{array}{ccc} K^\times & \xrightarrow{\mathrm{ord}} & \mathbf{Z} \\ {\scriptstyle \mathrm{rec}}\downarrow & & \downarrow{\scriptstyle 1 \mapsto \varphi_F} \\ G_K^{\mathrm{ab}} & \longrightarrow & G_F \end{array} \tag{2.2.7}$$

*is commutative.*

*Proof* (1) By Proposition 2.1.3(4) and Corollary 2.1.2(2), the morphism $\mathrm{Br}\, K \to X_F$ (2.1.9) is an isomorphism. Since $\mathrm{ev}_{\varphi_F} : X_F \to \mathbf{Q}/\mathbf{Z}$ is an isomorphism by Lemma 2.2.1(1), the assertion follows.

(2) Since $\{\ , t\} : X_F \to \mathrm{Br}\, K$ is a section of the surjection $\mathrm{Br}\, K \to X_F$ by Proposition 2.1.3(4), the assertion follows.

(3) The commutative diagram (2.2.5) and the equality (2.2.6) are rephrasings of the definition of $\mathrm{rec}_K : K^\times \to G_K^{\mathrm{ab}}$ in Definition 2.2.2(2). For an unramified character $\chi \in X_F \subset X_K$, the right-hand side of (2.2.6) is $\chi(\varphi_F^{\mathrm{ord}_K x})$ by (2.2.4). Hence, the diagram (2.2.7) is commutative. □

The isomorphism $\mathrm{inv}_K : \mathrm{Br}\, K \to \mathbf{Q}/\mathbf{Z}$ is characterized by (2.2.4) and the reciprocity morphism $\mathrm{rec}_K : K^\times \to G_K^{\mathrm{ab}}$ is characterized by (2.2.6). The definition of $\mathrm{inv}_K : \mathrm{Br}\, K \to \mathbf{Q}/\mathbf{Z}$ depends on the choice of generator $\varphi_F \in G_F \cong \widehat{\mathbf{Z}}$. In arithmetic geometry, it is common to take the inverse $\varphi_F^{-1}$, called the geometric Frobenius. This modifies the definition of the reciprocity mapping by $-1$.

**Proposition 2.2.4** *Let $K$ be a henselian discrete valuation field with finite residue field $F$, and let $L$ be a finite extension of degree $n = [L : K]$ of $K$.*

(1) *The diagram*

$$\begin{array}{ccc} \mathrm{Br}\, K & \xrightarrow{\mathrm{inv}_K} & \mathbf{Q}/\mathbf{Z} \\ {\scriptstyle \mathrm{Res}_{L/K}}\downarrow & & \downarrow{\scriptstyle n} \\ \mathrm{Br}\, L & \xrightarrow{\mathrm{inv}_L} & \mathbf{Q}/\mathbf{Z} \end{array} \tag{2.2.8}$$

*is commutative.*

(2) *The diagram*

$$\begin{array}{ccc} L^\times & \xrightarrow{\mathrm{rec}_L} & G_L^{\mathrm{ab}} \\ {\scriptstyle \mathrm{N}_{L/K}}\downarrow & & \downarrow \\ K^\times & \xrightarrow{\mathrm{rec}_K} & G_K^{\mathrm{ab}} \end{array} \tag{2.2.9}$$

*is commutative.*

*Proof* (1) If $L$ is an unramified extension of $K$ of residue field $E$, we have the commutative diagram

$$\begin{array}{ccccc} \mathrm{Br}\, K & \longrightarrow & X_F & \longrightarrow & \mathbf{Q}/\mathbf{Z} \\ \downarrow & & \downarrow & & \downarrow{\scriptstyle [E:F]} \\ \mathrm{Br}\, L & \longrightarrow & X_E & \longrightarrow & \mathbf{Q}/\mathbf{Z} \end{array}$$

If $L$ is a totally ramified extension of $K$ of ramification index $e$, we have the commutative diagram

$$\begin{array}{ccc} \mathrm{Br}\, K & \longrightarrow & X_F \\ \downarrow & & \downarrow{\scriptstyle e} \\ \mathrm{Br}\, L & \longrightarrow & X_F \end{array}$$

Hence, we obtain the commutative diagram (2.2.8).

(2) By the commutative diagram (2.2.8), the restriction $\mathrm{Res}_{L/K}\colon \mathrm{Br}\, K \to \mathrm{Br}\, L$ is a surjection. Hence, by $\mathrm{Cor}_{L/K} \circ \mathrm{Res}_{L/K} = [L : K]$, we obtain the commutative diagram

$$\begin{array}{ccc} \mathrm{Br}\, L & \xrightarrow{\mathrm{inv}_L} & \mathbf{Q}/\mathbf{Z} \\ {\scriptstyle \mathrm{Cor}_{L/K}}\downarrow & & \| \\ \mathrm{Br}\, K & \xrightarrow{\mathrm{inv}_K} & \mathbf{Q}/\mathbf{Z} \end{array} \tag{2.2.10}$$

Thus, we have $\mathrm{inv}_K\{\chi, \mathrm{N}_{L/K}x\} = \mathrm{inv}_K \mathrm{Cor}_{L/K}\{\chi, x\} = \mathrm{inv}_L\{\chi, x\}$ for $\chi \in X_K$ and $x \in L^\times$ by Lemma 2.1.5(2), and we obtain the commutative diagram (2.2.9). □

**Theorem 2.2.5** *Let $K$ be a henselian discrete valuation field with finite residue field $F$. Let $L$ be a finite abelian extension of $K$.*

(1) *The commutative diagram* (2.2.9) *defines an isomorphism*

$$K^\times/\mathrm{N}_{L/K}L^\times \to \mathrm{Gal}(L/K). \tag{2.2.11}$$

(2) *The mapping*

$$\{\text{subextensions } M \subset L\} \to \{\text{subgroups } \mathrm{N}_{L/K}L^\times \subset N \subset K^\times\}$$

*sending* $M$ *to* $\mathrm{N}_{M/K}M^\times$ *is a bijection reversing inclusions.*

*Proof* (1) First, we show the isomorphism (2.2.11) in the case where $L$ is the cyclic extension of degree $n$ corresponding to the kernel of a character $\chi \in X_K$. Then, by Propositions 2.1.1(2) and 2.2.4(1), we obtain isomorphisms

$$K^\times/\mathrm{N}_{L/K}L^\times \xrightarrow{\{\chi,\ \}} \mathrm{Br}(L/K) \xrightarrow{\mathrm{inv}_K} \frac{1}{n}\mathbf{Z}/\mathbf{Z}.$$

Hence, the assertion follows in this case.

We show the isomorphism (2.2.11) in the general case by induction on $[L:K]$. Let $K \subsetneqq M \subsetneqq L$ be a subextension. By the commutative diagram (2.2.9) for $M$ over $K$, we have the commutative diagram

$$\begin{array}{ccccccccc} & & M^\times/\mathrm{N}_{L/M}L^\times & \xrightarrow{\mathrm{N}_{M/K}} & K^\times/\mathrm{N}_{L/K}L^\times & \longrightarrow & K^\times/\mathrm{N}_{M/K}M^\times & \longrightarrow & 0 \\ & & \downarrow{\scriptstyle \mathrm{rec}_M} & & \downarrow{\scriptstyle \mathrm{rec}_K} & & \downarrow{\scriptstyle \mathrm{rec}_K} & & \\ 0 & \longrightarrow & \mathrm{Gal}(L/M) & \longrightarrow & \mathrm{Gal}(L/K) & \longrightarrow & \mathrm{Gal}(M/K) & \longrightarrow & 0 \end{array} \tag{2.2.12}$$

of exact sequences. Since the left and right vertical arrows are isomorphisms by the hypothesis of induction, the middle arrow is also an isomorphism.

(2) By Galois theory, we have a bijection

$$\{\text{subextensions } M \subset L\} \to \{\text{subgroups of } \mathrm{Gal}(L/K)\}$$

reversing inclusions. The isomorphism (2.2.11) defines a bijection {subgroups $\mathrm{N}_{L/K}L^\times \subset N \subset K^\times$} $\to$ {subgroups of $\mathrm{Gal}(L/K)$} preserving inclusions. By the commutative diagram (2.2.12), the subgroup $N \subset K^\times$ corresponding to $M \subset L$ equals $\mathrm{N}_{M/K}M^\times$. □

**Corollary 2.2.6** *Let* $K$ *be a henselian discrete valuation field with finite residue field* $F$ *of order* $q = \#F$. *Let* $K_u \subset K_t \subset K_s$ *and* $K_a$ *be respectively the maximal unramified extension, the maximal tamely ramified extension, and the maximal abelian extension in a separable closure of* $K$.

(1) *We say that a subgroup of* $K^\times$ *is an open subgroup if it contains* $1 + \mathfrak{m}_K^n$ *for some integer* $n \geqslant 1$. *Then, the mapping*

$$\begin{aligned} &\{\text{finite abelian extensions of } K \text{ in } K_a\} \\ &\qquad \to \{\text{open subgroups of } K^\times \text{ of finite index}\} \end{aligned} \tag{2.2.13}$$

*sending* $L$ *to* $\mathrm{N}_{L/K}L^\times$ *is an injection reversing inclusions.*

(2) *The maximal tamely ramified abelian extension* $K_t \cap K_a$ *is generated by a* $(q-1)$*th root of a uniformizer of* $K$ *over the maximal unramified extension* $K_u$.

(3) *The reciprocity morphism* $\mathrm{rec}\colon K^\times \to G_K^{\mathrm{ab}}$ *induces a commutative diagram*

$$\begin{array}{ccccccc} 0 \longrightarrow & F^\times & \longrightarrow & K^\times/(1+\mathfrak{m}_K) & \longrightarrow & \mathbf{Z} & \longrightarrow 0 \\ & \downarrow & & \downarrow & & \downarrow & \\ 0 \longrightarrow & \mathrm{Gal}((K_t \cap K_a)/K_u) & \longrightarrow & \mathrm{Gal}((K_t \cap K_a)/K) & \longrightarrow & G_F & \longrightarrow 0 \end{array} \tag{2.2.14}$$

*of exact sequences. The left vertical arrow is an isomorphism of cyclic groups of order* $q-1$*, and the right vertical arrow is an injection* $\mathbf{Z} \to G_F = \widehat{\mathbf{Z}}$ *sending* 1 *to the* $q$*th-power Frobenius* $\varphi_F$.

*Proof* (1) By Proposition 1.10.3(1), the image $\mathrm{N}_{L/K}L^\times \subset K^\times$ is an open subgroup. The remaining assertions follow from Theorem 2.2.5(2).

(2) and (3) Since $\{\chi, 1+\mathfrak{m}_K\} = 0$ for tamely ramified characters $\chi \in X_K$ by Corollary 2.1.2(3), the middle vertical arrow of (2.2.14) is defined. The right square of the diagram is obtained by (2.2.7). Hence, the middle vertical arrow induces the left vertical arrow and the induced arrow is a surjection by Theorem 2.2.5(1). Since $K^\times$ contains a primitive $(q-1)$th root of 1, for a uniformizer $t$ of $K$ we have an inclusion $K(t^{1/(q-1)}) \subset K_t \cap K_a$. Hence, the left vertical arrow induces a surjection $F^\times \to \mathrm{Gal}(K_u(t^{1/(q-1)})/K_u)$ of cyclic groups of the same order $q-1$. Thus, this left vertical arrow is an isomorphism and we have $K_u(t^{1/(q-1)}) = K_t \cap K_a$. □

We define a filtration on the Galois group of a finite abelian extension of a henselian discrete valuation field with finite residue field using the reciprocity mapping.

**Definition 2.2.7** Let $K$ be a henselian discrete valuation field with finite residue field, and let $L$ be a finite abelian extension of $K$. For integers $n \geqslant 1$, let $G^n_{\mathrm{rec}} \subset G = \mathrm{Gal}(L/K)$ be the image of $1 + \mathfrak{m}_K^n$ by the reciprocity mapping $K^\times \to \mathrm{Gal}(L/K)$.

We will remove the assumption that the residue field is finite in the next section.

The existence theorem of local class field theory asserts that the injection (2.2.13) is a bijection and that the reciprocity mapping $\mathrm{rec}_K\colon K^\times \to G_K^{\mathrm{ab}}$ is an injection. The construction of abelian extensions using Lubin–Tate formal groups is explained in [46]. We will see the injectivity of the reciprocity mapping for $K = \mathbf{Q}_p$ in Exercise 2.2.

**Exercise 2.2** Let $p$ be a prime number.

(1) Let $n \geqslant 1$ be an integer. Show that the reciprocity morphism induces an isomorphism $(\mathbf{Z}_p/p^n\mathbf{Z}_p)^\times \to \mathrm{Gal}(\mathbf{Q}_p(\zeta_{p^n})/\mathbf{Q}_p)$. Deduce the equality $\mathrm{N}_{\mathbf{Q}_p(\zeta_{p^n})/\mathbf{Q}_p}\mathbf{Q}_p(\zeta_{p^n})^\times = (1+p^n\mathbf{Z}_p)\cdot\langle p\rangle$.
(2) Show that the maximal abelian extension $\mathbf{Q}_p^{\mathrm{ab}}$ is the composition of the maximal unramified extension $\mathbf{Q}_p^{\mathrm{ur}}$ and $\mathbf{Q}_p(\zeta_{p^n}; n \geqslant 1)$.

## 2.3 Cohomological Filtration

**Definition 2.3.1** Let $K$ be a henselian discrete valuation field. Let $K_1$ be the fraction field of the henselization of the local ring of $O_K[T]$ at the prime ideal $\mathfrak{m}_K O_K[T]$.

(1) We define an increasing filtration $F_\bullet X_K \subset X_K$ indexed by integers $n \geqslant 0$ by

$$F_n X_K = \{\chi \in X_K \mid \{\chi, 1 + \mathfrak{m}_{K_1}^{n+1}\} = 0 \text{ in } \mathrm{Br}\, K_1\}. \tag{2.3.1}$$

We set $F_{-1}X_K = 0$.

For $n \geqslant 0$, let $\mathrm{Gr}_n^F X_K$ denote the graded quotient $F_n X_K / F_{n-1} X_K$.

(2) For $\chi \in \bigcup_{n\geqslant 0} F_n X_K \subset X_K$, we call the smallest integer $n \geqslant 0$ satisfying $\chi \in F_n X_K$ the Swan conductor $\mathrm{Sw}\,\chi$ of $\chi$.
(3) We say that $\chi \in X_K$ is unramified (resp. tamely ramified) if the corresponding cyclic extension is unramified (resp. tamely ramified). We say that $\chi$ is wildly ramified if $\chi$ is not tamely ramified.

Since the inclusion $\bigcup_{n\geqslant 0} F_n X_K \subset X_K$ is an equality, as we will see in Proposition 2.3.3(1), the Swan conductor $\mathrm{Sw}\,\chi \in \mathbf{N}$ is defined for every character $\chi \in X_K$. In the case where $K$ is of characteristic $p > 0$, we will give an explicit description of the filtration $F_\bullet X_K$ using Witt vectors in Section 16.3.

Since the canonical morphism $\mathrm{Br}\, K \to \mathrm{Br}\, K_1$ is known to be an injection [50, (1.8.1) and proof of Proposition 6.3 (iv) $\Rightarrow$ (i)] (see also Corollary 16.4.2(2)), the equality $\mathrm{Sw}\,\chi = n$ implies $\{\chi, 1 + \mathfrak{m}_K^{n+1}\} = 0$, although we will not use this fact. The reason we need to introduce an extension $K_1$ is that the Brauer group $\mathrm{Br}\, K$ can be too small. For example, if the residue field $F$ is algebraically closed we have $\mathrm{Br}\, K = 0$, as we have seen in Corollary 2.1.2(1). The following property on the construction of $K_1$ will be used repeatedly.

**Lemma 2.3.2** *Let $K$ be a henselian discrete valuation field and $L$ be a finite extension of $K$. Let $K_1$ and $L_1$ be the fraction fields of the henselizations of the local rings of $O_K[T]$ and of $O_L[T]$ at the prime ideals $\mathfrak{m}_K O_K[T]$ and*

$\mathfrak{m}_L O_L[T]$. *Then, the canonical morphisms induce an isomorphism* $O_L \otimes_{O_K} O_{K_1} \to O_{L_1}$.

*Proof* Since $O_K[T] \to O_L \otimes_{O_K} O_K[T] = O_L[T]$ is finite, we obtain the isomorphism by Corollary 1.3.7(2). □

Let $K$ be a henselian discrete valuation field and let $K_u \subset K_t \subset K_s$ be the maximal unramified extension and the maximal tamely ramified extension in a separable closure.

**Proposition 2.3.3** (1) *We have* $X_K = \bigcup_{n \geqslant 0} F_n X_K$. *More precisely, let* $\chi \in X_K$ *and* $L$ *be the corresponding cyclic extension. If* $\chi$ *is tamely ramified, we have* $\operatorname{Sw}\chi = 0$. *Assume that* $\chi$ *is wildly ramified. Let* $d_{L/K}$ *and* $e_{L/K}$ *denote the valuation of the logarithmic different and the ramification index, and let* $g_{L/K} \geqslant 1$ *denote the largest integer such that* $G_i \neq 1$. *Then*

$$\operatorname{Sw}\chi \leqslant \frac{d_{L/K} + g_{L/K}}{e_{L/K}} \leqslant \frac{1}{e_{L/K}} \left( \sum_{\sigma \in P - \{1\}} i_G(\sigma) + g_{L/K} \right). \tag{2.3.2}$$

(2) *For* $n \geqslant 1$, $p \cdot \operatorname{Gr}_n^F X_K = 0$.
(3) *Let* $K_t$ *be the maximal tamely ramified extension of* $K$. *Then we have an inclusion* $\operatorname{Ker}(X_K \to X_{K_t}) \subset F_0 X_K$.

We will give equivalent conditions for the equalities in (2.3.2) to hold as the Hasse–Arf theorem, Theorem 5.3.4. We will show in Corollary 5.2.3(1) that the inclusion in Proposition 2.3.3(3) is an equality.

*Proof* Let $K_1$ be the extension of $K$ in Definition 2.3.1 and, for a finite extension $L$ of $K$, let $L_1 = LK_1$ be as in Lemma 2.3.2.

(1) Let $\chi \in X_K$ be a character and $L$ be the cyclic extension corresponding to the kernel of $\chi$.

Let $\chi$ be a tamely ramified character. Then $\{\chi, 1 + \mathfrak{m}_{K_1}\} = 0$ by Corollary 2.1.2(3).

Assume that $\chi$ is wildly ramified. Let $j \leqslant g_{L/K}$ be the largest integer such that $e_{L/K}$ divides $d_{L/K}+j$, and set $0 \leqslant n = (d_{L/K}+j)/e_{L/K} \leqslant (d_{L/K}+g_{L/K})/e_{L/K}$. Since $d_{L/K} = d_{L_1/K_1}$ and so on by Lemma 2.3.2, and since $e_{L/K} + j > g_{L/K}$, by Corollary 1.10.6 we have

$$\mathrm{N}_{L_1/K_1}(1 + \mathfrak{m}_{L_1}^{e_{L/K}+j}) = 1 + \mathfrak{m}_{K_1}^{n+1} \subset \mathrm{N}_{L_1/K_1} L_1^{\times}.$$

Since $\{\chi, 1 + \mathfrak{m}_{K_1}^{n+1}\} \subset \{\chi, \mathrm{N}_{L_1/K_1} L_1^{\times}\} = 0 \subset \operatorname{Br} K_1$ by Proposition 2.1.1(2), we have $\chi \in F_n X_K$ and $\operatorname{Sw}\chi \leqslant n$. The second inequality follows from (1.9.4).

(2) Let $n \geqslant 1$ and $\chi \in F_n X_K$. Then, since $(1 + \mathfrak{m}_{K_1}^n)^p \subset 1 + p \cdot \mathfrak{m}_{K_1}^n + \mathfrak{m}_{K_1}^{2n} \subset 1 + \mathfrak{m}_{K_1}^{n+1}$, we have $\{p \cdot \chi, 1 + \mathfrak{m}_{K_1}^n\} = \{\chi, (1 + \mathfrak{m}_{K_1}^n)^p\} \subset \{\chi, 1 + \mathfrak{m}_{K_1}^{n+1}\} = 0$.

(3) This follows from $\operatorname{Sw}\chi = 0$ for tamely ramified characters $\chi$. □

For a finite abelian extension of henselian discrete valuation fields, the cohomological filtration of the Galois group is defined as follows.

**Definition 2.3.4** Let $L \subset K_s$ be an abelian extension of a henselian discrete valuation field $K$ of Galois group $G = \operatorname{Gal}(L/K)$. For an integer $n \geqslant 0$, let $G^n_{\mathrm{coh}} \subset G$ be the subgroup defined by

$$G^n_{\mathrm{coh}} = \{\sigma \in G \mid \chi(\sigma) = 0 \text{ for every } \chi \in F_{n-1}X_K \cap \operatorname{Hom}(G, \mathbf{Q}/\mathbf{Z})\}. \quad (2.3.3)$$

We define the graded quotient by $\operatorname{Gr}^n_{\mathrm{coh}} G = G^n_{\mathrm{coh}}/G^{n+1}_{\mathrm{coh}}$.

By the definition of $G^n_{\mathrm{coh}}$, the canonical pairing $X_K \times G_K^{\mathrm{ab}} \to \mathbf{Q}/\mathbf{Z}$ induces a perfect pairing $\operatorname{Gr}^F_n X_K \times \operatorname{Gr}^n_{\mathrm{coh}} G_K^{\mathrm{ab}} \to \mathbf{Q}/\mathbf{Z}$.

**Lemma 2.3.5** *Let $L$ be a finite abelian extension of $K$, and $G = \operatorname{Gal}(L/K)$ be the Galois group. Then there exists an integer $n \geqslant 0$ such that $G^n_{\mathrm{coh}} = 1$.*

*Proof* This follows from Proposition 2.3.3(1). □

We compute the cohomological filtration for cyclic extensions of degree $p$ in the next chapter. Using this, we prove a relation of the cohomological filtration with the lower ramification groups, first for cyclic extensions in Theorem 5.3.4. We will generalize the relation later in Corollary 6.3.9 for abelian extensions with separable extensions of residue fields.

**Lemma 2.3.6** *Let $K$ be a henselian discrete valuation field with finite residue field $F$. Let $L$ be a finite abelian extension of $K$ of Galois group $G$. Then, for each integer $n \geqslant 0$, we have an inclusion $G^n_{\mathrm{coh}} \supset G^n_{\mathrm{rec}}$.*

*Proof* Let $K_1$ be as in Definition 2.3.1. For $\chi \in \operatorname{Hom}(G, \mathbf{Q}/\mathbf{Z}) \subset X_F$, the condition $\chi(G^n_{\mathrm{rec}}) = 0$ is equivalent to $\{\chi, 1 + \mathfrak{m}_K^n\} = 0$ in $\operatorname{Br} K$ and the condition $\chi(G^n_{\mathrm{coh}}) = 0$ is equivalent to $\{\chi, 1 + \mathfrak{m}^n_{K_1}\} = 0$ in $\operatorname{Br} K_1$. Since $\operatorname{Br} F = 0$, the morphism $\operatorname{Br} K \to \operatorname{Br} K_1$ is an injection by Corollary 2.1.4. Hence, $\chi(G^n_{\mathrm{coh}}) = 0$ implies $\chi(G^n_{\mathrm{rec}}) = 0$. This means the inclusion $G^n_{\mathrm{coh}} \supset G^n_{\mathrm{rec}}$. □

We will prove that the equality holds in Corollary 5.3.8(2). In Exercise 2.3, we show the equality $G^m_{\mathrm{coh}} = G^m_{\mathrm{rec}}$ for $G = \operatorname{Gal}(\mathbf{Q}_p(\zeta_{p^n})/\mathbf{Q}_p)$.

**Exercise 2.3** Let $p$ be a prime number and $n \geqslant 1$ be an integer.

(1) Let $K$ be a discrete valuation field of characteristic 0 such that the residue field is of characteristic $p$. Show that $(1 + p\mathcal{O}_K)^{p^{n-1}} \subset 1 + p^n\mathcal{O}_K$ for $p \neq 2$, and that $(1 + 4\mathcal{O}_K)^{2^{n-2}} \subset 1 + 2^n\mathcal{O}_K$ for $n \geqslant 2$. Show that the equality holds if $K$ is henselian.
(2) Let $\chi \colon \operatorname{Gal}(\mathbf{Q}_p(\zeta_{p^n})/\mathbf{Q}_p) \to \mathbf{Q}/\mathbf{Z}$ be a character. Show that $\operatorname{Sw}\chi \leqslant n-1$.

(3) Assume that $\chi$ is primitive in the sense that $\chi$ does not factor through the quotient $\mathrm{Gal}(\mathbf{Q}_p(\zeta_{p^n})/\mathbf{Q}_p) \to \mathrm{Gal}(\mathbf{Q}_p(\zeta_{p^{n-1}})/\mathbf{Q}_p)$. Show that the equality $\mathrm{Sw}\,\chi = n - 1$ holds.
(4) Show the equality $G^m_{\mathrm{coh}} = G^m_{\mathrm{rec}}$ for $G = \mathrm{Gal}(\mathbf{Q}_p(\zeta_{p^n})/\mathbf{Q}_p)$ and $m \geqslant 1$.

(In fact, the composition $\mathbf{Z}_p^\times \subset \mathbf{Q}_p^\times \to \mathrm{Gal}(\mathbf{Q}_p(\zeta_{p^n})/\mathbf{Q}_p) \to (\mathbf{Z}/p^n\mathbf{Z})^\times$ of the restriction of the reciprocity mapping with the canonical isomorphism is known to be the *inverse* of the canonical surjection.)

## 2.4 Differential Forms with Logarithmic Poles

To study the cohomological filtration, we will introduce a canonical morphism $\beta\colon \Omega^1_F(\log) \to \mathrm{Br}\,K$. In this section, we define and study a vector space $\Omega^1_F(\log) \supset \Omega^1_F$ over the residue field $F$. In the next section we define the morphism $\beta\colon \Omega^1_F(\log) \to \mathrm{Br}\,K$.

First, we study differential forms of a field $F$ of characteristic $p > 0$. We use the roman letter F to write the Frobenius operator F and the italic letter $F$ for a field in order to distinguish them. For $x \in F^\times$, let $d\log x$ denote $x^{-1}dx \in \Omega^1_F$.

**Theorem 2.4.1** *Let $F$ be a field of characteristic $p > 0$.*

(1) [54, 7.2] *There exists a unique injection $\mathrm{F} = \mathrm{C}^{-1}\colon \Omega^1_F \to \Omega^1_F/dF$ such that, for $x \in F$ and $y \in F^\times$, $\mathrm{F}(x \cdot d\log y) = x^p \cdot d\log y$.*
(2) [42, (2.1.24)] *The sequence*

$$0 \longrightarrow F^\times/F^{\times p} \xrightarrow{d\log} \Omega^1_F \xrightarrow{\mathrm{F}-1} \Omega^1_F/dF \qquad (2.4.1)$$

*is exact. The last arrow is a surjection if $F$ is separably closed.*

The exact sequence (2.4.1) is a multiplicative variant of the exact sequence

$$0 \longrightarrow F \xrightarrow{\mathrm{F}} F \xrightarrow{d} \Omega^1_F \qquad (2.4.2)$$

and is an analogue of the Artin–Schreier sequence $0 \to \mathbf{F}_p \to \mathbf{G}_a \xrightarrow{\mathrm{F}-1} \mathbf{G}_a \to 0$ for a field $F$ of characteristic $p > 0$ and the additive group $\mathbf{G}_a = F_s$. We define a surjection $b\colon \Omega^1_F \to \mathrm{Br}\,F[p]$ similar to the surjection

$$a\colon F \to \mathrm{H}^1(F, \mathbf{F}_p) \qquad (2.4.3)$$

defined by the Artin–Schreier sequence.

**Proposition 2.4.2** *Let $F$ be a field of characteristic $p > 0$. Then there exists a unique morphism*

$$b\colon \Omega^1_F \to \mathrm{Br}\,F[p] \qquad (2.4.4)$$

*such that, for* $x \in F$ *and* $y \in F^\times$,

$$b(x \cdot d\log y) = \{a(x), y\}. \tag{2.4.5}$$

*The morphism* $b\colon \Omega^1_F \to \operatorname{Br} F[p]$ (2.4.4) *is a surjection and induces an isomorphism* $\operatorname{Coker}(\mathrm{F}-1\colon \Omega^1_F \to \Omega^1_F/dF) \to \operatorname{Br} F[p]$.

*Proof* Since the $F$-vector space $\Omega^1_F$ is generated by the image of $d\log\colon F^\times \to \Omega^1_F$, the uniqueness follows. By Theorem 2.4.1(2) for a separable closure $F_s$, we obtain an exact sequence $0 \to F_s^\times/F_s^{\times p} \xrightarrow{d\log} \Omega^1_{F_s} \xrightarrow{\mathrm{F}-1} \Omega^1_{F_s}/dF_s \to 0$. Since $F_s$ and $\Omega^1_{F_s}$ are $F_s$-vector spaces, we have an isomorphism $\Omega^1_F/dF \to \mathrm{H}^0(F, \Omega^1_{F_s}/dF_s)$ and $\mathrm{H}^1(F, \Omega^1_{F_s}) = 0$ by (2.1.1). The boundary mapping of Galois cohomology defines an isomorphism

$$\operatorname{Coker}(\mathrm{F}-1\colon \Omega^1_F \to \Omega^1_F/dF) \to \mathrm{H}^1(F, F_s^\times/F_s^{\times p}). \tag{2.4.6}$$

The boundary mapping of the exact sequence $0 \to F_s^\times \to F_s^\times \to F_s^\times/F_s^{\times p} \to 0$ defines an isomorphism

$$\mathrm{H}^1(F, F_s^\times/F_s^{\times p}) \to \operatorname{Br} F[p] \tag{2.4.7}$$

by Hilbert 90. Their composition defines a surjection $b\colon \Omega^1_F \to \operatorname{Br} F[p]$ inducing an isomorphism $\operatorname{Coker}(\mathrm{F}-1\colon \Omega^1_F \to \Omega^1_F/dF) \to \operatorname{Br} F[p]$.

We show that the morphism $b\colon \Omega^1_F \to \operatorname{Br} F[p]$ satisfies the equality (2.4.5). For $y \in F^\times$ we have the commutative diagrams

$$\begin{array}{ccccccccc}
0 & \longrightarrow & \mathbf{F}_p & \longrightarrow & F_s & \xrightarrow{\mathrm{F}-1} & F_s & \longrightarrow & 0 \\
 & & \big\downarrow{\scriptstyle y} & & \big\downarrow{\scriptstyle d\log y} & & \big\downarrow{\scriptstyle d\log y} & & \\
0 & \longrightarrow & F_s^\times/F_s^{\times p} & \xrightarrow{d\log} & \Omega^1_{F_s} & \xrightarrow{\mathrm{F}-1} & \Omega^1_{F_s}/dF_s & \longrightarrow & 0
\end{array}$$

$$\begin{array}{ccccccccc}
0 & \longrightarrow & \mathbf{Z} & \xrightarrow{p} & \mathbf{Z} & \longrightarrow & \mathbf{F}_p & \longrightarrow & 0 \\
 & & \big\downarrow{\scriptstyle y} & & \big\downarrow{\scriptstyle y} & & \big\downarrow{\scriptstyle y} & & \\
0 & \longrightarrow & F_s^\times & \longrightarrow & F_s^\times & \longrightarrow & F_s^\times/F_s^{\times p} & \longrightarrow & 0
\end{array}$$

of exact sequences of $G_F$-modules. They induce a commutative diagram

$$\begin{array}{ccccc}
F & \xrightarrow{a} & \mathrm{H}^1(F, \mathbf{F}_p) & \longrightarrow & \mathrm{H}^2(F, \mathbf{Z}) = X_F \\
\big\downarrow{\scriptstyle \cdot d\log y} & & \big\downarrow{\scriptstyle y} & & \big\downarrow{\scriptstyle \{\ ,y\}} \\
\Omega^1_F & \xrightarrow{(2.4.6)} & \mathrm{H}^1(F, F_s^\times/F_s^{\times p}) & \xrightarrow{(2.4.7)} & \mathrm{H}^2(F, F_s^\times) = \operatorname{Br} F
\end{array}$$

where the horizontal arrows are the boundary morphisms. The composition of the lower line is the morphism $b$. The images of $x \in F$ in $\mathrm{H}^2(F, F_s^\times) = \mathrm{Br}\, F$ through the upper right are $\{a(x), y\}$, and through the lower left are $b(x \cdot d\log y)$. Hence, we obtain (2.4.5). □

**Corollary 2.4.3** *Let $E$ be a finite separable extension of a field $F$ of characteristic $p > 0$. Then the diagram*

$$\begin{array}{ccc} \Omega^1_E & \xrightarrow{b_E} & \mathrm{Br}\, E \\ {\scriptstyle \mathrm{Tr}_{E/F}}\downarrow & & \downarrow{\scriptstyle \mathrm{Cor}_{E/F}} \\ \Omega^1_F & \xrightarrow{b_F} & \mathrm{Br}\, F \end{array} \tag{2.4.8}$$

*is commutative.*

*Proof* Since $E$ is a finite separable extension, the morphism $\Omega^1_F \otimes_F E \to \Omega^1_E$ is an isomorphism and $\mathrm{Tr}_{E/F} \colon \Omega^1_E \to \Omega^1_F$ is defined as $1 \otimes \mathrm{Tr}_{E/F}$. Hence, it suffices to show that $b_F(\mathrm{Tr}_{E/F}x \cdot d\log y) = \mathrm{Cor}_{E/F}(b_E(x \cdot d\log y))$ for $x \in E$ and $y \in F^\times$. Since the left-hand side is $\{a_F(\mathrm{Tr}_{E/F}x), y\}$ by (2.4.5) and the right-hand side is $\mathrm{Cor}_{E/F}\{a_E(x), y\} = \{\mathrm{Cor}_{E/F} a_E(x), y\}$ by (2.4.5) and Lemma 2.1.5(2), the assertion follows from Lemma 2.1.5(3). □

**Definition 2.4.4** Let $K$ be a discrete valuation field and let $F = O_K/\mathfrak{m}_K$ be the residue field. We define an extension $\Omega^1_F(\log)$ of $F$ by $\Omega^1_F$ as

$$\Omega^1_F(\log) = (\Omega^1_F \oplus (F \otimes_{\mathbf{Z}} K^\times))/\langle (d\bar{x}, -x \otimes x) \mid x \in O_K,\ x \neq 0\rangle. \tag{2.4.9}$$

For $x \in K^\times$, let $d\log x \in \Omega^1_F(\log)$ denote the image of $1 \otimes x \in F \otimes K^\times$.

Note that $\Omega^1_F(\log)$ does not depend only on $F$ but also on $K$, although $K$ does not appear in the notation.

**Lemma 2.4.5** *The canonical morphism $\Omega^1_F \to \Omega^1_F(\log)$ and the valuation $\mathrm{ord} \colon K^\times \to \mathbf{Z}$ define an exact sequence*

$$0 \longrightarrow \Omega^1_F \longrightarrow \Omega^1_F(\log) \xrightarrow{\mathrm{res}} F \longrightarrow 0 \tag{2.4.10}$$

*of $F$-vector spaces. The diagram*

$$\begin{array}{ccc} K^\times & \xrightarrow{\mathrm{ord}} & \mathbf{Z} \\ {\scriptstyle d\log}\downarrow & & \downarrow \\ \Omega^1_F(\log) & \xrightarrow{\mathrm{res}} & F \end{array} \tag{2.4.11}$$

*is commutative.*

*Proof* Since $K^\times/(1+\mathfrak{m}_K)$ is isomorphic to the direct sum $F^\times \oplus \mathbf{Z}$, and since $d\bar{x} = 0$ for $x \in 1+\mathfrak{m}_K$ and $x \in \mathfrak{m}_K$, we have an exact sequence

$$0 \to (\Omega^1_F \oplus (F \otimes_{\mathbf{Z}} F^\times))/\langle (dx, -x \otimes x) \mid x \in F^\times \rangle \to \Omega^1_F(\log) \to F \to 0.$$

Since the morphism $d\log\colon F^\times \to \Omega^1_F$ sending $x$ to $d\log x = dx/x$ defines the inverse of the canonical surjection $\Omega^1_F \to (\Omega^1_F \oplus (F \otimes_{\mathbf{Z}} F^\times))/\langle (dx, -x \otimes x) \mid x \in F^\times \rangle$, we obtain the exact sequence (2.4.10).

For a uniformizer $t \in K^\times$, we have $\mathrm{res}(d\log t) = \mathrm{ord}_K t = 1$. Since $K^\times$ is generated by uniformizers, the diagram (2.4.11) is commutative. □

We extend the constructions in Theorem 2.4.1 to differential forms with logarithmic poles.

**Proposition 2.4.6** *Let $K$ be a discrete valuation field with residue field $F$ of positive characteristic $p$.*

(1) *There exists a unique injection* $\mathrm{F}\colon \Omega^1_F(\log) \to \Omega^1_F(\log)/dF$ *such that, for $x \in F$ and $y \in K^\times$,* $\mathrm{F}(x \cdot d\log y) = x^p \cdot d\log y$.

(2) *We have an exact sequence*

$$0 \longrightarrow K^\times/(1+\mathfrak{m}_K)K^{\times p} \xrightarrow{d\log} \Omega^1_F(\log) \xrightarrow{\mathrm{F}-1} \Omega^1_F(\log)/dF. \tag{2.4.12}$$

*The last arrow is a surjection if $F$ is separably closed.*

*Proof* (1) Since the $F$-vector space $\Omega^1_F(\log)$ is generated by the image of $d\log\colon K^\times \to \Omega^1_F(\log)$, the uniqueness follows.

Let $t \in K^\times$ be a uniformizer and define a linear mapping

$$f\colon \Omega^1_F(\log) = \Omega^1_F \oplus F \cdot d\log t \to \Omega^1_F(\log)/dF = \Omega^1_F/dF \oplus F \cdot d\log t$$

by $f(\omega, x \cdot d\log t) = (\mathrm{F}\omega, x^p \cdot d\log t)$ for $\omega \in \Omega^1_F$, $x \in F$. For $x \in F$, $y = ut^n \in K^\times$, $u \in O_K^\times$, and $n \in \mathbf{Z}$, we have $f(x \cdot d\log y) = f(x \cdot d\log u, nx \cdot d\log t) = (x^p \cdot d\log u, nx^p \cdot d\log t) = x^p \cdot d\log y$. By the exact sequence (2.4.10) and the injectivity of $\mathrm{F}\colon \Omega^1_F \to \Omega^1_F/dF$ in Theorem 2.4.1(1), the injectivity follows.

(2) Since a uniformizer of $K$ defines a splitting of the exact sequence $0 \to F^\times \to K^\times/(1+\mathfrak{m}_K) \to \mathbf{Z} \to 0$, we obtain the top exact sequence in the

commutative diagram

$$\begin{array}{ccccccccc} & & 0 & & 0 & & 0 & & \\ & & \downarrow & & \downarrow & & \downarrow & & \\ 0 & \longrightarrow & F^\times/F^{\times p} & \longrightarrow & K^\times/(1+\mathfrak{m}_K)K^{\times p} & \longrightarrow & \mathbf{F}_p & \longrightarrow & 0 \\ & & \downarrow{\scriptstyle d\log} & & \downarrow{\scriptstyle d\log} & & \downarrow{\scriptstyle \cap} & & \\ 0 & \longrightarrow & \Omega^1_F & \longrightarrow & \Omega^1_F(\log) & \longrightarrow & F & \longrightarrow & 0 \\ & & \downarrow{\scriptstyle \mathrm{F}-1} & & \downarrow{\scriptstyle \mathrm{F}-1} & & \downarrow{\scriptstyle \mathrm{F}-1} & & \\ 0 & \longrightarrow & \Omega^1_F/dF & \longrightarrow & \Omega^1_F(\log)/dF & \longrightarrow & F & \longrightarrow & 0 \end{array} \tag{2.4.13}$$

The second line is the exact sequence (2.4.10), and the third line is an induced exact sequence. The left column is exact by Theorem 2.4.1(2). Since the right column is exact, the middle column is also exact.

If $F$ is separably closed, the lower left vertical arrow is surjective by Theorem 2.4.1(2). Since the lower right vertical arrow is also surjective, the middle arrow is surjective. □

The following injectivity will be used in the definition of the refined Swan conductor in the next section.

**Corollary 2.4.7** *Let $K_1$ be the fraction field of the henselization of the localization of the polynomial ring $O_K[T]$ at the prime ideal $\mathfrak{m}_K O_K[T]$, and let $F(T)$ be the residue field of $K_1$. Then the composition*

$$\Omega^1_F(\log) \xrightarrow{T\cdot} \Omega^1_{F(T)}(\log) \to \mathrm{Coker}(\mathrm{F}-1\colon \Omega^1_{F(T)}(\log) \to \Omega^1_{F(T)}(\log)/dF(T)) \tag{2.4.14}$$

*is an injection.*

*Proof* Set $S = 1/T$ and let $F[S]_{(S)} \subset F(T)$ be the local ring at infinity. We have an exact sequence $0 \to F[S]_{(S)} \to F(T) \to TF[T] \to 0$, and $TF[T] = T^pF[T^p] \oplus \bigoplus_{p\nmid n} F\cdot T^n$. Define an $F[S]_{(S)}$-submodule

$$\begin{aligned}\Omega^1_{F[S]_{(S)}}(\log) &= \Omega^1_F(\log)\otimes F[S]_{(S)} \oplus F[S]_{(S)}\cdot d\log T\\ &\subset \Omega^1_{F(T)}(\log) = \Omega^1_F(\log)\otimes F(T) \oplus F(T)\cdot d\log T.\end{aligned}$$

Since the diagram

$$\begin{array}{ccc} F[S]_{(S)} & \longrightarrow & F(T) \\ \downarrow{\scriptstyle d} & & \downarrow{\scriptstyle d} \\ \Omega^1_{F[S]_{(S)}}(\log) & \longrightarrow & \Omega^1_{F(T)}(\log) \end{array}$$

is commutative, we obtain a commutative diagram of exact sequences:

$$\begin{array}{ccccccccc} 0 \to & T^pF[T^p] & \longrightarrow & F(T)/F[S]_{(S)} & \longrightarrow & \bigoplus_{p\nmid n} F\cdot T^n & \to 0 \\ & \downarrow d & & \downarrow d & & \downarrow d & \\ 0 \to & \Omega^1_F(\log)\otimes_F TF[T] & \longrightarrow & \Omega^1_{F(T)}(\log)/\Omega^1_{F[S]_{(S)}}(\log) & \longrightarrow & TF[T]\cdot d\log T & \to 0 \end{array}$$

Since the right vertical arrow is an injection, the sequence of cokernels gives the second line of the commutative diagram

$$\begin{array}{ccccccccc} 0 \to & \Omega^1_F(\log)\otimes_F TF[T] & \longrightarrow & \Omega^1_{F(T)}(\log)/\Omega^1_{F[S]_{(S)}}(\log) & \longrightarrow & TF[T]\cdot d\log T & \to 0 \\ & \downarrow \mathrm{F}-1 & & \downarrow \mathrm{F}-1 & & \downarrow \mathrm{F}-1 & \\ 0 \to & \dfrac{\Omega^1_F(\log)\otimes_F TF[T]}{d(T^pF[T^p])} & \longrightarrow & \dfrac{\Omega^1_{F(T)}(\log)}{d(F(T))+\Omega^1_{F[S]_{(S)}}(\log)} & \longrightarrow & T^pF[T^p]\cdot d\log T & \to 0 \end{array}$$

of exact sequences. Since the right vertical arrow is an injection, we obtain the injection

$$\begin{aligned} A = \operatorname{Coker}\left(\mathrm{F}-1\colon \Omega^1_F(\log)\otimes_F TF[T] \to \frac{\Omega^1_F(\log)\otimes_F TF[T]}{d(T^pF[T^p])}\right) \qquad (2.4.15)\\ \to \operatorname{Coker}(\mathrm{F}-1\colon \Omega^1_{F(T)}(\log) \to \Omega^1_{F(T)}(\log)/(dF(T)+\Omega^1_{F[S]_{(S)}}(\log))). \end{aligned}$$

Thus, it suffices to show that the morphism $T\cdot\colon \Omega^1_F(\log) \to A$ is an injection.

By $TF[T] = \bigoplus_{p\nmid n}\bigoplus_{e\geqslant 0} F\cdot T^{p^e n}$, the cokernel

$$A = \bigoplus_{p\nmid n}\operatorname{Coker}\left(\mathrm{F}-1\colon \bigoplus_{e\geqslant 0}\Omega^1_F(\log)\cdot T^{np^e} \to \Omega^1_F(\log)\cdot T^n \oplus \bigoplus_{e\geqslant 1}\frac{\Omega^1_F(\log)}{dF}\cdot T^{np^e}\right)$$

has a direct summand

$$A_1 = \operatorname{Coker}\left(\mathrm{F}-1\colon \bigoplus_{e\geqslant 0}\Omega^1_F(\log)\cdot T^{p^e} \to \Omega^1_F(\log)\cdot T \oplus \bigoplus_{e\geqslant 1}\frac{\Omega^1_F(\log)}{dF}\cdot T^{p^e}\right)$$

for $n = 1$. Since $\mathrm{F}\colon \Omega^1_F(\log) \to \Omega^1_F(\log)/dF$ is an injection by Proposition 2.4.6(1), the morphism $T\cdot\colon \Omega^1_F(\log) \to A_1$ is an injection. □

**Exercise 2.4** Let $F$ be a finite field of characteristic $p$, and $K = F((t))$ be the field of Laurent power series. Define the residue morphism $\mathrm{Res}\colon \Omega^1_K \to F$ by sending $\sum_n a_n t^n \cdot d\log t$ to $a_0$. Show that $\mathrm{inv}\{a(x), y\} = (1/p)\mathrm{Tr}_{F/\mathbf{F}_p}\mathrm{Res}(x\cdot d\log y)$ for $x \in K$ and $y \in K^\times$.

## 2.5 Refined Swan Conductor

Let $K$ be a henselian discrete valuation field and $F$ be the residue field. First, we construct an analogue $\beta\colon \Omega^1_F(\log) \to \mathrm{Br}\,K$ for Brauer groups of the morphism $\alpha\colon F \to X_F \to X_K$ (2.5.1) defined below.

Let $K_s$ be a separable closure of $K$, and $G_K = \mathrm{Gal}(K_s/K)$ be the absolute Galois group. The residue field of the maximal unramified extension of $K_u \subset K_s$ over $K$ is a separable closure $F_s$ of $F$ and the Galois group $\mathrm{Gal}(K_u/K)$ is canonically identified with the absolute Galois group $\mathrm{Gal}(F_s/F) = G_F$ of the residue field. The canonical surjection $G_K \to \mathrm{Gal}(K_u/K) = G_F$ induces an injection $X_F \to X_K$.

We assume that the residue field $F$ is of positive characteristic $p > 0$. We consider the composition

$$\alpha\colon F \xrightarrow{a} X_F \to X_K, \tag{2.5.1}$$

where $a\colon F \to X_F$ (2.4.3) is defined by the Artin–Schreier sequence. To define an analogue $\beta\colon \Omega^1_F(\log) \to \mathrm{Br}\,K$ of $\alpha$, we compute the unramified part $\mathrm{Br}(K_u/K)$ and its $p$-torsion part $\mathrm{Br}(K_u/K)[p]$. Recall that the isomorphisms (2.1.6) and (2.1.7) induce an isomorphism

$$\mathrm{H}^2(F, K_u^\times/(1+\mathfrak{m}_{K_u})) \to \mathrm{Br}(K_u/K). \tag{2.5.2}$$

**Proposition 2.5.1** *Let $K$ be a henselian discrete valuation field, and let $K_u \subset K_s$ be the maximal unramified extension in a separable closure of $K$.*

(1) *The exact sequence $0 \to K_u^\times/(1+\mathfrak{m}_{K_u}) \xrightarrow{p\cdot} K_u^\times/(1+\mathfrak{m}_{K_u}) \to K_u^\times/(1+\mathfrak{m}_{K_u}) \cdot K_u^{\times p} \to 0$ defines an isomorphism*

$$\mathrm{H}^1(F, K_u^\times/(1+\mathfrak{m}_{K_u}) \cdot K_u^{\times p}) \to \mathrm{Br}(K_u/K)[p]. \tag{2.5.3}$$

(2) *The boundary mapping of Galois cohomology for the exact sequence (2.4.12)*

$$0 \longrightarrow K_u^\times/(1+\mathfrak{m}_{K_u})K_u^{\times p} \xrightarrow{d\log} \Omega^1_{F_s}(\log) \xrightarrow{\mathrm{F}-1} \Omega^1_{F_s}(\log)/dF_s \longrightarrow 0 \tag{2.5.4}$$

*for the maximal unramified extension $K_u$ defines an isomorphism*

$$\mathrm{Coker}(\mathrm{F}-1\colon \Omega^1_F(\log) \to \Omega^1_F(\log)/dF) \to \mathrm{H}^1(F, K_u^\times/(1+\mathfrak{m}_{K_u})K_u^{\times p}). \tag{2.5.5}$$

*Proof* (1) Since $K_u^\times/(1+\mathfrak{m}_{K_u})$ is an extension of $\mathbf{Z}$ by $F_s^\times$, the $p$th power morphism of $K_u^\times/(1+\mathfrak{m}_{K_u})$ is an injection. Hence, we have an exact sequence $0 \to K_u^\times/(1+\mathfrak{m}_{K_u}) \xrightarrow{p\cdot} K_u^\times/(1+\mathfrak{m}_{K_u}) \to K_u^\times/(1+\mathfrak{m}_{K_u}) \cdot K_u^{\times p} \to 0$. By the exact sequence $0 \to F_s^\times \to K_u^\times/(1+\mathfrak{m}_{K_u}) \to \mathbf{Z} \to 0$ of $G_F$-modules and by

Hilbert 90 ($\mathrm{H}^1(F, F_s^\times) = 0$ and $\mathrm{H}^1(F, \mathbf{Z}) = 0$), we have $\mathrm{H}^1(F, K_u^\times/(1+\mathfrak{m}_{K_s})) = 0$. Hence, the boundary mapping of Galois cohomology defines an isomorphism

$$\mathrm{H}^1(F, K_u^\times/(1+\mathfrak{m}_{K_u}) \cdot K_u^{\times p}) \to \mathrm{H}^2(F, K_u^\times/(1+\mathfrak{m}_{K_u}))[p]. \tag{2.5.6}$$

Thus, (2.5.2) induces an isomorphism (2.5.3).

(2) Similarly to the proof of Proposition 2.4.2, we have $\mathrm{H}^0(F, \Omega^1_{F_s}(\log)) = \Omega^1_F(\log)$ and $\mathrm{H}^0(F, \Omega^1_{F_s}(\log))/dF_s = \Omega^1_F(\log)/dF$. Since $\mathrm{F} - 1\colon \Omega^1_{F_s}(\log) \to \Omega^1_{F_s}(\log)/dF_s$ is a surjection for a separably closed field $F_s$, we obtain an isomorphism (2.5.5). □

We define a canonical isomorphism

$$\beta\colon \mathrm{Coker}(\mathrm{F}{-}1\colon \Omega^1_F(\log) \to \Omega^1_F(\log)/dF) \to \mathrm{Br}(K_u/K)[p] \subset \mathrm{Br}\,K \tag{2.5.7}$$

as the composition of the isomorphisms (2.5.5) and (2.5.3). The induced morphism $\Omega^1_F(\log) \to \mathrm{Br}(K_u/K)[p] \subset \mathrm{Br}\,K$ will also be denoted by $\beta$. Recall that the surjection $a_F\colon F \to X_F[p]$ (2.4.3) is defined by the Artin–Schreier sequence, and its analogue $b_F\colon \Omega^1_F \to \mathrm{Br}\,F[p]$ (2.4.4) is a surjection defined in Proposition 2.4.2.

**Corollary 2.5.2** *The commutative diagram* (2.4.13) *for the maximal unramified extension $K_u$ defines a commutative diagram*

$$\begin{array}{ccccccccc} 0 & \longrightarrow & \Omega^1_F & \longrightarrow & \Omega^1_F(\log) & \xrightarrow{\text{res}} & F & \longrightarrow & .0 \\ & & {\scriptstyle b_F\,(2.4.4)}\downarrow & & {\scriptstyle\beta}\downarrow & & \downarrow{\scriptstyle a_F\,(2.4.3)} & & \\ 0 & \longrightarrow & \mathrm{Br}\,F[p] & \longrightarrow & \mathrm{Br}(K_u/K)[p] & \longrightarrow & X_F[p] & \longrightarrow & 0 \end{array} \tag{2.5.8}$$

*of exact sequences, where the lower line is the $p$-torsion part of the split exact sequence* (2.1.8). *The vertical arrows are surjections.*

*Proof* The upper line of (2.5.8) is (2.4.10) and the lower line is the $p$-torsion part of the split exact sequence (2.1.8). The diagram (2.4.13) for the maximal unramified extension $K_u$ is a commutative diagram of $G_F$-modules and the lower vertical arrows are surjections. The residue field of the maximal unramified extension $K_u$ is a separable closure $F_s$ of $F$. Hence, the lower line of (2.5.8) is also obtained by applying the functor $\mathrm{H}^1(F, -)$ and identifying $\mathrm{H}^1(F, K_u^\times/(1+\mathfrak{m}_{K_u}) \cdot K_u^{\times p})$ with $\mathrm{Br}(K_u/K)[p]$ by the isomorphism (2.5.3). The boundary morphisms define the vertical arrows in (2.5.8), and the diagram is commutative.

The surjectivities are clear from the construction. □

**Theorem 2.5.3** *Let $K$ be a henselian discrete valuation field with residue field*

*$F$ of characteristic $p > 0$. For $x \in F$, let $\alpha_x \in X_K$ denote the image of $x \in F$ by $\alpha\colon F \to X_K$. Then*

$$\beta\colon \Omega^1_F(\log) \to \mathrm{Br}(K_u/K)[p] \subset \mathrm{Br}\, K \tag{2.5.9}$$

*is the unique morphism satisfying the condition that, for every $x \in F$ and $y \in K^\times$,*

$$\{\alpha_x, y\} = \beta(x \cdot d\log y). \tag{2.5.10}$$

*Proof* Since the $F$-vector space $\Omega^1_F(\log)$ is generated by the image of the morphism $d\log\colon K^\times \to \Omega^1_F(\log)$, the uniqueness follows.

The morphism $\beta\colon \Omega^1_F(\log) \to \mathrm{Br}\, K$ is defined as the composition of the morphism (2.5.7) with the surjection $\Omega^1_F(\log) \to \mathrm{Coker}(\mathrm{F} - 1\colon \Omega^1_F(\log) \to \Omega^1_F(\log)/dF)$ and the inclusion $\mathrm{Br}(K_u/K)[p] \subset \mathrm{Br}\, K$. To show the existence, it suffices to show that this satisfies the equality (2.5.10). Let $x \in F$ and $y \in K^\times$. Similarly to the proof of Proposition 2.4.2, the left and middle squares in the diagram

$$\begin{array}{ccccccc}
F & \xrightarrow{\alpha} & \mathrm{H}^1(F, \mathbf{F}_p) & \longrightarrow & \mathrm{H}^2(F, \mathbf{Z}) & \longrightarrow & \mathrm{H}^2(K, \mathbf{Z}) \\
\downarrow{\scriptstyle \cdot d\log y} & & {\scriptstyle y}\downarrow & & \downarrow{\scriptstyle \{\ ,y\}} & & \downarrow{\scriptstyle \{\ ,y\}} \\
\Omega^1_F(\log) & \xrightarrow{(2.5.5)} & \mathrm{H}^1\left(F, \dfrac{K_u^\times}{K_u^{\times p}(1+\mathfrak{m}_K)}\right) & \xrightarrow{(2.5.6)} & \mathrm{H}^2(F, K_u^\times/(1+\mathfrak{m}_{K_u})) & \xrightarrow{(2.5.2)} & \mathrm{Br}\, K
\end{array}$$

(where the horizontal arrows are the boundary morphisms) is commutative. The compositions of the upper and lower lines are $\alpha$ and $\beta$ respectively. Since the right square is also commutative, we obtain the equality (2.5.10) as required. □

**Corollary 2.5.4** *For $x \in O_K - \{0\}$, $\{\alpha_x, x\} = 0$ in* $\mathrm{Br}\, K$.

*Proof* We have $\{\alpha_x, x\} = \beta(x \cdot d\log x)$. Since $\beta$ factors through $\Omega^1_F(\log) \to \Omega^1_F(\log)/dF$, we have $\beta(dx) = 0$. □

**Definition 2.5.5** Let $K$ be a henselian discrete valuation field with residue field $F$ of characteristic $p > 0$. Let $\chi \in X_K$ be a character and assume that $n = \mathrm{Sw}\, \chi \geqslant 1$. Let $K_1$ be as in Definition 2.3.1, and identify $\mathfrak{m}^n_{K_1}/\mathfrak{m}^{n+1}_{K_1} = \mathfrak{m}^n_K/\mathfrak{m}^{n+1}_K \otimes_F F(T)$ and $\Omega^1_F(\log) \otimes_F F(T) \subset \Omega^1_{F(T)}(\log)$. We say that a morphism $f\colon \mathfrak{m}^n_K/\mathfrak{m}^{n+1}_K \to \Omega^1_F(\log)$ of $F$-vector spaces is the refined Swan conductor of $\chi$ if, for every $x \in \mathfrak{m}^n_K$ and $x_1 \in \mathfrak{m}^n_{K_1}$,

$$\{\chi, 1 + x\} = \beta_K(f(\bar{x})), \qquad \{\chi, 1 + x_1\} = \beta_{K_1}((f \otimes 1_{F(T)})(\overline{x_1})) \tag{2.5.11}$$

in $\mathrm{Br}\, K$ and in $\mathrm{Br}\, K_1$ respectively. If the refined Swan conductor $f$ exists, we will write $f = \mathrm{rsw}\, \chi$.

**Lemma 2.5.6** *Let $K$ be a henselian discrete valuation field with residue field $F$ of characteristic $p > 0$.*

(1) *Let $K_1$ be the fraction field of henselization of the local ring at the prime ideal $\mathfrak{m}_K O_K[T] \subset O_K[T]$ with the residue field $F_1 = F(T)$, as in Corollary 2.4.7. Then the composition*

$$\Omega^1_F(\log) \xrightarrow{T\cdot} \Omega^1_{F_1}(\log) \xrightarrow{\beta_{K_1}} \operatorname{Br} K_1 \tag{2.5.12}$$

*is an injection.*

(2) *Let $\chi \in X_K$ be a character such that the Swan conductor satisfies $n = \operatorname{Sw}\chi \geqslant 1$. Then the refined Swan conductor* $\operatorname{rsw}\chi$ *is unique if it exists.*

*Proof* (1) Since $\beta_{K_1}$ induces an injection

$$\operatorname{Coker}(\mathrm{F} - 1 \colon \Omega^1_{F_1}(\log) \to \Omega^1_{F_1}(\log)/dF_1) \to \operatorname{Br} K_1,$$

the assertion follows from Corollary 2.4.7.

(2) For $x \in \mathfrak{m}_K^n$, the second equality in (2.5.11) implies $\{\chi, 1 + xT\} = \beta(T \cdot f(\bar{x}))$. This determines $f$ uniquely by (1). □

The first equality in (2.5.11) implies $\{\chi, 1 + \mathfrak{m}_K^{n+1}\} = 0$. Since the morphism $\Omega^1_F(\log) \to \operatorname{Hom}(F, \operatorname{Br} K)$ sending $\omega$ to the morphism $x \mapsto \beta(x\omega)$ is not an injection in general, the first equality in (2.5.11) alone may not characterize the refined Swan conductor uniquely. This is why we introduce $K_1$ in Definition 2.3.1 and impose the second equality in (2.5.11). In fact, since the morphism $\operatorname{Br} K \to \operatorname{Br} K_1$ is known to be an injection, the first equality is redundant, although we will not use this fact. In the case where $F$ is finite, the morphism $\Omega^1_F(\log) = F \to \operatorname{Hom}(F, \operatorname{Br} K)$ is an isomorphism and the first equality in (2.5.11) is enough to characterize the refined Swan conductor, as we will see in Section 5.3.

The refined Swan conductor always exists and the second equality in (2.5.11) holds for every extension, as we will see in Theorem 16.3.1. We will prove the existence in the case where the corresponding cyclic extensions are log monogenic and compute them explicitly in the next chapters.

**Exercise 2.5** For a field $L$ of characteristic $\neq 2$ and $a \in L^\times$, let $\kappa_a \in X_L[2]$ be the character corresponding to the quadratic extension $L(\sqrt{a})$ if $a \notin L^{\times 2}$ and $\kappa_a = 0$ otherwise. Further, for $a, b \in L^\times$, let $(a, b)$ denote $\{\kappa_a, b\} \in \operatorname{Br} L[2]$.

(1) For $a \in L^\times - \{1\}$, show that $(a, 1 - a) = 0$ in $\operatorname{Br} L[2]$.
(2) Let $\chi \colon \operatorname{Gal}(\mathbf{Q}_2(\sqrt{-1})/\mathbf{Q}_2) \to \mathbf{Q}/\mathbf{Z}$ be the non-trivial character. Show that the refined Swan conductor $\operatorname{rsw}\chi \colon 2\mathbf{Z}_2/4\mathbf{Z}_2 \to \Omega^1_{\mathbf{F}_2}(\log) = \mathbf{F}_2$ exists.

## Historical Notes

The cohomological filtration and the refined Swan conductor are defined by Kato in [50]. The canonical morphism $\beta$ is also defined in [50].

The sheaf of differential forms with log poles is used in [19] to study the cohomology of the complement of a divisor with normal crossing in a smooth variety. It is defined in a more general setting of log geometry in [51].

# PART II

# CYCLIC EXTENSIONS

We study cyclic extensions in detail. Cyclic extensions of degree $p$ admit a classification by the Artin–Schreier theory in the equal characteristic case and by the Kummer theory in the mixed characteristic case under the existence of a primitive $p$th root of 1. For these extensions, we determine the lower ramification groups and the cohomological filtration explicitly.

The Hasse–Arf theorem connects the lower ramification groups for cyclic log monogenic extensions and the cohomological filtration. The study of cyclic extensions of degree $p$ makes the first step of induction in the proof. To make the induction work, we introduce the trace of differential forms with log poles and establish various compatibilities.

A prototype of the theory of cohomological filtration is the local class field theory where the residue field is finite. In this case, the Brauer group is canonically identified with $\mathbf{Q}/\mathbf{Z}$ and the Hasse–Arf theorem means equality of the filtrations on the multiplicative group and on the abelianized quotient of the absolute Galois group via the reciprocity mapping.

# 3

# Cyclic Extensions of Degree $p$

We compute explicitly the lower numbering filtration and the cohomological filtration for cyclic extensions of degree $p$. For fields of characteristic $p > 0$, such extensions are defined by Artin–Schreier equations. We show in Section 3.1 that the cohomological filtration is induced by the filtration on the additive group defined by negative powers of the maximal ideal. We also determine the lower numbering filtration. For a field of characteristic 0 containing a primitive $p$th root of unity, such extensions are defined by Kummer equations. In Section 3.2, we show that the cohomological filtration is induced by the filtration on the multiplicative group defined by the congruence modulo positive powers of the maximal ideal. We also determine the lower numbering filtration.

The relation between the lower ramification groups and the cohomological filtration studied in this section will serve as the first step of induction in the proof of the Hasse–Arf theorem in Chapter 5.

## 3.1 Artin–Schreier Extensions

We study Artin–Schreier extensions of a henselian discrete valuation field of characteristic $p > 0$. For a field $K$ of characteristic $p > 0$, let $\mathrm{F}\colon K \to K$ denote the absolute Frobenius morphism. The boundary mapping of the exact sequence $0 \to \mathbf{F}_p \to \mathbf{G}_a \overset{\mathrm{F}-1}{\to} \mathbf{G}_a \to 0$ defines an isomorphism

$$a\colon K/(\mathrm{F}-1)K \to \mathrm{H}^1(K, \mathbf{F}_p) = X_K[p]. \tag{3.1.1}$$

For a discrete valuation field $K$ of characteristic $p > 0$, we define an increasing filtration on $K$ by $F_n K = \mathfrak{m}_K^{-n}$ and the graded quotients $\mathrm{Gr}_n^F K = F_n K / F_{n-1} K$. We also define an increasing filtration $F_n^B X_K[p]$ to be the image of $F_n K$ by the surjection $a\colon K \to X_K[p]$ and the graded quotients $\mathrm{Gr}_n^B X_K[p] = F_n^B X_K[p]/F_{n-1}^B X_K[p]$. We will show in Corollary 3.1.7 that

$F_n^B X_K[p]$ equals the restriction $F_n X_K \cap X_K[p]$ of the cohomological filtration. The morphism $a$ (3.1.1) induces a canonical surjection

$$\mathrm{Gr}_n^F K = \mathfrak{m}_K^{-n}/\mathfrak{m}_K^{-n+1} \to \mathrm{Gr}_n^B X_K[p]. \tag{3.1.2}$$

**Lemma 3.1.1** *Let $K$ be a henselian discrete valuation field of characteristic $p > 0$ and $F$ be the residue field. We have $F_{-1}^B X_K[p] = 0$. For integers $n \geqslant 0$, the surjections* (3.1.2) *induce isomorphisms*

$$\begin{aligned} \mathfrak{m}_K^{-n}/\mathfrak{m}_K^{-n+1} &\to \mathrm{Gr}_n^B X_K[p] \quad \textit{for } n \geqslant 1 \textit{ prime to } p, \qquad (3.1.3)\\ \mathfrak{m}_K^{-n}/(\mathfrak{m}_K^{-n+1} + \mathrm{F}(\mathfrak{m}_K^{-m})) &\to \mathrm{Gr}_n^B X_K[p] \quad \textit{for } n = pm \geqslant p \textit{ divisible by } p,\\ F/(\mathrm{F}-1)F &\to \mathrm{Gr}_0^B X_K[p]. \end{aligned}$$

*Proof* Since $K$ is henselian, for $x \in \mathfrak{m}_K$, the Artin–Schreier equation $u^p - u = x$ has a unique solution in $\mathfrak{m}_K$ by Hensel's lemma. Hence, we have $F_{-1} X_K[p] = 0$.

We consider the morphism $\mathrm{F} - 1 \colon K \to K$. On the graded quotients this induces injections $\mathrm{F} \colon \mathfrak{m}_K^{-m}/\mathfrak{m}_K^{-m+1} \to \mathfrak{m}_K^{-n}/\mathfrak{m}_K^{-n+1}$ for $n = pm \geqslant p$ and $\mathrm{F}-1 \colon F \to F$ for $n = 0$. Since $K = \bigcup_{n \geqslant 0} F_n K$, we have $F_n K \cap \mathrm{Im}(\mathrm{F}-1 \colon K \to K) = \mathrm{Im}(\mathrm{F}-1 \colon F_m K \to K)$ for $n \geqslant 0$ and the integral part $m = [n/p]$. Further, the morphism $\mathrm{Gr}_n^F K \to \mathrm{Gr}_n^B X_K[p]$ (3.1.2) is an isomorphism for $n > 0$ prime to $p$ and induces an isomorphism $\mathrm{Coker}(\mathrm{Gr}_m^F K \to \mathrm{Gr}_n^F K) \to \mathrm{Gr}_n^B X_K[p]$ for $n = pm \geqslant 0$ divisible by $p$. Thus, we obtain the isomorphisms in (3.1.3). □

We compute the lower numbering filtration for Artin–Schreier extensions, according to the classification in Lemma 3.1.1.

**Proposition 3.1.2** *Let $K$ be a henselian discrete valuation field of characteristic $p > 0$ with residue field $F$. For $x \in K - (\mathrm{F} - 1)(K)$, let $L = K(u)$ be the cyclic extension of degree $p$ of $K$ generated by a solution $u$ of the Artin–Schreier equation $u^p - u = x$, and let $\mathbf{F}_p \to G = \mathrm{Gal}(L/K)$ be the isomorphism defined by the generator $\sigma \in G$ satisfying $\sigma(u) = u + 1$.*

(1) *Let $n \geqslant 1$ be an integer prime to $p$, and $x \in \mathfrak{m}_K^{-n} - \mathfrak{m}_K^{-n+1}$. Then $L$ is a totally ramified extension and $n$ is the largest integer satisfying $G = G_n$.*
(2) *Let $n = pm \geqslant 1$ be an integer divisible by $p$, and $x \in \mathfrak{m}_K^{-n} - (\mathfrak{m}_K^{-n+1} + \mathrm{F}(\mathfrak{m}_K^{-m}))$. Define $w \in O_L^\times$ by $u = wt^{-m}$ for a uniformizer $t$ of $K$. Then the residue field $E$ of $L$ is a purely inseparable extension $F(w)$ of degree $p$ generated by $w$ of the residue field $F$ of $K$, and $m$ is the largest integer satisfying $G = G_m$.*
(3) *Let $x \in O_K - (\mathrm{F} - 1)O_K$. Then $L$ is an unramified extension of $K$.*

*Proof* Let $u \in L = K(u)$ be a solution of $u^p - u = x$, and $t$ be a uniformizer of $K$. Let $e = e_{L/K}$ be the ramification index.

(1) Since $u \notin O_L$, we have $p \cdot \mathrm{ord}_L u = \mathrm{ord}_L(u^p - u) = e \cdot \mathrm{ord}_K x = -en$. Hence, $e = p$, $\mathrm{ord}_L u = -n$, and $L$ is totally ramified. Let $k, l$ be integers satisfying $kp - ln = 1$. Then $s = t^k u^l$ is a uniformizer of $O_L = O_K[s]$. We have $\sigma(s)/s = (1 + 1/u)^l \equiv 1 - 1/(nu)$, and $i_G(\sigma) = \mathrm{ord}_L(-1/(nu)) = n$ is the largest integer satisfying $G = G_n$.

(2) Set $x = v/t^n \in K$ for a unit $v \in O_K^\times$. Then $\bar{v} \in F - F^p$, and $w = t^m u$ satisfies $w^p - t^{(p-1)m} w = v$. Hence, the residue field $E = F(\bar{w})$ of $O_L = O_K[w]$ is a purely inseparable extension of $F$ generated by the $p$th root of $\bar{v}$.

We have $\sigma(w)/w = 1 + 1/u$, and $i_G(\sigma) = \mathrm{ord}_L(1/u) = \mathrm{ord}_L(t^m/w) = m$ is the largest integer satisfying $G = G_m$.

(3) If $x \in O_K$, then $O_K[u]/(u^p - u - x)$ is étale over $O_K$ by Lemma 1.1.6. Hence, $L$ is an unramified extension by Lemma 1.5.2(1). □

To study the refined Swan conductor, we also consider an increasing filtration on $\Omega^1_K$ by sub-$O_K$-modules by $F_n\Omega^1_K = \mathfrak{m}_K^{-n} \cdot d\log K^\times$, and set $\mathrm{Gr}^F_n \Omega^1_K = F_n\Omega^1_K / F_{n-1}\Omega^1_K$. The morphism $d \colon K \to \Omega^1_K$ sends $F_n K$ to $F_n\Omega^1_K$ and induces a morphism $\mathrm{Gr}^F_n K \to \mathrm{Gr}^F_n \Omega^1_K$, also denoted $d$. Let $a \colon K \to X_K$ and $b \colon \Omega^1_K \to \mathrm{Br}\, K$ be the morphisms (3.1.1) and (2.4.4).

**Lemma 3.1.3** *Let $K$ be a henselian discrete valuation field of characteristic $p > 0$.*

(1) *Let $n$ and $i \geqslant 1$ be integers. For $x \in \mathfrak{m}_K^{-n}$ and $y \in \mathfrak{m}_K^i$, we have $x \cdot d\log(1 + y) \in F_{n-i}\Omega^1_K$ and*

$$x \cdot d\log(1 + y) \equiv -y \cdot dx \bmod F_{n-2i}\Omega^1_K + dK. \tag{3.1.4}$$

(2) *The morphism $b \colon \Omega^1_K \to \mathrm{Br}\, K$ annihilates $F_{-1}\Omega^1_K$ and $dK$.*
(3) *There exists a unique isomorphism $\Omega^1_F(\log) \to \mathrm{Gr}^F_0 \Omega^1_K$ of $F$-vector spaces sending $d\log y$ to $y^{-1}dy$ for $y \in K^\times$.*
(4) *For every integer $n \geqslant 0$, $F^B_n X_K[p] \subset F_n X_K \cap X_K[p]$.*

*Proof* (1) We have $x \cdot d\log(1 + y) = x/(1 + y) \cdot dy = xy/(1 + y) \cdot d\log y \in F_{n-i}\Omega^1_K$. The difference $x \cdot (d\log(1 + y) - dy) = x(1/(1 + y) - 1) \cdot dy = -xy \cdot y/(1 + y) \cdot d\log y$ is contained in $F_{n-2i}\Omega^1_K$. Hence, (3.1.4) follows from the Leibniz rule $d(xy) = xdy + ydx$.

(2) For $x \in \mathfrak{m}_K$, we have $a(x) = 0$ by Hensel's lemma. Hence, $b$ annihilates $F^{-1}\Omega^1_K = \mathfrak{m}_K \cdot d\log K^\times$. By the definition of $b$ as the boundary mapping, $b$ also annihilates $dK$.

(3) The canonical morphism $\Omega^1_{O_K} \to \Omega^1_K$ induces $\Omega^1_{O_K} \otimes_{O_K} F \to \mathrm{Gr}^F_0 \Omega^1_K$. For a uniformizer $t$ of $K$, we have isomorphisms $(\Omega^1_{O_K} \otimes_{O_K} F)/F \cdot dt \to \Omega^1_F$ and $\Omega^1_F(\log) = \Omega^1_F \oplus F \cdot d\log t$. Since $dt = t \cdot d\log t \in F_{-1}\Omega^1_K$, the morphism

$\Omega^1_{O_K} \otimes_{O_K} F \to \mathrm{Gr}^F_0 \Omega^1_K$ induces a surjection $\Omega^1_F(\log) \to \mathrm{Gr}^F_0 \Omega^1_K$ independent of the choice of uniformizer $t$.

Let $K_1$ be the fraction field of the henselization of the localization of the polynomial ring $O_K[T]$ at the prime ideal $\mathfrak{m}_K O_K[T]$ as in Definition 2.3.1. By (2), the morphism $\Omega^1_K \to \mathrm{Br}\, K_1$ sending $\omega$ to $b(T\cdot\omega)$ induces $\mathrm{Gr}^F_0 \Omega^1_K \to \mathrm{Br}\, K_1$. Its image is the image of the injection $\Omega^1_F(\log) \to \mathrm{Br}\, K_1$ (2.5.12) sending $\omega$ to $\beta_{K_1}(T \cdot \omega)$. Hence, we obtain an inverse of $\Omega^1_F(\log) \to \mathrm{Gr}^F_0 \Omega^1_K$ and the assertion follows.

(4) By (1) and (2) applied to $K_1$ in the proof of (3), we have $\mathfrak{m}_K^{-n} \cdot d\log(1 + \mathfrak{m}_{K_1}^{n+1}) \subset F_{-1}\Omega^1_{K_1}$ and $\{a(\mathfrak{m}_K^{-n}), 1 + \mathfrak{m}_{K_1}^{n+1}\} \subset b(F_{-1}\Omega^1_{K_1}) = 0$. Hence, $a(F_n K) \subset F_n X_K$. □

**Lemma 3.1.4** *Let $K$ be a henselian discrete valuation field of characteristic $p > 0$. Let $n \geqslant 1$ be an integer. Then there exists a unique morphism*

$$r \colon \mathrm{Gr}^B_n X_K[p] \to \mathrm{Gr}^F_n \Omega^1_K$$

*such that the diagram*

$$\begin{array}{ccc} \mathrm{Gr}^F_n K & & \\ {\scriptstyle a}\downarrow & \searrow^{-d} & \\ \mathrm{Gr}^B_n X_K[p] & \xrightarrow{\ r\ } & \mathrm{Gr}^F_n \Omega^1_K \end{array} \tag{3.1.5}$$

*is commutative. The morphism $r \colon \mathrm{Gr}^B_n X_K[p] \to \mathrm{Gr}^F_n \Omega^1_K$ is an injection.*

*Proof* Assume $p \nmid n$. By Lemma 3.1.1, the surjection $a \colon \mathrm{Gr}^F_n K \to \mathrm{Gr}^B_n X_K$ is an isomorphism. Hence, there exists a unique morphism $r \colon \mathrm{Gr}^B_n X_K \to \mathrm{Gr}^F_n \Omega^1_K$ making the diagram (3.1.5) commutative. By Lemma 3.1.3(3), $\mathrm{Gr}^F_n K \to \mathrm{Gr}^F_n \Omega^1_K$ is injective for $p \nmid n$, and $r \colon \mathrm{Gr}^B_n X_K \to \mathrm{Gr}^F_n \Omega^1_K$ is also injective.

Assume $p \mid n = pm$. By Lemma 3.1.1, we have an exact sequence $0 \to \mathrm{Gr}^F_m K \xrightarrow{\mathrm{F}} \mathrm{Gr}^F_n K \xrightarrow{a} \mathrm{Gr}^B_n X_K \to 0$. By Lemma 3.1.3(3) and the exact sequence (2.4.2), we have an exact sequence $0 \to \mathrm{Gr}^F_m K \xrightarrow{\mathrm{F}} \mathrm{Gr}^F_n K \xrightarrow{d} \mathrm{Gr}^F_n \Omega^1_K$. Hence, there exists a unique injection $r \colon \mathrm{Gr}^B_n X_K \to \mathrm{Gr}^F_n \Omega^1_K$ making the diagram (3.1.5) commutative. □

We show the existence of the refined Swan conductor for Artin–Schreier extensions. We identify $\mathrm{Gr}^F_n \Omega^1_K = \mathrm{Hom}_F(\mathfrak{m}_K^n/\mathfrak{m}_K^{n+1}, \Omega^1_F(\log))$ by the isomorphism in Lemma 3.1.3(3).

**Proposition 3.1.5** *Let $K$ be a henselian discrete valuation field of characteristic $p > 0$. Let $n \geqslant 1$ be an integer, and $L$ be a cyclic extension corresponding to $\chi \in F^B_n X_K[p] - F^B_{n-1} X_K[p]$.*

(1) *The Swan conductor of* $\chi \in X_K$ *equals* $n$, *and*

$$r(\chi) \in \mathrm{Gr}_n^F \Omega_K^1 = \mathrm{Hom}_F(\mathfrak{m}_K^n/\mathfrak{m}_K^{n+1}, \Omega_F^1(\log))$$

*is the refined Swan conductor* $\mathrm{rsw}\,\chi$.

(2) *Let* $x \in F_n K$ *be an element such that* $a(x) = \chi$, *and let* $u \in L = K(u)$ *be a solution of the Artin–Schreier equation* $u^p - u = x$. *Let* $\sigma \in G = \mathrm{Gal}(L/K)$ *be the generator satisfying* $\sigma(u) = u + 1$, *and* $m = i_G(\sigma) \geqslant 1$. *Set* $v = \sigma(u)/u - 1 \in L$, $w = \mathrm{N}_{L/K} v$, *and* $z = \mathrm{N}_{L/K} u \in K$. *Then we have* $n = f_{L/K} \cdot m$ *and* $\mathrm{ord}_L(v) = m$, $\mathrm{ord}_K(w) = n$. *The refined Swan conductor* $\mathrm{rsw}\,\chi\colon \mathfrak{m}_K^n/\mathfrak{m}_K^{n+1} \to \Omega_F^1(\log)$ *maps a basis* $-w$ *to* $d\log z$.

*Proof* (1) We have $\mathrm{Sw}\,\chi \leqslant n$ by Lemma 3.1.3(4). Let $x \in F_n K$ be an element such that $a(x) = \chi$. For $y \in \mathfrak{m}_K^n$, we have $\{\chi, 1 + y\} = b(x \cdot d\log(1 + y)) = -b(y dx) = -\beta_K(y dx)$ by Lemma 3.1.3. Similarly, we have $\{\chi, 1 + y_1\} = -\beta_{K_1}(y_1 dx)$ for $y_1 \in \mathfrak{m}_{K_1}^n$. Since $r\colon \mathrm{Gr}_n^B X_K[p] \to \mathrm{Gr}_n^F \Omega_K^1$ is an injection by Lemma 3.1.4, we have $dx \neq 0$ in $\mathrm{Gr}_n^F \Omega_K^1$. Thus, we have $\mathrm{Sw}\,\chi = n$ and $\mathrm{rsw}\,\chi = -dx = r(\chi)$.

(2) We have $n = f_{L/K} \cdot m$ by Proposition 3.1.2. We have $v = 1/u$, $z = x$, $w = 1/x$, and $\mathrm{ord}_L v = -\mathrm{ord}_L u = -e_{L/K}/p \cdot \mathrm{ord}_K x = n/f_{L/K} = m$, $\mathrm{ord}_K(w) = f_{L/K} \mathrm{ord}_L v = n$. The refined Swan conductor $\mathrm{rsw}\,\chi = -dx$ maps $-w = -1/x$ to $d\log x = d\log z$. □

**Example 3.1.6** Let $K$ be a henselian discrete valuation field of characteristic $p > 0$, $F$ be the residue field, and $t \in K$ be a uniformizer.

(1) Let $L$ be the Artin–Schreier extension defined by $x^p - x = 1/t^n$ for $n = pm - 1 > 0$, as in Example 1.4.4(1). Let $\sigma \in G = \mathrm{Gal}(L/K)$ be the generator sending $x$ to $x + 1$, and $\chi\colon G = \mathrm{Gal}(L/K) \to \mathbf{Q}/\mathbf{Z}$ be the character sending $\sigma$ to $1/p$. By Proposition 3.1.5(1), $\mathrm{Sw}\,\chi = n$ and $\mathrm{rsw}\,\chi$ is the morphism $\mathfrak{m}_K^n/\mathfrak{m}_K^{n+1} \to \Omega_F^1(\log)$ sending the basis $t^n$ to $-d\log t$.

(2) Let $L$ be the Artin–Schreier extension defined by $x^p - x = v/t^n$ for $n = pm > 0$ and for $v \in O_K$ such that $\bar{v} \in F - F^p$. Let $u = t^m x$ be the generator of the residue field $E = F(u)$ as in Example 1.4.4(2). Let $\sigma \in G = \mathrm{Gal}(L/K)$ be the generator defined by $\sigma(x) = x + 1$, and $\chi\colon G = \mathrm{Gal}(L/K) \to \mathbf{Q}/\mathbf{Z}$ be the character sending $\sigma$ to $1/p$. By Proposition 3.1.5(1), $\mathrm{Sw}\,\chi = n$ and $\mathrm{rsw}\,\chi$ is the morphism $\mathfrak{m}_K^n/\mathfrak{m}_K^{n+1} \to \Omega_F^1(\log)$ sending the basis $t^n$ to $-du$.

**Corollary 3.1.7** *For every integer* $n \geqslant 0$, $F_n^B X_K[p] = F_n X_K \cap X_K[p]$. *For* $n = 0$, *we have* $\mathrm{Im}(X_F[p] \to X_K[p]) = F_0 X_K \cap X_K[p]$.

*Proof* By Proposition 3.1.5(1), the set $\{\chi \in X_K[p] \mid \mathrm{Sw}\,\chi = n\}$ equals the complement $F_n^B X_K[p] - F_{n-1}^B X_K[p]$ for $n \geqslant 1$. Hence, we have $\{\chi \in X_K[p] \mid \mathrm{Sw}\,\chi = 0\} = F_0^B X_K[p]$. Further, for $n = 0$ we have $\mathrm{Im}(X_F[p] \to X_K[p]) = F_0^B X_K[p]$. □

Let $L$ be a finite abelian extension over $K$ such that $G = \mathrm{Gal}(L/K)$ is an $\mathbf{F}_p$-vector space. Then, by Corollary 3.1.7, the cohomological filtration $G^n_{\mathrm{coh}} \subset G$ indexed by integers $n \geqslant 0$ is defined by

$$G^n_{\mathrm{coh}} = \{\sigma \in G \mid \chi_a(\sigma) = 0 \text{ for } a \in \mathfrak{m}_K^{-n+1}\}.$$

We will show in Corollary 6.3.9 that in the classical case where the residue field extension is separable, this filtration is the same as the lower numbering filtration up to renumbering. In the general case, Exercise 3.1 shows that there is no such relation. More precisely, the upper numbering filtrations are not compatible with subgroups even after modifying the numbering. Since the lower numbering filtrations are compatible with subgroups, this implies that the two filtrations are different even allowing modification of the numbering.

**Exercise 3.1** Let $k$ be a field of characteristic $p > 0$, and $K = k(X, Y)((t))$. Let $1 \leqslant n < m < l$ be integers, and let $M = K(u) \subset L = K(u, v, w)$ be the Galois extensions of $K$ defined by the Artin–Schreier equations: $u^p - u = X/t^{pn}$, $v^p - v = Y/t^{pm}$, $w^p - w = X/t^{pl}$. Then the Galois groups $H = \mathrm{Gal}(L/M) \subset G = \mathrm{Gal}(L/K)$ are isomorphic to $\mathbf{F}_p^2 \subset \mathbf{F}_p^3$. Let $\chi$ and $\chi'$ be the characters of $G$ corresponding to $K(v)$ over $K$ and to $K(w)$ over $K$ respectively.

(1) Show that $G = G^{pn}_{\mathrm{coh}} \supsetneqq G^{pn+1}_{\mathrm{coh}} = H = G^{pm}_{\mathrm{coh}} \supsetneqq G^{pm+1}_{\mathrm{coh}} = \mathrm{Gal}(L/M(v)) = G^{pl}_{\mathrm{coh}} \supsetneqq G^{pl+1}_{\mathrm{coh}} = 1$. Show that $\mathrm{Sw}\,\chi < \mathrm{Sw}\,\chi'$.

Assume that $n$ is divisible by $p$, and that $l < m + (p-1)n/p$.

(2) Let $\chi_M$ and $\chi'_M$ be the restrictions to $H \subset G$. Show that $\mathrm{Sw}\,\chi_M > \mathrm{Sw}\,\chi'_M$.
(3) Show that there is no integer $r$ such that $H^r_{\mathrm{coh}} = G^{pl}_{\mathrm{coh}}$.

## 3.2 Kummer Extensions

Assume that $K$ is a henselian discrete valuation field of characteristic 0 containing a $p$th root of 1 with residue field $F$ of characteristic $p > 0$. We study the ramification of Kummer extensions of degree $p$.

For a field $K$ of characteristic 0 and a prime number $p$, let $\mu_p \subset \mathbf{G}_m$ denote the subgroup of roots of 1 of order $p$ of the multiplicative group. By Hilbert 90 ($\mathrm{H}^1(K, \mathbf{G}_m) = 0$), the exact sequence $0 \to \mu_p \to \mathbf{G}_m \xrightarrow{p} \mathbf{G}_m \to 0$ defines an isomorphism

$$\kappa \colon K^\times/K^{\times p} \to \mathrm{H}^1(K, \mu_p). \tag{3.2.1}$$

If $K$ contains a primitive $p$th root of 1, the target $\mathrm{H}^1(K, \mu_p)$ is identified with $\mathrm{Hom}_{\mathrm{cont}}(G_K^{\mathrm{ab}}, \mu_p) = X_K[p] \otimes_{\mathbf{F}_p} \mu_p$. For $x \in K^\times$, let $\kappa_x \in X_K[p] \otimes_{\mathbf{F}_p} \mu_p$ denote the character defined as the image.

Let $K$ be a henselian discrete valuation field of characteristic 0 with residue field $F$ of characteristic $p > 0$. For integers $i \geqslant 1$, let $U^i K^\times \subset K^\times$ denote the subgroup $1 + \mathfrak{m}_K^i$ and set $U^0 K^\times = K^\times$. The graded quotient $\mathrm{Gr}_U^i K^\times$ is canonically identified with $\mathfrak{m}_K^i/\mathfrak{m}_K^{i+1}$ for $i \geqslant 1$, and $\mathrm{Gr}_U^0 K^\times = K^\times/(1 + \mathfrak{m}_K)$ is an extension of $\mathbf{Z}$ by $F^\times$.

We define a decreasing filtration $U^i\mathrm{H}^1(K, \mu_p)$ indexed by integers $i \geqslant 0$ to be the image of $U^i K^\times$ by the surjection $\kappa\colon K^\times \to \mathrm{H}^1(K, \mu_p)$ (3.2.1). For the graded quotients $\mathrm{Gr}_U^i \mathrm{H}^1(K, \mu_p) = U^i\mathrm{H}^1(K, \mu_p)/U^{i+1}\mathrm{H}^1(K, \mu_p)$, we have canonical surjections

$$\begin{aligned} \mathrm{Gr}_U^i K^\times &= \mathfrak{m}_K^i/\mathfrak{m}_K^{i+1} \to \mathrm{Gr}_U^i \mathrm{H}^1(K, \mu_p) \quad \text{for } i \geqslant 1, \\ \mathrm{Gr}_U^0 K^\times &= K^\times/(1 + \mathfrak{m}_K) \to \mathrm{Gr}_U^0 \mathrm{H}^1(K, \mu_p). \end{aligned} \tag{3.2.2}$$

**Lemma 3.2.1** *Let $K$ be a henselian discrete valuation field of characteristic 0 with residue field $F$ of characteristic $p > 0$. Assume that $K$ contains a primitive $p$th root $\zeta_p$ of 1, and let $e = \mathrm{ord}_K p$ be the absolute ramification index and $e' = ep/(p-1) = p \cdot \mathrm{ord}_K(\zeta_p - 1)$.*

*For integers $i > e' = ep/(p-1)$, $U^i\mathrm{H}^1(K, \mu_p) = 0$. For integers $0 \leqslant i < e'$, the surjections (3.2.2) induce isomorphisms*

$$\begin{aligned} K^\times/(1 + \mathfrak{m}_K)K^{\times p} &\to \mathrm{Gr}_U^0 \mathrm{H}^1(K, \mu_p), \\ \mathfrak{m}_K^i/\mathfrak{m}_K^{i+1} &\to \mathrm{Gr}_U^i \mathrm{H}^1(K, \mu_p) \\ &\qquad \text{for } 1 \leqslant i < e' \text{ prime to } p, \\ \mathfrak{m}_K^i/(\mathfrak{m}_K^{i+1} + \{x^p \mid x \in \mathfrak{m}_K^j\}) &\to \mathrm{Gr}_U^i \mathrm{H}^1(K, \mu_p) \\ &\qquad \text{for } 1 \leqslant i = pj < e' \text{ divisible by } p. \end{aligned} \tag{3.2.3}$$

*The mapping $O_K \to \mathrm{Gr}_U^{e'} \mathrm{H}^1(K, \mathbf{F}_p) = U^{e'}\mathrm{H}^1(K, \mathbf{F}_p)$ sending $x$ to $\{1 + (\zeta_p - 1)^p x\}$ induces an isomorphism*

$$F/(\mathrm{F} - 1)F \to \mathrm{Gr}_U^{e'} \mathrm{H}^1(K, \mathbf{F}_p). \tag{3.2.4}$$

*Proof* The proof is similar to that of Lemma 3.1.1. For $x \in \mathfrak{m}_K$, the equation $(1 + (\zeta_p - 1)X)^p = 1 + (\zeta_p - 1)^p x$ is equivalent to

$$X^p + \frac{p}{(\zeta_p - 1)^{p-1}} \sum_{i=1}^{p-1} \binom{p-1}{i-1} \frac{(\zeta_p - 1)^{i-1}}{i} X^i = x$$

and has a unique solution in $\mathfrak{m}_K$ by Hensel's lemma since the coefficient $p/(\zeta_p - 1)^{p-1}$ of $X$ is a unit. Thus, the $p$th power morphism $K^\times \to K^\times$

induces a surjection $U^{e/(p-1)+1}K^\times \to U^{e'+1}K^\times$ and we have $U^i\mathrm{H}^1(K,\mu_p)=0$ for integers $i>e'=\mathrm{ord}_K(\zeta_p-1)^p$.

We consider the $p$th power morphism $K^\times \to K^\times$. Since this induces injections $F^\times \xrightarrow{p} F^\times$ and $p\colon \mathbf{Z}\to\mathbf{Z}$, the induced morphism $\mathrm{Gr}^0_U K^\times \to \mathrm{Gr}^0_U K^\times$ is an injection and we obtain the first isomorphism in (3.2.3). For $i \geqslant 1$ and $v \in \mathfrak{m}_K^i$, we have

$$(1+v)^p \equiv 1+pv+v^p \bmod \mathfrak{m}_K^{2i+e} \tag{3.2.5}$$

by binomial expansion. Hence, for $1 \leqslant j < e/(p-1)$, $i=pj$, and a uniformizer $t \in K$, the $p$th power morphism $K^\times \to K^\times$ induces an injection $\mathrm{Gr}^j_U K^\times = F\cdot t^j \to \mathrm{Gr}^i_U K^\times = F\cdot t^i$. Thus, we have $U^iK^\times \cap K^{\times p} = (U^jK^\times)^p$ for $0 \leqslant i < e'$ and the round-up $j=\lceil i/p\rceil$ by induction on $i$. Further, $\mathrm{Gr}^i_U K^\times \to \mathrm{Gr}^i_U \mathrm{H}^1(K,\mu_p)$ is an isomorphism for $1 \leqslant i < e'$ prime to $p$ and induces an isomorphism $\mathrm{Coker}(\mathrm{Gr}^j_U K^\times \to \mathrm{Gr}^i_U K^\times) \to \mathrm{Gr}^i_U \mathrm{H}^1(K,\mu_p)$ for $0 \leqslant i = pj < e'$ divisible by $p$. Thus, we obtain the last two isomorphisms in (3.2.3).

Since $(\zeta_p-1)^{p-1} \equiv -p$, substituting $v=u(\zeta_p-1)$ in (3.2.5) we obtain $(1+u(\zeta_p-1))^p \equiv 1+(u^p-u)(\zeta_p-1)^p \bmod \mathfrak{m}_K^{e'+1}$. Hence, we obtain the isomorphism (3.2.4). □

**Proposition 3.2.2** *Let $K$ be a henselian discrete valuation field of characteristic 0 with residue field $F$ of characteristic $p>0$. Assume $K$ contains a primitive $p$th root of 1. Let $e=\mathrm{ord}_K p$ be the absolute ramification index and set $e'=ep/(p-1)$. For a cyclic extension $L$ of degree $p$ of $K$, let $G=\mathrm{Gal}(L/K)$ be the Galois group.*

(1) *Let $1 \leqslant i < e' = ep/(p-1)$ be an integer prime to $p$, and let $1+x \in (1+\mathfrak{m}_K^i) - (1+\mathfrak{m}_K^{i+1})$. Then the cyclic extension $L=K(u)$ of degree $p$ of $K$ generated by a solution $u$ of the Kummer equation $(1+u)^p = 1+x$ is a totally ramified extension, and $n=e'-i$ is the largest integer satisfying $G=G_n$.*
(2) *Let $1 \leqslant i = pj < e' = ep/(p-1)$ be an integer divisible by $p$, and let $1+x \in (1+\mathfrak{m}_K^i) - (1+\mathfrak{m}_K^{i+1} + \{z^p \mid z \in \mathfrak{m}_K^j\})$. Then, for the cyclic extension $L=K(u)$ defined by $(1+u)^p = 1+x$, the ramification index $e_{L/K}$ is 1 and the residue field $E$ of $L$ is a purely inseparable extension of degree $p$ of the residue field $F$ of $K$. Define $w \in O_L^\times$ by $u = wt^{-j}$ for a uniformizer $t$ of $K$. Then $E=F(w)$ and $m = e/(p-1)-j$ is the largest integer satisfying $G=G_m$.*
(3) *Let $x \in K^\times$ be an element such that $\mathrm{ord}_K x$ is prime to $p$. Then the cyclic extension $L=K(u)$ defined by $u^p=x$ is a totally ramified extension, and $n=e'$ is the largest integer satisfying $G=G_n$.*
(4) *Let $x \in O_K^\times$ be a unit such that $\bar{x} \in F^\times - F^{\times p}$. Then, for the cyclic*

*extension* $L = K(u)$ *defined by* $u^p = x$, *the ramification index* $e_{L/K}$ *is* 1 *and the residue field* $E = F(u)$ *is a purely inseparable extension of degree* $p$ *of* $F$. *Further,* $m = e/(p-1)$ *is the largest integer satisfying* $G = G_m$.

(5) *Let* $x \in \mathfrak{m}_K^{e'} - (\mathfrak{m}_K^{e'+1} + \{(\zeta_p - 1)^p(z^p - z) \mid z \in O_K\})$. *Then the cyclic extension* $L = K(u)$ *defined by* $(1+u)^p = 1+x$ *is an unramified extension of* $K$ *of degree* $p$ *defined by* $w^p - w = x/(\zeta_p - 1)^p$.

*Proof* (1) Since $u \in \mathfrak{m}_L$, we have $p \cdot \mathrm{ord}_L u = \mathrm{ord}_L u^p = \mathrm{ord}_L x = e_{L/K} \cdot i$ by (3.2.5). Since $i$ is prime to $p$, we have $e = e_{L/K} = p$, $\mathrm{ord}_L u = i$, and $L$ is totally ramified. Let $k, l$ be integers satisfying $kp + li = 1$, and let $t \in K$ be a uniformizer. Then $s = t^k u^l$ is a uniformizer of $L$. For the generator $\sigma \in G$ satisfying $\sigma(1+u) = \zeta_p \cdot (1+u)$, we have $\sigma(u)/u = 1 + (\zeta_p - 1)(1+u)/u$. Hence, $\sigma(s)/s = (1 + (\zeta_p - 1)(1+u)/u)^l \equiv 1 + (\zeta_p - 1)(1+u)/iu$ and $i_G(\sigma) = \mathrm{ord}_L((\zeta_p - 1)(1+u)/iu) = e' - i = n$.

(2) Define a unit $v$ by $x = vt^i \in O_K$ for a uniformizer $t$ of $K$. Then the coefficients of the polynomial $P = (1+Wt^j)^p - (1+vt^i) \in O_K[W]$ are divisible by $t^i$. A solution $w \in O_L^\times$ of $P/t^i = 0$ satisfies $w^p \equiv v$ and we have $u = wt^j$. Hence, the residue field $E = F(\overline{w})$ of $O_L = O_K[w]$ is a purely inseparable extension of $F$ generated by the $p$th root of $\overline{v}$. For the generator $\sigma \in G$ satisfying $\sigma(1+u) = \zeta_p \cdot (1+u)$, we have $\sigma(w)/w = \sigma(u)/u = 1 + (\zeta_p - 1)(1+u)/u$ and $i_G(\sigma) = \mathrm{ord}_L((\zeta_p - 1)(1+u)/u) = e/(p-1) - j = m$.

(3) Assume that $c = \mathrm{ord}_K x$ is prime to $p$. Then we have $p \cdot \mathrm{ord}_L u = e_{L/K} \cdot c$. Hence, we have $e = e_{L/K} = p$, $\mathrm{ord}_L u = c$, and $L$ is totally ramified.

Let $k, l$ be integers satisfying $kp + lc = 1$. Then $s = t^k u^l$ is a uniformizer of $L$. For the generator $\sigma \in G$ satisfying $\sigma(u) = \zeta_p \cdot u$, we have $\sigma(s)/s = \zeta_p^l \equiv 1 - (\zeta_p - 1)/c$ and $i_G(\sigma) = \mathrm{ord}_L((\zeta_p - 1)/c) = e'$.

(4) The residue field $E = F(\overline{u})$ of $O_L = O_K[u]$ is a purely inseparable extension of $F$ generated by the $p$th root of $\overline{x}$. For the generator $\sigma \in G$ satisfying $\sigma(u) = \zeta_p \cdot u$, we have $\sigma(u)/u = \zeta_p \equiv 1 + (\zeta_p - 1)$ and $i_G(\sigma) = \mathrm{ord}_L(\zeta_p - 1) = e/(p-1)$.

(5) Let $x = (\zeta_p - 1)^p \cdot v$ for $v \in O_K$. Since the coefficients of the polynomial $P = (1+(\zeta_p - 1)W)^p - (1+(\zeta_p - 1)^p \cdot v) \in O_K[W]$ are divisible by $(\zeta_p - 1)^p$, a solution $w \in O_L^\times$ of $P = 0$ satisfies $w^p - w \equiv v$ as in the proof of Lemma 3.2.1. Then $u = w(\zeta_p - 1)$ satisfies $(1+u)^p = 1+x$. Hence, $L$ is an unramified extension defined by $w^p - w = x/(\zeta_p - 1)^p$. □

**Lemma 3.2.3** *Let* $0 \leqslant i < e'$ *be an integer and let* $n = e' - i$. *Define morphisms*

$$-d\colon \mathrm{Gr}_U^i K^\times \to \mathrm{Hom}_F(\mathfrak{m}_K^n/\mathfrak{m}_K^{n+1}, \Omega_F^1(\log)) \quad \textit{for } i > 0,$$
$$-d\log\colon \mathrm{Gr}_U^0 K^\times \to \mathrm{Hom}_F(\mathfrak{m}_K^{e'}/\mathfrak{m}_K^{e'+1}, \Omega_F^1(\log))$$

*by sending the class of* $1+x$ *to the morphism* $y \mapsto -\overline{xy/(\zeta_p - 1)^p} \cdot d\log x$ *and by sending the class of* $x$ *to the morphism* $(\zeta_p - 1)^p \mapsto -d\log x$ *respectively. Then there exists a unique morphism*

$$r\colon \mathrm{Gr}_U^i \mathrm{H}^1(K, \mu_p) \to \mathrm{Hom}_F(\mathfrak{m}_K^j/\mathfrak{m}_K^{j+1}, \Omega_F^1(\log)) \tag{3.2.6}$$

*such that the diagram*

$$\begin{array}{ccc} \mathrm{Gr}_U^i K^\times & & \\ \downarrow{\scriptstyle \kappa} & \searrow^{-d \text{ or } -d\log} & \\ \mathrm{Gr}_U^i \mathrm{H}^1(K, \mu_p) & \xrightarrow{\ r\ } & \mathrm{Hom}_F(\mathfrak{m}_K^j/\mathfrak{m}_K^{j+1}, \Omega_F^1(\log)) \end{array} \tag{3.2.7}$$

*is commutative. The morphism*

$$r\colon \mathrm{Gr}_U^i \mathrm{H}^1(K, \mu_p) \to \mathrm{Hom}_F(\mathfrak{m}_K^j/\mathfrak{m}_K^{j+1}, \Omega_F^1(\log))$$

*is an injection.*

*Proof* The proof is similar to that of Lemma 3.1.4. Assume $p \nmid i$. By Lemma 3.2.1, the surjection $\mathrm{Gr}_U^i K^\times \to \mathrm{Gr}_U^i \mathrm{H}^1(K, \mu_p)$ is an isomorphism. Hence, there exists a unique morphism

$$r\colon \mathrm{Gr}_U^i \mathrm{H}^1(K, \mu_p) \to \mathrm{Hom}_F(\mathfrak{m}_K^j/\mathfrak{m}_K^{j+1}, \Omega_F^1(\log))$$

making the diagram (3.2.7) commutative. Since the morphism $\mathrm{Gr}_U^i K^\times \to \mathrm{Hom}_F(\mathfrak{m}_K^j/\mathfrak{m}_K^{j+1}, \Omega_F^1(\log))$ is an injection for $p \nmid n$, $r\colon \mathrm{Gr}_U^i \mathrm{H}^1(K, \mu_p) \to \mathrm{Hom}_F(\mathfrak{m}_K^j/\mathfrak{m}_K^{j+1}, \Omega_F^1(\log))$ is also an injection.

Assume $p \mid i = pk > 0$. By Lemma 3.2.1, we have an exact sequence $0 \to \mathrm{Gr}_U^k K^\times \to \mathrm{Gr}_U^i K^\times \to \mathrm{Gr}_U^i \mathrm{H}^1(K, \mu_p) \to 0$. By the exact sequence (2.4.2), we have an exact sequence

$$0 \to \mathrm{Gr}_U^k K^\times \to \mathrm{Gr}_U^i K^\times \to \mathrm{Hom}_F(\mathfrak{m}_K^j/\mathfrak{m}_K^{j+1}, \Omega_F^1(\log)).$$

Hence, there exists a unique injection $r$ (3.2.6) making the diagram (3.1.5) commutative.

For $i = 0$, $K^\times/K^{\times p}(1 + \mathfrak{m}_K) \to \mathrm{Gr}_U^0 \mathrm{H}^1(K, \mu_p)$ is an isomorphism by Lemma 3.2.1 and $d\log\colon K^\times/K^{\times p}(1 + \mathfrak{m}_K) \to \Omega_F^1(\log)$ is an injection by Proposition 2.4.6(2). Hence, the assertion follows. □

We introduce some notation on Galois cohomology, define filtrations on $\mathrm{H}^2(K, \mu_n^{\otimes 2})$ for integers $n \geqslant 1$ invertible in $K$, and study their properties.

Let $K$ be a field and $n \geqslant 1$ be an integer invertible in $K$. Let $\mu_n$ denote the group of $n$th roots of unity, and let $K^\times/K^{\times n} \to \mathrm{H}^1(K, \mu_n)$ denote the isomorphism defined by the Kummer sequence $0 \to \mu_n \to \mathbf{G}_m \to \mathbf{G}_m \to 0$.

For $x \in K^\times$, let $\{x\}$ denote its image in $\mathrm{H}^1(K, \mu_n)$. For $x, y \in K^\times$, let $\{x, y\} \in \mathrm{H}^2(K, \mu_n^{\otimes 2})$ denote the cup-product $\{x\} \cup \{y\}$.

**Lemma 3.2.4** *Let $K$ be a field and $n \geqslant 1$ be an integer invertible in $K$.*

(1) *For $x \in K^\times - \{1\}$, $\{x, 1-x\} = 0$ in $\mathrm{H}^2(K, \mu_n^{\otimes 2})$.*
(2) *For $x \in K^\times$, $\{x, -x\} = 0$.*
(3) *For $x, y \in K^\times$, $\{x, y\} = -\{y, x\}$.*

*Proof* (1) Since $A = K[u]/(u^n - x)$ is étale over $K$ by Lemma 1.1.6, $A$ is decomposed as a product $\prod_i L_i$ of finite separable extensions of $K$ by Proposition 1.1.8(2) (a) $\Rightarrow$ (b). Let $u_i \in L_i$ be the projection of $u \in A$. Since $1 - x = \mathrm{N}_{A/K}(1-u) = \prod_i \mathrm{N}_{L_i/K}(1-u_i)$, we have $\{x, 1-x\} = \sum_i \{x, \mathrm{N}_{L_i/K}(1-u_i)\} = \sum_i \mathrm{Cor}_{L_i/K}\{u_i^n, 1-u_i\} = n \cdot \sum_i \mathrm{Cor}_{L_i/K}\{u_i, 1-u_i\} = 0$ by Lemma 2.1.5(2).

(2) If $x \neq 1$, we have $\{x, -x\} = \{x, -x\} - \{x, 1-x\} = \{1/x, 1 - 1/x\} = 0$ by (1). If $x = 1$, we have $\{1, -1\} = 0$.

(3) For $x, y \in K^\times$, we have $\{xy, -xy\} = 0$ by (2). From this and (2), we obtain $\{x, y\} + \{y, x\} = 0$. □

Let $K$ be a discrete valuation field. For an integer $n \geqslant 1$ invertible in $K$, the subgroup $U^i\mathrm{H}^1(K, \mu_n)$ is defined as the image of $1 + \mathfrak{m}_K^i \subset K^\times$ for $i \geqslant 1$ by the boundary morphism $K^\times \to \mathrm{H}^1(K, \mu_n)$ of the Kummer sequence $0 \to \mu_n \to \mathbf{G}_m \to \mathbf{G}_m \to 0$, and $U^0\mathrm{H}^1(K, \mu_n) = \mathrm{H}^1(K, \mu_n)$. For an integer $n \geqslant 1$ invertible in $K$, we define a decreasing filtration $U^\bullet\mathrm{H}^2(K, \mu_n^{\otimes 2})$ by

$$U^i\mathrm{H}^2(K, \mu_n^{\otimes 2}) = \{U^i\mathrm{H}^1(K, \mu_n), \mathrm{H}^1(K, \mu_n)\}$$

for $i \geqslant 0$.

**Lemma 3.2.5** *Let $i, j \geqslant 1$ be integers.*

(1) $\{U^iK^\times, U^jK^\times\} \subset U^{i+j}\mathrm{H}^2(K, \mu_n^{\otimes 2})$.
(2) *For $1 + x \in U^iK^\times$ and $1 + y \in U^jK^\times$, we have*

$$\{1+x, 1+y\} \equiv \{1 - xy, -x\} \bmod U^{i+j+1}\mathrm{H}^2(K, \mu_n^{\otimes 2}).$$

*Proof* (1) For $1 + x \in U^iK^\times$ and $1 + y \in U^jK^\times$, by Lemma 3.2.4 we have

$$\begin{aligned}\{1+x, 1+y\} &= \{1+x, -x(1+y)\} = \{(1+x)/(1+x(1+y)), -x(1+y)\} \\ &= -\{1 + xy/(1+x), -x(1+y)\}.\end{aligned}$$

The right-hand side is an element of $U^{i+j}\mathrm{H}^2(K, \mu_n^{\otimes 2})$.

(2) Further, the right-hand side of the equality in the proof of (1) is congruent to $\{1 - xy, -x\} \bmod U^{i+j+1}\mathrm{H}^2(K, \mu_n^{\otimes 2})$ by (1). □

**Lemma 3.2.6** *Let $K$ be a henselian discrete valuation field of characteristic* 0 *with residue field $F$ of characteristic $p > 0$. Assume that $K$ contains a primitive pth root $\zeta_p$ of* 1. *We identify $\mathrm{H}^1(K, \mu_p) = X_K[p]$ and $\mathrm{H}^2(K, \mu_p^{\otimes 2}) = \mathrm{Br}\, K[p]$ by fixing a primitive pth root $\zeta_p \in \mu_p$ of* 1. *Let $e = \mathrm{ord}\, p$ be the absolute ramification index and set $e' = pe/(p-1)$.*

(1) *We have $U^{e'+1}\mathrm{H}^1(K, \mu_p) = 0$, and hence $U^{e'+1}\mathrm{H}^2(K, \mu_p^{\otimes 2}) = 0$.*
(2) *For $x \in \mathfrak{m}_K^{e'}$, the character $\kappa_{1+x} \in U^{e'}\mathrm{H}^1(K, \mu_p)$ equals*

$$\alpha\Big(\frac{x}{(\zeta_p - 1)^p}\Big) \in F_0 X_K[p] = F_0 X_K \cap X_K[p].$$

(3) *Let $0 \leqslant i < e' = pe/(p-1)$ be an integer and set $n = e' - i$. For $x \in \mathfrak{m}_K^i$ and $y \in \mathfrak{m}_K^n$, we have*

$$\{1 + x, 1 + y\} = -\beta\Big(\frac{xy}{(\zeta_p - 1)^p} \cdot d\log x\Big)$$

*in* $\mathrm{Br}\, K[p]$.
(4) *For $x \in K^\times$ and $y \in \mathfrak{m}_K^{e'}$, we have*

$$\{x, 1 + y\} = -\beta\Big(\frac{y}{(\zeta_p - 1)^p} \cdot d\log x\Big)$$

*in* $\mathrm{Br}\, K[p]$.
(5) *For integers $0 \leqslant i \leqslant e'$ and $n = e' - i$, we have $U^i\mathrm{H}^1(K, \mu_p) \subset F_n X_K \cap X_K[p]$.*

*Proof* (1) This is proved in Lemma 3.2.1.

(2) This is proved in Proposition 3.2.2(5).

(3) We have $\{1 + x, 1 + y\} \equiv \{1 - xy, -x\} \bmod U^{e'+1}\mathrm{H}^2(K_1, \mu_p^{\otimes 2}) = 0$ by Lemma 3.2.5(2) and (1). By (2), the right-hand side equals

$$\{\alpha\Big(\frac{-xy}{(\zeta_p - 1)^p}\Big), -x\} = -\beta\Big(\frac{xy}{(\zeta_p - 1)^p} \cdot d\log x\Big).$$

(4) We have $\{x, 1+y\} = -\{1+y, x\}$ by Lemma 3.2.4(3). By (2), the right-hand side equals

$$-\{\alpha\Big(\frac{y}{(\zeta_p - 1)^p}\Big), x\} = -\beta\Big(\frac{y}{(\zeta_p - 1)^p} \cdot d\log x\Big).$$

(5) By Lemma 3.2.5(1) and (1), we have an inclusion

$$\{U^i\mathrm{H}^1(K, \mu_p), U^{n+1}K_1^\times\} \subset U^{e'+1}\mathrm{H}^2(K_1, \mu_p^{\otimes 2}) = 0. \qquad \square$$

We study the refined Swan conductor of a wildly ramified character of order $p$ and the graded quotients of the filtration $U^\bullet\mathrm{H}^1(K, \mu_p)$.

**Proposition 3.2.7** *Assume that $K$ contains a primitive pth root $\zeta$ of 1, and identify $\mathrm{H}^1(K,\mu_p) = X_K[p]$ and $\mathrm{H}^2(K,\mu_p^{\otimes 2}) = \mathrm{Br}\, K[p]$ by fixing a primitive pth root $\zeta_p \in \mu_p$ of 1. Let $e = \mathrm{ord}_K p$ be the absolute ramification index and set $e' = pe/(p-1)$. For $x \in K^\times$, let $\kappa_x = \{x\} \in X_K[p] = \mathrm{H}^1(K,\mu_p)$ denote the associated character. For integers $n \geqslant 0$, let $F_n X_K[p]$ denote the intersection $F_n X_K \cap X_K[p]$.*

(1) *Let $0 \leqslant i < e'$, and let $\chi \in U^i\mathrm{H}^1(K,\mu_p) - U^{i+1}\mathrm{H}^1(K,\mu_p)$ be a character. Then we have $\mathrm{Sw}\,\chi = n = e' - i$. The image*

$$r(\chi) \in \mathrm{Hom}(\mathfrak{m}_K^n/\mathfrak{m}_K^{n+1}, \Omega_F^1(\log))$$

*by (3.2.6) is the refined Swan conductor $\mathrm{rsw}\,\chi$.*

(2) *For integers $0 \leqslant i \leqslant e'$ and $n = e' - i$, we have*

$$U^i\mathrm{H}^1(K,\mu_p) = F_n X_K[p] \tag{3.2.8}$$

*and*

$$F_0 X_K[p] = \mathrm{Im}(X_F[p] \to X_K[p]). \tag{3.2.9}$$

*Proof* (1) First, we consider the case $1 \leqslant i < e'$. Let $x \in \mathfrak{m}_K^i$ be an element such that $\kappa_{1+x} = \chi \in U^i\mathrm{H}^1(K,\mu_p)$. By Lemma 3.2.6(3), we have

$$\{1+x, 1+y\} = -\beta\Big(\frac{xy}{(\zeta_p - 1)^p} \cdot d\log x\Big)$$

for $y \in \mathfrak{m}_K^n$ in $\mathrm{Br}\, K[p]$. Similarly, we have

$$\{1+x, 1+y_1\} = -\beta\Big(\frac{xy_1}{(\zeta_p - 1)^p} \cdot d\log x\Big)$$

for $y_1 \in \mathfrak{m}_{K_1}^n$ in $\mathrm{Br}\, K_1[p]$. Since $r(\chi) \neq 0$ in $\Omega_F^1(\log)$ by Lemma 3.2.3, we have $\mathrm{Sw}\,\chi = n$ and $\mathrm{rsw}\,\chi = r(\chi)$.

Let $x \in K^\times$ be an element such that $\kappa_x = \chi \in \mathrm{H}^1(K,\mu_p)$. By Lemma 3.2.6(3), we have

$$\{x, 1+y\} = -\beta\Big(\frac{y}{(\zeta_p - 1)^p} \cdot d\log x\Big)$$

in $\mathrm{Br}\, K[p]$ for $y \in \mathfrak{m}_K^{e'}$. We also have

$$\{x, 1+y_1\} = -\beta\Big(\frac{y_1}{(\zeta_p - 1)^p} \cdot d\log x\Big)$$

in $\mathrm{Br}\, K_1[p]$ for $y_1 \in \mathfrak{m}_{K_1}^{e'}$. Since $r(\chi) \neq 0$ in $\Omega_F^1(\log)$ by Lemma 3.2.3, we have $\mathrm{Sw}\,\chi = e'$ and $\mathrm{rsw}\,\chi = r(\chi)$.

(2) By (1) and Lemma 3.2.6(5), we have $\{\chi \in \mathrm{H}^1(K,\mu_p) \mid \mathrm{Sw}\,\chi \leqslant n\} = \mathrm{Gr}_U^i\mathrm{H}^1(K,\mu_p)$ for $n = e' - i$. Hence, we have (3.2.8).

By Lemma 3.2.6(2), we have a commutative diagram

$$\begin{array}{ccc} F & \xrightarrow{a} & X_F[p] \\ \big\downarrow & & \big\downarrow \\ U^{e'}\mathrm{H}^1(K,\mu_p) & \longrightarrow & F_0X_K[p] \end{array} \tag{3.2.10}$$

where the left vertical arrow is the surjection sending the class of $x \in O_K$ to the class of $1 + x(\zeta_p - 1)^p$. Since the bottom horizontal arrow is an isomorphism, the right vertical injection is also an isomorphism and we obtain the equality (3.2.9). □

**Proposition 3.2.8** *Let $K$ be a henselian discrete valuation field of characteristic 0 with residue field $F$ of characteristic $p > 0$ containing a primitive pth root of 1. We fix a primitive pth root $\zeta_p$ of 1 and identify $X_K[p] = \mathrm{H}^1(K, \mu_p)$.*

(1) *Let $1 \leqslant i < e' = p/(p-1) \cdot \mathrm{ord}_K p$ be an integer and $x \in \mathfrak{m}_K^i$ such that $\mathrm{Sw}\,\kappa_{1+x} = n = e' - i$. Let $L = K(u)$ be the cyclic extension of $K$ generated by a solution $u \in L$ of the equation $(1+u)^p = 1+x$. Let $\sigma \in G = \mathrm{Gal}(L/K)$ be the generator satisfying $1+\sigma(u) = \zeta_p \cdot (1+u)$, and $m \geqslant 1$ be the largest integer satisfying $G_m = G$. Set $v = \sigma(u)/u - 1 \in L$, $w = \mathrm{N}_{L/K} v$, and $z = \mathrm{N}_{L/K} u \in K$.*

   *Then we have $n = f_{L/K} \cdot m$, $\mathrm{ord}_L(v) = m$, and $\mathrm{ord}_K(w) = n$. The refined Swan conductor $\mathrm{rsw}\,\kappa_{1+x}\colon \mathfrak{m}_K^n/\mathfrak{m}_K^{n+1} \to \Omega_F^1(\log)$ maps a basis $w$ to $-d\log z$.*

(2) *Let $x \in K^\times$ such that $\mathrm{Sw}\,\kappa_x = e'$. Let $L = K(u)$ be the cyclic extension of $K$ generated by a solution $u \in L$ of the equation $u^p = x$. Let $\sigma \in G = \mathrm{Gal}(L/K)$ be the generator satisfying $\sigma(u) = \zeta_p \cdot u$, and $m \geqslant 1$ be the largest integer satisfying $G_m = G$. Set $v = \sigma(u)/u - 1 \in L$, $w = \mathrm{N}_{L/K} v$, and $z = \mathrm{N}_{L/K} u \in K$.*

   *Then we have $e' = f_{L/K} \cdot m$, $\mathrm{ord}_L(v) = m$, and $\mathrm{ord}_K(w) = e'$. The refined Swan conductor $\mathrm{rsw}\,\kappa_x\colon \mathfrak{m}_K^{e'}/\mathfrak{m}_K^{e'+1} \to \Omega_F^1(\log)$ maps a basis $-w$ to $d\log z$.*

*Proof* (1) We have $n = f_{L/K} \cdot m$ by Proposition 3.2.2. We have $v = (\zeta_p - 1)(1+u)/u$, $w = (\zeta_p - 1)^p(1+x)/x$, $z = (-1)^{p-1}x$ and $\mathrm{ord}_L v = -\mathrm{ord}_L u = -e_{L/K}/p \cdot \mathrm{ord}_K x = n/f_{L/K} = m$, $\mathrm{ord}_K(w) = f_{L/K}\mathrm{ord}_L v = n$. The refined Swan conductor $\mathrm{rsw}\,\kappa_{1+x}$ maps $-w \equiv -(\zeta_p - 1)^p/x$ to $d\log x = d\log z$ by Proposition 3.2.7(1).

(2) We have $f_{L/K} \cdot m = e'$ by Proposition 3.2.2. We have $v = \zeta_p - 1$, $w = (\zeta_p - 1)^p$, $z = (-1)^{p-1}x$, and $\mathrm{ord}_L(\zeta_p - 1) = e_{L/K}/(p-1) \cdot \mathrm{ord}_K p = e'/f_{L/K} = m$, $\mathrm{ord}_K(w) = f_{L/K}\mathrm{ord}_L v = e'$. The refined Swan conductor $\mathrm{rsw}\,\kappa_x$ maps $-w = -(\zeta_p - 1)^p$ to $d\log x = d\log z$ by Proposition 3.2.7(1). □

**Exercise 3.2** Let $p$ be a prime number and $n \geqslant 2$ be an integer. Set $K = \mathbf{Q}_p(\zeta_{p^{n-1}})$ and $L = \mathbf{Q}_p(\zeta_{p^n})$. Let $\sigma$ be the generator of $G = \mathrm{Gal}(L/K)$ defined by $\sigma(\zeta_{p^n}) = \zeta_p \cdot \zeta_{p^n}$, and let $\chi\colon G \to \mathbf{Q}/\mathbf{Z}$ be the faithful character sending $\sigma$ to $1/p$. Compute $m = \mathrm{Sw}\,\chi \in \mathbf{N}$ and $\mathrm{rsw}\,\chi\colon \mathfrak{m}_K^m/\mathfrak{m}_K^{m+1} \to \Omega^1_{\mathbf{F}_p}(\log) = \mathbf{F}_p$.

## Historical Notes

The proofs of Corollary 3.1.7 for Artin–Schreier extensions and of Proposition 3.2.7 for Kummer extensions are taken from [50, §3] and [50, §4] respectively.

# 4

# Trace of Differential Forms

In the proof by induction on the degree of a relation between the cohomological filtration and the lower numbering filtration in the next chapter, a key tool, even in the classical situation where the residue field extension is separable, is the trace mapping of differential forms with log poles. In this preliminary chapter we prepare some generalities on the trace mapping.

We introduce logarithmic complete intersections in Section 4.1 and study their duality in Section 4.2. We introduce the trace mapping of differential forms with log poles in Section 4.3 and study the trace map for inseparable extensions of fields in Section 4.4. The reader who is ready to admit the existence of trace mapping satisfying Propositions 4.4.1 and 5.1.1 along with Lemma 5.1.2(2) may skip the first three sections.

## 4.1 Logarithmic Complete Intersection

**Definition 4.1.1** Let $A$ be a ring and $B$ be a finitely generated ring over $A$.

(1) We say that $B$ is quasi-finite over $A$ if, for every prime ideal $\mathfrak{p}$ of $A$ and the residue field $k(\mathfrak{p})$, the $k(\mathfrak{p})$-vector space $B \otimes_A k(\mathfrak{p})$ is of finite dimension.

(2) We say that $B$ is smooth of relative dimension $n$ over $A$ if, locally on Spec $B$, there exists an étale morphism $A[X_1, \ldots, X_n] \to B$ over $A$.

**Lemma 4.1.2** *Let $A$ be a henselian local ring with the maximal ideal $\mathfrak{m}$, and $B$ be a quasi-finite ring over $A$. Then there exists an isomorphism $B \to B_1 \times B_2$ such that $B_1$ is finite over $A$ and $B_2/\mathfrak{m}B_2 = 0$.*

*Proof* We prove the assertion by induction on the cardinality of Spec $B/\mathfrak{m}B$. Assume that Spec $B/\mathfrak{m}B \neq \varnothing$ and let $\mathfrak{n} \subset B$ be the inverse image of a maximal ideal of $B/\mathfrak{m}B$. Then, by Theorem 1.1.9, there exist a subring $B' \subset B$, finite over $A$, and an element $b \in B' - (B' \cap \mathfrak{n})$ such that $B'[1/b] \to B[1/b]$ is an isomorphism. Let $\mathfrak{n}' = \mathfrak{n} \cap B'$. Since $A$ is a henselian local ring, we have an isomorphism $B' \to B'_{\mathfrak{n}'} \times B'_3$ by Proposition 1.3.5 (1) $\Rightarrow$ (2). The isomorphism $B' \to B'_{\mathfrak{n}'} \times B'_3$ induces an isomorphism $B \to B_{\mathfrak{n}} \times B_3$. Hence, the assertion follows by induction. □

**Proposition 4.1.3** *Let $A$ be a ring and $B$ be a quasi-finite ring over $A$. Then the following conditions are equivalent:*

(1) *Let $\widetilde{A}$ be a smooth ring of relative dimension $n$ over $A$, and let $\widetilde{A} \to B[b^{-1}]$ for $b \in B$ be a surjection of rings over $A$. Then, locally on* Spec $\widetilde{A}$, *the kernel is generated by $n$ elements.*
(2) *There exists a surjection $A[X_1, \ldots, X_n] \to B$ of rings over $A$ such that, locally on* Spec $A[X_1, \ldots, X_n]$, *the kernel is generated by $n$ elements.*

Note that the number $n$ of relations equals the number of indeterminants in (2).

*Proof* Since there exists a surjection $A[X_1, \ldots, X_n] \to B$ over $A$, and since $A[X_1, \ldots, X_n]$ is smooth over $A$, the implication (1) $\Rightarrow$ (2) is clear.

We show (2) $\Rightarrow$ (1). Let $A_1 = A[Y_1, \ldots, Y_m] \to B$ be a surjection over $A$ such that the kernel $I_1 = \operatorname{Ker}(A_1 \to B)$ is locally generated by $m$ elements. Let $g \in A_1$ be an element lifting $b \in B$. Then the kernel of the surjection $A_1[Y_{m+1}] \to A_1[Y_{m+1}]/(gY_{m+1} - 1) = A_1[g^{-1}] \to B[b^{-1}]$ is locally generated by $m+1$ elements. Hence, by replacing the surjection $A_1 \to B$ by $A_1[Y_{m+1}] \to B[b^{-1}]$, we may assume $B[b^{-1}] = B$.

Set $\widetilde{A}_1 = \widetilde{A} \otimes_A A_1$ and let $\widetilde{A}_1 \to B$ be the induced morphism. By the assumption, locally on Spec $B$, the conormal module $N_{B/A_1} = I_1/I_1^2$ is generated by $m$ elements. By the exact sequence $0 \to N_{B/A_1} \to N_{B/\widetilde{A}_1} \to \Omega^1_{\widetilde{A}_1/A_1} \otimes_{\widetilde{A}_1} B \to 0$, locally on Spec $B$, the conormal module $N_{B/\widetilde{A}_1}$ is generated by $n+m$ elements. Hence, by the split exact sequence $0 \to N_{B/\widetilde{A}} \to N_{B/\widetilde{A}_1} \to \Omega^1_{\widetilde{A}_1/\widetilde{A}} \otimes_{\widetilde{A}_1} B \to 0$, the conormal module $N_{B/\widetilde{A}}$ is generated by $n$ elements locally on Spec $B$. Since $B$ is of finite presentation over $A$, the kernel $\widetilde{I}$ of $\widetilde{A} \to B$ is of finite type and hence, locally on Spec $\widetilde{A}$, is generated by $n$ elements by Nakayama's lemma. □

**Definition 4.1.4** Let $A$ be a ring and $B$ be a quasi-finite ring over $A$. We say that $B$ is locally of complete intersection over $A$ if the equivalent conditions in Proposition 4.1.3 are satisfied.

The composition of quasi-finite morphisms locally of complete intersection is quasi-finite locally of complete intersection.

**Proposition 4.1.5** *Let $A$ be a ring and $B$ be a quasi-finite ring locally of complete intersection over $A$.*

(1) *Let $\widetilde{A}$ be a smooth ring of relative dimension $n$ over $A$, and let $\widetilde{A} \to B$ be a surjection over $A$. Let $\mathfrak{q}$ be a prime ideal of $B$, and let $\mathfrak{p} \subset A$ and $\widetilde{\mathfrak{p}} \subset \widetilde{A}$ be the inverse images of $\mathfrak{q}$. Assume that the kernel of the surjection $\widetilde{A}_{\widetilde{\mathfrak{p}}} \to B_{\mathfrak{q}}$ is generated by $f_1, \ldots, f_n$. Then the image $f_i \in \widetilde{A}_{\widetilde{\mathfrak{p}}}/(f_1, \ldots, f_{i-1})$ is a non-zero divisor for every $i = 1, \ldots, n$, and $\widetilde{A}_{\widetilde{\mathfrak{p}}}/(f_1, \ldots, f_n) = B_{\mathfrak{q}}$ is flat over the local ring $A_{\mathfrak{p}}$.*

(2) *$B$ is flat over $A$.*

*Proof* (1) First, we assume that $A$ is noetherian. We may assume that $A$ is local by replacing $A$ by the localization $A_{\mathfrak{p}}$. Let $k$ be the residue field of $A$. Then, since the regular ring $\widetilde{A}_{\mathfrak{p}} \otimes_A k$ is Cohen–Macaulay by [31, Chapitre $0_{\text{IV}}$, Corollaire (17.1.3)], for every $i = 1, \ldots, n$ the image $f_i \in \widetilde{A}_{\mathfrak{p}}/(f_1, \ldots, f_{i-1}) \otimes_A k$ is a non-zero divisor by [31, Corollaire (16.5.6) b) $\Rightarrow$ a)]. Since $\widetilde{A}_{\mathfrak{p}}$ is flat over $A$, the quotient $\widetilde{A}_{\mathfrak{p}}/(f_1, \ldots, f_n)$ is flat over $A$ by Proposition 1.1.5 (2) $\Rightarrow$ (1).

We show the general case. The ring $A$ is the filtered inductive limit of subrings of finite type over $\mathbf{Z}$. Hence, it is reduced to the case where $A$ is of finite type over $\mathbf{Z}$ and hence is noetherian.

(2) It suffices to take a surjection $A[X_1, \ldots, X_n] \to B$ and apply (1). □

We introduce a logarithmic variant.

**Definition 4.1.6** (1) We call a pair $(A, I)$ of a ring $A$ and an invertible ideal $I$ a log ring.

For a log ring $(A, I)$, let $A[I^{-1}]$ be the inductive limit $\varinjlim I^{\otimes -n}$ with respect to the natural transition mappings $I^{\otimes -n} \to I^{\otimes -n-1}$.

(2) Let $(A, I)$ and $(B, J)$ be log rings. We say that a pair $(f, e)$ of a morphism $f\colon A \to B$ of rings and an integer $e \geqslant 1$ is a morphism of log rings $(A, I) \to (B, J)$ if $f(I)B = J^e$. If $e = 1$, we say that $f$ is strict.

Let $(f, e)\colon (A, I) \to (B, J)$ be a morphism of log rings, and let

$$\begin{array}{ccc} \mathbf{N} & \longrightarrow & \mathbf{N} \times \mathbf{Z} \\ \downarrow & & \downarrow \\ A & \longrightarrow & B \end{array} \tag{4.1.1}$$

be a commutative diagram of monoids where $A$ and $B$ are regarded as monoids by the multiplications. We say that (4.1.1) is a chart of $(f, e)\colon A \to B$ if the following conditions are satisfied:

(a) The images of $1 \in \mathbf{N}$ and $(1, 0) \in \mathbf{N} \times \mathbf{Z}$ in $A$ and $B$ are generators of the ideals $I$ and $J$ respectively.

(b) The image of $1 \in \mathbf{N}$ in $\mathbf{N} \times \mathbf{Z}$ is $(e, c)$ for an integer $c$ relatively prime to $e$.

If the image of $1 \in \mathbf{N}$ is $t \in A$ and the images of $(1,0), (0,1) \in \mathbf{N} \times \mathbf{Z}$ are $s, u \in B$, we say that the chart (4.1.1) is defined by $u^c s^e = t$.

We may consider every ring $A$ as a log ring $(A, A)$. For a log ring $(A, I)$, the complement $\operatorname{Spec} A - \operatorname{Spec} A/I$ is $\operatorname{Spec} A[I^{-1}]$. In the following, we often omit the invertible ideal $I$ to denote a log ring.

**Lemma 4.1.7** *Let $(f, e)\colon (A, I) \to (B, J)$ be a morphism of log rings such that $B$ is of finite presentation over $A$. Assume that $B[J^{-1}]$ is flat over $A[I^{-1}]$ and $B/J$ is flat over $A/I$. Then $B$ is flat over $A$.*

*Proof* Since $A$ is the inductive limit of subrings finitely generated over $\mathbf{Z}$, we may assume that $A$ is noetherian by [31, Chapitre IV, Corollaire (11.2.6.1)]. Since $B/J$ is flat over $A/I$, $B/J^e = B/IB$ is also flat over $A/I$. Since $I$ and $J$ are invertible ideals and $B[J^{-1}]$ is flat over $A[I^{-1}]$, $B$ is flat over $A$ by [14, Chapitre III, §5, Théorème 1]. □

**Lemma 4.1.8** *Let $(A, I)$ be a log ring and let $t \in I$ be a generator.*

(1) *Let $e \geqslant 1$ and $c$ be relatively prime integers. Then the ring*

$$\widetilde{A} = A[X_1, \ldots, X_{n-1}, S, U^{\pm 1}]/(U^c S^e - t)$$

*is flat over $A$ and $\widetilde{I} = S\widetilde{A}$ is an invertible ideal. The morphisms $A[t^{-1}] \to \widetilde{A}[S^{-1}]$ and $A/I[U^{\pm 1}] \to \widetilde{A}/\widetilde{I}$ are smooth.*

(2) *Let $(f, e)\colon (A, I) \to (B, J)$ be a morphism of log rings and assume that $B$ is of finite presentation over $A$. Then the following conditions are equivalent:*

(a) *Locally on* Spec $A$ *and on* Spec $B$, *there exists a chart* (4.1.1) *such that $A \otimes_{\mathbf{Z}[\mathbf{N}]} \mathbf{Z}[\mathbf{N} \times \mathbf{Z}] \to B$ is smooth.*

(b) *$B[J^{-1}]$ is smooth over $A[I^{-1}]$ and, locally on* Spec $A$ *and on* Spec $B$, *there exists a chart* (4.1.1) *such that $A/I[\{0\} \times \mathbf{Z}] \to B/J$ is smooth.*

*If these equivalent conditions are satisfied, $B$ is flat over $A$.*

(3) *Let $(f, e)\colon (A, I) \to (B, J)$ be a morphism of log rings such that $B$ is quasi-finite over $A$. Then, locally on* Spec $A$, *there exists a strict surjection*

$$\widetilde{A} = A[X_1, \ldots, X_{n-1}, S, U^{\pm 1}]/(US^e - t) \to B \tag{4.1.2}$$

*of log rings over $A$.*

*Proof* (1) We may assume $n = 1$. The morphism $\mathbf{Z}[T] \to \mathbf{Z}[S, U^{\pm 1}]$ sending $T$ to $U^c S^e$ is the composition of flat morphisms $\mathbf{Z}[T] \to \mathbf{Z}[T, U^{\pm 1}] \to \mathbf{Z}[S, U^{\pm 1}]$ and is flat. Hence, $\widetilde{A} = A \otimes_{\mathbf{Z}[T]} \mathbf{Z}[S, U^{\pm 1}]$ is flat over $A$. Since $\widetilde{A}$ is flat over $A$, the image of $t$ in $\widetilde{A}$ is a non-zero divisor and hence a divisor $S$ of $t$ is also a non-zero divisor.

Since $e$ and $c$ are relatively prime, the morphism $\mathbf{Z}[T^{\pm 1}] \to \mathbf{Z}[S^{\pm 1}, U^{\pm 1}]$ is smooth. Hence, $\widetilde{A}[S^{-1}] = A[t^{-1}] \otimes_{\mathbf{Z}[T^{-1}]} \mathbf{Z}[S^{-1}, U^{\pm 1}]$ is smooth over $A[t^{-1}]$. We have $\widetilde{A}/\widetilde{I} = A/I[U^{\pm 1}]$.

(2) The implication (a) $\Rightarrow$ (b) follows from (1).

(b) $\Rightarrow$ (a): By Lemma 4.1.7, condition (b) implies that $B$ is flat over $A$.

Since the assertion is clear on Spec $B[J^{-1}]$, it suffices to show the assertion on a neighborhood of Spec $B/J$. We may assume that there exist generators $t \in I, s \in J$, units $u \in B^\times$, and an integer $c \geqslant 1$ relatively prime to $e$ satisfying $s^e = u^c t$ such that $B/J$ is smooth over $A/I[u^{-1}]$. We consider the morphism $A[S, U^{\pm 1}]/(U^c S^e - t) \to B$ sending $S, U$ to $s, u$. Since $A[S, U^{\pm 1}]/(U^c S^e - t)/(S) = A/I[U^{\pm 1}] \to B/J$ is smooth and $B$ is flat over $A$, the morphism $A[S, U^{\pm 1}]/(U^c S^e - t) \to B$ is smooth on a neighborhood of $B/J$ by [31, IV, Chapitre 17, Théorème (17.5.1) b) $\Rightarrow$ a)].

(3) For every prime ideal $\mathfrak{p}$ of $A$, since $B$ is quasi-finite over $A$, the $B \otimes_A k(\mathfrak{p})$-module $J \otimes_A k(\mathfrak{p})$ is a free module of rank 1. Hence, by Nakayama's lemma, we may assume that the ideal $J$ is a principal ideal. Let $s \in J$ be a generator and define a unit $u \in B^\times$ by $us^e = t$. By taking a system $b_1, \ldots, b_{n-1} \in B$ of generators over $A$, we obtain a surjection (4.1.2). □

**Definition 4.1.9** Let $(f, e) \colon (A, I) \to (B, J)$ be a morphism of log rings. We say that $B$ is log smooth over $A$ if the equivalent conditions in Lemma 4.1.8(2) are satisfied.

The definition of log smoothness is more restrictive than the usual one. In the following, we consider $\widetilde{A} = A[X_1, \ldots, X_{n-1}, S, U^{\pm 1}]/(U^c S^e - t)$ as a log ring log smooth over $(A, I)$ with the invertible ideal $\widetilde{I} = S\widetilde{A}$. If a strict morphism $(f, 1) \colon (A, I) \to (B, J)$ is log smooth then $A \to B$ is smooth since $f$ is flat.

**Lemma 4.1.10** *If $(f, e) \colon (A, I) \to (B, J)$ and $(g, l) \colon (B, J) \to (C, K)$ are log smooth morphisms of log rings, then the composition $(g \circ f, el) \colon (A, I) \to (C, K)$ is log smooth.*

*Proof* Since $B[J^{-1}]$ and $C[K^{-1}]$ are smooth over $A[I^{-1}]$ and $B[J^{-1}]$ respectively, $C[K^{-1}]$ is smooth over $A[I^{-1}]$. We may assume that there exist charts defined by $u^a s^e = t$ and $v^b z^l = s'$ such that $B/J$ and $C/K$ are smooth over $A/I[u^{\pm 1}]$ and $B/J[v^{\pm 1}]$ respectively. Let $w \in B^\times$ be the unit satisfying $s = ws'$. Since $b$ and $l$ are relatively prime, there exist integers $m, n$ such that $bm + ln = 1$. By replacing $v, z$ by $w^m v, w^n z$, we may assume $s' = s$.

Let $c = (a, b)$ be the greatest common divisor and set $a = ca', b = cb'$. Since $(a, e) = (b, l) = 1$, $c$ and $el$ are relatively prime and $a'$ and $b'e$ are relatively prime. We set $w = u^{a'} v^{b'e}$. Then we have $w^c z^{el} = t$. By the assumption, $C/K$ is smooth over $B/J[v^{\pm1}]$ and over $A/I[u^{\pm1}, v^{\pm1}]$. Since $A/I[u^{\pm1}, v^{\pm1}]$ is smooth over $A/I[w^{\pm1}]$, $C/K$ is smooth over $A/I[w^{\pm1}]$. □

**Proposition 4.1.11** *Let $(A, I) \to (B, J)$ be a morphism of log rings. Assume that $B$ is quasi-finite of finite presentation over $A$. Then the following conditions are equivalent:*

(1) *Locally on* Spec $A$, *there exist an integer $n \geqslant 1$ and a strict surjection*

$$\widetilde{A} = A[X_1, \ldots, X_{n-1}, S, U^{\pm1}]/(US^e - t) \to B \tag{4.1.3}$$

*of log rings over $A$ such that $t \in A$ and the image $s \in B$ of $S$ are generators of $I$ and $J$ respectively, and the kernel $I_{\widetilde{A}} = \mathrm{Ker}(\widetilde{A} \to B)$ is generated by $n$ elements locally on* Spec $\widetilde{A}$.

(2) *Locally on* Spec $B$, *for every log ring $\widetilde{A}$ log smooth over $A$ and for every strict surjection $\widetilde{A} \to B$ of log rings over $A$, the kernel $I_{\widetilde{A}} = \mathrm{Ker}(\widetilde{A} \to B)$ is generated by $n$ elements locally on* Spec $\widetilde{A}$.

(3) *The ring $B[J^{-1}]$ is locally of complete intersection over $A[I^{-1}]$ and $B/J$ is locally of complete intersection over $A/I$.*

*If these equivalent conditions are satisfied, then $B$ is locally of complete intersection and flat over $A$.*

*Proof* (2) ⇒ (1): By Lemma 4.1.8(3), locally on Spec $A$, there exists a surjection of log rings (4.1.3). Hence (2) implies (1).

(1) ⇒ (3): Assume that the kernel of the surjection

$$A[X_1, \ldots, X_{n-1}, S, U^{\pm1}]/(US^e - t) \to B$$

is locally generated by $f_1, \ldots, f_n$. Then the kernel of the surjection

$$A[X_1, \ldots, X_{n-1}, S, U^{\pm1}]/(US^e - t)[S^{-1}] = A[X_1, \ldots, X_{n-1}, S^{-1}] \to B[s^{-1}]$$

is locally generated by $n$ elements $f_1, \ldots, f_n$. Hence, $B[J^{-1}]$ is locally of complete intersection over $A[I^{-1}]$. Further, the kernel of the surjection

$$A/I[X_1, \ldots, X_{n-1}, U^{\pm1}] \to B/J$$

is locally generated by $n$ elements $f_1, \ldots, f_n$. Hence, $B/J$ is locally of complete intersection over $A/I$.

(3) ⇒ (2): By Proposition 4.1.5(2) and Lemma 4.1.7, condition (3) implies that $B$ is flat over $A$.

Locally on Spec $B$, let $(\widetilde{A}, \widetilde{I})$ be a log ring log smooth over $A$ and let $(\widetilde{A}, \widetilde{I}) \to (B, J)$ be a surjection of log rings over $A$. It suffices to show that the kernel $I_{\widetilde{A}} = \mathrm{Ker}(\widetilde{A} \to B)$ is generated by $n$ elements locally on Spec $B$.

Since $B[J^{-1}]$ is locally of complete intersection over $A[I^{-1}]$, the kernel of the surjection $\widetilde{A}[t^{-1}] \to B[J^{-1}]$ is generated by $n$ elements locally on $\operatorname{Spec} \widetilde{A}[t^{-1}]$. Since $B/J$ is locally of complete intersection over $A/I$ and $\widetilde{A}/\widetilde{I}$ is smooth over $A/I$, the kernel $I'$ of the surjection $\widetilde{A}/\widetilde{I} \to B/J$ is also generated by $n$ elements locally on $\operatorname{Spec} B/J$. Since $I'$ equals the image of $I_{\widetilde{A}}$, after localizing $B$ if necessary, we may assume that $I'$ is generated by the images of $f_1, \ldots, f_n \in I_{\widetilde{A}}$. It suffices to show that the surjection $B' = \widetilde{A}/(f_1, \ldots, f_n) \to B$ is an isomorphism on a neighborhood of $\operatorname{Spec} B/IB \subset \operatorname{Spec} B'$.

The surjection $B' \to B$ induces an isomorphism $B'/\widetilde{I}B' = (\widetilde{A}/\widetilde{I})/I' \to B/J = B/sB$. Since the image of $\widetilde{I}$ in $B'/IB'$ is nilpotent, after localizing $B'$ we may assume that $B'$ is quasi-finite over $A$ and is flat over $A$ by Proposition 4.1.5(2). Then, locally on $B'$, a generator $S$ of $\widetilde{I}B'$ is a non-zero divisor of $B'$ and of $B$, and the surjection $B' \to B$ induces an isomorphism $B'/S^eB' = B'/IB' \to B/IB$ of rings over $A/I$. Since $B$ is also flat over $A$ by Proposition 4.1.5(2) and Lemma 4.1.7, the kernel $I''$ of the surjection $B' \to B$ satisfies $I''/II'' = 0$. Hence, by Nakayama's lemma, the surjection $B' \to B$ is an isomorphism on a neighborhood of $\operatorname{Spec} B'/IB'$.

Assume that the kernel of the surjection $A[X_1, \ldots, X_{n-1}, S, U^{\pm 1}]/(US^e - t) \to B$ is locally generated by $f_1, \ldots, f_n$. By multiplying by a power of $U$, we may assume that $f_1, \ldots, f_n \in A[X_1, \ldots, X_{n-1}, S, U]$. Then, the kernel of the surjection $A[X_1, \ldots, X_{n-1}, S, U^{\pm 1}] \to B$ is locally generated by $n+1$ elements $f_1, \ldots, f_n, US^e - t$ and $B$ is locally of complete intersection. □

**Definition 4.1.12** Let $(f, e)\colon (A, I) \to (B, J)$ be a morphism of log rings and assume that $B$ is quasi-finite over $A$. We say that $B$ is locally of log complete intersection over $A$ if the equivalent conditions (1)–(3) in Proposition 4.1.11 are satisfied.

Using (3), we see that the composition of quasi-finite morphisms locally of log complete intersection is quasi-finite locally of log complete intersection.

**Proposition 4.1.13** *Let $(f, e)\colon A \to B$ and $(g, m)\colon A \to A'$ be morphisms of log rings. Let $d = (e, m)$ be the greatest common divisor and set $e = de'$ and $m = dm'$. Assume that $A \to B$ is flat and that $A'/I'$ and $B/J$ are flat over $A/I$.*

(1) *There exists a commutative diagram*

$$\begin{array}{ccc} (A, I) & \xrightarrow{(f,e)} & (B, J) \\ {\scriptstyle (g,m)}\downarrow & & \downarrow{\scriptstyle (g',m')} \\ (A', I') & \xrightarrow{(f',e')} & (B', J') \end{array}$$

*of log rings satisfying the property that, for every commutative diagram*

$$\begin{array}{ccc} (A,I) & \xrightarrow{(f,e)} & (B,J) \\ {\scriptstyle (g,m)}\downarrow & & \downarrow{\scriptstyle (h',m')} \\ (A',I') & \xrightarrow{(h,e')} & (C,K) \end{array} \tag{4.1.4}$$

*of log rings, there exists a unique strict morphism* $(k,1)\colon (B',J') \to (C,K)$ *of log rings over* $(A,I)$ *such that the diagram*

$$\begin{array}{ccc} (B',J') & \xleftarrow{(g',m')} & (B,J) \\ {\scriptstyle (f',e')}\uparrow & \searrow^{(k,1)} & \downarrow{\scriptstyle (h',m')} \\ (A',I') & \xrightarrow{(h,e')} & (C,K) \end{array} \tag{4.1.5}$$

*is commutative.*

(2) *Let* $B'$ *be the ring in* (1). *Then the morphism* $B \otimes_A A' \to B'$ *is an injection. If* $B$ *is of finite presentation over* $A$*, then* $B'$ *is flat of finite presentation over* $A'$.

(3) *Further, if* $B$ *is log smooth (resp. quasi-finite and locally of log complete intersection) over* $A$*, then* $B'$ *in* (1) *is log smooth (resp. quasi-finite and locally of log complete intersection) over* $A'$.

We let $B \otimes_A^{\log} A'$ denote $B'$ in Proposition 4.1.13 and call it the log tensor product. To prove the proposition, we prepare a lemma.

**Lemma 4.1.14** *Let* $C$ *be a ring,* $s, t \in C$ *be non-zero divisors, and* $u, v \in C^\times$ *be units. Assume that integers* $e, m \geqslant 1$ *satisfy* $us^e = vt^m$*, and let* $d = (e,m)$ *be the greatest common divisor of* $e = de'$ *and* $m = dm'$. *Assume that, for every integer* $0 \leqslant i' \leqslant i \leqslant e$ *and* $0 \leqslant j' \leqslant j \leqslant m$*, the following conditions are satisfied:*

- *The multiplication induces an injection* $s^i t^j \cdot \colon C/(s^{e-i}, t^{m-j}) \to C/(t^m)$.
- *As ideals of* $C/(t^m)$*, the intersection* $(s^i t^{j'}) \cap (s^{i'} t^j)$ *equals* $(s^i t^j)$.

(1) *For* $0 \leqslant k \leqslant e'm$*, define index sets*

$$\begin{aligned} M_k &= \{(i,j) \in \mathbf{N}^2 \mid i < e, j < m, m'i + e'j = k + e'm\} \\ \subset \widetilde{M}_k &= \{(i,j) \in \mathbf{N}^2 \mid i < e, j < m, m'i + e'j \geqslant k + e'm\} \end{aligned}$$

*and an ideal* $F_k(C/t^mC) = (s^i t^j \mid (i,j) \in \widetilde{M}_k) \subset C/t^mC$. *Then, for every* $0 \leqslant k < e'm$*, the surjection*

$$\sum s^i t^j \cdot \colon \bigoplus_{(i,j) \in M_k} C/(s,t) \to F_k(C/t^mC)/F_{k+1}(C/t^mC)$$

*is an isomorphism.*

(2) *Let $a, b$ be integers satisfying $am' - be' = 1$ and define*

$$B = C[S', W]/(W^d - v/u, S'^{e'} - W^a t, S'^{m'} - W^b s).$$

*Then, the morphism $C \to B$ is an injection and we have $t^m B \subset C$. Further, $S' \in B$ is a non-zero divisor.*

(3) *The annihilator ideal $\mathrm{Ann}_C(B/C) \subset C$ equals $t^{m-1}s^{-1}s'B$.*

*Proof* (1) We prove the assertion by induction on $k$ from above. If $e'm - k < e' + m'$, then we have $\widetilde{M}_k = \varnothing$ and the assertion follows. Assume $e'm - k \geqslant e'$ and let $0 \leqslant i < e$ be the smallest integer satisfying $m'i + e'(m - 1) \geqslant k$. Then, since $(s^i t'^{m-1}) \subset (t'^{m-1}) \cap F_k \subset (t'^{m-1}) \cap (s^i) = (s^i t'^{m-1})$ by the assumption, we obtain an exact sequence $0 \to (s^i t'^{m-1}) \to F_k \xrightarrow{t'\cdot} F_{k+e'} \to 0$. Since we have a similar exact sequence for $F_{k+1}$, we have an exact sequence $0 \to (s^i t'^{m-1})/(s^{i'} t'^{m-1}) \to F_k/F_{k+1} \xrightarrow{t'\cdot} F_{k+e'}/F_{k+1+e'} \to 0$ if $i' = i$ or $i + 1$ denotes the smallest integer satisfying $m'i' + e'(m - 1) \geqslant k + 1$. Since $(s^i t'^{m-1})/(s^{i'} t'^{m-1})$ is isomorphic to $C/(s, t)$ if $i' = i + 1$ and is 0 if $i' = i$, the assertion follows by induction.

(2) We show that the multiplication $t^m : B \to B$ factors through $C \to B$. The $C$-module $\widetilde{B} = C[S', W]/(W^d - v/u, S'^{e'} - W^a t)$ is free with basis $S'^k W^l$, $(0 \leqslant k < e', 0 \leqslant l < d)$ and we have $B = \mathrm{Coker}((S'^{m'} - W^b s)\cdot : \widetilde{B} \to \widetilde{B})$. Let $M$ be the monoid $\mathbf{N}^2$ written multiplicatively, and let $s, t$ denote the basis $(1, 0), (0, 1)$. Then $S' = s^a/t^b$, $W = s^{e'}/t^{m'}$ is a basis of $M^{\mathrm{gp}} = \mathbf{Z}^2$ since $am' - be' = 1$. The morphism $M \to C$ of multiplicative monoids defined by sending $s, t \in M$ to $s, t \in C$ is extended to the submonoid $\widetilde{M} = \langle s, t, W^{\pm d} \rangle \subset M^{\mathrm{gp}}$ by $W^d = v/u$. For each $S'^k W^l$, $(0 \leqslant k < e', 0 \leqslant l < d)$, there exist integers $p, r$ and $0 \leqq q < e$ such that $S'^k W^{l+dp} = s^q/t^r$. Since $(m - r)e > mq - er = k \geqslant 0$, we have $r < m$ and $t^m \cdot S'^k W^l \in \widetilde{M}$. Thus, by sending the basis $S'^k W^l$ to the image of $t^m \cdot S'^k W^l$ in $C$, we define a $C$-linear mapping $\widetilde{f} : \widetilde{B} \to C$. Since the elements $S'^{m'} \cdot S'^k W^l, W^b s \cdot S'^k W^l$ are equal in $M^{\mathrm{gp}}$, the composition $\widetilde{B} \to \widetilde{B} \xrightarrow{\widetilde{f}} C$ is the 0-mapping and a $C$-linear mapping $f : B \to C$ is induced. Hence, $t^m \cdot : B \to B$ is the composition of $f$ with $C \to B$. Since $t^m$ is a non-zero divisor of $C$ and the multiplication $t^m \cdot : C \to C$ is the composition $C \to B \xrightarrow{f} C$, the mapping $C \to B$ is an injection.

We show that $S' \in B$ is a non-zero divisor. Since $S'$ is a divisor of $t^m$, by the

commutative diagram

$$\begin{array}{ccccccccc} 0 & \longrightarrow & C & \longrightarrow & B & \longrightarrow & B/C & \longrightarrow & 0 \\ & & \Vert & & \downarrow{\scriptstyle f} & & \downarrow{\scriptstyle \overline{f}} & & \\ 0 & \longrightarrow & C & \xrightarrow{t^m\cdot} & C & \longrightarrow & C/t^mC & \longrightarrow & 0 \end{array}$$

it suffices to show that the right vertical arrow $\overline{f}\colon B/C \to C/t^mC$ is an injection. Let $F_k(B/C)$ be the image of $S'^kB$ for $0 \leqslant k \leqslant e'm$. Then, $\overline{f}\colon B/C \to C/t^mC$ maps $F_k(B/C)$ to $F_k(C/t^mC)$ for $0 \leqslant k \leqslant e'm$. Since $\sum s^it^{j-m}\cdot\colon \bigoplus_{(i,j)\in M_k} C/(s,t) \to F_k(B/C)/F_{k+1}(B/C)$ is a surjection for $0 \leqslant k \leqslant e'm$, the morphism $B/C = F_0(B/C) \to F_0(C/t^mC) \subset C/t^mC$ is thus an isomorphism by (1).

(3) Since the ideal $t^mB/t^mC = F_0(C/t^mC) \subset C/t^mC$ is generated by $s^it^j, (i,j) \in \widetilde{M}_0$, we have

$$\mathrm{Ann}_C(B/C) = \mathrm{Ann}_C(t^mB/t^mC) = \bigcap_{(i,j)\in\widetilde{M}_0} \mathrm{Ann}_C(s^it^j) = \bigcap_{(i,j)\in\widetilde{M}_0} (s^{e-i}, t^{m-j}).$$

By repeatedly applying the equality $I_1 \cap (I_2 + I_3) = (I_1 \cap I_2) + I_3$ for ideals satisfying $I_1 \supset I_3$, we see that the right-hand side equals $\sum_{(i,j)\in P}(s^it^j)$ for some subset $P \subset \mathbf{N}^2$.

Set $P' = \{(i,j) \in \mathbf{N}^2 \mid m'i + e'j \geqslant e'm - (e' + m') + 1\}$; we show that $P = P'$. We have $P = \{(i,j) \in \mathbf{N}^2 \mid s^it^j \cdot t^mB/t^mC = 0\}$, and the condition $s^it^j \cdot t^mB/t^mC = 0$ means that either $i + i' \geqslant e$ or $j + j' \geqslant m$ for every $(i',j') \in \widetilde{M}_0 = \{(i',j') \in \mathbf{N}^2 \mid i < e, j < m, m'i + e'j \leqslant e'm\}$. For $(i,j) \in \mathbf{N}^2$ satisfying $m'i + e'j > e'm - (e' + m')$ and $(i',j') \in \widetilde{M}_0$, we have $m'(i + i') + e'(j + j') > m'(e - 1) + e'(m - 1)$ and hence either $i + i' \geqslant e$ or $j + j' \geqslant m$. Thus, we have $P' \subset P$. Conversely, if $(i,j) \in \mathbf{N}^2 - P'$, we have $i \leqslant e - 1, j \leqslant m - 1$ and, for $i' = (e-1) - i$ and $j' = (m-1) - j$, we have $(i',j') \in \widetilde{M}_0$ and $s^it^j \cdot s^{i'}t^{j'} \neq 0$ in $C/t^mC$. Thus, we have $P = P'$ and $\mathrm{Ann}_C(B/C) = (s^it^j \mid (i,j) \in P) = S'^{em'-e'-m'+1}B \subset C$. □

*Proof of Proposition* 4.1.13 Since the assertion is local on Spec $A$, Spec $A'$, and on Spec $B$, we may assume that $t \in I$, $t' \in I'$, and $s \in J$ are generators of the invertible ideals. Let $u \in B^\times$ and $v \in A'^\times$ be the units satisfying $us^e = t$, $vt'^m = t$, and let $a, b$ be integers satisfying $am' - be' = 1$. Since $B \otimes_A A'$ is flat over $A'$, $t'$ is a non-zero divisor of $B \otimes_A A'$. Since $us^e = vt'^m$ in $B \otimes_A A'$, both $s$ and $t'$ are non-zero divisors. Since $B/J$ and $A'/I'$ are flat over $A/I$, for $0 \leqslant i \leqslant e$ and $0 \leqslant j \leqslant m$, the quotient rings $B/J^i$ and $A'/I'^j$ are also flat over $A/I$. Hence, the exact sequences $0 \to B/J^{e-i} \to B/J^{e-i'} \to B/I^{i-i'} \to 0$ and $0 \to A'/I'^{m-j} \to A'/I'^{m-j'} \to A'/I'^{j-j'} \to 0$

for $0 \leqslant i' \leqslant i \leqslant e$ and $0 \leqslant j' \leqslant j \leqslant m$ define an exact sequence $0 \to (s^i t'^j) \to (s^i t'^{j'}) \oplus (s^{i'} t'^j) \to (s^{i'} t'^{j'}) \xrightarrow{s^{e-i} t^{m-j}} (s^{e-i+i'} t^{m-j+j'}) \to 0$ of ideals of $B \otimes_A A'/(t) = B/J^e \otimes_{A/I} A'/I'^m$ and $s, t' \in B \otimes_A A'$ satisfy the assumptions in Lemma 4.1.14.

(1) Let (4.1.4) be a commutative diagram of log rings. Then there exists a unique generator $s'$ of the invertible ideal $K$ and a unique unit $w$ of $C^\times$ satisfying $s' \cdot t'^b = s^a$ and $w \cdot t'^{m'} = s^{e'}$. Since $s'^{e'} = w^a t'$, $s'^{m'} = w^b s$, and $w^d = v/u$, there exists a unique morphism

$$B' = B \otimes_A A'[S', W]/(S'^{e'} - W^a t', S'^{m'} - W^b s, W^d - v/u) \to C.$$

The ideal $J' = S'B'$ is generated by a non-zero divisor $S'$ by Lemma 4.1.14. The diagram (4.1.5) of log rings is commutative.

(2) The morphism $B \otimes_A A' \to B'$ is an injection by Lemma 4.1.14(2). If $B$ is of finite presentation over $A$, then $B'$ is of finite presentation over $A'$. Since $B'/S'B' = B/J \otimes_{A/I} A'/I'[W]/(W^d - v/u)$ is flat over $A'/I'$ and $B'[S'^{-1}] = B \otimes_A A'[t'^{-1}]$ is flat over $A'[t'^{-1}]$, $B'$ is flat over $A'$ by Lemma 4.1.7.

(3) We show the assertion for log smoothness. We may assume that the chart (4.1.1) defined by $us^e = t$ satisfies the condition in Lemma 4.1.8(2). The morphism $A/I[U^{\pm 1}] \to B/J$ sending $U$ to $u$ is smooth. Hence, $B'/S'B' = B/J \otimes_{A/I} A'/I'[W]/(u - vW^{-d}) = B/J \otimes_{A/I[U^{\pm 1}]} A'/I'[W^{\pm 1}]$ with respect to the morphism sending $U$ to $vW^{-d}$ is smooth over $A'/I'[W^{\pm 1}]$. Thus, the chart for $A' \to B'$ defined by $S'^{e'} = W^a t'$ satisfies the condition in Lemma 4.1.8(2) and $B'$ is log smooth over $A'$.

We show the assertion for log complete intersection. We may assume that there exist a log ring $\widetilde{A}$, log smooth of relative dimension $n$ over $A$, and a strict surjection $\widetilde{A} \to B$ of log rings over $A$ such that the kernel is defined by $n$ elements. Then, $\widetilde{A}' = \widetilde{A} \otimes_A^{\log} A'$ is log smooth over $A'$ and $\widetilde{A}' \to B' = B \otimes_A^{\log} A'$ is a strict surjection such that the kernel is generated by the image of the kernel of $\widetilde{A} \to B$. Since $B \otimes_A A' \to B'$ is finite, $B'$ is quasi-finite over $A'$. □

Let $t$ be a generator of $I$ and let $A' = A[X_1, \ldots, X_{n-1}, S, U^{\pm 1}]/(US^e - t)$. Then the morphism $A' \to B[X_1, \ldots, X_{n-1}, X_n^{\pm 1}]$ sending $U, S$ to $uX_n^{-e}, sX_n$ induces an isomorphism $B' \to B[X_1, \ldots, X_{n-1}, X_n^{\pm 1}]$.

**Exercise 4.1** Let $K$ be a henselian discrete valuation field and $L$ be a finite separable extension of ramification index $e$. Let $(f, e)\colon (O_K, \mathfrak{m}_K) \to (O_L, \mathfrak{m}_L)$ be the morphism of log rings defined by the inclusion. Show that $(O_L, \mathfrak{m}_L)$ is locally of log complete intersection over $(O_K, \mathfrak{m}_K)$.

## 4.2 Logarithmic Residues and Duality

**Definition 4.2.1** Let $A$ be a ring and let $f_1, \dots, f_n \in A$. Define an $A$-linear form $f\colon L = A^{\oplus n} \to A$ by sending the standard basis $e_1, \dots, e_n$ to $f_1, \dots, f_n$. The Koszul complex $K(L, f)$ is the complex

$$K(L, f) = [\wedge_A^n L \to \wedge_A^{n-1} L \to \cdots \to L \to A]$$

where $\wedge_A^i L \to \wedge_A^{i-1} L$ is defined by sending $x_1 \wedge \cdots \wedge x_i$ to $\sum_{j=1}^i (-1)^{j-1} f(x_j) \cdot x_1 \wedge \cdots \wedge x_{j-1} \wedge x_{j+1} \wedge \cdots \wedge x_i$.

**Lemma 4.2.2** *Let $A$ be a ring and let $f_1, \dots, f_n \in A$.*

(1) *The following conditions are equivalent:*

  (a) *For every $i = 1, \dots, n$, the image $\overline{f}_i \in \overline{A}_i = A/(f_1, \dots, f_{i-1})$ is a non-zero divisor.*

  (b) *For each $i = 0, \dots, n$, the Koszul complex $K_i = K(A^{\oplus i}, (f_1, \dots, f_i))$ is a free resolution of $A/(f_1, \dots, f_i)$.*

(2) *If these equivalent conditions are satisfied, the conormal module*

$$N_{\overline{A}_n/A} = (f_1, \dots, f_n)/(f_1, \dots, f_n)^2$$

*is a free $\overline{A}_n$-module of rank $n$.*

*Proof* (1) We prove the equivalence by the induction on $i \geqslant 1$. The Koszul complex $K_i$ is the simple complex associated with the double complex $f_i\colon K_{i-1} \to K_{i-1}$. The condition that $\overline{f}_i \in A/(f_1, \dots, f_{i-1})$ is a non-zero divisor means that the sequence $0 \to \overline{A}_{i-1} \xrightarrow{\overline{f}_i} \overline{A}_{i-1} \to \overline{A}_i \to 0$ is exact. Hence, if $K_{i-1}$ is a resolution of $\overline{A}_{i-1}$, then $K_i$ is a resolution of $\overline{A}_i$ if and only if $\overline{f}_i \in A/(f_1, \dots, f_{i-1})$ is a non-zero divisor.

Assume that the Koszul complex $K(L, f)$ is a free resolution of the $A$-module $\overline{A}_n$. Then we have an exact sequence $\wedge_A^2 L \to L \to (f_1, \dots, f_n) \to 0$. This induces an isomorphism $L \otimes_A \overline{A}_n \to N_{\overline{A}_n/A}$ of free $\overline{A}_n$-modules of rank $n$.

(2) If condition (b) is satisfied, by removing the 0th component $A$ from the Koszul complex $K_n$, we obtain a free resolution of the ideal $(f_1, \dots, f_n) \subset A$. Hence, the $\overline{A}_n$-module $N_{\overline{A}_n/A} = (f_1, \dots, f_n) \otimes_A \overline{A}_n$ is free of rank $n$. □

**Proposition 4.2.3** (1) *Let $B$ be a quasi-finite ring locally of complete intersection over $A$. If $\widetilde{A}$ is a smooth ring over $A$ of relative dimension $n$ and if $\widetilde{A} \to B$ is a surjection of rings over $A$, then the conormal module $N_{B/\widetilde{A}}$ is a locally free $B$-module of rank $n$. Further, if the kernel of $\widetilde{A} \to B$ is generated by $f_1, \dots, f_n$, the Koszul complex $K(\widetilde{A}^n, f)$ is a free resolution of $B$.*

(2) *Let $(B, J)$ be a quasi-finite log ring locally of log complete intersection over $(A, I)$. If $\widetilde{A}$ is a log ring log smooth of relative dimension $n$ over $A$ and if $\widetilde{A} \to B$ is a strict surjection of log rings over $A$, then the conormal module $N_{B/\widetilde{A}}$ is a locally free $B$-module of rank $n$. Further, if the kernel of $\widetilde{A} \to B$ is generated by $f_1, \dots, f_n$, the Koszul complex $K(\widetilde{A}^n, f)$ is a free resolution of $B$.*

*Proof* (1) Since the question is local, we may assume that the kernel of $\widetilde{A} \to B$ is generated by $f_1, \dots, f_n$. Let $\mathfrak{p}$ be any prime ideal of $A$, and let $\mathfrak{m}$ be any maximal ideal of $\widetilde{A} \otimes_A k(\mathfrak{p})$ containing $f_1, \dots, f_n$. By abuse of notation, let $\mathfrak{m}$ denote the corresponding prime ideal of $\widetilde{A}$. Since $B$ is quasi-finite over $A$, the elements $f_1, \dots, f_n$ of $\widetilde{A}_{\mathfrak{m}}$ satisfy condition (a) in Lemma 4.2.2(1) by Proposition 4.1.5(1). Therefore, $K(\widetilde{A}^n, f)$ is a free resolution of $B$ and $N_{B/\widetilde{A}}$ is a free $B$-module of rank $n$ by Lemma 4.2.2.

(2) We may assume that there exist relatively prime integers $e \geqslant 1, c$ and an étale morphism $A[X_1, \dots, X_{n-1}, S, U^{\pm 1}]/(U^c S^e - t) \to \widetilde{A}$ over $A$, and that the kernel of $\widetilde{A} \to B$ is generated by $f_1, \dots, f_n$. There exist a ring $\widetilde{A}'$ étale over $A[X_1, \dots, X_{n-1}, U^{\pm 1}, S]$ and a surjection $\widetilde{A}' \to \widetilde{A}$ over $A$ by [31, Chapitre IV, Corollaire (18.4.7)]. Let $\widetilde{f}_1, \dots, \widetilde{f}_n \in \widetilde{A}'$ be liftings of $f_1, \dots, f_n$. By localizing if necessary, we may assume that the kernel of the surjection $\widetilde{A}' \to B$ is generated by $\widetilde{f}_1, \dots, \widetilde{f}_n, \widetilde{f}_{n+1} = U^c S^e - t$. Since $B$ is quasi-finite over $A$, the elements $\widetilde{f} = (\widetilde{f}_1, \dots, \widetilde{f}_{n+1})$ of $\widetilde{A}'$ satisfy the assumption of (1). Since the Koszul complex $K(\widetilde{A}', \widetilde{f}_{n+1})$ is a resolution of $\widetilde{A}$, we obtain a quasi-isomorphism $K(\widetilde{A}'^{n+1}, \widetilde{f}) \to K(\widetilde{A}^n, f)$ of Koszul complexes. Since $K(\widetilde{A}'^{n+1}, \widetilde{f})$ is a resolution of $B$ by (1), $K(\widetilde{A}^n, f)$ is also a resolution. Further, since $N_{B/\widetilde{A}'}$ is a free $B$-module of rank $n+1$ and the images of $\widetilde{f}_{n+1}$ are a part of a basis, the quotient $N_{B/\widetilde{A}} = N_{B/\widetilde{A}'}/(\widetilde{f}_{n+1})$ is free of rank $n$. □

Let $(A, I) \to (B, J)$ be a quasi-finite morphism locally of log complete intersection of log rings. Let $\widetilde{A}$ be a log smooth log ring over $A$, and $\widetilde{A} \to B$ be a strict surjection of log rings over $A$ as in Proposition 4.2.3(2). Let $\widetilde{B} = B \otimes_A^{\log} \widetilde{A}$ be the log tensor product and $\widetilde{B} \to B$ be the surjection induced by $\widetilde{A} \to B$. Since $B \to \widetilde{B}$ is smooth by Proposition 4.1.13(3), the $\widetilde{B}$-module $\Omega^1_{\widetilde{B}/B}$ is a free module of rank $n$ and its exterior power $\Omega^n_{\widetilde{B}/B} = \wedge^n_{\widetilde{B}} \Omega^1_{\widetilde{B}/B}$ is free of rank 1. We construct a canonical isomorphism

$$\operatorname{Hom}_A(J, I) \otimes_B \operatorname{Hom}_B(\Omega^n_{\widetilde{B}/B} \otimes_{\widetilde{B}} B, B) \to \operatorname{Hom}_B(\wedge^n_B N_{B/\widetilde{A}}, B) \tag{4.2.1}$$

of invertible $B$-modules.

**Proposition 4.2.4** *Let $(B, J)$ be a finite log ring locally of log complete intersection over $(A, I)$.*

(1) *The B-modules* $\mathrm{Hom}_A(B, A)$ *and* $\mathrm{Hom}_A(J, I)$ *are invertible B-modules.*

(2) *Let* $(A, I) \to (A', I')$ *be a morphism of log rings such that* $A'/I'$ *is flat over* $A/I$. *Let* $B' = B \otimes_A^{\log} A'$ *and* $J' \subset B'$ *be the invertible ideal. Then the morphism* $B \otimes_A A' \to B'$ *is an injection and induces an isomorphism*

$$B' \otimes_B \mathrm{Hom}_A(J, I) \to \mathrm{Hom}_{A'}(J', I') \tag{4.2.2}$$

*of invertible* $B'$*-modules.*

(3) *Let* $\widetilde{A}$ *be a log smooth log ring over* $A$, *and* $\widetilde{A} \to B$ *be a strict surjection of log rings over* $A$. *Let* $\widetilde{B} = B \otimes_A^{\log} \widetilde{A} \to B$ *be the induced surjection. Then, there exists a unique isomorphism*

$$\mathrm{Hom}_A(J, I) \otimes_B \mathrm{Hom}_B(\Omega^n_{\widetilde{B}/B} \otimes_{\widetilde{B}} B, B) \to \mathrm{Hom}_B(\wedge^n_B N_{B/\widetilde{A}}, B)$$

*(see* (4.2.1)*) of invertible B-modules satisfying the following condition.*

*Locally on* $\mathrm{Spec}\, A$, *let* $f_1, \ldots, f_n$ *and* $g_1, \ldots, g_n$ *be generators of the kernels* $I_{\widetilde{A}} = \mathrm{Ker}(\widetilde{A} \to B)$ *and* $I_{\widetilde{B}} = \mathrm{Ker}(\widetilde{B} \to B)$. *Define a matrix* $D = (D_{ij}) \in \mathrm{M}_n(\widetilde{B})$ *by*

$$f_j = \sum_{i=1}^{n} D_{ij} g_i. \tag{4.2.3}$$

*Let* $(f_1 \wedge \cdots \wedge f_n)^\vee : \wedge^n_B N_{B/\widetilde{A}} \to B$ *and* $(dg_1 \wedge \cdots \wedge dg_n)^\vee : \Omega^n_{\widetilde{B}/B} \otimes_{\widetilde{B}} B \to B$ *denote the dual bases. For a linear form* $h: J \to I$, *let* $\widetilde{h}: \widetilde{B} \to \widetilde{A}$ *denote the corresponding linear form by* (4.2.2) *for* $A' = \widetilde{A}$, *and let* $\widetilde{h}(\det D)|_B$ *denote the image of* $\widetilde{h}(\det D)$ *by the surjection* $\widetilde{A} \to B$. *Then* (4.2.1) *maps* $h \otimes (dg_1 \wedge \cdots \wedge dg_n)^\vee$ *to* $\widetilde{h}(\det D)|_B \cdot (f_1 \wedge \cdots \wedge f_n)^\vee$.

*Proof* (1) Let $\widetilde{A}$ be a log smooth log ring of relative dimension $n$ over $A$, and $\widetilde{A} \to B$ be a strict surjection of log rings over $A$. Locally on $\mathrm{Spec}\, A$, let $f_1, \ldots, f_n$ be generators of the kernel $I_{\widetilde{A}} = \mathrm{Ker}(\widetilde{A} \to B)$. Let $L = \widetilde{A}^{\oplus n}$ be the free $\widetilde{A}$-module of rank $n$ and define an $\widetilde{A}$-linear morphism $f: L \to \widetilde{A}$ by $f_1, \ldots, f_n$. By Proposition 4.2.3, the surjection $\widetilde{A} \to B$ defines a resolution $K_{\widetilde{A}}(L, f) \to B$ by free $\widetilde{A}$-modules. Similarly for the surjection $\widetilde{B} = B \otimes_A^{\log} \widetilde{A} \to B$, let $M = \widetilde{B}^{\oplus n}$ be a free module and define $g: M \to \widetilde{B}$ by local generators $g_1, \ldots, g_n$ of $I_{\widetilde{B}} = \mathrm{Ker}(\widetilde{B} \to B)$. Then the Koszul complex defines a resolution $K_{\widetilde{B}}(M, g) \to B$ by free $\widetilde{B}$-modules.

Since $\widetilde{B}$ is flat over $\widetilde{A}$ by Proposition 4.1.13(2), it is finite and flat over $\widetilde{A}$. Hence we may compute $\mathrm{Ext}^n_{\widetilde{A}}(B, \widetilde{A})$ by using the resolutions $K_{\widetilde{A}}(L, f)$ and $K_{\widetilde{B}}(M, g)$ of $B$. Define $D = (D_{ij}) \in \mathrm{M}_n(\widetilde{B})$ by (4.2.3). This defines an $\widetilde{A}$-linear morphism $D\colon L \to M$ such that the diagram

$$\begin{array}{ccc} L & \longrightarrow & \widetilde{A} \\ {\scriptstyle D}\downarrow & & \downarrow \\ M & \longrightarrow & \widetilde{B} \end{array}$$

is commutative. The morphism $D\colon L \to M$ induces a morphism

$$D_*\colon K_{\widetilde{A}}(L, f) \to K_{\widetilde{B}}(M, g)$$

of the Koszul complexes compatible with the identity of $B$. The morphism $D$ induces an isomorphism

$$D^*\colon \mathrm{Hom}_{\widetilde{A}}(\wedge^n_{\widetilde{B}} M, \widetilde{A}) \otimes_{\widetilde{B}} B \to \mathrm{Hom}_{\widetilde{A}}(\wedge^n_{\widetilde{A}} L, \widetilde{A}) \otimes_{\widetilde{A}} B. \tag{4.2.4}$$

We identify $\mathrm{Hom}_{\widetilde{A}}(\wedge^n_{\widetilde{B}} M, \widetilde{A}) \otimes_{\widetilde{B}} B = \mathrm{Hom}_{\widetilde{A}}(\widetilde{B}, \widetilde{A}) \otimes_{\widetilde{B}} \mathrm{Hom}_B(\wedge^n_B(M \otimes_{\widetilde{B}} B), B)$ and $\mathrm{Hom}_{\widetilde{A}}(\wedge^n_{\widetilde{A}} L, \widetilde{A}) \otimes_{\widetilde{A}} B = \mathrm{Hom}_B(\wedge^n_B(L \otimes_{\widetilde{A}} B), B)$. Further, we identify $L \otimes_{\widetilde{A}} B = N_{B/\widetilde{A}}$ and $M \otimes_{\widetilde{B}} B = N_{B/\widetilde{B}}$ with $\Omega^1_{\widetilde{B}/B} \otimes_{\widetilde{B}} B$ by the isomorphism $N_{B/\widetilde{B}} \to \Omega^1_{\widetilde{B}/B} \otimes_{\widetilde{B}} B$ of $B$-modules induced by the derivation $d$. Then (4.2.4) induces an isomorphism

$$\mathrm{Hom}_{\widetilde{A}}(\widetilde{B}, \widetilde{A}) \otimes_{\widetilde{B}} \mathrm{Hom}_B(\Omega^n_{\widetilde{B}/B} \otimes_{\widetilde{B}} B, B) \to \mathrm{Hom}_B(\wedge^n_B N_{B/\widetilde{A}}, B) \tag{4.2.5}$$

of invertible $B$-modules.

Since $B$ is locally of complete intersection over $A$, (4.2.5) for $(A, A) \to (B, B)$ is an isomorphism

$$\mathrm{Hom}_A(B, A) \otimes_B \mathrm{Hom}_B(\Omega^n_{\widetilde{B}/B} \otimes_{\widetilde{B}} B, B) \to \mathrm{Hom}_B(\wedge^n_B N_{B/\widetilde{A}}, B)$$

of $B$-modules. Hence, $\mathrm{Hom}_A(B, A)$ and $\mathrm{Hom}_A(J, I)$ are invertible $B$-modules.

(2) Since $B$ is flat over $A$ and $B/J$ is flat over $A/I$, the log tensor product $B' = B \otimes^{\log}_A A'$ is defined by Proposition 4.1.13(1). Set $C' = B \otimes_A A'$ and identify

$$\begin{aligned} \mathrm{Hom}_{A'}(C', A') = A' \otimes_A \mathrm{Hom}_A(B, A) &= A' \otimes_A B \otimes_B \mathrm{Hom}_A(B, A) \\ &= C' \otimes_B \mathrm{Hom}_A(B, A). \end{aligned}$$

Since $\mathrm{Hom}_A(B, A)$ is an invertible $B$-module, we may further identify

$$\begin{aligned} \mathrm{Hom}_{A'}(B', A') = \mathrm{Hom}_{A'}(B' \otimes_{C'} C', A') &= \mathrm{Hom}_{C'}(B', \mathrm{Hom}_{A'}(C', A')) \\ &= \mathrm{Hom}_{C'}(B', C') \otimes_B \mathrm{Hom}_A(B, A). \end{aligned} \tag{4.2.6}$$

Since $C' \to B'$ is an injection and a generator $t$ of $I$ is a non-zero divisor of $C'$ by Proposition 4.1.13(2), we may apply Lemma 4.2.5 to $C' \to B'$

and identify $\mathrm{Hom}_{C'}(B', C') = \mathrm{Ann}_{C'}(B'/C')$. By Lemma 4.1.14(3), we have $\mathrm{Ann}_{C'}(B'/C') = II'^{-1}J^{-1}J' \subset B'$ and we obtain the isomorphism (4.2.2) from (4.2.6).

(3) Since $\widetilde{A} \to \widetilde{B}$ is strict, we have $\widetilde{J} = \widetilde{I}\widetilde{B}$ and the right-hand side $\mathrm{Hom}_{\widetilde{A}}(\widetilde{J}, \widetilde{I})$ of (4.2.5) for $A' = \widetilde{A}$ is identified with $\mathrm{Hom}_{\widetilde{A}}(\widetilde{B}, \widetilde{A})$. Locally on Spec $A$, define an isomorphism (4.2.1) by (4.2.5) and (4.2.2). Since the linear mapping $L \to M$ is defined by the matrix $D = (D_{ij}) \in \mathrm{M}_n(\widetilde{B})$, the morphism (4.2.4) maps the class of the composition $\widetilde{h} \circ (dg_1 \wedge \cdots \wedge dg_n)^\vee$ to $(f_1 \wedge \cdots \wedge f_n)^\vee \otimes \widetilde{h}(\det D)|_B$. Hence, the isomorphism (4.2.1) satisfies the condition for the chosen $f_1, \ldots, f_n$, $g_1, \ldots, g_n$, and $D \in \mathrm{M}_n(\widetilde{B})$.

Let $f'_1, \ldots, f'_n$ and $g'_1, \ldots, g'_n$ be other systems of generators of $I_{\widetilde{A}}$ and $I_{\widetilde{B}}$, and define Koszul resolutions $K_{\widetilde{A}}(L') \to B$ and $K_{\widetilde{B}}(M') \to B$. Write $f'_j = \sum_{i=1}^n C_{ij} f_i$ as $\widetilde{A}$-linear combinations and $f'_j = \sum_{i=1}^n D'_{ij} g'_i$, $g'_j = \sum_{i=1}^n C'_{ij} g_i$ as $\widetilde{B}$-linear combinations, and define $\widetilde{A}$-linear morphisms $C \colon L' \to L$ and $D' \colon L' \to M'$ and a $\widetilde{B}$-linear morphism $C' \colon M' \to M$ by the matrices $C = (C_{ij}) \in \mathrm{M}_n(\widetilde{A})$ and $D' = (D'_{ij}) \in \mathrm{M}_n(\widetilde{B})$, $C' = (D'_{ij}) \in \mathrm{M}_n(\widetilde{B})$. Then $C$, $D'$, and $C'$ induce morphisms $C_* \colon K_{\widetilde{A}}(L) \to K_{\widetilde{A}}(L)$, $D'_* \colon K_{\widetilde{A}}(L') \to K_{\widetilde{B}}(M')$, and $D'_* \colon K_{\widetilde{B}}(M') \to K_{\widetilde{B}}(M)$ of Koszul complexes. Since the compositions $D \circ C$ and $C' \circ D'$ are both compatible with the identity of $B$, the morphisms $D_* \circ C_*$ and $C'_* \circ D'_*$ are homotopic to each other. Hence, the morphisms (4.2.5) defined as $C^* \circ D^*$ and $D'^* \circ C'^*$ are the same. Since $C$ induces the identity on $N_{B/\widetilde{A}}$ and $C'$ induces the identity on $\Omega^1_{\widetilde{B}/B} \otimes_{\widetilde{B}} B$, the isomorphism (4.2.1) is independent of the choice and is well-defined on Spec $A$. Further, it satisfies the condition for every $f_1, \ldots, f_n, g_1, \ldots, g_n$, and $D \in \mathrm{M}_n(\widetilde{B})$. □

**Lemma 4.2.5** *Let $C$ be a ring and let $C \to B$ be an injection of rings. Assume that there exists a non-zero divisor $t \in C$ such that its image in $B$ is a non-zero divisor and $tB \subset C$. Then the evaluation morphism $\mathrm{ev}_1 \colon \mathrm{Hom}_C(B, C) \to C$ is an injection and the image equals the annihilator ideal $\mathrm{Ann}_C B/C$.*

*Proof* Since the $C$-module $B/C$ is annihilated by $t$ and $C$ is $t$-torsion free, we have $\mathrm{Hom}_C(B/C, C) = 0$. Hence, by applying $\mathrm{Hom}_C(-, C)$ to the exact sequence $C \to B \to B/C \to 0$, we see that $\mathrm{Hom}_C(B, C) \to \mathrm{Hom}_C(C, C) = C$ is an injection. The image equals $\{b \in C \mid bB \subset C\} = \mathrm{Ann}_C B/C$. □

We make explicit the isomorphism (4.2.1) in the case where $A \to B$ is strict and $n = 1$.

**Lemma 4.2.6** *Let $P = X^n + a_1 X^{n-1} + \cdots + a_n \in \widetilde{A} = A[X]$ and $B = A[X]/P$. Let $b \in B$ be the image of $X$ and define $Q = c_1 X^{n-1} + c_2 X^{n-2} + \cdots + c_n \in B[X] = \widetilde{B}$ by $P = (X - b) \cdot Q$. Let $f_1, \ldots, f_n \in \mathrm{Hom}_B(B, A)$ be the dual* *basis* $c_1 = 1, c_2, \ldots, c_n$ *of the free* $A$*-module* $B$ *as in Lemma* 1.2.1. *Then the isomorphism* $\mathrm{Hom}_A(B, A) \otimes \mathrm{Hom}_B(\Omega^1_{\widetilde{B}/B} \otimes_{\widetilde{B}} B, B) \to \mathrm{Hom}_B(N_{B/\widetilde{A}}, B)$ (4.2.1) *maps the tensor product of the basis* $f_n$ *of a* $B$*-module* $\mathrm{Hom}_B(B, A)$ *with the dual basis* $dX^\vee$ *to the dual basis* $P^\vee$.

*Proof* Let $f \colon B \to A$ be an $A$-linear form and let $\widetilde{f} \colon \widetilde{B} \to \widetilde{A}$ denote the induced $\widetilde{A}$-linear form. Then (4.2.1) maps $f \otimes dX^\vee$ to $\widetilde{f}(Q)(b) \cdot P$. Since $\widetilde{f_n}(Q)(b) = 1$, the assertion follows. □

We study the transitivity of the isomorphism (4.2.1) and its independence of the choice of strict surjections. Let $(A, I) \to (B, J) \to (C, K)$ be morphisms of log rings. Assume that the morphisms are finite locally of log complete intersection. Let $A_1$ and $B_2$ be log ring log smooth over $A$ and over $B$ respectively, and let $A_1 \to B$ and $B_2 \to C$ be strict surjections of log rings.

After replacing Spec $B_2$ by an affine open subset, we construct a log smooth morphism $A_1 \to A_3$ and an isomorphism $B \otimes_{A_1} A_3 \to B_2 \times B'$ for some ring $B'$. We assume that $B_2$ is étale over $B[Y_1, \ldots, Y_{m-1}, Z, W^{\pm 1}]/(W^c Z^l - s)$ for relatively prime integers $l \geqslant 1, c$ and a generator $s \in J$. Then, further replacing by an affine open subset if necessary, there exists an étale ring $A_3$ over $A_1[Y_1, \ldots, Y_{m-1}, Z, W^{\pm 1}]/(W^c Z^l - s)$ for a lifting of $s$ denoted abusively $s \in I_1$ and a surjection $A_3 \to B_2$ extending the composition $A_1[Y_1, \ldots, Y_{m-1}, Z, W^{\pm 1}]/(W^c Z^l - s) \to B[Y_1, \ldots, Y_{m-1}, Z, W^{\pm 1}]/(W^c Z^l - s) \to B_2$ by [31, Chapitre IV, Corollaire (18.4.7)]. Since the surjection $B \otimes_{A_1} A_3 \to B_2$ is étale, we have an isomorphism $B \otimes_{A_1} A_3 \to B_2 \times B'$ for some ring $B'$.

We consider the commutative diagram

$$\begin{array}{ccccccc}
A & & & & & & \\
\downarrow & \searrow & & & & & \\
A_1 & \longrightarrow & A_3 & & & & \\
\downarrow & & \downarrow & \searrow & & & \\
B_1 = B \otimes^{\log}_A A_1 & \longrightarrow & B_3 = B \otimes^{\log}_A A_3 & \longrightarrow & C_3 = C \otimes^{\log}_A A_3 & & \\
\downarrow & & \downarrow & & \downarrow & \searrow & \\
B & \longrightarrow & B_2 & \longrightarrow & C_2 = C \otimes^{\log}_B B_2 & \longrightarrow & C.
\end{array} \tag{4.2.7}$$

The lower vertical arrows and the arrows in the lower right triangle are induced by the surjections $A_1 \to B$ and $A_3 \to B_2 \to C$. Since $A \to A_1$ and $A_1 \to A_3$ are log smooth, the composition $A \to A_3$ is log smooth by Lemma 4.1.10.

Assume that $A_1$ is of relative dimension $n$ over $A$. By applying the construction of (4.2.1) to the right column, the bottom line, and to the slant line, we obtain

$$\begin{aligned}
&\mathrm{Hom}_A(J,I)\otimes_B(\Omega^n_{B_1/B}\otimes_{B_1}B)^\vee \to (\wedge^n_B N_{B/A_1})^\vee,\\
&\mathrm{Hom}_B(K,J)\otimes_C(\Omega^m_{C_2/C}\otimes_{C_2}C)^\vee \to (\wedge^m_C N_{C/B_2})^\vee,\\
&\mathrm{Hom}_A(K,I)\otimes_C(\Omega^{n+m}_{C_3/C}\otimes_{C_3}C)^\vee \to (\wedge^{n+m}_C N_{C/A_3})^\vee.
\end{aligned}\tag{4.2.8}$$

Since the log smooth morphism $A_1 \to A_3$ is flat, and since $B\otimes_{A_1}A_3 \to B_2$ and $C\otimes_{C_1}C_3 \to C_2$ are localizations, we have exact sequences

$$\begin{aligned}
&0\to N_{B/A_1}\otimes_B C\to N_{C/A_3}\to N_{C/B_2}\to 0,\\
&0\to \Omega^1_{C_1/C}\otimes_{C_1}C_2\to \Omega^1_{C_3/C}\otimes_{C_3}C_2\to \Omega^1_{C_2/C}\to 0.
\end{aligned}\tag{4.2.9}$$

**Lemma 4.2.7** *We consider the diagram*

$$\begin{array}{ccc}
(\mathrm{Hom}_A(J,I)\otimes_B \mathrm{Hom}_B(K,J))\otimes_C & \longrightarrow & (\Omega^n_{B_1/B}\otimes_{B_1}B)^\vee\otimes_B\\
(\wedge^n_B N_{B/A_1})^\vee\otimes_B(\wedge^m_C N_{C/B_2})^\vee & & (\Omega^m_{C_2/C}\otimes_{C_2}C)^\vee\\
\downarrow & & \downarrow\\
\mathrm{Hom}_A(K,I)\otimes_C(\Omega^{n+m}_{C_3/C}\otimes_{C_3}C)^\vee & \longrightarrow & (\wedge^{n+m}_C N_{C/A_3})^\vee
\end{array}\tag{4.2.10}$$

*of invertible $C$-modules. The horizontal arrows are defined by* (4.2.8) *and the vertical arrows are defined by the composition* $\mathrm{Hom}_A(J,I)\otimes_B\mathrm{Hom}_B(K,J)\to \mathrm{Hom}_A(K,I)$ *and the exact sequences* (4.2.9). *Then the diagram* (4.2.10) *is commutative.*

*Proof* We may assume that the kernel $I_1 = \mathrm{Ker}(A_1 \to B)$ is generated by $n$ elements $f_1,\dots,f_n$, and $I_2 = \mathrm{Ker}(B_2\to C)$ is generated by $m$ elements $f'_1,\dots,f'_m$. Then the images $f_1,\dots,f_n \in I_3 = \mathrm{Ker}(A_3\to C)$ and liftings $\widetilde{f'_1},\dots,\widetilde{f'_m}\in I_3$ form generators of $I_3$. We may also assume that the kernel $I'_1 = \mathrm{Ker}(B_1\to B)$ is generated by $n$ elements $g_1,\dots,g_n$, and $I'_2 = \mathrm{Ker}(C_2\to C)$ is generated by $l$ elements $g'_1,\dots,g'_m$. Then the images $g_1,\dots,g_n\in I'_3 = \mathrm{Ker}(C_3\to C)$ and liftings $\widetilde{g}'_1,\dots,\widetilde{g}'_m\in I'_3$ form generators of $I'_3$.

Express $f_1,\dots,f_n$ as linear combinations of $g_1,\dots,g_n$, and define a matrix $D_1\in \mathrm{M}_n(B_1)$ by the coefficients. Express $f_1,\dots,f_n,\widetilde{f'_1},\dots\widetilde{f'_m}$ as linear combinations of $g_1,\dots,g_n,\widetilde{g}_1,\dots,\widetilde{g}_m$, and define a matrix $D_3\in \mathrm{M}_{n+m}(C_3)$ extending $D_1$. The lower right $m\times m$ block of $D_3$ defines a matrix $D_2\in \mathrm{M}_m(C_2)$ such that the linear combinations of $g'_1,\dots,g'_m$ with coefficients in $D_2$ express $f'_1,\dots,f'_m$. Set $L_1 = A_1^{\oplus n}$, $L_2 = B_2^{\oplus m}$, $L_3 = A_3^{\oplus n+m}$, $M_1 = B_1^{\oplus n}$, $M_2 = C_2^{\oplus m}$, and $M_3 = C_3^{\oplus n+m}$. Then the matrices $D_1, D_2, D_3$ define a commutative diagram

$$\begin{array}{ccccc} L_1 \otimes_{A_1} A_3 & \longrightarrow & L_3 & \longrightarrow & L_2 \\ {\scriptstyle D_1}\downarrow & & {\scriptstyle D_3}\downarrow & & \downarrow{\scriptstyle D_2} \\ M_1 \otimes_{A_1} A_3 & \longrightarrow & M_3 & \longrightarrow & M_2 \end{array}$$

Then, since (4.2.8) are defined by the vertical arrows and since (4.2.9) are defined by the horizontal arrows, the commutativity implies that the diagram (4.2.10) is commutative. □

Let $A_1$ and $A_2$ be log rings log smooth over $A$, and let $A_1 \to B$ and $A_2 \to B$ be two strict surjections of log rings over $A$. We show a compatibility of the isomorphism (4.2.1) for them. We define a commutative diagram

$$\begin{array}{ccccc} A_1 & \longrightarrow & A_3 = A_1 \otimes_A^{\log} A_2 & \longleftarrow & A_2 \\ \downarrow & & \downarrow & & \downarrow \\ B_1 = B \otimes_A^{\log} A_1 & \longrightarrow & B_3 = B_1 \otimes_B B_2 & \longleftarrow & B_2 = B \otimes_A^{\log} A_2 \end{array} \tag{4.2.11}$$

Applying Lemma 4.2.7 to the surjections $A_1 \to B$ and $B_2 \to B = C$, we see that the lower triangle of the diagram

$$\begin{array}{ccc} \mathrm{Hom}_A(J, I) & \longrightarrow & \mathrm{Hom}_B(\wedge_B^m N_{B/A_2}, \Omega^m_{B_2/B} \otimes_{B_2} B) \\ \downarrow & \searrow & \downarrow \\ \mathrm{Hom}_B(\wedge_B^n N_{B/A_1}, \Omega^n_{B_1/B} \otimes_{B_1} B) & \longrightarrow & \mathrm{Hom}_B(\wedge_B^{n+m} N_{B/A_3}, \Omega^{n+m}_{B_3/B} \otimes_{B_3} B) \end{array} \tag{4.2.12}$$

is commutative. Since the roles of $A_1$ and $A_2$ are symmetric, the upper triangle is also commutative. This commutative diagram expresses the independence of (4.2.1) of the choice of generators.

We study a compatibility of (4.2.1) with base change. Let $(f, e) \colon (A, I) \to (B, J)$ be a finite flat morphism of log rings locally of log complete intersection. Let $(g, m) \colon (A, I) \to (A', I')$ be a morphism of log rings and assume that $A'$ is flat over $A$ and $A'/I'$ is flat over $A/I$. Then, by Proposition 4.1.13(3), $B' = B \otimes_A^{\log} A'$ is finite and locally of log complete intersection. Let $\widetilde{A}$ be a log ring log smooth over $A$, and let $\widetilde{A} \to B$ be a strict surjection over $A$. Let $\widetilde{B} = \widetilde{A} \otimes_A^{\log} B$. We consider the commutative diagram

$$\begin{array}{ccccc} \widetilde{A} & \longrightarrow & \widetilde{B} & \longrightarrow & B \\ \downarrow & & \downarrow & & \downarrow \\ \widetilde{A}' = \widetilde{A} \otimes_A^{\log} A' & \longrightarrow & \widetilde{B}' = \widetilde{B} \otimes_B B' & \longrightarrow & B' \end{array}$$

of log rings. The horizontal arrows are strict. The lower line is the base change $\otimes_{\widetilde{A}}\widetilde{A}'$ of the upper line, and hence the morphism

$$\widetilde{B}' \otimes_{\widetilde{B}} \mathrm{Hom}_{\widetilde{A}}(\widetilde{B}, \widetilde{A}) \to \mathrm{Hom}_{\widetilde{A}'}(\widetilde{B}', \widetilde{A}') \tag{4.2.13}$$

is an isomorphism of invertible $\widetilde{B}'$-modules.

The isomorphisms (4.2.13) and (4.2.2) are compatible in the sense that the diagram

$$\begin{array}{ccc} \mathrm{Hom}_A(J, I) & \longrightarrow & \mathrm{Hom}_{\widetilde{A}}(\widetilde{B}, \widetilde{A}) \\ \downarrow & & \downarrow \\ \mathrm{Hom}_{A'}(J', I') & \longrightarrow & \mathrm{Hom}_{\widetilde{A}'}(\widetilde{B}', \widetilde{A}') \end{array} \tag{4.2.14}$$

is commutative. The commutative diagram (4.2.14) implies the commutative diagram

$$\begin{array}{ccc} \mathrm{Hom}_A(J, I) \otimes_B \mathrm{Hom}_B(\Omega^n_{\widetilde{B}/B} \otimes_{\widetilde{B}} B, B) & \longrightarrow & \mathrm{Hom}_B(\wedge^n_B N_{B/\widetilde{A}}, B) \\ \downarrow & & \downarrow \\ \mathrm{Hom}_{A'}(J', I') \otimes_{B'} \mathrm{Hom}_{B'}(\Omega^n_{\widetilde{B}'/B'} \otimes_{\widetilde{B}'} B', B') & \longrightarrow & \mathrm{Hom}_{B'}(\wedge^n_{B'} N_{B'/\widetilde{A}'}, B') \end{array} \tag{4.2.15}$$

of the isomorphisms (4.2.1).

**Exercise 4.2** Let $(A, I)$ be a log ring and $t \in I$ be a generator. Let $P = S^e + t(q_1 S^{e-1} + \cdots + q_e) \in A[S] = \widetilde{A}_0$. Let $B = A[S]/P$, and $s \in B$ be the image of $S$. Let $Q = q_1 S^{e-1} + \cdots + q_e \in A[S]$ and assume that $Q(s) \in B$ is invertible. Consider $B$ as a log ring with the invertible ideal $J = sB$. Define strict surjections $\widetilde{A} = A[U^{\pm 1}, S]/(S^e - Ut) \to B$ and $\widetilde{B} = B[X^{\pm 1}] \to B$ by sending $U, S$ to $-Q(s), s$ and $X$ to 1. They are compatible with the morphism $\widetilde{A} \to \widetilde{B}$ sending $U, S$ to $-Q(s)X^e, sX$.

Find a basis $f$ of the free $B$-module $\mathrm{Hom}_A(J, I)$ of rank 1 such that the isomorphism $\mathrm{Hom}_A(J, I) \otimes_B \mathrm{Hom}_B(\Omega^1_{\widetilde{B}/B} \otimes_{\widetilde{B}} B, B) \to \mathrm{Hom}_B(N_{B/\widetilde{A}}, B)$ (4.2.1) sends $f \otimes dX^\vee$ to the dual basis $(U + Q(S))^\vee$.

## 4.3 Trace of Differential Forms with Log Poles

**Definition 4.3.1** Let $(A, I)$ be a log ring and assume that $I$ is a principal ideal. Let $I^\times$ denote the $A^\times$-torsor of generators of $I$, and let $A^{(I^\times)}$ denote the free $A$-module with the set of basis $I^\times$. For $t \in I^\times$, let $t$ also abusively denote the corresponding member of the canonical basis of $A^{(I^\times)}$. Then, we define an

$A$-module $\Omega^1_A(\log)$ by

$$\Omega^1_A(\log) = (\Omega^1_A \oplus A^{(I^\times)})/(\,(0,t) + (u^{-1}du, 0) - (0, ut); t \in I^\times, u \in A^\times,\\ (dt, 0) - t \otimes (0,t); t \in I^\times)). \tag{4.3.1}$$

For $t \in I^\times$, let $d\log t \in \Omega^1_A(\log)$ denote the image of $(0,t)$. For a morphism $(A, I) \to (B, J)$ of log rings, define a $B$-module

$$\Omega^1_{B/A}(\log/\log) = \mathrm{Coker}(\Omega^1_A(\log) \otimes_A B \to \Omega^1_B(\log)).$$

In this section we assume that the ideal $I$ for a log ring $(A, I)$ is principal.

**Lemma 4.3.2** *Let $(A, I)$ be a log ring.*

(1) *Let $t$ be a generator of $I$. We have an exact sequence*

$$0 \to A \to \Omega^1_A \oplus A \cdot d\log t \to \Omega^1_A(\log) \to 0 \tag{4.3.2}$$

*where the first morphism is defined by $(dt, -t \cdot d\log t)$.*

(2) *If $A \to B$ is strict, the canonical morphism $\Omega^1_{B/A} \to \Omega^1_{B/A}(\log/\log)$ is an isomorphism.*

*Proof* (1) The set $I^\times$ is an $A^\times$-torsor and, for $u \in A^\times$, we have $(dut, 0) - ut \otimes (0, ut) = u((dt, 0) - t \otimes (0,t)) + ut((0,t) + (u^{-1}du, 0) - (0, ut))$. Hence, by definition (4.3.1), we have $\Omega^1_A(\log) = (\Omega^1_A \oplus A)/((dt, 0) - t \otimes (0,1))$. Since $t$ is a non-zero divisor, this means (4.3.2).

(2) The assertion follows from (4.3.2) for $A, t$ and for $B, t$ by the snake lemma. □

If $I = A$, we have $\Omega^1_A(\log) = \Omega^1_A$.

**Lemma 4.3.3** *Let $(A, I) \to (\widetilde{A}, \widetilde{I})$ be a log smooth morphism of log rings.*

(1) *The $\widetilde{A}$-module $\Omega^1_{\widetilde{A}/A}(\log/\log) = \mathrm{Coker}(\Omega^1_A(\log) \otimes_A \widetilde{A} \to \Omega^1_{\widetilde{A}}(\log))$ is locally free of rank $n$ and*

$$0 \to \Omega^1_A(\log) \otimes_A \widetilde{A} \to \Omega^1_{\widetilde{A}}(\log) \to \Omega^1_{\widetilde{A}/A}(\log/\log) \to 0$$

*is a split short exact sequence.*

(2) *Let $A \to A'$ be a morphism of log rings and let $\widetilde{A}' = \widetilde{A} \otimes^{\log}_A A'$. Then the canonical morphism*

$$\Omega^1_{\widetilde{A}/A}(\log/\log) \otimes_{\widetilde{A}} \widetilde{A}' \to \Omega^1_{\widetilde{A}'/A'}(\log/\log) \tag{4.3.3}$$

*is an isomorphism.*

*Proof* We may assume that $\widetilde{A} = A[S, U^{\pm 1}]/(U^c S^e - t)$ for relatively prime integers $e \geqslant 1, c$ and a generator $t$ of $I$.

(1) By (4.3.2), we have $\Omega^1_{\widetilde{A}}(\log) = \Omega^1_{\widetilde{A}} \oplus \widetilde{A} \cdot d\log S/(dS - S \cdot d\log S)$ and $\Omega^1_{\widetilde{A}} = \Omega^1_A \otimes_A \widetilde{A} \oplus \widetilde{A} \cdot dU \oplus \widetilde{A} \cdot dS/(d(U^c S^e - t))$. Hence, (4.3.2) further defines a commutative diagram

$$\begin{array}{ccccccc} 0 \to A & \longrightarrow & \Omega^1_A \oplus A \cdot d\log t & \longrightarrow & \Omega^1_A(\log) \to 0 \\ \downarrow & & \downarrow & & \downarrow \\ 0 \to \widetilde{A} & \longrightarrow & \Omega^1_A \otimes_A \widetilde{A} \oplus \widetilde{A} \cdot dU \oplus \widetilde{A} \cdot d\log S & \longrightarrow & \Omega^1_{\widetilde{A}}(\log) \to 0 \end{array}$$

of exact sequences where the upper left arrow maps $1 \in A$ to $dt - t \cdot d\log t$ and the lower one maps $1 \in \widetilde{A}$ to $dt - t \cdot (aU^{-1}dU + c d\log S)$. The middle vertical arrow is defined by sending $d\log t$ to $cU^{-1}dU + e \cdot d\log S$. Since $e$ and $c$ are relatively prime, by the snake lemma, the right vertical arrow induces an injection $\Omega^1_A(\log) \otimes_A \widetilde{A} \to \Omega^1_{\widetilde{A}}(\log)$ and the cokernel $\Omega^1_{\widetilde{A}/A}(\log/\log)$ is a free $\widetilde{A}$-module of rank 1.

(2) We may assume that $t'$ is a generator of $I'$ and $v \in A'^{\times}$ satisfies $vt'^m = t$. Let $d = (e, m)$, $e = de'$, $m = dm'$, and $a, b$ be integers satisfying $am' - be' = 1$. Then $\widetilde{A}' = \widetilde{A} \otimes_A A'[S', W^{\pm 1}]/(W^d - v/U^c, S'^{e'} - W^a t', S'^{m'} - W^b S) = A'[S, U^{\pm 1}, S', W^{\pm 1}]/(U^c S^e - vt'^m, W^d - v/U^c, S'^{e'} - W^a t', S'^{m'} - W^b S)$ is isomorphic to $A'[U^{\pm 1}, S', W^{\pm 1}]/(W^d - v/U^c, S'^{e'} - W^a t')$. Hence, we have a commutative diagram

$$\begin{array}{ccc} \widetilde{A}' \cdot d\log v \oplus \widetilde{A}' \cdot d\log t' & \longrightarrow & \widetilde{A}' \cdot U^{-1}dU \oplus \widetilde{A}' \cdot W^{-1}dW \oplus \widetilde{A}' \cdot d\log S' \\ \uparrow & & \uparrow \\ \widetilde{A} \cdot d\log t & \longrightarrow & \widetilde{A} \cdot U^{-1}dU \oplus \widetilde{A} \cdot d\log S \end{array}$$

of free modules with the arrows given by $d\log t \mapsto cU^{-1}dU + e d\log S$, $d\log t \mapsto d\log v + m d\log t'$, $d\log S \mapsto -bW^{-1}dW + m' d\log S'$, and $d\log t' \mapsto -aW^{-1}dW + e' d\log S'$. The horizontal arrows are injections and the cokernels are $\Omega^1_{\widetilde{A}'/A'}(\log/\log)$ and $\Omega^1_{\widetilde{A}/A}(\log/\log)$ respectively. The vertical arrows with $\otimes_{\widetilde{A}} \widetilde{A}'$ on the sources are injections. Since $am' - be' = 1$, the morphism of their cokernels is an isomorphism. Hence, $\Omega^1_{\widetilde{A}/A}(\log/\log) \otimes_{\widetilde{A}} \widetilde{A}' \to \Omega^1_{\widetilde{A}'/A'}(\log/\log)$ is an isomorphism. $\square$

The highest exterior power $\Omega^n_{\widetilde{A}/A}(\log/\log) = \wedge^n_{\widetilde{A}} \Omega^1_{\widetilde{A}/A}(\log/\log)$ is a free $\widetilde{A}$-module of rank 1. If $e = 1$, then $\widetilde{A}$ is smooth over $A$ and $\Omega^1_{\widetilde{A}/A}(\log/\log) = \Omega^1_{\widetilde{A}/A}$.

Let $B$ be a quasi-finite log ring locally of log complete intersection over $A$, and let $\widetilde{A}$ be a log ring log smooth of relative dimension $n$ over $A$. Let $\widetilde{A} \to B$

be a strict surjection and let $\widetilde{B} = B \otimes_A^{\log} \widetilde{A}$. We identify $\Omega^1_{\widetilde{A}/A}(\log/\log) \otimes_{\widetilde{A}} \widetilde{B}$ with $\Omega^1_{\widetilde{B}/B}$ by the isomorphism (4.3.3). Then (4.2.1) defines an isomorphism

$$\mathrm{Hom}_A(J, I) \to \mathrm{Hom}_B(\wedge^n_B N_{B/\widetilde{A}}, \Omega^n_{\widetilde{A}/A}(\log/\log) \otimes_{\widetilde{A}} B) \tag{4.3.4}$$

of invertible $B$-modules. By the commutative diagram (4.2.12), the isomorphism (4.3.4) is independent of the choice of strict surjection $\widetilde{A} \to B$. In the case that $A \to B$ is strict and $e = 1$, the morphism $\Omega^1_{\widetilde{A}/A} \to \Omega^1_{\widetilde{A}/A}(\log/\log)$ is an isomorphism and we recover the inverse

$$\mathrm{Hom}_A(B, A) \longrightarrow \mathrm{Hom}_B(\wedge^n_B N_{B/\widetilde{A}}, \Omega^n_{\widetilde{A}/A} \otimes_{\widetilde{A}} B) \tag{4.3.5}$$

of the residue symbol [36, §9].

Let $A$ be a ring and $B$ be a ring over $A$. Assume that $B$ is a free $A$-module of finite rank $n$ and consider the isomorphism

$$B \otimes_A B \longrightarrow \mathrm{Hom}_A(\mathrm{Hom}_A(B, A), B) \tag{4.3.6}$$

sending $x \otimes b$ to the morphism sending $f$ to $xf(b)$.

**Lemma 4.3.4** *Let $A$ be a ring and $B$ be a ring over $A$. Assume that $B$ is a free $A$-module of finite rank $n$, and that the $B$-module $\mathrm{Hom}_A(B, A)$ is free of rank 1. Then the restriction*

$$\mathrm{Hom}_B(\mathrm{Hom}_A(B, A), B) \subset \mathrm{Hom}_A(\mathrm{Hom}_A(B, A), B) \to B \otimes_A B \to B \tag{4.3.7}$$

*of the composition of the inverse of the isomorphism (4.3.6) and the morphism induced by the multiplication equals the evaluation mapping at the trace mapping $\mathrm{Tr}_{B/A} \in \mathrm{Hom}_A(B, A)$.*

*Proof* Take a basis $b_1, \ldots, b_n \in B$ of the free $A$-module and let $f_1, \ldots, f_n \in \mathrm{Hom}_A(B, A)$ be the dual basis of $b_1, \ldots, b_n$. Then, by Lemma 1.2.1, we have $\mathrm{Tr}_{B/A} = \sum_{i=1}^n b_i f_i$. Let $f \in \mathrm{Hom}_A(B, A)$ be a basis of the free $B$-module of rank 1 and define $x_1, \ldots, x_n \in B$ by $f_i = x_i f$. Then the dual basis $f^\vee \in \mathrm{Hom}_B(\mathrm{Hom}_A(B, A), B)$ is the image of $\sum_{i=1}^n x_i \otimes b_i \in B \otimes_A B$ by the isomorphism (4.3.6) since $f^\vee(f_j) = x_j = \sum_{i=1}^n x_i \otimes f_j(b_i)$ for $j = 1, \ldots, n$. Since $f^\vee(\mathrm{Tr}_{B/A}) = \sum_{i=1}^n b_i x_i$, the assertion follows. □

**Proposition 4.3.5** *Assume that $A \to B$ is strict and $\widetilde{A} = A[X_1, \ldots, X_n]$. Let $d\colon N_{B/\widetilde{A}} \to \Omega^1_{\widetilde{A}/A} \otimes_{\widetilde{A}} B$ denote the $B$-linear morphism induced by the derivation. Then the isomorphism (4.3.5) maps the linear form $\mathrm{Tr}_{B/A}\colon B \to A$ to the morphism $\wedge^n d\colon \wedge^n N_{B/\widetilde{A}} \to \Omega^n_{\widetilde{A}/A} \otimes_{\widetilde{A}} B$.*

*Proof* Let $g \in \mathrm{Hom}_A(B, A)$ be the basis corresponding to the isomorphism

$\wedge^n_B N_{B/\widetilde{A}} \to \Omega^n_{\widetilde{A}/A} \otimes_{\widetilde{A}} B = \Omega^n_{\widetilde{B}/B} \otimes_{\widetilde{B}} B$ sending $f_1 \wedge \cdots \wedge f_n$ to $dX_i \wedge \cdots \wedge dX_n$ by the isomorphism (4.3.5), and let $g^\vee$ be the dual basis. Let $b_1, \dots, b_n \in B$ be the elements defining the composition $\widetilde{B} = B \otimes_A \widetilde{A} = B[X_1, \dots, X_n] \to B \otimes_A B \to B$. Then we have $g^\vee(\wedge^n d) = \det D_{ij}(b_1, \dots, b_n)$. Hence, by Lemma 4.3.4, it suffices to show that (4.3.7) maps $g^\vee$ to $\det D_{ij}(b_1, \dots, b_n)$.

For $b \in B$ and $f \in \widetilde{A}$, the composition of the surjection $\widetilde{B} = B \otimes_A \widetilde{A} \to B \otimes_A B$ with the isomorphism (4.3.6) maps $b \otimes f \in \widetilde{B}$ to the morphism sending $h\colon B \to A$ to $h(b) f|_B$. Let $\widetilde{g}\colon \widetilde{B} = B \otimes_A \widetilde{A} \to \widetilde{A}$ be the morphism induced by $g$. Then, by Proposition 4.2.4(3), we have $\widetilde{g}(\det D)|_B = 1$. Hence, (4.3.6) maps $\det D|_B \in B \otimes_A B$ to the dual basis $g^\vee \in \mathrm{Hom}_B(\mathrm{Hom}_A(B, A), B)$. Since $D_{ij}(b_1, \dots, b_n) = (\partial f_i / \partial X_j)(b_1, \dots, b_n)$, the multiplication $B \otimes_A B \to B$ maps $\det D|_B$ to $\det D_{ij}(b_1, \dots, b_n)$. □

We study the transitivity of the isomorphism (4.3.4). We consider the commutative diagram (4.2.7) of log rings. By Lemma 4.3.3(1), the sequence

$$0 \to \Omega^1_{A_1/A}(\log/\log) \otimes_{A_1} A_3 \longrightarrow \Omega^1_{A_3/A}(\log/\log) \longrightarrow \Omega^1_{A_3/A_1}(\log/\log) \to 0 \tag{4.3.8}$$

of free $A_3$-modules is exact. By Lemma 4.3.3(2), the morphism

$$\Omega^1_{A_3/A_1}(\log/\log) \otimes_{A_3} B_2 \to \Omega^1_{B_2/B}(\log/\log) \tag{4.3.9}$$

is an isomorphism. Similarly to (4.2.10), we consider the diagram

$$\begin{array}{ccc}
\mathrm{Hom}_A(J, I) \otimes_B & \longrightarrow & \mathrm{Hom}_B(\wedge^n_B N_{B/A_1}, \Omega^n_{A_1/A}(\log/\log) \otimes_{A_1} B) \otimes_B \\
\mathrm{Hom}_B(K, J) & & \mathrm{Hom}_C(\wedge^m_C N_{C/B_2}, \Omega^m_{B_2/B}(\log/\log) \otimes_{B_2} C) \\
\downarrow & & \downarrow \\
\mathrm{Hom}_A(K, I) & \longrightarrow & \mathrm{Hom}_C(\wedge^{n+m}_C N_{C/A_3}, \Omega^{n+m}_{A_3/A}(\log/\log) \otimes_{A_3} C)
\end{array} \tag{4.3.10}$$

The horizontal arrows are defined by (4.3.4) and the left vertical arrow is the composition. The right vertical arrow is defined by the exact sequences (4.2.9) and (4.3.8), and by the isomorphism (4.3.9). In the commutative diagram

$$\begin{array}{ccccc}
\Omega^1_{A_1/A}(\log/\log) \otimes_{A_1} C_1 & \longrightarrow & \Omega^1_{A_3/A}(\log/\log) \otimes_{A_3} C_3 & \longrightarrow & \Omega^1_{B_2/B}(\log/\log) \otimes_{B_2} C_2 \\
\downarrow & & \downarrow & & \downarrow \\
\Omega^1_{C_1/C} & \longrightarrow & \Omega^1_{C_3/C} & \longrightarrow & \Omega^1_{C_2/C},
\end{array}$$

the vertical arrows are isomorphisms by Lemma 4.3.3(2). Hence, the diagram (4.3.10) is commutative by Lemma 4.2.7.

Let $B$ be a log ring locally of log complete intersection over $A$. We define the trace mapping

$$\mathrm{Tr}_{B/A}\colon \Omega^1_B(\log) \to \Omega^1_A(\log) \tag{4.3.11}$$

following the method in [59, Section 16]. Let $\widetilde{A}$ be a log ring log smooth of relative dimension $n$ over $A$, and let $\widetilde{A} \to B$ be a strict surjection of log rings over $A$. Then, since the $B$-module $N_{B/\widetilde{A}}$ is locally free of rank $n$, the exact sequence $N_{B/\widetilde{A}} \to \Omega^1_{\widetilde{A}}(\log) \otimes_{\widetilde{A}} B \to \Omega^1_B(\log) \to 0$ induces a morphism

$$\Omega^1_B(\log) \otimes_B \wedge^n_B N_{B/\widetilde{A}} \to \Omega^{n+1}_{\widetilde{A}}(\log) \otimes_{\widetilde{A}} B \tag{4.3.12}$$

of $B$-modules. Since the exact sequence $0 \to \Omega^1_A(\log) \otimes_A \widetilde{A} \to \Omega^1_{\widetilde{A}}(\log) \to \Omega^1_{\widetilde{A}/A}(\log/\log) \to 0$ splits by Lemma 4.3.3(1), we have a canonical projection

$$\Omega^{n+1}_{\widetilde{A}}(\log) \to \Omega^1_A(\log) \otimes_A \Omega^n_{\widetilde{A}/A}(\log/\log). \tag{4.3.13}$$

Combining (4.3.12) and (4.3.13), we obtain a morphism

$$\Omega^1_B(\log) \otimes_B \wedge^n_B N_{B/\widetilde{A}} \to \Omega^1_A(\log) \otimes_A (\Omega^n_{\widetilde{A}/A}(\log/\log) \otimes_{\widetilde{A}} B). \tag{4.3.14}$$

This, together with the isomorphism (4.3.4), induces a $B$-linear morphism

$$\begin{aligned}\Omega^1_B(\log) &\to \mathrm{Hom}_B(\wedge^n_B N_{B/\widetilde{A}}, \Omega^n_{\widetilde{A}/A}(\log/\log) \otimes_{\widetilde{A}} B) \otimes_A \Omega^1_A(\log)\\ &\to \mathrm{Hom}_A(J, I) \otimes_A \Omega^1_A(\log).\end{aligned} \tag{4.3.15}$$

Since $\mathrm{Hom}_A(J, I) \subset \mathrm{Hom}_A(IB, I) = \mathrm{Hom}_A(B, A)$, we have

$$\Omega^1_B(\log) \to \mathrm{Hom}_A(J, I) \otimes_A \Omega^1_A(\log) \to \mathrm{Hom}_A(B, A) \otimes_A \Omega^1_A(\log). \tag{4.3.16}$$

Since $B$ is a finitely generated projective $A$-module, the functor $\mathrm{Hom}_A(B, A)\otimes_A -$: ($A$-modules) $\to$ ($B$-modules) is isomorphic to the right adjoint $\mathrm{Hom}_A(B, -)$ of the forgetful functor ($B$-modules) $\to$ ($A$-modules). Hence, the morphisms (4.3.16) define a commutative diagram

$$\begin{array}{ccc}\Omega^1_B(\log) & \xrightarrow{\mathrm{Tr}_{B/A}} & \Omega^1_A(\log)\\ \uparrow & & \uparrow\\ J \otimes_B \Omega^1_B(\log) & \longrightarrow & I \otimes_A \Omega^1_A(\log)\end{array} \tag{4.3.17}$$

Concretely, if $g \in \mathrm{Hom}_A(J, I)$ is a basis of the $B$-module and if (4.3.16) maps $\omega \in \Omega^1_B(\log)$ to $g \otimes \alpha$ for $\alpha \in \Omega^1_A(\log)$, then we have $\mathrm{Tr}_{B/A}(b \cdot \omega) = g(b) \cdot \alpha$ for $b \in J$. If $A \to B$ is strict and $e = 1$, we obtain

$$\mathrm{Tr}_{B/A}\colon \Omega^1_B \to \Omega^1_A. \tag{4.3.18}$$

The diagram (4.3.17) means that $\mathrm{Tr}_{B/A}\colon \Omega^1_B(\log) \to \Omega^1_A(\log)$ induces a morphism

$$\Omega^1_B(\log) \otimes_B B/J \to \Omega^1_A(\log) \otimes_A A/I. \tag{4.3.19}$$

Since $\mathrm{Hom}_A(J, I) \subset \mathrm{Hom}_A(B, A)$, we have a morphism $\mathrm{Hom}_A(J, I) \to \mathrm{Hom}_{A/I}(B/J, A/I)$ and hence

$$\Omega^1_B(\log) \to \mathrm{Hom}_A(J, I) \otimes_A \Omega^1_A(\log) \to \mathrm{Hom}_{A/I}(B/J, A/I) \otimes_A \Omega^1_A(\log)$$

and $\Omega^1_B(\log) \otimes_B B/J \to \mathrm{Hom}_{A/I}(B/J, A/I) \otimes_{A/I} (\Omega^1_A(\log) \otimes_A A/I)$. This corresponds to (4.3.19).

We prove some basic properties of the morphisms (4.3.17) and (4.3.18).

**Proposition 4.3.6** *Let $B$ be a finite log ring locally of log complete intersection over $A$.*

(1) *The morphisms* (4.3.17) *and* (4.3.18) *are independent of the choice of surjection $\widetilde{A} \to B$. Let $C$ be a finite log ring locally of log complete intersection over $B$. Then*

$$\mathrm{Tr}_{C/A} = \mathrm{Tr}_{B/A} \circ \mathrm{Tr}_{C/B}. \tag{4.3.20}$$

(2) *The trace mappings* (4.3.17) *and* (4.3.18) *make a commutative diagram*

$$\begin{array}{ccc} \Omega^1_B & \longrightarrow & \Omega^1_B(\log) \\ {\scriptstyle \mathrm{Tr}_{B/A}}\downarrow & & \downarrow{\scriptstyle \mathrm{Tr}_{B/A}} \\ \Omega^1_A & \longrightarrow & \Omega^1_A(\log) \end{array} \tag{4.3.21}$$

(3) *Let $A \to A'$ be a strict morphism of log rings, and let $B' = B \otimes_A A'$. Then the diagram*

$$\begin{array}{ccc} \Omega^1_B(\log) & \longrightarrow & \Omega^1_{B'}(\log) \\ {\scriptstyle \mathrm{Tr}_{B/A}}\downarrow & & \downarrow{\scriptstyle \mathrm{Tr}_{B'/A'}} \\ \Omega^1_A(\log) & \longrightarrow & \Omega^1_{A'}(\log) \end{array} \tag{4.3.22}$$

*is commutative.*

(4) *For $b \in B$ and $\omega \in \Omega^1_A(\log)$, we have*

$$\mathrm{Tr}_{B/A} db = d\,\mathrm{Tr}_{B/A} b, \qquad \mathrm{Tr}_{B/A}(b\omega) = (\mathrm{Tr}_{B/A} b) \cdot \omega.$$

(5) *For a generator $s \in J^\times$ and a unit $u \in B^\times$, we have*

$$\mathrm{Tr}_{B/A} d\log s = d\log \mathrm{N}_{B/A} s, \quad \mathrm{Tr}_{B/A} d\log u = d\log \mathrm{N}_{B/A} u.$$

(6) *Assume that* $A \to B$ *is strict and that* $\mathrm{Tr}_{B/A}\colon B \to A$ *is the* 0*-mapping. Then the compositions* $\mathrm{can} \circ \mathrm{Tr}_{B/A}\colon \Omega^1_B \to \Omega^1_A \to \Omega^1_B$ *and* $\mathrm{Tr}_{B/A} \circ \mathrm{can}\colon B \otimes_A \Omega^1_A \to \Omega^1_B \to \Omega^1_A$ *are the* 0*-mappings.*

(7) *Let* $(A, I) \to (A', I')$ *be a morphism of log rings such that* $A'/I'$ *is flat over* $A/I$*, and let* $B' = B \otimes^{\log}_A A'$*. Then the diagrams*

$$\begin{array}{ccc} \Omega^1_B(\log) & \xrightarrow{\mathrm{Tr}_{B/A}} & \Omega^1_A(\log) \\ \downarrow & & \downarrow \\ \Omega^1_{B'}(\log) & \xrightarrow{\mathrm{Tr}_{B'/A'}} & \Omega^1_{A'}(\log), \end{array} \tag{4.3.23}$$

$$\begin{array}{ccc} \Omega^1_B(\log) \otimes_B B/J & \xrightarrow{\mathrm{Tr}_{B/A}} & \Omega^1_A(\log) \otimes_A A/I \\ \downarrow & & \downarrow \\ \Omega^1_{B'}(\log) \otimes_{B'} B'/J' & \xrightarrow{\mathrm{Tr}_{B'/A'}} & \Omega^1_{A'}(\log) \otimes_{A'} A'/I' \end{array}$$

*are commutative.*

*Proof* (1) First, we show the transitivity (4.3.20). We consider the commutative diagram (4.2.7) and

$$\begin{array}{ccc} \Omega^1_C(\log) \otimes_C \wedge^m_C N_{C/B_2} \otimes_{B_2} \wedge^n_{B_2} N_{B_2/B_3} & \longrightarrow & (\Omega^{m+1}_{B_2}(\log) \otimes_{B_2} C) \otimes_B \wedge^n_B N_{B/A_1} \\ \downarrow & & \downarrow \\ \Omega^1_C(\log) \otimes_C \wedge^{n+m}_C N_{C/A_3} & & (\Omega^1_B(\log) \otimes_B \Omega^m_{B_2/B}(\log/\log) \otimes_B \wedge^n_B N_{B/A_1}) \otimes_{B_2} C \\ \downarrow & & \downarrow \\ \Omega^{n+m+1}_{A_3}(\log) \otimes_{A_3} C & \longrightarrow & \Omega^{n+1}_{A_1}(\log) \otimes_{A_1} \Omega^m_{A_3/A_1}(\log/\log) \otimes_{A_3} C \\ \downarrow & & \downarrow \\ \Omega^1_A(\log) \otimes_A \Omega^{n+m}_{A_3/A}(\log/\log) \otimes_{A_3} C & \longrightarrow & \Omega^1_A(\log) \otimes_A \Omega^n_{A_1/A}(\log/\log) \otimes_{A_1} \Omega^m_{A_3/A_1}(\log/\log) \otimes_{A_3} C \end{array}$$

The upper left vertical arrow is defined by the exact sequence $0 \to N_{B_2/B_3} \otimes_{B_2} C \to N_{C/B_3} \to N_{C/B_3} \to 0$, and the lower left vertical arrows define $\mathrm{Tr}_{C/A}$ for the surjection $A_3 \to C$. In the top horizontal arrow, the isomorphism $N_{B/A_1} \otimes_B B_2 \to N_{B_2/B_3}$ is induced by the isomorphism $B \otimes_{A_1} B_3 \to B_2$

since $A_1 \to A_3 \to B_3$ is flat by Lemma 4.1.8(1) and Proposition 4.1.13(2). The top horizontal arrow and the upper right vertical arrow define $\mathrm{Tr}_{C/B}$ for the surjection $B_2 \to C$. In the right middle vertical arrow, the isomorphism $\Omega^1_{A_3/A_1} \otimes_{A_3} B_2 \to \Omega^1_{B_2/B}$ is induced by the isomorphism $A_1 \otimes_{A_3} B \to B_2$. The lower right vertical arrows define $\mathrm{Tr}_{B/A}$ for the surjection $A_1 \to B$. The lower horizontal arrows are defined by the split short exact sequence (4.3.8) and the induced one (4.3.8) $0 \to \Omega^1_{A_1/A}(\log/\log) \to \Omega^1_{A_3/A}(\log/\log) \to \Omega^1_{A_3/A_1}(\log/\log) \to 0$.

The diagram is commutative. From this and (4.3.10), the transitivity (4.3.20) follows.

To show that the morphism (4.3.15) is independent of the choice of $\widetilde{A} \to B$, we consider the diagram (4.2.11). Then, $\mathrm{Tr}_{B/B}$ defined by the surjection $B_3 \to B$ is the identity and the transitivity proved above implies that $\mathrm{Tr}_{B/A}$ defined by the surjections $A_1 \to B$ and $A_3 \to B$ are the same. Since the roles of $A_1$ and $A_2$ are symmetric, $\mathrm{Tr}_{B/A}$ defined by the surjections $A_2 \to B$ and $A_3 \to B$ are the same. Hence, $\mathrm{Tr}_{B/A}$ is independent of the choice of $\widetilde{A} \to B$.

(2) Since the assertion is local on Spec $A$, we may assume that there is a strict surjection $\widetilde{A} = A[X_1, \ldots, X_{n-1}, S, U^{\pm 1}]/(US^e - t) \to B$ by Lemma 4.1.8(3). Modifying $\widetilde{A} \to B$ if necessary, we may assume that the composition $\widetilde{A}_0 = A[X_1, \ldots, X_{n-1}, U^{\pm 1}] \to \widetilde{A} \to B$ is a surjection. Since $B$ is locally of complete intersection by Proposition 4.1.11 (1) $\Rightarrow$ (3), the kernel of $\widetilde{A}_0 \to B$ is locally generated by $n$ elements. Then the morphism $\widetilde{A}_0 \to \widetilde{A}$ defines a commutative diagram

$$\begin{array}{ccccc}
\Omega^1_B \otimes_B \wedge^n_B N_{B/\widetilde{A}_0} & \longrightarrow & \Omega^{n+1}_{\widetilde{A}_0} \otimes_{\widetilde{A}} B & \longrightarrow & \Omega^1_A \otimes_A (\Omega^n_{\widetilde{A}_0/A} \otimes_{\widetilde{A}_0} B) \\
\downarrow & & \downarrow & & \downarrow \\
\Omega^1_B(\log) \otimes_B \wedge^n_B N_{B/\widetilde{A}} & \longrightarrow & \Omega^{n+1}_{\widetilde{A}}(\log) \otimes_{\widetilde{A}} B & \longrightarrow & \Omega^1_A(\log) \otimes_{\widetilde{A}} (\Omega^n_{\widetilde{A}/A}(\log/\log) \otimes_{\widetilde{A}} B)
\end{array}$$

From this, we obtain a commutative diagram

$$\begin{array}{ccc}
\Omega^1_B & \longrightarrow & \mathrm{Hom}_A(B, A) \otimes_A \Omega^1_A \\
\downarrow & & \downarrow \\
\Omega^1_B(\log) & \longrightarrow & \mathrm{Hom}_A(J, I) \otimes_A \Omega^1_A(\log)
\end{array}$$

and the assertion follows.

(3) We may assume that there is a strict surjection $\widetilde{A} \to B$. Let $\widetilde{A}' = \widetilde{A} \otimes_A A' \to B'$ be the strict surjection. Then the morphism $\Omega^1_{B'}(\log) \otimes_{B'} \wedge^n_{B'} N_{B'/\widetilde{A}'} \to \Omega^1_{A'}(\log) \otimes_{A'} (\Omega^n_{\widetilde{A}'/A'}(\log/\log) \otimes_{\widetilde{A}'} B')$ defining $\mathrm{Tr}_{B'/A'}$ is ob-

tained from $\Omega^1_B(\log)\otimes_B \wedge^n_B N_{B/\widetilde{A}} \to \Omega^1_A(\log)\otimes_A (\Omega^n_{\widetilde{A}/A}(\log/\log)\otimes_{\widetilde{A}} B)$ defining $\mathrm{Tr}_{B/A}$ by $\otimes_A A'$. Hence, the diagram (4.3.22) is commutative.

(4) As in the proof of (2), we may assume that there is a strict surjection $\widetilde{A} \to B$. We may further assume that this induces a surjection $\widetilde{A}_0 = A[X_1, \ldots, X_{n-1}, U^{\pm 1}] \to B$. Then, as in the proof of (2), the morphism $\widetilde{A}_0 \to \widetilde{A}$ induces a commutative diagram

$$\begin{array}{ccccc}
B \otimes_B \wedge^n_B N_{B/\widetilde{A}_0} & \longrightarrow & \Omega^n_{\widetilde{A}_0} \otimes_{\widetilde{A}_0} B & \longrightarrow & A \otimes_{\widetilde{A}_0} (\Omega^n_{\widetilde{A}_0/A} \otimes_{\widetilde{A}_0} B) \\
{\scriptstyle d\otimes 1}\downarrow & & {\scriptstyle d}\downarrow & & \downarrow{\scriptstyle d\otimes 1} \\
\Omega^1_B(\log) \otimes_B \wedge^n_B N_{B/\widetilde{A}} & \longrightarrow & \Omega^{n+1}_{\widetilde{A}}(\log) \otimes_{\widetilde{A}} B & \longrightarrow & \Omega^1_A(\log)\otimes_{\widetilde{A}} \\
& & & & (\Omega^n_{\widetilde{A}/A}(\log/\log) \otimes_{\widetilde{A}} B)
\end{array}$$

By Proposition 4.3.5, the upper line defines the trace mapping $\mathrm{Tr}_{B/A}\colon B \to A$ in the same way as $\mathrm{Tr}_{B/A}\colon \Omega^1_B(\log) \to \Omega^1_A(\log)$ is defined. Hence, we obtain the first equality.

Similarly, we have a commutative diagram

$$\begin{array}{ccccc}
B \otimes_B \wedge^n_B N_{B/\widetilde{A}_0} & \longrightarrow & \Omega^n_{\widetilde{A}_0} \otimes_{\widetilde{A}_0} B & \longrightarrow & A \otimes_{\widetilde{A}_0} (\Omega^n_{\widetilde{A}_0/A} \otimes_{\widetilde{A}_0} B) \\
{\scriptstyle \omega\otimes 1}\downarrow & & {\scriptstyle \omega\wedge}\downarrow & & \downarrow{\scriptstyle \omega\otimes 1} \\
\Omega^1_B(\log) \otimes_B \wedge^n_B N_{B/\widetilde{A}} & \longrightarrow & \Omega^{n+1}_{\widetilde{A}}(\log) \otimes_{\widetilde{A}} B & \longrightarrow & \Omega^1_A(\log)\otimes_{\widetilde{A}} \\
& & & & (\Omega^n_{\widetilde{A}/A}(\log/\log) \otimes_{\widetilde{A}} B)
\end{array}$$

and the second equality follows.

(5) First, we show the case where $A$ is a field and $B$ is étale over $A$. Let $A'$ be a separable closure. Then, since $\Omega^1_A \to \Omega^1_{A'}$ is an injection, by (3) we may assume that $A$ is separably closed. Then $B$ is isomorphic to a product $\prod_{i=1}^n A$ and for $u = (u_i) \in B^\times = \prod_{i=1}^n A^\times$ we have $\mathrm{Tr}_{B/A} d\log u = \mathrm{Tr}_{B/A}(d\log u_i) = \sum_{i=1}^n d\log u_i = d\log \prod_{i=1}^n u_i = \mathrm{N}_{B/A} d\log u$.

Since $us$ is also a generator of $J$, it suffices to prove the first equality. We prove the equality by reducing it to a certain universal case. We may assume that $B = A[X_1, \ldots, X_{n-1}, U^{\pm 1}, S]/(US^e - t, f_1, \ldots, f_n)$, and that $s$ is the image of $S$. By multiplying by a power of $U$ if necessary, we may assume that $f_1, \ldots, f_n \in A[X_1, \ldots, X_{n-1}, U, S]$. Let $N \geqslant 1$ be an integer such that $f_1, \ldots, f_n$ are polynomials of degree at most $N$. We construct a universal case. Let $I = \{(d_1, \ldots, d_{n+1}) \in \mathbf{N}^{n+1} \mid \sum_i d_i \leqslant N\}$ and let

$$F = \sum_{d\in I} C_d X_1^{d_1} \cdots X_{n-1}^{d_{n-1}} U^{d_n} S^{d_{n+1}} \in \mathbf{Z}[C_d;\ d \in I, X_1, \ldots, X_{n-1}, U^{\pm 1}, S]$$

be the universal polynomial of degree at most $N$. Let $A_0 = \mathbf{Z}[C_{d,i}, T; d \in I, i =$

$1, \ldots, n]$ and $B_0 = \mathbf{Z}[X_1, \ldots, X_{n-1}, U^{\pm 1}, S, C_{d,i}; d \in I - \{0\}, i = 1, \ldots, n]$, and regard them as log rings with the invertible ideals $A_0 T \subset A_0$ and $B_0 S \subset B_0$ respectively. We identify

$$\begin{aligned} B_0 &= \mathbf{Z}[X_1, \ldots, X_{n-1}, U^{\pm 1}, S, C_{d,i}; d \in I, i = 1, \ldots, n]/(F_1, \ldots, F_n) \\ &= A_0[X_1, \ldots, X_{n-1}, U^{\pm 1}, S]/(US^e - T, F_1, \ldots, F_n) \end{aligned}$$

and define a morphism $A_0 \to B_0$ of log rings by sending $T$ to $US^e$. Then, the coefficients of $f_1, \ldots, f_{n+1}$ define a strict morphism $A_0 \to A$ of log rings and $B_0 \otimes_{A_0} A \to B$ is an isomorphism. By [31, Théorème (13.1.3)], $U_0 = \{\mathfrak{p} \in \operatorname{Spec} A_0 \mid \dim_{k(\mathfrak{p})} B_0 \otimes_{A_0} k(\mathfrak{p}) \text{ is finite}\}$ is a dense open subscheme of $\operatorname{Spec} A_0$. Then $V_0 = \operatorname{Spec} B_0 \times_{\operatorname{Spec} A_0} U_0$ is flat over $U_0$ by Proposition 4.1.5(2), and the morphism $\operatorname{Spec} A \to \operatorname{Spec} A_0$ factors through $U_0$.

To show the equality, it suffices to show it for the localization $A_{\mathfrak{m}}$ at every maximal ideal. Hence, we may assume that $A$ is local. Since the morphism $A \to A^{\mathrm{h}}$ to the henselization is faithfully flat, we may further assume that $A$ is henselian. Let $\mathfrak{p} \in U_0 \subset \operatorname{Spec} A_0$ be the image of the closed point of $\operatorname{Spec} A$, and let $A_1$ be the henselization of the localization $A_{0,\mathfrak{p}}$. Then, by Lemma 4.1.2, $B_0 \otimes_{A_0} A_1$ is the product $B_1 \times B_2$ of finite flat ring $B_1$ over $A_1$ and $B_2$ such that $B_2 \otimes_{A_1} k(\mathfrak{p}) = 0$. Since $B_1 \otimes_{A_1} A \to B$ is an isomorphism by Nakayama's lemma, by (3) it suffices to show the equaity for $B_1$ over $A_1$.

Let $K_1$ be the fraction field of $A_1$. Since $B_0$ is smooth over $\mathbf{Z}$, $B_1 \otimes_{A_1} K_1$ is étale over $K_1$. Since $\Omega^1_{A_0}(\log)$ is a free $A_0$-module, $\Omega^1_{A_1}(\log) = \Omega^1_{A_0}(\log) \otimes_{A_0} A_1$ is also a free $A_0$-module. Hence, the equality already proved in $\Omega^1_{K_1}(\log) = \Omega^1_{A_1}(\log) \otimes_{A_1} K_1$ implies the required equality in $\Omega^1_{A_1}(\log)$.

(6) Let $\widetilde{A} = A[X_1, \ldots, X_{n-1}, X_n^{\pm 1}] \to B$ be a surjection as in the proof of (2) and (4), and assume that the kernel of the surjections $\widetilde{A} \to B$ is generated by $n$ elements $f_1, \ldots, f_n$. Let $\widetilde{B} = \widetilde{A} \otimes_A B$ and $dX_1, \ldots, dX_n$ be the basis of $\Omega^1_{\widetilde{B}/B}$. We consider the commutative diagram

$$\begin{array}{ccccc} \Omega^1_B \otimes_B \wedge^n_B N_{B/\widetilde{A}} & \longrightarrow & \Omega^{n+1}_{\widetilde{A}} \otimes_{\widetilde{A}} B & \longrightarrow & \Omega^1_A \otimes_A (\Omega^n_{\widetilde{A}/A} \otimes_{\widetilde{A}} B) \\ \downarrow & & \downarrow & & \downarrow \\ \Omega^1_B \otimes_B \wedge^n_B N_{B/\widetilde{B}} & \longrightarrow & \Omega^{n+1}_{\widetilde{B}} \otimes_{\widetilde{B}} B & \longrightarrow & \Omega^1_B \otimes_B (\Omega^n_{\widetilde{B}/B} \otimes_{\widetilde{B}} B) \end{array}$$

The upper line defines $\operatorname{Tr}_{B/A} \colon \Omega^1_B \to \Omega^1_A$ and the right vertical arrow is the tensor product of $\mathrm{can} \colon \Omega^1_A \to \Omega^1_B$ and the canonical isomorphism $\Omega^n_{\widetilde{A}/A} \otimes_{\widetilde{A}} B \to \Omega^n_{\widetilde{B}/B} \otimes_{\widetilde{B}} B$. The composition through the lower left is the tensor product of the identity of $\Omega^1_B$ and the morphism induced by $d \colon N_{B/\widetilde{A}} \to \Omega^1_{\widetilde{B}/B} \otimes_{\widetilde{B}} B$.

This composition is 0 by the assumption that $\mathrm{Tr}_{B/A}\colon B \to A$ is 0 and by Proposition 4.3.5.

We consider a basis of the free $B$-module $\mathrm{Hom}_B(\wedge^n_B N_{B/\widetilde{A}}, \Omega^n_{\widetilde{B}/B})$ of rank 1 sending $f_1 \wedge \cdots \wedge f_n$ to $dX_1 \wedge \cdots \wedge dX_n$, and let $g \in \mathrm{Hom}_A(B, A)$ be the corresponding basis by the isomorphism (4.3.5). For $\omega \in \Omega^1_B$, let $\alpha \in \Omega^1_A$ be the element such that the image of $\omega \otimes f_1 \wedge \cdots \wedge f_n$ by the composition of the upper line is $\alpha \otimes dX_1 \wedge \cdots \wedge dX_n$. By the commutative diagram, the image of $\alpha$ in $\Omega^1_B$ is 0. Since $\mathrm{Tr}_{B/A}\,\omega = g(1)\alpha$, the first composition can $\circ\, \mathrm{Tr}_{B/A}\colon \Omega^1_B \to \Omega^1_A \to \Omega^1_B$ is 0.

By the second equality in (4), the vanishing of $\mathrm{Tr}_{B/A}\colon B \to A$ implies the vanishing of the second composition $\mathrm{Tr}_{B/A} \circ \mathrm{can}\colon \Omega^1_A \to \Omega^1_B \to \Omega^1_A$.

(7) The commutative diagram (4.2.15) implies that the left square of the diagram

$$\begin{array}{ccccc}
\Omega^1_B(\log) & \to & \mathrm{Hom}_A(J, I) \otimes_A \Omega^1_A(\log) & \to & \mathrm{Hom}_{A/I}(B/J, A/I) \otimes_A \Omega^1_A(\log) \\
\downarrow & & \downarrow & & \downarrow \\
\Omega^1_{B'}(\log) & \to & \mathrm{Hom}_{A'}(J', I') \otimes_{A'} \Omega^1_{A'}(\log) & \to & \mathrm{Hom}_{A'/I'}(B'/J', A'/I') \otimes_{A'} \Omega^1_{A'}(\log)
\end{array} \tag{4.3.24}$$

is commutative. The commutativity of the right square is obtained by the commutative diagram

$$\begin{array}{ccccccccc}
0 \to & \mathrm{Hom}_A(B, I) & \longrightarrow & \mathrm{Hom}_A(J, I) & \longrightarrow & \mathrm{Hom}_{A/I}(B/J, A/I) & \to 0 \\
& \downarrow & & \downarrow & & \downarrow & \\
0 \to & \mathrm{Hom}_{A'}(B', I') & \longrightarrow & \mathrm{Hom}_{A'}(J', I') & \longrightarrow & \mathrm{Hom}_{A'/I'}(B'/J', A'/I') & \to 0
\end{array}$$

of exact sequences. By (4.3.24), we obtain commutative diagrams (4.3.23). □

We make an explicit description of (4.3.18) in the case $n = 1$.

**Proposition 4.3.7** *Let $P = X^n + a_1X^{n-1} + \cdots + a_n \in \widetilde{A} = A[X]$ and $B = A[X]/P$. Let $b \in B$ be the image of $X$, and define $Q = c_1X^{n-1} + c_2X^{n-2} + \cdots + c_n \in B[X] = \widetilde{B}$ by $P = (X - b) \cdot Q$. Let $f_1, \ldots, f_n \in \mathrm{Hom}_A(B, A)$ be the dual basis of the basis $c_1 = 1, c_2, \ldots, c_n$ of the free $A$-module $B$, as in Lemma* 1.2.1. *Then, for $c \in B$, we have*

$$\mathrm{Tr}_{B/A}(c \cdot db) = -\sum_{i=1}^{n} f_i(c) \cdot da_i. \tag{4.3.25}$$

*Proof* The morphism (4.3.14) sends $db \otimes P$ to the projection $-(da_1 \wedge b^{n-1}dX + \cdots + da_n \wedge dX) \in \Omega^1_A \otimes_A (\Omega^1_{\widetilde{A}/A} \otimes_{\widetilde{A}} B)$ of $dX \wedge dP \in \Omega^1_{\widetilde{A}} \otimes_{\widetilde{A}} B$. By Lemma 4.2.6, the basis $f_n$ of the $B$-module $\mathrm{Hom}_B(B, A)$ corresponds to the isomorphism $N_{B/\widetilde{A}} \to \Omega^1_{\widetilde{A}/A} \otimes_{\widetilde{A}} B$ sending the basis $P$ to $dX$. Since $b^{n-i} f_n = f_i$ for $i = 1, \ldots, n$, the image of $db$ by (4.3.16) equals $-\sum_{i=1}^n b^{n-i} f_n \otimes da_i = -\sum_{i=1}^n f_i \otimes da_i$. This means (4.3.25). □

**Exercise 4.3** Let $(A, tA)$, $P = S^e + t(q_1 S^{e-1} + \cdots + q_e) = S^e + tQ \in A[S] = \widetilde{A}_0$, $s \in B = A[S]/P$, and the strict surjection $\widetilde{A} = A[U^{\pm 1}, S]/(S^e - Ut) \to B$ be as in Exercise 4.2. Compute $\mathrm{Tr}_{B/A}(sc \otimes d\log s) \in I \otimes_A \Omega^1_A(\log)$ and $\mathrm{Tr}_{B/A}(c \cdot d\log s) \in \Omega^1_A(\log)$ for $c \in B$.

## 4.4 Trace for Inseparable Extensions

Let $E$ be a finite extension of a field $F$. Since the polynomial ring of one variable over a field is a principal ideal domain, the kernel of a surjection $F[X_1, \ldots, X_n] \to E$ is generated by $n$ elements, by induction on $n$. Hence, an $F$-linear mapping $\mathrm{Tr}_{E/F} \colon \Omega^1_E \to \Omega^1_F$ is defined in (4.3.18). They satisfy the following properties by Propositions 4.3.6 and 4.3.7.

**Proposition 4.4.1** *Let $E$ be a finite extension over a field $F$.*

(1) *Let $E' \subset E$ be a subextension. Then* $\mathrm{Tr}_{E/F} = \mathrm{Tr}_{E'/F} \circ \mathrm{Tr}_{E/E'}$.

(2) *For $b \in E$ and $\omega \in \Omega^1_F$, we have* $\mathrm{Tr}_{E/F}(b\omega) = (\mathrm{Tr}_{E/F} b) \cdot \omega$ *and* $\mathrm{Tr}_{E/F}(db) = d\,\mathrm{Tr}_{E/F} b$.

(3) *The diagram*

$$\begin{array}{ccc} E^\times & \xrightarrow{d\log} & \Omega^1_E \\ {\scriptstyle \mathrm{N}_{E/F}}\downarrow & & \downarrow{\scriptstyle \mathrm{Tr}_{E/F}} \\ F^\times & \xrightarrow{d\log} & \Omega^1_F \end{array} \tag{4.4.1}$$

*is commutative.*

(4) *Assume that $E = F(b)$ is generated by a single element $b$ over $F$, and let $P = X^n + a_1 X^{n-1} + \cdots + a_n \in F[X]$ be its minimal polynomial. Define $c_1 = 1, c_2, \ldots, c_n \in E$ by $P = (X - b)(c_1 X^{n-1} + c_2 X_{n-2} + \cdots + c_n)$, and let $f_1, \ldots, f_n \in \mathrm{Hom}_F(E, F)$ be the dual basis of the basis $c_1, c_2, \ldots, c_n$ of the $F$-vector space $E$ as in Lemma* 1.2.1. *Then, for $c \in E$, we have*

$$\mathrm{Tr}_{B/A}(c \cdot db) = -\sum_{i=1}^n f_i(c) \cdot da_i. \tag{4.4.2}$$

(5) *If* $\mathrm{Tr}_{E/F} \colon E \to F$ *is the* 0*-mapping, the compositions* $\mathrm{can} \circ \mathrm{Tr}_{E/F} \colon \Omega^1_E \to \Omega^1_F \to \Omega^1_E$ *and* $\mathrm{Tr}_{E/F} \circ \mathrm{can} \colon E \otimes_F \Omega^1_F \to \Omega^1_E \to \Omega^1_F$ *are the* 0*-mappings.*

We study the condition for the trace mapping $\mathrm{Tr}_{E/F}\colon \Omega^1_E \to \Omega^1_F$ to be non-zero.

**Proposition 4.4.2** *Let $F$ be a field of characteristic $p > 0$ and $E$ be a finite extension.*

(1) *If $\mathrm{Tr}_{E/F}\colon \Omega^1_E \to \Omega^1_F$ is non-zero, for every subfield $F \subset F_1 \subset E_1 \subset E$ the trace mapping $\mathrm{Tr}_{E_1/F_1}\colon \Omega^1_{E_1} \to \Omega^1_{F_1}$ is non-zero.*
(2) *The following conditions* (a) *and* (b) *are equivalent:*

  (a) *The extension $E$ is generated by a single element over $F$.*
  (b) *The trace mapping $\mathrm{Tr}_{E/F}\colon \Omega^1_E \to \Omega^1_F$ is non-zero.*

Recall that by Lemma 1.8.4 (1) $\Leftrightarrow$ (2), $E$ is generated by a single element over $F$ if and only if $E$ is generated by a single element over the separable closure $F' \subset E$ of $F$.

*Proof* (1) Since $\mathrm{Tr}_{E/F} = \mathrm{Tr}_{F_1/F} \circ \mathrm{Tr}_{E_1/F_1} \circ \mathrm{Tr}_{E/E_1}$ by Proposition 4.4.1(1), the assertion follows.

(2) We may assume that $E$ is an inseparable extension of $F$.

(a) $\Rightarrow$ (b): First we show the case where $E$ is a purely inseparable extension. Let $[E : F] = q = p^e$ and assume that $E = F(u)$ for a solution of $u^q = v$. Then $\mathrm{Tr}_{E/F} d \log u = d \log v \neq 0$ by Proposition 4.4.1(3).

We show the general case. Let $F' \subset E$ be the separable closure. Then, since $E$ is a purely inseparable extension of $F'$, the image $\mathrm{Tr}_{E/F'}(\Omega^1_E) \subset \Omega^1_{F'} = F' \otimes_F \Omega^1_F$ is a non-zero $F'$-linear subspace. Hence, the assertion follows from (1) and Lemma 4.4.3.

(b) $\Rightarrow$ (a): By Lemma 1.8.4 (3) $\Rightarrow$ (1), it suffices to show that $\dim_E \Omega^1_{E/F} = 1$. Let $E_1 = F(E^p) \subset E$ be as in the proof of Lemma 1.8.4 (3) $\Rightarrow$ (2). By (1), $\mathrm{Tr}_{E/E_1}$ is non-zero. Hence, by replacing $F$ by $E_1$, we may assume that $E^p \subset F$.

By contradiction, we assume $\dim_E \Omega^1_{E/F} = d \geqq 2$ and $[E : F] = p^d$. Then we may assume that $E = F(u_i; i = 1, \ldots, d)$ and $\Omega^1_{E/F} = \bigoplus_{i=1}^d E \cdot du_i$. Since $E \supsetneqq F(u_i)$ for $i = 1, \ldots, d$, we have $\mathrm{Tr}_{E/F(u_i)}(E \cdot du_i) = (\mathrm{Tr}_{E/F(u_i)} E) \cdot du_i = 0$ by Proposition 4.4.1(2). Hence, $\mathrm{Tr}_{E/F}(\Omega^1_{E/F}) = \sum_{i=1}^d \mathrm{Tr}_{E/F}(E \cdot du_i) = 0$ by Proposition 4.4.1(1). □

**Lemma 4.4.3** *Let $E$ be a finite extension of a field $F$, and $T\colon E \to F$ be a non-zero $F$-linear form. Let $V$ be an $F$-vector space, and $W \subset E \otimes_F V$ a non-zero $E$-linear subspace. Then $(T \otimes 1)(W) \neq 0$.*

*Proof* Since $W = 0$ if and only if the image $(1 \otimes f)(W) \subset E$ is 0 for every linear form $f\colon V \to F$, it is reduced to the case where $V = F$. Then $W \neq 0$ means $W = E$ and the assertion follows since $T\colon E \to F$ is non-zero. □

We will show that if $\mathrm{Tr}_{E/F}\colon \Omega^1_E \to \Omega^1_F$ is non-zero for a finite inseparable extension $E$ over $F$, its image equals the subspace $Z_{F\backslash E} \subset \Omega^1_F$ defined by

$$Z_{F\backslash E} = \mathrm{Ker}(\Omega^1_F \to \Omega^1_E). \tag{4.4.3}$$

**Lemma 4.4.4** *Let $F$ be a field of characteristic $p > 0$ and $E$ be a finite extension of $F$.*

(1) *If $E$ is an inseparable extension of $F$, the morphism $\mathrm{Tr}_{E/F}$ is factorized as*

$$\Omega^1_E \to \Omega^1_{E/F} \to Z_{F\backslash E} \to \Omega^1_F. \tag{4.4.4}$$

(2) *We have $\dim_F Z_{F\backslash E} \leqslant \dim_E \Omega^1_{E/F}$ and the following conditions are equivalent:*

  (a) *The inclusion $Z_{F\backslash E} \to \Omega^1_F$ induces an isomorphism*

$$Z_{F\backslash E} \otimes_F E \to \mathrm{Ker}(\Omega^1_F \otimes_F E \to \Omega^1_E). \tag{4.4.5}$$

  (b) $\dim_F Z_{F\backslash E} = \dim_E \Omega^1_{E/F}$.

The complex $[\Omega^1_F \otimes_F E \to \Omega^1_E]$ is canonically quasi-isomorphic to the cotangent complex $L_{E/F}$ studied in Section 14.1, and hence $\mathrm{Ker}(\Omega^1_F \otimes_F E \to \Omega^1_E)$ is isomorphic to the cohomology group $\mathrm{H}_1(L_{E/F})$.

*Proof* (1) Since $\mathrm{Tr}_{E/F}\colon E \to F$ is 0, the morphism $\mathrm{Tr}_{E/F}\colon \Omega^1_E \to \Omega^1_F$ factors through the quotient $\Omega^1_{E/F} = \Omega^1_E/(E \otimes_F \Omega^1_F)$ by Proposition 4.4.1(5). Since the composition can $\circ\, \mathrm{Tr}_{E/F}\colon \Omega^1_E \to \Omega^1_F \to \Omega^1_E$ is 0 by Proposition 4.4.1(5), the image of $\mathrm{Tr}_{E/F}\colon \Omega^1_E \to \Omega^1_F$ is a subspace of $Z_{F\backslash E}$.

(2) We have $\dim_E \mathrm{Ker}(\Omega^1_F \otimes_F E \to \Omega^1_E) = \dim_E \Omega^1_{E/F}$ by [15, Chapter V, §16.6, Theorem 4]. Since the inclusion $Z_{F\backslash E} \to \Omega^1_F$ induces an injection $Z_{F\backslash E} \otimes_F E \to \mathrm{Ker}(\Omega^1_F \otimes_F E \to \Omega^1_E)$, the equality is equivalent to the isomorphism. □

**Proposition 4.4.5** *Let $F$ be a field of characteristic $p > 0$ and $E$ be a finite inseparable extension. Assume that $E$ is generated by a single element over $F$.*

(1) *The image of $\mathrm{Tr}_{E/F}\colon \Omega^1_E \to \Omega^1_F$ equals $Z_{F\backslash E}$ and we have $\dim Z_{F\backslash E} = 1$. Further, the morphism $Z_{F\backslash E} \otimes_F E \to \mathrm{Ker}(\Omega^1_F \otimes_F E \to \Omega^1_E)$ (4.4.5) is an isomorphism.*

(2) *Suppose that $E = F(u)$ is a purely inseparable extension of $F$ of degree $p$. Then $\mathrm{Tr}_{E/F}\colon \Omega^1_E \to \Omega^1_F$ induces an isomorphism $\Omega^1_{E/F}/F \cdot dE \to Z_{F\backslash E}$ of $F$-vector spaces of dimension 1.*

*Proof* (1) By Lemma 4.4.4(2), we have $\dim_F Z_{F\backslash E} \leqslant \dim_E \Omega^1_{E/F} = 1$. By Lemma 4.4.4(1), we have an inclusion $\mathrm{Tr}_{E/F}(\Omega^1_E) \subset Z_{F\backslash E}$. Since $\mathrm{Tr}_{E/F}\colon \Omega^1_E \to$

$\Omega^1_F$ is non-zero by Proposition 4.4.2(2) (a) $\Rightarrow$ (b), we have $\mathrm{Tr}_{E/F}(\Omega^1_E) = Z_{F\backslash E}$ and $\dim_F Z_{F\backslash E} = 1$.

Since $\dim_F Z_{F\backslash E} = \dim_E \Omega^1_{E/F} = 1$, the morphism (4.4.5) is an isomorphism by Lemma 4.4.4(2).

(2) The $E$-vector space $\Omega^1_{E/F} = F du$ and the $F$-vector space $\Omega^1_{E/F}/F \cdot dE$ are of dimension 1. Hence by Proposition 4.4.1(2) and (1), $\mathrm{Tr}_{E/F}\colon \Omega^1_E \to \Omega^1_F$ induces a non-zero morphism $\Omega^1_{E/F}/F \cdot dE \to Z_{F\backslash E}$ of $F$-vector spaces of dimension 1. □

**Corollary 4.4.6** *Let $E$ be a finite inseparable extension of a field $F$ of characteristic $p > 0$. Assume that $E$ is generated by a single element. Let $E' \subsetneq E$ be a subextension such that $E$ is an inseparable extension of $E'$. Then the canonical morphism $Z_{F\backslash E} \to Z_{F'\backslash E}$ induces an isomorphism*

$$Z_{F\backslash E} \otimes_F F' \to Z_{F'\backslash E} \tag{4.4.6}$$

*and we have $\mathrm{Tr}_{E'/F}(Z_{E'\backslash E}) = Z_{F\backslash E}$.*

*Proof* Since (4.4.6) is an injection of vector spaces of dimension 1, it is an isomorphism. We have $\mathrm{Tr}_{E'/F}(Z_{E'\backslash E}) = \mathrm{Tr}_{E'/F}(\mathrm{Tr}_{E/E'}(\Omega^1_E)) = \mathrm{Tr}_{E/F}(\Omega^1_E) = Z_{F\backslash E}$ by Proposition 4.4.5(1) and Proposition 4.4.1(1). □

**Proposition 4.4.7** *Let $E$ be a purely inseparable extension of $F$ generated by a single element, and let $[E : F] = q > 1$.*

(1) *The $F$-vector space $Z_{F\backslash E}$ is generated by the image of $\mathrm{Tr}_{E/F} \circ d\log\colon E^\times \to \Omega^1_F$ and the inclusion $E^q \to F$ induces an isomorphism*

$$\Omega^1_{E^q/F^q} \otimes_{E^q} F \to Z_{F\backslash E} \tag{4.4.7}$$

*of $F$-vector spaces of dimension 1.*

(2) *There exists a unique isomorphism $i = i_{E/F}\colon Z_{F\backslash E} \otimes_F E \to (\Omega^1_{E/F})^{\otimes q}$ of $E$-vector spaces of dimension 1 satisfying $i(\mathrm{Tr}_{E/F}(d\log x)) = (d\log x)^{\otimes q}$ for every $x \in E^\times$.*

*Proof* (1) Assume $E = F(u)$ and let $v = u^q \in F$. Then the image of $\Omega^1_{E^q/F^q} \otimes_{E^q} F \to \Omega^1_F$ is generated by $dv$. Since $dv = v \cdot d\log v$ is a non-zero element of $Z_{F\backslash E}$, the space $Z_{F\backslash E}$ is generated by $d\log v = \mathrm{Tr}_{E/F} d\log u \in \mathrm{Tr}_{E/F}(d\log E^\times) \subset \mathrm{Tr}_{E/F}(\Omega^1_E) = Z_{F\backslash E}$ by Proposition 4.4.1(3).

Since $\Omega^1_{E^q} \to \Omega^1_E$ is the 0-mapping and since $\Omega^1_{E^q/F^q}$ is generated by $dv$, (4.4.7) is an isomorphism.

(2) Since $Z_{F\backslash E}$ is generated by the image of $\mathrm{Tr}_{E/F'} \circ d\log\colon E^\times \to \Omega^1_{F'}$, the uniqueness follows.

For an $E$-vector space $V$ of dimension 1 and $V \otimes_E E$ with respect to the morphism $E \to E$ sending $x$ to $x^q$, the morphism $V \to V^{\otimes q}$ sending $v$ to $v^{\otimes q}$ induces an isomorphism $V \otimes_E E \to V^{\otimes q}$ of an $E$-vector space of dimension 1. Thus, if $V'$ denotes $V \otimes_E E^q$ with respect to the isomorphism $E \to E^q$ sending $x$ to $x^q$, we obtain an isomorphism $V' \otimes_{E^q} E = V \otimes_E E \to V^{\otimes q}$ of an $E$-vector space of dimension 1.

Applying this construction to $V = \Omega^1_{E/F}$ and $V' = \Omega^1_{E^q/F^q}$, we obtain a canonical isomorphism $\Omega^1_{E^q/F^q} \otimes_{E^q} E \to (\Omega^1_{E/F})^{\otimes q}$. Taking the composition with the isomorphism $Z_{F\backslash E} \otimes_F E \to \Omega^1_{E^q/F^q} \otimes_{E^q} E$ induced by the inverse of (4.4.7), we define an isomorphism $i\colon Z_{F\backslash E} \otimes_F E \to (\Omega^1_{E/F})^{\otimes q}$.

We show that this isomorphism satisfies the required condition. Since $\Omega^1_{E/F} = \Omega^1_{E/FE^p}$, it suffices to show that $i(\mathrm{Tr}_{E/F}\, d\log u) = (d\log u)^{\otimes q}$ for every generator $u$ of $E$ over $F$. For $v = u^q \in E^q \subset F$, we have $\mathrm{Tr}_{E/F}\, d\log u = d\log \mathrm{N}_{L/K}\, u = d\log v$ by Proposition 4.4.1(3). Since the isomorphism

$$\Omega^1_{E^q/F^q} \otimes_{E^q} E \to (\Omega^1_{E/F})^{\otimes q}$$

maps $d\log v$ to $(d\log u)^{\otimes q}$, we have $i(\mathrm{Tr}_{L/K}\, d\log u) = (d\log u)^{\otimes q}$. □

**Exercise 4.4** Let $E = F(u)$ be a purely inseparable extension of degree $p$ over a field $F$ of characteristic $p > 0$. Show that $\mathrm{Tr}_{E/F}\colon \Omega^1_{E/F} \to \Omega^1_F$ is the unique $F$-linear morphism such that the diagram (4.4.1) is commutative.

## Historical Notes

The trace mapping $\Omega^1_E \to \Omega^1_F$ for an inseparable extension is used in [49, (2.3)] and considered to be a morphism induced by the norm map of $K$ theory. The definition here follows the method in [59] using the computation of $\mathrm{Hom}_A(B, A)$ in [36].

# 5

# The Hasse–Arf Theorem

For cyclic extensions, a relation between the cohomological filtration and the lower numbering filtration is given by the Hasse–Arf theorem, Theorem 5.3.4. We give equivalent conditions for the equality of the Swan conductor using the trace mapping of the differential forms, and also for the refined Swan conductor. The first case of the theorem, where the character is of order $p$, has already been essentially established in Sections 3.1 and 3.2, by studying Artin–Schreier extensions and Kummer extensions.

We study the trace mappings of differential forms with log poles for extensions of discrete valuation rings in Section 5.1. We prove its compatibility with the corestriction morphism of Brauer groups in Section 5.2. This is a crucial ingredient of the proof given in Section 5.3 of the induction step of the Hasse–Arf theorem. In the classical case where the residue field is finite, we deduce the equality of the cohomological filtration with that defined by the reciprocity mapping from the Hasse–Arf theorem.

## 5.1 Trace for Discrete Valuation Rings

We consider a discrete valuation ring $O_K$ as a log ring $(O_K, \mathfrak{m}_K)$ with the maximal ideal. Let $L$ be a finite extension of $K$. Assume that the integral closure $O_L$ of $O_K$ is a discrete valuation ring and that $O_L$ is of finite type over $O_K$. Let $F$ and $E$ be the residue fields of $K$ and $L$. Let $t$, $s$ be uniformizers of $O_K$ and $O_L$, and let $e = e_{L/K}$ be the ramification index. Define a unit $u \in O_L^\times$ by $t = us^e$, and a surjection $O_K[X_1, \ldots, X_{n-1}, S, U^{\pm 1}]/(US^e - t) \to O_L$ sending $U, S$ to $u, s$. Since a discrete valuation ring and $O_K[X_1, \ldots, X_{n-1}, S, U^{\pm 1}]/(US^e - t)$ are regular, the kernel is locally generated by $n$ elements. Hence, $O_L$ is locally of log complete intersection over $O_K$ and $O_K$-linear mappings $\mathrm{Tr}_{O_L/O_K} : \Omega^1_{O_L}(\log) \to$

$\Omega^1_{O_K}(\log)$ and $\mathrm{Tr}_{O_L/O_K}\colon \Omega^1_{O_L} \to \Omega^1_{O_K}$ are defined in (4.3.17) and (4.3.18). They satisfy the following properties by Proposition 4.3.6.

**Proposition 5.1.1** *Let $L$ be a finite extension of a discrete valuation field $K$. Assume that the integral closure $O_L$ of $O_K$ is a discrete valuation ring and that $O_L$ is finite over $O_K$. Let $F$ and $E$ be the residue fields of $K$ and $L$.*

(1) $\mathrm{Tr}_{O_L/O_K}(\mathfrak{m}_L\Omega^1_{O_L}(\log)) \subset \mathfrak{m}_K\Omega^1_{O_K}(\log)$.
(2) *Let $M \subset L$ be a subextension. Then* $\mathrm{Tr}_{O_L/O_K} = \mathrm{Tr}_{O_M/O_K} \circ \mathrm{Tr}_{O_L/O_M}$.
(3) *For $b \in O_L$ and $\omega \in \Omega^1_{O_K}(\log)$,* $\mathrm{Tr}_{O_L/O_K}(b\omega) = (\mathrm{Tr}_{O_L/O_K} b)\cdot\omega$.
(4) *The diagrams*

$$\begin{array}{ccc} \Omega^1_{O_L} & \longrightarrow & \Omega^1_{O_L}(\log) \\ {\scriptstyle \mathrm{Tr}_{O_L/O_K}}\downarrow & & \downarrow{\scriptstyle \mathrm{Tr}_{O_L/O_K}} \\ \Omega^1_{O_K} & \longrightarrow & \Omega^1_{O_K}(\log), \end{array} \qquad \begin{array}{ccc} L^\times & \xrightarrow{d\log} & \Omega^1_{O_L}(\log) \\ {\scriptstyle \mathrm{N}_{L/K}}\downarrow & & \downarrow{\scriptstyle \mathrm{Tr}_{O_L/O_K}} \\ K^\times & \xrightarrow{d\log} & \Omega^1_{O_K}(\log) \end{array} \tag{5.1.1}$$

*are commutative.*

**Lemma 5.1.2** *Let $K$ be a henselian discrete valuation field and $K'$ be a tamely ramified extension.*

(1) *The morphism* $\Omega^1_{O_K}(\log) \otimes_{O_K} O_{K'} \to \Omega^1_{O_{K'}}(\log)$ *is an isomorphism.*
(2) *Let $L$ be an extension of $K$. Then the log tensor product $O_L \otimes^{\log}_{O_K} O_{K'}$ is the integral closure $\prod_i O_{L'_i}$ of $O_L$ in $L \otimes_K K' = \prod_i L'_i$.*
*Let* $\mathrm{Tr}_{O_L\otimes^{\log}_{O_K} O_{K'}/O_{K'}}\colon \Omega^1_{O_L\otimes^{\log}_{O_K} O_{K'}}(\log) \to \Omega^1_{O_{K'}}(\log)$ *denote the direct sum of* $\mathrm{Tr}_{O_{L'_i}/O_{K'}}\colon \Omega^1_{O_{L'_i}}(\log) \to \Omega^1_{O_{K'}}(\log)$. *Then the diagram*

$$\begin{array}{ccc} \Omega^1_{O_L}(\log) & \longrightarrow & \Omega^1_{O_L\otimes^{\log}_{O_K} O_{K'}}(\log) \\ {\scriptstyle \mathrm{Tr}_{O_L/O_K}}\downarrow & & \downarrow{\scriptstyle \mathrm{Tr}_{O_L\otimes^{\log}_{O_K} O_{K'}/O_{K'}}} \\ \Omega^1_{O_K}(\log) & \longrightarrow & \Omega^1_{O_{K'}}(\log) \end{array} \tag{5.1.2}$$

*is commutative.*

*Proof* (1) By considering the maximal unramified extension of $K$ in $K'$, it suffices to prove the cases where $K'$ is an unramified extension and $K'$ is a totally and tamely ramified extension respectively. Let $t$ be a uniformizer of $K$. Then we have an isomorphism $(\Omega^1_{O_K} \oplus O_K \cdot d\log t)/(dt - t\cdot d\log t) \to \Omega^1_{O_K}(\log)$. If $K'$ is an unramified extension, then $O_K \to O_{K'}$ is étale by Lemma 1.5.2(1) and $\Omega^1_{O_K} \otimes_{O_K} O_{K'} \to \Omega^1_{O_{K'}}$ is an isomorphism. Since $t$ is also a uniformizer of $K'$, this is extended to an isomorphism $\Omega^1_{O_K}(\log) \otimes_{O_K} O_{K'} \to \Omega^1_{O_{K'}}(\log)$.

Assume that $K'$ is a totally and tamely ramified extension. Then, by Proposition 1.5.5(2), by replacing $t$ if necessary, we may assume that there exists a uniformizer $t'$ of $K'$ such that $t = t'^m$ for $m = [K' : K]$. Then we have $O_{K'} = O_K[t']/(t'^m - t)$ and $\Omega^1_{O_{K'}} = (\Omega^1_{O_K} \otimes_{O_K} O_{K'} \oplus O_{K'} \cdot dt')/(dt - mt'^{m-1}dt')$. Hence, we have $\Omega^1_{O_{K'}}(\log) = (\Omega^1_{O_K} \otimes_{O_K} O_{K'} \oplus O_{K'} \cdot d\log t')/(dt - mt'^m d\log t')$ and the morphism $\Omega^1_{O_K}(\log) \otimes_{O_K} O_{K'} = (\Omega^1_{O_K} \otimes_{O_K} O_{K'} \oplus O_{K'} \cdot d\log t)/(dt - td\log t) \to \Omega^1_{O_{K'}}(\log)$ sending $d\log t$ to $md\log t'$ is an isomorphism.

(2) Let $t$ be a uniformizer of $O_K$ and define charts for $O_K \to O_L$ and $O_K \to O_{K'}$ by $t = us^e$ and $t = vt'^m$. Then, in the notation of the proof of Proposition 4.1.13, $s'$ is a non-zero divisor of $O_L \otimes^{\log}_{O_K} O_{K'}$ and $O_L \otimes^{\log}_{O_K} O_{K'}/(s')$ is $E \otimes_F F'[w]/(w^d - u/v)$. Since $d \mid m$ is invertible in $F$, this is étale over $E$ and hence $O_L \otimes^{\log}_{O_K} O_{K'}$ is a regular ring of dimension 1.

The commutative diagram (5.1.2) is the first of (4.3.23). □

Recall that $\Omega^1_{O_K}(\log) \otimes_{O_K} F$ is denoted by $\Omega^1_F(\log)$. By Proposition 5.1.1(1), the morphism $\mathrm{Tr}_{O_L/O_K} : \Omega^1_{O_L}(\log) \to \Omega^1_{O_K}(\log)$ induces

$$\mathrm{Tr}_{L/K} : \Omega^1_E(\log) \to \Omega^1_F(\log). \tag{5.1.3}$$

This is a special case of (4.3.19). If the ramification index is 1, the diagram

$$\begin{array}{ccc} \Omega^1_E & \longrightarrow & \Omega^1_E(\log) \\ {\scriptstyle \mathrm{Tr}_{E/F}}\big\downarrow & & \big\downarrow{\scriptstyle \mathrm{Tr}_{L/K}} \\ \Omega^1_F & \longrightarrow & \Omega^1_F(\log). \end{array} \tag{5.1.4}$$

is commutative by the left commutative diagram of (5.1.1) and Proposition 4.3.6(7). We will generalize this in Proposition 5.1.6(1).

Define

$$\Omega^1_{E/F}(\log/\log) = \mathrm{Coker}(\Omega^1_F(\log) \otimes_F E \to \Omega^1_E(\log)), \tag{5.1.5}$$

$$Z_{F\backslash E}(\log) = \mathrm{Ker}(\Omega^1_F(\log) \to \Omega^1_E(\log)). \tag{5.1.6}$$

Notice that although $L$ and $K$ do not appear in the notation, these vector spaces depend on $K \to L$, not just on $F \to E$.

**Lemma 5.1.3** *Let $L$ be a finite separable extension of a henselian discrete valuation field $K$.*

(1) *We have*

$$\dim_E \mathrm{Ker}(\Omega^1_F(\log) \otimes_F E \to \Omega^1_E(\log)) = \dim_E \Omega^1_{E/F}(\log/\log). \tag{5.1.7}$$

(2) *We have* $\dim_F Z_{F\backslash E}(\log) \leqslant \dim_E \Omega^1_{E/F}(\log/\log)$ *and the following conditions are equivalent:*

(a) *The inclusion* $Z_{F\setminus E}(\log) \to \Omega^1_F(\log)$ *induces an isomorphism*

$$Z_{F\setminus E}(\log) \otimes_F E \to \operatorname{Ker}(\Omega^1_F(\log) \otimes_F E \to \Omega^1_E(\log)). \quad (5.1.8)$$

(b) *We have* $\dim_F Z_{F\setminus E}(\log) = \dim_E \Omega^1_{E/F}(\log/\log)$.

(3) *If the ramification index* $e_{L/K}$ *is prime to* $p$, *the inclusion* $\Omega^1_E \to \Omega^1_E(\log)$ *induces an isomorphism* $\Omega^1_{E/F} \to \Omega^1_{E/F}(\log/\log)$.

*If* $E$ *is a separable extension of* $F$ *and if* $e_{L/K}$ *is divisible by* $p$, *then the residue mapping* $\operatorname{res}\colon \Omega^1_E(\log) \to E$ *induces an isomorphism* $\operatorname{res}\colon \Omega^1_{E/F}(\log/\log) \to E$.

(4) *The following conditions are equivalent:*

(a) $\Omega^1_{E/F}(\log/\log) = 0$.

(b) *The canonical morphism*

$$E \otimes_F \Omega^1_F(\log) \to \Omega^1_E(\log) \quad (5.1.9)$$

*is an isomorphism.*

(c) $L$ *is tamely ramified.*

(5) *If* $L$ *is not tamely ramified over* $K$, *then* $\operatorname{Tr}_{L/K}\colon \Omega^1_E(\log) \to \Omega^1_F(\log)$ *factors through* $\Omega^1_{E/F}(\log/\log)$.

*Proof* We consider the commutative diagram

$$\begin{array}{ccccccccc} 0 & \longrightarrow & \Omega^1_F \otimes_F E & \longrightarrow & \Omega^1_F(\log) \otimes_F E & \longrightarrow & E & \longrightarrow & 0 \\ & & \downarrow & & \downarrow & & \downarrow{\scriptstyle e_{L/K}} & & \\ 0 & \longrightarrow & \Omega^1_E & \longrightarrow & \Omega^1_E(\log) & \longrightarrow & E & \longrightarrow & 0. \end{array} \quad (5.1.10)$$

(1) By the snake lemma and by the equality $\dim_E \operatorname{Ker}(\Omega^1_F \otimes_F E \to \Omega^1_E) = \dim_E \Omega^1_{E/F}$ ([15, Chapter V, §16.6, Theorem 4]), we obtain (5.1.7).

(2) Similarly to the proof of Lemma 4.4.4(2), the inclusion $Z_{F\setminus E}(\log) \to \Omega^1_F(\log)$ induces an injection $Z_{F\setminus E}(\log) \otimes_F E \to \operatorname{Ker}(\Omega^1_F(\log) \otimes_F E \to \Omega^1_E(\log))$ and we obtain $\dim_F Z_{F\setminus E}(\log) \leqslant \dim_E \Omega^1_{E/F}(\log/\log)$ by (5.1.7). Further, the equality is equivalent to the isomorphism (5.1.8).

(3) If the ramification index $e_{L/K}$ is invertible in $E$, the right vertical arrow in (5.1.10) is an isomorphism and $\Omega^1_{E/F} \to \Omega^1_{E/F}(\log/\log)$ is an isomorphism.

If $E$ is a separable extension of $F$, the left vertical arrow in (5.1.10) is an isomorphism. Further, if the right vertical arrow is the 0-mapping, the residue mapping induces an isomorphism $\Omega^1_{E/F}(\log/\log) \to E$.

(4) We obtain (a) $\Leftrightarrow$ (b) by (1) and (c) $\Rightarrow$ (b) by Lemma 5.1.2(1).

(b) $\Rightarrow$ (c): Assume that the middle arrow in (5.1.10) is an isomorphism.

Then the multiplication by $e_{L/K}$ induces a surjection and $e_{L/K}$ is invertible in $E$. Hence, we have $\Omega^1_{E/F} = 0$ by (3) and $E$ is a separable extension of $F$.

(5) If $L$ is not a tamely ramified extension, then we have $\mathrm{Tr}_{L/K}(O_L) \subset \mathfrak{m}_K$ by Proposition 1.9.1(2). Hence, the restriction of $\mathrm{Tr}_{L/K}\colon \Omega^1_E(\log) \to \Omega^1_F(\log)$ on the image of $E \otimes_F \Omega^1_F(\log)$ is 0 by Proposition 5.1.1(3). □

**Lemma 5.1.4** *Let $L$ be a tamely ramified extension of $K$ and let $e_{L/K}$ be the ramification index. Then the composition of the isomorphism $E \otimes_F \Omega^1_F(\log) \to \Omega^1_E(\log)$ (5.1.9) with $\mathrm{Tr}_{L/K}\colon \Omega^1_E(\log) \to \Omega^1_F(\log)$ equals $e_{L/K} \cdot \mathrm{Tr}_{E/F} \otimes 1$ and is a surjection. The diagram*

$$\begin{array}{ccccccccc} 0 & \longrightarrow & \Omega^1_E & \longrightarrow & \Omega^1_E(\log) & \xrightarrow{\ \mathrm{res}\ } & E & \longrightarrow & 0 \\ & & {\scriptstyle e_{L/K}\cdot \mathrm{Tr}_{E/F}}\big\downarrow & & {\scriptstyle \mathrm{Tr}_{L/K}}\big\downarrow & & \big\downarrow{\scriptstyle \mathrm{Tr}_{E/F}} & & \\ 0 & \longrightarrow & \Omega^1_F & \longrightarrow & \Omega^1_F(\log) & \xrightarrow{\ \mathrm{res}\ } & F & \longrightarrow & 0 \end{array} \tag{5.1.11}$$

*is commutative.*

*Proof* For $b \in O_L$ and $\omega \in \Omega^1_{O_K}(\log)$, we have $\mathrm{Tr}_{L/K}(b\,\omega) = (\mathrm{Tr}_{L/K} b)\omega = e_{L/K} \cdot (\mathrm{Tr}_{E/F} \bar{b})\omega$ by Proposition 5.1.1(3). Hence, the composition equals $e_{L/K} \cdot \mathrm{Tr}_{E/F} \otimes 1$. Since $e_{L/K}$ is invertible and $E$ is a separable extension, $e_{L/K} \cdot \mathrm{Tr}_{E/F}$ is a surjection.

If we identify the upper line of (5.1.11) with $\otimes_F E$ of the lower line, then the vertical arrows are $e_{L/K} \cdot \mathrm{Tr}_{E/F}$ and the diagram is commutative. □

We give a characterization of log monogenic extensions in terms of the dimension of $\Omega^1_{E/F}(\log/\log)$.

**Proposition 5.1.5** *Let $L$ be a finite separable extension of a henselian discrete valuation field $K$.*

(1) *The following conditions are equivalent:*
   (a) *The $E$-vector space $\Omega^1_{E/F}(\log/\log)$ is of dimension 1.*
   (b) *The extension $L$ is a log monogenic extension of $K$ and is a wildly ramified extension of $K$.*

(2) *Assume that $L$ is a log monogenic extension of $K$. Then $\mathrm{Tr}_{L/K}\colon \Omega^1_E(\log) \to \Omega^1_F(\log)$ is non-zero and $\dim_F Z_{F\setminus E}(\log) \leqslant 1$.*

We will later prove a partial converse of Proposition 5.1.5(2).

*Proof* (1) Assume first that $e_{L/K}$ is prime to $p$. Then, by Lemma 5.1.3(3), condition (a) is equivalent to $\dim_E \Omega^1_{E/F} = 1$. Hence by Lemma 1.8.4 (1) $\Leftrightarrow$ (3), this is further equivalent to the condition that $E$ is an inseparable extension generated by a single element over $F$. Thus, this is equivalent to (b) by Lemma 1.8.3(2) in this case.

We may assume that $K$ equals the maximal tamely ramified extension $M \subset L$ by Lemma 5.1.3(4). We may assume that $e_{L/K} > 1$ is a power of $p$. We consider the commutative diagram (5.1.10). If $\Omega^1_{E/F} = 0$, then $E$ is a separable extension and the equivalence follows from Lemma 1.8.3(1) in this case. Assume otherwise. Then condition (a) is equivalent to the boundary mapping $E \to \Omega^1_{E/F}$ being an isomorphism by the snake lemma. Let $t, s$ be uniformizers of $M, L$ respectively, and define $u \in O_L^\times$ by $u = t/s^{e_{L/M}}$. Since $p \mid e_{L/M}$, we have $d\log u \equiv d\log t$. Hence, the condition is further equivalent to the purely inseparable extension $E$ being generated by $u$ over $F_M$ by Lemma 1.8.4. This means that $L$ is a log monogenic extension of $K$ and the equivalence is proved.

(2) First, we show that $\mathrm{Tr}_{L/K}\colon \Omega^1_E(\log) \to \Omega^1_F(\log)$ is non-zero in the case where $L$ is a purely wildly ramified extension of $K$. The morphism $d\log\colon K^\times/(1+\mathfrak{m}_K)K^{\times p} \to \Omega^1_F(\log)$ is an injection by Proposition 2.4.6(2). Hence, by the second commutative diagram in (5.1.1), if $\mathrm{N}_{L/K}\colon L^\times/(1+\mathfrak{m}_L)L^{\times p} \to K^\times/(1+\mathfrak{m}_K)K^{\times p}$ is non-zero, so $\mathrm{Tr}_{L/K}\colon \Omega^1_E(\log) \to \Omega^1_F(\log)$ is non-zero.

We prove the general case. Let $M \subset L$ be the maximal tamely ramified extension and $F_M$ be the residue field. Then, since $L$ is a purely wildly ramified extension of $M$, the image $\mathrm{Tr}_{L/M}(\Omega^1_E(\log)) \subset \Omega^1_{F_M}(\log) = F_M \otimes_F \Omega^1_F(\log)$ is a non-zero $F'$-linear subspace. Hence, the assertion follows from the chain rule Proposition 5.1.1(2), Lemma 5.1.4, and Lemma 4.4.3.

We have $\dim_F Z_{F\backslash E}(\log) \leqslant \dim_E \Omega^1_{E/F}(\log/\log) = 1$ by Lemma 5.1.3(2) and (1). □

In this section, we prove the following.

**Proposition 5.1.6** *Let $L$ be a finite separable extension of a henselian discrete valuation field $K$.*

(1) *The diagram*

$$\begin{array}{ccccccccc} 0 & \longrightarrow & \Omega^1_E & \longrightarrow & \Omega^1_E(\log) & \xrightarrow{\mathrm{res}} & E & \longrightarrow & 0 \\ & & \Big\downarrow{\scriptstyle e_{L/K}\cdot\mathrm{Tr}_{E/F}} & & \Big\downarrow{\scriptstyle \mathrm{Tr}_{L/K}} & & \Big\downarrow{\scriptstyle \mathrm{Tr}_{E/F}} & & \\ 0 & \longrightarrow & \Omega^1_F & \longrightarrow & \Omega^1_F(\log) & \xrightarrow{\mathrm{res}} & F & \longrightarrow & 0 \end{array} \tag{5.1.12}$$

*is commutative.*

(2) *The following conditions are equivalent:*

(a) *$L$ is a log monogenic extension of $K$.*

(b) $\mathrm{Tr}_{L/K}\colon \Omega^1_E(\log) \to \Omega^1_F(\log)$ *is non-zero.*

(3) *Assume that L is log monogenic and is not tamely ramified. Then the image of* $\mathrm{Tr}_{L/K}\colon \Omega^1_E(\log) \to \Omega^1_F(\log)$ *equals* $Z_{F\backslash E}(\log)$, *and* $\dim Z_{F\backslash E}(\log) = 1$. *Further, the inclusion* $Z_{F\backslash E}(\log) \to \Omega^1_F(\log)$ *induces an isomorphism*

$$Z_{F\backslash E}(\log) \otimes_F E \to \mathrm{Ker}(\Omega^1_F(\log) \otimes_F E \to \Omega^1_E(\log)). \tag{5.1.13}$$

We will prove Proposition 5.1.6 after studying the totally and wildly ramified cases and, separately, the case where the ramification index is 1 and the residue field extension is inseparable. First, we consider the totally ramified case.

**Proposition 5.1.7** *Let L be a totally ramified extension of K of degree n (divisible by p).*

(1) *The kernel* $Z_{F\backslash E}(\log)$ *is an F-vector space of dimension 1 generated by* $d\log \mathrm{N}_{L/K} s$ *for a uniformizer* $s \in L$. *The restriction of the residue mapping* $\mathrm{res}\colon \Omega^1_F(\log) \to F$ *defines an isomorphism* $Z_{F\backslash E}(\log) \to F$.
(2) *The morphism* $\mathrm{Tr}_{L/K}\colon \Omega^1_E(\log) \to \Omega^1_F(\log)$ *is factorized as*

$$\Omega^1_E(\log) \to \Omega^1_{E/F}(\log/\log) \to Z_{F\backslash E}(\log) \to \Omega^1_F(\log)$$

*and the middle arrow* $\Omega^1_{E/F}(\log/\log) \to Z_{F\backslash E}(\log)$ *is an isomorphism. The diagram*

$$\begin{array}{ccccccccc} 0 & \longrightarrow & \Omega^1_E & \longrightarrow & \Omega^1_E(\log) & \xrightarrow{\mathrm{res}} & E & \longrightarrow & 0 \\ & & \downarrow{\scriptstyle 0} & & \downarrow{\scriptstyle \mathrm{Tr}_{L/K}} & & \downarrow{\scriptstyle \mathrm{Tr}_{E/F}=1_F} & & \\ 0 & \longrightarrow & \Omega^1_F & \longrightarrow & \Omega^1_F(\log) & \xrightarrow{\mathrm{res}} & F & \longrightarrow & 0. \end{array} \tag{5.1.14}$$

*is commutative.*

*Proof* (1) Let $s \in L$ be a uniformizer. Then $t = d\log \mathrm{N}_{L/K} s \in K$ is also a uniformizer and we have $t \equiv (-1)^{n-1}s^n \bmod \mathfrak{m}_K\mathfrak{m}_L$ by Proposition 1.4.2(3). Hence, $d\log t = n \cdot d\log s = 0$ in $\Omega^1_E(\log)$ and $d\log t \in Z_{F\backslash E}(\log)$. Since $\mathrm{res}\, d\log t = 1$, we have $d\log t \neq 0$ and $d\log t$ is a basis of $Z_{F\backslash E}(\log)$ by Proposition 5.1.5(2). Further, the residue mapping defines an isomorphism $Z_{F\backslash E}(\log) \to F$ of $F$-vector spaces of dimension 1.

(2) By Lemma 5.1.3(5), $\mathrm{Tr}_{L/K}\colon \Omega^1_E(\log) \to \Omega^1_F(\log)$ factors through the quotient $\Omega^1_{E/F}(\log/\log)$. The $E = F$-vector space $\Omega^1_{E/F}(\log/\log)$ of dimension 1 is generated by $d\log s$ for a uniformizer $s \in L^\times$ by Lemma 5.1.3(3). Since $\mathrm{Tr}_{L/K} d\log s = d\log \mathrm{N}_{L/K} s \in \Omega^1_F(\log)$ by Proposition 5.1.1(4), the first assertion follows from (1).

Since $\Omega^1_E \subset \Omega^1_E(\log)$ equals the image of $\Omega^1_F(\log)$, the restriction of $\mathrm{Tr}_{L/K}$ on $\Omega^1_E$ is 0 by Lemma 5.1.3(5) and the left square of (5.1.14) is commutative.

Since $\operatorname{res} d\log t = 1$ and $\operatorname{res} d\log \mathrm{N}_{L/K}t = f_{L/K}\cdot 1 = 1$, the right square in (5.1.14) is commutative. □

Next, we consider the case where the ramification index is 1 and $E$ is an inseparable extension of $F$.

**Proposition 5.1.8** *Assume that the ramification index $e_{L/K}$ is 1 and $E$ is an inseparable extension of $F$.*

(1) *The inclusion $\Omega^1_F \to \Omega^1_F(\log)$ induces an equality $Z_{F\backslash E} = Z_{F\backslash E}(\log)$.*
(2) *The morphism $\mathrm{Tr}_{L/K}\colon \Omega^1_E(\log) \to \Omega^1_F(\log)$ is factorized as*

$$\Omega^1_E(\log) \to \Omega^1_{E/F}(\log/\log) \to Z_{F\backslash E}(\log) \to \Omega^1_F(\log)$$

*and the diagram*

$$\begin{array}{ccccccccc} 0 & \longrightarrow & \Omega^1_E & \longrightarrow & \Omega^1_E(\log) & \xrightarrow{\mathrm{res}} & E & \longrightarrow & 0 \\ & & \downarrow{\scriptstyle \mathrm{Tr}_{E/F}} & & \downarrow{\scriptstyle \mathrm{Tr}_{L/K}} & & \downarrow{\scriptstyle 0} & & \\ 0 & \longrightarrow & \Omega^1_F & \longrightarrow & \Omega^1_F(\log) & \xrightarrow{\mathrm{res}} & F & \longrightarrow & 0. \end{array} \tag{5.1.15}$$

*is commutative. If $[L:K] = p$, the middle arrow induces an isomorphism $\Omega^1_{E/F}(\log/\log)/dE \to Z_{F\backslash E}(\log)$.*

*Proof* (1) Since the ramification index is 1, the right vertical arrow of the commutative diagram

$$\begin{array}{ccccccccc} 0 & \longrightarrow & \Omega^1_F & \longrightarrow & \Omega^1_F(\log) & \xrightarrow{\mathrm{res}} & F & \longrightarrow & 0 \\ & & \downarrow & & \downarrow & & \downarrow & & \\ 0 & \longrightarrow & \Omega^1_E & \longrightarrow & \Omega^1_E(\log) & \xrightarrow{\mathrm{res}} & E & \longrightarrow & 0, \end{array} \tag{5.1.16}$$

is the inclusion. Hence, the assertion follows.

(2) By Lemma 5.1.3(5), $\mathrm{Tr}_{L/K}$ factors through the surjection $\Omega^1_E(\log) \to \Omega^1_{E/F}(\log/\log)$. Since, by Lemma 5.1.3(3), $\Omega^1_{E/F} \to \Omega^1_{E/F}(\log/\log)$ is an isomorphism, the factorization follows from Lemma 4.4.4(1) and (1). The left square of (5.1.15) is the commutative diagram (5.1.4). Let $t$ be a uniformizer of $K$. Then, since $\mathrm{Tr}_{L/K}(b\, d\log t) = (\mathrm{Tr}_{E/F} b) d\log t = 0$ by Proposition 5.1.1(4), the right square is also commutative.

If $[L : K] = [E : F] = p$, the isomorphism $\Omega^1_{E/F}(\log/\log)/F\cdot dE \to Z_{F\backslash E}(\log)$ follows from Proposition 4.4.5(2). □

*Proof of Proposition* 5.1.6 Let $N$ be a Galois closure of $L$ over $K$, and let $M' \subset N$ be the maximal tamely ramified extension over $K$. Then $L' = LM'$ is a tamely ramified extension of $L$ and, by Corollary 1.5.11, there exists a sequence

$M' = L'_0 \subsetneqq L'_1 \subsetneqq \cdots \subsetneqq L'_e = L'$ of extensions such that, for $i = 1, \ldots, e$, $L'_i$ is a wildly ramified extension of $L'_{i-1}$ of degree $p$. Let $E'$ denote the residue field of $L'$.

(1) Since $L'$ is a tamely ramified extension of $L$, $\mathrm{Tr}_{L'/L}\colon \Omega^1_{E'}(\log) \to \Omega^1_E(\log)$ and its restriction $\Omega^1_{E'} \to \Omega^1_E$ are surjections and the diagram (5.1.12) for $L'$ over $L$ is commutative by Lemma 5.1.4. Since $M'$ is a tamely ramified extension of $K$, the diagram (5.1.12) for $M'$ over $K$ is also commutative by Lemma 5.1.4. Hence, it suffices to show that the diagram (5.1.12) for $L'_i$ over $L'_{i-1}$ is commutative for $i = 1, \ldots, e$. Thus, we may assume that $L$ is a wildly ramified extension of degree $p$. In this case, the assertion holds by Proposition 5.1.7(2) and Proposition 5.1.8(2).

(2) Since (a) $\Rightarrow$ (b) is already proved in Proposition 5.1.5(2), it suffices to prove (b) $\Rightarrow$ (a). Since $L'$ is a tamely ramified extension of $L$, the morphism $\mathrm{Tr}_{L'/L}\colon \Omega^1_{E'}(\log) \to \Omega^1_E(\log)$ is a surjection by Lemma 5.1.4. Hence, by Proposition 5.1.1(2), the condition $\mathrm{Tr}_{L/K} \neq 0$ implies $\mathrm{Tr}_{L'/M'} \neq 0$ and $\mathrm{Tr}_{L'_i/L'_{i-2}} \neq 0$ for every $i = 2, \ldots, e$. Since $L'$ is a tamely ramified extension of $L$, it suffices to show that $L'$ is a log monogenic extension of $M'$ by Proposition 1.8.5(1). Further, by Corollary 1.8.6(2), it suffices to show that $L'_i$ is a log monogenic extension of $L'_{i-2}$ for $i = 2, \ldots, e$.

Hence, we may assume that $L$ is a purely wildly ramified extension of $K$ of degree $p^2$, and there exists a subextension $K \subsetneqq K' \subsetneqq L$. By Propositions 5.1.1(2), 5.1.7(2), and 5.1.8(2), if $e_{L/K'} = 1$ then we have $e_{K'/K} = 1$. Further, by Proposition 4.4.2(2), in this case the residue field $E$ is a purely inseparable extension of $F$ generated by a single element and $L$ is a log monogenic extension of $K$. Assume that $L$ is a totally ramified extension of $K'$. It suffices to consider the case where $e_{K'/K} = 1$. Let $t, s$ be uniformizers of $K$ and $L$, let $u = \mathrm{N}_{L/K'}s/t \in O_{K'}$, and let $F'$ denote the residue field of $F'$. Then the image of $\mathrm{Tr}_{L/K'}\colon \Omega^1_E(\log) \to \Omega^1_{F'}(\log)$ is the $F'$-vector space generated by $\mathrm{Tr}_{L/K'} d\log s = d\log \mathrm{N}_{L/K'}s = d\log u + d\log t$ by Propositions 5.1.7(2) and 5.1.1(4). By the assumption that $\mathrm{Tr}_{L/K} = \mathrm{Tr}_{K'/K} \circ \mathrm{Tr}_{L/K'} \neq 0$, we have $\mathrm{Tr}_{K'/K}(F' \cdot (d\log u + d\log t)) \neq 0$. Since $\mathrm{Tr}_{K'/K}(F' \cdot d\log t) = (\mathrm{Tr}_{F'/F}F') \cdot d\log t = 0$ and since $\mathrm{Tr}_{F'/F}\colon \Omega^1_{F'} \to \Omega^1_F$ factors through $\Omega^1_{F'/F}$, we have $d\log u \neq 0$ in $\Omega^1_{F'/F}$. Hence, $u$ is a generator of a purely inseparable extension $F'$ over $F$, and $L$ is a log monogenic extension of $K$.

(3) We first show $\mathrm{Im}(\mathrm{Tr}_{L/K}\colon \Omega^1_E(\log) \to \Omega^1_F(\log)) \subset Z_{F\backslash E}(\log)$. Since $L \cap M' = M$ and $M \otimes_K M'$ is isomorphic to the product of copies of $M'$, the tensor product $L \otimes_K M'$ is also isomorphic to the product of copies of $L'$. By Lemma 5.1.2(2), the log tensor product $O_L \otimes^{\log}_{O_K} O'_M$ is isomorphic to the integral closure $O_{L\otimes_K M'} = \prod_i O_{L'_i}$ of $O_K$ in $L \otimes_K M' = \prod_i L'_i$. Let $F'$ and $E'$

denote the residue field of $M'$ and of $L'$, and let $E'_i$ denote the residue fields of factors $L'_i$ of $L \otimes_K M'$. Then the left square in the diagram

$$\begin{array}{ccccc} \Omega^1_E(\log) & \xrightarrow{\mathrm{Tr}_{L/K}} & \Omega^1_F(\log) & \longrightarrow & \Omega^1_E(\log) \\ \downarrow & & \downarrow & & \downarrow \\ \bigoplus_i \Omega^1_{E'_i}(\log) & \xrightarrow{\sum_i \mathrm{Tr}_{L'_i/M'}} & \Omega^1_{F_{M'}}(\log) & \longrightarrow & \Omega^1_{E'}(\log) \end{array}$$

is commutative. By the functoriality, the right square is commutative. By Propositions 5.1.7(2) and 5.1.8(2), the composition of lower horizontal arrows is 0. Since the right vertical arrow is injective by Lemma 5.1.2(1), the composition of upper horizontal arrows is 0 and we have $\mathrm{Im}(\mathrm{Tr}_{L/K} : \Omega^1_E(\log) \to \Omega^1_F(\log)) \subset Z_{F\backslash E}(\log)$.

Since $\mathrm{Im}(\mathrm{Tr}_{L/K} : \Omega^1_E(\log) \to \Omega^1_F(\log)) \neq 0$ by Proposition 5.1.5(2), and $\dim_F Z_{F\backslash E}(\log) \leqslant \dim_E \Omega^1_{E/F}(\log/\log) = 1$ by Lemma 5.1.3(2) and Proposition 5.1.5(1), the inclusion $\mathrm{Im}(\mathrm{Tr}_{L/K} : \Omega^1_E(\log) \to \Omega^1_F(\log)) \subset Z_{F\backslash E}(\log)$ is an equality and we have $\dim_F Z_{F\backslash E}(\log) = 1$.

Since $\dim_F Z_{F\backslash E}(\log) = \dim_E \Omega^1_{E/F}(\log/\log) = 1$, the morphism (5.1.13) is an isomorphism by Lemma 5.1.3(2). □

**Corollary 5.1.9** *Let $L$ be a log monogenic extension of a henselian discrete valuation field $K$, and $M \subset L$ be a subextension. Assume that $L$ over $K$ and $L$ over $M$ are wildly ramified. Let $F_M$ denote the residue field.*

(1) *Assume that $M$ is wildly ramified over $K$. Then the composition*

$$Z_{F_M\backslash E}(\log) \to \Omega^1_{F_M}(\log) \to \Omega^1_{F_M/F}(\log/\log)$$

*is an isomorphism of $F_M$-vector spaces of dimension* 1.

(2) *The trace morphism* $\mathrm{Tr}_{M/K} : \Omega^1_{F_M}(\log) \to \Omega^1_F(\log)$ *induces a surjection* $Z_{F_M\backslash E}(\log) \to Z_{F\backslash E}(\log)$.

(3) *If $M$ is tamely ramified over $K$, the inclusion $\Omega^1_F(\log) \to \Omega^1_{F_M}(\log)$ induces an isomorphism $Z_{F\backslash E}(\log) \otimes_F F_M \to Z_{F_M\backslash E}(\log)$.*

*Proof* (1) Since $\mathrm{Tr}_{L/K} : \Omega^1_E(\log) \to \Omega^1_F(\log)$ is non-zero and is factorized as $\Omega^1_E(\log) \to Z_{F_M\backslash E}(\log) \to \Omega^1_{F_M}(\log) \to \Omega^1_{F_M/F}(\log/\log) \to \Omega^1_F(\log)$ by Proposition 5.1.6(3), the assertion follows.

(2) Since $Z_{F_M\backslash E}(\log)$ is the image of $\mathrm{Tr}_{L/M} : \Omega^1_E(\log) \to \Omega^1_{F_M}(\log)$ and $Z_{F\backslash E}(\log)$ is the image of $\mathrm{Tr}_{L/K} = \mathrm{Tr}_{M/K} \circ \mathrm{Tr}_{L/M}$, the morphism $\mathrm{Tr}_{M/K}$ defines a surjection $Z_{F_M\backslash E}(\log) \to Z_{F\backslash E}(\log)$.

(3) Since the morphism $Z_{F\backslash E}(\log) \otimes_F F_M \to Z_{F_M\backslash E}(\log)$ is an injection of vector spaces of dimension 1, it is an isomorphism. □

The following elementary properties will be used in a reduction step in the proof of Theorem 5.3.4.

**Lemma 5.1.10** *Let $L$ be a finite separable extension of $K$, and $K'$ be a finite tamely ramified extension of $K$. Set $L' = LK'$.*

(1) *Let $d = \mathrm{ord}_L D^{\log}_{L/K}$ and $d' = \mathrm{ord}_{L'} D^{\log}_{L'/K'}$ be the valuations of the logarithmic differents. Let $n$ be an integer and set $m = e_{L/K}n - d$, $n' = e_{K'/K}n$, and $m' = e_{L'/K'}n' - d'$. Then $d' = e_{L'/L}d$, $m' = e_{L'/L}m$, and the diagram*

$$\begin{array}{ccc} \mathfrak{m}_L^m/\mathfrak{m}_L^{m+1} & \xleftarrow{e_{L'/L}\cdot \mathrm{Tr}_{E'/E}} & \mathfrak{m}_{L'}^{m'}/\mathfrak{m}_{L'}^{m'+1} \\ {\scriptstyle \mathrm{Tr}_{L/K}}\downarrow & & \downarrow{\scriptstyle \mathrm{Tr}_{L'/K'}} \\ \mathfrak{m}_K^n/\mathfrak{m}_K^{n+1} & \xleftarrow{e_{K'/K}\cdot \mathrm{Tr}_{F'/F}} & \mathfrak{m}_{K'}^{n'}/\mathfrak{m}_{K'}^{n'+1}, \end{array}$$

*is commutative.*

(2) *The diagram*

$$\begin{array}{ccc} \Omega^1_E(\log) & \xleftarrow{e_{L'/L}\cdot \mathrm{Tr}_{E'/E}} & \Omega^1_{E'}(\log) \\ {\scriptstyle \mathrm{Tr}_{L/K}}\downarrow & & \downarrow{\scriptstyle \mathrm{Tr}_{L'/K'}} \\ \Omega^1_F(\log) & \xleftarrow{e_{K'/K}\cdot \mathrm{Tr}_{F'/F}} & \Omega^1_{F'}(\log) \end{array}$$

*is commutative.*

*Proof* (1) Since $K'$ over $K$ and $L'$ over $L$ are tamely ramified, we have $\mathrm{ord}_{K'} D^{\log}_{K'/K} = \mathrm{ord}_{L'} D^{\log}_{L'/L} = 0$. Hence, by the chain rule (1.9.3), we have $e_{L'/L}d = \mathrm{ord}_{L'} D^{\log}_{L'/K} = d'$. The equality $m' = e_{L'/L}m$ follows from this and the chain rule for the ramification indices. By the commutative diagram (1.9.1), the horizontal arrows are induced by $\mathrm{Tr}_{L'/L}$ and $\mathrm{Tr}_{K'/K}$. Hence, the diagram is commutative.

(2) Since the horizontal arrows are $\mathrm{Tr}_{L'/L}$ and $\mathrm{Tr}_{K'/K}$ by Lemma 5.1.4, the assertion follows from Proposition 5.1.1(3). □

**Exercise 5.1** Let $L$ be a totally ramified extension of $K$ of degree $n$. Let $s \in L$ be a uniformizer and $P = X^n + a_1X^{n-1} + \cdots + a_n \in O_K[X]$ be the minimal polynomial. Using Exercise 4.3, compute $\mathrm{Tr}_{L/K} d\log s \in \Omega^1_F(\log) = \Omega^1_{O_K}(\log) \otimes_{O_K} F$.

## 5.2 Trace and Corestriction

We prove the compatibility of the morphism $\beta\colon \Omega^1_F(\log) \to \operatorname{Br} K$ (2.5.9) with the trace mappings.

**Proposition 5.2.1** *Let $L$ be a finite separable extension of a henselian discrete valuation field $K$. Then the diagram*

$$\begin{array}{ccc} \Omega^1_E(\log) & \xrightarrow{\beta_L} & \operatorname{Br} L \\ {\scriptstyle \operatorname{Tr}_{L/K}}\downarrow & & \downarrow{\scriptstyle \operatorname{Cor}_{L/K}} \\ \Omega^1_F(\log) & \xrightarrow{\beta_K} & \operatorname{Br} K \end{array} \tag{5.2.1}$$

*is commutative.*

*Proof* We divide the proof into the following steps.

(1) $L$ is a totally ramified extension of degree divisible by $p$.

(2) The degree $[L : K]$ is $p$, the ramification index $e_{L/K}$ is 1, and the residue field $E$ is an inseparable extension.

(3) $L$ is a tamely ramified extension.

(4) The general case.

(1) For $x \in E = F$ and $y \in L^\times$, we have $\operatorname{Cor}_{L/K}\beta_L(x \cdot d\log y) = \operatorname{Cor}_{L/K}\{\alpha_x, y\}_L = \{\alpha_x, \mathrm{N}_{L/K}y\}_K \in \operatorname{Br} K$ by Lemma 2.1.5(2). On the other hand, we have $\operatorname{Tr}_{L/K}(x \cdot d\log y) = x \cdot d\log \mathrm{N}_{L/K}y$ by Proposition 5.1.1. Hence, $\beta_K(\operatorname{Tr}_{L/K}(x \cdot d\log y)) = \beta_K(x \cdot d\log \mathrm{N}_{L/K}y) = \{\alpha_x, \mathrm{N}_{L/K}y\}_K$ and the diagram (5.2.1) is commutative.

(2) The trace mapping $\operatorname{Tr}_{L/K}\colon \Omega^1_E(\log) \to \Omega^1_F(\log)$ factors through the quotient $\Omega^1_{E/F}(\log/\log)$ by Lemma 5.1.3(5). Further, it factors through the quotient $\Omega^1_{E/F}(\log/\log)/F \cdot dE = \Omega^1_{E/F}/F \cdot dE$ by Proposition 5.1.8(2).

We show that the composition through the upper right $\operatorname{Br} L$ also annihilates the images of $F \cdot dE$ and $\Omega^1_F(\log) \otimes_F E$. For $a \in F$ and $b \in E$, we have $a \cdot db = d(ab) - b \cdot da$. Since $\beta_L$ annihilates $dE$ by the construction (2.5.7), it suffices to show that the composition annihilates the image of $\Omega^1_F(\log) \otimes_F E$. For $x \in E$ and $y \in K^\times$, we have $z = x^p \in F$ and $\operatorname{Cor}_{L/K}\beta_L(x \cdot d\log y) = \operatorname{Cor}_{L/K}\{\alpha_x, y\}_L = \operatorname{Cor}_{L/K}\{\alpha_z, y\}_L = p\{\alpha_z, y\}_K = 0$ by Lemma 2.1.5(1).

Since $\Omega^1_{E/F}(\log/\log)/F \cdot dE = \Omega^1_{E/F}/F \cdot dE$ is an $F$-vector space of dimension 1 generated by $d\log y$ for $y \in E$ such that $E = F(y)$, the same computation as in (1) shows the assertion.

(3) The $E$-vector space $\Omega^1_E(\log)$ is generated by $x \cdot d\log y$ for $x \in E$ and $y \in K^\times$ by Lemma 5.1.4. Since $\operatorname{Cor}_{L/K}\beta_L(x \cdot d\log y) = \operatorname{Cor}_{L/K}\{\alpha_x, y\} =$

$e_{L/K}\cdot\{\alpha_{\mathrm{Tr}_{E/F}x}, y\} = \beta_K(e_{L/K}\cdot \mathrm{Tr}_{E/F}x\cdot d\log y)$ by Lemma 2.1.5(2), the assertion follows from $\mathrm{Tr}_{L/K} = e_{L/K}\cdot\mathrm{Tr}_{E/F}\otimes 1$.

(4) Since the morphism $\mathrm{Tr}_{L'/L}\colon \Omega^1_{E'}(\log)\to\Omega^1_E(\log)$ is a surjection for a tamely ramified extension $L'$ over $L$ by Lemma 5.1.4, by the same argument as in the proof of Proposition 5.1.6 using Corollary 1.5.11, the assertion is reduced to the cases (1), (2), and (3). □

We show that to prove the existence of the refined Swan conductor, it suffices to do so after taking the tamely ramified extension. This will be used in the first reduction step of the proof of Theorem 5.3.4.

**Proposition 5.2.2** *Let $\chi\in X_K$ be a character, and let $K'$ be a finite tamely ramified Galois extension of $K$ of ramification index $e = e_{K'/K}$. Let $\chi'\in X_{K'}$ be the restriction of $\chi$. Assume that* $\mathrm{Sw}\,\chi' = n'\geqslant 1$ *and that the refined Swan conductor* $\mathrm{rsw}\,\chi'\colon \mathfrak{m}^{n'}_{K'}/\mathfrak{m}^{n'+1}_{K'}\to\Omega^1_{F'}(\log)$ *exists. Set* $\mathrm{Sw}\,\chi = n\geqslant 0$.

(1) $n' = en$.
(2) *The refined Swan conductor* $\mathrm{rsw}\,\chi\colon \mathfrak{m}^n_K/\mathfrak{m}^{n+1}_K\to\Omega^1_F(\log)$ *exists and* $\mathrm{rsw}\,\chi'$ *is the scalar extension of* $\mathrm{rsw}\,\chi$.

*Proof* (1) Let $K_1$ be as in Definition 2.3.1 and let $K'_1 = K'K_1$. Since the refined Swan conductor $\mathrm{rsw}\,\chi'\colon \mathfrak{m}^{n'}_{K'}/\mathfrak{m}^{n'+1}_{K'}\to\Omega^1_{F'}(\log)$ is an injection invariant under the action of the Galois group $\mathrm{Gal}(K'/K)$, and the inertia group $I\subset\mathrm{Gal}(K'/K)$ acts trivially on the target, the Swan conductor $n' = \mathrm{Sw}\,\chi$ is divisible by $e$ by Proposition 1.6.2(2). Let $n_1 = n'/e\geqslant 1$. Then there exists a unique $F$-linear mapping $f\colon \mathfrak{m}^{n_1}_K/\mathfrak{m}^{n_1+1}_K\to\Omega^1_F(\log)$ such that the diagrams

$$\begin{array}{ccc} \mathfrak{m}^{en_1}_{K'}/\mathfrak{m}^{en_1+1}_{K'} & \xrightarrow{\mathrm{rsw}\,\chi'} & \Omega^1_{F'}(\log) \\ {\scriptstyle \mathrm{Tr}_{K'/K}}\downarrow & & \downarrow{\scriptstyle \mathrm{Tr}_{K'/K}} \\ \mathfrak{m}^{n_1}_K/\mathfrak{m}^{n_1+1}_K & \xrightarrow{f} & \Omega^1_F(\log), \end{array} \qquad \begin{array}{ccc} \mathfrak{m}^{en_1}_{K'_1}/\mathfrak{m}^{en_1+1}_{K'_1} & \xrightarrow{\mathrm{rsw}\,\chi'\otimes 1} & \Omega^1_{F'_1}(\log) \\ {\scriptstyle \mathrm{Tr}_{K'_1/K_1}}\downarrow & & \downarrow{\scriptstyle \mathrm{Tr}_{K'_1/K_1}} \\ \mathfrak{m}^{n_1}_{K_1}/\mathfrak{m}^{n_1+1}_{K_1} & \xrightarrow{f\otimes 1} & \Omega^1_{F_1}(\log) \end{array} \tag{5.2.2}$$

are commutative. We have

$$\{\chi, 1+\mathfrak{m}^{n_1+1}_{K_1}\} = \{\chi, \mathrm{N}_{K'_1/K_1}(1+\mathfrak{m}^{n'+1}_{K'_1})\} = \mathrm{Cor}_{K'_1/K_1}\{\chi', 1+\mathfrak{m}^{n'+1}_{K'_1}\} = 0$$

by Proposition 1.10.3(2) (a) $\Rightarrow$ (b) and by Lemma 2.1.5(2), and we have $n\leqslant n_1$. Further, we have

$$\begin{aligned} \{\chi, 1+\mathfrak{m}^{n_1}_{K_1}\} &= \{\chi, \mathrm{N}_{K'_1/K_1}(1+\mathfrak{m}^{n'}_{K'_1})\} = \mathrm{Cor}_{K'_1/K_1}\{\chi, 1+\mathfrak{m}^{n'}_{K'_1}\} \\ &= \mathrm{Cor}_{K'_1/K_1}\beta_{K'_1}((\mathrm{rsw}\,\chi'\otimes 1)(\mathfrak{m}^{n'}_{K'_1}/\mathfrak{m}^{n'+1}_{K'_1})) \\ &= \beta_{K_1}(\mathrm{Tr}_{K'_1/K_1}(\mathrm{rsw}\,\chi'\otimes 1)(\mathfrak{m}^{n'}_{K'_1}/\mathfrak{m}^{n'+1}_{K'_1})) \end{aligned}$$

by Proposition 1.10.3(2) (a) $\Rightarrow$ (b), Lemma 2.1.5(2), and by Proposition 5.2.1. Similarly, we have $\{\chi, 1+\mathfrak{m}_K^{n_1+1}\} = 0$. By the right commutative square in (5.2.2) and by the surjectivity of its left vertical arrow, the right-hand side equals $\beta_{K_1}((f\otimes 1)(\mathfrak{m}_{K_1}^n/\mathfrak{m}_{K_1}^{n+1}))$ and is non-zero by Lemma 2.5.6(1). Thus, we have $n = n_1 = n'/e$.

(2) The commutative diagrams (5.2.2) together with Lemma 1.10.1(2), Lemma 2.1.5(2), and Proposition 5.2.1 imply that, for $x \in \mathfrak{m}_{K'}^{en}$ and for $x_1 \in \mathfrak{m}_{K_1'}^{en}$, we have

$$\begin{aligned}\{\chi, 1+\mathrm{Tr}_{K'/K}x\} &= \{\chi, \mathrm{N}_{K'/K}(1+x)\} = \mathrm{Cor}_{K'/K}\{\chi', 1+x\}\\ &= \mathrm{Cor}_{K'/K}\beta_{K'}(\mathrm{rsw}\chi'(\bar{x})) = \beta_K(\mathrm{Tr}_{K'/K}\mathrm{rsw}\chi'(\bar{x})) = \beta_K(f(\mathrm{Tr}_{K'/K}(\bar{x}))),\end{aligned}$$

and $\{\chi, 1+\mathrm{Tr}_{K_1'/K_1}x_1\} = \beta_{K_1}((f\times 1)(\mathrm{Tr}_{K_1'/K_1}(\bar{x}_1)))$ similarly. Since the left vertical arrows in (5.2.2) are surjections, by Lemma 2.5.6(1) we have $f = \mathrm{rsw}\,\chi$. □

Let $K_u \subset K_t \subset K_s$ be the maximal unramified extension and the maximal tamely ramified extension in a separable closure $K_s$ of $K$, and let $K_a \subset K_s$ be the maximal abelian extension. Recall that we say that $\chi \in X_K$ is unramified (resp. tamely ramified) if the corresponding cyclic extension is unramified (resp. tamely ramified).

**Corollary 5.2.3** *Let $K$ be a henselian discrete valuation field with residue field $F$ of characteristic $p > 0$.*

(1) *We have $F_0X_K = \mathrm{Ker}(X_K \to X_{K_t})$. The abelian extension of $K$ corresponding to $G_K^{\mathrm{ab}\,1} \subset G_K^{\mathrm{ab}} = \mathrm{Gal}(K_a/K)$ equals the maximal tamely ramified abelian extension $K_t \cap K_a$.*
(2) *The image of $\alpha\colon F \to X_F \to X_K$ (2.5.1) equals the $p$-torsion part $F_0X_K[p]$ of $F_0X_K$.*
(3) *Let $\mu_F \subset F^\times$ be the subgroup of roots of 1, and define the inductive limit $\mu_F^\vee = \varinjlim_{\mu_m\subset\mu_F} \mathrm{Hom}(\mu_m, \mathbf{Q}/\mathbf{Z})$ with respect to the dual of the $(m'/m)$th power morphisms $\mu_{m'} \to \mu_m$. Then we have an exact sequence*

$$0 \to X_F \to F_0X_K \to \mu_F^\vee \to 0. \tag{5.2.3}$$

*Proof* (1) The inclusion $F_0X_K \subset \mathrm{Ker}(X_K \to X_{K_t})$ is proved in Proposition 2.3.3(3). For the other inclusion $F_0X_K \supset \mathrm{Ker}(X_K \to X_{K_t})$, it suffices to show that the condition that $\chi$ is wildly ramified implies $n = \mathrm{Sw}\,\chi \geqslant 1$. Let $L$ be the cyclic extension of $K$ corresponding to the kernel of $\chi$, and let $M \subsetneq L$ be the maximal tamely ramified extension of $K$. Let $K'$ be $M(\zeta_p)$ if $K$ is of characteristic 0 and let $K' = M$ if otherwise.

Let $p^e > 1$ be the order of the restriction $\chi_{K'}$, and let $\chi' = p^{e-1}\chi_{K'} \in X_{K'}$

be the character of order $p$. Then $\operatorname{Sw}\chi' \geqslant 1$ and the refined Swan conductor $\operatorname{rsw}\chi'$ exists by Proposition 3.1.5(1) if $K$ is of characteristic $p$ and by Proposition 3.2.8(2) if $K$ is of characteristic 0. Hence, we have $\operatorname{Sw} p^{e-1}\chi \geqslant 1$ by Proposition 5.2.2, and consequently $\operatorname{Sw}\chi \geqslant 1$.

The equality $F_0X_K = \operatorname{Ker}(X_K \to X_{K_t})$ means exactly that the extension corresponding to $G_K^{\mathrm{ab}\,1}$ equals the maximal tamely ramified abelian extension $K_t \cap K_a$.

(2) Since $\operatorname{Ker}(X_{K_u} \to X_{K_t})$ is a $\mathbf{Z}[1/p]$-module, we have $F_0X_K[p] = \operatorname{Ker}(X_K \to X_{K_u})[p]$ by (1). By the exact sequence $0 \to X_F \to X_K \to X_{K_u}$, we further have $F_0X_K[p] = \operatorname{Im}(X_F \to X_K)[p]$. Since $a\colon F \to X_F[p]$ is a surjection, the assertion follows.

(3) Identify $\operatorname{Gal}(K_u/K)$ with $G_F = \operatorname{Gal}(F_s/F)$ and $I_t = \operatorname{Gal}(K_t/K_u)$ with $\varprojlim_{p\nmid m} \mu_m$ as in Corollary 1.5.10. Since the group $\operatorname{Gal}(K_t/K)$ is isomorphic to the semi-direct product $G_F \ltimes I_t$ by Corollary 1.5.10, $F_0X_K = \operatorname{Ker}(X_K \to X_{K_t}) = \operatorname{Hom}_{\mathrm{cont}}(\operatorname{Gal}(K_t/K), \mathbf{Q}/\mathbf{Z})$ fits in an exact sequence $0 \to X_F \to F_0X_K \to \operatorname{Hom}_{G_F}(I_t, \mathbf{Q}/\mathbf{Z}) \to 0$ by (1). Since

$$\operatorname{Hom}_{G_F}(I_t, \mathbf{Q}/\mathbf{Z}) = \operatorname{Hom}_{\mathrm{cont}}(\varprojlim_{p\nmid m} \mu_m, \mathbf{Q}/\mathbf{Z})^{G_F}$$

is identified with $\mu_F^\vee$, we obtain (5.2.3). □

**Exercise 5.2** Let $p$ be a prime number. We identify $G = \operatorname{Gal}(\mathbf{Q}_p(\zeta_{p^2})/\mathbf{Q}_p)$ with $(\mathbf{Z}/p^2\mathbf{Z})^\times$ and let $\sigma \in G$ be the element corresponding to $1 + p$. Let $\chi\colon G \to \mathbf{Q}/\mathbf{Z}$ be a primitive character such that $\chi(\sigma) = 1/p$. Compute $m = \operatorname{Sw}\chi$ and $\operatorname{rsw}\chi\colon \mathfrak{m}_{\mathbf{Z}_p}^m/\mathfrak{m}_{\mathbf{Z}_p}^{m+1} \to \mathbf{F}_p$.

## 5.3 The Hasse–Arf Theorem

In this section, we prove a relation between the lower numbering filtration and the Swan conductor, called the Hasse–Arf theorem, Theorem 5.3.4, for cyclic extensions. First, we give an equivalent condition for an equality on the Swan conductor and also a procedure to construct the refined Swan conductor.

**Proposition 5.3.1** *Let $K$ be a henselian discrete valuation field and $\chi \in X_K$. Let $M$ be a cyclic extension of $K$ of Galois group $G$, and $\chi_M \in X_M$ be the restriction. Let $g = g_{M/K}$ be the largest integer such that $G_g \neq 1$, and $d_{M/K}$ be the valuation $\operatorname{ord}_M D^{\log}_{M/K}$ of the logarithmic different. Set $n = \operatorname{Sw}\chi$ and suppose $n' = \operatorname{Sw}\chi_M > g_{M/K}$.*

(1) *We have*

$$n \leqslant \frac{d_{M/K} + n'}{e_{M/K}}. \tag{5.3.1}$$

(2) *Let $F_M$ denote the residue field of $M$, and assume that the refined Swan conductor* $\operatorname{rsw}\chi_M\colon \mathfrak{m}_M^{n'}/\mathfrak{m}_M^{n'+1} \to \Omega^1_{F_M}(\log)$ *exists. Then the following conditions are equivalent:*

(a) *The equality in* (5.3.1) *holds.*

(b) *The composition*

$$\mathfrak{m}_M^{n'}/\mathfrak{m}_M^{n'+1} \xrightarrow{\operatorname{rsw}\chi_M} \Omega^1_{F_M}(\log) \xrightarrow{\operatorname{Tr}_{M/K}} \Omega^1_F(\log) \tag{5.3.2}$$

*is non-zero.*

*If these equivalent conditions are satisfied, there exists a unique $F$-linear morphism $f\colon \mathfrak{m}_K^n/\mathfrak{m}_K^{n+1} \to \Omega^1_F(\log)$ such that the diagram*

$$\begin{array}{ccc} \mathfrak{m}_M^{n'}/\mathfrak{m}_M^{n'+1} & \xrightarrow{\operatorname{rsw}\chi_M} & \Omega^1_{F_M}(\log) \\ {\scriptstyle \operatorname{Tr}_{M/K}}\downarrow & & \downarrow{\scriptstyle \operatorname{Tr}_{M/K}} \\ \mathfrak{m}_K^n/\mathfrak{m}_K^{n+1} & \xrightarrow{f} & \Omega^1_F(\log) \end{array} \tag{5.3.3}$$

*is commutative. Further, the morphism $f\colon \mathfrak{m}_K^n/\mathfrak{m}_K^{n+1} \to \Omega^1_F(\log)$ is the refined Swan conductor* $\operatorname{rsw}\chi$.

*Proof* (1) Let $K_1$ be as in Definition 2.3.1. Then $M_1 = MK_1$ is also constructed from $M$ as in Definition 2.3.1, and the canonical morphism $O_M \otimes_{O_K} O_{K_1} \to O_{M_1}$ is an isomorphism by Lemma 2.3.2. Let $j \leqslant n'$ be the largest integer such that $d_{M/K} + j$ is divisible by $e_{M/K}$, and set $i = (d_{M/K} + j)/e_{M/K} \leqslant (d_{M/K} + n')/e_{M/K}$. Since $j$ is the largest, we have $j + e_{M/K} > n' > g_{M/K}$ and $d_{M/K} + j + e_{M/K} = e_{M/K} \cdot (i + 1)$. Hence, by Corollary 1.10.6, we have

$$\mathrm{N}_{M_1/K_1}(1 + \mathfrak{m}_{M_1}^{j+e_{M/K}}) = 1 + \mathfrak{m}_{K_1}^{i+1}. \tag{5.3.4}$$

Since $\{\chi, 1+\mathfrak{m}_{K_1}^{i+1}\} = \{\chi, \mathrm{N}_{M_1/K_1}(1+\mathfrak{m}_{M_1}^{j+e_{M/K}})\} \subset \operatorname{Cor}_{M_1/K_1}\{\chi_M, 1+\mathfrak{m}_{M_1}^{n'+1}\} = 0$ by (5.3.4) and Lemma 2.1.5(2), $\operatorname{Sw}\chi \leqslant i \leqslant (d_{M/K} + n')/e_{M/K}$.

(2) (a) $\Rightarrow$ (b): Assume that $n' = e_{M/K} \cdot n - d_{M/K}$. Since $n' > g_{M/K}$, by Corollary 1.10.6 we have

$$\mathrm{N}_{M_1/K_1}(1 + \mathfrak{m}_{M_1}^{n'}) = 1 + \mathfrak{m}_{K_1}^{n}. \tag{5.3.5}$$

By (5.3.5), by Lemma 2.1.5(2), and by Proposition 5.2.1, we have

$$\begin{aligned}\{\chi, 1+\mathfrak{m}^n_{K_1}\} &= \{\chi, \mathrm{N}_{M_1/K_1}(1+\mathfrak{m}^{n'}_{M_1})\}\\ &= \mathrm{Cor}_{M_1/K_1}\{\chi_M, 1+\mathfrak{m}^{n'}_{M_1}\}\\ &= \mathrm{Cor}_{M_1/K_1}\beta_{M_1}((\mathrm{rsw}\,\chi_M(\mathfrak{m}^{n'}_{M_1}/\mathfrak{m}^{n'+1}_{M_1})))\\ &= \beta_{K_1}(\mathrm{Tr}_{M_1/K_1}((\mathrm{rsw}\,\chi_M\otimes 1)(\mathfrak{m}^{n'}_{M_1}/\mathfrak{m}^{n'+1}_{M_1}))).\end{aligned}$$

Since the left-hand side is non-zero, the right-hand side is also non-zero. Since $\mathrm{Tr}_{M_1/K_1}\circ(\mathrm{rsw}\,\chi_M\otimes 1)$ is the base change of the composition of (5.3.2), condition (b) follows.

(b) $\Rightarrow$ (a): Set $\varphi(j) = (d_{M/K}+j)/e_{M/K}$ and let $i = \lceil\varphi(n')\rceil$ be the ceiling. Since $n' > g_{M/K}$, by Corollary 1.10.6, by Lemma 2.1.5(2), and by Proposition 5.2.1, we have

$$\begin{aligned}\{\chi, 1+\mathfrak{m}^i_{K_1}\} &= \{\chi, \mathrm{N}_{M_1/K_1}(1+\mathfrak{m}^{n'}_{M_1})\}\\ &= \mathrm{Cor}_{M_1/K_1}\{\chi_M, 1+\mathfrak{m}^{n'}_{M_1}\}\\ &\supset \mathrm{Cor}_{M_1/K_1}\beta_{M_1}(T\cdot(\mathrm{rsw}\,\chi_M(\mathfrak{m}^{n'}_{M}/\mathfrak{m}^{n'+1}_{M})))\\ &= \beta_{K_1}(T\cdot\mathrm{Tr}_{M/K}(\mathrm{rsw}\,\chi_M(\mathfrak{m}^{n'}_{M}/\mathfrak{m}^{n'+1}_{M}))).\end{aligned}$$

By (b) and Lemma 2.5.6(1), the last term $\beta_{K_1}(T\cdot\mathrm{Tr}_{M/K}(\mathrm{rsw}\,\chi_M(\mathfrak{m}^{n'}_{M}/\mathfrak{m}^{n'+1}_{M})))$ is non-zero. Thus, we have $\varphi(n')\leqslant i\leqslant n$. Since $n\leqslant\varphi(n')$ is proved in (1), we have the equality and (a) holds.

We show the existence of $f\colon \mathfrak{m}^n_K/\mathfrak{m}^{n+1}_K \to \Omega^1_F(\log)$, which makes the diagram (5.3.3) commutative. It suffices to show that the composition $\mathrm{Tr}_{M/K}\circ\mathrm{rsw}\,\chi_M$ through $\Omega^1_{F_M}(\log)$ in (5.3.3) annihilates the kernel of the surjection $\mathrm{Tr}_{M/K}\colon \mathfrak{m}^{n'}_M/\mathfrak{m}^{n'+1}_M \to \mathfrak{m}^n_K/\mathfrak{m}^{n+1}_K$. For $x\in\mathfrak{m}^{n'}_{M_1}$, by Proposition 5.2.1 and by Lemma 2.1.5(2), we have

$$\begin{aligned}\beta_{K_1}(\mathrm{Tr}_{M_1/K_1}(\mathrm{rsw}\,\chi_M\otimes 1)(\bar{x})) &= \mathrm{Cor}_{M_1/K_1}\beta_{M_1}((\mathrm{rsw}\,\chi_M\otimes 1)(\bar{x})) \quad (5.3.6)\\ &= \mathrm{Cor}_{M_1/K_1}\{\chi_M, 1+x\}\\ &= \{\chi, \mathrm{N}_{M_1/K_1}(1+x)\}.\end{aligned}$$

Since $n' > g_{M/K}$, we have $\mathrm{N}_{M_1/K_1}(1+x)\equiv 1+\mathrm{Tr}_{M_1/K_1}x \bmod \mathfrak{m}^{n+1}_{K_1}$ by Corollary 1.10.6. Hence, the right-hand side in (5.3.6) equals $\{\chi, 1+\mathrm{Tr}_{M_1/K_1}x\}$ by $\{\chi, 1+\mathfrak{m}^{n+1}_{K_1}\} = 0$. For $x\in\mathfrak{m}^{n'}_M$, this means $\beta_{K_1}(T\cdot\mathrm{Tr}_{M/K}(\mathrm{rsw}\,\chi_M(\bar{x})) = \{\chi, 1+T\cdot\mathrm{Tr}_{M/K}x\}$. Hence, by Lemma 2.5.6(1), the composition $\mathrm{Tr}_{M/K}\circ\mathrm{rsw}\,\chi_M$ through $\Omega^1_{F_M}(\log)$ in (5.3.3) annihilates the kernel of the surjection $\mathrm{Tr}_{M/K}\colon \mathfrak{m}^{n'}_M/\mathfrak{m}^{n'+1}_M \to \mathfrak{m}^n_K/\mathfrak{m}^{n+1}_K$. Thus, there exists a unique $F$-linear mapping $f\colon \mathfrak{m}^n_K/\mathfrak{m}^{n+1}_K \to \Omega^1_F(\log)$ that makes the diagram (5.3.3) commutative.

The equality (5.3.6) means that we have an equality

$$\beta_{K_1}((f \otimes 1)(\mathrm{Tr}_{M_1/K_1}(\bar{x})) = \{\chi, 1 + \mathrm{Tr}_{M_1/K_1} x\} \tag{5.3.7}$$

for $x \in \mathfrak{m}_{M_1}^{n'}$. Since $\mathrm{Tr}_{M_1/K_1} \colon \mathfrak{m}_{M_1}^{n'}/\mathfrak{m}_{M_1}^{n'+1} \to \mathfrak{m}_{K_1}^{n}/\mathfrak{m}_{K_1}^{n+1}$ is a surjection, we have the second equality in (2.5.11) for $x \in \mathfrak{m}_{K_1}^{n}$. Similarly to (5.3.6) and (5.3.7), for $x \in \mathfrak{m}_{M}^{n'}$, we have $\beta_K(\mathrm{Tr}_{M/K}\,\mathrm{rsw}\,\chi_M(\bar{x})) = \{\chi, \mathrm{N}_{M/K}(1+x)\} = \{\chi, 1 + \mathrm{Tr}_{M/K} x\}$ by Proposition 5.2.1 and Corollary 1.10.6. Thus, we have the first equality in (2.5.11) for $x \in \mathfrak{m}_K^n$ and $f = \mathrm{rsw}\,\chi$. □

In order to realize the Hasse–Arf theorem, Theorem 5.3.4, as a statement on linear mappings, we introduce a refinement of the morphism $\delta_\sigma \colon L^\times/(1 + \mathfrak{m}_L) \to \mathfrak{m}_L^i/\mathfrak{m}_L^{i+1}$ (1.6.3) for a finite Galois extension. For an extension $L$ of a discrete valuation field $K$ with residue field $E$, recall that $\Omega^1_{E/F}(\log/\log)$ denotes the cokernel of $\Omega^1_F(\log) \otimes_F E \to \Omega^1_E(\log)$. Let $L$ be a Galois extension of $K$ and let $G$ be the Galois group. We rephrase the injection $\delta \colon \mathrm{Gr}_i G \to \mathrm{Hom}(L^\times/(1 + \mathfrak{m}_L), \mathfrak{m}_L^i/\mathfrak{m}_L^{i+1})$ (1.6.3) using $\Omega^1_{E/F}(\log/\log)$.

**Proposition 5.3.2** *Assume that the residue field $F$ is of characteristic $p > 0$. Let $i \geqslant 1$ be an integer.*

(1) *For $\sigma \in G_i$, the morphism $\delta_\sigma \colon L^\times/(1+\mathfrak{m}_L) \to \mathfrak{m}_L^i/\mathfrak{m}_L^{i+1}$ (1.6.3) induces an $E$-linear morphism $d_\sigma \colon \Omega^1_{E/F}(\log/\log) \to \mathfrak{m}_L^i/\mathfrak{m}_L^{i+1}$.*
(2) *The mapping $G_i \to \mathrm{Hom}_E(\Omega^1_{E/F}(\log/\log), \mathfrak{m}_L^i/\mathfrak{m}_L^{i+1})$ sending $\sigma$ to $d_\sigma$ defines an injection*

$$d \colon \mathrm{Gr}_i G \to \mathrm{Hom}_E(\Omega^1_{E/F}(\log/\log), \mathfrak{m}_L^i/\mathfrak{m}_L^{i+1}) \tag{5.3.8}$$

*compatible with the actions of $G$.*

*Proof* (1) Since $\sigma(bb') - bb' = b(\sigma(b') - b') + b'((\sigma(b) - b) + (\sigma(b) - b)(\sigma(b') - b')$ for $b, b \in O_L$, the mapping $O_L \to \mathfrak{m}_L^i/\mathfrak{m}_L^{i+1}$ sending $b$ to $\sigma(b) - b$ is a derivation. Since $\sigma(b) - b = b \cdot ((\sigma(b)/b) - 1)$ for $b \in O_K \cap K^\times$, the morphism $\delta_\sigma \colon L^\times/(1 + \mathfrak{m}_L) \to \mathfrak{m}_L^i/\mathfrak{m}_L^{i+1}$ induces an $E$-linear morphism $d_\sigma \colon \Omega^1_E(\log) \to \mathfrak{m}_L^i/\mathfrak{m}_L^{i+1}$. Since its restriction to the image of $\Omega^1_F(\log)$ is trivial, $d_\sigma$ further induces $d_\sigma \colon \Omega^1_{E/F}(\log/\log) \to \mathfrak{m}_L^i/\mathfrak{m}_L^{i+1}$.

(2) Since $\delta \colon \mathrm{Gr}_i G \to \mathrm{Hom}(L^\times/(1 + \mathfrak{m}_L), \mathfrak{m}_L^i/\mathfrak{m}_L^{i+1})$ is an injection by Corollary 1.6.4, $d \colon \mathrm{Gr}_i G \to \mathrm{Hom}_E(\Omega^1_{E/F}(\log/\log), \mathfrak{m}_L^i/\mathfrak{m}_L^{i+1})$ (5.3.8) is also an injection.

For $\sigma \in G_i$ and $\tau \in G$, the morphism $d_{\tau\sigma\tau^{-1}} \colon \Omega^1_{E/F}(\log/\log) \to \mathfrak{m}_L^i/\mathfrak{m}_L^{i+1}$ maps $\tau(x \cdot d\log y) = \tau(x) \cdot d\log \tau(y)$ to $\tau(x) \cdot (\tau\sigma\tau^{-1}\tau(y)/\tau(y) - 1)) = \tau(x \cdot (\sigma(y)/y - 1)) = \tau(d_\sigma(x \cdot d\log y))$. □

We show a compatibility with tamely ramified base extension.

**Lemma 5.3.3** *Let $L$ be a Galois extension of a henselian discrete valuation field $K$, and let $K'$ be a tamely ramified extension. Let $E, F, E', F'$ be the residue fields. Let $G = \mathrm{Gal}(L/K)$ be the Galois group and identify $G' = \mathrm{Gal}(L'/K')$ for $L' = LK'$ as a subgroup of $G'$. Let $i \geqslant 1$ and $i' = e_{L'/L}i$ be integers, and let $\sigma \in G_i$ be the image of $\sigma' \in G'_{i'}$. Then the diagram*

$$\begin{array}{ccc} \Omega^1_{E'}(\log) & \xrightarrow{d_{\sigma'}} & \mathfrak{m}^{i'}_{L'}/\mathfrak{m}^{i'+1}_{L'} \\ {\scriptstyle \mathrm{Tr}_{\mathcal{O}_{L'}/\mathcal{O}_L}}\downarrow & & \downarrow{\scriptstyle \mathrm{Tr}_{L'/L}} \\ \Omega^1_E(\log) & \xrightarrow{d_\sigma} & \mathfrak{m}^i_L/\mathfrak{m}^{i+1}_L \end{array} \tag{5.3.9}$$

*is commutative.*

*Proof* By Corollary 1.7.3, the inclusion $G' \to G$ induces an isomorphism $\mathrm{Gr}_{i'}G' \to \mathrm{Gr}_i G$. Since the diagram

$$\begin{array}{ccc} \Omega^1_{E'}(\log) & \xrightarrow{d_{\sigma'}} & \mathfrak{m}^{i'}_{L'}/\mathfrak{m}^{i'+1}_{L'} \\ \uparrow & & \uparrow \\ \Omega^1_E(\log) & \xrightarrow{d_\sigma} & \mathfrak{m}^i_L/\mathfrak{m}^{i+1}_L \end{array} \tag{5.3.10}$$

is commutative, the upper arrow is identified with the scalar extension of the left lower arrow by Lemma 5.1.4. Hence, the diagram (5.3.9) is commutative. □

Now we are ready to prove a relation between the lower numbering filtration and the cohomological filtration for cyclic extensions.

**Theorem 5.3.4** (Hasse–Arf theorem) *Let $\chi \in X_K$ be a character and let $L$ be the corresponding cyclic extension of $K$. Let $F$ and $E$ be the residue fields of $K$ and $L$ respectively. Assume that $\mathrm{Sw}\,\chi = n_{L/K} \geqslant 1$, and let $d_{L/K} > 0$ be the valuation of the logarithmic different $D^{\log}_{L/K}$. Let $G = \mathrm{Gal}(L/K)$ be the Galois group and $g_{L/K} > 0$ be the largest integer $i$ such that $G_i \neq 1$.*

(1) *The following conditions are equivalent:*

(a) *We have an equality*

$$n_{L/K} = \frac{d_{L/K} + g_{L/K}}{e_{L/K}}. \tag{5.3.11}$$

(a′) *We have an equality*

$$n_{L/K} = \frac{1}{e_{L/K}}\Big(\sum_{\sigma \in P-\{1\}} i_G(\sigma) + g_{L/K}\Big). \tag{5.3.12}$$

(b) *The extension of $L$ over $K$ is log monogenic.*

(c) *The trace mapping* $\mathrm{Tr}_{L/K}\colon \Omega^1_E(\log) \to \Omega^1_F(\log)$ *is non-zero.*

(2) *Assume that the equivalent conditions in (1) are satisfied. Let* $\sigma \in G$ *be the generator of the subgroup of order* $p$ *such that* $\chi(\sigma) = 1/p$, *and let* $d_\sigma\colon \Omega^1_E(\log) \to \mathfrak{m}_L^{g_{L/K}}/\mathfrak{m}_L^{g_{L/K}+1}$ *be the morphism* (5.3.8) *sending* $d\log b$ *to* $\sigma(b)/b-1$. *Then the refined Swan conductor* $\mathrm{rsw}\,\chi\colon \mathfrak{m}_K^{n_{L/K}}/\mathfrak{m}_K^{n_{L/K}+1} \to Z_{F\setminus E}(\log) \subset \Omega^1_F(\log)$ *exists and the diagram*

$$\begin{array}{ccc} \mathfrak{m}_L^{g_{L/K}}/\mathfrak{m}_L^{g_{L/K}+1} & \xleftarrow{\ d_\sigma\ } & \Omega^1_{E/F}(\log/\log) \\ {\scriptstyle \mathrm{Tr}_{L/K}}\downarrow & & \downarrow{\scriptstyle \mathrm{Tr}_{L/K}} \\ \mathfrak{m}_K^{n_{L/K}}/\mathfrak{m}_K^{n_{L/K}+1} & \xrightarrow{\ \mathrm{rsw}\,\chi\ } & Z_{F\setminus E}(\log) \end{array} \tag{5.3.13}$$

*is commutative.*

The refined Swan conductor $\mathrm{rsw}\chi$ is uniquely determined by the commutative diagram (5.3.13). In the proof of the integrality of the conductor in Part III and its geometric applications in Part IV, we will use the refined Swan conductor only for log monogenic extensions.

*Proof* Conditions (b) and (c) are equivalent by Proposition 5.1.6(2). By the inequalities (2.3.2), the equality (5.3.12) implies (5.3.11). If (b) is satisfied, then (1.9.5) implies that (5.3.11) and (5.3.12) are equivalent. Hence, for the equivalences in (1), it suffices to show that (a) and (b) are equivalent. We divide the proof into the following three steps:

(i) Reduction to the case where $\chi$ is purely wildly ramified and $K$ contains a primitive root of 1 if $K$ is of characteristic 0.
(ii) The case where $\chi$ is purely wildly ramified and is of order $p$.
(iii) The case where $\chi$ is purely wildly ramified and is of order $p^e > p$.

(i) We show that we may replace $K$ by a tamely ramified Galois extension $K'$ and $\chi \in X_K$ by the restriction $\chi' \in X_{K'}$. Let $L' = LK'$, $G' = \mathrm{Gal}(L'/K')$, and $n_{L'/K'}, \ldots$ be the corresponding invariants. Since $n_{L'/K'} = e_{K'/K} \cdot n_{L/K}$ by Proposition 5.2.2(1) and $g_{L'/K'} = e_{L'/L} \cdot g_{L/K}$, $d_{L'/K'} = e_{L'/L} \cdot d_{L/K}$, the first equality (5.3.11) for $\chi'$ implies the equality for $\chi$. Since $i_{G'}(\sigma) = e_{L'/L} \cdot i_G(\sigma)$ for $\sigma \in P'$ by Corollary 1.7.3, the second equality (5.3.11) for $\chi'$ implies the equality for $\chi$. By Proposition 1.8.5(1), the extension $L$ over $K$ is log monogenic if and only if so is $L'$ over $K'$.

By Proposition 5.2.2(2), the existence of $\mathrm{rsw}\,\chi'$ implies the existence of $\mathrm{rsw}\,\chi$, and then $\mathrm{rsw}\,\chi'$ is the scalar extension of $\mathrm{rsw}\,\chi$. The composition $L' = LK'$ is a tamely ramified extension of $L$. By Lemma 5.1.10, Lemma 5.3.3, and by the surjectivity of $\Omega_{E'}(\log) \to \Omega_E(\log)$, the commutative diagram (5.3.13) for $L'$ over $K'$ with the bottom horizontal arrow rsw $\chi'$ implies the diagram for $L$ over $K$ with the bottom horizontal arrow rsw $\chi$.

Hence, if $K$ is of characteristic 0, by replacing $K$ by $K(\zeta_p)$, we may assume that $K$ contains a primitive $p$th root $\zeta_p$ of 1. Further, by replacing $K$ by the maximal tamely ramified extension $M \subset L$ over $K$, we may assume that $\chi$ is of order a power $p^e$ of $p$ and purely wildly ramified.

(ii) We show the case $[L : K] = p$ assuming that $K$ contains a primitive $p$th root $\zeta_p$ of 1 if $K$ is of characteristic 0. In this case, (b) holds. We show (a). Assume that $L$ is a ramified cyclic extension of degree $p$. By Proposition 3.1.5(2) if $K$ is of characteristic $p > 0$, and by Proposition 3.2.8 if $K$ is of characteristic 0 and contains $\zeta_p$, we have $\mathrm{Sw}\,\chi = p \cdot g_{L/K}/e_{L/K} = (d_{L/K} + g_{L/K})/e_{L/K}$ (5.3.11) by (1.9.5), and the refined Swan conductor exists.

We prove the commutativity of the diagram (5.3.13). Assume that $\chi \in X_K$ is a wildly ramified character of order $p$ and that $K$ contains a primitive $p$th root of 1 if $K$ is of characteristic 0. Let $u \in L^\times$ be as in Proposition 3.1.5(2) if $K$ is of characteristic $p$ or as in Proposition 3.2.8 if $K$ is of characteristic 0. Set $v = \sigma(u)/u - 1$, $w = \mathrm{N}_{L/K}v$, and $z = \mathrm{N}_{L/K}u$ be as in in Proposition 3.2.8. We have $\mathrm{Tr}_{L/K} d\log u = d\log \mathrm{N}_{L/K}u = d\log z$ by Proposition 5.1.1(4). We show that the image of $d\log u \in \Omega^1_{E/F}(\log/\log)$ in $Z_{F\backslash E}(\log)$ by the composition is also $d\log z$. We have $d_\sigma(d\log u) = \sigma(u)/u - 1 = v$. For $j = m = g_{L/K}$, $i = n = pm/e$, and $d = (p-1)m$, the inequalities $j - 1 \geqslant e \cdot (i-1) - d$ and $2j - 1 \geqslant e \cdot i - d$ are satisfied. Since $\mathrm{N}_{L/K}(1+v) = \mathrm{N}_{L/K}(\sigma(u)/u) = 1$, we have $1 + \mathrm{Tr}_{L/K}v + \mathrm{N}_{L/K}v \equiv 1 \bmod \mathfrak{m}_K^{n+1}$ by Proposition 1.10.5(3). Hence, we have $\mathrm{Tr}_{L/K}v \equiv -\mathrm{N}_{L/K}v = -w \bmod \mathfrak{m}_K^{n+1}$ and the image is $d\log z$ by Proposition 3.1.5(2) and Proposition 3.2.8.

Thus, in the case $E = F$, the diagram (5.3.13) of $F$-vector spaces of dimension 1 is commutative.

We show the case $e_{L/K} = 1$. Since the kernel of the right vertical arrow $\mathrm{Tr}_{L/K} \colon \Omega^1_{E/F}(\log/\log) = \Omega^1_{E/F} \to Z_{F\backslash E} = Z_{F\backslash E}(\log)$ is generated by $dE$ by Proposition 4.4.5(2), and since $\Omega^1_{E/F}(\log/\log) = F \cdot d\log u \oplus F \cdot dE$, it suffices to show that the composition through the left vertical arrow also annihilates $F \cdot dE$. Since $\beta_{K_1}(T \cdot -) \colon \Omega^1_F(\log) \to \mathrm{Br}\, K_1$ is an injection by Lemma 2.5.6(1), it suffices to show $\beta_{K_1}(T \cdot \mathrm{rsw}\,\chi(\mathrm{Tr}_{L/K} d_\sigma\omega)) = \{\chi, 1 + T \cdot \mathrm{Tr}_{L/K} d_\sigma\omega\}$ is 0 for $\omega \in F \cdot dE$. By Proposition 1.10.5(3), for $a \in \mathfrak{m}_L^g$ we have $\mathrm{N}_{L/K}(1 + T \cdot a) \equiv 1 + T \cdot \mathrm{Tr}_{L/K}a + T^p \cdot \mathrm{N}_{L/K}a \bmod 1 + \mathfrak{m}_{K_1}^{n+1}$. Hence, by Proposition 2.1.1(2), we have $\{\chi, 1 + T \cdot \mathrm{Tr}_{L/K}(d_\sigma\omega)\} = \{\chi, 1 - T^p \cdot \mathrm{N}_{L/K}(d_\sigma\omega)\} = \beta_{K_1}(T^p \cdot \mathrm{rsw}\,\chi(-\mathrm{N}_{L/K}(d_\sigma\omega)))$.

To prove $\beta_{K_1}(T^p \cdot \mathrm{rsw}\,\chi(-\mathrm{N}_{L/K}(d_\sigma\omega))) = 0$ for $\omega \in F \cdot dE$, we claim that the composition

$$\Omega^1_{E/F}(\log/\log) = \Omega^1_{E/F} \xrightarrow{d_\sigma} \mathfrak{m}_L^g/\mathfrak{m}_L^{g+1} \xrightarrow{-\mathrm{N}_{L/K}} \mathfrak{m}_K^n/\mathfrak{m}_K^{n+1} \xrightarrow{\mathrm{rsw}\,\chi} Z_{F\backslash E}(\log) = Z_{F\backslash E}$$

maps $a \cdot d\log u \in \Omega^1_{E/F}(\log/\log)$ to $a^p \cdot d\log z$ and hence $FdE$ to $dF \cap Z_{F\backslash E}$. For $a \cdot d\log u \in \Omega^1_{E/F}(\log/\log)$, we have $d_\sigma(a \cdot d\log u) = av$, $-\mathrm{N}_{L/K}(av) = -a^p w$, and $\mathrm{rsw}\,\chi(-a^p w) = a^p \cdot d\log z$ by Proposition 3.1.5(2) and Proposition 3.2.8. Let $t$ be a uniformizer of $K$ and define units $u_1 \in O_L^\times$ and $z_1 \in O_K^\times$ by $u = u_1 t^{-m_1}$ and $z = z_1 t^{-n_1}$ for $n_1 = pm_1$. Then $O_L = O_K[u_1]$ by Proposition 3.1.2(2) and Proposition 3.2.2(2) and (4). Hence, $F \cdot dE \subset \Omega^1_{E/F}(\log/\log) = E \cdot d\log u = F(u_1) \cdot du_1$ is generated by $du_1^i = iu_1^i \cdot d\log u_1 = iu_1^i \cdot d\log u$ over $F$. Since $(iu_1^i)^p \cdot d\log z = iz_1^i \cdot d\log z_1 = d(z_1^i) \in dF \cap Z_{F\backslash E}$, the claim follows. Since $T^p \cdot \mathrm{rsw}\,\chi(-\mathrm{N}_{L/K}(d_\sigma\omega)) \in T^p \cdot dF \subset dF_1$ for $\omega \in F \cdot dE$ as we have shown, it is annihilated by $\beta_{K_1} : \Omega^1_{F_1}(\log) \to \mathrm{Br}\,K_1$ by the definition (2.5.7), as required.

(iii) Assume that $[L : K] = p^e$ for $e \geqslant 2$. Let $M$ be the subextension such that $[L : M] = p$, and let $\chi_M \in X_M$ be the restriction of $\chi$. By (1.7.11) and step (ii), we have $g_{M/K} < f_{L/M} \cdot g_{L/K} = \mathrm{Sw}\,\chi_M$ and the assumption $\mathrm{Sw}\,\chi_M > g_{M/K}$ in Proposition 5.3.1 is satisfied. By the chain rule of the different, we have

$$\frac{d_{M/K} + f_{L/M} \cdot g_{L/K}}{e_{M/K}} = \frac{e_{L/M} \cdot d_{M/K} + (p-1) \cdot g_{L/K} + g_{L/K}}{e_{L/K}} = \frac{d_{L/K} + g_{L/K}}{e_{L/K}}.$$

Hence, the equality (5.3.11) is equivalent to (5.3.1). By Proposition 5.3.1(2) (a) $\Leftrightarrow$ (b), this is further equivalent to the condition that the composition (5.3.2) is non-zero. We consider the commutative diagram (5.3.13) for $\chi_M$ and $L$ over $M$. Then, since the composition $\Omega^1_E(\log) \to \mathfrak{m}_L^{g_{L/K}}/\mathfrak{m}_L^{g_{L/K}+1} \to \mathfrak{m}_M^{n_{L/M}}/\mathfrak{m}_M^{n_{L/M}+1}$ is a surjection, by the transitivity of Tr, it is further equivalent to (c), and hence to (b).

Assume that these equivalent conditions are satisfied. Then, by Proposition 5.3.1(2), the refined Swan conductor $\mathrm{rsw}\,\chi : \mathfrak{m}_K^{n_{L/K}}/\mathfrak{m}_K^{n_{L/K}+1} \to \Omega^1_F(\log)$ exists and the diagram (5.3.3) with $f = \mathrm{rsw}\,\chi$ is commutative. Combining this with (5.3.13) for $\chi_M$, we obtain the commutative diagram (5.3.13) for $\chi$. □

**Example 5.3.5** Let $K$ be a henselian discrete valuation field of characteristic $p > 0$, $F$ be the residue field, and $t \in K$ be a uniformizer.

(1) Let $L$ be the Artin–Schreier extension defined by $x^p - x = 1/t^n$ for

$n = pm - 1 > 0$, and $s = t^m x \in L$ be a uniformizer as in Example 1.4.4(1). Since the minimal polynomial of $s$ is $S^p - t^{(p-1)m}S - t$, we have $\mathrm{N}_{L/K}s = t$ and the right vertical arrow $\mathrm{Tr}_{L/K}$ in (5.3.13) maps the basis $d\log s \in \Omega^1_{L/K}(\log/\log)$ to $d\log t = d\log \mathrm{N}_{L/K}s \in Z_{F\backslash E}(\log)$. Let $\sigma \in G = \mathrm{Gal}(L/K)$ be the generator defined by $\sigma(x) = x + 1$. The top horizontal arrow $d_\sigma$ maps $d\log s$ to $\sigma(s)/s - 1 = 1/x$. Since the minimal polynomial of $1/x$ is $Y^p + t^n Y^{p-1} - t^n$, we have $\mathrm{Tr}_{L/K} 1/x = -t^n$. Let $\chi\colon G \to \mathbf{Q}/\mathbf{Z}$ be the character sending $\sigma$ to $1/p$. Since the refined Swan conductor $\mathrm{rsw}\,\chi$ maps the basis $-t^n \in \mathfrak{m}_K^n/\mathfrak{m}_K^{n+1}$ to $d\log t \in Z_{F\backslash E}(\log)$ as in Example 3.1.6(1), the diagram (5.3.13) is commutative.

(2) Let $L$ be the Artin–Schreier extension defined by $x^p - x = v/t^n$ for $n = pm > 0$ and for $v \in O_K$ such that $\bar{v} \in F - F^p$. Let $u = t^m x$ be the generator of the residue field $E = F(u)$ as in Example 1.4.4(2). Since the minimal polynomial of $u$ is $U^p - t^{(p-1)m}U - v$, we have $\mathrm{N}_{L/K}u = v$ and the right vertical arrow $\mathrm{Tr}_{L/K}$ in (5.3.13) maps the basis $d\log u \in \Omega^1_{L/K}(\log/\log)/F \cdot dE$ to $d\log v = d\log \mathrm{N}_{L/K}u \in Z_{F\backslash E}(\log)$. Let $\sigma \in G = \mathrm{Gal}(L/K)$ be the generator defined by $\sigma(x) = x+1$. The top horizontal arrow $d_\sigma$ maps $d\log u$ to $\sigma(u)/u - 1 = 1/x$. Since the minimal polynomial of $1/x$ is $Z^p + t^n/v \cdot Z^{p-1} - t^n/v$, we have $\mathrm{Tr}_{L/K} 1/x = -t^n/v$. Let $\chi\colon G \to \mathbf{Q}/\mathbf{Z}$ be the character sending $\sigma$ to $1/p$. Since the refined Swan conductor $\mathrm{rsw}\,\chi$ maps the basis $-t^n \in \mathfrak{m}_K^n/\mathfrak{m}_K^{n+1}$ to $dv = v \cdot d\log v \in Z_{F\backslash E}(\log)$ as in Example 3.1.6(2), the diagram (5.3.13) is commutative.

In the classical case where the residue field is perfect, the refined Swan conductor always exists.

**Proposition 5.3.6** *Let $K$ be a henselian discrete valuation field, and assume that the residue field $F$ is a perfect field of characteristic $p > 0$. We identify $\Omega^1_F(\log)$ with $F$ by the residue mapping $\Omega^1_F(\log) \to F$.*

(1) *The composition of $\beta\colon \Omega^1_F(\log) = F \to \mathrm{Br}\,K$ with $\mathrm{Br}\,K \to X_F$ equals $a\colon F \to X_F$.*

(2) *Let $\chi \in X_K$ be a character such that $\mathrm{Sw}\,\chi = n \geqslant 1$. Then the refined Swan conductor $\mathrm{rsw}\,\chi\colon \mathfrak{m}_K^n/\mathfrak{m}_K^{n+1} \to \Omega^1_F(\log) = F$ exists. For $x \in \mathfrak{m}_K^n$, the image of $\{\chi, 1+x\} \in \mathrm{Br}\,K$ in $X_F$ satisfies*

$$\overline{\{\chi, 1+x\}} = a(\mathrm{rsw}\,\chi(\bar{x})). \tag{5.3.14}$$

(3) *Let $n \geqslant 1$ be an integer. Then the refined Swan conductor defines an injection*

$$\mathrm{rsw} : \mathrm{Gr}^F_n X_K \to \mathrm{Hom}_F(\mathfrak{m}_K^n/\mathfrak{m}_K^{n+1}, F). \tag{5.3.15}$$

(4) *Assume that $F$ is algebraically closed. Let $L$ be a cyclic extension corresponding to $\chi \in X_K$ and $G = \mathrm{Gal}(L/K)$ be the Galois group. Assume $n = \mathrm{Sw}\,\chi \geqslant 1$ and let $i = g_{L/K} \geqslant 1$ be the largest integer such that $G_i \neq 1$. Then we have a commutative diagram of exact sequences*

$$\begin{array}{ccccccc} 0 \to G_i & \xrightarrow{(1.6.6)} & (1+\mathfrak{m}_L^i)/(1+\mathfrak{m}_L^{i+1}) & \xrightarrow{\mathrm{N}_{L/K}} & (1+\mathfrak{m}_K^n)/(1+\mathfrak{m}_K^{n+1}) & \to 0 \\ \downarrow & & \downarrow & & \downarrow{\scriptstyle -\mathrm{rsw}\,\chi} & \\ 0 \to \mathbf{F}_p & \longrightarrow & F & \xrightarrow{x \mapsto x^p - x} & F & \to 0 \end{array} \tag{5.3.16}$$

*The morphism $G_i \to (1+\mathfrak{m}_L^i)/(1+\mathfrak{m}_L^{i+1})$ (1.6.6) maps $\sigma$ to $\sigma(s)/s$ for a uniformizer $s$ of $L$, and $G_i \to \mathbf{F}_p$ maps the generator $\sigma$ such that $\chi(\sigma) = 1/p \in \mathbf{Q}/\mathbf{Z}$ to $1 \in \mathbf{F}_p$.*

*Proof* (1) This follows from the right square in (2.5.8).

(2) For every finite extension $L$ of $K$, the residue field $E$ is a separable extension of $F$. The refined Swan conductor $\mathrm{rsw}\,\chi$ exists by Theorem 5.3.4. Since $\{\chi, 1+x\} = \beta(\mathrm{rsw}\,\chi(\bar{x}))$, the equality (5.3.14) follows from (1).

(3) Since the refined Swan conductor exists by (2) and is compatible with the addition, we obtain a morphism (5.3.15). Since $\mathrm{Sw}\,\chi = n$ implies $\mathrm{rsw}\,\chi \neq 0$ in $\mathrm{Hom}_F(\mathfrak{m}_K^n/\mathfrak{m}_K^{n+1}, F)$, the morphism (5.3.15) is an injection.

(4) If $[L : K] = p$, the commutative diagram (5.3.13) and the commutative diagrams in Corollary 1.10.7(1) give (5.3.16). The general case follows from this and (5.3.3). □

**Proposition 5.3.7** *Let $K$ be a henselian discrete valuation field and assume that the residue field $F$ is finite. Let $n \geqslant 1$ be an integer.*

(1) *The reciprocity mapping sends $1 + \mathfrak{m}_K^n$ to $G_K^{\mathrm{ab}\,n}$.*
(2) *For $\chi \in X_K$ such that $\mathrm{Sw}\,\chi = n \geqslant 1$ and for $x \in \mathfrak{m}_K^n$, we have*

$$\mathrm{inv}_K\{\chi, 1+x\} = \mathrm{Tr}_{F/\mathbf{F}_p}(\mathrm{rsw}\,\chi(\bar{x})) \tag{5.3.17}$$

*under the identification of $\mathbf{F}_p$ with $(1/p)\mathbf{Z}/\mathbf{Z}$ sending 1 to $1/p$. This property characterizes the refined Swan conductor* $\mathrm{rsw}\,\chi$.

(3) *For every $n \geqslant 1$, we have a surjection $\mathfrak{m}_K^n/\mathfrak{m}_K^{n+1} \to \mathrm{Gr}^n G_K^{\mathrm{ab}}$. If $K$ is complete, the reciprocity morphism induces a surjection $1 + \mathfrak{m}_K^n \to G_K^{\mathrm{ab}\,n}$.*

*Proof* (1) Since the pairing $\mathrm{inv}_K\{\ ,\ \}\colon X_K \times K^\times \to \mathrm{Br}\,K \to \mathbf{Q}/\mathbf{Z}$ annihilates $F_{n-1}X_K \times (1+\mathfrak{m}_K^n)$, the reciprocity morphism sends $1+\mathfrak{m}_K^n$ to $G_K^{\mathrm{ab}\,n} = \{\sigma \in G_K^{\mathrm{ab}} \mid \chi(\sigma) = 0$ for every $\chi \in F_{n-1}X_K\}$.

(2) Since $a(\mathrm{rsw}\,\chi(x)) \in X_F$ equals the image of $\{\chi, 1+x\} \in \mathrm{Br}\,K$ by Propo-

sition 5.3.6(2), the equality (5.3.17) follows from the commutative diagram (2.2.1).

Since the mapping $F \to \mathrm{Hom}(F, \mathbf{F}_p)$ sending 1 to $\mathrm{Tr}_{F/\mathbf{F}_p}$ is an isomorphism of $F$-vector spaces, $\mathrm{rsw}\,\chi$ is characterized by (5.3.17).

(3) By (5.3.17), the injection $\mathrm{rsw} : \mathrm{Gr}^F_n X_K \to \mathrm{Hom}_F(\mathfrak{m}^n_K/\mathfrak{m}^{n+1}_K, F)$ (5.3.15) defines an injection $\mathrm{rsw} : \mathrm{Gr}^F_n X_K \to \mathrm{Hom}(\mathfrak{m}^n_K/\mathfrak{m}^{n+1}_K, (1/p)\mathbf{Z}/\mathbf{Z})$. Taking the dual, we obtain a surjection $\mathfrak{m}^n_K/\mathfrak{m}^{n+1}_K \to \mathrm{Gr}^n G^{\mathrm{ab}}_K$ for every $n \geqslant 1$.

If $K$ is complete, by taking the limit we obtain a surjection $1 + \mathfrak{m}^n_K \to G^{\mathrm{ab}\,n}_K$ for every $n \geqslant 1$. □

**Corollary 5.3.8** *Let $K$ be a henselian discrete valuation field with finite residue field $F$.*

(1) *Let $\chi \in X_F$ and $n \geqslant 1$ be an integer. Then the following conditions are equivalent:*
  (a) $\mathrm{Sw}\,\chi = n$.
  (b) $\mathrm{inv}_K\{\chi, 1 + \mathfrak{m}^n_K\} \neq 0$ *and* $\mathrm{inv}_K\{\chi, 1 + \mathfrak{m}^{n+1}_K\} = 0$.

(2) *Let $L$ be a finite abelian extension of $K$, and $G = \mathrm{Gal}(L/K)$. Then, for every $n \geqslant 1$, we have $G^n_{\mathrm{coh}} = G^n_{\mathrm{rec}}$.*

*Proof* (1) By the commutative diagram (2.2.5), the assertion follows from the surjectivity of $\mathfrak{m}^n_K/\mathfrak{m}^{n+1}_K \to \mathrm{Gr}^n G^{\mathrm{ab}}_K$ in Proposition 5.3.7(3).

(2) This is also an immediate consequence of the surjectivity of $\mathfrak{m}^n_K/\mathfrak{m}^{n+1}_K \to \mathrm{Gr}^n G^{\mathrm{ab}}_K$ in Proposition 5.3.7(3). □

**Exercise 5.3** Let $p$ be a prime number and $n \geqslant 2$ be an integer. We identify $G = \mathrm{Gal}(\mathbf{Q}_p(\zeta_{p^n})/\mathbf{Q}_p)$ with $(\mathbf{Z}/p^n\mathbf{Z})^\times$ and let $\sigma \in G$ be the element corresponding to $1 + p^{n-1}$. Let $\chi : G \to \mathbf{Q}/\mathbf{Z}$ be a primitive character such that $\chi(\sigma) = 1/p$.

Compute $m = \mathrm{Sw}\,\chi$ and $\mathrm{rsw}\,\chi : \mathfrak{m}^m_{\mathbf{Z}_p}/\mathfrak{m}^{m+1}_{\mathbf{Z}_p} \to \mathbf{F}_p$.

## Historical Notes

The Hasse–Arf theorem, Theorem 5.3.4, is proved by Kato in [49, Theorem (3.6)] in a slightly weaker form. The classical case where the residue field is finite, stated in the form of Corollary 6.3.9(2), is proved by Hasse [37]. It is generalized to the algebraically closed residue field case by Arf [7]. A more conceptual proof by Serre is given in [78, Chapitre V, §7]. See a comment at the end of Serre's letter dated "9 novembre 1958" recorded in [18].

# PART III

# CONDUCTOR AND REFINEMENTS

The Swan conductor is a numerical invariant of Galois representations measuring the wild ramification. In Chapter 6, we define the Swan conductor under the classical assumption that the residue field extension is separable, using the break decomposition. The Swan conductor will be applied in Chapter 8 to state the Grothendieck–Ogg–Shafarevich formula computing the Euler–Poincaré characteristic of an étale sheaf on a smooth curve over an algebraically closed field.

We introduce a refinement of the Swan conductor taking values in the quotient group of the multiplicative group modulo the units congruent to 1.

We prove the integrality of the Swan conductor. In the rank 1 case, this is a consequence of the Hasse–Arf theorem. The general case is proved by reducing it to the rank 1 case by the induction formula computing the Swan conductor of induced representations. A description of the Swan conductor in terms of the Swan character implies the induction formula, and also the compatibility with the quotient groups of the Swan conductor. We state the description using Brauer traces as we consider Galois representations with mod $\ell$ coefficients in geometric applications.

In Chapter 7, we generalize the Swan conductor to log monogenic extensions. A refinement is then defined as a certain twisted differential form of the residue field. This refinement will be applied in Chapter 10 to state the Deligne–Kato formula computing the dimension of the space of nearby cycles of an étale sheaf on a curve over a discrete valuation ring with algebraically closed residue field.

# 6

# Swan Conductor

A main application of filtration by ramification groups is to the theory of conductors of Galois representations of local fields. After recalling some basic facts on representations of finite groups in Section 6.1, we study the conductor under the classical assumption that the extension of residue fields is separable.

Under this assumption, we first introduce a rescaling function and its refinement in Section 6.2. A geometric structure behind the definition of the rescaling function and the refinement will be clarified in Section 11.5. Using them, in Section 6.3 we define the Swan conductor as a rational number and a refinement of the Swan conductor defined as an element of the quotient group of the multiplicative group of a Galois extension by the subgroup of units congruent to 1 modulo the maximal ideal.

In Section 6.3, we give another description of the Swan conductor and its refinement using the Swan character. From this description, we derive the induction formula in Section 6.4 and deduce the integrality of Swan conductor from the Hasse–Arf theorem by using the induction formula.

## 6.1 Brauer Trace

In this section $G$ denotes a finite group.

**Lemma 6.1.1** *Let $\Lambda$ be a field. Let $M$ be a finitely generated $\Lambda[G]$-module and $P$ be a finitely generated projective $\Lambda[G]$-module. Then $P \otimes_\Lambda M$ with the diagonal action of $G$ is a finitely generated projective $\Lambda[G]$-module.*

*Proof* We may assume that $P$ is a free $\Lambda[G]$-module and then $P = \Lambda[G]$. Let $\Lambda[G] \otimes'_\Lambda M$ denote the tensor product with the action of $G$ on the first factor $\Lambda[G]$. Then the morphism $\Lambda[G] \otimes'_\Lambda M \to \Lambda[G] \otimes_\Lambda M$ defined by sending

$g \otimes x$ to $g \otimes gx$ is an isomorphism of $\Lambda[G]$-modules. Since $\Lambda[G] \otimes'_\Lambda M$ is a free $\Lambda[G]$-module of finite rank, the assertion follows. □

Let $K_\Lambda(G)$ and $P_\Lambda(G)$ denote the Grothendieck groups of the categories of finitely generated $\Lambda[G]$-modules and finitely generated projective $\Lambda[G]$-modules respectively. The tensor products over $\Lambda$ with the diagonal actions define a ring structure on $K_\Lambda(G)$. Lemma 6.1.1 implies that $P_\Lambda(G)$ is a $K_\Lambda(G)$-module and that the canonical morphism $P_\Lambda(G) \to K_\Lambda(G)$ is $K_\Lambda(G)$-linear.

Let $\ell$ be a prime number and $E$ be a finite extension of $\mathbf{Q}_\ell$. Let $O_E$ be the valuation ring and $\Lambda$ be the residue field. Let $P_{O_E}(G)$ be the Grothendieck group of the category of finitely generated projective $O_E[G]$-modules. Tensor products and the inclusion of categories define morphisms

$$K_E(G) \leftarrow P_{O_E}(G) \to P_\Lambda(G) \to K_\Lambda(G).$$

**Proposition 6.1.2** (1) [79, Partie III, 1.5] *The morphism $P_{O_E}(G) \to P_\Lambda(G)$ defined by tensor product $\otimes_{O_E}\Lambda$ is an isomorphism.*

(2) [79, Partie III, 1.4] *The groups $P_{O_E}(G)$ and $P_\Lambda(G)$ are free abelian groups of finite rank generated by the isomorphism classes of indecomposable projective modules.*

(3) [79, Partie III, 3.1, Théorème 1] *For any finite extension $E'$ of $E$, the morphism $P_{O_E}(G) \to K_{E'}(G)$ defined by the tensor product $\otimes_{O_E} E'$ is an injection and its cokernel is torsion free.*

(4) [79, Partie III, 2.5] *If the order of $G$ is prime to $\ell$, the morphisms $P_\Lambda(G) \to K_\Lambda(G)$ and $P_{O_E}(G) \to K_E(G)$ are isomorphisms.*

We recall the definition of the Brauer trace of a semi-simple endomorphism of a vector space of finite dimension over a perfect field of positive characteristic. Let $\Lambda$ be a perfect field of characteristic $\ell > 0$, and $O_E = W(\Lambda) \subset E = W(\Lambda)[1/\ell]$ be the ring of Witt vectors and its fraction field.

Let $M$ be a $\Lambda$-vector space of finite dimension and let $\sigma$ be a semi-simple endomorphism of $M$. Let $A = \Lambda[\sigma] \subset \mathrm{End}_\Lambda(M)$ be the semi-simple commutative $\Lambda$-algebra generated by $\sigma$. Since $A$ is finite étale over $\Lambda$, we have a decomposition $A = \prod_i A_i$ into a product of finite separable extensions of $\Lambda$. Let $\sigma = (\sigma_i)_i$ and $M = \prod_i M_i$ be the corresponding decompositions, and let $m_i = \dim_{A_i} M_i$ be the multiplicity. Then the restriction of the trace mapping $\mathrm{Tr}(- : M)\colon \mathrm{End}_\Lambda(M) \to \Lambda$ on $A = \prod_i A_i \subset \prod_i \mathrm{End}_\Lambda(M_i) \subset \mathrm{End}_\Lambda(M)$ equals the sum $\sum_i m_i \cdot \mathrm{Tr}_{A_i/\Lambda}$.

Let $W(A) = \prod_i W(A_i)$ be the product of the rings of Witt vectors, and let $\widetilde{\sigma} = (\widetilde{\sigma}_i)_i$ be the Teichmüller lifting. Then the Brauer trace $\mathrm{Tr}^{\mathrm{Br}}(\sigma : M) \in O_E$

is defined by

$$\mathrm{Tr}^{\mathrm{Br}}(\sigma : M) = \sum_i m_i \cdot \mathrm{Tr}_{W(A_i)/W(\Lambda)}(\widetilde{\sigma}_i). \tag{6.1.1}$$

**Lemma 6.1.3** *Let $M$ and $N$ be $\Lambda$-vector spaces of finite dimension, and let $f$ and $g$ be semi-simple endomorphisms of $M$ and $N$ respectively. Then*

$$\mathrm{Tr}^{\mathrm{Br}}(f \otimes g : M \otimes_\Lambda N) = \mathrm{Tr}^{\mathrm{Br}}(f : M) \cdot \mathrm{Tr}^{\mathrm{Br}}(g : N).$$

*Proof* Extending the scalar, we may assume that $f$ and $g$ are diagonalizable. Then the assertion is clear. □

**Definition 6.1.4** Let $G$ be a finite group and $\Lambda$ be a perfect field of characteristic $\ell \neq 0$. Let $E$ be a complete discrete valuation field of characterstic 0 with residue field $\Lambda$.

(1) We say that an element $\sigma \in G$ is $\ell$-regular if the order of $\sigma$ is prime to $\ell$. We let $G_{\ell\text{-reg}} \subset G$ denote the subset consisting of $\ell$-regular elements. If $\ell$ is clear from the context, we drop it in the notation and write $G_{\mathrm{reg}} = G_{\ell\text{-reg}}$.
(2) Let $M$ be a representation of $G$ on a $\Lambda$-vector space of finite dimension. Then the action of an $\ell$-regular element $\sigma \in G_{\mathrm{reg}}$ is semi-simple and we define the Brauer trace $\mathrm{Tr}^{\mathrm{Br}}(\sigma : M) \in W(\Lambda) \subset O_E \subset E = O_E[1/l]$.

**Proposition 6.1.5** *Let $P$ be a finitely generated projective $O_E[G]$-module.*

(1) *We have*

$$\mathrm{Tr}(\sigma : P) = \begin{cases} \mathrm{Tr}^{\mathrm{Br}}(\sigma : P \otimes_{O_E} \Lambda) & \text{if } \sigma \in G_{\mathrm{reg}}, \\ 0 & \text{otherwise.} \end{cases}$$

(2) *Let $M$ be a finitely generated $\Lambda[G]$-module. Then*

$$\dim_\Lambda (P \otimes_{O_E} M)^G = \frac{1}{\#G} \sum_{\sigma \in G_{\mathrm{reg}}} \mathrm{Tr}(\sigma : P) \cdot \mathrm{Tr}^{\mathrm{Br}}(\sigma : M). \tag{6.1.2}$$

*Proof* (1) If $\sigma \in G_{\mathrm{reg}}$, the subring $O_E[\sigma] \subset \mathrm{End}_{O_E}(P)$ is finite étale over $O_E$ and the equality follows in this case. Let $\sigma \in G - G_{\mathrm{reg}}$ and let $C = \langle \sigma \rangle \subset G$ be the subgroup generated by $\sigma$. Decompose $C = C_1 \times C_2$ into the product of the prime-to-$\ell$-part and the $\ell$-part, and set $\sigma = (\sigma_1, \sigma_2)$. Extending the scalar, we may assume that $\sigma_1$ is diagonalizable. Since $P$ is a projective $O_E[G]$-module, it is also projective as an $O_E[C_2]$-module and hence is a free module. By decomposing $P$ by the characters of $C_1$, we may assume that $\sigma_1$ acts on $P$ as a scalar $a$. For the regular representation, we have $\mathrm{Tr}(\sigma_2 : O_E[C_2]) = 0$ for $\sigma_2 \neq 1$. Since $P$ is a free $O_E[C_2]$-module, we have $\mathrm{Tr}(\sigma : P) = a \cdot \mathrm{Tr}(\sigma_2 : P) = 0$.

(2) Let $a = [M] \in K_\Lambda(G)$ and $b = [P] \in P_{O_E}(G)$. Then the product

$ab$ defines an element of $P_{O_E}(G)$ by Lemma 6.1.1 and Proposition 6.1.2(1). Since $(1/\#G)\sum_{\sigma\in G}\sigma$ is the projector to the $G$-fixed part in characteristic 0, the left-hand side equals $(1/\#G)\sum_{\sigma\in G}\mathrm{Tr}(\sigma : ab)$. By (1) and Lemma 6.1.3, this equals the right-hand side. □

**Lemma 6.1.6** *Let $G$ be a finite group. Let $\Lambda$ be a field of characteristic $\ell$, different from a prime number $p$, and let $V$ be a $\Lambda$-vector space of finite dimension with an action of $G$.*

(1) *Let $(G_i)_{i\geqslant 1}$ be a decreasing separated filtration by normal subgroups of $G$ such that $G_1$ is a $p$-group. Then there exists a unique decomposition*

$$V = \bigoplus_{i\in\mathbf{N}} V_i \tag{6.1.3}$$

*by sub-$G$-representations such that for every $i \geqslant 1$ the $G_i$-fixed part $V^{G_i}$ is the sum $\bigoplus_{j<i} V_j$ of $V_j$ for $0 \leqslant j < i$.*

(2) *Assume that $G$ is an $\mathbf{F}_p$-vector space and that $\Lambda^\times$ contains a primitive $p$th root of 1, and let $G^\vee = \mathrm{Hom}(G, \Lambda^\times)$ be the character group. There exists a unique decomposition*

$$V = \bigoplus_{\chi\in G^\vee} V_\chi \tag{6.1.4}$$

*by sub-$G$-representations such that $G$ acts on $V_\chi$ via $\chi : G \to \Lambda^\times$.*

*Proof* (1) The $G_i$-fixed part $V^{G_i}$ defines increasing and exhaustive filtration. We construct a splitting. For each integer $i \geqslant 1$, the operator $e_i = (1/\#G_i)\sum_{\sigma\in G_i}\sigma$ is a projector to the $G_i$-fixed part, and they are mutually commutative. We set $V_i = \mathrm{Im}(1 - e_i)e_{i+1}$ for $i \geqslant 1$ and $V_0 = \mathrm{Im}\, e_1$. Then $V_0 = V^{G_1}$ and $V^{G_{i+1}} = V^{G_i} \oplus V_i$ for $i \geqslant 1$. Hence, the subspaces $V_i$ satisfy the required property.

(2) The group algebra $\Lambda[G]$ is semi-simple and the decomposition $\Lambda[G] = \prod_{\chi\in G^\vee}\Lambda$ defines the required decomposition. □

**Proposition 6.1.7** ([79, Partie II, 9.5, Théorème 20]) *Let $G$ be a $p$-group and $\Lambda$ be a field of characteristic $\neq p$ containing the $\#G$th roots of 1. Then every irreducible representation of $G$ is isomorphic to a monomial representation.*

The induced representation $\mathrm{Ind}_H^G\chi$ of a character $\chi$ of degree 1 of a subgroup $H \subset G$ is called a monomial representation.

**Exercise 6.1** Let $\Lambda$ be a finite field of characteristic $\ell \neq 0$ and $V$ be a $\Lambda$-vector space of finite dimension. Let $\sigma$ be an endomorphism of $V$ of order $n$ prime to $\ell$, and regard the Brauer trace $\mathrm{Tr}^{\mathrm{Br}}(\sigma : M)$ as an algebraic integer in $\mathbf{Z}[\zeta_n]$.

Define the rational Brauer trace by

$$\mathrm{Tr}^{\mathrm{Br}}_{\mathbf{Q}}(\sigma : M) = \frac{1}{[\mathbf{Q}(\zeta_n) : \mathbf{Q}]}\mathrm{Tr}_{\mathbf{Q}(\zeta_n)/\mathbf{Q}}\mathrm{Tr}^{\mathrm{Br}}(\sigma : M). \tag{6.1.5}$$

(1) Assume that $n = p^e > 1$ for a prime number $p \neq \ell$, and let $M^{\sigma} \subset M^{\sigma^p} \subset M$ denote the fixed subspaces. Show that

$$\begin{aligned}\mathrm{Tr}^{\mathrm{Br}}_{\mathbf{Q}}(\sigma : M) &= \frac{1}{(p-1)p^{e-1}} \sum_{i \in \mathbf{Z}/p^e\mathbf{Z} - p\mathbf{Z}/p^e\mathbf{Z}} \mathrm{Tr}^{\mathrm{Br}}(\sigma^i : M) \qquad (6.1.6)\\ &= \frac{1}{p-1}(p \cdot \dim M^{\sigma} - \dim M^{\sigma^p}).\end{aligned}$$

(2) Assume in Proposition 6.1.5 that $\mathrm{Tr}(\sigma : P)$ is a rational integer for any $\sigma \in G_{\mathrm{reg}}$. Show that we may replace the Brauer trace $\mathrm{Tr}^{\mathrm{Br}}(\sigma : M)$ in (6.1.2) by $\mathrm{Tr}^{\mathrm{Br}}_{\mathbf{Q}}(\sigma : M)$.

## 6.2 A Rescaling Function

In this section we introduce a rescaling function and its refinement.

Let $K$ be a henselian discrete valuation field such that the residue field $F$ is of characteristic $p > 0$. Let $L$ be a finite Galois extension of $K$ of Galois group $G = \mathrm{Gal}(L/K)$. In this chapter, we assume that the residue field $E$ of $L$ is a *separable* extension of $F$, unless otherwise stated. The essential case is where $L$ is totally ramified over $K$. Recall that the filtration $(G_i)_{i \geqslant 0}$ of $G$ by the lower numbering ramification subgroups is defined by $G_i = \mathrm{Ker}(G \to \mathrm{Aut}(L^{\times}/1 + \mathfrak{m}_L^i))$ for integers $i \geqslant 1$, and $G_0 = \mathrm{Ker}(G \to \mathrm{Gal}(E/F)) = I$. The subgroup $G_1$ is the $p$-Sylow subgroup $P$ of the inertia group $I = G_0$.

**Definition 6.2.1** Let $L$ be a finite Galois extension of a henselian discrete valuation field $K$ of Galois group $G$. Assume that the residue field $E$ of $L$ is a separable extension of the residue field $F$ of $K$. We define a function $\varphi = \varphi_{L/K} : \mathbf{N} \to \mathbf{Q}$ by

$$\varphi(i) = \sum_{j=1}^{i} \frac{1}{[G_0 : G_j]}. \tag{6.2.1}$$

More canonically, the source $\mathbf{N}$ of the rescaling function $\varphi_{L/K}$ should be regarded as a submonoid of $L^{\times}/O_L^{\times} \cong \mathbf{Z}$, and the target $\mathbf{Q}$ should be regarded as $(K^{\times}/O_K^{\times}) \otimes_{\mathbf{Z}} \mathbf{Q}$. This will be clarified by the introduction of groups $R_K \subset R_L$ later in this section.

We give descriptions of the rescaling function $\varphi_{L/K}$ in terms of the logarithmic different and the function $i_G(\sigma)$.

**Proposition 6.2.2** *Let $L$ be a finite Galois extension of a henselian discrete valuation field $K$ of Galois group $G$ such that the residue field $E$ of $L$ is a separable extension of the residue field $F$ of $K$.*

(1) *For each integer $i \geqslant 0$, let $M_i \subset L$ be the extension corresponding to $G_i \subset G$ and let $d_{M_i/K} = \operatorname{ord}_{M_i} D^{\log}_{M_i/K}$ be the valuation of the logarithmic different. Then*

$$\varphi(i) = \frac{d_{M_i/K} + i}{e_{M_i/K}}. \tag{6.2.2}$$

(2) *For $i \geqslant 0$, we have*

$$\varphi(i) = \frac{1}{\#G_0}\Big( \sum_{\sigma \in I - G_i} i_G(\sigma) + \sum_{\sigma \in G_i} i \Big). \tag{6.2.3}$$

*Proof* (1) Let $f(i)$ denote the right-hand side of (6.2.2). Since $M_0 \subset L$ is the maximal unramified extension over $K$, the logarithmic different $d_{M_0/K}$ is 0 and we have $f(0) = 0$. By the assumption that $E$ is a separable extension of $F$, the index $[G_0 : G_i]$ equals the ramification index $e_{M_i/K}$. Hence, to show (6.2.2) by induction on $i \geqslant 0$, it suffices to show the equality

$$f(i) = \frac{d_{M_{i+1}/K} + i}{e_{M_{i+1}/K}}. \tag{6.2.4}$$

By the chain rule, Lemma 1.9.3(2), we have $d_{M_{i+1}/K} = e_{M_{i+1}/M_i} \cdot d_{M_i/K} + d_{M_{i+1}/M_i}$. Hence, the equality (6.2.4) follows from $d_{M_{i+1}/M_i} = (e_{M_{i+1}/M_i} - 1) \cdot i$ (1.9.5).

(2) By the definition of $i_G(\sigma)$, the content of the parentheses on the right-hand side of (6.2.3) equals $\#\{(\sigma, j) \in G \times \mathbf{N} \mid 1 \leqslant j \leqslant i, \sigma \in G_j\}$. Hence, the equality follows. □

We prepare to introduce a refinement of the rescaling function in the sense of (6.2.2). First, we define an extension of $\mathbf{Z}$ by the multiplicative group of the residue field $F^\times$ in which the refinement takes values. Until the paragraph before Definition 6.2.4, we drop the assumption that the residue field extension be separable.

Let $K$ be a discrete valuation field and let $F$ be the residue field of $K$. The group

$$R_K = K^\times/(1 + \mathfrak{m}_K) \tag{6.2.5}$$

is an extension of $\mathbf{Z}$ by $F^\times$. We identify $F^\times$ as the subgroup $O_K^\times/(1 + \mathfrak{m}_K)$ of $R_K$ and let $[a]$ denote $a \in F^\times$ regarded as an element of $R_K$. For an $F$-vector space $V$ of dimension 1, let $V^\times = V - \{0\}$ denote the set of bases of $V$. We identify $R_K$ with the disjoint union $\coprod_{i \in \mathbf{Z}} (\mathfrak{m}_K^i/\mathfrak{m}_K^{i+1})^\times$. For an extension $L$ of the discrete valuation field of $K$ with residue field $E$ of ramification index $e_{L/K}$, the morphism $R_K \to R_L$ induced by the inclusion $K^\times \to L^\times$ is an injection and fits in the commutative diagram

$$\begin{array}{ccccccccc} 0 & \longrightarrow & F^\times & \longrightarrow & R_K & \longrightarrow & \mathbf{Z} & \longrightarrow & 0 \\ & & \downarrow & & \downarrow & & \downarrow{\scriptstyle e_{L/K}} & & \\ 0 & \longrightarrow & E^\times & \longrightarrow & R_L & \longrightarrow & \mathbf{Z} & \longrightarrow & 0 \end{array} \tag{6.2.6}$$

of exact sequences. The norm mapping $L^\times \to K^\times$ induces $\mathrm{N}_{L/K} \colon R_L \to R_K$. If $L$ is a Galois extension of Galois group $G$, the composition $\mathrm{N}_{L/K} \colon R_L \to R_K \to R_L$ with the inclusion is the same as $\mathrm{N}_G = \sum_{\sigma \in G} \sigma$.

**Lemma 6.2.3** *Let $L$ be a tamely ramified finite Galois extension of $K$ of degree $n$ of Galois group $G$. Then the canonical morphism $R_K \to R_L$ induces an isomorphism $R_K \to R_L^G$ to the $G$-fixed part.*

*Proof* We identify $R_L$ with the disjoint union $\coprod_{i \in \mathbf{Z}} (\mathfrak{m}_L^i/\mathfrak{m}_L^{i+1})^\times$. Since $L$ is tamely ramified, the ramification index $m = e_{L/K}$ is prime to $p$ and the action of the inertia subgroup $I$ on $\mathfrak{m}_L/\mathfrak{m}_L^2$ is through a surjection $I \to \mu_m \subset E^\times$. Since the action on $\mathfrak{m}_L^i/\mathfrak{m}_L^{i+1}$ is by its $i$th power, the inertia fixed part of $\mathfrak{m}_L^i/\mathfrak{m}_L^{i+1}$ is 0 unless $i$ is divisible by $m$. For any multiple $i = mj$ of $m$, we have a canonical isomorphism $\mathfrak{m}_K^j/\mathfrak{m}_K^{j+1} \otimes_F E \to \mathfrak{m}_L^i/\mathfrak{m}_L^{i+1}$. This induces a bijection $\mathfrak{m}_K^j/\mathfrak{m}_K^{j+1} \to (\mathfrak{m}_L^i/\mathfrak{m}_L^{i+1})^G$ on the $G$-fixed part. Thus, we obtain an isomorphism $R_K = \coprod_{j \in \mathbf{Z}} (\mathfrak{m}_K^j/\mathfrak{m}_K^{j+1})^\times \to R_L^G = \coprod_{i \in \mathbf{Z}} (\mathfrak{m}_L^i/\mathfrak{m}_L^{i+1})^{\times G}$ as required. □

From here on, we assume that the residue field $E$ of $L$ is a separable extension of the residue field $F$ of $K$. We define a refinement of the valuation of the logarithmic different as an element of $R_L$. Let $L$ be a finite separable extension of $K$ and let $D_{L/K}^{\log} \subset O_L$ denote the logarithmic different. Then $(D_{L/K}^{\log})^{-1} \cdot \mathrm{Tr}_{L/K} = \mathrm{Hom}_{O_K}(\mathfrak{m}_L, \mathfrak{m}_K)$ and $\mathrm{Tr}_{L/K} \colon (D_{L/K}^{\log})^{-1} \to O_K$ induces a surjection $(D_{L/K}^{\log})^{-1} \otimes_{O_L} E \to F$. Since the $E$-vector space $\mathrm{Hom}_F(E, F)$ is of dimension 1, there exists a unique isomorphism $(D_{L/K}^{\log})^{-1} \otimes_{O_L} E \to E$ such that the surjection $(D_{L/K}^{\log})^{-1} \otimes_{O_L} E \to F$ induced by $\mathrm{Tr}_{L/K}$ is its composition with $\mathrm{Tr}_{E/F} \colon E \to F$.

**Definition 6.2.4** Let $L$ be a finite separable extension of a henselian discrete valuation field $K$ such that the residue field $E$ of $L$ is a separable extension of $F$. Let $d_{L/K}$ denote the valuation of the logarithmic different $D_{L/K}^{\log}$. We call the unique element

$$\Delta_{L/K} \in (D^{\log}_{L/K} \otimes_{O_L} E)^\times = (\mathfrak{m}_L^{d_{L/K}}/\mathfrak{m}_L^{d_{L/K}+1})^\times \subset R_L$$

satisfying the following property the logarithmic different: For every integer $i \in \mathbf{Z}$ and $j = e_{L/K} \cdot i - d_{L/K}$, the diagram

$$\begin{array}{ccc} \mathfrak{m}_L^j/\mathfrak{m}_L^{j+1} & \xrightarrow{\Delta_{L/K}\cdot} & \mathfrak{m}_L^{e_{L/K}i}/\mathfrak{m}_L^{e_{L/K}i+1} \\ {\scriptstyle \mathrm{Tr}_{L/K}}\downarrow & & \| \\ \mathfrak{m}_K^i/\mathfrak{m}_K^{i+1} & \xleftarrow{1\otimes \mathrm{Tr}_{E/F}} & \mathfrak{m}_K^i/\mathfrak{m}_K^{i+1} \otimes_F E \end{array}$$

is commutative.

The image of $\Delta_{L/K} \in R_L$ in $\mathbf{Z}$ is the valuation $\mathrm{ord}_L D^{\log}_{L/K}$ of the logarithmic different. In an abuse of terminology we call both the element $\Delta_{L/K}$ and the ideal $D^{\log}_{L/K}$ the logarithmic different.

**Lemma 6.2.5** *Let $L$ be a finite separable extension of a henselian discrete valuation field $K$ such that the residue field $E$ of $L$ is a separable extension of $F$.*

(1) *Let $M$ be a subextension of $L$. Then*

$$\Delta_{L/K} = \Delta_{L/M} + \Delta_{M/K} \tag{6.2.7}$$

*in $R_L$.*

(2) *Assume that $L$ is tamely ramified and let $e_{L/K}$ be the ramification index. Then*

$$\Delta_{L/K} = [e_{L/K}]. \tag{6.2.8}$$

(3) *Assume that $L$ is a totally ramified extension of degree $n$. Let $s$ be a uniformizer of $L$ and $P \in O_K[T]$ be the minimal polynomial of $t$. Then $\Delta_{L/K} = P'(s)/s^{n-1}$ in $R_L$.*

*Proof* (1) Since the trace mappings satisfy the chain rule, the assertion follows from the isomorphism $\mathrm{Hom}_{O_L}(\mathfrak{m}_L, \mathfrak{m}_M) \otimes_{O_M} \mathrm{Hom}_{O_M}(\mathfrak{m}_M, \mathfrak{m}_K) \to \mathrm{Hom}_{O_L}(\mathfrak{m}_L, \mathfrak{m}_K)$ defined by the composition.

(2) We have $D^{\log}_{L/K} = O_L$ by Lemma 1.9.4(1) (a) $\Rightarrow$ (b). Since $\mathrm{Tr}_{L/K} \colon O_L \to O_K$ induces $e_{L/K} \cdot \mathrm{Tr}_{E/F} \colon E \to F$ by (1.9.1), we have $\Delta_{L/K} = [e_{L/K}]$.

(3) In the notation of Proposition 1.2.2, $\mathrm{Tr}_{L/K}$ induces an isomorphism $D^{\log}_{L/K} \otimes_{O_L} F \to F$ sending $s^{n-1}/P'(s)$ to 1 by (1.2.2). Hence, the assertion follows. □

Let $L$ be a finite Galois extension of $K$ of Galois group $G$. Assume that the extension $E$ of the residue field $F$ is separable. For $i \geqslant 1$, a canonical injection

$$\delta\colon \mathrm{Gr}_i G = G_i/G_{i+1} \to \mathfrak{m}_L^i/\mathfrak{m}_L^{i+1} \tag{6.2.9}$$

sending $\sigma$ to $\delta(\sigma) = \sigma(s)/s - 1$ for a uniformizer $s$ of $L$ is defined as (1.6.6) in Corollary 1.6.4(1). For $\sigma \in G_i - G_{i+1}$, we consider $\delta(\sigma) \in \mathfrak{m}_L^i/\mathfrak{m}_L^{i+1} - \{0\}$ as an element of $R_L = \coprod_{n\in\mathbf{Z}}(\mathfrak{m}_L^n/\mathfrak{m}_L^{n+1})^\times$. If $K'$ is a tamely ramified extension of $K$ and $L' = LK'$, for $\sigma' \in P' \subset \mathrm{Gal}(L'/K')$ corresponding to $\sigma \in P \subset \mathrm{Gal}(L/K)$, the morphism $R_L \to R_{L'}$ maps $\delta(\sigma)$ to $\delta(\sigma') + [e_{L'/L}]$ by (5.3.10).

We also have an injection $G_0/G_1 \to E^\times$ sending $\sigma$ to $\sigma(s)/s$ for a uniformizer $s$ of $L$ by Proposition 1.6.2(2). We extend $\delta\colon G_1 - \{1\} \to R_L$ to

$$\delta\colon G_0 - \{1\} \to R_L \tag{6.2.10}$$

by sending $\sigma$ to $\delta(\sigma) = \sigma(s)/s - 1 \bmod \mathfrak{m}_L$. If $L$ is tamely ramified over $K$ and $m = e_{L/K}$, for a primitive $m$th root $\zeta$ of 1 we have $\sum_{\sigma\in I-\{1\}} \delta(\sigma) = \prod_{i=1}^{m-1}(\zeta^i - 1) = [(-1)^{m-1}m]$.

Similarly to Propositions 1.8.7 and 1.9.5(3) and Corollary 1.9.6, we have the following equalities.

**Proposition 6.2.6** *Let $L$ be a finite Galois extension of $K$ of ramification index $n$ of Galois group $G$. Assume that the residue extension $E$ over $F$ is separable. Let $M_0 \subset M_1 \subset L$ be the maximal unramified extension and the maximal tamely ramified extension corresponding to $G_1 = P \subset G_0 = I \subset G$, and set $m_{L/K} = [G_0 : G_1]$ to be the prime-to-$p$ part $e_{M_1/M_0}$ of the ramification index $e_{L/K}$.*

(1) *Let $\sigma \in P - \{1\}$. For an integer $j$ prime to $p$, we have $\delta(\sigma^j) = \delta(\sigma) + [j]$. For $\tau \in G$, we have $\delta(\tau\sigma\tau^{-1}) = \tau\delta(\sigma)$.*

(2) *We have*

$$\Delta_{L/K} = \Delta_{L/M_0} = \sum_{\sigma\in I-\{1\}} \delta(\sigma) + [(-1)^{n-1}], \tag{6.2.11}$$

$$\Delta_{L/M_1} = \sum_{\sigma\in P-\{1\}} \delta(\sigma) \tag{6.2.12}$$

*in $R_L$.*

(3) *Let $N \subset G$ be a normal subgroup and let $M$ be the corresponding extension. We identify $\overline{G} = \mathrm{Gal}(M/K)$ with $G/N$ and $\overline{G}_1 = \overline{P} \subset \overline{G}$ with the image of*

*$G_1 = P$. Then*

$$\Delta_{M/K} = \Delta_{L/K} - \sum_{\tau \in N\cap P-\{1\}} \delta(\tau) - [m_{L/M}], \tag{6.2.13}$$

$$\delta(\sigma) = \sum_{\tau\in\sigma\cap P} \delta(\tau) + [m_{L/M}] \tag{6.2.14}$$

*for $\sigma \in \overline{P} - \{1\}$ in $R_M \subset R_L$.*

The fact that the same correction term $[m_{L/M}]$ appears in both (6.2.13) and (6.2.14) will be used in the proof of the compatibility with the quotient group of the refined Swan conductor, Corollary 6.3.8.

*Proof* (1) Let $i \geqslant 1$ and assume $\sigma \in G_i - G_{i+1}$. Then, since $\delta\colon G_i/G_{i+1} \to \mathfrak{m}_L^i/\mathfrak{m}_L^{i+1}$ is a morphism, we have $\delta(\sigma^j) = [j]+\delta(\sigma)$. The equality $\delta(\tau\sigma\tau^{-1}) = \tau\delta(\sigma)$ follows from the compatibility in Corollary 1.6.4(1).

(2) Since $\Delta_{M_0/K} = 1$, we have $\Delta_{L/K} = \Delta_{L/M_0}$ by the chain rule (6.2.7). Hence, we may assume that $L$ is a totally ramified extension of $M_0 = K$ of degree $n$. Let $s \in O_L$ be a uniformizer and $Q \in O_{M_0}[T]$ be the minimal polynomial of $s$. Then $\Delta_{L/M_0} = Q'(s)/s^{n-1}$ by Lemma 6.2.5(3). Similarly to the proof of Proposition 1.9.5(3), the right-hand side equals $\prod_{\sigma\in I-\{1\}}(1-\sigma(s)/s)$ and the second equality in (6.2.11) follows.

Since $[L : K_1]$ is a power of $p$, we have $(-1)^{[L:K_1]-1} = 1$ in $E^\times \subset R_L$ and the equality (6.2.12) follows from (6.2.11).

(3) Let $M_1 \subset L$ be the maximal tamely ramified extension of $M$ corresponding to the subgroup $P \cap N \subset N$. By the chain rule (6.2.7) and (6.2.8), we have $\Delta_{L/K} - \Delta_{M/K} = \Delta_{L/M} = \Delta_{L/M_1} + [m_{L/M}]$. Since $\Delta_{L/M_1} = \sum_{\tau\in P\cap N-\{1\}} \delta(\tau)$ by (6.2.12), the equality (6.2.13) follows.

To show the equality (6.2.14), by considering the extension corresponding to $N \cap P$, the assertion is reduced to showing the following two cases:

(i) $L$ is a tamely ramified extension of $M$;
(ii) $N$ is a subgroup of $P$,

respectively.

(i) Let $e_{L/M} = m_{L/M}$ be the ramification index. Let $\tau \in P$ be the unique inverse image of $\sigma \in \overline{P}$. Since $\delta_\tau\colon L^\times/O_L^\times \to \mathfrak{m}_L^i/\mathfrak{m}_L^{i+1}$ is a morphism of groups, the image of $\delta(\sigma) \in R_M$ in $R_L$ equals $\delta(\tau) + [m_{L/M}]$.

(ii) In this case, $L$ is a totally ramified extension of $M$ of degree a power of $p$. Let $s$ be a uniformizer of $L$. Then the constant term $t = a_n \in O_M$ of the minimal polynomial $X^n + a_1X^{n-1} + \cdots + a_n \in O_M[X]$ of $s$ over $M$ is a uniformizer of $M$ and we have $t \equiv -s^n$. By (1.7.3), we have $1-\sigma(t)/t = \prod_{\tau\in\sigma}(1-\tau(s)/s)$ in

$R_L$. Since $\#\sigma = \#P$ is a power of $p$, we obtain $\sigma(t)/t - 1 = \prod_{\tau\in\sigma}(\tau(s)/s - 1)$ and the equality (6.2.14) follows. □

We define a refinement of the rescaling function.

**Definition 6.2.7** Let $i \geqslant 1$ be an integer and let $\chi\colon \mathrm{Gr}_i G = G_i/G_{i+1} \to \mathbf{F}_p$ be a non-trivial character. Let $M_{i,\chi} \subset L$ be the extension corresponding to the inverse image $G_{i,\chi} \subset G_i$ of $\mathrm{Ker}(\chi\colon G_i/G_{i+1} \to \mathbf{F}_p)$, and let $\Delta_{M_{i,\chi}/K} \in R_{M_{i,\chi}} \subset R_L$ be the logarithmic different. We define $\varphi(i,\chi) \in R_L$ by

$$\varphi(i,\chi) = \Delta_{M_{i,\chi}/K} + \sum_{\sigma\in G_i, \chi(\bar{\sigma})=1} \delta(\sigma). \tag{6.2.15}$$

We give a geometric interpretation of $\varphi(i,\chi)$ in Section 15.3. The following lemma confirms that $\varphi(i,\chi)$ is a refinement of $\varphi(i)$.

**Lemma 6.2.8** *Let $i \geqslant 1$ be an integer and let $\chi\colon \mathrm{Gr}_i G = G_i/G_{i+1} \to \mathbf{F}_p$ be a non-trivial character. Let $G_{i,\chi} \subset G_i$ be the inverse image of the kernel* $\mathrm{Ker}(\chi\colon G_i/G_{i+1} \to \mathbf{F}_p)$.

(1) *We have*

$$e_{L/K} \cdot \varphi(i) = \sum_{\sigma\in G_1 - G_i} \mathrm{ord}_L \delta(\sigma) + i \cdot \#G_i \tag{6.2.16}$$

*in* $\mathbf{Z}$ *and*

$$\varphi(i,\chi) = \Delta_{L/K} - \sum_{\sigma\in G_{i,\chi}-\{1\}} \delta(\sigma) + \sum_{\sigma\in G_i,\chi(\bar{\sigma})=1} \delta(\sigma) \tag{6.2.17}$$

$$= \sum_{\sigma\in I - G_{i,\chi}} \delta(\sigma) + \sum_{\sigma\in G_i,\chi(\bar{\sigma})=1} \delta(\sigma) + [(-1)^{n-1}] \tag{6.2.18}$$

*in* $R_L$.

(2) *The element $\varphi(i,\chi) \in R_L$ is a lifting of $e_{L/K} \cdot \varphi(i) \in \mathbf{Z}$.*

*Proof* (1) Since $\mathrm{ord}_L\delta(\sigma) = i(\sigma)$ and $e_{L/K} = \#I$, the equality (6.2.16) is equivalent to (6.2.3). Applying (6.2.12) to $\Delta_{L/M_{i,\chi}} = \Delta_{L/K} - \Delta_{M_{i,\chi}/K}$, we obtain the first equality in (6.2.17). The second equality follows from the first and (6.2.11).

(2) Since $\mathrm{ord}_L\delta(\sigma) = i$ for $\sigma \in G_i - G_{i+1}$, substituting (6.2.12) and taking the valuation of (6.2.17), we obtain (6.2.16). □

In the fundamental case where $L/K$ is the cyclic extension defined by a character $\chi \in X_K$, the Hasse–Arf theorem, Theorem 5.3.4, means that the rescaling function and its refinement are determined by the Swan conductor $\mathrm{Sw}\,\chi$ and the refined Swan conductor $\mathrm{rsw}\,\chi$ as follows.

**Proposition 6.2.9** *Let $\chi \in X_K$ be a character, L be the corresponding cyclic extension of K, and let $G = \mathrm{Gal}(L/K)$ be the Galois group. Assume that $\mathrm{Sw}\,\chi = n_{L/K} \geqslant 1$ and let $g = g_{L/K} > 0$ be the largest integer i such that $G_i \neq 1$. Assume that the residue field E of L is a separable extension of F.*

(1) $n_{L/K} = \varphi(g_{L/K})$.

(2) *The composition* $\mathrm{res} \circ \mathrm{rsw}\,\chi \colon \mathfrak{m}_K^n/\mathfrak{m}_K^{n+1} \to \Omega_F^1(\log) \to F$ *of the refined Swan conductor with the residue mapping is non-zero and defines an element of $R_K$. For the restriction $\chi_0$ of $\chi$ on $G_m$, the element $\varphi(g_{L/K}, \chi_0) \in R_L$ equals the* inverse *of* $\mathrm{res} \circ \mathrm{rsw}\,\chi \in R_K \subset R_L$.

*Proof* (1) Since $G_{g+1} = 1$, we have

$$\varphi(g_{L/K}) = \frac{d_{L/K} + g_{L/K}}{e_{L/K}}$$

by Proposition 6.2.2(1) and (6.2.4). Hence, the assertion follows from (5.3.11).

(2) Let $\sigma \in G_g$ be the generator such that $\chi(\sigma) = 1$. Let $d_\sigma \colon E \to \mathfrak{m}_L^i/\mathfrak{m}_L^{i+1}$ denote the morphism sending 1 to $\sigma(s)/s - 1$ for a uniformizer $s \in L$. Then, since $L = M_{m,\chi_0}$, we have $\varphi(g_{L/K}, \chi_0) = \Delta_{L/K} + \delta(\sigma)$ in $R_L$. By the commutative diagram (5.3.13) and Proposition 5.1.7, the composition $\mathrm{res} \circ \mathrm{rsw}\,\chi \colon \mathfrak{m}_K^n/\mathfrak{m}_K^{n+1} \to F$ is an isomorphism and equals the inverse of $\Delta_{L/K} + \delta(\sigma)$ in $R_L$. □

**Example 6.2.10** Let $K$ be a henselian discrete valuation field of characteristic $p > 0$, and $t \in K$ be a uniformizer. Let $L$ be the Artin–Schreier extension defined by $x^p - x = 1/t^n$ for $n = pm - 1 > 0$, and $s = t^m x \in L$ be a uniformizer as in Example 1.4.4(1). For $\sigma_i \in G = \mathrm{Gal}(L/K)$ such that $\sigma_i(x) = x + i$, $i \in \mathbf{F}_p^\times$, we have $\delta(\sigma_i) = \sigma_i(s)/s - 1 = i/x \in \mathfrak{m}_L^n/\mathfrak{m}_L^{n+1} - \{0\} \subset R_L$. Let $\chi \colon G \to \mathbf{Q}/\mathbf{Z}$ be the character sending $\sigma_1$ to $1/p$. Then $\varphi(n, \chi) = \sum_{\sigma \in G, \neq 1} \delta(\sigma) + \delta(\sigma_1)$ equals $(p-1)!x^{-p} = -t^n \in R_K \subset R_L$ by (6.2.18). Since the refined Swan conductor $\mathrm{rsw}\,\chi$ maps the basis $-t^n \in \mathfrak{m}_K^n/\mathfrak{m}_K^{n+1}$ to $d\log t \in Z_{F\backslash E}(\log)$ as in Example 3.1.6(1) and $\mathrm{res}\, d\log t = 1$, $\varphi(n, \chi) = -t^n$ is the inverse of $\mathrm{res} \circ \mathrm{rsw}\,\chi$.

**Exercise 6.2** Let $p$ be a prime number and $n \geqslant 2$ be an integer. Let $G = \mathrm{Gal}(\mathbf{Q}_p(\zeta_{p^n})/\mathbf{Q}_p)$ and $i \geqslant 1$ be the largest integer such that $G_i \neq 0$. Let $\chi \colon (1 + p^{n-1}\mathbf{Z}_p)/(1 + p^n\mathbf{Z}_p) \to \mathbf{F}_p$ be the faithful character sending $1 + p^{n-1}$ to $1 \in \mathbf{F}_p$. Compute the value $\varphi(i) \in \mathbf{N}$ of the rescaling function and $\varphi(i, \chi) \in R_{\mathbf{Q}_p} = \mathbf{Q}_p^\times/(1 + p\mathbf{Z}_p)$, and verify the equality in Proposition 6.2.9.

## 6.3 Swan Conductor and its Refinement

We define the Swan conductor and its refinement using the break decomposition, the rescaling function, and its refinement.

**Definition 6.3.1** Let $L$ be a finite Galois extension of a henselian discrete valuation field $K$ of Galois group $G$. Assume that the residue field $E$ of $L$ is a separable extension of the residue field $F$ of $K$, and let $\varphi = \varphi_{L/K}$ be the rescaling function defined in Definition 6.2.1. Let $\Lambda$ be a field of characteristic $\ell$ different from $p = \operatorname{char} F$, and let $V$ be a $\Lambda$-vector space of finite dimension with an action of $G$.

(1) We call the unique decomposition $V = \bigoplus_{i \in \mathbf{N}} V_i$ in Lemma 6.1.6(1) such that for any $i \in \mathbf{N}$ the $G_i$-fixed part $V^{G_i}$ is the sum $\bigoplus_{j<i} V_j$ of $V_j$ for $j < i$ the break decomposition.
(2) Let $V = \bigoplus_{i \in \mathbf{N}} V_i$ be the break decomposition. We define the Swan conductor $\mathrm{Sw} V \in \mathbf{Q}$ by

$$\mathrm{Sw} V = \sum_i \dim V_i \cdot \varphi(i) \geqslant 0. \tag{6.3.1}$$

(3) Fix a primitive $p$th root $\zeta \in \Lambda^\times$ of 1. For $i \geqslant 1$, we identify a character $\chi\colon \mathrm{Gr}_i G = G_i/G_{i+1} \to \Lambda^\times$ with $\chi\colon \mathrm{Gr}_i G \to \mathbf{F}_p$ by $\zeta$. Let $V_i = \bigoplus_{\chi \in G^\vee} V_\chi$ (6.1.4) be the decomposition by non-trivial characters. We define the refined Swan conductor $\mathrm{rsw}_\zeta V \in R_L$ of $V$ with respect to $\zeta$ by

$$\mathrm{rsw}_\zeta V = \sum_{i=1}^{\infty} \sum_{\chi\colon \mathrm{Gr}_i G \to \Lambda^\times,\, \chi \neq 1} \dim V_\chi \cdot \varphi(i, \chi). \tag{6.3.2}$$

Some authors call the break decomposition the slope decomposition. When we want to make the group $G$ explicit, we write $\mathrm{Sw}_G V = \mathrm{Sw}\, V$ and $\mathrm{rsw}_{\zeta,G} V = \mathrm{rsw}_\zeta V$. If the choice of $\zeta$ is not important, we drop it and abusively write $\mathrm{rsw}\, V$. The refined Swan conductor is a refinement of the Swan conductor by Lemma 6.3.2(1). The Swan conductor $\mathrm{Sw}\, V \geqslant 0$ measures the wild ramification of $V$ in the sense that $\mathrm{Sw}\, V = 0$ is equivalent to the condition that $G_1 = P$ acts trivially on $V$, as we see in Lemma 6.3.2(2). We will derive in Theorem 6.4.2 that the Swan conductor $\mathrm{Sw}\, V$ is in fact an integer as a consequence (Lemma 6.3.3) of Theorem 5.3.4.

A variant of the Swan conductor called the Artin conductor is defined by

$$\mathrm{Art}\, V = \dim V - \dim V^I + \mathrm{Sw}\, V.$$

**Lemma 6.3.2** (1) *The refined Swan conductor* $\mathrm{rsw}_\zeta V \in R_L$ *is a lifting of* $e_{L/K} \cdot \mathrm{Sw}\, V \in \mathbf{Z}$.

(2) *The following conditions are equivalent:*

   (a) *$V$ is tamely ramified; namely, the action of $P = G_1$ on $V$ is trivial.*
   (b) $\operatorname{Sw} V = 0$.
   (c) $\operatorname{rsw} V = 0$.

(3) *Let $P = G_1 \subset G$ be the wild inertia subgroup and let $m_{L/K}$ be the prime-to-$p$ part of the ramification index $e_{L/K}$. Then*

$$\operatorname{rsw}_{\zeta,G} V = \operatorname{rsw}_{\zeta,P} V + \dim V/V^P \cdot [m_{L/K}]. \tag{6.3.3}$$

(4) $\operatorname{rsw}_\zeta V \in R_L$ *lies in the $G$-fixed part $R_L^G$.*

*Proof* (1) The assertion follows from (6.3.1), (6.3.2), and Lemma 6.2.8.

(2) (a) $\Rightarrow$ (c): The condition $V^P = V_0$ implies $\operatorname{rsw} V = 0$.

(c) $\Rightarrow$ (b): This follows from (1).

(b) $\Rightarrow$ (a): Since $\varphi(i) \geqslant 0$ and the equality is equivalent to $i = 0$, the vanishing $\operatorname{Sw} V = 0$ implies $V_i = 0$ for all $i > 0$.

(3) The extension $M_1 \subset L$ corresponding to $P = G_1$ is the maximal tamely ramified extension over $K$. By the chain rule (6.2.7), we have $\operatorname{rsw}_{\zeta,G} V = \operatorname{rsw}_{\zeta,P} V + \dim V/V^P \cdot \Delta_{M_1/K}$. Hence, the assertion follows from (6.2.8).

(4) For the conjugate $\chi^\sigma$ of $\chi\colon \operatorname{Gr}_i G \to \Lambda^\times$ by $\sigma \in G$, we have $\dim V_{\chi^\sigma} = \dim V_\chi$ and $\varphi(i, \chi^\sigma) = \sigma(\varphi(i, \chi))$. □

In the rank-1 case, the Swan conductor and its refinement are the same as those defined by using the cohomological filtration.

**Lemma 6.3.3** *Let $L$ be a cyclic extension of $K$ such that the residue field $E$ of $L$ is a separable extension of $F$. Let $\rho\colon G \to \Lambda^\times$ and $\chi\colon G \to \mathbf{Q}/\mathbf{Z}$ be faithful characters of the Galois group $G = \operatorname{Gal}(L/K)$, and let $V$ denote the representation of degre 1 of $G$ defined by $\rho$.*

(1) $\operatorname{Sw} V = \operatorname{Sw} \chi$.
(2) *Assume that* $\operatorname{Sw} \chi = n = n_{L/K} \geqslant 1$. *Let $\sigma \in G$ be a unique element of order $p$ satisfying $\chi(\sigma) = 1/p$, and let $\zeta = \rho(\sigma) \in \Lambda^\times$ be a primitive $p$th root of 1. Then $\operatorname{rsw}_\zeta V \in R_L$ equals the* inverse *of* $\operatorname{res} \circ \operatorname{rsw} \chi\colon \mathfrak{m}_K^n/\mathfrak{m}_K^{n+1} \to \Omega_F^1(\log) \to F$ *regarded as an element of $R_K \subset R_L$.*

*Proof* This is a rephrasing of Proposition 6.2.9. □

We will compute the Swan conductor in Proposition 6.3.7 using the Swan character $s\colon G \to \mathbf{Z}$ defined by

$$s(\sigma) = \begin{cases} -[G : I] \cdot i_G(\sigma) & \text{if } \sigma \neq 1, \\ \ [G : I] \cdot \operatorname{ord}_{O_L} D_{L/K}^{\log} & \text{if } \sigma = 1. \end{cases} \tag{6.3.4}$$

Recall that for $\sigma \neq 1$ and for a uniformizer $s \in O_L$, we have $i_G(\sigma) = \mathrm{ord}_L(\sigma(s)/s - 1)$ if $\sigma \in I$ and $i_G(\sigma) = 0$ if $\sigma$ is not contained in the $p$-Sylow subgroup $P$ of $I$. We have $\sum_{\sigma \in G} s(\sigma) = 0$ by Proposition 1.9.5(3). If it is necessary to indicate the group $G$, we write $s = s_G$.

We will show in Corollary 6.4.3 that the Swan character is in fact a character of $G$. By Proposition 6.3.7, this means that for any representation $V$ of $G$, the Swan conductor $\mathrm{Sw}\, V$ is an integer.

We introduce a refinement of the Swan character.

**Definition 6.3.4** (1) Let $L$ be a finite Galois extension of a henselian discrete valuation field $K$ such that the residue field $E$ of $L$ is a separable extension of the residue field $F$ of $K$. Let $M_1 \subset L$ be the maximal tamely ramified extension over $K$. We define $\widetilde{s}(\sigma) \in R_L$ for $\sigma \in P = G_1$ by

$$\widetilde{s}(\sigma) = \begin{cases} -\delta(\sigma) & \text{if } \sigma \in G_i - G_{i+1} \text{ for } i \geqslant 1, \\ \Delta_{L/K} & \text{if } \sigma = 1. \end{cases} \tag{6.3.5}$$

(2) Let $\zeta_p \in \mathbf{Z}[\zeta_p]$ be a primitive $p$th root of 1. We define an element $\varepsilon = \varepsilon(\zeta_p) \in \mathbf{F}_p^\times \otimes_{\mathbf{Z}} \mathbf{Z}[\zeta_p]$ by

$$\varepsilon = \sum_{l=1}^{p-1} l \otimes \zeta_p^l. \tag{6.3.6}$$

The minus sign in (6.3.5) indicates the *inverse* in $R_L$. The non-zero element $\delta(\sigma) \in \mathfrak{m}_L^i/\mathfrak{m}_L^{i+1}$ and the logarithmic different $\Delta_{L/K}$ are regarded as elements of $R_L$. The definition of $\widetilde{s}(\sigma)$ here is modified from the original definition in [49, (1.8)] by $[-1] \in E^\times \subset R_L$. By Proposition 6.2.6(3), we have

$$\sum_{\sigma \in P} \widetilde{s}(\sigma) = \Delta_{M_1/K} \tag{6.3.7}$$

in $R_L$. Since $\varepsilon$ is an element of $\mathbf{F}_p^\times \otimes_{\mathbf{Z}} \mathbf{Z}[\zeta]$, we have $p \cdot \varepsilon = \varepsilon$. The element $\varepsilon$ appears as a correction term through the following lemma.

**Lemma 6.3.5** *Let $G$ be a finite $p$-group and $\rho\colon G \to \mathbf{Z}[\zeta_p]^\times$ be a non-trivial character of order $p$. Let $\varepsilon = \varepsilon(\zeta_p) \in \mathbf{F}_p^\times \otimes_{\mathbf{Z}} \mathbf{Z}[\zeta_p]$ be $\varepsilon = \sum_{l=1}^{p-1} l \otimes \zeta_p^l$ as in* (6.3.6)*. Let $R$ be a commutative group written additively, and let $[\ ]\colon \mathbf{F}_p^\times \to R$ be a morphism. Let $\widetilde{s}\colon G \to R$ be a function satisfying $\widetilde{s}(\sigma^i) = \widetilde{s}(\sigma) - [i]$ for any $\sigma \in G - \{1\}$ and integer $i$ prime to $p$. Then*

$$\sum_{\sigma \in G, \rho(\sigma)=1} \widetilde{s}(\sigma) - \sum_{\sigma \in G, \rho(\sigma)=\zeta_p} \widetilde{s}(\sigma) = \sum_{\sigma \in G} \widetilde{s}(\sigma) \otimes \rho(\sigma) + \varepsilon(\zeta_p)$$

*in $R \otimes_{\mathbf{Z}} \mathbf{Z}[\zeta_p]$.*

*Proof* For $\sigma \neq 1$, we have $\sum_{i=1}^{p-1} \widetilde{s}(\sigma^i) \otimes \zeta_p^i = \widetilde{s}(\sigma) \otimes \sum_{i=1}^{p-1} \zeta_p^i - \sum_{i=1}^{p-1} [i] \otimes \zeta_p^i$ by the assumption on $s$. Hence, we have

$$\sum_{i=1}^{p-1} \widetilde{s}(\sigma^i) \otimes \zeta_p^i = -\widetilde{s}(\sigma) - \varepsilon(\zeta_p) \tag{6.3.8}$$

in $R \otimes_{\mathbf{Z}} \mathbf{Z}[\zeta_p]$ by $\sum_{i=0}^{p-1} \zeta_p^i = 0$ and the definition of $\varepsilon(\zeta_p)$.

For $i \in \mathbf{F}_p$, let $G(i) \subset G$ denote the inverse image $s^{-1}(\zeta_p^i) = \{\sigma \in G \mid \rho(\sigma) = \zeta_p^i\}$ and decompose $G = \coprod_{i \in \mathbf{F}_p} G(i)$. Then

$$\sum_{\sigma \in G} \widetilde{s}(\sigma) \otimes \rho(\sigma) - \sum_{\sigma \in G, \rho(\sigma)=1} \widetilde{s}(\sigma) = \sum_{i=1}^{p-1} \sum_{\sigma \in G(i)} \widetilde{s}(\sigma) \otimes \zeta^i. \tag{6.3.9}$$

Since $G$ is a $p$-group, for any integer $i$ prime to $p$, the mapping $G \to G$ sending $\sigma$ to $\sigma^i$ is a bijection and induces a bijection $G(1) \to G(i)$. Hence, the right-hand side of (6.3.9) equals $\sum_{\sigma \in G(1)} \sum_{i=1}^{p-1} \widetilde{s}(\sigma^i) \otimes \zeta^i$. Thus, by (6.3.8), the right-hand side of (6.3.9) is further equal to

$$-\sum_{\sigma \in G, \rho(\sigma)=\zeta_p} \widetilde{s}(\sigma) - \#G(1) \cdot \varepsilon(\zeta_p).$$

Since $\#G(1) = \#G(0)$ is a power of $p$ and since $p \cdot \varepsilon(\zeta_p) = \varepsilon(\zeta_p)$, the assertion follows. □

**Lemma 6.3.6** *Let $i \geqslant 1$ be an integer and let $\chi \colon \mathrm{Gr}_i G = G_i/G_{i+1} \to \mathbf{F}_p$ be a non-trivial character. Let $\zeta = \zeta_p \in \mathbf{Z}[\zeta_p]$ be a primitive pth root of 1 and define a character $\rho \colon G_i \to \mathbf{Z}[\zeta_p]^\times$ by $\rho(\sigma) = \zeta^{\chi(\overline{\sigma})}$. Then*

$$\varphi(i, \chi) = \sum_{\sigma \in G_i} \widetilde{s}(\sigma) \cdot \rho(\sigma) + \varepsilon(\zeta) \tag{6.3.10}$$

*in $R_L \otimes_{\mathbf{Z}} \mathbf{Z}[\zeta_p]$.*

*Proof* By Proposition 6.2.6(3) and (6.2.17), we have $\Delta_{M_{i,\chi}/K} = \sum_{s \in G_{i,\chi}} \widetilde{s}(\sigma)$. Further, by (6.2.17) we have

$$\varphi(i, \chi) = \sum_{\sigma \in G_{i,\chi}} \widetilde{s}(\sigma) - \sum_{\sigma \in G_i, \chi(\sigma)=1} \widetilde{s}(\sigma).$$

Since $\delta(\sigma^i) = \delta(\sigma) + [i]$ for $\sigma \in G_1 - \{1\}$ and $i$ prime to $p$, by taking the inverse we obtain $\widetilde{s}(\sigma^i) = \widetilde{s}(\sigma) - [i]$. Thus, we may apply Lemma 6.3.5 to obtain (6.3.10). □

**Proposition 6.3.7** *Let $L$ be a finite Galois extension of a henselian discrete valuation field $K$ of Galois group $G$ such that the residue field $E$ of $L$ is a* *separable extension of the residue field $F$ of $K$. Let $\Lambda$ be a field of characteristic $\ell \neq p$, and let $V$ be a representation of $G$ on a $\Lambda$-vector space of finite dimension.*

(1) *We have*

$$\operatorname{Sw} V = \frac{1}{\# G} \sum_{\sigma \in P} s(\sigma) \cdot \operatorname{Tr}^{\mathrm{Br}}(\sigma : V). \tag{6.3.11}$$

(2) *Let $\zeta = \zeta_p \in \Lambda^\times$ be a primitive $p$th root of 1 and $\widetilde{\zeta} \in W(\Lambda)$ be the Teichmüller lifting. Then*

$$\operatorname{rsw}_\zeta V = \sum_{\sigma \in P} \widetilde{s}(\sigma) \cdot \operatorname{Tr}^{\mathrm{Br}}(\sigma : V) + \dim V/V^P \cdot \varepsilon(\zeta) - \dim V^P \cdot [m_{L/K}] \tag{6.3.12}$$

*in $R_L \otimes_{\mathbf{Z}} \mathbf{Z}[\widetilde{\zeta}]$.*

As a consequence of the modification of the definition of $\widetilde{s}(\sigma)$, $\operatorname{rsw}_\zeta V$ is different from that in [49, (3.1)] by $\dim V/V^P \cdot [-1]$. This explains the minus sign in $\{\chi, 1 - cz\}$ in [49, Theorem (3.6)]. If $\ell \neq 0$, since $P = G_1$ is a subset of $G_{\ell\text{-reg}}$, the Brauer traces on the right-hand side are defined. If $\ell = 0$, we understand that $\operatorname{Tr}^{\mathrm{Br}}(\sigma : V) = \operatorname{Tr}(\sigma : V)$. By (6.2.8), the prime-to-$p$ part $m_{L/K}$ of the ramification index $e_{L/K}$ in the last term of (6.3.12) equals $\Delta_{M_1/K}$.

*Proof* By Lemma 6.3.2(1), the equality (6.3.12) implies (6.3.11). Hence, it suffices to show (2).

By (6.3.3), by the chain rule $\widetilde{s}(1) = \Delta_{L/K} = \Delta_{L/M_1} + \Delta_{M_1/K}$, and $\Delta_{M_1/K} = [m_{L/K}]$ (6.2.8), we may assume $G = P$. Since both sides are additive in $V$, we may assume $V = V_i$ by Lemma 6.1.6(1). By (6.2.12), we have $\sum_{\sigma \in P} \widetilde{s}(\sigma) = 0$. Hence, if $i = 0$, both sides are 0.

Assume $i \geqslant 1$. Replacing $K$ by its maximal tamely ramified extension in $L$, we may assume that $G = P$. Then, since $[P, G_i] \subset G_{i+1}$ and $\operatorname{Gr}_i G$ is an $\mathbf{F}_p$-vector space by Corollary 1.6.4, we may further assume that $V = V_\chi$ for a non-trivial character $\chi \colon \operatorname{Gr}_i G \to \Lambda^\times$ by Lemma 6.1.6(2). Let $G_{i,\chi} \supset G_{i+1}$ denote the kernel $\operatorname{Ker}(\chi \colon G_i \to \Lambda^\times)$ and let $\widetilde{\chi} \colon \operatorname{Gr}_i G \to W(\Lambda)^\times$ be the Teichmüller lifting of $\chi$.

For $j < i$, the complement $G_j - G_{j+1}$ is decomposed into orbits by $G_i$. On each orbit, $\widetilde{s}(\sigma)$ is constant and $\operatorname{Tr}^{\mathrm{Br}}(\sigma\tau : V) = \operatorname{Tr}^{\mathrm{Br}}(\sigma : V) \cdot \widetilde{\chi}(\tau)$ for $\sigma \in G_j - G_{j+1}$ and $\tau \in G_i$. For $\sigma \in G_i$, we have $\operatorname{Tr}^{\mathrm{Br}}(\sigma : V) = \dim V \cdot \widetilde{\chi}(\sigma)$. Since $\sum_{\tau \in G_i} \widetilde{\chi}(\tau) = 0$, the first summation on the right-hand side of (6.3.12) equals $\dim V \cdot \sum_{\sigma \in G_i} \widetilde{s}(\sigma) \cdot \widetilde{\chi}(\sigma)$. Thus, the equality (6.3.12) follows from (6.3.10). □

The Swan conductor and the refined Swan conductor are compatible with quotient groups.

**Corollary 6.3.8** *Let $L$ be a finite Galois extension of $K$ such that the residue field $E$ of $L$ is a separable extension of the residue field $F$ of $K$, and let $G = \mathrm{Gal}(L/K)$ be the Galois group. Let $N \subset G$ be a normal subgroup and let $M$ be the corresponding extension. We identify $\overline{G} = \mathrm{Gal}(M/K)$ with $G/N$. Let $V$ be a representation of the quotient $\overline{G}$, and we also regard $V$ as a representation of $G$.*

(1) $\mathrm{Sw}_G V = \mathrm{Sw}_{\overline{G}} V$.
(2) $\mathrm{rsw}_{\zeta,G} V = \mathrm{rsw}_{\zeta,\overline{G}} V$ *in* $R_M \subset R_L$.

*Proof* It suffices to show (2). The wild inertia subgroup $\overline{P} \subset \overline{G}$ is the image of $P \subset G$. We may assume that $\Lambda$ contains a primitive $p$th root of 1. Let $\widetilde{\zeta} \in W(\Lambda)$ be the Teichmüller lifting. Since $\mathbf{Z}[\widetilde{\zeta}]$ is faithfully flat over $\mathbf{Z}$, it suffices to show the equality in $R_L \otimes_{\mathbf{Z}} \mathbf{Z}[\widetilde{\zeta}]$. By (6.3.12), it suffices to show that

$$\sum_{\tau\in P} \widetilde{s}_G(\tau)\cdot \mathrm{Tr}^{\mathrm{Br}}(\tau : V) + \dim V/V^P \cdot \varepsilon(\zeta) - \dim V^P \cdot [m_{L/K}]$$
$$= \sum_{\sigma\in \overline{P}} \widetilde{s}_{\overline{G}}(\sigma)\cdot \mathrm{Tr}^{\mathrm{Br}}(\sigma : V) + \dim V/V^{\overline{P}} \cdot \varepsilon(\zeta) - \dim V^{\overline{P}} \cdot [m_{M/K}]$$

in $R_L \otimes \mathbf{Z}[\zeta]$. For $\sigma \in \overline{P}$, the image of $\widetilde{s}_{\overline{G}}(\sigma) \in R_M$ in $R_L$ by the inclusion equals $\sum_{\tau\in\sigma\cap P} \widetilde{s}_G(\tau) - [m_{L/M}]$ by Proposition 6.2.6(3). Thus, it is reduced to showing the equality $\dim V^P \cdot ([m_{L/K}] - [m_{M/K}]) = \sum_{\sigma\in\overline{P}} \mathrm{Tr}^{\mathrm{Br}}(\sigma : V)\cdot [m_{L/M}]$. By the chain rule $m_{L/K} = m_{L/M}\cdot m_{M/K}$ and $\sum_{\sigma\in\overline{P}} \mathrm{Tr}^{\mathrm{Br}}(\sigma : V) = \#\overline{P}\cdot \dim V^P$, the assertion follows from $\#\overline{P}\cdot[m_{L/M}] = [m_{L/M}]$. □

**Corollary 6.3.9** *Let $L$ be a finite Galois extension of $K$ of Galois group $G = \mathrm{Gal}(L/K)$, and assume that the residue field $E$ of $L$ is a separable extension of the residue field $F$ of $K$.*

(1) *Let $\Lambda$ be a field of characteristic $\neq p$, and let $V$ be a representation of $G$ on a $\Lambda$-vector space of dimension 1. Then the Swan conductor* $\mathrm{Sw}\, V$ *is an integer and the refined Swan conductor* $\mathrm{rsw}\, V$ *is an element of $R_K \subset R_L$.*
(2) (Hasse–Arf theorem) *Assume that $L$ is an abelian extension. For every integer $i \geqslant 1$ such that $G_i \supsetneqq G_{i+1}$, the value $n = \varphi(i)$ of the rescaling function is an integer and we have $G_i = G^n_{\mathrm{coh}}$ and $G_{i+1} = G^{n+1}_{\mathrm{coh}}$.*

*Proof* (1) By Corollary 6.3.8, we may assume that $V$ is defined by a faithful character $G \to \Lambda^\times$. Then the assertion follows from Proposition 6.2.9.

(2) Let $N$ be the order of $G$. Taking a finite field $\Lambda$ containing a primitive $N$th root of 1, we identify characters $G \to \Lambda^\times$ with characters $G \to \mathbf{Q}/\mathbf{Z}$. Since the rescaling function $\varphi$ is strictly increasing, for any character $\chi\colon G \to (1/N)\mathbf{Z}/\mathbf{Z} \subset \mathbf{Q}/\mathbf{Z}$, the inequality $\mathrm{Sw}\,\chi \leqslant \varphi(i)$ is equivalent to $\chi(G_{i+1}) = 0$

and then the equality holds if and only if $\chi(G_i) \neq 0$. Hence, $n = \varphi(i)$ is an integer by Theorem 5.3.4, and we have $G_i = G^n_{\mathrm{coh}}$ and $G_{i+1} = G^{n+1}_{\mathrm{coh}}$. □

The relation between the lower numbering filtration and the upper numbering filtration will be generalized in Chapter 12.

**Exercise 6.3** Let $p$ be a prime number and $n \geqslant 2$ be an integer. We identify $\mathrm{Gal}(\mathbf{Q}_p(\zeta_{p^n})/\mathbf{Q}_p) = (\mathbf{Z}/p^n\mathbf{Z})^\times$. Let $\rho\colon (\mathbf{Z}/p^n\mathbf{Z})^\times \to \Lambda^\times$ be a primitive character, and let $\zeta = \rho(1+p^{n-1}) \in \Lambda^\times$ be a primitive $p$th root of 1. Compute the Swan conductor $\mathrm{Sw}\,\rho$ and its refinement $\mathrm{rsw}_\zeta\,\rho$, and verify that the equalities (6.3.11) and (6.3.12) hold.

## 6.4 Induction Formula and Integrality

**Proposition 6.4.1** *Let $L$ be a finite Galois extension of $K$ such that the residue field $E$ of $L$ is a separable extension of the residue field $F$ of $K$, and let $G = \mathrm{Gal}(L/K)$ be the Galois group. Let $\Lambda$ be a field of characteristic $\ell \neq p$, and let $W$ be a representation of a subgroup $H \subset G$ on a $\Lambda$-vector space of finite dimension. Let $V = \mathrm{Ind}_H^G W$ be the induced representation. Let $M$ be the extension corresponding to $H$, and set $d_{M/K} = \mathrm{ord}_M D^{\log}_{M/K}$; let $f_{M/K}$ be the residual degree. Let $\zeta = \zeta_p \in \Lambda^\times$ be a primitive $p$th root of* 1.

(1) *We have*

$$\mathrm{Sw}_G\, V = f_{M/K} \cdot (\mathrm{Sw}_H\, W + \dim W \cdot d_{M/K}). \tag{6.4.1}$$

*More precisely, we have*

$$\mathrm{Sw}_G\, V = f_{M/K} \cdot \mathrm{Sw}_H\, W + \dim W \cdot \mathrm{Sw}_G\, \mathrm{Ind}_H^G \Lambda, \tag{6.4.2}$$

$$\mathrm{Sw}_G\, \mathrm{Ind}_H^G \Lambda = f_{M/K} \cdot d_{M/K}. \tag{6.4.3}$$

(2) *Let $\mathrm{N}_{G/H}\colon R_L^H \to R_L^G$ be the norm mapping from the $H$-fixed part to the $G$-fixed part. Then*

$$\mathrm{rsw}_{\zeta,G}\, V = \mathrm{N}_{G/H}\big(\mathrm{rsw}_{\zeta,H}\, W + \dim W \cdot \Delta_{M/K} - \dim W^{P\cap H} \cdot [m_{M/K}]\big) \tag{6.4.4}$$

*in $R_L^G$. More precisely, we have*

$$\mathrm{rsw}_{\zeta,G}\, V = \mathrm{N}_{G/H}(\mathrm{rsw}_{\zeta,H}\, W) + \dim W \cdot \mathrm{rsw}_{\zeta,G}\, \mathrm{Ind}_H^G \Lambda + \dim W/W^{P\cap H} \cdot [M:K] \cdot [m_{M/K}], \tag{6.4.5}$$

$$\mathrm{rsw}_{\zeta,G}\, \mathrm{Ind}_H^G \Lambda = \mathrm{N}_{G/H}(\Delta_{M/K} - [m_{M/K}]). \tag{6.4.6}$$

The formulas (6.4.1) and (6.4.2) are called the induction formulas. Their refinements (6.4.4) and (6.4.5) are also called the induction formulas. The conductor-discriminant formula (1.9.5) is so called because it is equivalent to the equality (6.4.3) that shows that the conductor of the permutation representation is computed by the discriminant.

*Proof* By Lemma 6.3.2(4), the content of the parentheses in (6.4.4) is an element of the $H$-fixed part $R_L^H$ and hence the right-hand side is defined. By Lemma 6.3.2(1), the equality (6.4.4) implies (6.4.1). Hence, it suffices to show (2).

Let $\widetilde{\zeta} \in W(\Lambda)$ be the Teichmüller lifting. Since $\mathbf{Z}[\widetilde{\zeta}]$ is faithfully flat over $\mathbf{Z}$, it suffices to show the equality in $R_L \otimes_{\mathbf{Z}} \mathbf{Z}[\widetilde{\zeta}]$. By Proposition 6.3.7(2), we have

$$\mathrm{rsw}_\zeta V - \dim V/V^P \cdot \varepsilon(\zeta) + \dim V^P \cdot [m_{L/K}] = \sum_{\sigma \in P} s_G(\sigma) \cdot \mathrm{Tr}^{\mathrm{Br}}(\sigma : V) \quad (6.4.7)$$

in $R_L \otimes_{\mathbf{Z}} \mathbf{Z}[\widetilde{\zeta}]$. Let $S \subset G$ be a complete set of representatives of $G/H$. Then, since $\mathrm{Tr}^{\mathrm{Br}}(\sigma : V) = \sum_{\tau \in S, \tau^{-1}\sigma\tau \in H} \mathrm{Tr}^{\mathrm{Br}}(\tau^{-1}\sigma\tau : W)$ and $s_G(\tau\sigma\tau^{-1}) = \tau s_G(\sigma)$, the right-hand side of (6.4.7) equals

$$\sum_{\sigma \in H \cap P} \sum_{\tau \in S} \tau s_G(\sigma) \cdot \mathrm{Tr}^{\mathrm{Br}}(\sigma : W).$$

Since $s_G(\sigma) = s_H(\sigma)$ for $\sigma \neq 1$ and $s_G(1) - s_H(1) = \Delta_{M/K}$ by the chain rule, Lemma 6.2.5(1), this is further equal to

$$\begin{aligned} &\sum_{\tau \in S} \tau \left( \sum_{\sigma \in H \cap P} s_H(\sigma) \cdot \mathrm{Tr}^{\mathrm{Br}}(\sigma : W) + \dim W \cdot \Delta_{M/K} \right) \qquad (6.4.8) \\ &= \mathrm{N}_{G/H}\big(\mathrm{rsw}_H W - \dim W/W^{P\cap H} \cdot \varepsilon(\zeta) \\ &\qquad + \dim W^{P\cap H} \cdot [m_{L/M}] + \dim W \cdot \Delta_{M/K}\big) \end{aligned}$$

by Proposition 6.3.7(2) for $W$.

Since $\dim V = [G : H] \cdot \dim W$, we have $\dim V \cdot \varepsilon(\zeta) = \mathrm{N}_{G/H}(\dim W \cdot \varepsilon(\zeta))$. Since $\dim V^P = [G : HP] \cdot \dim W^{P\cap H}$ and $[G : H]$ equals $[G : HP]$ times a power of $p$, we have $\dim V^P \cdot \varepsilon(\zeta) = \mathrm{N}_{G/H}(\dim W^{P\cap H} \cdot \varepsilon(\zeta))$ and $\dim V^P \cdot [m_{L/K}] = \mathrm{N}_{G/H}(\dim W^{P\cap H} \cdot ([m_{L/M}] + [m_{M/K}]))$. Hence, comparing (6.4.7) and (6.4.8), we obtain (6.4.4). □

We prove the integrality of the Swan conductor and its refinement.

**Theorem 6.4.2** *Let $L$ be a finite Galois extension of $K$, and let $G = \mathrm{Gal}(L/K)$ be the Galois group. Let $\Lambda$ be a perfect field of characteristic $\neq p$, and let $V$ be a representation of $G$ on a $\Lambda$-vector space of finite dimension.*

(1) *The Swan conductor* $\mathrm{Sw}\, V$ *is an integer.*

(2) *The refined Swan conductor* $\mathrm{rsw}_\zeta V \in R_L$ *is in the image of the injection* $R_K \to R_L$. *Its image by* $R_K \to \mathbf{Z}$ *is* $\mathrm{Sw}\, V$.

*Proof* If $\dim V = 1$, the assertion is Corollary 6.3.9(1). By Lemma 6.3.2(1), it suffices to show (2).

First, we show the case where $L$ is a totally ramified extension of $K$ and $[L : K]$ is a power of $p$. We may assume that $V$ is irreducible. Then, since $G$ is a $p$-group, by Proposition 6.1.7, we may assume that $V = \mathrm{Ind}_H^G W$ is a monomial representation for a rank-1 representation $W$ of $H \subset G$, and let $M \subset L$ be the subextension corresponding to $H$. On the right-hand side of (6.4.4), the content $\mathrm{rsw}_H W + \dim W \cdot \Delta_{M/K} - \dim W^{P\cap H} \cdot [m_{M/K}]$ in the parentheses is in $R_M$ by the rank-1 case. By Proposition 6.4.1(2), its norm $\mathrm{rsw}_G V$ is an element of $R_K$ and the assertion follows in this case.

We show the general case. Let $M_1 \subset L$ be the maximal tamely ramified extension over $K$. Then, since $L$ is a totally ramified extension of $M_1$ and $[L : M_1]$ is a power of $p$, we have $\mathrm{rsw}_{\zeta,P} V \in R_{M_1}$. Since this is $G$-invariant, we obtain $\mathrm{rsw}_{\zeta,G} V = \mathrm{rsw}_{\zeta,P} V + \dim V/V^P \cdot \Delta_{M_1/K} \in R_{M_1}^G = R_K$ by Lemma 6.2.3. □

**Corollary 6.4.3** *There exists a finitely generated projective* $\mathbf{Z}_\ell[G]$*-module* $P$*, unique up to isomorphism, whose character equals the central function* $s_G \colon G \to \mathbf{Z}$.

*Proof* By Theorem 6.4.2, the Swan conductor $\mathrm{Sw}\, V$ is an integer for any representation $V$ of $G$. Hence, by Proposition 6.3.7(1), there exists a finite extension $E$ of $\mathbf{Q}_\ell$ and an element $a \in K_E(G)$ whose character is $s_G$. Let $m \geqslant 0$ be the largest integer such that $G_m \neq 1$. Then $\#G \cdot a = \sum_{i=1}^m \#G_i \cdot \mathrm{Ind}_{G_i}^G (r_{G_i} - 1_{G_i})$, where $r_{G_i}$ and $1_{G_i}$ denote the regular representation and the unit representation respectively. Since the order of $G_i$ for $i \geqslant 1$ is a power of $p$ and prime to $\ell$, the injection $P_{\mathbf{Z}_\ell}(G_i) \to K_{\mathbf{Q}_\ell}(G_i)$ is an isomorphism by Proposition 6.1.2(4). Hence, $\#G \cdot a$ defines an element of $P_{\mathbf{Z}_\ell}(G)$ and the assertion follows from Proposition 6.1.2(3) and (2). □

**Exercise 6.4** Let $p$ be a prime number and $n \geqslant 2$ be an integer. Let $\rho$ be the regular representation of $G = \mathrm{Gal}(\mathbf{Q}_p(\zeta_{p^n})/\mathbf{Q}_p) = (\mathbf{Z}/p^n\mathbf{Z})^\times$. Compute the Swan conductor $\mathrm{Sw}\, \rho$ and its refinement $\mathrm{rsw}_\zeta\, \rho$ by applying the induction formula to $\rho = \mathrm{Ind}_1^G \Lambda$ and by decomposing it as a sum of characters of degree 1, and verify that the equalities (6.4.3) and (6.4.6) hold.

## Historical Notes

The Swan conductor is defined by (6.3.11) in [79], and also in [34] and [66]. Its genesis can be traced in the letters of Grothendieck dated "10.04.63" and of Serre dated "April 63" recorded in [18]. The definition using the break decomposition presented here is closer to that in [55, Definition 1.6]. The Artin conductor, slightly different from the Swan conductor, is studied in [78, Chapitre VI]. The rescaling function $\varphi_{L/K}$ is also studied in [78, Chapitre IV, §3]. The theory of Brauer traces is described in [43].

The refined Swan conductor was introduced and studied by Kato in [49]. The relation in Proposition 6.2.9 in the rank-1 case with the Swan conductor and its refinements defined cohomologically are given in [50, Proposition (6.8)] and the following remark.

# 7

# Conductor and Differential Forms

We study a generalization of the refined Swan conductor to log monogenic extensions. First, in Section 7.1, we introduce the group where the generalization takes values as an extension of $\mathbf{Z}^2$ by the multiplicative group of the residue field involving differential forms. We define a refinement of the logarithmic different using the trace mapping of the spaces of differential forms studied in Section 5.1.

We define the Swan conductor and its refinement in Section 7.2. They share the same properties as in the classical case proved by essentially the same arguments as in the previous chapter. In Section 7.3, assuming that the residue field is a function field of one variable over a perfect field, we study the valuation of the refined Swan conductor. This will become a crucial ingredient in the proof in Section 10.5 of the Deligne–Kato formula computing the dimension of the space of nearby cycles of sheaves on a curve over a local field.

## 7.1 Different and Differential Forms

Let $F$ be a field. For an $F$-vector space $L$ of dimension 1, let $L^\times$ denote the set of non-zero elements of $L$ regarded as an $F^\times$-torsor. Let $L_1$ and $L_2$ be $F$-vector spaces of dimension 1. We introduce a group $F^\times\langle L_1^\times, L_2^\times\rangle$ generated by $L_1^\times, L_2^\times$ as follows.

Let $F[L_1, L_2]$ denote the symmetric algebra $\mathrm{S}_F(L_1 \oplus L_2)$. Bases $S$ and $T$ of $L_1$ and $L_2$ define an isomorphism $F[S,T] \to F[L_1, L_2]$ from the polynomial ring. By inverting non-zero elements of $L_1$ and $L_2$, we obtain the localization $F[L_1, L_2][L_1^{\times -1}, L_2^{\times -1}]$, isomorphic to $F[S^{\pm 1}, T^{\pm 1}]$. We define $F^\times\langle L_1^\times, L_2^\times\rangle$ to be the multiplicative group $F[L_1, L_2][L_1^{\times -1}, L_2^{\times -1}]^\times$. This is non-canonically isomorphic to $F[S^{\pm 1}, T^{\pm 1}]^\times = F^\times \times S^{\mathbf{Z}} \times T^{\mathbf{Z}}$. In the following, we use the additive notation to write the multiplication in $F^\times\langle L_1^\times, L_2^\times\rangle$.

The underlying set $F^\times\langle L_1^\times, L_2^\times\rangle$ is identified with the disjoint union

$$F^\times\langle L_1^\times, L_2^\times\rangle = \coprod_{(m,n)\in\mathbf{Z}^2} \mathrm{Isom}_F(L_1^{\otimes m}, L_2^{\otimes n}) \tag{7.1.1}$$

of the sets of isomorphisms. Under this identification, the group structure on $F^\times\langle L_1^\times, L_2^\times\rangle$ is given by tensor product. Namely, for isomorphisms $f\colon L_1^{\otimes m} \to L_2^{\otimes n}$ and $g\colon L_1^{\otimes m'} \to L_2^{\otimes n'}$, the sum $f+g$ is defined as the tensor product $f\otimes g\colon L_1^{\otimes m+m'} \to L_2^{\otimes n+n'}$. The kernel of the surjection $F^\times\langle L_1^\times, L_2^\times\rangle \to \mathbf{Z}^2$ sending $f\colon L_1^{\otimes m} \to L_2^{\otimes n}$ to $(-m,n)\in\mathbf{Z}^2$ is $F^\times = \mathrm{Isom}_F(L_1^{\otimes 0}, L_2^{\otimes 0})$, and we obtain an exact sequence $0\to F^\times \to F^\times\langle L_1^\times, L_2^\times\rangle \to \mathbf{Z}^2 \to 0$.

Let $L_1, L_2$ be $F$-vector spaces of dimension 1, and let $E$ be an extension of $F$. Let $M_1, M_2$ be $E$-vector spaces of dimension 1. Let $e, w$ be integers and let $f_1\colon L_1\otimes_F E \to M_1^{\otimes e}$ and $f_2\colon L_2\otimes_F E\to M_2^{\otimes w}$ be isomorphisms of $E$-vector spaces of dimension 1. Then they induce a morphism $F[L_1, L_2] \to E[M_1, M_2]$ of rings. This induces a morphism $F^\times\langle L_1^\times, L_2^\times\rangle \to E^\times\langle M_1^\times, M_2^\times\rangle$ sending an isomorphism $f\colon L_1^{\otimes m}\to L_2^{\otimes n}$ to the isomorphism $f_2^{\otimes n}\circ(f\otimes 1_E)\circ(f_1^{\otimes m})^{-1}\colon M_1^{\otimes me}\to M_2^{\otimes nq}$. The diagram

$$\begin{array}{ccccccccc}
0 & \longrightarrow & F^\times & \longrightarrow & F^\times\langle L_1^\times, L_2^\times\rangle & \longrightarrow & \mathbf{Z}^2 & \longrightarrow & 0\\
 & & \big\downarrow & & \big\downarrow & & \big\downarrow{\scriptstyle (e,w)} & & \\
0 & \longrightarrow & E^\times & \longrightarrow & E^\times\langle M_1^\times, M_2^\times\rangle & \longrightarrow & \mathbf{Z}^2 & \longrightarrow & 0
\end{array}$$

is commutative.

Assume that $E$ is a finite extension of degree $d$. Then the ring morphism $F[L_1, L_2]\to E[M_1, M_2]$ is finite flat of degree $dew$ and the norm mapping

$$\mathrm{N}_{E/F}\colon E^\times\langle M_1^\times, M_2^\times\rangle \to F^\times\langle L_1^\times, L_2^\times\rangle \tag{7.1.2}$$

is defined. Under the identification (7.1.1), the norm mapping (7.1.2) sends $f\colon M_1^{\otimes m}\to M_2^{\otimes n}$ to $(-1)^{d(m(e-1)w+ne(w-1))}\mathrm{N}_{E/F}b\cdot g^{\otimes d}\colon L_1^{\otimes mwd}\to L_2^{\otimes ned}$ for $b\in E^\times$ and $g\colon L_1^{\otimes mw}\to L_2^{\otimes ne}$ such that the diagram

$$\begin{array}{ccc}
M_1^{\otimes mew} & \xrightarrow{f^{\otimes ew}} & M_2^{\otimes new}\\
{\scriptstyle f_1^{\otimes mw}}\big\uparrow & & \big\uparrow{\scriptstyle f_2^{\otimes ne}}\\
L_1^{\otimes mw}\otimes_F E & \xrightarrow{g\otimes b} & L_2^{\otimes ne}\otimes_F E
\end{array}$$

is commutative.

Let $K$ be a discrete valuation field with residue field $F$ and let $L$ be a wildly ramified log monogenic extension of $K$. Assume that the residue field $E$ of $L$ is an inseparable extension of $F$. Then the $F$-vector space

$$Z_{F\backslash E}(\log) = \mathrm{Ker}(\Omega^1_F(\log) \to \Omega^1_E(\log))$$

is of dimension 1 by Proposition 5.1.6. The $E$-vector space

$$\Omega^1_{E/F}(\log/\log) = \mathrm{Coker}(\Omega^1_F(\log) \otimes_F E \to \Omega^1_E(\log))$$

is of dimension 1 by Proposition 5.1.5(1). We define extensions

$$S_{L/K} = F^\times\langle(\mathfrak{m}_K/\mathfrak{m}_K^2)^\times, Z_{F\backslash E}(\log)^\times\rangle$$

of $\mathbf{Z}^2$ by $F^\times$ and

$$T_{L/K} = E^\times\langle(\mathfrak{m}_L/\mathfrak{m}_L^2)^\times, \Omega^1_{E/F}(\log/\log)^\times\rangle$$

of $\mathbf{Z}^2$ by $E^\times$.

By the canonical isomorphisms $(\mathfrak{m}_K/\mathfrak{m}_K^2)^{\otimes m} \to \mathfrak{m}_K^m/\mathfrak{m}_K^{m+1}$, we identify

$$S_{L/K} = \coprod_{m,n} \mathrm{Isom}_F(\mathfrak{m}_K^m/\mathfrak{m}_K^{m+1}, Z_{F\backslash E}(\log)^{\otimes n})$$

and similarly for $T_{L/K}$. If $L$ is a Galois extension of Galois group $G$, since the actions of the wild inertia subgroup $P \subset G$ on $E, \Omega^1_{E/F}(\log/\log)$ and $\mathfrak{m}_L/\mathfrak{m}_L^2$ are trivial, the group $G$ acts on $T_{L/K}$ through the quotient $G/P$. If $K$ and $L$ are clear from the context, we drop them in the notation and we write $S_K = S_{L/K}$ and $T_L = T_{L/K}$.

Sending $x \in K^\times$ such that $\mathrm{ord}_K x = m$ to the isomorphism $\mathfrak{m}_K^{-m}/\mathfrak{m}_K^{-m+1} \to F = (Z_{F\backslash E}(\log))^{\otimes 0}$ defined by the multiplication by $x$, we define an injection $R_K \to S_{L/K}$. Similarly, we define an injection $R_L \to T_{L/K}$.

Let $M \subset L$ be the maximal tamely ramified extension of $K$. The wild degree $w = [L : M]$ is a power of $p$. We construct an injection $i\colon S_{L/K} \to T_{L/K}$ fitting in the commutative diagram

$$\begin{array}{ccccccccc} 0 & \longrightarrow & R_K & \longrightarrow & S_{L/K} & \longrightarrow & \mathbf{Z} & \longrightarrow & 0 \\ & & \cap\big\downarrow & & i\big\downarrow & & \big\downarrow w & & \\ 0 & \longrightarrow & R_L & \longrightarrow & T_{L/K} & \longrightarrow & \mathbf{Z} & \longrightarrow & 0 \end{array} \tag{7.1.3}$$

of exact sequences extending (6.2.6). To define this, we define a canonical isomorphism

$$i\colon Z_{F\backslash E}(\log) \otimes_F E \to \Omega^1_{E/F}(\log/\log)^{\otimes w} \tag{7.1.4}$$

of $E$-vector spaces of dimension 1.

**Proposition 7.1.1** *Let $L$ be a log monogenic extension of a henselian discrete valuation field $K$. Assume that $L$ is wildly ramified and $\dim_F Z_{F\backslash E}(\log) = 1$. Let $M \subset L$ be the maximal tamely ramified extension of $K$, and let $w = [L : M] > 1$ be the wild degree.*

(1) *There exists a unique isomorphism*

$$i = i_{L/K} : Z_{F\backslash E}(\log) \otimes_F E \to \Omega^1_{E/F}(\log/\log)^{\otimes w}$$

*of $E$-vector spaces of dimension 1 satisfying the following condition:*

*Define an isomorphism $Z_{F_M\backslash E}(\log) \otimes_F E \to Z_{F\backslash E}(\log) \otimes_F E$ to be the inverse of the isomorphism induced by the inclusion $Z_{F\backslash E}(\log) \to Z_{F_M\backslash E}(\log)$. Then the composition*

$$Z_{F_M\backslash E}(\log) \otimes_F E \longrightarrow Z_{F\backslash E}(\log) \otimes_F E \xrightarrow{\ i\ } \Omega^1_{E/F}(\log/\log)^{\otimes w}$$

*maps $\mathrm{Tr}_{L/M}(d\log x)$ to $(d\log x)^{\otimes w}$ for any $x \in L^\times$.*

(2) *Let $K' \subset L$ be a subextension such that $L$ is wildly ramified over $K'$ and $K'$ is wildly ramified over $K$, and let $F'$ be the residue field of $K'$. Let $M_1 = M \cap K' \subset K'$ and $M_2 = MK' \subset L$ be the maximal tamely ramified extensions over $K$ and over $K'$ respectively, and set $w_1 = [K' : M_1]$ and $w_2 = [L : M_2]$. Then we have the commutative diagram*

$$\begin{array}{ccc} Z_{F\backslash E}(\log) \otimes_F E & \xrightarrow{i_{L/K}} & \Omega^1_{E/F}(\log/\log)^{\otimes w} \\ {\scriptstyle i_{K'/K}\otimes 1}\downarrow & & \uparrow {\scriptstyle i_{L/K'}^{\otimes w_1}} \\ \Omega^1_{F'/F}(\log/\log)^{\otimes w_1} \otimes_{F'} E & \longleftarrow & Z_{F'\backslash E}(\log)^{\otimes w_1} \otimes_{F'} E \end{array} \tag{7.1.5}$$

*of isomorphisms of $E$-vector spaces of dimension 1, where the bottom horizontal arrow is the isomorphism induced by the composition $Z_{F'\backslash E}(\log) \overset{\subset}{\to} \Omega^1_{F'}(\log) \to \Omega^1_{F'/F}(\log/\log)$ in Corollary 5.1.9(1).*

*Proof* (1) Since $Z_{F_M\backslash E}(\log)$ is of dimension 1, and since $d\log : M^\times/(1 + \mathfrak{m}_M)M^{\times p} \to Z_{F_M\backslash E}(\log)$ is an injection, by the second commutative diagram (5.1.1) the uniqueness follows. We may assume that $K = M$.

We consider the commutative diagram of exact sequences

$$\begin{array}{ccccccccc} 0 & \longrightarrow & \Omega^1_F & \longrightarrow & \Omega^1_F(\log) & \xrightarrow{\text{res}} & F & \longrightarrow & 0 \\ & & \downarrow & & \downarrow & & \downarrow & & \\ 0 & \longrightarrow & \Omega^1_E & \longrightarrow & \Omega^1_E(\log) & \xrightarrow{\text{res}} & E & \longrightarrow & 0 \end{array}$$

First, we consider the case where the right vertical arrow is 0 and the left vertical arrow is non-zero. Then $L$ is totally ramified over $K$, and the residue morphisms define isomorphisms $Z_{F\backslash E}(\log) \to F$ and $\Omega^1_{E/F}(\log/\log) \to E$ by Proposition 5.1.7(1) and Lemma 5.1.3(3). By identifying them, we define $i$ to be the identity. For $x \in L^\times$, $\mathrm{Tr}_{L/M}(d\log x) = d\log(\mathrm{N}_{L/M}x)$ and $(d\log x)^{\otimes w}$ are both identified with $\mathrm{ord}_L(x) \in \mathbf{F}_p \subset F \subset E$. Hence, the required property is satisfied.

Next, we consider the case where the left vertical arrow is not an injection and the right vertical arrow is non-zero. Then $e_{L/K} = 1$ and $E$ is a non-trivial, purely inseparable extension of $F$ generated by a single element, and the inclusions define isomorphisms $Z_{F\backslash E} \to Z_{F\backslash E}(\log)$ and $\Omega^1_{E/F} \to \Omega^1_{E/F}(\log/\log)$ by Proposition 5.1.8(1) and Lemma 5.1.3(3). We define $i\colon Z_{F\backslash E}(\log) \otimes_F E \to \Omega^1_{E/F}(\log/\log)^{\otimes w}$ to be the morphism $i\colon Z_{F\backslash E} \otimes_F E \to (\Omega^1_{E/F})^{\otimes w}$ constructed in Proposition 4.4.7(2). For $x \in K^\times$, $\mathrm{Tr}_{L/M}(d\log x)$ and $(d\log x)^{\otimes w}$ are both 0. Since $L^\times/(1+\mathfrak{m}_L)L^{\times p}$ is generated by the image of $K^\times/(1+\mathfrak{m}_K)K^{\times p}$ and $E^\times/E^{\times p}$, the required property is satisfied by Proposition 4.4.7(2).

Finally, assume that the left and right vertical arrows are 0. Then $p \mid e_{L/K}$ and $E$ is a purely inseparable extension of $F$ of degree $f$ divisible by $p$ generated by a single element. Then, we have isomorphisms $Z_{F\backslash E} \to Z_{F\backslash E}(\log)$ and $\mathrm{res}\colon \Omega^1_{E/F}(\log/\log) \to E$ by Proposition 4.4.5(2) (b), Proposition 5.1.5(2), and Lemma 5.1.3(3). Let $t, s$ be uniformizers of $K, L$ respectively, and $u = s^{e_{L/K}}/t$. Since the middle vertical arrow is non-zero, $(-1)$ times the boundary morphism induces an isomorphism $E \to \Omega^1_{E/F}$ sending 1 to $d\log u$.

Then, we define $i$ to be the composition of the morphism $i\colon Z_{F\backslash E} \to (\Omega^1_{E/F})^{\otimes f}$ constructed in Proposition 4.4.7(2) with the identity of $E$ by identifying with the isomorphisms $E \to \Omega^1_{E/F}$ and $\Omega^1_{E/F}(\log/\log) \to E$ above. We show that this satisfies the required property. Since $L^\times$ is generated by uniformizers, we may assume $x = s$ is a uniformizer of $L$. Let $t$ be a uniformizer of $K$. Then, by Lemma 1.8.3(3), $u = s^{e_{L/K}}/t$ is a generator of $E$ over $F$. For $v = u^f$, we have $\mathrm{Tr}_{L/K}(d\log s) = d\log v$. The mapping $i\colon Z_{F\backslash E} \to (\Omega^1_{E/F})^{\otimes f}$ maps $d\log v$ to $(d\log u)^{\otimes f}$, and $d\log u$ is the image of $-1 = \mathrm{res}(-d\log t)$ by the boundary mapping. Hence, the required property is satisfied.

(2) The morphism $i_{M_2/M}\colon Z_{F_M\backslash E}(\log) \otimes_{F_M} F_{M_2} \to \Omega^1_{F_{M_2}/F_M}(\log/\log)$ is canonically identified with the base change of $i_{K'/K}\colon Z_{F\backslash E}(\log) \otimes_F F' \to \Omega^1_{F'/F}(\log/\log)$ since $K' \otimes_{M_1} M \to M_2$ is an isomorphism. Hence, by replacing $K$ by $M$, we may assume that $K' = M_2$.

We compute the images of $\mathrm{Tr}_{L/K}(d\log x) \in Z_{F\backslash E}(\log)$ on the upper left corner for $x \in L^\times$. The image in the upper right corner $\Omega^1_{E/F}(\log/\log)^{\otimes w}$ is $(d\log x)^{\otimes w}$, and that in the lower left corner $(\Omega^1_{F'/F}(\log/\log))^{\otimes w_1} \otimes_{F'} E$ is $(\mathrm{Tr}_{L/K'} d\log x)^{\otimes w_1}$. Similarly, we compute the images of $(\mathrm{Tr}_{L/K'} \circ d\log x)^{\otimes w_1} \in Z_{F'\backslash E}(\log)^{\otimes w_1} \otimes_{F'} E$ on the lower right corner. The image in $\Omega^1_{E/F}(\log/\log)^{\otimes w}$

is also $((d\log x)^{\otimes w_2})^{\otimes w_1} = (d\log x)^{\otimes w}$, and that in $(\Omega^1_{F'/F}(\log/\log))^{\otimes w_1} \otimes_{F'} E$ is $(\mathrm{Tr}_{L/K'} d\log x)^{\otimes w_1}$. Hence, the diagram is commutative. □

The canonical isomorphisms $(\mathfrak{m}_K/\mathfrak{m}_K^2) \otimes_F E \to \mathfrak{m}_L^e/\mathfrak{m}_L^{e+1}$ and $Z_{F\backslash E} \otimes_F E \to (\Omega^1_{E/F})^{\otimes m}$ (7.1.4) define an injection $i\colon S_{L/K} \to T_{L/K}$ fitting in (7.1.3).

Let $K' \subset L$ be any subextension and let $F'$ be the residue field. If $L$ is wildly ramified over $K'$ and if the residue field $E$ of $L$ is an inseparable extension of the residue field $F'$ of $K'$, the group $S_{K'} = S_{L/K'}$ and the morphism $i_{L/K'}\colon S_{K'} \to T_{L/K'} = T_{L/K}$ are defined. If $K'$ is wildly ramified over $K$ and if the residue field $F'$ of $K'$ is an inseparable extension of the residue field $F$ of $K$, the group $T_{K'} = T_{K'/K}$ and the morphism $i_{K'/K}\colon S_{L/K} = S_{K'/K} \to T_{K'}$ are defined. When they are defined, the groups are extensions of $\mathbf{Z}^2$ by $F'^{\times}$. The composition $Z_{F'\backslash E}(\log) \to \Omega^1_{F'}(\log) \to \Omega^1_{F'/F}(\log/\log)$ is an isomorphism by Corollary 5.1.9(1) and induces an isomorphism $i_{K'/K}\colon S_{K'} \to T_{K'}$. If $L$ is wildly ramified over $K'$ and if $K'$ is wildly ramified over $K$, then the commutative diagram (7.1.5) implies that the diagram

$$\begin{array}{ccc} S_{L/K} & \xrightarrow{i_{L/K}} & T_{L/K} \\ {\scriptstyle i_{K'/K}}\big\downarrow & & \big\uparrow{\scriptstyle i_{L/K'}} \\ T_{K'/K} & \xleftarrow{i_{K'/K'}} & S_{L/K'} \end{array} \tag{7.1.6}$$

is commutative, where the bottom horizontal arrow is an isomorphism.

We define a norm mapping $\mathrm{N}_{L/K}\colon T_{L/K} \to S_{L/K}$.

**Lemma 7.1.2** *Let $L$ be a wildly ramified log monogenic extension of a henselian discrete valuation field $K$. Let $M \subsetneq L$ be the maximal tamely ramified extension and $w = [L : M]$. Then, there exists a unique morphism*

$$N_{L/M}\colon T_{L/K} \to S_{L/M} \tag{7.1.7}$$

*such that the composition $T_{L/K} \to S_{L/M} \to T_{L/M} = T_{L/K}$ maps $x$ to $x^w$.*

*Proof* Since $i\colon S_{L/M} \to T_{L/M}$ is an injection, it suffices to show the inclusion $(T_{L/M})^w \subset \mathrm{Im}(i\colon S_{L/M} \to T_{L/M})$. Since $T_{L/K}$ is generated by $R_L$ and $\Omega^1_{E/F}(\log/\log)^{\times}$, it suffices to show the inclusion for each of them. Since the composition $R_L \to R_K \to R_L$ maps $x$ to $x^w$, we have an inclusion for $R_L$. The inclusion for $\Omega^1_{E/F}(\log/\log)^{\times}$ is clear from the definition of the isomorphism $i\colon Z_{F\backslash E}(\log) \otimes_F E \to \Omega^1_{E/F}(\log/\log)^{\otimes w}$ (7.1.4). □

Let $M \subset L$ be the maximal tamely ramified extension of $K$, and let $F_M \subset E$ be the residue field of $M$. We define the norm mapping $\mathrm{N}_{L/K}\colon T_{L/K} \to S_{L/K}$ to be the composition of the morphism $\mathrm{N}_{L/M}\colon T_{L/K} \to S_{L/M}$ (7.1.7) and

$\mathrm{N}_{F_M/F}\colon S_{L/M} \to S_{L/K}$ defined by applying the construction of (7.1.2) to the isomorphisms $\mathfrak{m}_K/\mathfrak{m}_K^2 \otimes_F F_M \to \mathfrak{m}_M^{e_{M/K}}/\mathfrak{m}_M^{e_{M/K}+1}$ and $Z_{F\backslash E}(\log) \otimes_F F_M \to Z_{F_M\backslash E}(\log)$ in Corollary 5.1.9(3).

**Lemma 7.1.3** *Let $L$ be a wildly ramified log monogenic Galois extension of Galois group $G$. Then the composition $i_{L/K} \circ \mathrm{N}_{L/K}\colon T_{L/K} \to S_{L/K} \to T_{L/K}$ equals $\mathrm{N}_G = \sum_{\sigma \in G} \sigma$.*

*Proof* Let $M \subset L$ be the maximal tamely ramified extension and let $\overline{G}$ be the Galois group $\mathrm{Gal}(M/K)$. Then the composition $S_{L/M} \to S_{L/K} \to S_{L/M}$ equals $\mathrm{N}_{\overline{G}} = \sum_{\sigma \in \overline{G}} \sigma$. Since the wild inertia group $P = \mathrm{Gal}(L/M) \subset G$ acts trivially on $T_{L/K}$, the composition $T_{L/K} \to S_{L/M} \to T_{L/K}$ equals $\mathrm{N}_P = \sum_{\sigma \in P} \sigma$ by Lemma 7.1.2. Since the injection $S_{L/M} \to T_{L/K}$ is compatible with the action of $G$, the assertion follows. □

We define a refinement $\Delta_{L/K}$ of the logarithmic different as an element of $T_{L/K}$ as follows. Let $d = d_{L/K}$ denote the valuation of the logarithmic different. Then $\mathrm{Tr}_{L/K}\colon L \to K$ induces a surjection $\mathfrak{m}_L^{-d}/\mathfrak{m}_L^{-d+1} \to F$ of $F$-vector spaces. We compare this with the trace mapping $\mathrm{Tr}_{L/K}\colon \Omega^1_{E/F}(\log/\log) \to Z_{F\backslash E}(\log)$, which is also a surjection of $F$-vector spaces.

**Lemma 7.1.4** *There exists a unique element $\Delta_{L/K} \in T_{L/K}$ satisfying the following condition: For any $a \in \mathrm{Isom}_E(\mathfrak{m}_L^{-d}/\mathfrak{m}_L^{-d+1}, \Omega^1_{E/F}(\log/\log)) \subset T_{L/K}$ and $b \in Z_{F\backslash E}(\log)^\times \subset S_{L/K}$ such that the diagram*

$$\begin{array}{ccc} \mathfrak{m}_L^{-d}/\mathfrak{m}_L^{-d+1} & \xrightarrow{\ a\ } & \Omega^1_{E/F}(\log/\log) \\ {\scriptstyle \mathrm{Tr}_{L/K}}\downarrow & & \downarrow{\scriptstyle \mathrm{Tr}_{L/K}} \\ F & \xrightarrow{\ b\ } & Z_{F\backslash E}(\log) \end{array} \tag{7.1.8}$$

*is commutative, we have*

$$\Delta_{L/K} = [a] - i[b]. \tag{7.1.9}$$

*Proof* Let $a' = ua$ and $b' = vb$ for $u \in E^\times$ and $v = F^\times$ be other isomorphisms such that the diagram (7.1.8) is commutative. Then $u = v$ since the $E$-vector space $\mathrm{Hom}_F(E, F)$ is of dimension 1, and the assertion follows. □

We call $\Delta_{L/K}$ characterized by (7.1.9) the logarithmic different of $L$ over $K$. The image of $\Delta_{L/K}$ by $T_{L/K} \to \mathbf{Z}^2$ equals $(\mathrm{ord}_L D^{\log}_{L/K}, -(w-1))$.

**Lemma 7.1.5** *Let $L$ be a log monogenic extension of a henselian discrete valuation field $K$.*

(1) *Let $K' \subset L$ be a subextension and let $F' \subset E$ be the residue fields. Then $\Delta_{L/K} = \Delta_{L/K'} + i_{L/K'}\Delta_{K'/K}$ in $T_{L/K}$.*

(2) *Let $M \subset L$ be the maximal tamely ramified extension. Let $e = e_{L/M}$ and $m_{L/K} = e_{M/K}$ be the ramification indices, $q = f_{L/M}$ be the residual degree, and $w = [L : M] = qe$. Let $t, s$ be uniformizers of $M$ and of $L$. Let $M \subset N \subset L$ be a subextension such that $L$ is a totally ramified extension of $N$ of degree $e_{L/M}$ and that the residue field $E$ of $L$ is a purely inseparable extension of the separable closure $F_M \subset E$ generated by the unit $u = \mathrm{N}_{L/N} s/t \in O_N^\times$. Let $P \in O_N[X]$ be the minimal polynomial of $s$ over $N$, and $Q \in O_M[X]$ be the minimal polynomial of $u$ over $M$. Then*

$$\begin{aligned}\Delta_{L/K} &= [m_{L/K}] + \frac{P'(s)}{s^{e-1}} - (d\log s)^{e-1} + \frac{Q'(u)}{u^{q-1}} - (d\log u)^{q-1}\\ &= [m_{L/K}] + \frac{P'(s)}{s^{e-1}} + \frac{Q'(u)}{u^{q-1}} - (d\log s)^{w-1}\end{aligned} \tag{7.1.10}$$

*in $T_{L/K}$.*

*Proof* (1) By the chain rule, Proposition 5.1.1(2), we have a commutative diagram

$$\begin{array}{ccc}\Omega^1_{E/F}(\log/\log) & \xrightarrow{\mathrm{Tr}_{L/K}} & Z_{F\backslash E}(\log)\\ {\scriptstyle \mathrm{Tr}_{L/K'}}\downarrow & & \uparrow{\scriptstyle \mathrm{Tr}_{K'/K}}\\ Z_{F'\backslash E}(\log) & \longrightarrow & \Omega^1_{F'/F}(\log/\log)\end{array} \tag{7.1.11}$$

Hence, the assertion follows from the chain rule $\mathrm{Tr}_{L/K} = \mathrm{Tr}_{M/K} \circ \mathrm{Tr}_{L/M}$ for the additive groups and the defining commutative diagram (7.1.8).

(2) First, we show the case where $L = M$ is tamely ramified over $K$. Then, by the commutative diagram (5.1.12), $a\colon \mathfrak{m}_L^{-d}/\mathfrak{m}_L^{-d+1} = E \to \Omega^1_{E/F}(\log/\log)$ defined by $e_{M/K} \cdot d\log s$ and $b\colon F \to Z_{F\backslash E}(\log)$ defined by $d\log \mathrm{N}_{L/K} s$ make the diagram (7.1.8) commutative. Hence, $\Delta_{M/K} = [m_{L/K}]$.

Since $d\log u = d\log z$ in $\Omega^1_{F_N/F}(\log/\log)$ and since $i\colon Z_{F_N\backslash E}(\log) \to \Omega^1_{E/F}(\log/\log)^{\otimes e}$ maps $d\log z = \mathrm{Tr}_{L/N} d\log s$ to $(d\log s)^{\otimes e}$, it suffices to show the first equality. By the chain rule (1), we may assume that $K = M$ and it suffices to consider the two cases $K = N$ and $L = N$.

First assume that $L$ is a totally ramified extension of degree a power $e > 1$ of $p$. Let $t = \mathrm{N}_{L/K} s$. Then $a\colon \mathfrak{m}_L^{-d}/\mathfrak{m}_L^{-d+1} \to \Omega^1_{E/F}(\log/\log)$, sending $s^{e-1}/P'(s)$ to $d\log s$, and $b\colon F \to Z_{F\backslash E}(\log)$ defined by $d\log t$ make the diagram (7.1.8) commutative. This implies the first case $K = N$.

We show the second case. Set $v = \mathrm{N}_{L/K} u \in O_K^\times$. It suffices to show that the isomorphism $a\colon \mathfrak{m}_L^{-d}/\mathfrak{m}_L^{-d+1} \to \Omega^1_{E/F}(\log/\log)$ sending $u^{q-1}/Q'(u)$ to $d\log u$ and the isomorphism $b\colon F \to Z_{F\backslash E}(\log)$ defined by $d\log v$ make the diagram (7.1.8) commutative. We have $\mathrm{Tr}_{L/K}(u^i/Q'(u)) = 0$ for $i =$

$0, \ldots, q-2$ and $\mathrm{Tr}_{L/K}(u^{q-1}/Q'(u)) = 1$ by Proposition 1.2.2. By (4.4.2), we have $\mathrm{Tr}_{L/K}(d\log u) = d\log v$ and $\mathrm{Tr}_{L/K}(u^{-i} d\log u) = 0$ for $i = 1, \ldots, q-1$. Hence the diagram (7.1.8) is commutative as required. □

Assume further that $L$ is a Galois extension and let $G = \mathrm{Gal}(L/K)$ be the Galois group. Recall that, for $i \geqslant 1$, we defined a canonical injection

$$d\colon \mathrm{Gr}_i G = G_i/G_{i+1} \to \mathrm{Hom}(\Omega^1_{E/F}(\log/\log), \mathfrak{m}_L^i/\mathfrak{m}_L^{i+1}) \tag{7.1.12}$$

in Proposition 5.3.2. On the source $\mathrm{Gr}_i G = G_i/G_{i+1}$ of (7.1.12), the group $G$ acts by the conjugate. On the target $\mathrm{Hom}_E(\Omega^1_{E/F}(\log/\log), \mathfrak{m}_L^i/\mathfrak{m}_L^{i+1})$, we consider the natural action. For $i \geqslant 1$ and $\sigma \in G_i - G_{i+1}$, the morphism $d(\sigma)\colon \Omega^1_{E/F}(\log/\log) \to \mathfrak{m}_L^i/\mathfrak{m}_L^{i+1}$ is an isomorphism and defines an element $d(\sigma) \in T_{L/K}$. If $K'$ is a tamely ramified extension of $K$ and $L' = LK'$, for $\sigma' \in P' \subset \mathrm{Gal}(L'/K')$ corresponding to $\sigma \in P \subset \mathrm{Gal}(L/K)$, the morphism $T_{L/K} \to T_{L'/K'}$ maps $d(\sigma)$ to $d(\sigma')$ by (5.3.10).

The following lemma is proved similarly to Proposition 6.2.6.

**Lemma 7.1.6** *Let $L$ be a log monogenic Galois extension of $K$ of Galois group $G$.*

(1) *Let $M \subset L$ be the maximal tamely ramified extension of $K$, $m_{L/K} = e_{M/K}$ be the ramification index, and $P \subset G$ be the corresponding wild inertia subgroup. Then*

$$\Delta_{L/K} = [m_{L/K}] + \sum_{\sigma \in P - \{1\}} d(\sigma) \tag{7.1.13}$$

*in $T_{L/K}$.*

(2) *Let $N \subset G$ be a normal subgroup, $\overline{G} = G/N$ be the quotient group, and let $L'$ be the corresponding extension. Then*

$$\Delta_{L/L'} = \Delta_{L/K} - \sum_{\tau \in N \cap P - \{1\}} d(\tau) - [m_{L'/K}],$$

$$d(\sigma) = \sum_{\tau \in \sigma \cap P} d(\tau) + [m_{L/M}]$$

*for $\sigma \in \overline{G}$, $\sigma \neq 1$, in $T_{L/K}$.*

*Proof* (1) By the chain rule, Lemma 7.1.5(1), and Lemma 6.2.5(2), we have $\Delta_{L/K} = [m_{L/K}] + \Delta_{L/M}$. Since $P = \mathrm{Gal}(L/M)$ is a $p$-group, there exists a subgroup $H \subset P$ of order $e_{L/M}$. Let $K = M \subset M' \subset L$ be the corresponding subextension. Let $t, s$ be uniformizers of $M, L$, $z = \mathrm{N}_{L/M'} s$ be a uniformizer of $M'$, and $u = z/t \in O^\times_{M'}$ be a unit generating $O_{M'}$ over $O_K$. Let $P \in O_{M'}[X]$ and $Q \in O_M[X]$ be the minimal polynomials of $s$ over $M'$ and of $u$ over $K = M$ as in Lemma 7.1.5(2). For $\sigma \in G - \{1\}$,

$d(\sigma) \in \mathrm{Isom}_E(\Omega^1_{E/F}(\log/\log), \mathfrak{m}_L^{i(\sigma)}/\mathfrak{m}_L^{i(\sigma)+1}) \subset T_{L/K}$ is an isomorphism sending $d\log s$ to $\sigma(s)/s - 1$.

We have $P'(s)/s^{e-1} = \prod_{\sigma\in H-\{1\}}(1-\sigma(s)/s)$. Let $S \subset G$ be a complete set of representatives of $(G - H)/H$. Then $Q'(u)/u^{q-1} = \prod_{\sigma\in S}(1-\sigma(u)/u)$. As in the proof of Proposition 1.8.7(2) by Lemma 1.7.4, we have $\sigma(u)/u - 1 = \sigma(z)/z - 1 = \prod_{\tau\in H}(\sigma\tau(s)/s - 1)$. Hence, the assertion follows from the first equality in (7.1.10) and $(-1)^{w-1} = 1$ in $F^\times$.

(2) By the chain rule, Lemma 7.1.5(1), we have $\Delta_{L/K} - \Delta_{L/L'} = i_{L/L'}\Delta_{L'/K}$. Hence, the first equality follows from (7.1.13) for $L'$ over $K$. We show the second equality. By replacing $K \subset L'$ by $M \subset ML'$, we may assume $K = M$. Let $K = M \subset M' \subset L$ and $s, t, z, u$ and so on be as in the proof of (1). Then, for $\tau \in G - \{1\}$, $d(\tau) \in \mathrm{Isom}_E(\Omega^1_{E/F}(\log/\log), \mathfrak{m}_L^{i(\tau)}/\mathfrak{m}_L^{i(\tau)+1}) \subset T_{L/K}$ is the isomorphism sending $d\log s$ to $\tau(s)/s - 1$.

First, we consider the case $[L : L'] \leqslant [L : M']$; by replacing $M'$, we may assume $M' \subset L'$. Since $L$ is a totally ramified extension of $L'$, the norm $s' = \mathrm{N}_{L/L'}s \in L'$ is a uniformizer and, for $\sigma \in \overline{G} - \{1\}$, $d(\sigma) \in \mathrm{Isom}_{E'}(\Omega^1_{E'/F}(\log/\log), \mathfrak{m}_{L'}^{i(\sigma)}/\mathfrak{m}_{L'}^{i(\sigma)+1}) \subset T_{L'/K}$ is the isomorphism sending $d\log s'$ to $\sigma(s')/s' - 1$. Since $\mathrm{Tr}_{L/L'} d\log s = d\log s'$, we have $i_{L/L'} d\log s' = (d\log s)^{\#N}$. Hence, the second equality follows from (1.7.4) and $(-1)^{\#N-1} = 1 \bmod p$.

Next, we assume that $[L : L'] \geqslant [L : M']$. In this case, by replacing $M'$, we may assume $L' \subset M'$. Since $E$ is a purely inseparable extension of $F$ generated by $u$, the norm $v = \mathrm{N}_{M'/L'}u \in E'$ is a generator over $F$ and, for $\sigma \in \overline{G} - \{1\}$, $d(\sigma) \in \mathrm{Isom}_{E'}(\Omega^1_{E'/F}(\log/\log), \mathfrak{m}_{L'}^{i(\sigma)}/\mathfrak{m}_{L'}^{i(\sigma)+1}) \subset T_{L'/K}$ is the isomorphism sending $d\log v$ to $\sigma(v)/v - 1$. As in the proof of (1), we have $\sigma(v)/v - 1 = \prod_{\tau\in N}(\sigma\tau(s)/s - 1)$. By the equalities $\mathrm{Tr}_{M'/L'} d\log u = d\log v$ in $Z_{E'\backslash E}(\log)$, $d\log u = -d\log z$ in $\Omega^1_{F_{M'}/F}(\log/\log)$, and $\mathrm{Tr}_{L/N} d\log s = d\log z$ in $Z_{F_{M'}\backslash E}(\log)$, we have $i_{L/L'} d\log v = (d\log s)^{\#N}$ in $T_{L/K}$. Hence, the second equality follows in this case as well. □

**Exercise 7.1** Let $K$ be a henselian discrete valuation field of characteristic $p > 0$ and $F$ be the residue field. Let $y$ be a uniformizer and let $u \in O_K$ be an element such that $du \in \Omega^1_F$ is non-zero. Let $L$ be the cyclic extension of $K$ of degree $p$ defined by the Artin–Schreier equation $t^p - t = u/y^p$. Compute the logarithmic different $\Delta_{L/K} \in T_{L/K}$ and verify the equality (7.1.13).

## 7.2 Refined Swan Conductor and Differential Forms

We keep the setting from the last section and introduce the rescaling function and its refinement.

**Definition 7.2.1** Let $L$ be a log monogenic Galois extension of a henselian discrete valuation field $K$ of Galois group $G$. Let $i \geqslant 1$ be an integer, and $M_i \subset L$ be the extension corresponding to the ramification subgroup $G_i \subset G$. Let $E$ be the residue field of $L$ and $F_i$ be the residue field of $M_i$. For $\sigma \in G_i - G_{i+1}$, $d(\sigma) \in \operatorname{Isom}(\Omega^1_{E/F}(\log/\log), \mathfrak{m}_L^i/\mathfrak{m}_L^{i+1}) \subset T_{L/K}$ (7.1.12) is defined.

(1) We define a rational number $\varphi(i) \in \mathbf{Q}$ by

$$\varphi(i) = \frac{\operatorname{ord}_{M_i} D_{M_i/K} + i \cdot [E : F_i]}{e_{M_i/K}}. \tag{7.2.1}$$

(2) Let $\chi\colon \operatorname{Gr}_i G = G_i/G_{i+1} \to \mathbf{F}_p$ be a non-trivial character. Let $G_{i,\chi} \subset G_i$ be the inverse image of $\operatorname{Ker}(\chi\colon G_i/G_{i+1} \to \mathbf{F}_p)$, and let $M_i \subset M_{i,\chi} \subset L$ be the extension corresponding to $G_{i,\chi} \subset G$. Then, we define an element $\varphi(i,\chi) \in T_{L/K}$ by

$$\varphi(i,\chi) = \Delta_{M_{i,\chi}/K} + \sum_{\sigma \in G_i, \chi(\overline{\sigma})=1} d(\sigma). \tag{7.2.2}$$

Similarly to Proposition 6.2.2(1), we have

$$\varphi(i) = \frac{\operatorname{ord}_{M_{i+1}} D_{M_{i+1}/K} + i \cdot [E : F_{i+1}]}{e_{M_{i+1}/K}}. \tag{7.2.3}$$

As in Definition 6.2.1, the function $\varphi$ is characterized by $\varphi(0) = 0$ and $\varphi(i) = \varphi(i-1) + [E : F_i]/e_{M_i/K}$ for $i \geqslant 1$. Similarly to Lemma 6.2.8, we have the following.

**Lemma 7.2.2** *Let $i \geqslant 1$ be an integer, and let $\chi\colon \operatorname{Gr}_i G = G_i/G_{i+1} \to \mathbf{F}_p$ be a non-trivial character. Let $G_{i,\chi} \subset G_i$ be the inverse image of the kernel* $\operatorname{Ker}(\chi\colon G_i/G_{i+1} \to \mathbf{F}_p)$.

(1) *For $\sigma \in G_i - G_{i+1}$, the image of $d(\sigma) \in T_{L/K}$ by the canonical morphism $T_{L/K} \to \mathbf{Z}^2$ is $(i,-1)$.*
(2) *Let $\operatorname{ord}_L\colon T_{L/K} \to \mathbf{Z}$ denote the composition of $T_{L/K} \to \mathbf{Z}^2$ with the first projection. Then*

$$e_{L/K} \cdot \varphi(i) = \sum_{\sigma \in G_1 - G_i} \operatorname{ord}_L d(\sigma) + i \cdot \# G_i \tag{7.2.4}$$

*and*

$$\varphi(i,\chi) = [m_{L/K}] + \sum_{\sigma \in G_1 - G_{i,\chi}} d(\sigma) + \sum_{\sigma \in G_i, \chi(\overline{\sigma})=1} d(\sigma) \tag{7.2.5}$$

*in* $T_{L/K}$.

(3) *The image of* $\varphi(i,\chi) \in T_{L/K}$ *is* $(e_{L/K} \cdot \varphi(i), -\#P) \in \mathbf{Z}^2$.

We give a concrete description of $\varphi(i,\chi) \in T_{L/K}$ as an isomorphism $\Omega^1_{E/F}(\log/\log)^{\otimes w} = Z_{F\backslash E}(\log) \otimes_F E \to \mathfrak{m}_L^{e_{L/K}\cdot\varphi(i)}/\mathfrak{m}_L^{e_{L/K}\cdot\varphi(i)+1}$, where $w = [L : M]$ denotes the wild degree for the maximal tamely ramified extension $M \subset L$ corresponding to $P = G_1 \subset G$. The residue field $F_M \subset E$ of $M$ is the separable closure of $F$ in $E$. Let $t, s$ be uniformizers of $L, M$ such that $u \equiv s^{e_{L/M}}/t \in O_L$ is a generator of the residue field $E$ over $F_M$, and let $v = u^q \in F_M$ for $q = [E : F_M]$. Then, for $i \geqslant 1$ and $\sigma \in G_i - G_{i+1}$, we have $d(\sigma)(d\log s) = \sigma(s)/s - 1 \in (\mathfrak{m}_L^i/\mathfrak{m}_L^{i+1})^\times \subset R_L$. Since $\#G_{i,\chi} = \#\{\sigma \in G_i, \chi(\bar{\sigma}) = 1\}$ and $d\log v = (d\log u)^q = (d\log s)^{qe} \in Z_{F_M\backslash E}(\log)^\times$ for $w = qe = \#P$, we have

$$\varphi(i,\chi) = [m_{L/K}] + \sum_{\sigma\in G_1 - G_{i,\chi}} d(\sigma)(d\log s) + \sum_{\sigma\in G_i, \chi(\bar{\sigma})=1} d(\sigma)(d\log s) - d\log v \tag{7.2.6}$$

as an element of $\mathrm{Isom}_E(Z_{F\backslash E}(\log) \otimes_F E, \mathfrak{m}_L^{e_{L/K}\cdot\varphi(i)}/\mathfrak{m}_L^{e_{L/K}\cdot\varphi(i)+1}) \subset T_{L/K}$. In other words, $\varphi(i,\chi)\colon Z_{F\backslash E} \otimes_F E \to \mathfrak{m}_L^{e_{L/K}\cdot\varphi(i)}/\mathfrak{m}_L^{e_{L/K}\cdot\varphi(i)+1}$ is the isomorphism sending $d\log v \in Z_{F\backslash E} \otimes_F E$ to $e_{L/M}$ times the product of $d(\sigma)(d\log s)$ for $\sigma \in G_1 - G_{i,\chi}$ and for $\sigma \in G_i$ such that $\chi(\bar{\sigma}) = 1$.

**Definition 7.2.3** Let $K$ be a henselian discrete valuation field with residue field $F$ of characteristic $p > 0$. Let $L$ be a log monogenic Galois extension of Galois group $G$. Let $\Lambda$ be a field of characteristic $\neq p$, and fix a primitive $p$th root $\zeta \in \Lambda^\times$ of 1. For $i \geqslant 1$, we identify $\chi\colon \mathrm{Gr}_i G = G_i/G_{i+1} \to \Lambda^\times$ with $\chi\colon \mathrm{Gr}_i G \to \mathbf{F}_p$.

Let $V$ be a $\Lambda$-vector space of finite dimension with an action of $G$. Let $V = \bigoplus_{i\in\mathbf{N}} V_i$ be the break decomposition (6.1.3), and $V_i = \bigoplus_{\chi\in G^\vee} V_\chi$ (6.1.4) be the decomposition by non-trivial characters as in Lemma 6.1.6. We define the refined Swan conductor $\mathrm{rsw}_\zeta V \in T_{L/K}$ of $V$ with respect to $\zeta$ by

$$\mathrm{rsw}_\zeta V = -\sum_{i=1}^{\infty} \sum_{\chi\colon \mathrm{Gr}_i G\to\Lambda^\times, \chi\neq 1} \dim V_\chi \cdot \varphi(i,\chi). \tag{7.2.7}$$

We also define the Swan conductor $\mathrm{Sw}V \in \mathbf{Q}$ by

$$\mathrm{Sw}V = \sum_{i=1}^{\infty} \dim V_i \cdot \varphi(i). \tag{7.2.8}$$

The definition in (7.2.7) of the refined Swan conductor is the same as the original one by Kato in [49] except for the sign. Here, we put the minus sign in

(7.2.7) so that the second component $\#P \cdot \dim V/V^P$ in the next lemma can be positive.

**Lemma 7.2.4** *Let $V$ be a $\Lambda$-vector space of finite dimension with an action of $G$.*

(1) *The image of* $\mathrm{rsw}_\zeta V$ *by the canonical morphism* $T_{L/K} \to \mathbf{Z}^2$ *is* $(-e_{L/K} \cdot \mathrm{Sw}V, \#P \cdot \dim V/V^P)$.
(2) *The following conditions are equivalent:*
   (a) *$V$ is tamely ramified. Namely, the action of $P = G_1$ on $V$ is trivial.*
   (b) $\mathrm{Sw}\, V = 0$.
   (c) $\mathrm{rsw}\, V = 0$.

*Proof* (1) The assertion follows from Lemma 7.2.2, (7.2.8), and $\sum_{i=1}^{\infty} \dim V_i = \dim V/V^P$.

(2) The same argument as in the proof of Lemma 6.3.2(2) also works here. □

Similarly to Proposition 6.2.9, the Hasse–Arf theorem, Theorem 5.3.4, implies the following for wildly ramified abelian characters.

**Proposition 7.2.5** *Let $L$ be a log monogenic cyclic extension of $K$, and let $G = \mathrm{Gal}(L/K)$ be the Galois group. Let $\chi \colon G \to \mathbf{Q}/\mathbf{Z}$ be a faithful character and $\rho \colon G \to \Lambda^\times$ be a faithful character.*

(1) $\mathrm{Sw}\, \chi = \mathrm{Sw}\, \rho$.
(2) *Assume that* $\mathrm{Sw}\, \chi = n \geqslant 1$. *Let $\sigma \in G$ be the element of order $p$ such that $\chi(\sigma) = 1/p$, and $\zeta = \rho(\sigma) \in \Lambda^\times$ be a primitive $p$th root of* 1. *Let $i \geqslant 1$ be the largest integer such that $G_i \neq 1$, and let $\chi_0$ be the restriction to $G_i$. We regard the refined Swan conductor* $\mathrm{rsw}\, \chi \colon \mathfrak{m}_K^n/\mathfrak{m}_K^{n+1} \to Z_{F\backslash E}(\log)$ *as an element of $S_{L/K}$. Then, its image by the injection $i_{L/K} \colon S_{L/K} \to T_{L/K}$ equals the inverse of $\varphi(i, \chi_0) \in T_{L/K}$ and we have an equality* $\mathrm{rsw}\, \chi = \mathrm{rsw}_\zeta\, \rho$ *of isomorphisms* $\mathfrak{m}_K^n/\mathfrak{m}_K^{n+1} \to Z_{F\backslash E}(\log) \subset \Omega_F^1(\log)$.

*Proof* (1) Since $L = M_{i+1}$, we have $E = F_{i+1}$ and the equality follows from (7.2.3) and (5.3.11).

(2) Similarly to the proof of Proposition 6.2.9(2), by comparing the commutative diagrams (7.1.8) and (5.3.13), we obtain $\Delta_{L/K} = -d(\sigma) - i_{L/K}(\mathrm{rsw}\, \chi)$. Hence, we obtain $\mathrm{rsw}_\zeta\, \rho = -\varphi(i, \chi_0) = -(\Delta_{L/K} + d(\sigma)) = i_{L/K}(\mathrm{rsw}\, \chi)$. □

Contrary to Proposition 6.2.9(2), we do not have the inverse because we put the minus sign in (7.2.7).

**Example 7.2.6** Let $K$ be a henselian discrete valuation field of characteristic $p > 0$, $F$ be the residue field, and $t \in K$ be a uniformizer. Let $L$ be the

Artin–Schreier extension defined by $x^p - x = v/t^n$ for $n = pm > 0$ and for $v \in O_K$ such that $\bar{v} \in F - F^p$. Let $u = t^m x$ be the generator of the residue field $E = F(u)$ as in Example 1.4.4(2). For $\sigma_i \in G = \mathrm{Gal}(L/K)$ such that $\sigma_i(x) = x + i$, $i \in \mathbf{F}_p^\times$, the isomorphism $d(\sigma_i) \in \mathrm{Isom}_E(\Omega^1_{E/F}, \mathfrak{m}_L^m/\mathfrak{m}_L^{m+1})$ maps $d\log u$ to $\sigma_i(u)/u - 1 = i/x$. Let $\chi\colon G \to \mathbf{Q}/\mathbf{Z}$ be the character sending $\sigma_1$ to $1/p$. Then the isomorphism $\varphi(m,\chi) = \sum_{\sigma\in G,\neq 1} d(\sigma) + d(\sigma_1) \in \mathrm{Isom}_E(\Omega^{1\otimes p}_{E/F}, \mathfrak{m}_L^n/\mathfrak{m}_L^{n+1}) \subset T_{L/K}$ maps $d\log u^{\otimes p}$ to $(p-1)!x^{-p} = -t^n/v \in \mathfrak{m}_L^n/\mathfrak{m}_L^{n+1} - \{0\} \subset R_L$ by (7.2.5). Since $\mathrm{Tr}_{L/K}(d\log u) = d\log v$, $\varphi(m,\chi)$ is the image of the isomorphism in $\mathrm{Isom}_F(Z_{F\backslash E}, \mathfrak{m}_K^n/\mathfrak{m}_K^{n+1}) \subset S_{L/K}$ sending the basis $d\log v$ to $-t^n/v \in \mathfrak{m}_K^n/\mathfrak{m}_K^{n+1} - \{0\} \subset R_K$.

Since the refined Swan conductor $\mathrm{rsw}\,\chi$ maps the basis $-t^n \in \mathfrak{m}_K^n/\mathfrak{m}_K^{n+1}$ to $dv \in Z_{F\backslash E}(\log)$ as in Example 3.1.6(2), $\varphi(m,\chi)$ is the inverse of $\mathrm{rsw}\,\chi$.

For an integer $i \geqslant 1$ and $\sigma \in G_i - G_{i+1}$, we define $\widetilde{s}(\sigma) \in T_{L/K}$ to be the *inverse* of $d(\sigma) \in \mathrm{Isom}_E(\Omega^1_{E/F}(\log/\log), \mathfrak{m}_L^i/\mathfrak{m}_L^{i+1}) \subset T_{L/K}$ (7.1.12) and $s(\sigma) = i \in \mathbf{Z}$. For $\sigma = 1$, define $\widetilde{s}(1) \in T_{L/K}$ to be the logarithmic different $\Delta_{L/K}$ (7.1.9) and let $s(1) \in \mathbf{Z}$ be $\mathrm{length}_{O_L}\Omega^1_{O_L/O_K}(\log/\log) = \mathrm{ord}_L D^{\log}_{L/K}$. Similarly to Lemma 6.3.6, for a primitive $p$th root $\zeta \in \Lambda^\times$ of 1 and the character $\rho\colon G_i \to \Lambda^\times$ defined by $\rho(\sigma) = \widetilde{\zeta}^{\chi(\sigma)}$, we have

$$\varphi(i,\chi) = \sum_{\sigma\in G_i} \widetilde{s}(\sigma)\cdot\rho(\sigma) + \varepsilon(\zeta) \tag{7.2.9}$$

in $T_{L/K} \otimes_{\mathbf{Z}} \mathbf{Z}[\widetilde{\zeta}]$, where $\varepsilon = \varepsilon(\zeta) \in R_L \otimes_{\mathbf{Z}} \mathbf{Z}[\widetilde{\zeta}] \subset T_{L/K} \otimes_{\mathbf{Z}} \mathbf{Z}[\widetilde{\zeta}]$ is defined by (6.3.6).

**Proposition 7.2.7** *We have*

$$\mathrm{rsw}_\zeta V = -\sum_{\sigma\in P} \widetilde{s}(\sigma)\cdot \mathrm{Tr}^{\mathrm{Br}}(\sigma : V) - \dim V/V^P \cdot \varepsilon(\zeta) + \dim V^P \cdot [m_{L/K}] \tag{7.2.10}$$

*in* $T_{L/K} \otimes_{\mathbf{Z}} \mathbf{Z}[\widetilde{\zeta}]$ *and*

$$\mathrm{Sw} V = \frac{1}{\#G}\sum_{\sigma\in P} s(\sigma)\cdot \mathrm{Tr}^{\mathrm{Br}}(\sigma : V). \tag{7.2.11}$$

*Proof* The proof is identical to that of Proposition 6.3.7 using (7.2.9), except for the signs in (7.2.10) inherited from (7.2.7). □

**Corollary 7.2.8** *Let L be a log monogenic Galois extension of K, and let* $G = \mathrm{Gal}(L/K)$ *be the Galois group.*

(1) *Let V be a representation of G, and* $N \subset G$ *be a normal subgroup. Assume that the restriction of V on N is trivial and regard V as a representation of the quotient group* $\overline{G} = G/N$ *of G. Then* $\mathrm{rsw}_{\zeta,G} V = \mathrm{rsw}_{\zeta,\overline{G}} V$.

(2) *Let $W$ be a representation of a subgroup $H \subset G$, and $M$ be the extension corresponding to $H$. For the induced representation $V = \mathrm{Ind}_H^G W$, we have an equality*

$$\mathrm{rsw}_{\zeta,G}\, V = \mathrm{N}_{G/H}\big(\mathrm{rsw}_{\zeta,H}\, W - \dim W \cdot \Delta_{M/K} + \dim W^{P\cap H} \cdot [m_{M/K}]\big) \tag{7.2.12}$$

*in the $G$-fixed part $T_{L/K}^G$. More precisely, we have*

$$\begin{aligned} \mathrm{rsw}_{\zeta,G}\, V = {} & \mathrm{N}_{G/H}(\mathrm{rsw}_{\zeta,H}\, W) + \dim W \cdot \mathrm{rsw}_{\zeta,G}\, \mathrm{Ind}_H^G \Lambda \quad (7.2.13)\\ & - \dim W/W^{P\cap H} \cdot [M : K] \cdot [m_{M/K}], \end{aligned}$$

$$\mathrm{rsw}_{\zeta,G}\, \mathrm{Ind}_H^G \Lambda = -\mathrm{N}_{G/H}(\Delta_{M/K} - [m_{M/K}]). \tag{7.2.14}$$

We have a minus sign in the induction formula (7.2.12) since we put the minus sign in the definition (7.2.7) of the refined Swan conductor. The integrality of the conductor also holds in the following sense.

**Theorem 7.2.9** *Let $\zeta \in \Lambda$ be a primitive pth root of* 1.

(1) *We have $\mathrm{rsw}_\zeta\, V \in S_{L/K}$. The image of $\mathrm{rsw}_\zeta\, V$ by $S_{L/K} \to \mathbf{Z}^2$ equals $(-\mathrm{Sw}\, V, \dim V/V^P)$.*
(2) *Let $n = \mathrm{Sw}\, V$ and $d = \dim V/V^P$. Then the refined Swan conductor $\mathrm{rsw}_\zeta\, V$ defines an isomorphism $\mathfrak{m}_K^n/\mathfrak{m}_K^{n+1} \to Z_{F\setminus E}(\log)^{\otimes d}$ of $F$-vector spaces of dimension* 1.

*Proof* (1) The same proof as for Theorem 6.4.2, replacing Proposition 6.4.1(2) by Corollary 7.2.8(2), proves the first assertion. The second assertion follows from Lemma 7.2.4(2) and the commutative diagram (7.1.3).

(2) The second assertion in (1) means that the refined Swan conductor $\mathrm{rsw}_\zeta\, V$ is an element of $\mathrm{Isom}_F(\mathfrak{m}_K^n/\mathfrak{m}_K^{n+1}, Z_{F\setminus E}(\log)^{\otimes d}) \subset S_{L/K}$. □

For later use, we state and prove the compatibility of the construction of the refined Swan conductor with totally ramified base extension.

**Lemma 7.2.10** *Assume that $L$ is a log monogenic Galois extension of $K$ with ramification index $e_{L/K} = 1$. Let $K'$ be a totally ramified extension of $K$, and $L' = LK'$ be the composition field. Let $G'$ denote the Galois group $\mathrm{Gal}(L'/K')$. Then the canonical morphism $S_{L/K} \to S_{L'/K'}$ sends the refined Swan conductor $\mathrm{rsw}_G V$ to $\mathrm{rsw}_{G'} V$.*

*Proof* Let $e = e_{K'/K}$ be the ramification index. Since $O_L \otimes_{O_K} O_{K'} \to O_{L'}$ is an isomorphism, the canonical morphism $G' \to G$ induces isomorphisms $G'_{ei} \to G_i$ and $G'_{ei+1} \to G_{i+1}$ for integers $i \geqslant 0$. Further, the canonical morphism $T_{L/K} \to T_{L'/K'}$ sends $\varphi(i,\chi)$ to $\varphi(ei,\chi')$ for characters $\chi\colon \mathrm{Gr}_i G \to \Lambda^\times$ and the composition with $\mathrm{Gr}_{ei} G' \to \mathrm{Gr}_i G$. Hence, the assertion follows. □

**Exercise 7.2** Let $K$ be a henselian discrete valuation field of characteristic $p > 0$, and $F$ be the residue field. Let $y$ be a uniformizer and let $u \in O_K$ be an element such that $du \in \Omega^1_F$ is non-zero. Let $L$ be the cyclic extension of $K$ of degree $p$ defined by the Artin–Schreier equation $t^p - t = u/y^p$ as in Exercise 7.1. Let $\chi\colon G = \mathrm{Gal}(L/K) \to \mathbf{F}_p$ be the faithful character sending the generator $\sigma$ satisfying $\sigma(t) = t + 1$ to 1. Compute $\varphi(1,\chi) \in T_{L/K}$ and verify the equality $\varphi(1,\chi) = -\mathrm{rsw}\,\chi$ in Proposition 7.2.5(2).

## 7.3 Valuation of the Refined Swan Conductor

In this section we study the valuation of the refined Swan conductor defined in Section 7.2 assuming that the residue field is a function field of one variable over a perfect field. First, we prepare some generalities on valuation. Recall that for a field $F$ and $F$-vector spaces $L_1, L_2$ of dimension 1, an extension $F^\times\langle L_1^\times, L_2^\times\rangle$ of $\mathbf{Z}^2$ by $F^\times$ is defined as the disjoint union of the $F^\times$-torsors of isomorphisms $L_1^{\otimes m} \to L_2^{\otimes n}$.

Let $v$ be a discrete valuation of $F$, and $A \subset F$ be the valuation ring. Let $L_{1A} \subset L_1$ and $L_{2A} \subset L_2$ be invertible $A$-submodules. Then, we define a morphism

$$v\colon F^\times\langle L_1^\times, L_2^\times\rangle \to \mathbf{Z} \tag{7.3.1}$$

as follows: For $m, n \in \mathbf{Z}^2$ and $f \in \mathrm{Isom}_F(L_1^{\otimes m}, L_2^{\otimes n})$, let $v(f) \in \mathbf{Z}$ be the unique integer $l$ such that $f(L_{1A}^{\otimes m}) = \mathfrak{m}_K^l L_{2A}^{\otimes n} \subset L_2^{\otimes n}$. The morphism $v\colon F^\times\langle L_1^\times, L_2^\times\rangle \to \mathbf{Z}$ extends the valuation $v\colon F^\times \to \mathbf{Z}$. For $l_1 \in L_1^\times \cap L_{1A} \subset F^\times\langle L_1^\times, L_2^\times\rangle$ and $l_2 \in L_2^\times \cap L_{2A} \subset F^\times\langle L_1^\times, L_2^\times\rangle$, we have $v(l_1) = \mathrm{length}_A(L_{1A}/Al_1)$ and $v(l_2) = \mathrm{length}_A(L_{2A}/Al_2)$.

**Lemma 7.3.1** *Let $v$ be a discrete valuation of $F$, and $w$ be a discrete valuation of $E$ extending $v$. Let $A \subset F$ and $B \subset E$ be the valuation rings and let $e = e_{w/v}$ be the ramification index. Let $L_{1A} \subset L_1$ and $L_{2A} \subset L_2$ be invertible $A$-submodules, and $M_{1B} \subset M_1$ and $M_{2B} \subset M_2$ be invertible $B$-submodules, and define morphisms $v\colon F^\times\langle L_1^\times, L_2^\times\rangle \to \mathbf{Z}$ and $w\colon E^\times\langle M_1^\times, M_2^\times\rangle \to \mathbf{Z}$ as above. Let $f_1\colon L_1 \otimes_F E \to M_1$ and $f_2\colon L_2 \otimes_F E \to M_2$ be isomorphisms of $E$-vector spaces of dimension 1, and define integers $m_1$ and $m_2$ by $f_1(L_{1A}) = \mathfrak{m}_B^{m_1} M_{1B}$ and $f_2(L_{2A}) = \mathfrak{m}_B^{m_2} M_{2B}$. Then the following diagram is commutative:*

$$\begin{array}{ccc} F^\times\langle L_1^\times, L_2^\times\rangle & \xrightarrow{(v,\mathrm{can})} & \mathbf{Z}\times\mathbf{Z}^2 \\ \downarrow & & \downarrow{\scriptstyle (e,m_1,m_2)} \\ E^\times\langle M_1^\times, M_2^\times\rangle & \xrightarrow{\ w\ } & \mathbf{Z} \end{array} \tag{7.3.2}$$

*Proof* Let $m, n \in \mathbf{Z}^2$, $f \in \mathrm{Isom}_F(L_1^{\otimes m}, L_2^{\otimes n})$, and $l = v(f) \in \mathbf{Z}$. Then, for the isomorphism $g \in \mathrm{Isom}_E(M_1^{\otimes m}, M_2^{\otimes n})$ such that $g \circ f_1 = f_2 \circ (f \otimes 1)$, we have $\mathfrak{m}_B^{m_1 m} g(M_{1B}^{\otimes m}) = \mathfrak{m}_B^{m_2 n} \mathfrak{m}_A^l M_{2B}^{\otimes n}$. Since the image of $f \in F^\times\langle L_1^\times, L_2^\times\rangle$ in $\mathbf{Z} \times \mathbf{Z}^2$ is $(l, -m, n)$, the assertion follows. □

Let $K$ be a henselian discrete valuation field, and let $k$ be a perfect field of characteristic $p > 0$. Assume that the residue field $F$ of $K$ is a function field of one variable over $k$. Define an extension $S_K$ of $\mathbf{Z}^2$ by $F^\times$ by

$$S_K = F^\times\langle(\mathfrak{m}_K/\mathfrak{m}_K^2)^\times, (\Omega^1_{F/k})^\times\rangle = \coprod_{d,n \in \mathbf{Z}} \mathrm{Isom}_F(\mathfrak{m}_K^n/\mathfrak{m}_K^{n+1}, \Omega^{1\otimes d}_{F/k}). \quad (7.3.3)$$

Let $C$ be a smooth curve over $k$ such that the residue field $F$ is the function field of $C$. Fix a closed point of $C$ and let $v$ be the corresponding discrete valuation of $F$. Let $A \subset F$ be the valuation ring of $v$. Fix an $A$-submodule $(\mathfrak{m}_K/\mathfrak{m}_K^2)_A \subset \mathfrak{m}_K/\mathfrak{m}_K^2$ free of rank 1 and, for an integer $n$, let $(\mathfrak{m}_K^n/\mathfrak{m}_K^{n+1})_A \subset \mathfrak{m}_K^n/\mathfrak{m}_K^{n+1}$ denote the tensor product $(\mathfrak{m}_K/\mathfrak{m}_K^2)_A^{\otimes n}$. We identify the free $A$-module $\Omega^1_{A/k}$ of rank 1 as a submodule of $\Omega^1_{F/k}$. Then, the valuation $v$ together with these submodules define a morphism

$$\mathrm{ord}_v \colon S_K \to \mathbf{Z} \quad (7.3.4)$$

as (7.3.1). For integers $d$ and $n$, and for an isomorphism

$$\alpha \in \mathrm{Isom}_F(\mathfrak{m}_K^n/\mathfrak{m}_K^{n+1}, \Omega^{1\otimes d}_{F/k}) \subset S_K,$$

the valuation $\mathrm{ord}_v\alpha \in \mathbf{Z}$ is the unique integer $m$ such that $\alpha((\mathfrak{m}_K^n/\mathfrak{m}_K^{n+1})_A) = \mathfrak{m}_A^m \cdot \Omega^{1\otimes d}_{A/k}$. We define a modification $\mathrm{ord}'_v \colon S_K \to \mathbf{Z}$ of $\mathrm{ord}_v \colon S_K \to \mathbf{Z}$ by replacing $\Omega^1_A$ by $\Omega^1_A(\log v)$.

Let $L$ be a finite separable extension of $K$ of ramification index 1, and let $E$ be the residue field. By the assumption that $F$ is a function field of one variable over a perfect field $k$, the extension $E$ is isomorphic to the tensor product $F' \otimes_F E'$ of the separable closure $F' \subset E$ of $F$ in $E$ with the largest purely inseparable extension $E' \subset E$ of $F$ in $E$. If $E \supsetneqq F'$, we have $S_K = S_{L/K}$ and $S_L = T_{L/K}$. By the assumption that $F$ is a function field of one variable over a perfect field $k$, we have $\dim \Omega^1_E = 1$ and $E$ is generated by a single element over $F$ by Lemma 1.8.4 (c) ⇒ (a).

Assume that $L$ is a Galois extension of $K$, and let $G$ be the Galois group. The inertia subgroup $P = \mathrm{Ker}(G \to \mathrm{Aut}_F(E)) \subset G$ corresponding to the maximal unramified extension $K' \subset L$ is a $p$-group since the ramification index $e_{L/K}$ is assumed to be 1. We identify the quotient group $\overline{G} = G/P$ with $\mathrm{Gal}(F'/F)$. Since $\mathfrak{m}_K^n/\mathfrak{m}_K^{n+1} \otimes_F F' \to \mathfrak{m}_{K'}^n/\mathfrak{m}_{K'}^{n+1}$ and $\Omega^1_F \otimes_F F' \to \Omega^1_{F'}$ are isomorphisms, the canonical injection $S_K \to S_{K'}$ induces an isomorphism $S_K \to S_{K'}^{\overline{G}}$ to the $\overline{G}$-fixed part.

Let $w$ be a discrete valuation of $E$ extending $v$, and let $B \subset E$ be the valuation ring of $w$. Define a free $B$-module $(\mathfrak{m}_L/\mathfrak{m}_L^2)_B$ of rank 1 by $(\mathfrak{m}_L/\mathfrak{m}_L^2)_B = B \cdot (\mathfrak{m}_K/\mathfrak{m}_K^2)_A \subset \mathfrak{m}_L/\mathfrak{m}_L^2$, and consider the free $B$-module $\Omega^1_{B/k}(\log w)$ of rank 1 as a submodule of $\Omega^1_{E/k}$. Then the valuation $w$ together with these submodules define a morphism

$$\mathrm{ord}'_w \colon S_L \to \mathbf{Z}. \tag{7.3.5}$$

**Lemma 7.3.2** *Let $K' \subset L$ be the maximal unramified extension over $K$, and let $F'$ be the residue field of $K'$. Let $q = [E : F'] = [E : F]_{\mathrm{insep}}$ be the inseparable degree. Let $v'$ be the restriction of $w$ to $F' \subset E$, and let $A' \subset F'$ be the valuation ring. Define $\mathrm{ord}'_{v'} \colon S_{K'} \to \mathbf{Z}$ by $(\mathfrak{m}_{K'}/\mathfrak{m}_{K'}^2)_{A'} = A' \otimes_A (\mathfrak{m}_K/\mathfrak{m}_K^2)_A$ and let $d_{v'/v} = \mathrm{ord}_{v'} D^{\log}_{A'/A}$ be the valuation of the logarithmic different. Then the diagram*

$$\begin{array}{ccccc} S_K & \longrightarrow & S_{K'} & \longrightarrow & S_L \\ {\scriptstyle(\mathrm{ord}'_v,\mathrm{pr}_2)}\downarrow & & {\scriptstyle(\mathrm{ord}'_{v'},\mathrm{pr}_2)}\downarrow & & \downarrow{\scriptstyle(\mathrm{ord}'_w,\mathrm{pr}_2)} \\ \mathbf{Z}^2 & \xrightarrow{\begin{pmatrix} e_{v'/v} & d_{v'/v} \\ 0 & 1 \end{pmatrix}} & \mathbf{Z}^2 & \xrightarrow{(e_{w/v'},q)} & \mathbf{Z}^2 \end{array} \tag{7.3.6}$$

*is commutative.*

*Proof* Since the canonical isomorphism $\Omega_E^{1\otimes q} = \Omega_{E/F}^{1\otimes q} \to Z_{F'\backslash E} \otimes_{F'} E = \Omega^1_{F'} \otimes_{F'} E$ induces an isomorphism $\Omega^1_B(\log w)^{\otimes q} \to \Omega^1_{A'}(\log v') \otimes_{A'} B$, and $(\mathfrak{m}_L/\mathfrak{m}_L^2)_B = B \otimes_{A'} (\mathfrak{m}_{K'}/\mathfrak{m}_{K'}^2)_{A'}$, the right square is commutative by Lemma 7.3.1. Since the length of the cokernel $A' \otimes_A \Omega^1_A(\log v) \to \Omega^1_{A'}(\log v')$ equals $d_{v'/v}$, and $(\mathfrak{m}_{K'}/\mathfrak{m}_{K'}^2)_{A'} = A' \otimes_A (\mathfrak{m}_K/\mathfrak{m}_K^2)_A$, the left square is also commutative by Lemma 7.3.1. □

For $\sigma \in P$, let $\mathrm{ord}'_w \widetilde{s}(\sigma) \in \mathbf{Z}$ be the image of $\widetilde{s}(\sigma) \in T_{L/K} = S_L$ by $\mathrm{ord}'_w \colon S_L \to \mathbf{Z}$. Let $\overline{I} \subset \overline{G} = \mathrm{Gal}(F'/F)$ denote the inertia subgroup of the restriction $v'$ to $F'$ of $w$. We extend the function $\mathrm{ord}'_w s \colon P \to \mathbf{Z}$ to the inverse image $I \subset G$ of $\overline{I} \subset \overline{G}$ as follows. The $G$-fixed part $E' = E^G \subset E$ is the largest purely inseparable extension of $F$, and the diagram

$$\begin{array}{ccc} F' & \longrightarrow & E \\ \uparrow & & \uparrow \\ F & \longrightarrow & E' \end{array}$$

is cocartesian. The extension $E$ of $E'$ is a Galois extension and the Galois group $\mathrm{Gal}(E/E')$ is canonically identified with $\mathrm{Gal}(F'/F) = G/P$. The subgroup

$\bar{I} \subset \mathrm{Gal}(F'/F)$ is identified with the inertia subgroup of $\mathrm{Gal}(E/E')$ at $w$. Let $s_w : \bar{I} \to \mathbf{Z}$ denote the Swan character. For $\sigma \in I - P$ and $\bar{\sigma}$ in $\bar{I} \subset \mathrm{Gal}(E/E')$, define

$$\mathrm{ord}'_w \widetilde{s}(\sigma) = s_w(\bar{\sigma}). \tag{7.3.7}$$

We have $\mathrm{ord}'_w \widetilde{s}(\sigma) = -\mathrm{ord}_w(\sigma(t) - t)/t$ for a uniformizer $t$ of $B$.

Let $\Lambda$ be a perfect field of characteristic $\ell \neq p$, and fix a primitive $p$th root $\zeta \in \Lambda$ of 1. Let $V$ be a $\Lambda$-vector space of finite dimension with an action of $G$. The restriction of $V$ to $P \subset G$ defines the refined Swan conductor $\mathrm{rsw}_\zeta V = \mathrm{rsw}\, V \in S_L$. By Theorem 7.2.9(1), we have $\mathrm{rsw} V \in S_K$. For $n = \mathrm{Sw}\, V$ and $d = \dim V/V^P$, the refined Swan conductor $\mathrm{rsw} V$ defines an isomorphism $\mathfrak{m}_K^n/\mathfrak{m}_K^{n+1} \to \Omega^{1\otimes d}_{F/k}$ of $F$-vector spaces of dimension 1 by Theorem 7.2.9(2). Its order, $\mathrm{ord}_v(\mathrm{rsw}\, V)$, is the unique integer $m$ satisfying $(\mathrm{rsw}\, V)((\mathfrak{m}_K^n/\mathfrak{m}_K^{n+1})_A) = \mathfrak{m}_A^m \cdot \Omega^{1\otimes d}_{A/k}$. A variant $\mathrm{ord}'_v(\mathrm{rsw}\, V)$ is defined to be the unique integer $m'$ satisfying $(\mathrm{rsw}\, V)((\mathfrak{m}_K^n/\mathfrak{m}_K^{n+1})_A) = \mathfrak{m}_A^{m'} \cdot \Omega^1_{A/k}(\log v)^{\otimes d}$.

**Lemma 7.3.3** *Let $L$ be a Galois extension of $K$, and let $V$ be a representation of $G = \mathrm{Gal}(L/K)$ on a $\Lambda$-vector space of finite dimension. Then*

$$\mathrm{ord}'_v(\mathrm{rsw}\, V) = \mathrm{ord}_v(\mathrm{rsw}\, V) + \dim V/V^P. \tag{7.3.8}$$

*Proof* Since $d = \dim V/V^P$, the assertion follows from $\Omega^1_{A/k}(\log v) = \mathfrak{m}_A^{-1} \cdot \Omega^1_{A/k}$. □

The $P$-fixed part $V^P$ defines a representation of $G/P = \mathrm{Gal}(F'/F)$, and hence the Swan conductor $\mathrm{Sw}_v V^P \in \mathbf{N}$ is defined.

**Proposition 7.3.4** *Let $V$ be a $\Lambda$-vector space of finite dimension with an action of $G$. Let $I \subset G$ be the inverse image of the inertia subgroup $\bar{I} \subset \bar{G} = G/P$ as above. Then*

$$\begin{aligned} &\mathrm{ord}_v(\mathrm{rsw}\, V) + \dim V/V^P - \mathrm{Sw}_v V^P \\ &\quad = -\frac{1}{\# I} \sum_{\sigma \in I, \neq 1} \mathrm{ord}'_w \widetilde{s}(\sigma) \cdot (\mathrm{Tr}^{\mathrm{Br}}(\sigma : V) - \dim V). \end{aligned} \tag{7.3.9}$$

*Proof* It suffices to prove the cases where $V = V^P$ and $V^P = 0$.

First, we show the case where $V^P = 0$. Since $\mathrm{Sw}_v V^P = 0$ in this case, the left-hand side $\mathrm{ord}_v(\mathrm{rsw}\, V) + \dim V/V^P$ equals $\mathrm{ord}'_v(\mathrm{rsw}\, V)$ by (7.3.8). By Lemma 7.2.4, the image of $\mathrm{rsw}\, V \in S_K$ by the left vertical arrow of (7.3.6) is $(\mathrm{ord}'_v(\mathrm{rsw}\, V), \dim V) \in \mathbf{Z}^2$. Hence, by the commutative diagram (7.3.6),

(7.2.10), and $\sum_{\sigma\in P}\widetilde{s}(\sigma)=0$, we obtain

$$\begin{aligned} &e_{w/v}\cdot \mathrm{ord}'_v(\mathrm{rsw}\,V)+e_{w/v'}d_{v'/v}\cdot \dim V\\ &\quad=-\sum_{\sigma\in P,\neq 1}\mathrm{ord}'_w\widetilde{s}(\sigma)\cdot(\mathrm{Tr}^{\mathrm{Br}}(\sigma:V)-\dim V). \end{aligned} \tag{7.3.10}$$

For a uniformizer $t$ of $B\subset E$, its power $t'=t^{e_{w/v'}}$ is a uniformizer of $A'=B\cap F'$. Hence, by Proposition 1.9.5(3), we have $d_{v'/v}=-\sum_{\sigma\in\bar{I}-\{1\}}s_{v'}(\sigma)$ and $e_{w/v'}d_{v'/v}=-\sum_{\sigma\in I-P}\mathrm{ord}'_w\widetilde{s}(\sigma)$. By the assumption $V^P=0$, for $\sigma\in I-P$ we have $\sum_{\tau\in P}\mathrm{Tr}^{\mathrm{Br}}(\sigma\tau:V)=0$. Since $\widetilde{s}(\overline{\sigma})=\widetilde{s}(\overline{\sigma\tau})$ for $\tau\in P$, we have $\sum_{\sigma\in I-P}\mathrm{ord}'_w\widetilde{s}(\sigma)\cdot\mathrm{Tr}^{\mathrm{Br}}(\sigma:V)=0$. Thus, we obtain

$$e_{w/v'}d_{v'/v}\cdot\dim V=\sum_{\sigma\in I-P}\mathrm{ord}'_w\widetilde{s}(\sigma)\cdot(\mathrm{Tr}^{\mathrm{Br}}(\sigma:V)-\dim V). \tag{7.3.11}$$

Subtracting (7.3.11) from (7.3.10) and dividing the result by $e_{w/v}=\#I$, we obtain (7.3.9) in the case $V^P=0$.

Next, we show the case where $V=V^P$ and hence $\mathrm{rsw}\,V=0$. In this case, we may assume $P=1$ and then the assertion follows from Proposition 6.3.7(1) and (7.3.7). □

**Exercise 7.3** Let $K$ be a henselian discrete valuation field of characteristic $p>0$ such that the residue field $F$ is a function field of one variable over a perfect field $k$ of characteristic $p$. Let $y$ be a uniformizer, and let $u\in O_K$ be an element such that $du\in\Omega^1_F$ is non-zero. Let $L$ be the cyclic extension of $K$ of degree $p$ defined by the Artin–Schreier equation $t^p-t=u/y^p$ as in Exercise 7.1. Let $\chi\colon G=\mathrm{Gal}(L/K)\to\Lambda^\times$ be the faithful character sending the generator $\sigma$ satisfying $\sigma(t)=t+1$ to a primitive $p$th root $\zeta$ of 1, and let $V$ be the representation of $G$ defined by $\chi$. Let $A$ be the local ring at a closed point $x$ of the smooth curve over $k$ with function field $F$, and let $(\mathfrak{m}_K/\mathfrak{m}_K^2)_A\subset\mathfrak{m}_K/\mathfrak{m}_K^2$ be the $A$-module generated by $y$. Compute the valuation of the refined Swan conductor $\mathrm{rsw}_\zeta V$ and verify the equality (7.3.9).

## Historical Notes

The refined Swan conductor involving differential forms was introduced and studied by Kato in [49]. The definition (7.2.7) here is modified from the original one by the sign. The valuation of the refined Swan conductor, which is the subject of Section 7.3, is studied in [48] in the context of valuation rings with valuation group $\mathbf{Z}^2$ with the lexicographic order.

# PART IV

# GEOMETRIC APPLICATIONS

The Grothendieck–Ogg–Shafarevich formula computes the Euler–Poincaré characteristic of cohomology of an étale sheaf on a smooth curve over an algebraically closed field of positive characteristic. The Swan conductor of Galois representations of local fields appears in the formula as the contribution of the ramification at the boundary.

For an étale sheaf on the generic fiber of a scheme over a discrete valuation ring, the sheaf of nearby cycles is defined on the closed fiber. The Deligne–Kato formula computes the dimension of the space of nearby cycles for curves over a discrete valuation ring. The variant of the refined Swan conductor with values as differential forms appears as the contribution of the ramification along the closed fiber.

The Deligne–Kato formula requires the assumption that the closed fibers of the curve itself, and also of a covering trivializing the sheaf, be reduced. The reduced fiber theorem ensures that, after taking a sufficiently large base extension, the condition is satisfied. The proof of the reduced fiber theorem is reduced to a property of the local fields at the generic points of irreducible components of the closed fiber. This property is proved as an application of the study of cyclic extensions of degree $p$ in Chapter 3.

# 8

# Grothendieck–Ogg–Shafarevich Formula

The Swan conductor was invented to formulate the Grothendieck–Ogg–Shafarevich formula computing the Euler–Poincaré characteristic of an étale sheaf on an algebraic curve over an algebraically closed field of positive characteristic. This formula is derived from the Lefschetz trace formula for étale cohomology. After recalling some basic terminology on étale cohomology, we prove the Lefschetz trace formula in Section 8.1. We prove the Grothendieck–Ogg–Shafarevich formula in Section 8.2.

## 8.1 Lefschetz Trace Formula

Let $X$ be a scheme, and let $n \geqslant 1$ and $\ell$ be an integer and a prime number invertible on $X$. The Picard group $\mathrm{Pic}(X)$ is defined to be the group of isomorphism classes of invertible $\mathcal{O}_X$-modules.

Let $\mathbf{G}_m$ denote the sheaf of multiplicative groups on the étale site of $X$. By the étale descent, the Picard group $\mathrm{Pic}(X)$ is canonically identified with $\mathrm{H}^1(X, \mathbf{G}_m)$ by identifying the class of an invertible $\mathcal{O}_X$-module $\mathcal{L}$ with that of the $\mathbf{G}_m$-torsor $\mathcal{I}som(\mathcal{O}_X, \mathcal{L})$. The Kummer sequence $0 \to \mu_n \to \mathbf{G}_m \to \mathbf{G}_m \to 0$ defines the boundary morphism $\mathrm{H}^1(X, \mathbf{G}_m) \to \mathrm{H}^2(X, \mu_n)$. Setting $n = \ell^e$ and passing to the limit, we obtain the cycle class mapping

$$\mathrm{cl}\colon \mathrm{Pic}(X) = \mathrm{H}^1(X, \mathbf{G}_m) \to \varprojlim_e \mathrm{H}^2(X, \mu_{\ell^e}) = \mathrm{H}^2(X, \mathbf{Z}_\ell(1)). \tag{8.1.1}$$

For a Cartier divisor $D$ of $X$, let $[D] \in \mathrm{Pic}(X)$ denote the class of the invertible sheaf $O(D)$. For a morphism $f\colon X \to Y$ of schemes, the diagram

$$\begin{array}{ccc} \mathrm{Pic}(X) & \xrightarrow{f^*} & \mathrm{Pic}(Y) \\ {\scriptstyle \mathrm{cl}}\downarrow & & \downarrow{\scriptstyle \mathrm{cl}} \\ \mathrm{H}^2(X, \mathbf{Z}_\ell(1)) & \xrightarrow{f^*} & \mathrm{H}^2(Y, \mathbf{Z}_\ell(1)) \end{array}$$

is commutative by the functoriality.

**Lemma 8.1.1** *Let $X$ be a regular noetherian scheme and $D \subset X$ be a reduced effective Cartier divisor. Let $j_D\colon Y \to X$ be the morphism from the disjoint union $Y$ of the generic points of irreducible components of $D$, and let $j\colon U = X - D \to X$ be the open immersion of the complement.*

(1) *The sequence*

$$0 \to \mathbf{G}_{m,X} \to j_*\mathbf{G}_{m,U} \to j_{D*}\mathbf{Z}_Y \to 0 \tag{8.1.2}$$

*is exact, and $R^1 j_*\mathbf{G}_{m,U} = 0$ and $R^1 j_{D*}\mathbf{Z}_Y = 0$.*

(2) *For an integer $n$ invertible on $X$, the Kummer sequence induces an isomorphism $j_{D*}(\mathbf{Z}/n\mathbf{Z})_Y \to R^1 j_*\mu_n$.*

*Proof* (1) At every geometric point $\bar{x}$ of $X$, the strict local ring $O_{X,\bar{x}}$ is a regular local ring and hence is a unique factorization domain. Thus, the stalk $0 \to O_{X,\bar{x}}^\times \to \mathbf{G}_m(U \times_X \operatorname{Spec} O_{X,\bar{x}}) \to \bigoplus_{\eta \in Y \times_X \operatorname{Spec} O_{X,\bar{x}}} \mathbf{Z} \to 0$ at $\bar{x}$ of (8.1.2) is exact, and consequently (8.1.2) is exact.

Similarly, the stalk of $R^1 j_*\mathbf{G}_{m,U}$ at $\bar{x}$ is $\mathrm{Pic}(U \times_X \operatorname{Spec} O_{X,\bar{x}}) = 0$ and we obtain $R^1 j_*\mathbf{G}_{m,U} = 0$. Since the stalk of $R^1 j_{D*}\mathbf{Z}_Y$ at $\bar{y}$ is the Galois cohomology $\mathrm{H}^1(Y \times_X \operatorname{Spec} O_{X,\bar{y}}, \mathbf{Z}) = 0$, we also have $R^1 j_{D*}\mathbf{Z}_Y = 0$.

(2) Since the étale morphism $\mathbf{Z}[\frac{1}{n}, T^{\pm 1}] \to \mathbf{Z}[\frac{1}{n}, T^{\pm 1}]$ defined by sending $T$ to $T^n$ is faithfully flat, the Kummer sequence $0 \to \mu_{n,X} \to \mathbf{G}_{m,X} \to \mathbf{G}_{m,X} \to 0$ is exact and the exact sequence (8.1.2) induces an isomorphism $j_*\mathbf{G}_{m,U}/(j_*\mathbf{G}_{m,U})^n \to j_{D*}\mathbf{Z}_Y/n j_{D*}\mathbf{Z}_Y$. The Kummer sequence on $U$ induces an isomorphism $j_*\mathbf{G}_{m,U}/(j_*\mathbf{G}_{m,U})^n \to R^1 j_*\mu_n$ by the vanishing $R^1 j_*\mathbf{G}_{m,U} = 0$ in (1). We obtain an isomorphism $j_{D*}\mathbf{Z}_Y/n j_{D*}\mathbf{Z}_Y \to j_{D*}(\mathbf{Z}/n\mathbf{Z})_Y$ by the vanishing $R^1 j_{D*}\mathbf{Z}_Y = 0$ in (1). □

Let $X$ be a regular noetherian scheme and $\ell$ be a prime number invertible on $X$. Let $i\colon D \to X$ be the closed immersion of a regular divisor, and $j\colon U = X - D \to X$ be the open immersion of the complement. Then, since $D$ is normal, the exact sequence (8.1.2) gives an exact sequence

$$0 \to \mathbf{G}_{m,X} \to j_*\mathbf{G}_{m,U} \to i_*\mathbf{Z}_D \to 0 \tag{8.1.3}$$

and an isomorphism $\mathbf{Z}_D \to R^1 i^! \mathbf{G}_{m,X}$. This morphism and the boundary morphism $R^1 i^! \mathbf{G}_{m,X} \to R^2 i^! \mu_n$ of the Kummer sequence induce an isomorphism $\mathbf{Z}/n\mathbf{Z} \to R^2 i^! \mu_n$ by Lemma 8.1.1. Since the adjunction morphism $\mu_n \to j_* \mu_n$ is an isomorphism, we have $R^q i^! \mu_n = 0$ for $q = 0, 1$ and the edge morphism $\mathrm{H}^q(D, R^2 i^! \mu_n) \to \mathrm{H}^{q+2}_D(X, \mu_n)$ is defined. Thus, we obtain the morphism $i_* \colon \mathrm{H}^q(D, \mathbf{Z}/n\mathbf{Z}) \to \mathrm{H}^q(D, R^2 i^! \mu_n) \to \mathrm{H}^{q+2}_D(X, \mu_n) \to \mathrm{H}^{q+2}(X, \mu_n)$. By passing to the limit, we obtain the Gysin morphism

$$i_* \colon \mathrm{H}^q(D, \mathbf{Z}_\ell) \to \mathrm{H}^{q+2}(X, \mathbf{Z}_\ell(1)). \tag{8.1.4}$$

**Lemma 8.1.2** (1) *The diagram*

$$\begin{array}{ccc} \mathbf{Z} & \xrightarrow{\cdot [D]} & \mathrm{Pic}(X) \\ \downarrow & & \downarrow \mathrm{cl} \\ \mathrm{H}^0(D, \mathbf{Z}_\ell) & \xrightarrow{i_*} & \mathrm{H}^2(X, \mathbf{Z}_\ell(1)) \end{array} \tag{8.1.5}$$

*is commutative.*

(2) *The composition*

$$\mathrm{H}^q(X, \mathbf{Z}_\ell) \xrightarrow{i^*} \mathrm{H}^q(D, \mathbf{Z}_\ell) \xrightarrow{i_*} \mathrm{H}^{q+2}(X, \mathbf{Z}_\ell(1)) \tag{8.1.6}$$

*equals the cup-product* $\cup \mathrm{cl}[D] \colon \mathrm{H}^q(X, \mathbf{Z}_\ell) \to \mathrm{H}^{q+2}(X, \mathbf{Z}_\ell(1))$ *with the cycle class of D.*

*Proof* (1) Since $i^! \mathbf{G}_m = 0$, the edge morphism $\Gamma(D, R^1 i^! \mathbf{G}_m) \to \mathrm{H}^1_D(X, \mathbf{G}_m)$ is defined and is an isomorphism. We consider the diagram

$$\begin{array}{ccccccc} \Gamma(D, \mathbf{Z}) & \longrightarrow & \Gamma(D, R^1 i^! \mathbf{G}_m) & \longrightarrow & \mathrm{H}^1_D(X, \mathbf{G}_m) & \longrightarrow & \mathrm{H}^1(X, \mathbf{G}_m) \\ \downarrow & & \downarrow & & \downarrow & & \downarrow \\ \Gamma(D, \mathbf{Z}/n\mathbf{Z}) & \longrightarrow & \Gamma(D, R^2 i^! \mu_n) & \longrightarrow & \mathrm{H}^2_D(X, \mu_n) & \longrightarrow & \mathrm{H}^2(X, \mu_n). \end{array} \tag{8.1.7}$$

The left vertical arrow is the canonical surjection and the other vertical arrows are induced by the Kummer sequence. The left horizontal arrows are induced by the isomorphisms $\mathbf{Z}_D \to R^1 i^! \mathbf{G}_{m,X}$ and $\mathbf{Z}/n\mathbf{Z}_D \to R^2 i^! \mu_n$ defined by the exact sequence (8.1.3). Hence, the diagram is commutative.

The class $[D] \in \mathrm{Pic}(X) = \mathrm{H}^1(X, \mathbf{G}_m)$ is the image of $1 \in \Gamma(D, \mathbf{Z})$ by the composition of the upper line, and the composition of the lower line is $i_*$. Hence, by taking the limit we obtain the assertion.

(2) The composition is induced by the morphism $\mathbf{Z}/n\mathbf{Z} \to i_* \mathbf{Z}/n\mathbf{Z} \to i_* R^2 i^! \mu_n$ of sheaves on $X$ and by the morphism $\mathrm{H}^{q+2}_D(X, \mu_n) \to \mathrm{H}^{q+2}(X, \mu_n)$, forgetting the support, for powers $n$ of $\ell$. Since these are compatible with the cup-product, the assertion follows from the commutative diagram (8.1.5). □

In the rest of this section, $X$ denotes a proper smooth geometrically connected curve of genus $g$ over a field $k$. If $X$ has a $k$-rational point, the functor sending a scheme $S$ over $k$ to the group $\mathrm{Ker}(\deg\colon \mathrm{Pic}(X\times S) \to \mathbf{Z}(S))/\mathrm{Im}(\mathrm{pr}_2^*\colon \mathrm{Pic}(S) \to \mathrm{Pic}(X \times S))$ is representable by an abelian variety of dimension $g$ over $k$ called the Jacobian of $X$. We will not prove the following standard facts on $\mathrm{H}^q(X, \mathbf{Z}_\ell) = \varprojlim_n \mathrm{H}^q(X, \mathbf{Z}/\ell\mathbf{Z})$.

**Theorem 8.1.3** *Let $X$ be a projective smooth irreducible curve of genus $g$ over a separably closed field $k$, and let $\ell$ be a prime number invertible in $k$.*

(1) *The canonical morphism $\mathbf{Z}_\ell \to \mathrm{H}^0(X, \mathbf{Z}_\ell)$ is an isomorphism.*
(2) *The isomorphism $\mathrm{H}^1(X, \mu_{\ell^e}) \to \mathrm{Pic}(X)[\ell^e]$ induces an isomorphism $\mathrm{H}^1(X, \mathbf{Z}_\ell(1)) \to T_\ell J = \varprojlim_e J[\ell^e]$ to the Tate module of the Jacobian, and $\mathrm{H}^1(X, \mathbf{Z}_\ell(1))$ is a free $\mathbf{Z}_\ell$-module of rank $2g$.*
(3) *There exists a unique isomorphism*

$$\mathrm{Tr}\colon \mathrm{H}^2(X, \mathbf{Z}_\ell(1)) \to \mathbf{Z}_\ell \tag{8.1.8}$$

*called the trace mapping sending the class $\mathrm{cl}([x]) \in \mathrm{H}^2(X, \mathbf{Z}_\ell(1))$ of a $k$-rational point $x \in X$ to 1. The diagram*

$$\begin{array}{ccc} \mathrm{Pic}(X) & \xrightarrow{\mathrm{cl}} & \mathrm{H}^2(X, \mathbf{Z}_\ell(1)) \\ {\scriptstyle \deg}\downarrow & & \downarrow{\scriptstyle \mathrm{Tr}} \\ \mathbf{Z} & \longrightarrow & \mathbf{Z}_\ell \end{array}$$

*is commutative.*

By the Künneth formula, the morphism

$$\sum_{q=0}^{2} \mathrm{pr}_1^* \cup \mathrm{pr}_2^*\colon \bigoplus_{q=0}^{2} \mathrm{H}^q(X, \mathbf{Z}_\ell) \otimes \mathrm{H}^{r-q}(X, \mathbf{Z}_\ell(1)) \to \mathrm{H}^r(X\times X, \mathbf{Z}_\ell(1)) \tag{8.1.9}$$

is an isomorphism. For $q = 0, 1, 2$, define $\mathrm{pr}_{1*}\colon \mathrm{H}^{q+2}(X \times X, \mathbf{Z}_\ell(1)) \to \mathrm{H}^q(X, \mathbf{Z}_\ell)$ to be the composition of

$$\mathrm{H}^{q+2}(X \times X, \mathbf{Z}_\ell(1)) \to \mathrm{H}^q(X, \mathbf{Z}_\ell) \otimes \mathrm{H}^2(X, \mathbf{Z}_\ell(1)) \xrightarrow{1\otimes \mathrm{Tr}} \mathrm{H}^q(X, \mathbf{Z}_\ell).$$

The first arrow is the composition of the inverse of (8.1.9) with the projection.

For $q = 0, 1, 2$, we define a canonical morphism

$$\mathrm{H}^2(X \times X, \mathbf{Z}_\ell(1)) \to \mathrm{End}\, \mathrm{H}^q(X, \mathbf{Z}_\ell) \tag{8.1.10}$$

by sending $\gamma \in \mathrm{H}^2(X \times X, \mathbf{Z}_\ell(1))$ to the endomorphism sending $\alpha$ to $\mathrm{pr}_{1*}(\gamma \cup \mathrm{pr}_2^* \alpha)$.

**Lemma 8.1.4** *Let $f: X \to X$ be an endomorphism. Let $\gamma = (1, f): X \to X \times X$ be the graph morphism and regard $\Gamma_f = \gamma(X) \subset X \times X$ as a smooth divisor.*

(1) *For $q = 0, 1, 2$, the composition $\mathrm{pr}_{1*} \circ \gamma_*: \mathrm{H}^q(X, \mathbf{Z}_\ell) \to \mathrm{H}^{q+2}(X \times X, \mathbf{Z}_\ell(1)) \to \mathrm{H}^q(X, \mathbf{Z}_\ell)$ is the identity.*

(2) *The morphism $f^*$ equals the morphism defined by the cycle class $[\Gamma_f]$.*

*Proof* (1) It suffices to show that the composition $\mathbf{Z}/n\mathbf{Z} \to R^2\gamma^!\mu_n \to R^2\mathrm{pr}_{2*}\mu_n \to \mathbf{Z}/n\mathbf{Z}$ of morphisms of locally constant sheaves on $X$ is the identity. It suffices to check this for the fiber at a closed point $x \in X$. This means that the composition $\mathbf{Z}/n\mathbf{Z} \to \mathrm{H}^2_x(X, \mu_n) \to \mathrm{H}^2(X, \mu_n) \to \mathbf{Z}/n\mathbf{Z}$ is the identity. The trace mapping sends the class $\mathrm{cl}[x]$ to 1 by Theorem 8.1.3(3), and the assertion follows.

(2) The morphism $f^*$ equals the composition $\mathrm{pr}_{1*} \circ \gamma_* \circ \gamma^* \circ \mathrm{pr}_2^*$ by (1). Since $\gamma_* \circ \gamma^*$ equals the cup-product with $\Gamma$ by Lemma 8.1.2, the assertion follows. □

**Lemma 8.1.5** *The diagram*

$$\begin{array}{ccc} \mathrm{H}^2(X \times X, \mathbf{Z}_\ell(1)) & \xrightarrow{\delta^*} & \mathrm{H}^2(X, \mathbf{Z}_\ell(1)) \\ {\scriptstyle (8.1.10)}\downarrow & & \downarrow{\scriptstyle \mathrm{Tr}} \\ \bigoplus_{q=0}^2 \mathrm{End}\, \mathrm{H}^q(X, \mathbf{Z}_\ell) & \xrightarrow{\sum_q (-1)^q \mathrm{Tr}} & \mathbf{Z}_\ell \end{array}$$

*is commutative.*

*Proof* By the Künneth formula, it suffices to show that, for the composition with $\mathrm{H}^q(X, \mathbf{Z}_\ell) \otimes \mathrm{H}^{2-q}(X, \mathbf{Z}_\ell(1)) \to \mathrm{H}^2(X \times X, \mathbf{Z}_\ell(1))$, the diagram is commutative. By the Poincaré duality, the morphism

$$\begin{aligned} \mathrm{H}^{2-q}(X, \mathbf{Z}_\ell(1)) &\to \mathrm{Hom}_{\mathbf{Z}_\ell}(\mathrm{H}^q(X, \mathbf{Z}_\ell), \mathrm{H}^2(X, \mathbf{Z}_\ell(1))) \\ &\xrightarrow{\mathrm{Tr}} \mathrm{Hom}_{\mathbf{Z}_\ell}(\mathrm{H}^q(X, \mathbf{Z}_\ell), \mathbf{Z}_\ell) \end{aligned}$$

is an isomorphism. Since the cup-product is antisymmetric, the assertion follows from the following lemma. □

**Lemma 8.1.6** *For a commutative ring $A$ and a free $A$-module $M$ of finite rank, the diagram*

$$\begin{array}{ccc} M \otimes M^\vee & \longrightarrow & A \\ \downarrow & & \| \\ \mathrm{End}_A(M) & \xrightarrow{\mathrm{Tr}} & A \end{array}$$

*is commutative. The upper horizontal arrow maps $x \otimes f$ to $f(x)$ and the left vertical arrow sends $x \otimes f$ to the endomorphism sending $y$ to $f(y)x$.*

*Proof* Since the assertion is compatible with the direct sum, it is reduced to the case where $M = A$, and the assertion follows. □

We prove the Lefschetz trace formula. For a divisor $D$ of $X \times X$, let $[D] \in \mathrm{Pic}(X \times X)$ denote the class of $D$ and define the intersection product $(D, \Delta_X)$ by $(D, \Delta_X) = \deg \delta^*[D]$.

**Theorem 8.1.7** (Lefschetz trace formula) *Let $X$ be a proper smooth geometrically connected curve over a separably closed field $k$. For an endomorphism $f\colon X \to X$, we have*

$$\sum_{q=0}^{2}(-1)^q \mathrm{Tr}(f^*\colon \mathrm{H}^q(X, \mathbf{Z}_\ell)) = (\Gamma_f, \Delta_X). \tag{8.1.11}$$

*Proof* Let $\gamma \in \mathrm{H}^2(X \times X, \mathbf{Z}_\ell(1))$ denote $\mathrm{cl}([\Gamma_f])$. By the commutative diagram

$$\begin{array}{ccccc} \mathrm{Pic}(X \times X) & \xrightarrow{\delta^*} & \mathrm{Pic}(X) & \xrightarrow{\deg} & \mathbf{Z} \\ {\scriptstyle \mathrm{cl}}\downarrow & & {\scriptstyle \mathrm{cl}}\downarrow & & \downarrow \\ \mathrm{H}^2(X \times X, \mathbf{Z}_\ell(1)) & \xrightarrow{\delta^*} & \mathrm{H}^2(X, \mathbf{Z}_\ell(1)) & \xrightarrow{\mathrm{Tr}} & \mathbf{Z}_\ell \end{array}$$

the right-hand side equals $\mathrm{Tr}(\delta^*\gamma)$. The endomorphism $f^*$ equals the endomorphism defined by $\gamma$ by Lemma 8.1.4(2). Hence, the assertion follows from Lemma 8.1.5. □

The Euler–Poincaré characteristic $\chi(X) = \chi(X, \mathbf{Z}_\ell) \in \mathbf{Z}$ is defined as the alternating sum

$$\chi(X, \mathbf{Z}_\ell) = \sum_{q=0}^{2}(-1)^q \mathrm{rank}\, \mathrm{H}^q(X, \mathbf{Z}_\ell).$$

If $X$ is irreducible of genus $g$, then we have $\chi(X, \mathbf{Z}_\ell) = 1 - 2g - 1 = 2 - 2g$. For the open immersion $j\colon U \to X$ of the complement $U = X - D$ of a reduced divisor $D$, the compact support cohomology $\mathrm{H}^q_c(U, \mathbf{Z}_\ell)$ is defined as $\mathrm{H}^q(X, j_!\mathbf{Z}_\ell)$ and its Euler–Poincaré characteristic is defined by $\chi_c(U) = \chi_c(U, \mathbf{Z}_\ell) = \sum_{q=0}^{2}(-1)^q \mathrm{rank}\, \mathrm{H}^q_c(U, \mathbf{Z}_\ell)$. They fit in a long exact sequence $0 \to \mathrm{H}^0_c(U, \mathbf{Z}_\ell) \to \mathrm{H}^0(X, \mathbf{Z}_\ell) \to \bigoplus_{x \in D} \mathbf{Z}_\ell \to \mathrm{H}^1_c(U, \mathbf{Z}_\ell) \to \mathrm{H}^1(X, \mathbf{Z}_\ell) \to 0$, and the canonical morphisms $\mathrm{H}^q_c(U, \mathbf{Z}_\ell) \to \mathrm{H}^q(X, \mathbf{Z}_\ell)$ are isomorphisms for $q \geqslant 2$. Hence, we have $\chi_c(U, \mathbf{Z}_\ell) = 2 - 2g - \#D$.

**Corollary 8.1.8** *Let $X$ be a proper smooth geometrically connected curve over an algebraically closed field.*

(1) *We have*

$$\chi(X, \mathbf{Z}_\ell) = -\deg \Omega^1_{X/k} = 2 \cdot \chi(X, \mathcal{O}_X). \tag{8.1.12}$$

(2) *Let $f\colon W \to X$ be a finite morphism of degree $n$ of proper smooth curves over $k$ such that the restriction $V = f^{-1}(U) \to U \subset X$ on dense open subsets are étale. For $w \in W - V$, let $d_w$ denote the valuation of the logarithmic different $D^{\log}_{W/X,w}$. Then*

$$\chi_c(V, \mathbf{Z}_\ell) = n \cdot \chi_c(U, \mathbf{Z}_\ell) - \sum_{w \in W - V} d_w. \tag{8.1.13}$$

*Proof* (1) Applying Theorem 8.1.7 to $f = 1_X$, we obtain $\chi(X, \Lambda) = (\Delta_X, \Delta_X)$. Since $\mathcal{O}_X(-\Delta_X) = \Omega^1_{X/k}$, the first equality follows. The second equality follows from Riemann–Roch.

(2) Let $D = X - U$ and $E = W - V$ be the reduced divisors on the complements. We have $\chi_c(U) = -\deg \Omega^1_X(\log D)$ and $\chi_c(V) = -\deg \Omega^1_W(\log E)$ by (8.1.12). Hence, $\chi_c(V) - n \cdot \chi_c(U)$ equals minus the length of the cokernel of $f^*\Omega^1_X(\log D) \to \Omega^1_W(\log E)$, and the assertion follows. □

**Exercise 8.1** Let $k$ be an algebraically closed field of characteristic $p > 0$, and let $n \geqslant 1$ be an integer prime to $p$. Let $C$ be the projective completion of the Artin–Schreier covering $C^\circ = \operatorname{Spec} k[t,u]/(u^p - u - t^n)$ of $\mathbf{A}^1_k = \operatorname{Spec} k[t]$, and $\sigma$ be the automorphism of $C$ defined by $\sigma(t) = t$ and $\sigma(u) = u + 1$. Compute $\operatorname{Tr}(\sigma^*\colon \mathrm{H}^q(C, \mathbf{Z}_\ell))$ for $q = 0, 1, 2$ using the Lefschetz trace formula.

## 8.2 Grothendieck–Ogg–Shafarevich Formula

We recall some basic facts on perfect complexes of representations of finite groups. Let $A$ be a noetherian ring and $G$ be a finite group. We say that a complex $K$ of $A[G]$-modules is a perfect complex if there exists a finite complex $P$ of finitely generated projective $A[G]$-modules and a quasi-isomorphism $P \to K$ of complexes of $A[G]$-modules.

**Lemma 8.2.1** *For a complex $K$ of $A[G]$-modules and integers $a \leqslant b$, the following conditions are equivalent:*

(1) *There exist a complex $P$ of finitely generated projective $A[G]$-modules such that $P^q = 0$ unless $a \leqslant q \leqslant b$, and a quasi-isomorphism $P \to K$ of complexes of $A[G]$-modules.*

(2) *For every finitely generated right $A[G]$-module $M$, the cohomology groups $H^q(M \otimes^L_{A[G]} K)$ are of finite type for all $q$ and are 0 unless $a \leqslant q \leqslant b$.*

*Proof* (1) $\Rightarrow$ (2): Let $P$ be a complex of finitely generated projective $A[G]$-modules such that $P^q = 0$ unless $a \leqslant q \leqslant b$, and let $P \to K$ be a quasi-isomorphism of complexes of $A[G]$-modules. Then $P$ satisfies condition (2) and hence $K$ does.

(2) $\Rightarrow$ (1): Let $K$ be a complex of $A[G]$-modules and $a \leqslant b$ be integers satisfying (2). Then (2) for $M = A[G]$ implies that the cohomology groups $H^q(K)$ are of finite type for all $q$ and are 0 unless $a \leqslant q \leqslant b$. Hence, there exist a complex $P$ of finitely generated projective $A[G]$-modules such that $P^q = 0$ unless $a < q \leqslant b$, and a morphism $P \to K$ of complexes of $A[G]$-modules such that the induced morphisms $H^q(P) \to H^q(K)$ are isomorphisms for $a + 1 < q \leqslant b$ and a surjection for $q = a + 1$. Then the mapping cone $C = \mathrm{Cone}[P \to K]$ is acyclic except at degree $a$. Hence, it suffices to show that the $A[G]$-module $L = \mathrm{H}^a(C)$ of finite type is projective.

Since $K$ satisfies (2) and $P$ satisfies (2) for $a + 1$ and $b$, the mapping cone $C$ also satisfies (2). In other words, for any finitely generated right $A[G]$-module $M$, the cohomology group $H^q(M \otimes^L_{A[G]} L)$ vanishes for any $q \notin [0, b-a]$. This means that the $A[G]$-module $L$ of finite presentation is flat. Hence, the canonical morphism $\mathrm{Hom}_{A[G]}(L, A[G]) \otimes_{A[G]} L \to \mathrm{End}_{A[G]} L$ is an isomorphism and and $L$ is projective. □

The Grothendieck groups $P_\Lambda(G)$ and $P_{O_E}(G)$ are isomorphic to the Grothendieck groups of the categories of perfect complexes of $\Lambda[G]$-modules and $O_E[G]$-modules respectively.

**Lemma 8.2.2** *Let $U$ be a separated scheme of finite type over a separably closed field $k$, and let $\pi: V \to U$ be an étale $G$-torsor for a finite group $G$. Let $\ell$ be a prime number different from the characteristic of $k$.*

(1) *The complex $R\Gamma_c(V, \mathbf{Z}_\ell)$ of $\mathbf{Z}_\ell[G]$-modules is a perfect complex.*
(2) *Assume that $k$ is algebraically closed, and that $U \subset X$ is a dense open subset of a proper smooth curve over $k$. Let $r_G: G \to \mathbf{Z}$ be the character of the regular representation of $G$ and, for each point $x \in X - U$, let $s_x: D_x \to \mathbf{Z}$ be the Swan character of the decomposition group at $x$. Then we have an equality*

$$\mathrm{Tr}(- : R\Gamma_c(V, \mathbf{Z}_\ell)) = \chi_c(U) \cdot r_G - \sum_{x \in X - U} \mathrm{Ind}_{D_x}^G s_x$$

*of characters of the group $G$.*

*Proof* (1) Let $M$ be a finitely generated right $\mathbf{Z}_\ell[G]$-module. Then, since

$\pi_*\mathbf{Z}_\ell$ is a free $\mathbf{Z}_\ell[G]$-module, the cohomology group $\mathrm{H}^q_c(U, M \otimes_{\mathbf{Z}_\ell[G]} \pi_*\mathbf{Z}_\ell)$ of the complex $M \otimes^L_{\mathbf{Z}_\ell[G]} R\Gamma_c(V, \mathbf{Z}_\ell) = R\Gamma_c(U, M \otimes_{\mathbf{Z}_\ell[G]} \pi_*\mathbf{Z}_\ell)$ are finitely generated $\mathbf{Z}_\ell$-modules and are 0 unless $0 \leqslant q \leqslant 2\dim U$. Hence, the assertion follows from Lemma 8.2.1.

(2) Let $W$ be the smooth compactification of $V$, and $D = W - V$ be the complement. Then, by the long exact sequence $\to \mathrm{H}^q_c(V, \mathbf{Z}_\ell) \to \mathrm{H}^q(W, \mathbf{Z}_\ell) \to \mathrm{H}^q(D, \mathbf{Z}_\ell) \to$, we have $\mathrm{Tr}(\sigma : R\Gamma_c(V, \mathbf{Z}_\ell)) = \mathrm{Tr}(\sigma : R\Gamma(W, \mathbf{Z}_\ell)) - \# D^\sigma$. Hence, the assertion for $\sigma \neq 1$ follows from the Lefschetz trace formula, Theorem 8.1.7. For $\sigma = 1$, the assertion follows from Corollary 8.1.8(2). □

As a preparation for the definition of the Swan conductor at a boundary point of a locally constant sheaf on a smooth curve over a perfect field, we briefly recall the dictionary between locally constant sheaves and Galois representations. We say that an étale sheaf $\mathcal{F}$ on a scheme $S$ is locally constant constructible if, étale locally on $S$, the sheaf $\mathcal{F}$ is a constant sheaf defined by a finite set. By étale descent, the functor

$$(\text{finite étale schemes on } S) \longrightarrow (\text{locally constant constructible sheaves on } S) \tag{8.2.1}$$

sending a finite étale scheme on $S$ to the sheaf represented by the scheme is an equivalence of categories.

Assume $S$ is normal and connected, and let $\bar{s}$ be a geometric point. Then the fundamental group $\pi_1(S, \bar{s})$ is a pro-finite group characterized by the property that the functor

$$(\text{finite étale schemes on } S) \longrightarrow (\text{finite sets with continuous action of } \pi_1(S, \bar{s})) \tag{8.2.2}$$

sending a finite étale scheme $X$ over $S$ to the fiber $X_{\bar{s}}$ is an équivalence of categories. If $S = \operatorname{Spec} k$ for a field and $\bar{s} = \operatorname{Spec} k_s$ for a separable closure, the fundamental group $\pi_1(S, \bar{s})$ is canonically identified with the absolute Galois group $G_k = \mathrm{Gal}(k_s/k)$.

The composition

$$\begin{aligned}(\text{locally constant constructible sheaves on } S)& \\ \longrightarrow (\text{finite sets with continuous action of } \pi_1(S, \bar{s}))&\end{aligned} \tag{8.2.3}$$

of the functors (8.2.1) and (8.2.2) is an equivalence of categories sending a sheaf $\mathcal{F}$ to the fiber $\mathcal{F}_{\bar{s}}$. For a finite field $\Lambda$, the functor (8.2.3) induces an equivalence of categories

$$\begin{aligned}&(\text{locally constant constructible sheaves of } \Lambda\text{-modules on } S) \\ &\longrightarrow (\text{finite dimensional } \Lambda\text{-vector spaces with continuous action of } \pi_1(S, \bar{s})).\end{aligned} \tag{8.2.4}$$

Let $N \subset \pi_1(S, \bar{s})$ be an open normal subgroup and let $U$ be a finite étale Galois scheme over $S$ corresponding to the finite $\pi_1(S, \bar{s})$-set $\pi_1(S, \bar{s})/N$. Then $U$ is a Galois covering of $S$ of Galois group $G = \pi_1(S, \bar{s})/N$. The functor (8.2.3) induces an equivalence of categories

$$\begin{gathered}(\text{locally constant constructible sheaves on } S \\ \text{such that the pull-back to } U \text{ is constant}) \to (\text{finite } G\text{-sets})\end{gathered} \tag{8.2.5}$$

on subcategories. We also have a variant for $\Lambda$-modules.

Let $K$ be a field and $\operatorname{Spec} K \to S$ be a morphism of schemes. Let $\mathcal{F}$ be a locally constant constructible sheaf of $\Lambda$-modules on $S$. Then the pull-back of $\mathcal{F}$ by $\operatorname{Spec} K \to S$ is a locally constant constructible sheaf on $\operatorname{Spec} K$. Hence, for a separable closure $K_s$ of $K$, the pull-back defines a representation of the absolute Galois group $G_K = \operatorname{Gal}(K_s/K)$.

Let $X$ be a noetherian normal scheme and $x \in X$ be a point of codimension 1. Then the local ring $\mathcal{O}_{X,x}$ is a discrete valuation ring. We call the fraction field $K_x$ of the completion $\widehat{\mathcal{O}}_{X,x}$ the local field at $x$. Let $U \subset X$ be a dense open subset, and let $\bar{\eta}$ be a geometric point above $\operatorname{Spec} K_x$. Let $\mathcal{F}$ be a locally constant constructible sheaf of $\Lambda$-modules on $U$. Then the stalk $\mathcal{F}_{\bar{\eta}}$ defines a representation of the absolute Galois group $G_{K_x}$ of the local field.

Let $V \to U$ be a finite étale Galois covering such that the pull-back $\mathcal{F}_V$ is a constant sheaf. Let $G = \operatorname{Gal}(V/U)$ and $M$ be the representation of $G$ corresponding to $\mathcal{F}$. Take a lifting of the geometric point $\bar{\eta}$ of $U$ to the Galois covering $V$ and let $D \subset G$ be the stabilizer of the lifting. Then the Galois representation $\mathcal{F}_{\bar{\eta}}$ of $G_{K_x}$ is the pull-back of the restriction of $M$ to $D$ by the surjection $G_{K_x} \to D$ defined by the lifting.

Assume that the residue field $F = k(x)$ of the discrete valuation field $K_x$ is perfect, and that $\Lambda$ is a finite field of characteristic different from that of $F$. Then the residue field $E$ of the finite Galois extension $L$ corresponding to $D = \operatorname{Gal}(L/K_x)$ is a separable extension of $F$, and consequently the Swan conductor $\operatorname{Sw} M$ is defined as an integer for the representation of $D$. Further, by Proposition 6.4.1(1), the Swan conductor $\operatorname{Sw} M$ is independent of the choice of $V$. Thus, the Swan conductor $n = \operatorname{Sw}_x \mathcal{F} \in \mathbf{N}$ is defined.

Let $k$ be an algebraically closed field, and let $U$ be a separated scheme of finite type over $k$. Let $\Lambda$ be a finite field of characteristic invertible in $k$, and let $\mathcal{F}$ be a locally constant constructible sheaf of $\Lambda$-modules on $U$. The compact support cohomology $\mathrm{H}^q_c(U, \mathcal{F})$ is defined as $\mathrm{H}^q(X, j_!\mathcal{F})$ using an open immersion $j\colon U \to X$ to a proper scheme $X$ over $k$. It is known to be independent of the choice of open immersion $j$, to be finite, and to vanish unless $0 \leqslant q \leqslant 2 \dim U$. The Euler–Poincaré characteristic $\chi_c(U, \mathcal{F})$ is defined as the alternating sum $\sum_{q=0}^{2\dim U}(-1)^q \dim_\Lambda \mathrm{H}_c^q(U,\mathcal{F})$. For the constant coefficient, $\chi_c(U,\Lambda)$ is denoted by $\chi_c(U)$.

**Theorem 8.2.3** (Grothendieck–Ogg–Shafarevich formula) *Let $U$ be a smooth connected curve over an algebraically closed field $k$, and let $X$ be a smooth compactification of $U$. Let $\Lambda$ be a finite field of characteristic invertible in $k$, and let $\mathcal{F}$ be a locally constant constructible sheaf of $\Lambda$-modules on $U$. Then*

$$\chi_c(U,\mathcal{F}) = \chi_c(U)\cdot \operatorname{rank}\mathcal{F} - \sum_{x\in X-U} \mathrm{Sw}_x\mathcal{F}. \tag{8.2.6}$$

*Proof* Let $V \to U$ be a $G$-torsor for a finite group $G$ trivializing $\mathcal{F}$, and let $M$ be the corresponding $\Lambda[G]$-module. By Lemma 8.2.2(1), the complex $R\Gamma_c(V,\Lambda)$ is a perfect complex of $\Lambda[G]$-modules. Let $G_{\mathrm{reg}} \subset G$ be the set of $\ell$-regular elements for $\ell = \operatorname{char}\Lambda$. Since $R\Gamma_c(U,\mathcal{F}) = (M\otimes R\Gamma_c(V,\Lambda))^G$, we have

$$\dim R\Gamma_c(U,\mathcal{F}) = \frac{1}{\#G}\sum_{\sigma\in G_{\mathrm{reg}}} \mathrm{Tr}(\sigma : R\Gamma_c(V,\mathbf{Z}_\ell))\cdot \mathrm{Tr}^{\mathrm{Br}}(\sigma : M)$$

by Proposition 6.1.5(2). By Lemma 8.2.2(2), the right-hand side equals the right-hand side of (8.2.6). □

**Exercise 8.2** Let $k$ be an algebraically closed field of characteristic $p > 0$, and $n \geqslant 1$ be an integer prime to $p$. Compute the genus $g$ of the proper smooth curve $X$ over $k$ defined by $u^p - u = t^n$ by applying the Grothendieck–Ogg–Shafarevich formula.

## Historical Notes

The Lefschetz trace formula (8.1.11) for endomorphisms of curves is called the Weil formula in [34, 5]. The genesis of the Grothendieck–Ogg–Shafarevich formula (8.2.6) is traced in the exchange of letters of Grothendieck dated “10.04.63” and of Serre dated “April 63” recorded in [18]. The Grothendieck–Ogg–Shafarevich formula was explained by Raynaud at the Bourbaki seminar [66], and also in [34].

# 9

# Reduced Fiber Theorem

For finite extensions of henselian discrete valuation fields, those without ramifications are unramified extensions. We study infinite extensions without ramification defined as formally smooth extensions in Section 9.1.

We prove in Section 9.2 that extensions of henselian discrete valuation fields become formally smooth after a suitable finite extension of the base field, under some assumption on the residue field extension. The crucial case is proved by applying the study of ramification of cyclic extensions of degree $p$ in Chapter 3.

For a flat scheme of finite type with smooth generic fiber over a henselian discrete valuation ring with perfect field, we show in Section 9.3 that, after a suitable finite separable extension of the base field, the geometric closed fiber of the normalization is reduced.

## 9.1 Formally Smooth Extensions

**Definition 9.1.1** Let $K_0 \to K$ be a morphism of discrete valuation fields. We say that $K$ is a formally smooth extension of $K_0$ if the residue field $F$ of $K$ is a separable extension of the residue field $k$ of $K_0$, and if the ramification index $e_{K/K_0}$ is 1.

If $F$ is a perfect field of characteristic $p > 0$, the fraction field $K$ of the ring $W(F)$ of Witt vectors is a formally smooth extension of $\mathbf{Q}_p$.

**Proposition 9.1.2** *Let $K_0$ be a discrete valuation field, and let $k$ be the residue field of $K_0$. Let $K$ be a formally smooth extension of $K_0$ with residue field $F$, and let $L$ be another extension of $K_0$ with residue field $E$.*

(1) *If $F \to E$ is a morphism over $k$, and if $L$ is complete, then there exists a morphism $K \to L$ over $K_0$ inducing $F \to E$.*

(2) *Let $K \to L$ be a morphism over $K_0$, and assume that $L$ is also formally smooth over $K_0$ and that $K$ and $L$ are complete. Then $K \to L$ is an isomorphism if and only if $F \to E$ is an isomorphism.*

*Proof* (1) By [31, Théorèmes (19.6.1), (19.7.1)], the condition that $K$ is a formally smooth extension of $K_0$ means that $\mathcal{O}_{K_0} \to \mathcal{O}_K$ is formally smooth. Since $\mathcal{O}_L$ is complete, the assertion follows from [31, Proposition (19.3.10)].

(2) It suffices to show the if part. Since the ramification index $e_{L/K}$ is 1, if $F \to E$ is an isomorphism, the morphism $K \to L$ of complete discrete valuation fields is an isomorphism. □

**Corollary 9.1.3** *Let $K$ be a henselian discrete valuation field with perfect residue field $k$, and $\widehat{K}$ be the completion. Then the functor*

$$\text{(finite extensions of } K) \to \text{(finite extensions of } \widehat{K}) \tag{9.1.1}$$

*defined by completion is essentially surjective.*

*Proof* By Proposition 1.4.8, it remains to consider the case where $K$ is of characteristic $p > 0$. Let $\widehat{L}$ be a finite extension of $\widehat{K}$. Then, by Proposition 1.4.8, there exists a finite separable extension $K'$ of $K$ such that $\widehat{K}'$ is the separable closure of $\widehat{K}$ in $\widehat{L}$. Let $[\widehat{L} : \widehat{K}'] = p^e$, $k'$ be the residue field of $K'$, and $t \in K'$ be a uniformizer. Then we have an isomorphism $k'((t)) \to \widehat{K}'$ by Proposition 9.1.2. Hence, $[\widehat{K}' : \widehat{K}'^p] = p$ and if we set $L = K'(t^{1/p^e})$, its completion is the unique purely inseparable extension $\widehat{L}$ of degree $p^e$. □

One can construct formally smooth extensions as inductive limits of essentially smooth extensions.

**Definition 9.1.4** Let $K_0 \to K$ be a morphism of henselian discrete valuation fields.

(1) We say that $\mathcal{O}_K$ is essentially of finite type over $\mathcal{O}_{K_0}$ if there exist a scheme $X$ of finite type over $\mathcal{O}_{K_0}$, a point $x \in X$ of the closed fiber, and an isomorphism $\mathcal{O}^{\mathrm{h}}_{X,x} \to \mathcal{O}_K$ of henselian local rings over $\mathcal{O}_{K_0}$.
(2) We say that $\mathcal{O}_K$ is essentially smooth over $\mathcal{O}_{K_0}$ if there exist a smooth scheme $X$ over $\mathcal{O}_{K_0}$, a point $x \in X$ of the closed fiber, and an isomorphism $\mathcal{O}^{\mathrm{h}}_{X,x} \to \mathcal{O}_K$ of henselian local rings over $\mathcal{O}_{K_0}$.

**Proposition 9.1.5** *Let $K_0$ be a henselian discrete valuation field, and let $F$ be a separable extension of the residue field $k$. Then there exists an inductive system $(\mathcal{O}_\lambda)_{\lambda \in \Lambda}$ of essentially smooth extensions of henselian discrete valuation rings of $\mathcal{O}_{K_0}$ such that the residue field of $\varinjlim_{\lambda \in \Lambda} \mathcal{O}_\lambda$ is $F$.*

*Proof* The construction is similar to that in [30, (10.3.1), Chapitre 0]. Let

$(x_i)_{i \in I}$ be a transcendental basis of $F$ over $k$, and let $E_0 \subset F$ be the separable closure of the subfield $F_I = k(x_i; i \in I)$. We have that $F = E_0$ if $k$ is of characteristic 0. If $k$ is of characteristic $p > 0$, we define an increasing sequence of subfields $E_n \subset F$ by $E_n = \{x \in F \mid x^{p^n} \in E_0\}$. We have $F = \bigcup_n E_n$ and $E_n^p \subset E_{n-1}$.

Let $K_I$ be the fraction field of the henselization $O_{K_I}$ of the localization of the polynomial ring $O_{K_0}[x_i; i \in I]$ at the prime ideal $\mathfrak{m}_{K_0} O_{K_0}[x_i; i \in I]$. Then the residue field of $K_I$ is $F_I$. Let $L_0$ be an unramified extension of $K_I$ with residue field $E_0 \supset F_I$. We have that $K = L_0$ if $k$ is of characteristic 0. In the following, we assume that $k$ is of characteristic $p > 0$, as the other case is similar and easier. Since $E_n^p \subset E_{n-1}$, there exists a family $(u_{n,i})_{i \in I_n}$ of elements of $E_{n-1}$ such that $du_{n,i} \in \Omega^1_{E_{n-1}}$, $i \in I_n$ is linearly independent, and that $E_n$ is generated by the $p$th roots of $u_{n,i}$, $i \in I_n$ over $E_{n-1}$. Take inductively liftings of $u_{n,i}$, $i \in I_n$, in $O_{L_{n-1}}$, also denoted abusively by $u_{n,i}$, $i \in I_n$, and define $O_{L_n} = O_{L_{n-1}}[X_i,\ i \in I_n]/(X_i^p - u_{n,i},\ i \in I_n)$. Then $(O_{L_n})_{n \in \mathbf{N}}$ is an inductive system of henselian discrete valuation rings, and the residue field of $K = \varinjlim_n L_n$ is $F = \varinjlim_n E_n$.

For a finite subset $J \subset I$, let $F_J = k(x_i,\ i \in J) \subset F_I$ and $K_J \subset K_I$ be the corresponding subextension. For a finite separable extension $E_0' \subset E_0$ of $E_J$, let $L_0' \subset L_0$ be the corresponding unramified extension. For a sequence $J_j \subset I_j$, $j = 1, \ldots, m$, of finite subsets, if subextensions $E_n' = E_{n-1}'(u_{n,i}^{1/p}, i \in J_n) \subset E_n$ for $n = 1, \ldots, m$ are defined inductively, let $L_n' \subset L_n$ denote the corresponding extensions. Then $K$ is the inductive limit of these $L_m'$.

By the construction, for each $L_m'$ there exist a flat ring $A$ of finite type over $O_{K_0}$, a prime ideal $\mathfrak{p}$ of $A$, and an isomorphism $A_{\mathfrak{p}}^{\mathrm{h}} \to O_{L_m'}$. Since the residue field $E_m'$ is a subfield of a separable extension $F$ of $k$, the extension $E_m'$ is itself a separable extension of $k$ and hence $A$ is smooth over $O_{K_0}$ at $\mathfrak{p}$. Namely, $O_{L_m'}$ is essentially smooth over $O_{K_0}$ as required. □

**Corollary 9.1.6** *Let $K$ be a complete discrete valuation field with residue field $F$ of characteristic $p > 0$. Let $k \subset F$ be a perfect subfield.*

(1) *Assume that $K$ is of characteristic 0. Let $O_{K_0} = W(k) \subset O_K$ be the ring of Witt vectors, and let $K_0 \to K$ be a morphism of discrete valuation fields. Then there exist an inductive system $(O_{K_\lambda})_{\lambda \in \Lambda}$ of discrete valuation rings, essentially of finite type over $O_{K_0}$, and an isomorphism from the completion of $\varinjlim_{\lambda \in \Lambda} O_{K_\lambda}$ to $O_K$ over $O_{K_0}$.*
(2) *Assume that $K$ is of characteristic $p > 0$, and let $t \in K$ be a uniformizer. Then there exists an isomorphism $F((t)) \to K$ of complete discrete valuation fields. Further, there exist an inductive system $(O_{K_\lambda})_{\lambda \in \Lambda}$ of discrete valuation rings, essentially smooth over the henselization $O_{K_0} = k[t]^{\mathrm{h}}_{(t)}$ of*

*the local ring of the affine line* $\mathbf{A}^1_k$ *at the origin, and an isomorphism from the completion of* $\varinjlim_{\lambda \in \Lambda} O_{K_\lambda}$ *to* $O_K$ *over* $O_{K_0}$.

*Proof* (1) Let $O_{K_1} = \varinjlim_{\lambda \in \Lambda} A_\lambda$ be a formally smooth extension of $K_0$ with residue field $F$, and $(A_\lambda)_{\lambda \in \Lambda}$ be an inductive system of henselian discrete valuation rings, essentially smooth over $O_{K_0}$ as in Proposition 9.1.5. Since $K$ is a finite separable extension of the completion $\widehat{K}_1$ by Proposition 9.1.2, there exist a finite separable extension $L_1$ of $K_1$ and an isomorphism $\widehat{L}_1 \to K$ by Proposition 1.4.8. Since $O_{L_1}$ is of finite presentation over $O_{K_1}$, there exist $\lambda \in \Lambda$, a finite separable extension $K_\lambda$ of the fraction field of $A_\lambda$, and an isomorphism $O_{K_\lambda} \otimes_{A_\lambda} O_{K_1} \to O_K$. Hence, the assertion follows.

(2) Since $F$ is formally smooth over $k$, there exists a morphism $F \to O_K$ such that the composition with $O_K \to F$ is an isomorphism. This induces an isomorphism $F((t)) \to K$.

Let $O_{K_1} = \varinjlim_{\lambda \in \Lambda} O_{K_\lambda}$ be a formally smooth extension of $K_0$ with residue field $F$, and $(O_{K_\lambda})_{\lambda \in \Lambda}$ be an inductive system of henselian discrete valuation rings, essentially smooth over $O_{K_0}$ as in Proposition 9.1.5. Then, since $O_{K_1}$ is formally smooth over $O_{K_0}$ and $K$ is complete, we have a morphism $K_1 \to K$ over $K_0$ inducing an isomorphism $\widehat{K}_1 \to K$ by Proposition 9.1.2. □

**Exercise 9.1** Let $K_0$ be a discrete valuation field and $t$ be a uniformizer. Let $m, n \geqslant 1$ be integers, and $L$ be the fraction field of the local ring $O_L = A_{\mathfrak{p}}$ of $A = O_{K_0}[x, y]/(x^m y^n - t)$ at the prime ideal $\mathfrak{p} = xA$. Find a formally smooth extension $K$ over $K_0$ and a finite morphism $K \to L$ over $K_0$.

## 9.2 Eliminating the Ramification Index

The goal of this section is to prove the following theorem asserting that an extension of henselian discrete valuation fields is potentially formally smooth under a certain mild assumption on the extension of the residue fields.

**Theorem 9.2.1** *Let* $K_0$ *be a henselian discrete valuation field with perfect residue field* $k$ *of characteristic* $p > 0$. *Let* $L$ *be an extension of the henselian discrete valuation field* $K_0$ *such that the largest perfect subfield* $k_E = \bigcap_{e \geqslant 0} E^{p^e}$ *of the residue field* $E$ *is a finite extension of* $k$.

(1) *There exists a finite extension* $K'_0$ *of* $K_0$ *such that* $L' = LK'_0$ *is formally smooth over* $K'_0$.
(2) *Assume further that* $L$ *is a separable extension of* $K_0$. *Then there exists a finite separable extension* $K'_0$ *of* $K_0$ *such that* $L' = LK'_0$ *is formally smooth over* $K'_0$.

This theorem will be used to prove the reduced fiber theorem in the next section.

First, we study the assumption on the residue fields. For a field $F$ of characteristic $p > 0$, let $k_F = \bigcap_{e \geqslant 0} F^{p^e} \subset F$ denote the largest perfect subfield.

**Lemma 9.2.2** *Let $E$ be a finite extension of a field $F$ of characteristic $p > 0$. Then $k_E$ is a finite extension of $k_F$.*

*Proof* Let $F' \subset E$ be the separable closure of $F$ in $E$. Since $E$ is a finite extension of $F'$, there exists an integer $e \geqslant 0$ such that $E^{p^e} \subset F'$. Hence, the inclusion $k_{F'} \subset k_E$ is an equality. Thus, it suffices to prove the assertion assuming that $E$ is a finite separable extension of $F$. Since $E$ is a separable extension of $F$, the tensor product $E \otimes_F F$ with respect to the absolute Frobenius morphism $\mathrm{F}\colon F \to F$ is reduced and has a unique prime ideal. Hence, $E \otimes_F F$ is an extension of $F$ of degree $[E : F]$, and $\mathrm{F}\colon E \to E$ induces an isomorphism $E \otimes_F F \to E$.

Let $x \in k_E$ be any element and $f \in F[X]$ be the minimal polynomial of $x$. For an integer $e \geqslant 0$, let $x = y_e^{p^e}$ for $y_e \in E$, and $g_e \in F[X]$ be the minimal polynomial of $y_e$. Since $F(y_e)$ is a purely inseparable extension of $F(x)$, we have $F(y_e) = F(x)$ and $f = \mathrm{F}^e g_e \in F^{p^e}[X]$. Hence, we have $f \in k_F[X]$ and $k_E$ is an algebraic extension of $k_F$. Further, for any element $x \in k_E$, we have $[k_F(x) : k_F] \leqslant [E : F]$. Since an algebraic extension $k_E$ of a perfect field $k_F$ is separable, we conclude that $[k_E : k_F] \leqslant [E : F]$. □

In the following three lemmas, we study the critical cases where the extension is a cyclic extension of degree $p$ of a formally smooth extension. We begin with the case where the fields are of characteristic $p > 0$.

**Lemma 9.2.3** *Let $F$ be a field of characteristic $p > 0$, and $k$ be the largest perfect subfield $k_F = \bigcap_{e \geqslant 0} F^{p^e}$. Let $K_0 = k((t)) \subset K = F((t))$ be the fields of Laurent power series. Let*

$$a = \sum_{n \in I} c_n t^{-n} + b \tag{9.2.1}$$

*be an element of $K$ where $I \subset \{n \in \mathbf{N} \mid n \neq 0\}$ is a finite set of indices, $c_n \in F^\times$ for any $n \in I$, and $b \in O_K = F[[t]]$. Assume that, for each $n \in I$, we have either $p \nmid n$ or $c_n \in F - F^p$. Set $I_\infty = \{n \in I \mid c_n \in k\}$ and, for $n \in J = I - I_\infty$, let $r(n)$ be the largest power of $p$ such that $c_n \in F^{r(n)}$. Let $L$ be a cyclic extension of degree $p$ defined by the Artin–Schreier equation $x^p - x = a$.*

(1) *If $I = I_\infty$, let $K_0'$ be the Artin–Schreier extension of $K_0$ defined by $x^p - x = \sum_{n \in I} c_n t^{-n}$. Then $L' = LK_0'$ is a formally smooth extension of $K_0'$.*
(2) *The rational numbers $n/r(n)$ for $n \in J$ are distinct from each other.*

(3) *Assume that $J$ is not empty. Let $R$ be a power of $p$ divisible by $p \cdot r(n)$ for any $n \in J$ such that* $\max_{n\in J} nR/r(n) > \max_{n\in I_\infty} n$, *and let* $K_0' = k((t^{1/R}))$ *be the purely inseparable extension of $K_0$ of degree $R$. Then $L' = LK_0'$ is a formally smooth extension of $K_0'$.*

*Proof* (1) Since $K$ is a formally smooth extension of $K_0$, the composition $K' = KK_0'$ is a formally smooth extension of $K_0'$. Since the Artin–Schreier extension $L'$ of $K'$ is defined by $x^p - x = b$ and is an unramified extension by Proposition 3.1.2(2), the assertion follows.

(2) Since we have either $p \nmid n$ or $c_n \in F - F^p$, we have either $p \nmid n$ or $r(n) = 1$. Consequently, we have $n = n/r(n)$ if $n/r(n)$ is an integer and $n$ is the numerator, or $n/r(n)$ otherwise. Hence, the assertion follows.

(3) Let $u \in K_0'$ be the $R$th root of $t$. The Artin–Schreier extension $L'$ of $K' = KK_0'$ is defined by $x^p - x = a'$ for

$$a' = \sum_{n\in J} c_n^{1/r(n)} u^{-nR/r(n)} + \sum_{n\in I_\infty} c_n^{1/R} u^{-n} + b.$$

By the definition of $r(n)$, for each $n \in J$ we have $c_n^{1/r(n)} \in F - F^p$. Since $nR/r(n)$ for $n \in J$ are divisible by $p$ and are distinct from each other by 2, by the assumption $\max_{n\in J} nR/r(n) > \max_{n\in I_\infty} n$ and Proposition 3.1.2(2), the ramification index $e_{L'/K'}$ is 1. Hence, $L'$ is formally smooth over $K_0'$. □

Next, we study fields of characteristic 0.

**Lemma 9.2.4** *Let $K$ be a henselian discrete valuation field of characteristic* 0 *with residue field $F$ of characteristic $p > 0$, and assume that $K$ contains a primitive $p$th root of* 1. *Let $t$ be a uniformizer of $K$, and let $K'$ be the totally ramified Kummer extension of $K$ defined by $x^p = t$. Let $a \in K^\times$ and $L$ be the Kummer extension of $K$ defined by $x^p = a$. Then, for $L' = LK'$, either we have $e_{L'/K'} = 1$ or $L'$ is the Kummer extension of $K'$ defined by $x^p = a'$ for $a' \in 1 + pO_{K'}$.*

*Proof* Let $k$ be the largest perfect subfield $k_F = \bigcap_{n\geqslant 0} F^{p^n}$, and let $e = \operatorname{ord}_K p$ be the absolute ramification index. Since $F$ is a separable extension of a perfect field $k$, the surjection $O_K/pO_K \to F$ admits a section $F \to O_K/pO_K$ over $k$ by [31, Théorème (19.6.1)]. We fix such a section and identify $O_K/pO_K$ with $F[t]/(t^e)$. Since $K^\times/(1 + pO_K)$ is an extension of $\mathbf{Z}$ by $(O_K/pO_K)^\times = (F[t]/(t^e))^\times$, there exist an integer $l$, $b \in F^\times$, and $c_i \in F$ satisfying

$$a \equiv t^l \cdot b \cdot \prod_{i=1}^{e-1} (1 + c_i t^i) \mod 1 + pO_K. \tag{9.2.2}$$

Let $s \in K_0'$ be a $p$th root of $t$. Then, the right-hand side of (9.2.2) is $s^{pl} \cdot$

$b \cdot \prod_{i=1}^{e-1}(1 + c_i s^{pi})$ and if $c_i \in F^p$ then $1 + c_i s^{pi}$ is a $p$th power. Hence, if $b \in F^\times - F^{\times p}$ or if $c_i \in F - F^p$ for some $1 \leqslant i < e$, we have $e_{L'/K'} = 1$ by Proposition 3.2.2(2). Otherwise, the right-hand side of (9.2.2) is a $p$th power in $(O_{K'}/pO_{K'})^\times = (F[s]/(s^{pe}))^\times$. □

**Lemma 9.2.5** *Let $\zeta_p \in K$ and $F$ be as in Lemma 9.2.4, and let $t \in K$ be a uniformizer. Assume that the absolute ramification index $e = \operatorname{ord}_K p$ is divisible by $p$. Let $E = F((t))$ be the fields of Laurent power series of characteristic $p > 0$. We fix a section $F \to O_K/pO_K$ over the largest perfect subfield $k = k_F = \bigcap_{n\geqslant 0} F^{p^n} \subset F$ of the canonical surjection $O_K/pO_K \to F$, and extend it to an isomorphism $F[t]/(t^e) = O_E/\mathfrak{m}_E^e \to O_K/pO_K$ sending $t$ to $t$.*

(1) *We have a commutative diagram*

$$\begin{array}{ccc} \mathfrak{m}_E^{-e/p(p-1)}/O_E & \longrightarrow & (1 + t^{e/p}O_K)/(1 + (\zeta_p - 1)O_K) \\ \Big\downarrow{\scriptstyle F-1} & & \Big\downarrow{\scriptstyle p\text{th power}} \\ \mathfrak{m}_E^{-e/(p-1)}/O_E & \longrightarrow & (1 + pO_K)/(1 + (\zeta_p - 1)^p O_K) \end{array} \tag{9.2.3}$$

*where the horizontal arrows are the isomorphisms sending $x \in t^{-e/p(p-1)} \cdot O_E/\mathfrak{m}_E^{e/p(p-1)} = t^{-e/p(p-1)} \cdot O_K/\mathfrak{m}_K^{e/p(p-1)}$ to $1 + x(\zeta_p - 1)$ and $y \in t^{-e/(p-1)} \cdot O_E/\mathfrak{m}_E^{e/(p-1)} = t^{-e/(p-1)} \cdot O_K/\mathfrak{m}_K^{e/(p-1)}$ to $1 + y(\zeta_p - 1)^p$ respectively.*

*Let*

$$\sum_{n\in I} c_n t^{-n} \tag{9.2.4}$$

*be an element of $\mathfrak{m}_E^{-e/(p-1)}/O_E$, where $I \subset \{n \in \mathbf{N} \mid 1 \leqslant n \leqslant e/(p-1)\}$ is a set of indices and $c_n \in F^\times$ for any $n \in I$. Assume that, for each $n \in I$, we have either $p \nmid n$ or $c_n \in F - F^p$. Set $I_\infty = \{n \in I \mid c_n \in k\}$ and, for $n \in J = I - I_\infty$, let $r(n)$ be the largest power of $p$ such that $c_n \in F^{r(n)}$.*

*Let $a \in 1 + pO_K \subset K^\times$ be an element congruent to the image by the lower horizontal arrow in (9.2.3) of (9.2.4) modulo $1 + (\zeta_p - 1)^p O_K$. Let $L$ be a cyclic extension of degree $p$ defined by the Kummer equation $x^p = a$.*

(2) *Assume $I = I_\infty \neq \varnothing$, and let $a' \in 1 + pO_K \subset K^\times$ be an element congruent modulo $1 + (\zeta_p - 1)^p O_K$ to the image of (9.2.4) by the lower horizontal arrow in (9.2.3); let $K'$ be the Kummer extension of $K$ defined by $x^p = a'$. Then $K'$ is a totally ramified extension of $K$, and $L' = LK'$ is an unramified extension of $K'$.*
(3) *Assume that $J$ is not empty. Then $n/r(n)$ for $n \in J$ are distinct from each other.*

(4) *Let $R$ be a power of $p$ divisible by $p \cdot r(n)$ for any $n \in J$ such that $\max_{n\in J} nR/r(n) > \max_{n\in I_\infty} n$, and let $K' = K(t^{1/R})$ be a totally ramified extension of $K$ defined by joining an $R$th root of the uniformizer $t$. Then, for $L' = LK'$, we have $e_{L'/K'} = 1$.*

*Proof* (1) Since $2e/p \geqslant e/(p-1)$ and $2e \geqslant ep/(p-1)$, the horizontal arrows are isomorphisms. Since the minimal polynomial of $\zeta_p - 1$ is $((X+1)^p - 1)/X \equiv X^{p-1} + p \bmod pX$, we have $(\zeta_p - 1)^{p-1} \equiv -p \bmod (\zeta_p - 1)^p$. Since $2/p \geqslant 1/(p-1)$, for $x \in \mathfrak{m}_K^{-e/p(p-1)}$, we have $(1 + x(\zeta_p - 1))^p \equiv 1 + px(\zeta_p - 1) + x^p(\zeta_p - 1)^p \equiv 1 + (x^p - x)(\zeta_p - 1)^p \bmod (\zeta_p - 1)^p$.

(2) We have $p \nmid n$ for any $n \in I = I_\infty \neq \emptyset$. Hence, $K'$ is a totally ramified extension of $K$ by Proposition 3.2.2(1). Since the Kummer extension $L'$ of $K'$ is defined by $x^p = 1 + b$ for $b \in (\zeta_p - 1)^p O_K$, it is an unramified extension by Proposition 3.2.2(2).

(3) This is the same as Lemma 9.2.3(2).

(4) Using (1), this is also proved in the same way as Lemma 9.2.3(3). □

As the end of preliminaries, we prove a lemma to reduce Theorem 9.2.1(2) for separable extensions to Theorem 9.2.1(1) for general extensions.

**Lemma 9.2.6** *Let $K$ be a henselian discrete valuation field of characteristic $p > 0$, and let $K'$ be a finite extension.*

(1) *Let $N \geqslant 1$ be an integer. Then there exists a finite separable extension $K_1$ of $K$ such that, for $K_1' = K'K_1$, we have $O_{K'} + \mathfrak{m}_K^N O_{K_1'} \subset O_{K_1} + \mathfrak{m}_K^N O_{K_1'}$.*

(2) *Let $A$ be a flat ring over $O_K$, and set $A_K = A \otimes_{O_K} K$. Let $A' \subset A_{K'} = A_K \otimes_K K'$ be a subring containing $A_{O_{K'}} = A \otimes_{O_K} O_{K'}$ as a subring. Assume that $A'$ is finite over $A_{O_{K'}}$, and let $N \geqslant 1$ be an integer satisfying $\mathfrak{m}_K^N A' \subset A_{O_{K'}}$. Let $K_1$ be a finite separable extension of $K$ satisfying the condition in (1). Then there exists a subring $A_1 \subset A_{K_1} = A_K \otimes_K K_1$, finite over $A_{O_{K_1}} = A \otimes_{O_K} O_{K_1}$, satisfying an equality $A_1 \otimes_{O_{K_1}} O_{K_1'} = A' \otimes_{O_{K'}} O_{K_1'}$ of subrings of $A_{K_1'} = A_K \otimes_K K_1'$ for $K_1' = K'K_1$.*

*Proof* (1) Let $M \subset K'$ be the separable closure of $K$ in $K'$, and let $x_1, \dots, x_n \in O_{K'}$ be liftings of a system of generators of the $O_M$-module $O_{K'}/\mathfrak{m}_K^N O_{K'}$ of finite type. The inseparable degree $q = [K' : M]$ is a power of $p$. Set $a_i = x_i^q \in O_M$ for $i = 1, \dots, n$. Let $t \in \mathfrak{m}_K$ be a uniformizer, and let $K_1$ be the extension of $M$ generated by solutions $y_1, \dots, y_n$ of the separable equations $y^q + t^{qN} y = a_i$. Then, since $(x_i - y_i)^q = t^{qN} y_i$, the elements $z_i = (x_i - y_i)/t^N \in K_1' = K'K_1$ satisfy $z_i^q = y_i \in O_{K_1}$ and hence are contained in $O_{K_1'}$. Thus, we have $x_i = y_i + t^N z_i \in O_{K_1} + \mathfrak{m}_K^N O_{K_1'}$ and $O_{K'} \subset O_{K_1} + \mathfrak{m}_K^N O_{K_1'}$.

(2) Let $x_1, \dots, x_n \in A'$ be a system of generators of an $A_{O_{K'}}$-module $A'$. Let

$t$ be a uniformizer of $K$, and write $t^N x_i = \sum_j s_{ij} a_{ij}$, $s_{ij} \in O_{K'}$, $a_{ij} \in A$. Let $t_{ij} \in O_{K_1}$ be elements such that $s_{ij} - t_{ij} \in t^N O_{K'_1}$, and set $y_i = t^{-N} \sum_j t_{ij} a_{ij} \in A_{K_1}$. Then, since $x_i - y_i = \sum_j t^{-N}(s_{ij} - t_{ij}) a_{ij} \in A_{O_{K'_1}} = A \otimes_{O_K} O_{K'_1}$, we have $y_i = (y_i - x_i) + x_i \in A'_{O_{K'_1}}$. Hence, $y_i$ are integral over $A_{O_{K_1}}$ and $A_1 = A_{O_{K_1}}[y_1, \ldots, y_n] \subset A_{K_1}$ is finite over $A_{O_{K_1}}$. Since $A_1$ is flat over $O_{K_1}$, the tensor product $A_1 \otimes_{O_{K_1}} O_{K'_1}$ is a subring of $A_{K'_1}$ generated by $y_1, \ldots, y_n$ over $A_{O_{K'_1}}$. Hence, this is equal to the subring $A' \otimes_{O_{K'}} O_{K'_1} \subset A_{K'_1}$ generated by $x_1, \ldots, x_n$ over $A_{O_{K'_1}}$. □

*Proof of Theorem* 9.2.1 (1) First, we show that the assertion is reduced to the case where $K_0$ and $L$ are complete. Let $\widehat{K}_0$ and $\widehat{L}$ be the completions, and let $\widehat{K}'_0$ be a finite extension of $\widehat{K}_0$ such that $\widehat{L}' = \widehat{L}\widehat{K}'_0$ is formally smooth over $\widehat{K}'_0$. Then, by Corollary 9.1.3, there exists a finite extension $K'_0$ of $K_0$ such that $\widehat{K}'_0 = \widehat{K}_0 K'_0$. Since $\widehat{L}' = \widehat{L}\widehat{K}'_0$ is the completion of $L' = LK'_0$, the extension $K'_0$ satisfies the required condition.

In the following, we assume that $K_0$ and $L$ are complete. There exists a formally smooth extension $K$ of $K_0$ with residue field $E$ by Proposition 9.1.5. Then, by Proposition 9.1.2, $L$ is a finite extension of $K$. By replacing $L$ by a normal closure over $K$, we may assume that $L$ is a normal extension over $K$. Let $K \subset M_t \subset M_s \subset L$ be the maximal tamely ramified extension and the separable closure of $K$ in $L$. Let $e = e_{M_t/K}$ be the ramification index, and let $K_1$ be a totally ramified extension of $K$ of degree $e$. Then $M_t K_1$ is an unramified extension of $M_t$ by Proposition 1.5.5(1) and hence is a formally smooth extension of $K_1$. Thus, by Lemma 9.2.2, by replacing $K_0$ by $K_1$, we may assume $K = M_t$.

Since the Galois group $\mathrm{Gal}(M_s/M_t)$ is a $p$-group, the Galois extension $M_s$ of $K = M_t$ is a successive extension of cyclic extensions of degree $p$ by Lemma 1.5.12(1). Further, the inseparable extension $L$ of $M_s$ is a successive extension of purely inseparable extensions of degree $p$. Hence, by induction on $[L : K]$ and by Lemma 9.2.2, it suffices to prove the assertion in the following two cases:

(i) $L$ is a cyclic extension of $K$ of degree $p$.

(ii) $L$ is a purely inseparable extension of $K$ of degree $p$.

We show case (i). First assume that the characteristic of $K_0$ is $p > 0$. Then a cyclic extension $L$ of $K$ of degree $p$ is an Artin–Schreier extension. We may assume that $K_0 = k((t))$ and $K = F((t))$. By Lemma 3.1.2(1), we may assume that $L$ is defined by an equation $x^p - x = a$, where $a \in K = F((t))$ satisfies the condition as in Lemma 9.2.3. Hence, the assertion follows from Lemma 9.2.3 in this case.

Next, we consider the case where $K_0$ is of characteristic 0. By joining a primitive $p$th root $\zeta_p$ of 1 to $K_0$, we may assume that $K_0$ contains $\zeta_p$. Then, a cyclic extension $L$ of $K$ of degree $p$ is a Kummer extension defined by $x^p = a$. In Lemma 9.2.4, we may take a uniformizer $t$ in $K_0$ and then we have $K' = KK'_0$ for the Kummer extension $K'_0$ of $K_0$ defined by $x^p = t$. Since $K'$ is formally smooth over $K'_0$, by Lemma 9.2.4, we may assume that $a \in 1 + p\mathcal{O}_K$ and that the absolute ramification index $e = \operatorname{ord}_{K_0} p$ is divisible by $p$.

Assume $a \in 1 + p\mathcal{O}_K$. By Lemma 3.2.1, we may assume that $a$ is the image of an element (9.2.4) satisfying the condition as in Lemma 9.2.5. Since we may take a uniformizer $t$ in $K_0$ and an element $a'$ in Lemma 9.2.5(2) from $1 + p\mathcal{O}_{K_0}$, the assertion follows from Lemma 9.2.5 in this case.

Finally, consider case (ii). In this case, $K$ is of characteristic $p > 0$ and we have $L \subset K^{1/p}$. Since $K^{1/p}$ is formally smooth over $K_0^{1/p}$, the purely inseparable extension $K_0^{1/p}$ of $K_0$ of degree $p$ satisfies the condition.

(2) First, we show that the assertion is reduced to the case where $K_0$ and $L$ are complete. The argument is the same as in the first paragraph of the proof of (1), except that $\widehat{K}'_0$ is a finite separable extension of $\widehat{K}_0$ and that we apply Proposition 1.4.8 to find a finite separable extension $K'_0$.

We assume that $K_0$ and $L$ are complete. Let $K''_0$ be a finite, not necessarily separable, extension of $K_0$ such that $e_{L''/K''_0} = 1$ for $L'' = LK''_0$. By replacing $K_0$ by its separable closure in $K''_0$, we may assume that $K''_0$ is a purely inseparable extension of $K_0$ and hence $LK''_0 = L \otimes_{K_0} K''_0$. We apply Lemma 9.2.6(2) to $A = \mathcal{O}_L$ and to the integral closure $A''$ of $A$ in $A \otimes_{\mathcal{O}_{K_0}} \mathcal{O}_{K''_0}$. Since $L$ is assumed to be complete, $A''$ is finite over $A \otimes_{\mathcal{O}_{K_0}} \mathcal{O}_{K''_0}$ by Corollary 1.4.6(2). Hence, by Lemma 9.2.6(2), there exist a finite separable extension $K'_0$ of $K_0$ and a subring $A' \subset L \otimes_{K_0} K'_0$ satisfying the condition in that lemma. Then, since $A''$ is formally smooth over $\mathcal{O}_{K''_0}$, for $K'''_0 = K'_0 K''_0$, the tensor product $A'' \otimes_{\mathcal{O}_{K''_0}} \mathcal{O}_{K'''_0} = A' \otimes_{\mathcal{O}_{K'_0}} \mathcal{O}_{K'''_0}$ is also formally smooth over $\mathcal{O}_{K'''_0}$. Thus, $A'$ is formally smooth over $\mathcal{O}_{K'_0}$ and the assertion follows. □

**Proposition 9.2.7** *Let $K$ be a henselian discrete valuation field, and let $L$ be a finite separable extension of $K$. Then there exists a finite separable extension $K'$ of $K$ satisfying the following conditions:*

(1) *For $L' = LK'$, we have $e_{L'/K'} = 1$.*
(2) *The residue field $F'$ of $K'$ is a separable extension of the residue field $F$ of $K$.*

Corollary 9.2.8 which follows this variant of Theorem 9.2.1, will be used in the proof of the equality of the cohomological filtration with the upper numbering filtration for abelian extensions.

*Proof* First, we show the case where $K$ is of characteristic $p > 0$. We first assume that $K$ is complete. Let $F$ be the residue field of $K$, and let $k = k_F = \bigcap_{e\geqslant 0} F^{p^e} \subset F$ be the largest perfect subfield. Then we may assume that $K = F((t))$ is a formally smooth extension of $K_0 = k((t))$ by Proposition 9.1.2(1). Let $E$ be the residue field of $L$. Then, by Lemma 9.2.2, the largest perfect subfield $k_E = \bigcap_{e\geqslant 0} E^{p^e} \subset E$ is a finite extension of $k = k_F$. Hence, by Theorem 9.2.1(2) applied to $L$ over $K_0$, there exists a finite separable extension $K'_0$ such that $L' = LK'_0$ is formally smooth over $K'_0$. Then, $K' = KK'_0$ is a totally ramified extension of $K$ and the ramification index is $e_{L'/K'} = 1$.

We show the general case in char $K = p$. Let $\widehat{K}'$ be a finite separable extension of the completion $\widehat{K}$ satisfying the conditions (1) and (2) for the extension $\widehat{L}$ over $\widehat{K}$. By Proposition 1.4.8, there exists a finite separable extension $K'$ of $K$ such that $\widehat{K}' = \widehat{K}K'$. Then $K'$ satisfies the conditions for $L$ over $K$.

We show the case where $K$ is of characteristic 0. By replacing $L$ by a Galois closure, we may assume that $L$ is a Galois extension of $K$. By replacing $K$ by its maximal tamely ramified extension in $L$, we may assume that the Galois group $G$ equals the wild inertia subgroup. Hence, by Lemma 1.5.12(1) and by induction on the order of $G$, we may assume that $L$ is a wildly ramified cyclic extension of degree $p$. By replacing $K$ by a tamely ramified extension $K(\zeta_p)$ if necessary, we may assume that $\zeta_p \in K$ and $L$ is a Kummer extension of $K$.

By Lemma 9.2.4, by replacing $K$ by $K(t^{1/p})$ for a uniformizer $t$ if necessary, we may assume that $L$ is defined by $x^p = a$ for $a \in 1 + pO_K$. Then the assertion follows from Lemma 9.2.5. □

**Corollary 9.2.8** *Let $K$ be a henselian discrete valuation field, and let $L$ be a finite separable extension of $K$. Then there exists an extension $K'$ of $K$ satisfying the following properties: Let $\overline{F} \to \overline{F}'$ be a morphism of algebraic closures compatible with the morphism $F \to F'$ of the residue fields. Then the morphism $\Omega^1_F \otimes_F \overline{F} \to \Omega^1_{F'} \otimes_{F'} \overline{F}'$ is an injection. The degree $[F' : F'^p]$ is $p$. For $L' = LK'$, the ramification index $e_{L'/K'}$ is 1.*

*Proof* Let $(u_i)_{i\in I}$ be a family of elements of $O_K$ such that $(d\bar{u}_i)_{i\in I}$ is a basis of the $F$-vector space $\Omega^1_F$. Let $K_1$ be the fraction field of the completion of the localization of the polynomial ring $O_K[v]$ at the prime ideal generated by $\mathfrak{m}_K$, and set $w_i = u_i/v \in O_{K_1}$ for $i \in I$. Then $d\bar{v}, d\bar{w}_i, i \in I$ form a basis of $\Omega^1_{F_1}$ for the residue field $F_1 = F(v)$ of $K_1$. Let $K_2$ be the completion of the extension of $K_1$ obtained by joining $p$-power-th roots of $w_i$. Then $O_{K_2}$ is a complete discrete valuation ring with residue field $F_2 = F(v, w_i^{1/p^\infty}; i \in I)$. Hence, $dv$ is a basis of $\Omega^1_{F_2}$ and $[F_2 : F_2^p] = p$. Since the image of $d\bar{u}_i$ is $w_i d\bar{v}$, the morphism $\Omega^1_F \otimes_F \overline{F} \to \Omega^1_{F_2} \otimes_{F_2} \overline{F}_2$ is an injection.

By Proposition 9.2.7 applied to $LK_2$ over $K_2$, there exists a finite separable extension $K'$ of $K_2$ such that the residue field $F'$ is a separable extension of $F_2$ and $e_{L'/K'} = 1$ for $L' = LK'$. Since the residue field $F'$ of $K'$ is a finite separable extension of $F_2$, we have $[F' : F'^p] = [F_2 : F_2^p] = p$ and the morphism $\Omega^1_F \otimes_F \overline{F} \to \Omega^1_{F'} \otimes_{F'} \overline{F}'$ is an injection. □

**Exercise 9.2** Let $k$ be a perfect field of characteristic $p > 0$ and $K_0 = k((y))$, $K = K_0(x)$. Define a cyclic extension $L$ of $K$ by the Artin–Schreier equation $t^p - t = x/y$.

(1) Let $K_0' = k((y'))$ be the purely inseparable extension of $K_0$ of degree $p$ generated by a $p$th root $y'$ of $y$. Show that $L' = LK_0'$ is a formally smooth extension of $K_0'$.
(2) Find a cyclic extension $K_1'$ of $K_0$ of degree $p$ such that $L_1' = LK_1'$ is a formally smooth extension of $K_1'$.

## 9.3 Reduced Fiber Theorem

**Definition 9.3.1** ([31, Définitions (4.6.2), (6.7.6)]) Let $k$ be a field and $X$ be a scheme of finite type over $k$. We say that $X$ is geometrically reduced (resp. geometrically normal) over $k$ if, for any finite extension $k'$ of $k$, the base change $X \times_k k'$ is reduced (resp. normal).

**Lemma 9.3.2** *Let $k$ be a field, and $X$ be a scheme of finite type over $k$.*

(1) *The following conditions are equivalent:*
  (a) *$X$ is geometrically reduced over $k$.*
  (b) *$X$ is reduced and, for any irreducible component $C$, the function field $k(C)$ is a separable extension of $k$.*
(2) *There exists a finite extension $k'$ of $k$ such that the normalization $X'$ of $X \times_k k'$ is geometrically normal.*

*Proof* (1) (a) $\Rightarrow$ (b): If $X$ is geometrically reduced, for any irreducible component $C$, the tensor product $k(C) \otimes_k k'$ is reduced for any finite extension $k'$ of $k$. Hence, $k(C)$ is a separable extension of $k$.

(b) $\Rightarrow$ (a): We may assume that $X = \operatorname{Spec} A$ is affine and $A$ is a subring of the product $\prod_i K_i$ of the residue field at minimal prime ideals. If $K_i \otimes_k k'$ is reduced for any finite extension $k'$ of $k$, then $A \otimes_k k'$ is also reduced.

(2) Let $\bar{k}$ be an algebraic closure of $k$, and $X^\natural_{\bar{k}} \to X_{\bar{k}} = X \times_k \bar{k}$ be the

normalization. Then, since $X^{\natural}_{\overline{k}}$ is of finite type over $\overline{k}$, there exist a finite extension $k' \subset \overline{k}$ of $k$ and a finite morphism $X' \to X_{k'}$ of schemes such that the base change to $\overline{k}$ defines $X^{\natural}_{\overline{k}} \to X_{\overline{k}}$. For any subextension $k' \subset k'' \subset \overline{k}$, the base change $X'_{k''}$ is normal by [31, Corollaire (6.5.4)(i)], since $X^{\natural}_{\overline{k}} \to X'_{k''}$ is faithfully flat and $X^{\natural}_{\overline{k}}$ is normal. □

For a discrete valuation ring $O_K$ with residue field $F$ and a scheme $X$ over $S = \operatorname{Spec} O_K$, let $X_\eta = X \times_S \operatorname{Spec} K$ and $X_s = X \times_S \operatorname{Spec} F$ denote the generic fiber and the closed fiber.

**Proposition 9.3.3** *Let $K$ be a discrete valuation field and $X$ be a flat scheme of finite type over $S = \operatorname{Spec} O_K$.*

(1) *Assume that $X$ is normal. Then the closed fiber $X_s$ is reduced (resp. geometrically reduced) if and only if, for any irreducible component $C$ of the closed fiber $X_s$, the local ring $O_{X_s,\xi}$ at the generic point $\xi$ of $C$ is a field (resp. a separable extension of $k$).*
(2) *Assume that $X_s$ is reduced and that the generic fiber $X_\eta$ is normal. Then $X$ is normal.*

*Proof* (1) By Serre's criterion [31, Théorème (5.8.6)], $X$ satisfies the condition ($S_2$). Since $X$ is flat over $S$, the closed fiber $X_s$ satisfies the condition ($S_1$) by [31, Corollaire (5.7.6)]. Hence, $X_s$ is reduced if and only if $X_s$ satisfies the condition ($R_0$) by [31, Proposition (5.8.5)]. This means that, for any irreducible component $C$, the local ring $O_{X_s,\xi}$ at the generic point $\xi$ of $C$ is a field. (The equivalence of the conditions in parentheses follows from Lemma 9.3.2(1).)

(2) By Serre's criterion [31, Théorème (5.8.6)], it suffices to show that $X$ satisfies the conditions ($S_2$) and ($R_1$). By the assumption that $X_\eta$ is normal, it suffices to verify the conditions ($S_2$) and ($R_1$) at each point of the closed fiber $X_s$. Since $X_s$ is reduced, $X_s$ satisfies the conditions ($S_1$) and ($R_0$) by [31, Proposition (5.8.5)]. Since $X$ is flat over $S$, $X$ satisfies the conditions ($S_2$) and ($R_1$) at each point of the closed fiber $X_s$ by [31, Chapitre 0, Proposition (16.4.6)(ii)] and [31, Chapitre 0, Corollaire (17.1.8) a) $\Rightarrow$ b)]. □

**Corollary 9.3.4** *Let $K$ be a discrete valuation field, and $X$ be a flat scheme of finite type over $S = \operatorname{Spec} O_K$. Assume that the generic fiber $X_\eta$ is geometrically normal, and that the closed fiber $X_s$ is geometrically reduced. Then, for any extension $K'$ of discrete valuation field, the base change $X' = X \times_S S'$ by $S' = \operatorname{Spec} O_{K'} \to S$ is normal.*

*Proof* The base change $X' \to S'$ of $X \to S$ is flat and of finite type. By the assumption that $X_\eta$ is geometrically normal and $X_s$ is geometrically reduced, the generic fiber $X'_{\eta'}$ is normal and the closed fiber $X'_{s'}$ is reduced. Hence, by Proposition 9.3.3(2), $X'$ is normal. □

The following is the reduced fiber theorem in the case where the base scheme is the spectrum of a discrete valuation ring.

**Proposition 9.3.5** *Let $K$ be a henselian discrete valuation field with perfect residue field, and $X$ be a flat normal scheme of finite type over $S = \operatorname{Spec} O_K$. Assume that the generic fiber $X_K$ over $K$ is geometrically normal. Then there exists a finite separable extension $K'$ of $K$ such that the normalization $X'$ of $X \times_S S'$ is of finite type over $S' = \operatorname{Spec} O_{K'}$ and has geometrically reduced fibers.*

*Proof* Applying Theorem 9.2.1(2) to the fraction fields $L_\xi$ of the henselization of the local rings $O_{X,\xi}$ at the generic points $\xi$ of irreducible components of the closed fiber $X_s$, we find a finite separable extension $K'$ of $K$ such that the composition fields $L_\xi K'$ are formally smooth over $K'$.

Since $K'$ is a finite separable extension of $K$, the normalization $X'$ of the base change $X \times_S S'$ is flat of finite type over $S'$. The generic fiber of $X'$ is normal. For any irreducible component $C$ of the closed fiber $X'_{s'}$, the local ring $O_{X'_{s'},\xi}$ at the generic point $\xi$ of $C$ is a separable extension of the residue field of $K'$. Hence, the assertion follows from Proposition 9.3.3(1). □

Proposition 9.3.5 implies the following finiteness theorem.

**Corollary 9.3.6** *The normalization of $X \times_{\operatorname{Spec} O_K} \operatorname{Spec} O_{\overline{K}}$ is of finite type over $\operatorname{Spec} O_{\overline{K}}$.*

*Proof* In the notation of Proposition 9.3.5, the normalization of $X \times_{\operatorname{Spec} O_K} \operatorname{Spec} O_{\overline{K}}$ equals $X' \times_{S'} \operatorname{Spec} O_{\overline{K}}$. □

**Exercise 9.3** Let $k$ and $K = k((y))$ be as in Exercise 9.2. Let $X$ be the normalization of $\mathbf{A}^1_S = \operatorname{Spec} O_K[x]$ in the Artin–Schreier covering $X_K$ of $\mathbf{A}^1_K$ defined by $t^p - t = x/y$.

(1) Let $K' = k((y'))$ be the purely inseparable extension of $K$ of degree $p$ generated by a $p$th root $y'$ of $y$. Show that the normalization $X'$ of the base change $X \times_S S'$ is smooth over $S' = \operatorname{Spec} O_{K'}$.

(2) Find a cyclic extension $K'_1$ of $K$ of degree $p$ such that the normalization $X'_1$ of the base change $X \times_S S'_1$ is smooth over $S' = \operatorname{Spec} O_{K'}$.

## Historical Notes

As pointed out in [58], there is an error in the proof of Theorem 9.2.1 in the original article [25]. To correct it, it suffices to require $p \nmid n$ or $c_n \in F - F^p$ in (9.2.1) as in Lemma 9.2.3. The author learned the relation between the Artin–Schreier extensions and the Kummer extensions in Lemma 9.2.5 from Kato in writing [52].

Corollary 9.3.6 is called the finiteness theorem of Grauert–Remmart in the context of rigid geometry.

# 10

# Nearby Cycles on Curves

As an application of the variant of the refined Swan conductor involving differential forms introduced in Chapter 7, we give, in Section 10.5, a proof, independent of Deligne's original proof, of the Deligne–Kato formula computing the dimension of the space of nearby cycles of an étale sheaf on a curve over a local field with perfect residue field. This formula is proved by combining the dimension formula in the constant coefficient case and a trace formula, proved in Sections 10.3 and 10.4 respectively. We recall basic facts on nearby cycles, including their computation in the semi-stable case, in Section 10.2.

The formulations of the dimension formula in the constant coefficient case and the trace formula both require the assumption that the closed fiber be reduced. The reduced fiber theorem proved in Section 9.3 ensures that the condition is satisfied after a sufficiently large base change. Both formulas are proved by reduction to the case where the curve is semi-stable. The reduction is based on a local version of the stable reduction theorem for curves recalled in Section 10.1.

## 10.1 Nodal Curves

**Definition 10.1.1** Let $O_K$ be a discrete valuation ring and let $X$ be a flat scheme of finite type over $S = \operatorname{Spec} O_K$.

(1) We say that $X$ is a curve over $S$ if, for any point $s$ of $S$, every irreducible component of the fiber $X_s$ is of dimension 1.

(2) Let $X$ be a curve over $S$. We say that a closed point $x$ of the closed fiber $X_s$ is an isolated singular point if there exists an open neighborhood $U \subset X$ of $x$ such that $U - \{x\}$ is smooth over $S$.

**Lemma 10.1.2** *Let $O_K$ be a discrete valuation ring, and let $X$ be a normal curve over $S = \operatorname{Spec} O_K$. The following conditions are equivalent:*

(1) *The generic fiber $X_\eta$ is geometrically normal and the closed fiber $X_s$ is geometrically reduced.*
(2) *There exists a finite set of closed points of the closed fiber such that the complement $U \subset X$ is smooth over $S$.*

*Proof* Let $U \subset X$ be the largest open subset smooth over $S$. Then condition (2) is equivalent to $X_\eta$ being a subset of $U$ and, for any irreducible component $C$ of the closed fiber $X_s$, the generic point $\xi$ of $C$ being contained in $U$. Since the generic fiber $X_\eta$ is of dimension 1, the first condition is equivalent to the condition that $X_\eta$ is geometrically normal. The second condition is equivalent to the condition that the closed fiber $X_s$ is geometrically reduced by Proposition 9.3.3(1). □

**Definition 10.1.3** Let $O_K$ be a discrete valuation ring and let $X$ be a curve over $S = \operatorname{Spec} O_K$. Let $F$ be the residue field of $O_K$ and $\overline{F}$ be an algebraic closure.

(1) Let $x \in X$ be an isolated singular point. We say that $x$ is a node of the closed fiber $X_s$ if the completion of the local ring $O_{X_{\bar{s}},\bar{x}}$ at a geometric point $\bar{x}$ above $x$ of the geometric closed fiber $X_{\bar{s}} = X \times_S \operatorname{Spec} \overline{F}$ is isomorphic to $\overline{F}[[u, v]]/(uv)$.
(2) We say that $X$ is a nodal curve over $S$ if $X$ is smooth over $S$ except at finitely many nodes $x_1, \ldots, x_n$ in the closed fiber. We say that a nodal curve $X$ over $S$ is semi-stable if $X$ is regular.

We give an étale local description of nodes.

**Lemma 10.1.4** *Let $X$ be a curve over a field $k$, and let $x$ be a closed point such that $X - \{x\}$ is smooth over $k$. Let $\bar{k}$ be an algebraic closure, and let $\bar{x}$ be a geometric point above $x$ of the geometric fiber $X_{\bar{k}}$. Then the following conditions are equivalent:*

(1) *The completion of $O_{X_{\bar{k}},\bar{x}}$ is isomorphic to $\bar{k}[[u, v]]/(uv)$.*
(2) *There exist an étale neighborhood $U \to X$ of $\bar{x}$ and an étale morphism $U \to \operatorname{Spec} k[u, v]/(uv)$.*

*Proof* (1) ⇒ (2): Let $A = O_{X,x}$ and $B = O_{X_{\bar{k}},\bar{x}}$ be the local rings, and let $A^\natural$ and $B^\natural$ be the normalizations. Since the completion $\widehat{B^\natural}$ is the normalization of the completion $\widehat{B}$ of $B$ and $\dim_{\bar{k}} \widehat{B^\natural}/\widehat{B} = 1$, we have $\dim_{\bar{k}} B^\natural/B = 1$. Since $B$ is faithfully flat over $A$, the tensor product $A^\natural \otimes_A B$ is reduced and $B^\natural$ is the normalization of $A^\natural \otimes_A B$. Since $\dim_{\bar{k}} B^\natural/B = 1$, we have $A^\natural \otimes_A B = B^\natural$ and

$\dim_{\bar{k}}(A^{\natural}/A) \otimes_A B = 1$. Since $\dim_{\bar{k}}(A^{\natural}/A) \otimes_A B = \text{length}\, A^{\natural}/A \cdot [k(x) : k]_{\text{insep}}$, the residue field $k(x)$ is a separable extension of $k$, and $A^{\natural}/A$ is a $k(x)$-vector space of dimension 1.

Since the normalization of $\bar{k}[[u, v]]/(uv)$ is $\bar{k}[[u]] \times \bar{k}[[v]]$, by replacing $X$ by $X \times_{\text{Spec}\, k} \text{Spec}\, k'$ for a separable quadratic extension $k'$ of $k(x)$, we may assume that $k(x) = k$ and $A^{\natural} \otimes_A k(x)$ is isomorphic to $k \times k$.

Let $A^{\text{h}}$ be the henselization of $A$. Since $A^{\natural}$ is finite over $A$, the tensor product $A^{\natural} \otimes_A A^{\text{h}}$ is isomorphic to the product $A_1^{\text{h}} \times A_2^{\text{h}}$ of henselian discrete valuation rings. Hence, by replacing $X$ by an étale neighborhood of $\bar{x}$, we may assume that the normalization $A^{\natural}$ is isomorphic to the product $A_1 \times A_2$ of discrete valuation rings.

Then, $A$ is identified with the inverse image of the diagonal $k \subset k \times k$ by the surjection $A^{\natural} = A_1 \times A_2 \to k \times k$. Let $u \in A_1$ and $v \in A_2$ be uniformizers. Let $u, v$ abusively denote the elements of $A$ corresponding to the pairs $(u, 0), (0, v)$. Since $uv = 0$, after shrinking $X$ if necessary, they define a morphism $X \to \text{Spec}\, k[u, v]/(uv)$ over $k$. Since this induces an isomorphism on henselizations, it is étale.

(2) $\Rightarrow$ (1): The assertion is clear if $X = \text{Spec}\, k[u, v]/(uv)$. Since taking an étale neighborhood does not change the completion after $\otimes_k \bar{k}$, the assertion follows. □

To give an étale local description of nodes of curves over a discrete valuation ring, we recall a basic fact on unramified morphisms.

**Lemma 10.1.5** ([31, Corollaire (18.4.7)]) *Let $f : X \to Y$ be a morphism of finite type of schemes, and $x \in X$. Then the following conditions are equivalent:*

(1) $\Omega^1_{X/Y,x} = 0$.
(2) *There exists a commutative diagram*

$$\begin{array}{ccc} U & \longrightarrow & P \\ \downarrow & & \downarrow \\ X & \longrightarrow & Y \end{array}$$

*where $U$ is an open neighborhood of $x$, the right vertical arrow is étale, and the top horizontal arrow is a closed immersion.*

**Proposition 10.1.6** *Let $X$ be a curve over a discrete valuation ring $O_K$, and let $x$ be an isolated singular point. Let $t \in O_K$ be a uniformizer. Then the following conditions are equivalent:*

(1) *$x$ is a node (resp. and $O_{X,x}$ is regular).*
(2) *Let $\bar{x}$ be a geometric point above $x$. Then there exist an étale neighborhood*

*$U \to X$ of $\bar{x}$, an integer $n \geqslant 1$ (resp. $n = 1$), and an étale morphism $U \to \operatorname{Spec} O_K[u, v]/(uv - t^n)$ sending $\bar{x}$ to the closed point $(u, v, t)$.*

*Proof* (1) $\Rightarrow$ (2): By Lemma 10.1.4, after replacing $X$ by an étale neighborhood there exists an étale morphism $X_F \to \operatorname{Spec} F[u, v]/(uv)$. Since the composition $X_F \to \operatorname{Spec} F[u, v]/(uv) \to \mathbf{A}^2_S = \operatorname{Spec} O_K[u, v]$ is unramified at $x$, after replacing $X$ by an open neighborhood, by Lemma 10.1.5 there exists a scheme $P$, étale over $\mathbf{A}^2_S$, and a closed immersion $X \to P$ such that the closed fiber $X_F \subset P_F$ is defined by $uv = 0$.

Let $I_X \subset O_P$ and $I_{X_F} \subset O_{P_F}$ be the ideal sheaves. Since $X$ is flat over $S$, the canonical morphism $I_X \otimes_{O_K} F \to I_{X_F}$ is an isomorphism. Hence, by Nakayama's lemma, after replacing $X$ by an open neighborhood of $x$ the ideal sheaf $I_X$ is an invertible $O_P$-module generated by $uv - a$ for a section $a \in \mathfrak{m}_K O_P$. If $a = t^n$ for $n \geqslant 1$, the immersion $X \to P$ defines an étale morphism $X \to \operatorname{Spec} O_K[u, v]/(uv - t^n)$ on a neighborhood of $x$.

Since $x$ is an isolated singular point, the $O_X$-module $\Omega^2_{X/S}$ is supported on $\{x\}$ on a neighborhood of $x$. Let $n \geqslant 1$ be the smallest integer satisfying $\mathfrak{m}^n_K \Omega^2_{X/S} = 0$. We show that $a \notin \mathfrak{m}^{n+1}_K O_P$. Otherwise, $X \times_S \operatorname{Spec} O_K/\mathfrak{m}^{n+1}_K$ would be étale over $\operatorname{Spec} O_K/\mathfrak{m}^{n+1}_K[u, v]/(uv)$ and $\Omega^2_{X/S} \otimes_{O_K} O_K/\mathfrak{m}^{n+1}_K$ is isomorphic to $O_K/\mathfrak{m}^{n+1}_K$.

Therefore, there exists a largest integer $n \geqslant 1$ such that $a \in \mathfrak{m}^n_K O_P$ even after changing the coordinates $u, v$. Set $a = wt^n$ for $w \in O_{P,x}$. If $w$ is a unit, it suffices to replace $u$ by $w^{-1}u$. Otherwise, we have $w = bu + cv \bmod \mathfrak{m}_K O_{P,x}$ for some $b, c \in O_{P,x}$ since the maximal ideal $\mathfrak{m}_x$ is generated by $u, v, t$. Then, since $(u - ct^n)(v - bt^n) = (w - bu - cv)t^n + bct^{2n} \in \mathfrak{m}^{n+1}_K O_{P,x}$, by replacing $u, v$ by $u - ct^n, v - bt^n$, we get a contradiction to the assumption that $n$ is the largest.

Further, if $O_{X,x}$ is regular, $uv - t^n$ is not contained in $\mathfrak{m}^2_x$ and we have $n = 1$.

(2) $\Rightarrow$ (1): The assertion is clear if $X = \operatorname{Spec} O_K[u, v]/(uv - t^n)$. Since taking an étale neighborhood does not change the completion after $\otimes_{O_K} \bar{k}$ and the regularity, the assertion follows. □

**Corollary 10.1.7** *Assume that $X$ is a nodal curve over $S = \operatorname{Spec} O_K$. Then $X$ is normal and the minimal resolution $X'$ of $X$ is a semi-stable curve over $S$.*

*Proof* By Proposition 9.3.3(2), a nodal curve is normal. Since the assertion is étale local on $X$, we may assume that $X = \operatorname{Spec} O_K[u, v]/(uv - t^n)$ for a uniformizer $t$ of $O_K$ and an integer $n \geqslant 1$. We show the assertion by induction on $n \geqslant 1$. If $n = 1$, $X$ is regular.

Suppose $n \geqslant 2$ and let $W \to X$ be the blow-up by the ideal $(u, v, t)$ of $X$. If $n = 2, 3$, $W$ is a semi-stable curve over $S$. Suppose $n \geqslant 4$. Then, $W$ is a nodal curve over $S$ and has three isolated singular points. There exists a unique singular point $w \in W$ at which the local ring is not regular. Further, there exists an affine open neighborhood of $w \in W$ isomorphic to $\operatorname{Spec} O_K[u, v]/(uv - t^{n-2})$. Hence, the assertion follows by induction on $n$. □

We recall a local variant of the stable reduction theorem.

**Theorem 10.1.8** ([80]) *Let $K$ be a discrete valuation field, and $X$ be a flat scheme of finite type over $S = \operatorname{Spec} O_K$. Let $U \subset X$ be the complement of a finite set of closed points of the closed fiber such that $U$ is a smooth curve over $S$. Then there exist a finite separable extension $K'$ of $K$, a semi-stable curve $X'$ over $S' = \operatorname{Spec} O_{K'}$, and a projective morphism $X' \to X \times_S S'$ that is an isomorphism on the inverse images of $U$.*

**Exercise 10.1** Let $k$ be a field of characteristic $p > 0$ and $O_K = k((y))$. Let $n > 1$ be a divisor of $p - 1$ and $X$ be the normalization of the affine line $\mathbf{A}^1_S = \operatorname{Spec} O_K[t]$ in the Artin–Schreier covering $X_K \to \mathbf{A}^1_K$ defined by $u^p - u = t^n/y^p$.

(1) Show that $X$ has a unique isolated singular point.
(2) Let $K' = k((y'))$ be a totally ramified extension of $K$ of degree $n$ defined by $y'^n = y$, and $S' = \operatorname{Spec} O_{K'}$. Show that a resolution $X'$ of singularities of the base change $X_{S'} = X \times_S S'$ is a nodal curve over $S'$.

## 10.2 Nearby Cycles

In this section we study the complex of nearby cycles. After recalling the definition, we state that the formation of the nearby cycles complex commutes with finite base change. Nearby cycles complexes are compatible with proper push forward and smooth pull-back, and satisfy finiteness. Some proofs are omitted or given only in basic cases. Finally, we compute the nearby cycles complex for semi-stable curves.

A henselian local ring is called strictly henselian if the residue field is separably closed. For a geometric point $\overline{x}$ of a scheme $X$, the strict henselization $X_{(\overline{x})}$ is defined to be the limit of étale neighborhoods, similarly to the definition of henselization, Definition 1.3.1.

**Definition 10.2.1** Let $O_K$ be a strict henselian discrete valuation ring, let $s$ and $\eta = \operatorname{Spec} K$ be the closed point and the generic point of $S = \operatorname{Spec} O_K$, and let $\overline{\eta} = \operatorname{Spec} K_s$ for a separable closure $K_s$ of $K$. Let $X$ be a scheme over $S$, and let

$$
\begin{array}{ccccc}
 & & & & X_{\overline{\eta}} = X \times_S \overline{\eta} \\
 & & & \overset{\overline{j}}{\swarrow} & \Big\downarrow g \\
X_s = X \times_S s & \xrightarrow{\ i\ } & X & \xleftarrow[\ j\ ]{} & X_\eta = X \times_S \eta
\end{array}
$$

be the canonical morphisms. For a sheaf $\mathcal{F}$ on the generic fiber $X_\eta$, define the complex of nearby cycles by $R\psi\mathcal{F} = i^* R\overline{j}_* g^* \mathcal{F}$.

Let $\overline{x}$ be a geometric point of $X$, and $X_{(\overline{x})}$ be the strict henselization. Then the stalk $(R\psi\mathcal{F})_{\overline{x}}$ is canonically identified with $R\Gamma(X_{(\overline{x})} \times_S \overline{\eta}, \mathcal{F})$. The scheme $X_{(\overline{x})} \times_S \overline{\eta}$ is called the Milnor fiber.

**Lemma 10.2.2** *Let $j \colon X_\eta \to X$ be the open immersion, $g \colon X_{\overline{\eta}} \to X_\eta$ be the canonical morphism, and let $I = \mathrm{Gal}(K_s/K)$ be the inertia group. Then the adjunction $\mathcal{F} \to g_* g^* \mathcal{F}$ induces an isomorphism $i^* Rj_* \mathcal{F} \to R\Gamma(I, R\psi\mathcal{F})$.*

*Proof* The adjunction induces an isomorphism $\mathcal{F} \to R\Gamma(I, g_* g^* \mathcal{F})$. This induces an isomorphism $Rj_*\mathcal{F} \to Rj_* R\Gamma(I, g_* g^* \mathcal{F}) \to R\Gamma(I, R\overline{j}_* g^* \mathcal{F})$. By applying $i^*$, we obtain the required isomorphism. □

**Proposition 10.2.3** *Let $O_K$ be a henselian discrete valuation ring. Let $X \to S = \mathrm{Spec}\, O_K$ be a morphism of finite type, and $\mathcal{F}$ be a torsion sheaf on $X_\eta$. Let $f \colon S' = \mathrm{Spec}\, O_{K'} \to S$ be the finite morphism for a finite extension $K'$ of $K$. Then the base change morphism $f^* R\psi\mathcal{F} \to R\psi f^* \mathcal{F}$ is an isomorphism.*

We say that the formation of nearby cycles commutes with base change if the base change morphism $f^* R\psi\mathcal{F} \to R\psi f^* \mathcal{F}$ is an isomorphism.

Although the finiteness assumption on $S' \to S$ is known to be redundant [21, Proposition 3.7], we give an elementary proof under this simplifying assumption.

*Proof* We consider the cartesian diagram

$$
\begin{array}{ccccc}
X'_{s'} & \longrightarrow & X' = X \times_S S' & \xleftarrow{\ \overline{j}'\ } & X'_{\overline{\eta}'} \\
\Big\downarrow & & \Big\downarrow f & & \Big\downarrow \\
X_s & \longrightarrow & X & \xleftarrow{\quad \overline{j} \quad} & X_{\overline{\eta}}
\end{array}
$$

Since the right vertical arrow induces a homeomorphism on the étale site, we have an isomorphism $R\overline{j}_*\mathcal{F} \to Rf_* R\overline{j}'_*\mathcal{F}$. Since $f$ is finite and the left vertical arrow also induces a homeomorphism, the assertion follows. □

**Theorem 10.2.4** (1) *Let $f: X \to Y$ be a proper morphism of schemes over $S$. If $\mathcal{F}$ is a torsion sheaf on $X$, the base change morphism $R\psi Rf_*\mathcal{F} \to Rf_*R\psi\mathcal{F}$ is an isomorphism.*

(2) *Let $f: X \to Y$ be a smooth morphism of schemes over $S$. If $\mathcal{F}$ is a torsion sheaf on $Y_\eta$ annihilated by an integer invertible on $S$, the base change morphism $f^*R\psi\mathcal{F} \to R\psi f^*\mathcal{F}$ is an isomorphism.*

(3) *Assume that $X$ is of finite type over $S$, and $\dim X_\eta = n$. Then $R^q\psi\mathcal{F} = 0$ for $q > n$.*

*Proof* (1) and (2): It suffices to apply the proper base change theorem [26, Theorem 7.3.1] and the smooth base change theorem [26, Theorem 7.7.2] to the left and right squares respectively of the diagram

$$\begin{array}{ccccc} Y_s & \longrightarrow & Y & \longleftarrow & Y_{\overline{\eta}} \\ \downarrow & & \downarrow & & \downarrow \\ X_s & \longrightarrow & X & \longleftarrow & X_{\overline{\eta}} \end{array}$$

(3): Since the Milnor fiber $X_{(\overline{x})} \times_S \overline{\eta}$ is a limit of affine schemes of finite type of dimension $n$ over $\overline{\eta}$, assertion (3) is a consequence of the cohomological dimension of affine schemes [26, Corollary 7.5.2]. □

In the rest of this section, $\ell$ denotes a prime number invertible on $S$ and $\Lambda$ denotes a local ring finite over $\mathbf{Z}_\ell$. The following property is a special case of the local acyclicity of smooth morphisms.

**Corollary 10.2.5** *Let $X$ be a smooth scheme over $S$, and $\mathcal{F}$ be a smooth sheaf of $\Lambda$-modules on $X$. Then the canonical morphism $\mathcal{F}|_{X_s} \to R\psi\mathcal{F}$ is an isomorphism.*

*Proof* We may assume that $\Lambda$ is finite. Since the assertion is étale local, we may assume that $\mathcal{F}$ is a constant sheaf. By Theorem 10.2.4(2), the assertion follows from the case $X = S$. □

As an application of Corollary 10.2.5, we show the vanishing of the nearby cycles at infinity of the étale sheaf defined by the Artin–Schreier covering $t^p - t = xy$ of $\mathbf{A}^2 = \operatorname{Spec} k[x, y]$.

**Example 10.2.6** Let $k$ be a field of characteristic $p > 0$, and fix a non-trivial additive character $\mathbf{F}_p \to \Lambda^\times$. Let $\mathcal{L}(xy)$ be the locally constant sheaf of $\Lambda$-modules of rank 1 on $\mathbf{A}^2 = \operatorname{Spec} k[x, y]$ defined by the Artin–Schreier covering $t^p - t = xy$. Let $S$ denote the strict henselization of the projective line $\mathbf{P}^1_k$ at $\infty$, and $X = \mathbf{A}^1_S$. Also, let $\mathcal{L}(xy)$ denote the pull-back by the base change $X_\eta = \mathbf{A}^1_\eta \to \mathbf{A}^2_k$ of $\eta \to \mathbf{A}^1_k = \operatorname{Spec} k[y]$. Then we have $R\psi\mathcal{L}(xy) = 0$.

In fact, let $F\colon S \to S$ be the morphism of degree $p$ defined by the Frobenius morphism $F\colon \mathbf{P}^1_k \to \mathbf{P}^1_k$. Let the domain $S$ be denoted by $S'$. Let $F\colon X' = \mathbf{A}^1_{S'} \to X$ denote the base change. By Proposition 10.2.3, it suffices to show that $R\psi F^*\mathcal{L} = 0$ for $X' \to S'$.

Let $z$ be the uniformizer of $O_{S'}$ satisfying $z^p = 1/y$. Then the pull-back $F^*\mathcal{L}$ is defined by the Artin–Schreier equation $t^p - t = x/z^p$. If we set $s = zt$, the normalization $f\colon W' \to X'$ of the Artin–Schreier covering is defined by $s^p - z^{p-1}s = x$, and we obtain an isomorphism $W' \to \mathbf{A}^1_{S'} = \operatorname{Spec} O_{S'}[s]$. Since $F^*\mathcal{L}$ is a direct summand of $f_*\Lambda$, the nearby cycles complex $R\psi F^*\mathcal{L}$ is a direct summand of $R\psi f_*\Lambda = f_*R\psi\Lambda$. Since $W' \to S'$ is smooth and since the morphism $W'_{s'} \to X'_{s'}$ on the closed fiber is purely inseparable, the assertion follows from Corollary 10.2.5.

The argument using the base change in Example 10.2.6 is a variant of the proof of Lemma 9.2.3 eliminating the ramification index.

**Proposition 10.2.7** *Let $X \to S$ be a morphism of finite type.*

(1) *Assume that $X$ is flat, the generic fiber $X_\eta$ is geometrically normal, and the closed fiber is $X_s$ is geometrically reduced. Then the canonical morphism $\Lambda \to R^0\psi\Lambda$ is an isomorphism.*
(2) *Assume that $X$ is a curve over $S$. Let $\mathcal{F}$ be a constructible sheaf of $\Lambda$-modules on $X_\eta$. Then $R\psi\mathcal{F}$ is constructible. The support of $R^1\psi\mathcal{F}$ is finite, and $R^q\psi\mathcal{F} = 0$ for $q > 1$.*

On the constructibility in assertion (2), the assumption on the dimension is known to be redundant [26, Theorem 9.4.1]. We give a proof by the global-to-local argument in this basic case.

*Proof* (1) For any finite extension $K' \subset K_s$, the base change $X' = X \times_S \operatorname{Spec} O_{K'}$ is normal by Corollary 9.3.4. Hence, the base change $X \times_S \operatorname{Spec} O_{K_s}$ is also normal. Therefore, the Milnor fiber $X_{(x)} \times_S \bar\eta$ is a dense open subscheme of a local normal scheme and hence is irreducible. Thus the assertion follows.

(2) By dévissage using Theorem 10.2.4(1), the proof of the constructibility is reduced to the case $\mathcal{F} = \Lambda$. By base change and normalization, we may assume that $X$ is normal and is smooth over $S$ except at finitely many closed points by Proposition 9.3.5 and Lemma 10.1.2(1). Then, by (1), we have an isomorphism $\Lambda \to R^0\psi\Lambda$. By Corollary 10.2.5, the morphism $\Lambda \to R\psi\Lambda$ is an isomorphism outside the finitely many singular points. By Theorem 10.2.4(3), we have $R^q\psi\mathcal{F} = 0$ for $q \neq 0, 1$. Thus, it suffices to show that the stalks of $R^1\psi\mathcal{F}$ at singular points are finite.

Since the assertion is local, we may assume that $X$ is affine and then projective. Then, by Theorem 10.2.4(1), we have a spectral sequence $E_2^{p,q} =$

$H^p(X_s, R^q\psi\mathcal{F}) \Rightarrow H^{p+q}(X_{\bar\eta}, \mathcal{F})$. We have $E_2^{p,q} = 0$ except for $(p,q) = (0,0)$, $(1,0)$, $(2,0)$, and $(0,1)$, and $E_2^{p,0}$ are finite. Since the abutments are finite, $E_2^{0,1} = H^0(X_s, R^1\psi\mathcal{F})$ is also finite and the assertion follows. □

Similarly to Lemma 8.2.2(1), finiteness implies perfection.

**Corollary 10.2.8** *Let $X$ be a curve over $S = \operatorname{Spec} O_K$ and $x$ be a geometric point of the closed fiber.*

(1) *Let $j\colon U \to X_K$ be an open immersion, and let $\pi\colon V \to U$ be an étale $G$-torsor for a finite group $G$. Let $\ell$ be a prime number different from the characteristic of the residue field $F$. Then the complex $R\psi_x j_!\pi_*\mathbf{Z}_\ell$ of $\mathbf{Z}_\ell[G]$-modules is a perfect complex.*
(2) *Assume further that $X$ is normal and that $X$ is smooth over $S$ except at a closed point $x$ of the closed fiber. Then the stalk $R^1\psi_x\mathbf{Z}_\ell$ is a free $\mathbf{Z}_\ell$-module of finite rank.*

*Proof* (1) Let $M$ be a finitely generated right $\mathbf{Z}_\ell[G]$-module. Then, since $\pi_*\mathbf{Z}_\ell$ is a locally constant sheaf of free $\mathbf{Z}_\ell[G]$-modules, the cohomology groups $R^q\psi j_!(M \otimes_{\mathbf{Z}_\ell[G]} \pi_*\mathbf{Z}_\ell)$ of $M \otimes^L_{\mathbf{Z}_\ell[G]} R\psi_x j_!\pi_*\mathbf{Z}_\ell = R\psi j_!(M \otimes_{\mathbf{Z}_\ell[G]} \pi_*\mathbf{Z}_\ell)$ are finitely generated $\mathbf{Z}_\ell$-modules and are 0 unless $q = 0, 1$, by Proposition 10.2.7(2). Hence, the assertion follows from Lemma 8.2.1.

(2) The complex $R\psi\mathbf{Z}_\ell$ is a perfect complex of $\mathbf{Z}_\ell$-modules by (1) applied to $G = 1$ and $U = X$. Hence, there exists a morphism $L_0 \to L_1$ of free $\mathbf{Z}_\ell$-modules of finite rank such that, for any $\mathbf{Z}_\ell$-module $M$ of finite type, there exist isomorphisms $R^0\psi M \to \operatorname{Ker}(L_0 \otimes_{\mathbf{Z}_\ell} M \to L_1 \otimes_{\mathbf{Z}_\ell} M)$ and $R^1\psi M \to \operatorname{Coker}(L_0 \otimes_{\mathbf{Z}_\ell} M \to L_1 \otimes_{\mathbf{Z}_\ell} M)$. Since $\mathbf{Z}_\ell \to R^0\psi\mathbf{Z}_\ell$ and $\mathbf{F}_\ell \to R^0\psi\mathbf{F}_\ell$ are isomorphisms by (1), $\operatorname{Coker}(L_0 \to L_1) = R^1\psi\mathbf{Z}_\ell$ is a free $\mathbf{Z}_\ell$-module of finite rank. □

**Theorem 10.2.9** *Let $f\colon X \to S$ be a semi-stable curve.*

(1) *For geometric points $x$ above singular points of $X_s$, the stalk $R^1\psi\Lambda_x$ is a free $\Lambda$-module of rank* 1.
(2) *Let*

$$\begin{array}{ccccc} X_s & \xrightarrow{\ i\ } & X & \xleftarrow{\ j\ } & X_\eta \\ {\scriptstyle f_s}\downarrow & & {\scriptstyle f}\downarrow & & \downarrow \\ s & \xrightarrow{\ i_S\ } & S & \xleftarrow{\ j_S\ } & \eta \end{array}$$

*be the cartesian diagram. Then the following sequence is exact:*

$$0 \to f_s^* i_S^* R^1 j_{S*}\Lambda \to i^* R^1 j_*\Lambda \to R^1\psi\Lambda \to 0.$$

*Proof* (1) Since the assertion is étale local, we may assume that $X$ is the blow-up of $\mathbf{A}^1_S$ at the origin of the closed fiber, and that $x$ is above the singular point of the closed fiber $X_s$. Let $g\colon X \to \mathbf{A}^1_S$ be the canonical morphism. By Theorem 10.2.4(1), the canonical morphism $R\psi\Lambda \to Rg_* R\psi\Lambda$ is an isomorphism. Hence, we have a spectral sequence $E_2^{p,q} = R^p g_* R^q\psi\Lambda \Rightarrow R\psi\Lambda$. Since the canonical morphism $\Lambda \to R^0\psi\Lambda$ is an isomorphism and $R^q\psi\Lambda = 0$ for $q > 1$ by Proposition 10.2.7, we obtain an isomorphism

$$R^1\psi\Lambda_x \to \mathrm{H}^2(\mathbf{P}^1, \Lambda) = \Lambda(-1). \tag{10.2.1}$$

(2) By Corollary 10.2.5, the assertion holds except at singular points of the closed fiber. By applying the functor $R\Gamma(I, -)$ to the distinguished triangle $f^*R\psi_S\Lambda \to R\psi_X\Lambda \to R^1\psi\Lambda[-1] \to$, we obtain a distinguished triangle

$$f^*i_S^*Rj_{S*}\Lambda \to i^*Rj_*\Lambda \to R\Gamma(I, R^1\psi\Lambda)[-1] \to .$$

Since $i_S^*R^2j_{S*}\Lambda = \mathrm{H}^2(I, \Lambda) = 0$ and $\mathrm{H}^0(I, R^1\psi\Lambda) = R^1\psi\Lambda$ by (10.2.1), the assertion follows. □

**Exercise 10.2** Let $k$ be a field of characteristic $p > 0$, and $S' = \operatorname{Spec} O_K$ for $K = k((y))$. Let $n > 1$ be a divisor of $p - 1$, and $X$ be the normalization of the affine line $\mathbf{A}^1_{S'} = \operatorname{Spec} O_K[t]$ in the Artin–Schreier covering defined by $u^p - u = t^n/y^{pn}$. Let $f\colon X' \to X$ be the minimal resolution, and $C = f^{-1}(x)$ be the fiber of the unique singular point $x \in X$. Construct a canonical isomorphism $\mathrm{H}^1(C_{\bar{k}}, \Lambda) \to R^1\psi\Lambda_x$.

## 10.3 Dimension at Isolated Singular Points

In the rest of this chapter, let $O_K$ be a henselian discrete valuation ring with algebraically closed residue field $k$ of characteristic $p \geqslant 0$, and let $S = \operatorname{Spec} O_K$. Let $X$ be a normal curve over $S$, and $x$ be a closed point of the closed fiber such that $X - \{x\}$ is smooth over $S$. By Proposition 9.3.3(1), this implies that the closed fiber $X_s$ is reduced.

Let $\Lambda$ be a finite field of characteristic $\ell$, different from the characteristic of the residue field $k$, and define an integer $\psi_x \in \mathbf{N}$ by

$$\psi_x = \dim_\Lambda R^1\psi\Lambda_x. \tag{10.3.1}$$

Let $\pi\colon X_s^\natural \to X_s$ be the normalization of the closed fiber and set $P_x = \pi^{-1}(x) \subset X_s^\natural$. Define integers $\gamma_x, \delta_x \in \mathbf{N}$ by

$$\gamma_x = \dim_\Lambda \operatorname{Coker}(\Lambda_{X_s} \to \pi_*\Lambda_{X_s^\natural})_x = \#P_x - 1, \tag{10.3.2}$$

$$\delta_x = \dim_k \operatorname{Coker}(O_{X_s,x} \to \pi_*O_{X_s^\natural,x}). \tag{10.3.3}$$

**Lemma 10.3.1** *The following conditions are equivalent:*

(1) *$X \to S$ is smooth.*

(2) *$X_s$ is smooth over $k$.*

(3) *$\delta_x = 0$.*

*These equivalent conditions imply that $\psi_x = \gamma_x = 0$.*

We will show in Corollary 10.3.5 that conversely, $\psi_x = 0$ implies the equivalent conditions (1)–(3).

*Proof* The vanishing $\delta_x = 0$ means that $X_s^\natural \to X_s$ is an isomorphism. This means that $X_s$ is normal and hence is equivalent to (2). Since $X \to S$ is flat, conditions (1) and (2) are equivalent to each other.

If $X_s^\natural \to X_s$ is an isomorphism, we have $\gamma_x = 0$. By Corollary 10.2.5, we have $\psi_x = 0$ if $X \to S$ is smooth. □

**Proposition 10.3.2** *Let $X$ be a normal curve over $S$, and let $x \in X_s$ be a closed point such that $X - \{x\}$ is smooth over $S$. Then*

$$\delta_x \geqslant \gamma_x. \tag{10.3.4}$$

*Proof* For the normalization $\pi\colon X_s^\natural \to X_s$, we have a commutative diagram of exact sequences

$$\begin{array}{ccccccccc} 0 & \longrightarrow & \mathfrak{m}_x & \longrightarrow & \mathcal{O}_{X_s,x} & \longrightarrow & k & \longrightarrow & 0\\ & & \downarrow & & \downarrow & & \downarrow & & \\ 0 & \longrightarrow & \mathrm{Ker} & \longrightarrow & \mathcal{O}_{X_s^\natural}\otimes_{\mathcal{O}_{X_s}}\mathcal{O}_{X_s,x} & \longrightarrow & \bigoplus_{x'\in P_x} k & \longrightarrow & 0 \end{array}$$

The invariants $\delta_x$ and $\gamma_x$ are the dimensions of the cokernels of the middle and right vertical arrows respectively. Since the right vertical arrow is an injection, the inequality (10.3.4) follows from the snake lemma. □

**Lemma 10.3.3** *Let $X$ be a normal curve over $S$, and let $x \in X_s$ be a closed point such that $X - \{x\}$ is smooth over $S$.*

(1) *Let $K'$ be a finite extension such that $S' = \operatorname{Spec} \mathcal{O}_{K'} \to S$ is a finite morphism. Then $X' = X \times_S S'$ is normal and, for the unique point $x' \in X'$ above $x$, we have $\psi_x = \psi_{x'}$, $\gamma_x = \gamma_{x'}$, and $\delta_x = \delta_{x'}$.*

(2) *Let $f\colon W \to X$ be a proper morphism of normal curves over $S$ such that $W$ is smooth over $S$ on the complement of a finite set $Z$ of closed points of the closed fiber $W_s$ and that $f$ induces an isomorphism $W - f^{-1}(x) \to X - \{x\}$. We identify the normalization $W_s^\natural$ of $W_s$ with the disjoint union*

*$X_s^\natural \amalg E$ of the normalization $X_s^\natural$ of $X_s$ and a proper smooth curve $E$. Then*

$$\psi_x + \gamma_x = \sum_{w\in Z}(\psi_w + \gamma_w) - \chi(E,\Lambda), \tag{10.3.5}$$

$$\delta_x = \sum_{w\in Z}\delta_w - \chi(E,\mathcal{O}_E). \tag{10.3.6}$$

*Proof* (1) The base change $X' = X\times_S S'$ is normal by Corollary 9.3.4. Since the formation of nearby cycles commutes with base change by Proposition 10.2.3, we have $\psi_x = \psi_{x'}$. Since $X_{s'} \to X_s$ is an isomorphism, we have $\gamma_x = \gamma_{x'}$ and $\delta_x = \delta_{x'}$.

(2) Since the canonical morphism $\Lambda \to R^0\psi\Lambda$ is an isomorphism by Proposition 10.2.7(1), we have $\psi_x = -\dim[\Lambda \to R\psi\Lambda]$, where $[\ \ ]$ denotes the mapping cone and dim denotes the dimension of the stalk at $x$; similarly for $\psi_w$. We also have $\gamma_x = \dim[\Lambda \to \pi_*\Lambda]$, and similarly for $\gamma_w$. Let $f_s\colon W_s \to X_s$ be the morphism on the closed fiber, and $\pi'\colon W_s^\natural \to W_s$ denote the normalization. Then $\psi_x - \sum_{w\in Z}\psi_w = -\dim[\Lambda \to Rf_{s*}\Lambda] + \dim[R\psi\Lambda \to Rf_{s*}R\psi\Lambda]$ and $\gamma_x - \sum_{w\in Z}\gamma_w = \dim[\Lambda \to Rf_{s*}\Lambda] - \dim[\pi_*\Lambda \to Rf_{s*}\pi'_*\Lambda]$. Thus, we obtain $(\psi_x+\gamma_x) - \sum_{w\in Z}(\psi_w+\gamma_w) = \dim[R\psi\Lambda \to Rf_{s*}R\psi\Lambda] - \dim[\pi_*\Lambda \to Rf_{s*}\pi'_*\Lambda]$.

Since, by Theorem 10.2.4(1), the morphism $R\psi\Lambda \to Rf_{s*}R\psi\Lambda$ is an isomorphism, we have $[R\psi\Lambda \to Rf_{s*}R\psi\Lambda] = 0$. The composition $f_s^\natural = f_s\circ\pi'\colon W_s^\natural = X_s^\natural \amalg E \to W_s \to X_s$ induces the morphism $\pi\colon X_s^\natural \to X_s$ of normalization on $X_s^\natural$ and contracts $E$ to the point $x$. Hence, we have $[\pi_*\Lambda \to Rf_{s*}\pi'_*\Lambda] = R\Gamma(E,\Lambda)$ and we obtain the first equality (10.3.5).

Since $Rf^\natural_{s*}\mathcal{O}_{W_s^\natural}$ is the direct sum of $\pi_*\mathcal{O}_{X_s^\natural}$ and $R\Gamma(E,\mathcal{O}_E)$, to show the second equality (10.3.6), it suffices to show that $\chi(\mathcal{O}_{X_s} \to Rf_*\mathcal{O}_{W_s}) = 0$. Since $W$ and $X$ are normal, the canonical morphism $\mathcal{O}_X \to f_*\mathcal{O}_W$ is an isomorphism. Hence, the exact sequence $0 \to \mathcal{O}_W \xrightarrow{\cdot t} \mathcal{O}_W \to \mathcal{O}_{W_s} \to 0$ for a uniformizer $t\in\mathcal{O}_K$ induces an exact sequence $0 \to \mathcal{O}_{X_s} \to f_*\mathcal{O}_{W_s} \to R^1f_*\mathcal{O}_W \to R^1f_*\mathcal{O}_W \to R^1f_*\mathcal{O}_{W_s} \to 0$. Since $R^1f_*\mathcal{O}_W$ is of finite length, we have $\chi(\mathcal{O}_{X_s} \to Rf_*\mathcal{O}_{W_s}) = \chi(R^1f_*\mathcal{O}_W) - \chi(R^1f_*\mathcal{O}_W) = 0$. □

**Proposition 10.3.4** *Let $X$ be a normal curve over $S = \operatorname{Spec}\mathcal{O}_K$, and $x$ be a closed point of the closed fiber such that $X - \{x\}$ is smooth over $S$. Then*

$$\psi_x + \gamma_x = 2\delta_x. \tag{10.3.7}$$

*Proof* First, we show the case where $X$ is a semi-stable curve. Let $\pi\colon X_s^\natural \to X_s$ be the canonical morphism of normalization. We have an exact sequence $0 \to \Lambda_{X_s} \to \pi_*\Lambda_{X_s^\natural} \to R^1\psi\Lambda \to 0$ by Theorem 10.2.9. This implies the equality $\psi_x = \gamma_x$ and hence the left-hand side equals $2\gamma_x$. Since $\delta_x = \gamma_x = 0$

or 1 depending on whether $X \to S$ is smooth at $x$ or not, the assertion follows in this case.

We deduce the general case from the semi-stable case by using Lemma 10.3.3. By Theorem 10.1.8 and Lemma 10.3.3(1), after replacing $K$ by a finite separable extension and $X$ by the base change if necessary, we may assume that there exist a semi-stable curve $W$ over $S$ and a proper morphism $f\colon W \to X$ over $S$ such that $f$ induces an isomorphism $W - f^{-1}(x) \to X - \{x\}$.

Let $E$ be the normalization of $f^{-1}(x)$. By $\chi(E, \Lambda) = 2\chi(E, \mathcal{O}_E)$ (Corollary 8.1.8), the equality (10.3.7) for any closed point $w \in f^{-1}(x)$ implies that the right-hand side of (10.3.5) equals twice the right-hand side of (10.3.6). Hence, the equality (10.3.7) for $x$ follows. □

**Corollary 10.3.5** *Let $X$ be a normal curve over $S$, and let $x \in X_s$ be a closed point such that $X - \{x\}$ is smooth over $S$.*

(1) *We have*

$$\psi_x \geqslant \delta_x \geqslant \gamma_x \geqslant 0. \tag{10.3.8}$$

(2) *The following conditions are equivalent:*

(a) *$X$ is smooth over $S$.*
(b) *$\psi_x = 0$.*
(c) *$\delta_x = 0$.*

*Proof* (1) Since $\delta_x$ is the additive mean of $\psi_x$ and $\gamma_x$ by (10.3.7), the inequality (10.3.4) implies (10.3.8).

(2) (a) $\Rightarrow$ (b): This follows from Corollary 10.2.5.

(b) $\Rightarrow$ (c): By (1), $\psi_x = 0$ implies $\delta_x = 0$.

(c) $\Leftrightarrow$ (a): This is proved in Lemma 10.3.1. □

**Theorem 10.3.6** *Let $X$ be a normal flat curve over $S$, and $x \in X$ be a closed point of the closed fiber such that $X - \{x\}$ is smooth over $S$. Let $W \to X$ be the minimal resolution such that the reduced closed fiber $W_{F,\mathrm{red}}$ is a divisor with normal crossings. Then the following conditions are equivalent.*

(1) *The curve $W$ over $S$ is semi-stable. Namely, the closed fiber $W_F$ is reduced.*
(2) *The action of the inertia group $I_K$ on $R^1\psi\mathbf{Z}_\ell$ is unipotent.*

The implication (1) $\Rightarrow$ (2) follows from Theorem 10.2.9 applied to the semi-stable curve $W$ and the spectral sequence $E_2^{p,q} = H^p(W_{\bar{F}}, R\psi\mathbf{Z}_\ell) \Rightarrow R\psi_x\mathbf{Z}_\ell$. The implication (2) $\Rightarrow$ (1) is proved in [69, Theorem (4.8)] using Proposition 10.3.4. Proposition 10.3.4 is proved by applying the semi-stable reduction theorem, Theorem 10.1.8. By Grothendieck's monodromy theorem, the implication (2) ⇒ (1) implies Theorem 10.1.8. This logical dependence is summarized in the following diagram:

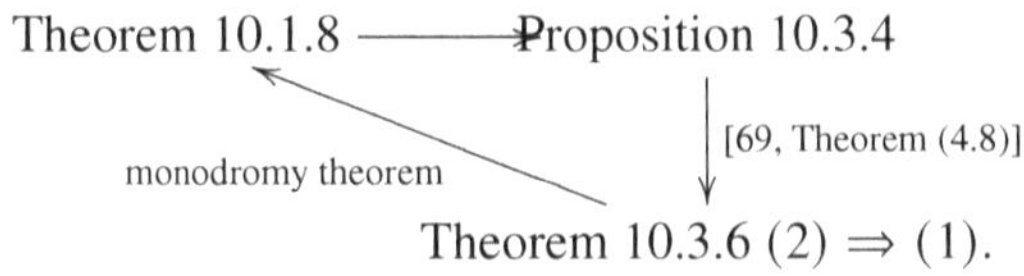


**Exercise 10.3** Let $k$ be a field of characteristic $p > 0$, and $S = \operatorname{Spec} \mathcal{O}_K$ for $K = k((y))$. Let $n > 1$ be a divisor of $p - 1$, and $X$ be the normalization of the affine line $\mathbf{A}^1_S = \operatorname{Spec} \mathcal{O}_K[t]$ in the Artin–Schreier covering defined by $u^p - u = t^n/y^p$ as in Exercise 10.1. Compute the space $R^1\psi\Lambda$ of nearby cycles and verify the formula (10.3.7).

## 10.4 Trace Formula

Let $X$ be a normal curve over $S = \operatorname{Spec} \mathcal{O}_K$, and let $x$ be a closed point of the closed fiber such that $X - \{x\}$ is smooth over $S$. Let $\sigma$ be an automorphism of $X$ over $S$ fixing $x$ such that the induced automorphism of $\mathcal{O}_{X,x}$ is non-trivial. Let $\ell \neq p$ be a prime number. Then the stalk $R^1\psi\mathbf{Z}_{\ell,x}$ is a free $\mathbf{Z}_\ell$-module of finite rank by Corollary 10.2.8(2). We define $\psi_x(\sigma) \in \mathbf{Z}_\ell$ by

$$\psi_x(\sigma) = \operatorname{Tr}(\sigma : R^1\psi\mathbf{Z}_{\ell,x}). \tag{10.4.1}$$

Let $\pi\colon X_s^\natural \to X_s$ be the normalization, and let

$$P_x^\sigma = P_x \cap X_s^{\natural\sigma} \tag{10.4.2}$$

be the fixed part of the inverse image $P_x = \pi^{-1}(x)$. Define an integer $\gamma_x(\sigma) \in \mathbf{N}$ by

$$\gamma_x(\sigma) = \operatorname{Tr}(\sigma : \operatorname{Coker}(\mathbf{Z}_{\ell,X_s} \to \pi_*\mathbf{Z}_{\ell,X_s^\natural})_x) = \#P_x^\sigma - 1. \tag{10.4.3}$$

We define an integer $\alpha_x(\sigma) \in \mathbf{Z}$. Let $x' \in P_x^\sigma \subset X_s^{\natural\sigma}$ be a closed fixed point of the normalization, and let $\xi \in X_s^\natural$ be the generic point of the irreducible component $C_\xi$ of $X_s^\natural$ containing $x'$. If $x'$ is an isolated fixed point of $C_\xi$, then the local ring $\mathcal{O}_{C_\xi^\sigma,x'}$ of the fixed part $C_\xi^\sigma$ is of finite length and we set $\alpha_{x'}(\sigma) = \dim_k \mathcal{O}_{C_\xi^\sigma,x'} \in \mathbf{N}$. Assume otherwise. Then $\mathcal{O}_{X^\sigma,\xi}$ is of finite length and we set $n = \operatorname{length} \mathcal{O}_{X^\sigma,\xi}$. The residue field $k(\xi)$ equals the function field $k(C_\xi)$ of the curve $C_\xi$ over $k$. Let $d_\sigma \in \operatorname{Hom}_{k(C_\xi)}(\Omega^1_{k(C_\xi)/k}, \mathfrak{m}_K^n/\mathfrak{m}_K^{n+1} \otimes_k k(C_\xi))$ denote the morphism sending $du$ to $\sigma(u) - u$, and define $\alpha_{x'}(\sigma) =$

$\mathrm{ord}_{x'} d_\sigma \in \mathbf{Z}$ to be the valuation at $x'$ with respect to the free $O_{C_\xi,x'}$-submodule $\mathrm{Hom}_{O_{C_\xi,x'}}(\Omega^1_{O_{C_\xi,x'}/k}, \mathfrak{m}_K^n/\mathfrak{m}_K^{n+1} \otimes_k O_{C_\xi,x'})$ of rank 1. Set

$$\alpha_x(\sigma) = \sum_{x' \in P_x^\sigma} \alpha_{x'}(\sigma). \tag{10.4.4}$$

We define an integer $\beta_x(\sigma) \in \mathbf{N}$. We set

$$Q_x = \mathrm{Spec}\, O_{X,x} \cap X_\eta. \tag{10.4.5}$$

For a closed fixed point $\xi \in Q_x^\sigma = Q_x \cap X_\eta^\sigma$ of the generic fiber, the local ring $O_{X^\sigma_{\eta,\xi}}$ of the fixed part $X^\sigma_{\eta,\xi}$ is of finite length and we set $\beta_\xi(\sigma) = \dim_K O_{X^\sigma_{\eta,\xi}}$. We define

$$\beta_x(\sigma) = \sum_{\xi \in Q_x^\sigma} \beta_\xi(\sigma). \tag{10.4.6}$$

**Lemma 10.4.1** *Let $X$ be a normal curve over $S = \mathrm{Spec}\, O_K$, and $x$ be a closed point of the closed fiber $X_s$ such that $X - \{x\}$ is smooth over $S$. Let $\sigma$ be an automorphism of $X$ over $S$ fixing $x$ such that the action on the local ring $O_{X,x}$ is non-trivial.*

(1) *Let $K'$ be a finite extension such that the morphism $S' = \mathrm{Spec}\, O_{K'} \to S$ is finite. Then, for the unique point $x' \in X' = X \times_S S'$ above $x$, we have $\psi_x(\sigma) = \psi_{x'}(\sigma)$, $\gamma_x(\sigma) = \gamma_{x'}(\sigma)$, $\alpha_x(\sigma) = \alpha_{x'}(\sigma)$ and $\beta_x(\sigma) = \beta_{x'}(\sigma)$.*
(2) *Let $f : W \to X$ be a proper morphism of normal curves over $S$, and $Z \subset f^{-1}(x)$ be a finite set of closed points such that $W - Z$ is smooth over $S$. Assume that $f$ induces an isomorphism $W - f^{-1}(x) \to X - \{x\}$, and that $\sigma$ induces an automorphism of $W$ over $S$ compatible with $f$. We identify the normalization $W_s^\natural$ of $W_s$ with the disjoint union $X_s^\natural \amalg E$ of the normalization $X_s^\natural$ of $X_s$ and a proper smooth curve $E$. Then*

$$\begin{aligned}\psi_x(\sigma) + \gamma_x(\sigma) &= \sum_{w \in Z^\sigma} (\psi_w(\sigma) + \gamma_w(\sigma)) \\ &\qquad - \mathrm{Tr}(\sigma : R\Gamma(E, \mathbf{Z}_\ell)), \end{aligned} \tag{10.4.7}$$

$$\alpha_x(\sigma) = \sum_{w \in f^{-1}(x) \cap W^\sigma} \alpha_w(\sigma) - \mathrm{Tr}(\sigma : R\Gamma(E, \mathbf{Z}_\ell)). \tag{10.4.8}$$

*Proof* Since the canonical morphism $R\psi_X \mathbf{Z}_\ell \to Rf_* R\psi_W \mathbf{Z}_\ell$ is an isomorphism by Theorem 10.2.4(1) and the canonical morphism $\mathbf{Z}_\ell \to R^0\psi \mathbf{Z}_\ell$ is an isomorphism by Proposition 10.2.7(1), the first equality (10.4.7) follows similarly to the proof of Lemma 10.3.3(2).

We show the second equality (10.4.8). Note that the sum on the right-hand side is a finite sum since $d_\sigma$ is a basis on a dense open subset of each irreducible component. The difference $\sum_{w \in f^{-1}(x) \cap W^\sigma} \alpha_w(\sigma) - \alpha_x(\sigma)$ equals the sum of

$\sum_{w'\in C\cap W^\sigma} \alpha_{w'}(\sigma)$ for irreducible components $C$ of $E$. If $\sigma(C) \neq C$, we have $\mathrm{Tr}(\sigma : R\Gamma(E_C, \mathbf{Z}_\ell)) = 0$ for the union $E_C \subset E$ of the orbit of $C$, and $E_C \cap W_s^{\natural\sigma} = \varnothing$. Suppose $\sigma(C) = C$. If $C^\sigma = C \cap W_s^{\natural\sigma}$ is finite, we have $\sum_{w'\in C^\sigma} \alpha_{w'}(\sigma) = \mathrm{Tr}(\sigma : R\Gamma(C, \mathbf{Z}_\ell))$ by the Lefschetz trace formula Theorem 8.1.7. If $C^\sigma = C$, and hence $\sigma|_C = 1_C$, we have $\sum_{w'\in C} \alpha_{w'}(\sigma) = -\deg \Omega^1_C$ and $\mathrm{Tr}(1_C : R\Gamma(C, \mathbf{Z}_\ell)) = -\deg \Omega^1_C$, also by the Lefschetz trace formula Theorem 8.1.7. □

**Proposition 10.4.2** *Let $X$ be a normal curve over $S = \operatorname{Spec} O_K$, and $x$ be a closed point of the closed fiber $X_s$ such that $X - \{x\}$ is smooth over $S$. Let $\sigma$ be an automorphism of $X$ over $S$ fixing $x$ such that the action on the local ring $O_{X,x}$ is non-trivial. Then $\psi_x(\sigma)$ is an integer independent of $\ell \neq p$, and we have*

$$\psi_x(\sigma) + \gamma_x(\sigma) = \alpha_x(\sigma) - \beta_x(\sigma). \tag{10.4.9}$$

*Proof* The proof is similar to that of Proposition 10.3.4. First, we show the case where $X$ is a semi-stable curve. Let $\pi : X_s^\natural \to X_s$ be the canonical morphism of normalization. We have an exact sequence $0 \to \Lambda_{X_s} \to \pi_*\Lambda_{X_s^\natural} \to R^1\psi\Lambda \to 0$ by Theorem 10.2.9. This implies the equality $\psi_x(\sigma) = \gamma_x(\sigma)$. Hence, the left-hand side equals $2\gamma_x(\sigma)$. If $X \to S$ is smooth at $x$, we have $\gamma_x(\sigma) = 0$. If $X \to S$ is not smooth at $x$, we may assume that $x$ is contained in two irreducible components of the closed fiber $X_s$ since the assertion is étale local. We have $\gamma_x(\sigma) = 1$ if $\sigma$ stabilizes the two components of $X_s$ containing $x$, and $\gamma_x(\sigma) = -1$ otherwise.

We compute the right-hand side. By Nakayama's lemma, if $u, v \in \mathfrak{m}_x$ is a system of generators, the ideal $I_\sigma \subset O_X$ defining the $\sigma$-fixed part $X^\sigma$ is generated by $\sigma(u) - u, \sigma(v) - v$ at $x$. We first consider the case where $X$ is smooth at $x$, and let $X \to \operatorname{Spec} O_K[u]$ be an étale morphism. Let $\xi \in X_s$ be the generic point, $t \in O_K$ be a uniformizer, and $n = \operatorname{ord}_\xi(\sigma(u) - u)$. Then $\alpha_x(\sigma) = \operatorname{ord}_x((\sigma(u) - u)/t^n)|_{X_s}$. Since $I_\sigma$ is generated by $\sigma(u) - u$ at $x$, we have $\alpha_x(\sigma) = \beta_x(\sigma)$ by Lemma 10.4.3.

Assume that $X \to S$ is not smooth at $x$, and let $X \to \operatorname{Spec} O_K[u, v]/(uv - t)$ be an étale morphism. Then $I_\sigma$ is generated by $\sigma(u) - u, \sigma(v) - v$ at $x$ and we have $v \cdot (\sigma(u) - u) = -\sigma(u)(\sigma(v) - v)$. Let $C_1$ and $C_2$ be the two components of $X_s$ defined by $u = 0$ and $v = 0$ respectively, and let $\xi_1$ and $\xi_2$ be the generic points.

First, we consider the case where $\sigma$ stabilizes the two components $C_1$ and $C_2$. Set $n_1 = \operatorname{ord}_{\xi_1}(\sigma(v) - v)$ and $n_2 = \operatorname{ord}_{\xi_2}(\sigma(u) - u)$. Then $\alpha_x(\sigma) = \operatorname{ord}_x((\sigma(v) - v)/t^{n_1})|_{C_1}) + \operatorname{ord}_x((\sigma(u) - u)/t^{n_2})|_{C_2})$. If we set $w = v \cdot (\sigma(u) - u)/(u^{n_1+1}v^{n_2+1}) = -\sigma(u)(\sigma(v) - v)/(u^{n_1+1}v^{n_2+1})$, we have $\alpha_x(\sigma) =$

$\mathrm{ord}_x(w_{C_1}) + (-n_1 + n_2 + 1) + \mathrm{ord}_x(w_{C_2}) + (-n_2 + n_1 + 1)$. Thus, we have $\alpha_x(\sigma) = \beta_x(\sigma) + 2$ by Lemma 10.4.3.

Assume that $\sigma$ switches the two components $C_1$ and $C_2$. Then we have $\alpha_x(\sigma) = 0$. The ideal $\mathcal{I}_\sigma$ is generated by $\sigma(u) - u = -\sigma(u)/v \cdot (\sigma(v) - v)$ at $x$. Since $\mathrm{ord}_x(\sigma(u) - u)|_{C_1}) = \mathrm{ord}_x(\sigma(u) - u)|_{C_2}) = 1$, we have $\beta_x(\sigma) = 2$ by Lemma 10.4.3. Thus, the equality (10.4.9) is proved in the semi-stable case.

We reduce the general case to the semi-stable case by using Lemma 10.4.1. By Theorem 10.1.8 and Lemma 10.3.3(1), after replacing $K$ by a finite separable extension and $X$ by the base change if necessary, we may assume that there exist a semi-stable curve $W$ over $S$, an automorphism $\sigma$ over $S$, and a proper morphism $f: W \to X$ over $S$ compatible with the action of $\sigma$ such that $f$ induces an isomorphism $W - f^{-1}(x) \to X - \{x\}$.

Since $\beta_x(\sigma) = \sum_{w \in f^{-1}(x)} \beta_w(\sigma)$, the equality (10.4.9) for any closed point $w \in W^\sigma \cap f^{-1}(x)$ implies the equality (10.4.9) for $x$ by Lemma 10.4.1.

Since $\alpha_x(\sigma)$, $\beta_x(\sigma)$, and $\gamma_x(\sigma)$ are integers independent of $\ell \neq p$, $\psi_x(\sigma)$ is also an integer independent of $\ell \neq p$ by (10.4.9). □

**Lemma 10.4.3** *Let $K$ be a henselian discrete valuation field with residue field $F$, and $X$ be a regular flat curve over $S = \mathrm{Spec}\, O_K$. Write the closed fiber $X_s$ as a linear combination $\sum_i m_i C_i$ of irreducible components as a Cartier divisor. Let $D \subset X$ be a Cartier divisor quasi-finite flat over $S$. Let $x \in D_s$ be a closed point in the closed fiber, and $(D, C_i)_x = \dim_F O_{D \cap C_i, x}$ be the intersection number at $x$. Let $D_x$ be the localization at $x$ and let $\deg D_{x,\eta} = \dim_K O_{D_x} \otimes K$. Then $\sum_i m_i (D, C_i)_x = \deg D_{x,\eta}$.*

*Proof* Since $O_K$ is henselian, by Lemma 4.1.2 the localization $D_x$ is finite over $S$. Hence, the $O_K$-module $O_{D_x}$ is free of finite rank and we have $\deg D_{x,s} = \deg D_{x,\eta}$. The left-hand side $\sum_i m_i (D, C_i)_x$ equals the degree $\deg D_{x,s}$, and the assertion follows. □

Let $D \subset X_\eta$ be a finite closed subset and let $U = X_\eta - D$ be the complement. Let $\sigma$ be a non-trivial automorphism of $X$ over $S$. Assume that $U$ is stable under the action of $\sigma$, and that $U^\sigma = \varnothing$. We define modifications $\alpha'_x(\sigma) \in \mathbf{Z}$ and $\beta'_x(\sigma) \in \mathbf{N}$ of the integers $\alpha_x(\sigma) \in \mathbf{Z}$ and $\beta_x(\sigma) \in \mathbf{N}$. Set

$$\alpha'_x(\sigma) = \sum_{x' \in P_x^\sigma} (\alpha_{x'}(\sigma) - 1) = \alpha_x(\sigma) - \gamma_x(\sigma) + 1. \tag{10.4.10}$$

For each fixed point $\xi \in D^\sigma$, let

$$\deg_s \xi = [K(\xi) : K]_s \tag{10.4.11}$$

denote the separable degree, and set

$$\beta'_\xi(\sigma) = \beta_\xi(\sigma) - \deg_s \xi = \dim_K \mathcal{O}_{X^\sigma_{\eta,\xi}} - \deg_s \xi. \tag{10.4.12}$$

We define

$$\beta'_x(\sigma) = \sum_{\xi \in Q_x \cap D^\sigma} \beta'_\xi(\sigma). \tag{10.4.13}$$

**Corollary 10.4.4** *Let $X$ be a normal curve over $S = \operatorname{Spec} \mathcal{O}_K$, and $x$ be a closed point of the closed fiber $X_s$ such that $X - \{x\}$ is smooth over $S$. Let $D \subset X_\eta$ be a finite closed subset, and let $j : U = X_\eta - D \to X$ be the open immersion of the complement.*

(1) *Let $\Lambda$ be a finite field of characteristic $\ell$ different from that of the residue field $k$. Then $\dim_\Lambda R\psi_x j_! \Lambda$ is independent of $\ell$ and we have*

$$-\dim_\Lambda R\psi_x j_! \Lambda = 2 \cdot \delta_x - \# P_x + \sum_{\xi \in Q_x \cap D^\sigma} \deg_s \xi. \tag{10.4.14}$$

(2) *Let $\sigma$ be an automorphism of $X$ over $S$ fixing $x$ such that the action on the local ring $\mathcal{O}_{X,x}$ is non-trivial. Assume that $U$ is stable under the action of $\sigma$, and that $U^\sigma = \varnothing$. Then the difference $\operatorname{Tr}(\sigma : R\psi_x j_! \mathbf{Z}_\ell) = \operatorname{Tr}(\sigma : R^0\psi_x j_! \mathbf{Z}_\ell) - \operatorname{Tr}(\sigma : R^1\psi_x j_! \mathbf{Z}_\ell)$ is a rational integer independent of $\ell \neq \operatorname{char} k$, and is $0$ unless $\sigma$ is of order a power of $p$. We have*

$$-\operatorname{Tr}(\sigma : R\psi_x j_! \mathbf{Z}_\ell) = \alpha'_x(\sigma) - \beta'_x(\sigma). \tag{10.4.15}$$

*Proof* (1) By the distinguished triangle $R\psi_x j_! \Lambda \to R\psi_x \Lambda \to R\psi_x \Lambda_D \to$, we have

$$\dim R\psi_x j_! \Lambda = \dim R\psi_x \Lambda - \dim R\psi_x \Lambda_D.$$

Since $\dim R\psi_x \Lambda_D = \sum_{\xi \in Q_x} \deg_s \xi$, the assertion follows by Proposition 10.3.4.

(2) By the distinguished triangle $R\psi_x j_! \mathbf{Z}_\ell \to R\psi_x \mathbf{Z}_\ell \to R\psi_x \mathbf{Z}_{\ell,D} \to$, we have

$$\operatorname{Tr}(\sigma : R\psi_x j_! \mathbf{Z}_\ell) = \operatorname{Tr}(\sigma : R\psi_x \mathbf{Z}_\ell) - \sum_{\xi \in Q_x \cap D^\sigma} \deg \xi.$$

Hence, the equality (10.4.15) follows from Proposition 10.4.2. The right-hand side of (10.4.15) is a rational integer independent of $\ell$. Since the complex $R\psi_x j_! \pi_* \mathbf{Z}_\ell$ of $\mathbf{Z}_\ell[G]$-modules is a perfect complex by Corollary 10.2.8(1), we have $\operatorname{Tr}(\sigma : R\psi_x j_! \mathbf{Z}_\ell) = 0$ if the order of $\sigma$ is divisible by $\ell \neq p$ by Proposition 6.1.5(1). Since $\operatorname{Tr}(\sigma : R\psi_x j_! \mathbf{Z}_\ell)$ is independent of $\ell$, we have $\operatorname{Tr}(\sigma : R\psi_x j_! \mathbf{Z}_\ell) = 0$ unless the order of $\sigma$ is a power of $p$. □

**Exercise 10.4** Let $k$ be a field of characteristic $p > 0$, and $S = \operatorname{Spec} \mathcal{O}_K$ for $K = k((y))$. Let $n > 1$ be a divisor of $p - 1$, and $X$ be the normalization of the affine line $\mathbf{A}^1_S = \operatorname{Spec} O_K[t]$ in the Artin–Schreier covering defined by $u^p - u = t^n/y^p$ as in Exercise 10.3. Let $\sigma$ be the automorphism of $X$ over $\mathbf{A}^1_S$ defined by $\sigma(u) = u + 1$. Compute the trace $\operatorname{Tr}(\sigma\colon R^1\psi\Lambda_x)$ at the singular point $x$ of the closed fiber of $X$, and verify the formula (10.4.9).

## 10.5 Deligne–Kato Formula

Let $X$ be a normal curve over $S = \operatorname{Spec} O_K$, and $x \in X_s$ be a closed point of the closed fiber such that the complement $X - \{x\}$ is smooth over $S$. Recall that $\pi\colon X^\natural_s \to X_s$ denotes the normalization and

$$P_x = \pi^{-1}(x), \qquad Q_x = \operatorname{Spec} O_{X,x} \cap X_\eta.$$

Let $\Lambda$ be a finite field of characteristic $\ell$ different from the characteristic of the residue field $F$. Let $\mathcal{F}$ be a locally constant constructible sheaf of $\Lambda$-modules on the complement $U = X_\eta - D \subset X_\eta$ of a finite closed subset $D$ of the generic fiber, and let $j\colon U \to X$ be the open immersion. Let $V \to U$ be a finite étale Galois covering of Galois group $G$ such that the pull-back $\mathcal{F}_V$ is constant, and let $f\colon W \to X$ be the normalization of $X$ in $V$. After replacing $K$ by a finite extension and $X$ by an open neighborhood of $x$, we may assume that $W - f^{-1}(x)$ is smooth over $S$ by Lemma 9.3.2 and Proposition 9.3.5.

The normalization $\pi\colon X^\natural_s \to X_s$ induces an isomorphism $X^\natural_s - P_x \to X_s - \{x\}$. By this isomorphism, we identify the generic point of an irreducible component of $X^\natural_s$ with the generic point of its image in $X_s$. Let $x' \in P_x$, and let $\xi \in X^\natural_s$ be the generic point of the irreducible component $C_\xi$ of $X^\natural_s$ containing $x'$. Let $K_\xi$ be the local field at $\xi$ and let $D_\xi \subset G$ denote the decomposition subgroup for $\xi \in X$. The residue field $F_\xi$ of $K_\xi$ is the function field of the curve $C_\xi$ over $k$.

Since, by Proposition 10.2.7(2), the sheaf $R^0\psi\mathcal{F}$ of nearby cycles is constructible, there exists a dense open subset $U_\xi$ of $C_\xi$ on which the restriction of $R^0\psi\mathcal{F}$ is a locally constant sheaf. Let

$$\operatorname{dimtot}_{x'} R^0\psi\mathcal{F}|_{U_\xi} = \operatorname{rank} R^0\psi\mathcal{F}|_{U_\xi} + \operatorname{Sw}_{x'} R^0\psi\mathcal{F}|_{U_\xi} \in \mathbf{N}$$

denote the sum of the rank and the Swan conductor at $x'$. Since the formation of nearby cycles commutes with base change by Proposition 10.2.3, the total dimension $\operatorname{dimtot}_{x'} R^0\psi\mathcal{F}|_{U_\xi} \in \mathbf{N}$ is invariant under the base change.

For the generic point $\xi$ of each irreducible component $C$ of the closed fiber $X_s$, we define the Swan conductor $n = \operatorname{Sw}_\xi \mathcal{F} \in \mathbf{N}$, $d = \operatorname{rank} \mathcal{F} - \operatorname{rank}_\xi R^0\psi\mathcal{F} \in \mathbf{N}$ and the refined Swan conductor $\operatorname{rsw}_\xi \mathcal{F} \in \operatorname{Hom}_F(\mathfrak{m}^n_K/\mathfrak{m}^{n+1}_K, \Omega^{1\otimes d}_{F_\xi})$ as follows. Let $V$ be a finite étale Galois covering of $U$ trivializing $\mathcal{F}$, and let $W$ be the normalization of $X$ in $V$. By Proposition 9.3.5, after replacing $K$ by a finite separable extension we may assume that the closed fiber $W_s$ is reduced.

Let $K_\xi$ be the local field at $\xi$ and let $L$ be the local field at a point of $W$ above $\xi$. Then the residue field $F_\xi$ of $K_\xi$ is the function field of a curve $C$ over $k$, and $L$ is a Galois extension of $K_\xi$ of ramification index 1. Let $M$ be the representation of the decomposition group $D = \mathrm{Gal}(L/K_\xi)$ defined by $\mathcal{F}$. Then, by the construction in Section 7.3, the Swan conductor $n = \mathrm{Sw}_\xi M \in \mathbf{N}$, the codimension $d = \dim_\Lambda M/M^{D_\xi} \in \mathbf{N}$, and the refined Swan conductor $\mathrm{rsw}_\xi M \in \mathrm{Hom}_F(\mathfrak{m}_K^n/\mathfrak{m}_K^{n+1}, \Omega^{1\otimes d}_{F_\xi}) - \{0\} \subset S_{K_\xi}$ are defined.

Further, by Corollary 7.2.8(1), the Swan conductor $n = \mathrm{Sw}_\xi M \in \mathbf{N}$ and the refined Swan conductor $\mathrm{rsw}\, M$ are independent of the choice of $V$. Thus, the Swan conductor $\mathrm{Sw}_x \mathcal{F}$ is defined and the refined Swan conductor

$$\mathrm{rsw}_\xi \mathcal{F} \in \mathrm{Hom}_k(\mathfrak{m}_K^n/\mathfrak{m}_K^{n+1}, \Omega^{1\otimes d}_{F_\xi/k})$$

is defined as a non-zero element of an $F_\xi$-vector space of dimension 1. Further, by Lemma 7.2.10, if we replace $K$ by a totally ramified extension $K'$, the Swan conductor $\mathrm{Sw}_\xi M$ is multiplied by the ramification index $e$ and the construction of the refined Swan conductor $\mathrm{rsw}_\xi \mathcal{F}$ commutes with base change.

The closed point $x' \in C_\xi$ defines a discrete valuation of the residue field $F_\xi$ and a free sub-$\mathcal{O}_{C_\xi,x'}$-module $\mathrm{Hom}_k(\mathfrak{m}_K^n/\mathfrak{m}_K^{n+1}, \Omega^{1\otimes d}_{\mathcal{O}_{C_\xi,x'}/k})$ of rank 1 of $\mathrm{Hom}_k(\mathfrak{m}_K^n/\mathfrak{m}_K^{n+1}, \Omega^{1\otimes d}_{F_\xi/k})$. Thus, the valuation

$$\mathrm{ord}_{x'} \mathrm{rsw}_\xi \mathcal{F} \in \mathbf{Z}$$

is defined as in (7.3.1). In the case $n = d = 0$, the refined Swan conductor $\mathrm{rsw}_\xi \mathcal{F}$ is $1 \in \mathrm{Hom}_k(\mathcal{O}_K/\mathfrak{m}_K, \Omega^{1\otimes 0}_{F_\xi/k}) = F_\xi$ and the valuation $\mathrm{ord}_{x'} \mathrm{rsw}_\xi \mathcal{F}$ vanishes.

For each $x' \in P_x$, let $\xi$ be the generic point of the irreducible component $C_\xi$ of $X_s^\natural$ containing $x'$, and we define an integer $s_{x'}(\mathcal{F}) \in \mathbf{Z}$ by

$$s_{x'}(\mathcal{F}) = \mathrm{dimtot}_{x'} R^0\psi\mathcal{F}|_{U_\xi} - \mathrm{ord}_{x'} \mathrm{rsw}_\xi \mathcal{F}. \tag{10.5.1}$$

By Lemma 7.2.10, the integer $s_{x'}(\mathcal{F}) \in \mathbf{Z}$ is invariant under finite extension of $K$.

**Lemma 10.5.1** *Let $V \to U$ be a finite étale Galois covering of Galois group $G$ such that the pull-back $\mathcal{F}_V$ is constant. Let $f: W \to X$ be the normalization in $V$, and assume that the complement $W - f^{-1}(x)$ is smooth over $S$. Let $x' \in P_x$. Let $w' \in P_w$ be a point above $x'$, and let $I \subset G$ be the inverse image of the inertia subgroup of $x$ by the surjection $G \to \overline{G}$. Then, in the notation of*

*Corollary* 10.4.4(2), *we have*

$$-(s_{x'}(\mathcal{F}) - \operatorname{rank}\mathcal{F}\cdot s_{x'}(\Lambda)) = \frac{1}{\#I}\sum_{\sigma\in I}\alpha'_{w'}(\sigma)\cdot(\operatorname{Tr}^{\mathrm{Br}}(\sigma\colon M) - \dim M). \tag{10.5.2}$$

*Proof* Let $C \subset W_s^\natural$ be the connected component containing $w'$, and let $\zeta \in W_s$ be the generic point of $C$. Let $\xi \in X_s$ be the image of $\zeta$, and let $\mathbf{K}$ and $\mathbf{L}$ be the local fields at $\xi$ and $\zeta$ respectively. Then $\mathbf{L}$ is a Galois extension of $\mathbf{K}$, and the Galois group is identified with the decomposition subgroup $D \subset G$ of $\zeta$. Let $P \subset D$ be the wild inertia subgroup.

We apply Proposition 7.3.4 to the representation $M$ of $D$ defined by $\mathcal{F}$. Then the $P$-fixed part $M^P$ is identified with the stalk $(R^0\psi\mathcal{F})_{\overline{\zeta}}$ at a geometric point above $\zeta$. Hence, the content in parentheses on the left-hand side equals $\dim M^P + \mathrm{Sw}_{x'} M^P - \mathrm{ord}_{x'}\mathrm{rsw}_\xi M - \dim M$. Since $\alpha'_{w'}(\sigma)$ on the right-hand side equals $-\mathrm{ord}'_w\widetilde{s}(\sigma)$ in (7.3.9), we obtain (10.5.2) by applying Proposition 7.3.4. □

By taking the sum over points $x' \in P_x$, we define an integer $s_x(\mathcal{F}) \in \mathbf{Z}$ by

$$s_x(\mathcal{F}) = \sum_{x'\in P_x} s_{x'}(\mathcal{F}). \tag{10.5.3}$$

In the case without ramification along the closed fiber, the invariant $s_x(\mathcal{F})$ is the sum of the total dimensions of the extension, as follows.

**Lemma 10.5.2** *Let $j\colon U \subset X_\eta \to X$ be the open immersion. Assume that $j_*\mathcal{F}$ is locally constant on the complement $X - T$ of a closed subset $T \subset X$, quasi-finite over $S$. For the generic point $\xi \in X_s^\natural$ of each irreducible component $C_\xi \subset X_s^\natural$, let $\widetilde{\mathcal{F}}|_{U_\xi}$ denote the pull-back of $j_*\mathcal{F}$ on the inverse image $U_\xi \subset C_\xi$ of $X - T$. Then, for each $x' \in P_x$ and the generic point $\xi$ of the irreducible component $C_\xi$ of $X_s^\natural$ containing $x'$, we have $s_{x'}(\mathcal{F}) = \dim\mathrm{tot}_{x'}\widetilde{\mathcal{F}}|_{U_\xi}$ and consequently*

$$s_x(\mathcal{F}) = \sum_{x'\in P_x} \dim\mathrm{tot}_{x'}\widetilde{\mathcal{F}}|_{U_\xi}. \tag{10.5.4}$$

*Proof* We may assume that $X - T$ is smooth over $S$. Then, since $j_*\mathcal{F}$ is locally constant on $X - T$, the canonical morphism $j_*\mathcal{F}|_{X_s} \to R^0\psi\mathcal{F}$ is an isomorphism on $X - T$ by Corollary 10.2.5, and we have $\mathrm{rsw}_\xi\mathcal{F} = 0$ by (10.5.1). □

We also define an integer $t_x(\mathcal{F}) \in \mathbf{N}$. After replacing $K$ by a finite extension, we may further assume that the residue fields at points of $Q_x \cap D$ and $f^{-1}(Q_x \cap D)$ are finite separable extensions of $K$. Then, for each point $\xi \in Q_x \cap D$, the

Swan conductor $\mathrm{Sw}_\xi \mathcal{F} \in \mathbf{N}$ is defined. For $\xi \in Q_x \cap D$, let $\deg \xi \in \mathbf{N}$ denote the degree $[K(\xi) : K]$. We define

$$t_x(\mathcal{F}) = \sum_{\xi \in Q_x \cap D} \deg \xi \cdot (\operatorname{rank} \mathcal{F} + \mathrm{Sw}_\xi \mathcal{F}). \tag{10.5.5}$$

The integer $t_x(\mathcal{F}) \in \mathbf{N}$ is also invariant under finite extension of $K$.

**Theorem 10.5.3** (Deligne–Kato formula) *Let $X$ be a normal curve over $S$, and $x \in X_s$ be a closed point of the closed fiber such that the complement $X - \{x\}$ is smooth over $S$. Let $j : U \to X_\eta$ be the open immersion of a dense open subset, and let $\mathcal{F}$ be a locally constant constructible sheaf of $\Lambda$-modules on $U$. Then*

$$-\dim_\Lambda R\psi_x j_! \mathcal{F} = 2 \operatorname{rank} \mathcal{F} \cdot \delta_x - s_x(\mathcal{F}) + t_x(\mathcal{F}). \tag{10.5.6}$$

By Lemma 10.5.2, if $\mathcal{F}$ has no ramification along the closed fiber, $s_x(\mathcal{F})$ is given by (10.5.4). Further, if $X$ is smooth over $S$, we have $\delta_x = 0$. This is the case originally proved by Deligne in [60].

*Proof* First, we show the case where $\mathcal{F} = \Lambda$ is a constant sheaf. In this case, the equality (10.5.6) is (10.4.14). Hence, the equality in this case follows from Corollary 10.4.4(1).

We show the general case. Since the case $\mathcal{F} = \Lambda$ is already proved, it suffices to show

$$\begin{aligned} &-(\dim_\Lambda R\psi_x j_! \mathcal{F} - \operatorname{rank} \mathcal{F} \cdot \dim_\Lambda R\psi_x j_! \Lambda) \\ &\quad = -(s_x(\mathcal{F}) - \operatorname{rank} \mathcal{F} \cdot s_x(\Lambda)) + (t_x(\mathcal{F}) - \operatorname{rank} \mathcal{F} \cdot t_x(\Lambda)). \end{aligned} \tag{10.5.7}$$

Let $V \to U$ be a finite étale Galois covering of Galois group $G$ such that the pull-back $\mathcal{F}_V$ is constant. Let $f : W \to X$ be the normalization of $X$ in $V$. Since the assertion is étale local, we may assume that $f^{-1}(x)$ consists of a single closed point $w$.

Let $M = \Gamma(V, \mathcal{F})$ be the corresponding representation of $G$, and let $G_p \subset G$ denote the subset consisting of elements $\sigma \neq 1$ of order a power of $p$. Then, since the complex $R\psi_x j_! \pi_* \mathbf{Z}_\ell$ of $\mathbf{Z}_\ell[G]$-modules is a perfect complex by Corollary 10.2.8(1), the left-hand side of (10.5.7) equals

$$-\frac{1}{\# G} \sum_{\sigma \in G_p} \operatorname{Tr}(\sigma : R\psi_w j_! \mathbf{Z}_\ell) \cdot (\operatorname{Tr}^{\mathrm{Br}}(\sigma : M) - \dim M) \tag{10.5.8}$$

by Proposition 6.1.5(2) and Corollary 10.4.4(2).

Let $Q_w = \operatorname{Spec} \mathcal{O}_{W,w} \cap W_\eta$. By Proposition 9.3.5, we may assume that $W - \{w\}$ is smooth over $S$ and that every point of $Q_w - (Q_w \cap V)$ is $K$-rational. Let $\overline{f} : W_s^\natural \to X_s^\natural$ denote the morphism induced by $f$. For $x' \in P_x$, set

$$A_{x'} = \frac{1}{\#G} \sum_{\sigma \in G_p} \sum_{\substack{w' \in P_w^\sigma, \\ \overline{f}(w') = x'}} \alpha'_{w'}(\sigma) \cdot (\mathrm{Tr}^{\mathrm{Br}}(\sigma\colon M) - \dim M). \tag{10.5.9}$$

For $\xi \in Q_x - (Q_x \cap U)$, set

$$B_\xi = \frac{1}{\#G} \sum_{\sigma \in G_p} \sum_{\substack{\zeta \in Q_w^\sigma, \\ f(\zeta) = \xi}} \beta'_\zeta(\sigma) \cdot (\mathrm{Tr}^{\mathrm{Br}}(\sigma\colon M) - \dim M). \tag{10.5.10}$$

Then, by Corollary 10.4.4(2) and (10.4.13), the number (10.5.8) equals

$$\sum_{x' \in P_x} A_{x'} - \sum_{\xi \in Q_x - (Q_x \cap U)} B_\xi.$$

Let $x' \in P_x$ and let $w' \in \overline{f}^{-1}(x')$. Then the stabilizer $I \subset G$ of $w'$ equals the inverse image of the inertia subgroup of $\overline{G}$ at $w'$. Hence, $\overline{f}^{-1}(x')$ is identified with $I/P$ and $A_{x'}$ equals the right-hand side of (10.5.2). Hence, by Lemma 10.5.1, we have $A_{x'} = -(s_{x'}(\mathcal{F}) - \operatorname{rank} \mathcal{F} \cdot s_{x'}(\Lambda))$.

Let $\xi \in Q_x - (Q_x \cap U)$ and $\zeta \in f^{-1}(\xi)$. Let $J \subset G$ be the inertia subgroup at $\zeta$. By replacing $K$ by a finite separable extension, we may assume $k(\xi) = k(\zeta)$. Then the inverse image $f^{-1}(\xi)$ is identified with $G/J$ and we have

$$B_\xi = \frac{1}{\#J} \sum_{\sigma \in J \cap G_p} \beta'_\zeta(\sigma) \cdot (\mathrm{Tr}^{\mathrm{Br}}(\sigma\colon M) - \dim M) = -(t_\xi(\mathcal{F}) - \operatorname{rank} \mathcal{F} \cdot t_\xi(\Lambda)).$$

Hence, the equality (10.5.7) follows. □

**Exercise 10.5** Let $\mathcal{F}$ be a locally constant constructible sheaf of $\Lambda$-modules on a dense open subset $U \subset \mathbf{A}^1_k = \operatorname{Spec} k[x]$ of an affine line over an algebraically closed field $k$ of characteristic $p > 0$, and $j_! \mathcal{F}$ be the extension by 0 on $\mathbf{A}^1_k$. Let $\mathcal{L}$ be the locally constant constructible sheaf of $\Lambda$-modules on the complement $\mathbf{A}^2_k - (y = 0) = \operatorname{Spec} k[x, y^{\pm 1}]$ of the $y$-axis in an affine plane defined by the Artin–Schreier equation $t^p - t = x/y$, and $j_{2!}\mathcal{L}$ be the extension by 0 on $\mathbf{A}^2_k$. Let $p_1, p_2\colon \mathbf{A}^2_k \to \mathbf{A}^1_k$ be the projections. Compute the dimension of the stalk at $(0, 0)$ of the sheaf $R\psi(p_1^* j_! \mathcal{F} \otimes j_{2!}\mathcal{L})$ of nearby cycles with respect to the second projection $p_2$ by applying the Deligne–Kato formula to the base change by the relative Frobenius $\mathbf{A}^1_k = \operatorname{Spec} k[y'] \to \mathbf{A}^1_k$ sending $y$ to $y'^p$.

## Historical Notes

The vanishing in Example 10.2.6 is an essential ingredient in the theory of $\ell$-adic Fourier transforms and is proved in [56, Théorème (2.4.4)] by a different method. The computation in Theorem 10.2.9 of the nearby cycles at singular points of a semi-stable curve is a special case of the Picard–Lefschetz formula in [20]. The proof given here using blow-up was found by the author during a discussion with Osamu Hyodo in a student room of the department of mathematics in the Hongo campus of the University of Tokyo, if memory serves (see [40]). Its generalization to higher dimensions reproves a computation by Rapoport and Zink [65] of nearby cycles of semi-stable schemes in arbitrary dimensions, and is documented by Illusie in [44].

The dimension formula (10.3.7) for the space of nearby cycles for the constant sheaf is proved in [48, Proposition (5.9)] and another proof is given in [69, Proposition (4.2)]. The trace formula (10.4.9) is proved in [70]. The Deligne–Kato formula (10.5.6) in the case where $X$ is smooth over $S$ and $\mathcal{F}$ has no ramification along the closed fiber was proved by Deligne and the detail is written up by Laumon in [60]. This is a crucial step in the proof of the semi-continuity of the Swan conductor proved in [60]. This property is later used in the construction of the characteristic cycle of an étale sheaf in [74]. The letter of Deligne to Illusie dated "October 28, 1976" stating the formula and giving an outline of the proof is described in [45] as Letter 2. The general case of the Deligne–Kato formula (10.5.6) is [48, Theorem (6.7)]. The dimension formula for the local Fourier transform in Exercise 10.5 is proved by Laumon in [61, Théorème (2.4.3)(i)b)] for $\overline{\mathbf{Q}}_\ell$-sheaves. Its deduction from the Deligne–Kato formula (10.5.6) was given in [5, Proposition B.6].

The proof of (10.5.6) in the case without ramification along the closed fiber in [60] uses deformation and compactification. The proof of (10.3.7) in [69] and the proof of (10.4.9) in [70] are similar to each other and both use compactification and replace deformation by the semi-stable reduction theorem, originally proved by Deligne and Mumford [23]. The proof of (10.3.7) and of (10.5.6) in the general case in [48] is by reduction to the case of (10.5.6) without ramification along the closed fiber as proved in [60] earlier.

The proofs given here of (10.3.7) and (10.4.9) are new, replacing compactification in the proofs of [70] and [69] by blow-up based on a local version, Theorem 10.1.8, of the semi-stable reduction theorem proved by Temkin [80]. As the proof of [69, Theorem (4.8)] shows that the semi-stable reduction theorem, Theorem 10.1.8, may be conversely deduced from (10.4.9), these two are equivalent. The proof of (10.5.6) given here, deriving it from (10.3.7) and (10.4.9), is mentioned in [48].

The proof of Proposition 10.3.2 and its Corollary 10.3.5 is also recorded in [11, Proposition 5.1]. A formulation of (10.5.6) using the logarithmic upper numbering filtration is given in [39].

# PART V

# UPPER RAMIFICATION SUBGROUPS

## Heuristics: Eliminating Rigid Geometry

In this preliminary section we present some observations leading to the definition of upper numbering ramification groups. As this serves only for heuristic purposes and is logically independent of the rest of the book, we omit the proof. We will also freely use some elementary terminology from rigid geometry when necessary to describe the ideas.

We begin with a geometric interpretation of the definition of lower numbering ramification groups. In this section, $K$ denotes a complete discrete valuation field. Let $L$ be a finite Galois extension of $K$ of Galois group $G$. Modifying the definition from Section 1.6, for an integer $i > 0$, define a normal subgroup $G_i$ of $G$ by

$$G_i = \mathrm{Ker}(G \to \mathrm{Aut}(O_L/\mathfrak{m}_L^i)).$$

For simplicity, we will assume in this section that the ring $O_L$ of integers is generated by a single element $a \in O_L$ over $O_K$. Even if we drop this assumption, the essential part of what follows still holds for higher-dimensional varieties, except for one point we will mention specifically.

Let $f \in O_K[X]$ be the minimal polynomial of a generator $a \in O_L$. Then we have an isomorphism $O_K[X]/(f) \to O_L$. Let $K_s$ be a separable closure of $K$ and let $O_{K_s}$ be the ring of integers. We regard $D = O_{K_s}$ as a rigid analytic unit disk. Fixing a morphism $L \to K_s$ over $K$, we identify the Galois group $G$ as a subset $\{x \in D \mid f(x) = 0\}$, and the identity $1 \in G$ with $a \in D$. Then the lower ramification subgroup $G_i$ is identified with the intersection $G_i = G \cap D_i$ with the subdisk $D_i = \{x \in D \mid \mathrm{ord}_L(x - a) \geqslant i\}$ with center at $1 \in G$, where $\mathrm{ord}_L$ is the normalized valuation of $L$ extended to $K_s$.

For the upper ramification subgroups, we define a morphism $f\colon D \to D$

of disks by sending $x$ to $f(x)$. Then the Galois group $G$ is identified with the inverse image $f^{-1}(0)$. Let $r > 0$ be a rational number, and let $D^r = \{y \in D \mid \operatorname{ord}_K y \geqslant r\}$ be the subdisk of the disk $D$ on the target of $f$, where $\operatorname{ord}_K$ is the normalized valuation of $K$ extended to $K_s$. Then the inverse image $f^{-1}(D^r) \subset D$ is an open neighborhood of $G = f^{-1}(0)$ and is the disjoint union of finitely many subdisks. This is the point requiring the assumption that $O_L$ is generated by a single element over $O_K$. In general, we need to replace subdisks by rational subdomains.

Let $U^r \subset f^{-1}(D^r)$ be the connected component containing the unit $1 \in G$. Then the upper numbering ramification group $G^r$ is defined as the intersection

$$G^r = G \cap U^r.$$

If $U^r = D_i$, we have $G^r = G_i$ and $r = \varphi(i)$ is the value of a rescaling function defined similarly to (6.2.1). As mentioned, this requires the assumption that $O_L$ is generated by a single element over $O_K$.

We give an algebraic interpretation of this analytic description involving an inequality on valuation. We regard $D^r = \{y \in D \mid \operatorname{ord}_K y \geqslant r\}$ as an ideal $\mathfrak{m}^r_{K_s} \subset O_{K_s}$. We consider the set $\operatorname{Mor}_{O_K}(O_L, O_{K_s}/\mathfrak{m}^r_{K_s})$ of morphisms of rings over $O_K$. Then a mapping $f^{-1}(D^r) \to \operatorname{Mor}_{O_K}(O_L, O_{K_s}/\mathfrak{m}^r_{K_s})$ is defined by sending a point $x$ to the morphism $\varphi_x\colon O_L \to O_{K_s}/\mathfrak{m}^r_{K_s}$ sending $a$ to the class of $x$. This is a surjection and the fiber of each morphism is a translation of the subdisk $D^r \subset D$. Thus, although we don't define a topology on the target set, we may define the set $\pi_0(\operatorname{Mor}_{O_K}(O_L, O_{K_s}/\mathfrak{m}^r_{K_s}))$ of connected components to be $\pi_0(f^{-1}(D^r))$. Then the upper ramification subgroup $G^r$ is defined by requiring the inclusion $G \to f^{-1}(D^r)$ to induce a bijection

$$G/G^r \to \pi_0(f^{-1}(D^r)) = \pi_0(\operatorname{Mor}_{O_K}(O_L, O_{K_s}/\mathfrak{m}^r_{K_s}))$$

to the set of connected components. Thus, the essential ingredient of the definition of upper ramification groups is the construction of the surjection

$$G \to \pi_0(f^{-1}(D^r))$$

of finite sets.

We replace rigid geometry by the language of schemes in the construction of the surjection $G \to \pi_0(f^{-1}(D^r))$. Besides its elementary nature, an advantage of using scheme theory instead of rigid geometry is the geometry of closed fibers. This is essential, for example, in the proof of the semi-continuity of upper numbering filtration and in the study of graded quotients.

Let $K' \subset K_s$ be a finite separable extension of $K$ containing an element $t$ satisfying $\operatorname{ord}_K t = r$. Then the open immersion $f^{-1}(D^r) \subset D$ is defined by the morphism $K'\langle X\rangle \to K'\langle X, f/t\rangle$ of Tate algebras. The Tate algebras are defined by $K'\langle X\rangle = O_{K'}\langle X\rangle \otimes_{O_{K'}} K'$, $K'\langle X, f/t\rangle = O_{K'}\langle X, f/t\rangle \otimes_{O_{K'}} K'$ for the $\mathfrak{m}_{K'}$-adic completions $O_{K'}\langle X\rangle$, $O_{K'}\langle X, f/t\rangle$ of $O_{K'}[X]$ and $A = O_{K'}[X, f/t] \subset K'[X]$.

We assume that the geometric connected components of the rational subdomain $f^{-1}(D_r) \subset D$ are defined over $K'$. Then they correspond bijectively to primitive idempotents of the Tate algebra $K'\langle X, f/t\rangle$. The idempotents are contained in the normalization $B \subset K'\langle X, f/t\rangle$ of $O_{K'}\langle X, f/t\rangle$. The subring $B \subset K'\langle X, f/t\rangle$ is called the unit ball. The set $\pi_0(f^{-1}(D^r))$ of connected components is canonically identified with $\pi_0(\operatorname{Spec} B)$. Further, this is identified with $\pi_0(\operatorname{Spec} B \otimes_{O_{K'}} F')$ for the closed fiber, where $F'$ denotes the residue field of $K'$ (cf. Remark 12.4.7).

Rational subdomains in rigid geometry correspond to partial blow-ups or dilatations for schemes. Set $S = \operatorname{Spec} O_K$ and $S' = \operatorname{Spec} O_{K'}$, and define a closed immersion $T = \operatorname{Spec} O_L \to Q = \mathbf{A}^1_S = \operatorname{Spec} O_K[X]$ by the surjection $O_K[X] \to O_L$. Then the ring $A = O_{K'}[X, f/t]$ is the coordinate ring of a dilatation $Q^{[r]}_{S'} = \operatorname{Spec} A \to Q_{S'} = Q \times_S S'$ at the closed subscheme of $T_{S'} = T \times_S S' \subset Q_{S'}$ defined by the ideal $(t, f)$.

Let $A^\natural$ be the normalization of the integral domain $A \subset K'[X]$, and set $Q^{(r)}_{S'} = \operatorname{Spec} A^\natural$. Since the unit ball $B$ is the $\mathfrak{m}_{K'}$-adic completion of $A^\natural$, and the canonical morphism $A^\natural \to B$ induces an isomorphism $A^\natural \otimes_{O_{K'}} F' \to B \otimes_{O_{K'}} F'$, we may further identify $\pi_0(f^{-1}(D^r))$ with the set $\pi_0(Q^{(r)}_{F'})$ of connected components of the closed fiber $Q^{(r)}_{F'} = \operatorname{Spec} A^\natural \otimes_{O_{K'}} F'$.

Assume further that $K'$ is an extension of $L$. Then the normalization $T^\natural_{S'}$ of $T_{S'}$ is the disjoint union of copies of $S'$ indexed by $G$. The morphism $T \to Q$ induces a morphism $T^\natural_{S'} \to Q^{(r)}_{S'}$ on the normalizations. The mapping $G \to \pi_0(Q^{(r)}_{F'})$ defined by this morphism is compatible with $G \to \pi_0(f^{-1}(D^r))$ under the identification $\pi_0(f^{-1}(D^r)) = \pi_0(Q^{(r)}_{F'})$. Thus, we may replace $G \to \pi_0(f^{-1}(D^r))$ defined using rigid geometry by

$$G = T^\natural_{F'} \to \pi_0(Q^{(r)}_{F'})$$

defined in the language of schemes without introducing completions. By the reduced fiber theorem, Proposition 9.3.5, for $K'$ sufficiently large, the formation of $Q^{(r)}_{F'}$ commutes with extension of $K'$. Thus, the set $\pi_0(Q^{(r)}_{F'})$ and the mapping $G \to \pi_0(Q^{(r)}_{F'})$ are independent of the choice of a sufficiently large separable extension $K'$.

We will define the upper ramification group $G^r$ to be the subgroup of $G$ such that $G \to \pi_0(Q^{(r)}_{F'})$ induces a bijection

$$G/G^r \to \pi_0(Q^{(r)}_{F'}).$$

We describe briefly the contents of each section in the following two chapters. As we don't need completions, we work with henselian discrete valuation rings rather than complete discrete valuation rings.

After preparatory discussions on dilatations and normalizations in Sections 11.1 and 11.2, we study immersions $T \to Q$ to smooth schemes in Section 11.3 and introduce the dilatation $Q^{[r]}_{S'}$ and its normalization $Q^{(r)}_{S'}$. In Section 11.4 we study their properties that are necessary to define and establish properties of upper ramification groups. We compute $Q^{(r)}_{S'}$ explicitly in the case where the ring $O_L$ of integers is generated by a single element over $O_K$ in Section 11.5.

Let $K_s$ be a separable closure, and let $G_K = \mathrm{Gal}(K_s/K)$ denote the absolute Galois group. Let $F$ denote the functor sending a finite separable extension $L$ over $K$ to the finite set $\mathrm{Mor}_K(L, K_s)$ with continuous $G_K$-action of morphisms over $K$. For each rational number $r > 0$, we define in Section 12.1 a quotient functor $F^r$ of the fiber functor $F$ such that $F^r(L)$ has a bijection to $\pi_0(Q^{(r)}_{F'})$ if $Q^{(r)}_{F'}$ is geometrically reduced. The fact that the functor $F^r$ is well-defined is a consequence of the property of $Q^{(r)}_{F'}$ that, for a smooth morphism $Q' \to Q$ compatible with immersions of $T$, the induced morphism $Q'^{(r)}_{F'} \to Q^{(r)}_{F'}$ is a vector bundle. We also introduce a variant $F^{r+}$ using the image of $T^{\natural}_{F'} \to Q^{(r)}_{F'}$.

We show in Section 12.1 that the quotient functor $F^r$ satisfies a certain axiom to be defined as the quotient of $F$ by a normal subgroup, as a consequence of the going down theorem for finite morphisms of normal rings. In Section 12.3 we define the upper ramification groups to be the subgroups defining the functors $F^r$. We prove a criterion for extension to be unramified or tamely ramified in terms of the functor $F^r$ in Section 12.2.

Finally, in Section 12.5 we prove the semi-continuity of the filtrations $G^r$ and $G^{r+}$. The proof relies on the reduced fiber theorem, the semi-stable reduction theorem, and a connectedness theorem of dense open subsets for flat and reduced morphisms, studied in Section 12.4.

**Exercise V.1** Let $K$ be a henselian discrete valuation field, and $L$ be a finite separable extension of degree $n$ of $K$. Assume that $O_L$ is generated by a single element $a$ over $O_K$. Let $a_1, \ldots, a_n = a$ be the conjugates of $a$ in a Galois closure of $L$, and let the minimal polynomial $f = \prod_{i=1}^{n}(X - a_i) \in O_K[X]$ of $a$ define a mapping $f\colon D = O_{K_s} \to D$. Set $s_i = \mathrm{ord}_L(a_i - a_n)$ for $i = 1, \ldots, n-1$, and define the rescaling function $\varphi\colon [0, \infty) \to [0, \infty)$ by

$$\varphi(s) = \frac{1}{e_{L/K}}\Big(\sum_{i=1}^{n-1} \min(s_i, s) + s\Big).$$

Let $s > 0$ be a rational number and set $r = \varphi(s) > 0$. Show that the inverse image $f^{-1}(D^r)$ of $D^r = \{y \in D \mid \mathrm{ord}_K y \geqslant r\}$ equals the union of disks $\{x \in D \mid \mathrm{ord}_L(x - a_i) \geqslant s\}$ for $i = 1, \ldots, n$.

## Historical Notes

The original definition in [1] of upper ramification subgroups uses rigid geometry. It was inspired by [22, A.6]. The approach using algebraic geometry rather than rigid geometry was initiated in [2].

The upper numbering filtration in the classical setting was introduced implicitly, without numbering, in a posthumous article of Herbrand edited by Chevalley in 1933 [38]. The upper numbering appears formally in the proceedings of the Artin–Tate seminar on class field theory published in 1968 [8, p. 119]. It is likely that Artin was aware of it during the seminar in 1951 or even earlier. The first published reference explicitly describing the filtration with its upper numbering appears to be due to Kawada in 1953 [57]. The author thanks Jean-Pierre Serre for his answers concerning the history of this filtration.

# 11

# Stable Integral Models

## 11.1 Dilatations

**Definition 11.1.1** Let $P$ be a scheme, $X \subset P$ be a closed subscheme, and $D \subset P$ be an effective Cartier divisor. Let $P' \to P$ be the blow-up at the intersection $X \cap D = X \times_P D \subset P$, and define the dilatation $P^{[D \cdot X]} \subset P'$ to be the largest open subscheme where the pull-back of the subscheme $X \cap D \subset P$ equals the pull-back of $D \subset P$.

If $\mathcal{I}_X \subset \mathcal{O}_P$ and $\mathcal{I}_D \subset \mathcal{O}_P$ denote the ideals defining the subschemes, and $p \colon P' \to P$ denotes the canonical morphism, the open subscheme $P^{[D \cdot X]} \subset P'$ is the largest open subset where $p^*\mathcal{I}_X \subset p^*\mathcal{I}_D$ holds. In other words, $P^{[D \cdot X]} \subset P'$ is the complement of the proper transform of $D$. If the pull-back $X \cap D \subset X$ of $D \subset P$ is a Cartier divisor, by the universality of blow-up the immersion $X \to P$ is lifted to $X \to P'$ and induces $X \to P^{[D \cdot X]}$. The inverse image $D \times_P P^{[D \cdot X]} \subset P^{[D \cdot X]}$ is a closed subscheme of the inverse image $X \times_P P^{[D \cdot X]}$.

If $P = \operatorname{Spec} A$ is affine and if $D$ is defined by a non-zero divisor $f \in A$, the dilatation is described as follows. Assume that $X = \operatorname{Spec} A/I$ for an ideal $I \subset A$. Then $P^{[D \cdot X]} = \operatorname{Spec} A[I/f]$ for the subring $A[I/f] \subset A[1/f]$. If the image of $f$ in $A/I$ is a non-zero divisor, the morphism $X \to P^{[D \cdot X]}$ is induced by the isomorphism $A/I \to A[I/f]/((I/f)A[I/f])$.

The construction of dilatation has the following functoriality.

**Lemma 11.1.2** *Let*

$$\begin{array}{ccc} X' & \longrightarrow & P' \\ \downarrow & & \downarrow \\ X & \longrightarrow & P \end{array} \tag{11.1.1}$$

*be a commutative diagram of schemes such that the horizontal arrows are* *closed immersions. Let $D \subset P$ be an effective Cartier divisor such that the pull-back $D \times_P P' \subset P'$ is a Cartier divisor, and let $D' \subset P'$ be an effective Cartier divisor divisible by the pull-back $D \times_P P'$.*

(1) *There exists a unique morphism $P'^{[D' \cdot X']} \to P^{[D \cdot X]}$ such that the diagram*

$$\begin{array}{ccc} P'^{[D' \cdot X']} & \longrightarrow & P' \\ \downarrow & & \downarrow \\ P^{[D \cdot X]} & \longrightarrow & P \end{array} \tag{11.1.2}$$

*is commutative.*

(2) *Assume that the diagram* (11.1.1) *is cartesian and $D' = D \times_P P'$. Then the diagram* (11.1.2) *defines an immersion $P'^{[D' \cdot X']} \to P^{[D \cdot X]} \times_P P'$. Further, if $P' \to P$ is flat, then the diagram* (11.1.2) *is cartesian.*

*Proof* (1) First, we consider the case where $P = \operatorname{Spec} A$ and $P' = \operatorname{Spec} B$ are affine and that $D, D'$ and $X, X'$ are defined by $f \in A$, $g \in B$, $I \subset A$, $J \subset B$ as above. Then the commutative diagram (11.1.1) means that $A \to B$ maps $I$ into $J$ and the assumption that $D'$ is divisible by $D \times_P P'$ means $g = fh$ in $B$ for a non-zero divisor $h$. Thus, $A \to B$ induces $A[1/f] \to B[1/g]$ sending $f$ to $h/g$ and $A[I/f] \to B[J/g]$. This defines the required morphism $P'^{[D' \cdot X']} \to P^{[D \cdot X]}$ in this case. The general case follows by patching.

(2) We may assume that $P = \operatorname{Spec} A$ and $P' = \operatorname{Spec} B$ are affine, and let the notation be as in the proof of (1). Then $A[I/f] \otimes_A B \to B[(IB)/f]$ is a surjection.

Further, if $A \to B$ is flat, then the injection $A[I/f] \to A[1/f]$ remains an injection after $\otimes_A B$, and the diagram (11.1.2) is cartesian. □

**Lemma 11.1.3** *Let $P$ be a scheme, $X \subset P$ be a closed subscheme, and $D \subset P$ be an effective Cartier divisor.*

(1) *Let $D' \subset P$ be another effective Cartier divisor and $D'' = D + D'$ be the sum as Cartier divisors. Assume that $D \cap X \subset X$ is a Cartier divisor and regard $X$ as a closed subscheme of the dilatation $P' = P^{[D \cdot X]}$. Then the pull-back $D'_{P'} \subset P'$ of $D'$ is a Cartier divisor and there exists a unique isomorphism $P'^{[D'_{P'} \cdot X]} \to P^{[D'' \cdot X]}$ lifting $P' \to P$.*

(2) *Let $E = P'$ be a vector bundle over $P$, and $D' = E \times_P D$ be the inverse image. Regard $X$ as a closed subscheme of $P'$ by the 0-section $X \subset E \times_P X \subset E = P'$. Then the commutative diagram*

$$\begin{array}{ccc} P'^{[D' \cdot X]} & \longrightarrow & E = P' \\ \downarrow & & \downarrow \\ P^{[D \cdot X]} & \longrightarrow & P \end{array}$$

*induces an isomorphism* $P'^{[D' \cdot X]} \to E(-D) \times_P P^{[D \cdot X]}$ *to the pull-back of the twist* $E(-D)$ *of* $E$.

*Proof* (1) First, we consider the case where $P = \operatorname{Spec} A$ is affine and $D, D' \subset P$ are defined by non-zero divisors $f, g \in A$. Assume that $X \subset P$ is defined by an ideal $I \subset A$. Then, since $X \to P'$ is defined by the surjection $B = A[I/f] \to A/I$ sending $J = I/f$ to 0, the immersion $X \to P'$ is defined by the ideal $J \subset B$. Since $A \to A[1/f]$ is flat, $g \in A[I/f]$ is a non-zero divisor. Since we have an equality $B[J/g] = A[I/(fg)]$ of subrings of $A[1/(fg)]$, the assertion follows in this case. The general case follows by patching.

(2) First, we consider the case where $P = \operatorname{Spec} A$ is affine, $D \subset P$ is defined by a non-zero divisor $f \in A$, and $P' = \operatorname{Spec} A[T_1, \ldots, T_n]$. Assume that $X \subset P$ is defined by an ideal $I \subset A$. Then, since $D', X \subset P'$ are defined by $f \in A[T_1, \ldots, T_n]$ and $I + (T_1, \ldots, T_n) \subset A[T_1, \ldots, T_n]$, we have $P^{[D \cdot X]} = \operatorname{Spec} A[I/f]$ and $P'^{[D' \cdot X]} = \operatorname{Spec} A[T_1, \ldots, T_n][(I + (T_1, \ldots, T_n))/f]$. We set $U_i = T_i/f \in A[T_1, \ldots, T_n][1/f] = A[1/f][U_1, \ldots, U_n]$. Then we have $A[T_1, \ldots, T_n][(I + (T_1, \ldots, T_n))/f] = A[I/f][U_1, \ldots, U_n]$ as its subrings, and the assertion is proved in this case. The general case follows by patching. □

**Lemma 11.1.4** *Let* $X$ *be a scheme and* $P$ *be a separated smooth scheme over* $X$. *Let* $D \subset X$ *be a Cartier divisor and* $D_P \subset P$ *be the pull-back. Let* $s \colon X \to P$ *be a section and consider* $X$ *as a closed subscheme of* $P$ *by* $s$.

(1) *The morphism* $P' = P^{[D_P \cdot X]} \to X$ *is smooth.*
(2) *The conormal bundle* $N_{X/P'}$ *with respect to the immersion* $X \to P'$ *lifting* $X \to P$ *is canonically identified with* $N_{X/P}(D)$.

*Proof* (1) Since the assertion is local on $P$, we may assume that there exists an étale morphism $P \to \mathbf{A}^n_X$ such that the composition $X \to P \to \mathbf{A}^n_X$ is the 0-section. Further, shrinking $P$ if necessary, we may assume that $X \to P \times_{\mathbf{A}^n_X} X$ is an isomorphism. Then, by Lemma 11.1.2(2), we may assume that $P = \mathbf{A}^n_X$. In this case, by Lemma 11.1.3(2), $P^{[D_P \cdot X]}$ is a vector bundle over $X^{[D \cdot X]} = X$ and is smooth.

(2) First, assume that $X = \operatorname{Spec} A$ and $P = \operatorname{Spec} B$, and that the immersions $D \subset X$ and $X \to P$ are defined by $f \in A$ and by an isomorphism $B/I \to A$. Then $P' = \operatorname{Spec} B'$ for $B' = B[I/f]$ and $N_{X/P} = I/I^2$, and $N_{X/P'} = I'/I'^2$ for $I' = I/f$. Hence, the assertion follows in this case. The general case follows by patching. □

**Lemma 11.1.5** *Let* $P$ *be a scheme,* $X \subset P$ *be a closed subscheme, and* $D \subset P$ *be an effective Cartier divisor. Assume that* $X \to P$ *is a regular immersion and that* $D \cap X \subset X$ *is a Cartier divisor. Let* $N_{X/P} = \mathcal{I}_X/\mathcal{I}_X^2$, *and* $N_{D/P}$ *be the* *conormal modules regarded as a vector bundle over X and a line bundle over D. Then there exists a cartesian diagram*

$$\begin{array}{ccc} \mathrm{Hom}_{X\cap D}(N_{X/P}|_{X\cap D}, N_{D/P}|_{X\cap D}) & \longrightarrow & P^{[D\cdot X]} \\ \downarrow & & \downarrow \\ D & \longrightarrow & P \end{array} \tag{11.1.3}$$

*Proof* Since the inverse image in $P^{[D\cdot X]}$ of $D$ equals the inverse image of $X\cap D$, we may replace $D$ by $X\cap D$ in (11.1.3). First, we consider the case where $P = \mathrm{Spec}\, A$ is affine, $X$ is defined by an ideal $I \subset A$ generated by a regular sequence $g_1, \ldots, g_n \in A$, and $D$ is defined by a non-zero divisor $f \in A$. Set $J = I + fA \subset A$, $\overline{A} = A/J$, and $\widetilde{A} = A[I/f]$ so that $X \cap D = \mathrm{Spec}\, \overline{A}$ and $P^{[D\cdot X]} = \mathrm{Spec}\, \widetilde{A}$. Since $J$ is generated by the regular sequence $f, g_1, \ldots, g_n$, we have $J/J^2 = \overline{A}\cdot f \oplus (I/I^2 \otimes_{A/I} \overline{A})$. It suffices to define a canonical isomorphism $\mathrm{S}_{\overline{A}}(f^{-1} \otimes (I/I^2 \otimes_{A/I} \overline{A})) \to \widetilde{A}/J\widetilde{A}$ from the symmetric algebra.

Since $\widetilde{A}$ is the inductive limit $\varinjlim f^{-n}J^n$ with respect to the inclusion, the quotient $\widetilde{A}/J\widetilde{A}$ is $\varinjlim f^{-n} \otimes J^n/J^{n+1}$. Since $J$ is generated by the regular sequence $f, g_1, \ldots, g_n$, we have a canonical isomorphism $\mathrm{S}^n_{\overline{A}}(f^{-1} \otimes J/J^2) \to f^{-n} \otimes J^n/J^{n+1}$. Since the transition mapping is the morphism induced by the inclusion $\overline{A} \to \overline{A} \oplus f^{-1} \otimes (I/I^2 \otimes_{A/I} \overline{A}) = f^{-1} \otimes J/J^2$, the inductive limit $\varinjlim f^{-n} \otimes J^n/J^{n+1} = \varinjlim \mathrm{S}^n_{\overline{A}}(f^{-1} \otimes J/J^2) = \varinjlim \bigoplus_{m=0}^{n} \mathrm{S}^m_{\overline{A}}(f^{-1} \otimes (I/I^2 \otimes_{A/I} \overline{A}))$ is canonically identified with $\mathrm{S}_{\overline{A}}(f^{-1} \otimes (I/I^2 \otimes_{A/I} \overline{A}))$.

The general case follows by patching. □

**Exercise 11.1** Let $O_K$ be a discrete valuation ring with residue field $F$ of characteristic $p > 0$, $a \in \mathfrak{m}_K$ be a non-zero element of the maximal ideal such that the valuation $\mathrm{ord}_K a$ is divisible by $p - 1$, and $n = pm \geqslant 1$ be an integer divisible by $p$. Let $X = \mathrm{Spec}\, O_K[x]/(x^p - ax) \subset P = \mathbf{A}^1_S = \mathrm{Spec}\, O_K[x]$ be a closed subscheme, and $D \subset P$ be a Cartier divisor defined by $t^n$ for a uniformizer $t$ of $K$. Show that the normalization $P^{(n)}$ of the dilatation $P^{[n]} = P^{[D\cdot X]}$ is smooth over $S = \mathrm{Spec}\, O_K$ and determine if the closed fiber $P^{(n)}_F$ is connected or not.

## 11.2 Going Down Theorem

**Definition 11.2.1** Let $X$ be a reduced scheme and $U \subset X$ be a dense open subscheme. Assume that $U$ is normal and let $V \to U$ be a finite flat morphism of normal schemes. The normalization of $X$ in $V$ is a normal scheme $Y$ affine over $X$ fitting in a cartesian diagram

$$\begin{array}{ccc} Y & \longleftarrow & V \\ {\scriptstyle \pi}\big\downarrow & & \big\downarrow{\scriptstyle \pi_U} \\ X & \overset{j}{\longleftarrow} & U \end{array}$$

and satisfying that the direct image $\pi_* \mathcal{O}_Y$ is the integral closure of $\mathcal{O}_X$ in $j_* \pi_{U*} \mathcal{O}_V$.

In the case $V = U$, the normalization in $U$ is called the normalization of $X$. If $X = \operatorname{Spec} A$ is affine and if $U = \operatorname{Spec} A[1/f]$ for a non-zero divisor $f \in A$, then the normalization of $X$ is $\operatorname{Spec} A^\natural$ for the integral closure $A^\natural$ of $A$ in $A[1/f]$.

The normalization has the following functoriality.

**Lemma 11.2.2** *Let $f : Y \to X$ be a morphism of reduced schemes, and let $U \subset X$ and $V \subset Y$ be normal dense open subschemes satisfying $f(V) \subset U$. Then the restriction $V \to U$ of $f$ is uniquely extended to a morphism $Y^\natural \to X^\natural$ of the normalizations $X^\natural \to X$ and $Y^\natural \to Y$.*

*Proof* We may assume that $X = \operatorname{Spec} A$, $Y = \operatorname{Spec} B$, $X^\natural = \operatorname{Spec} A^\natural$, $Y^\natural = \operatorname{Spec} B^\natural$, $U = \operatorname{Spec} A_1$, and $V = \operatorname{Spec} B_1$ are affine. Since $A^\natural$ is integral over $A$, the image of $A^\natural \subset A_1$ by $A_1 \to B_1$ is a subring of the integral closure $B^\natural$ of $B$ in $B_1$. Hence, the assertion follows. □

We recall the going down theorem in commutative algebra and its consequences.

**Proposition 11.2.3** *Let*

$$\begin{array}{ccc} Y & \longleftarrow & V \\ {\scriptstyle f}\big\downarrow & & \big\downarrow \\ X & \longleftarrow & U \end{array} \qquad (11.2.1)$$

*be a cartesian diagram of normal noetherian schemes such that $f : Y \to X$ is finite, $V \to U$ is faithfully flat, and the horizontal arrows are dense open immersions.*

(1) *Let $Z \subset X$ be a closed irreducible subset and let $W$ be an irreducible component of $f^{-1}(Z)$. Then $W \to Z$ is surjective.*
(2) *Let $Z \subset X$ be a closed connected subset, and let $W$ be a connected component of $f^{-1}(Z)$. Then $W \to Z$ is surjective.*

*Proof* (1) Since $f$ is a closed mapping, it suffices to show that the generic point $z$ of $Z$ is contained in the image of $W$. Let $w$ be a point of $W$. Replacing $X$ by an affine neighborhood of $x = f(w) \in Z$, we may assume that $X = \operatorname{Spec} A$

and $Y = \operatorname{Spec} B$ are affine. Then the assumption implies that $A \to B$ is an injection and $B$ is finite over $A$. Since $x$ is a point of the closure $Z$ of $z$, the assertion follows from [14, Chapter V, Section 2.4, Theorem 3].

(2) Let $Z' \subset Z$ be an irreducible component meeting $f(W)$. Then there exists an irreducible component $W'$ of $f^{-1}(Z')$ meeting $W$. Since $W$ is a connected component, we have $W' \subset W$. By (1), we have $Z' = f(W') \subset f(W)$.

By the assumption that $X$ is noetherian, the closed subset $Z$ is the disjoint union of the finite union $f(W)$ of irreducible components of $Z$ meeting $f(W)$ and the finite union $Z - f(W)$ of irreducible components of $Z$ not meeting $f(W)$. Since $Z$ is connected and $f(W)$ is not empty, $Z - f(W)$ is empty. □

**Corollary 11.2.4** *Let*

$$\begin{array}{ccccc} T^{\natural} & \longrightarrow & Y & \longleftarrow & V \\ \big\downarrow & & {\scriptstyle f}\big\downarrow & & \big\downarrow \\ S^{\natural} & \longrightarrow & X & \longleftarrow & U \end{array} \tag{11.2.2}$$

*be a commutative diagram of normal noetherian schemes such that the right square is cartesian. Assume that the right horizontal arrows are open immersions of dense open subsets, and that the restriction of the left horizontal arrows on each irreducible component is a closed immersion. Assume that the middle vertical arrow is finite, and the right vertical arrow is finite and faithfully flat. Assume that the open subsets $S^{\natural} \times_X U \subset S^{\natural}$ and $T^{\natural} \times_Y V \subset T^{\natural}$ are dense, that $S^{\natural} \times_X U \to U$ and $T^{\natural} \times_Y V \to V$ are closed immersions, and that $T^{\natural} \times_Y V \to S^{\natural} \times_X V$ is an isomorphism.*

(1) *Let $S \subset X$ and $T \subset Y$ be the images of $S^{\natural} \to X$ and $T^{\natural} \to Y$. Then $f^{-1}(S) = T$.*
(2) *Let $D \subset X$ be a closed subset and $E = f^{-1}(D) \subset Y$ be the inverse image. We consider the commutative diagram*

$$\begin{array}{ccc} T \cap E & \longrightarrow & \pi_0(E) \\ \big\downarrow & & \big\downarrow \\ S \cap D & \longrightarrow & \pi_0(D) \end{array} \tag{11.2.3}$$

*where the horizontal arrows send a point to the connected component containing it, and the vertical arrows are induced by $f$. Then the mapping $T \cap E \to (S \cap D) \times_{\pi_0(D)} \pi_0(E)$ defined by (11.2.3) is a surjection.*
(3) *Assume that every irreducible component of $S$ meets $D$ at exactly one point, and that similarly every irreducible component of $T$ meets $E$ at exactly one*

*point. We consider the commutative diagram*

$$\begin{array}{ccc} \pi_0(T^\natural) & \longrightarrow & T \cap E \\ \downarrow & & \downarrow \\ \pi_0(S^\natural) & \longrightarrow & S \cap D \end{array} \qquad (11.2.4)$$

*defined as follows: We define the horizontal arrows by sending an irreducible component to the unique point of intersection. The vertical arrows are induced by* $f$. *Then, the mapping* $\pi_0(T^\natural) \to \pi_0(S^\natural) \times_{(S\cap D)} (T \cap E)$ *defined by* (11.2.4) *is a surjection.*

(4) *The mapping* $\pi_0(T^\natural) \to \pi_0(S^\natural) \times_{\pi_0(D)} \pi_0(E)$ *defined by the combination of* (11.2.3) *and* (11.2.4) *is a surjection.*

*Proof* (1) By the left square in (11.2.2), we have $T \subset f^{-1}(S)$. Since $T$ is closed, we have $\overline{T \cap V} \subset T$. It suffices to show that $T \cap V = f^{-1}(S \cap U)$ is dense in $f^{-1}(S)$. Let $W$ be any irreducible component of $f^{-1}(S)$, and let $Z \subset S$ be an irreducible component containing $f(W) \subset S$ as a subset. Then $W \subset f^{-1}(Z)$ is an irreducible component and hence, by Proposition 11.2.3(1), we have $f(W) = Z$. Since $S \cap U$ is dense in $S$, the intersection $Z \cap U$ is not empty. Hence, $W \cap f^{-1}(S \cap U)$ is also not empty.

(2) By (1), we have $f^{-1}(S \cap D) = T \cap E$. Let $C' \subset E$ be a connected component, and let $C \subset D$ be the connected component satisfying $f(C') \subset C$. Since the mapping $C' \to C$ is a surjection by Proposition 11.2.3(2), we have that $C' \cap (T \cap E) = C' \cap f^{-1}(S \cap D) \to C \cap (S \cap D)$ is also a surjection and the assertion follows.

(3) Since $S \cap U = S^\natural \times_X U$ is dense in $S$ and $S^\natural$, we may identify irreducible components of $S$ with connected components of $S^\natural$. Similarly, we identify irreducible components of $T$ with connected components of $T^\natural$. Let $w \in T \cap E$ and let $Z$ be an irreducible component of $S$ such that $z = f(w)$ is the unique point of the intersection $Z \cap D$. Let $Z'$ be an irreducible component of $f^{-1}(Z)$ containing $w$. Since $Z' \to Z$ is a surjection by Proposition 11.2.3(1), and since $Z$ is a connected component of $S^\natural$, $Z'$ is a connected component of $T^\natural$ and $w$ is the unique point of $Z' \cap E$.

(4) By (2), the mapping $\pi_0(S^\natural) \times_{(S\cap D)} (T \cap E) \to \pi_0(S^\natural) \times_{(S\cap D)} ((S \cap D) \times_{\pi_0(D)} \pi_0(E)) = \pi_0(S^\natural) \times_{\pi_0(D)} \pi_0(E)$ is a surjection. Hence, the assertion follows from (3). □

**Exercise 11.2** As in Exercise 11.1, let $O_K$ be a henselian discrete valuation ring with residue field $F$ of characteristic $p > 0$, and let $l, m \geqq 1$ be integers. Let $t$ be a uniformizer of $K$, and set $a = t^{(p-1)l}$ and $n = pm$. Let $T = \operatorname{Spec} O_K[x]/(x^p - ax) \to Q = \mathbf{A}^1_S = \operatorname{Spec} O_K[x]$ be a closed immersion, and $E \subset Q$ be the Cartier divisor defined by $t^n$. Let $Q^{[n]} = Q^{[E \cdot T]}$ be the dilatation and $Q^{(n)}$ be the normalization. Let $T^\natural \to Q^{(n)}$ be the morphism of normalizations induced by $T \to Q^{[n]}$, and let $T_n \subset Q^{(n)}$ be the reduced closed subscheme supported on the image of $T^\natural \to Q^{(n)}$. Let $E_n = Q_F^{(n)} \subset Q^{(n)}$ be the closed fiber and $V = Q_K = Q_K^{(n)} \subset Q^{(n)}$ be the generic fiber.

Show that every irreducible component of $T_n$ meets $E_n$ at a single point, compute the finite sets $\pi_0(T_n \cap V)$, $T_n \cap E_n$, and $\pi_0(E_n)$, and verify that the mappings $\pi_0(T_n \cap V) \to T_n \cap E_n \to \pi_0(E_n)$ are surjections.

## 11.3 Immersions to Smooth Schemes

Let $K$ be a henselian discrete valuation field and let $L$ be a finite separable extension. We consider an immersion $T = \operatorname{Spec} O_L \to Q$ to a smooth scheme over $S = \operatorname{Spec} O_K$.

**Example 11.3.1** (1) Let $K$ be a henselian discrete valuation field and $L$ be a finite separable extension. Let $x_1, \ldots, x_n \in O_L$ be a system of generators over $O_K$. Then the surjection $O_K[X_1, \ldots, X_n] \to O_L$ sending $X_1, \ldots, X_n$ to $x_1, \ldots, x_n$ defines a closed immersion

$$T = \operatorname{Spec} O_L \to Q = \mathbf{A}^n_S = \operatorname{Spec} O_K[X_1, \ldots, X_n]$$

to a smooth scheme over $S$.

(2) Let $k$ be a perfect field and $X$ be a smooth scheme over $k$. Let $D \subset X$ be a smooth divisor and assume that $K$ is the fraction field of the henselization of the local ring $O_{X,\xi}$ at the generic point $\xi$ of $D$. Let $V \to U = X - D$ be a finite étale morphism on the complement, and let $Y$ be the normalization of $X$ in $V$. Let $L$ be the fraction field of the henselization of the local ring $O_{Y,\eta}$ at a point $\eta$ above $\xi$. After replacing $X$ by a neighborhood of $\xi$, we may assume that $Y$ is smooth over $k$. Then the morphism $T = \operatorname{Spec} O_L \to Q = Y \times_k S$ defined by the canonical morphism $T \to Y$ is an immersion to a smooth scheme over $S = \operatorname{Spec} O_K$.

**Lemma 11.3.2** *Let $K \to K'$ be a morphism of henselian discrete valuation fields, and let $L$ and $L'$ be finite separable extensions of $K$ and of $K'$. Let $L \to L'$ be a morphism of fields compatible with $K \to K'$. Set $S = \operatorname{Spec} O_K$, $S' = \operatorname{Spec} O_{K'}$, $T = \operatorname{Spec} O_L$, and $T' = \operatorname{Spec} O_{L'}$. Let $T \to Q$ and $T' \to Q'$ be immersions to smooth schemes over $S$ and $S'$. Then, after replacing $Q'$ by an étale neighborhood of $T'$, there exists a morphism $Q' \to Q$ that makes the* *diagram*

$$\begin{array}{ccccc} T' & \longrightarrow & Q' & \longrightarrow & S' \\ \downarrow & & \downarrow & & \downarrow \\ T & \longrightarrow & Q & \longrightarrow & S \end{array} \tag{11.3.1}$$

*commutative.*

*Proof* The composition $T' \to T \to Q$ defines a section of the smooth morphism $\mathrm{pr}_2\colon Q \times_S T' \to T'$. Hence, there exist a neighborhood $V \subset Q \times_S T'$ of the image of the section $T' \to Q \times_S T'$, an integer $n \geqslant 0$, and an étale morphism $V \to \mathbf{A}^n_{T'}$ such that the inverse image of the 0-section equals the image of $T'$. We may lift $V \to \mathbf{A}^n_{T'}$ to an étale morphism $W \to \mathbf{A}^n_{Q'}$ on a neighborhood of the image of $T' \to Q \times_S T' \to Q \times_S Q'$. After replacing $Q'$ by the inverse image $Q'_1 = W \times_{\mathbf{A}^n_{Q'}} Q'$ of the 0-section, the restriction $Q'_1 \to Q$ of $\mathrm{pr}_1|_W\colon W \subset Q \times_S Q' \to Q$ defines the middle vertical arrow of the diagram. □

**Lemma 11.3.3** *Let $L \subset L'$ be finite separable extensions of $K$, and set $T = \operatorname{Spec} O_L$ and $T' = \operatorname{Spec} O_{L'}$. Let*

$$\begin{array}{ccc} T' & \longrightarrow & Q' \\ \downarrow & & \downarrow \\ T & \longrightarrow & Q \end{array} \tag{11.3.2}$$

*be a commutative diagram of schemes over $S = \operatorname{Spec} O_K$. Assume that the horizontal arrows are immersions to smooth schemes over $S$.*

(1) *Assume that $Q' \to Q$ is smooth of relative dimension $n$. Then, after replacing $Q'$ by an open neighborhood of $T'$, there exists a cartesian diagram*

$$\begin{array}{ccc} T' & \longrightarrow & Q' \\ \downarrow & & \downarrow \\ T & \longrightarrow & \mathbf{A}^n_Q \end{array} \tag{11.3.3}$$

*over $S$ where $T \to \mathbf{A}^n_Q$ is the composition of $T \to Q$ with the 0-section $Q \to \mathbf{A}^n_Q$.*

(2) *Assume that* (11.3.2) *is cartesian. Then, on a neighborhood of $T'$, the right vertical arrow is quasi-finite and flat. There exist an étale neighborhood $Q_1$ of $T \subset Q$, an open neighborhood of $Q'_1$ of $T' \subset Q' \times_Q Q_1$, and a*

*commutative diagram*

$$\begin{array}{ccccc} T' & \longrightarrow & Q'_1 & \longrightarrow & Q' \\ \downarrow & & \downarrow & & \downarrow \\ T & \longrightarrow & Q_1 & \longrightarrow & Q \end{array}$$

*such that* $Q'_1 \to Q_1$ *is finite flat and that the left square is cartesian.*

*Proof* (1) Since $Q' \times_Q T$ is smooth of relative dimension $n$ over $T$, the immersion $T' \to Q' \times_Q T$ is a regular immersion of codimension $n$. Hence, there exist an affine open neighborhood $U$ of $T'$ in $Q' \times_Q T$ and a cartesian diagram

$$\begin{array}{ccc} T' & \longrightarrow & U \\ \downarrow & & \downarrow \\ T & \longrightarrow & \mathbf{A}^n_T \end{array}$$

By lifting $U \to \mathbf{A}^n_T$ to $Q' \to \mathbf{A}^n_Q$ on a neighborhood of $T'$, we obtain a cartesian diagram (11.3.3).

(2) Since $T' \to T$ is finite, the right vertical arrow $Q' \to Q$ is quasi-finite on a neighborhood of $T'$ [31, Théorème (13.1.3)]. Since $T \to Q$ and $T' \to Q'$ are regular immersions of the same codimension and since $T' \to T$ is flat, the right vertical arrow $Q' \to Q$ is flat on a neighborhood of $T'$ by [31, Chapitre 0, Proposition (15.1.21)] and [31, Théorème (11.1.1)].

For the second assertion, we may assume that $Q' \to Q$ is quasi-finite. Then it suffices to apply Theorem 11.3.4 to the morphism $Q' \to Q$ and the closed points $t \in T \subset Q$ and $t' \in T' \subset Q'$. □

**Theorem 11.3.4** ([31, Théorème (18.12.1), Remarque (18.12.2)]) *Let* $f: X \to Y$ *be a quasi-finite morphism of schemes and* $x \in X, y = f(x) \in Y$. *Then there exists an étale neighborhood* $V$ *of* $y$ *and an open neighborhood* $U \subset X \times_Y V$ *of the point above* $x$ *such that* $U \to V$ *is finite.*

**Proposition 11.3.5** *Let* $L \subset L'$ *be finite separable extensions of a henselian discrete valuation field* $K$. *Then there exists a cartesian diagram*

$$\begin{array}{ccc} \operatorname{Spec} O_{L'} = T' & \longrightarrow & Q' \\ \downarrow & & \downarrow \\ \operatorname{Spec} O_L = T & \longrightarrow & Q \end{array} \qquad (11.3.4)$$

*of schemes over* $S = \operatorname{Spec} O_K$ *such that the horizontal arrows are closed immersion to smooth schemes over* $S$ *and the vertical arrows are finite and flat.*

*Proof* We take closed immersions $T \to Q = \mathbf{A}^n_S$ and $T' \to Q'_1 = \mathbf{A}^{n'}_S$ as in Example 11.3.1(1). We consider the commutative diagram

$$\begin{array}{ccc} T' & \longrightarrow & Q'_1 \times_S Q \\ \big\downarrow & & \big\downarrow {\scriptstyle \mathrm{pr}_2} \\ T & \longrightarrow & Q \end{array}$$

where $T' \to Q'_1 \times_S Q$ is the product of the immersion $T' \to Q'_1$ and the composition $T' \to T \to Q$, and the right vertical arrow is the second projection $\mathrm{pr}_2 \colon Q'_1 \times_S Q \to Q$. Since the second projection is smooth, it suffices to apply Lemma 11.3.3. □

**Exercise 11.3** Let $L$ be a finite extension of a discrete valuation field $K$. Assume that $O_L$ is generated over $O_K$ by a single element $a \in O_L$, and define closed immersions $S = \operatorname{Spec} O_K \to P = \mathbf{A}^1_S = \operatorname{Spec} O_K[X]$ and $T = \operatorname{Spec} O_L \to Q = \mathbf{A}^1_S = \operatorname{Spec} O_K[Y]$ by sending $X$ to 0 and $Y$ to $a$. Construct a cartesian diagram

$$\begin{array}{ccc} T & \longrightarrow & Q \\ \big\downarrow & & \big\downarrow \\ S & \longrightarrow & P \end{array}$$

such that the vertical arrows are finite flat.

## 11.4 Geometric Construction

Let $L$ be a finite separable extension of a henselian discrete valuation field $K$, and let $T = \operatorname{Spec} O_L \to Q$ be a closed immersion to a smooth scheme over $S = \operatorname{Spec} O_K$. Let $r \geqslant 0$ be a rational number and let $K'$ over $K$ be an extension of henselian discrete valuation fields of ramification index $e$ such that $r' = er$ is an integer. Let $T_{F'} = T \times_S \operatorname{Spec} F' \to T_{S'} = T \times_S S' \to Q_{S'} = Q \times_S S'$ be the base change of $T \to S$ by $\operatorname{Spec} F' \to S' = \operatorname{Spec} O_{K'} \to S$, and let $D'_r \subset Q_{S'}$ denote the Cartier divisor defined by the principal ideal $\mathfrak{m}_{K'}^{r'} O_{Q_{S'}}$. Define the dilatation $Q_{S'}^{[r]} \to Q_{S'}$ to be $Q_{S'}^{[D'_r \cdot T_{S'}]} \to Q_{S'}$ and its normalization $Q_{S'}^{(r)}$.

Assume $r > 0$. Since the closed subscheme of $Q_{S'}^{[r]}$ defined by the ideal $\mathfrak{m}_{K'}^{r'} O_{Q_{S'}^{[r]}}$ is a closed subscheme of the inverse image $T_{S'} \times_{Q_{S'}} Q_{S'}^{[r]}$, the closed fiber $Q_{F'}^{[r]} = Q_{S'}^{[r]} \times_{S'} \operatorname{Spec} F'$ equals the inverse image $T_{F'} \times_{Q_{S'}} Q_{S'}^{[r]}$.

Let $T^\natural_{S'}$ be the normalization of $T_{S'} = T \times_S S'$. Then the morphism $T_{S'} \to$

$Q_{S'}$ is lifted to the dilatation and, by the functoriality of normalizations, Lemma 11.2.2, we have a commutative diagram

$$\begin{array}{ccc} T^{\natural}_{S'} & \longrightarrow & Q^{(r)}_{S'} \\ \downarrow & & \downarrow \\ T_{S'} & \longrightarrow & Q^{[r]}_{S'} \end{array} \tag{11.4.1}$$

If $K'$ contains a Galois closure of $L$ as a subfield, then the normalization $T^{\natural}_{S'}$ is the disjoint union of copies of $S'$ indexed by the finite set $\mathrm{Mor}_K(L, K')$ of morphisms over $K$.

**Lemma 11.4.1** *Let $K'$ be an extension of a henselian discrete valuation field $K$ containing a Galois closure of $L$ as a subfield, and let $e = e_{K'/K}$ be the ramification index. Let $m > 0$ be an integer satisfying $\mathfrak{m}^m_{K'}(O_L \otimes_{O_K} O_{K'})^{\natural} \subset O_L \otimes_{O_K} O_{K'}$ for the normalization $(O_L \otimes_{O_K} O_{K'})^{\natural}$ of $O_L \otimes_{O_K} O_{K'}$. Then, for $r \geqslant 2m/e > 0$, the morphism $T^{\natural}_{S'} \to Q^{(r)}_{S'}$ is an immersion.*

*Proof* We may assume that $Q = \mathrm{Spec}\, A$ is affine. Let $I$ be the kernel of the surjection $A_{O_{K'}} = A \otimes_{O_K} O_{K'} \to O_L \otimes_{O_K} O_{K'}$, and let $a \in O_{K'}$ be an element of valuation $m$. By the assumption that $K'$ contains a Galois closure of $L$ as a subfield, the normalization $(O_L \otimes_{O_K} O_{K'})^{\natural}$ is isomorphic to the product of copies of $O_{K'}$ and hence is generated over $O_{K'}$ by primitive idempotents $e_1, \ldots, e_n \in (O_L \otimes_{O_K} O_{K'})^{\natural}$. Hence, it suffices to show that the primitive idempotents are in the image of the normalization of $A_{O_{K'}}[I/a^2]$.

Let $e_i \in (O_L \otimes_{O_K} O_{K'})^{\natural}$ be an idempotent. By the assumption $a \cdot (O_L \otimes_{O_K} O_{K'})^{\natural} \subset O_L \otimes_{O_K} O_{K'}$, we have $b_i = a \cdot e_i \in O_L \otimes_{O_K} O_{K'}$. Since $e_i$ is an idempotent, we have $b_i^2 - ab_i = 0$. Let $c_i \in A_{O_{K'}}$ be an element lifting $b_i$. Then $b_i^2 - ab_i = 0$ means $c_i^2 - ac_i \in I$. Hence, $(c_i/a) \in A_{O_{K'}}[1/a]$ is a root of the polynomial

$$X^2 - X - \frac{c_i^2 - ac_i}{a^2} \in A_{O_{K'}}[I/a^2][X]$$

and is integral over $A_{O_{K'}}[I/a^2]$. Since the image of $c_i/a$ is $b_i/a = e_i$, the assertion follows. □

The construction of (11.4.1) has the following functoriality. Let $K'$ be an extension of a henselian discrete valuation field $K$ of ramification index $e$. Let $L$ and $L'$ be finite separable extensions of $K$ and of $K'$ respectively, and let $L \to L'$ be a morphism of fields compatible with $K \to K'$. Set $S = \mathrm{Spec}\, O_K$, $S' = \mathrm{Spec}\, O_{K'}$, $T = \mathrm{Spec}\, O_L$, $T' = \mathrm{Spec}\, O_{L'}$, and let $T \to Q$ and $T' \to Q'$ be immersions to smooth schemes over $S$ and over $S'$ respectively. Then, after replacing $Q'$ by an étale neighborhood of $T'$, we have a commutative diagram

$$\begin{array}{ccc} T' & \longrightarrow & Q' \\ \downarrow & & \downarrow \\ T & \longrightarrow & Q \end{array} \tag{11.4.2}$$

of schemes compatible with $S' \to S$ by Lemma 11.3.2.

Let (11.4.2) be a commutative diagram of schemes such that the horizontal arrows are immersions to smooth schemes over $S$ and over $S'$ respectively and the vertical arrows are compatible with $S' \to S$. Let $r > 0$ and $r' > er$ be rational numbers. Let $K_1$ and $K'_1$ be extensions of discrete valuation fields $K$ and $K'$ of ramification indices $e_1$ and $e'_1$ such that $e_1 r$ and $e'_1 r'$ are integers. Let $K_1 \to K'_1$ be morphisms of fields compatible with $K \to K'$ and set $S_1 = \operatorname{Spec} O_{K_1}$, $S'_1 = \operatorname{Spec} O_{K'_1}$. Then the diagram (11.4.2) defines a commutative diagram

$$\begin{array}{ccccccc} T'^{\natural}_{S'_1} & \longrightarrow & Q'^{(r')}_{S'_1} & \longrightarrow & Q'^{[r']}_{S'_1} & \longrightarrow & Q' \times_{S'} S'_1 \\ \downarrow & & \downarrow & & \downarrow & & \downarrow \\ T^{\natural}_{S_1} & \longrightarrow & Q^{(r)}_{S_1} & \longrightarrow & Q^{[r]}_{S_1} & \longrightarrow & Q \times_S S_1 \end{array} \tag{11.4.3}$$

compatible with $S'_1 \to S_1$, by the funtoriality of dilatations, Lemma 11.1.2, and normalizations, Lemma 11.2.2.

In particular, if $S' = S$ in the commutative diagram (11.4.2) and if $K_1 \to K_2$ is a morphism of extensions of $K$, we have a commutative diagram

$$\begin{array}{ccc} Q'^{(r')}_{S_1} & \longleftarrow & Q'^{(r')}_{S_2} \\ \downarrow & & \downarrow \\ Q^{(r)}_{S_1} & \longleftarrow & Q^{(r)}_{S_2} \end{array} \tag{11.4.4}$$

compatible with the other morphisms in (11.4.3).

In the rest of this section we study properties of $Q^{(r)}_{S'}$. Let $K'$ be an extension of henselian discrete valuation fields of $K$ of ramification index $e = e_{K'/K}$, and $S' = \operatorname{Spec} O_{K'} \to S = \operatorname{Spec} O_K$ be the canonical morphism. Let $r > 0$ be a rational number and assume that $r' = er$ is an integer.

**Lemma 11.4.2** *Let $L \subset L'$ be finite separable extensions of $K$ and set*

*$T = \operatorname{Spec} O_L$ and $T' = \operatorname{Spec} O_{L'}$. Let*

$$\begin{array}{ccc} T' & \longrightarrow & Q' \\ \downarrow & & \downarrow \\ T & \longrightarrow & Q \end{array} \tag{11.4.5}$$

*be a commutative diagram of schemes over $S$ such that the horizontal arrows are immersions to smooth schemes over $S$.*

(1) *Assume that $Q' \to Q$ and $T' \to T$ are étale. Then $Q'^{(r)}_{S'} \to Q^{(r)}_{S'} \times_Q Q'$ is an isomorphism on the inverse image of an open neighborhood of $T' \subset Q'$ and induces an isomorphism $Q'^{(r)}_{F'} \to Q^{(r)}_{F'} \times_T T'$.*

(2) *Assume that $Q' \to Q$ is smooth of relative dimension $n$, and $T' = T$. Then $Q'^{(r)}_{S'} \to Q^{(r)}_{S'}$ is smooth of relative dimension $n$. There exists an isomorphism $Q'^{(r)}_{F'} \to \mathbf{A}^n_{F'} \times_{F'} Q^{(r)}_{F'}$ such that the morphism $T^{\natural}_{F'} \to Q'^{(r)}_{F'}$ on the closed fiber is the lifting of $T^{\natural}_{F'} \to Q^{(r)}_{F'}$ to the 0-section.*

*Proof* (1) Since the immersion $T' \to T \times_Q Q'$ of étale schemes over $T$ is étale, this is an open immersion. Hence, replacing $Q'$ by an open neighborhood of $T'$, we may assume that the diagram (11.4.5) is cartesian. Then, since $Q' \to Q$ is flat, the right square in the diagram

$$\begin{array}{ccccc} Q'^{(r)}_{S'} & \longrightarrow & Q'^{[r]}_{S'} & \longrightarrow & Q' \\ \downarrow & & \downarrow & & \downarrow \\ Q^{(r)}_{S'} & \longrightarrow & Q^{[r]}_{S'} & \longrightarrow & Q \end{array}$$

is cartesian by Lemma 11.1.2(2). Since the right vertical arrow is étale, the left square is also cartesian by Corollary 1.2.4.

Since the closed fiber $Q^{[r]}_{F'}$ equals the inverse image $T_{F'} \times_{Q_{S'}} Q^{[r]}_{S'}$, the last assertion follows.

(2) By Lemma 11.3.3, there exists an étale morphism $Q' \to \mathbf{A}^n_Q$ such that the composition $T \to \mathbf{A}^n_Q$ is the lifting of $T \to Q$ to the 0-section. Hence, by (1), we may assume that $Q' = \mathbf{A}^n_Q$. Then the assertion follows from Lemma 11.1.3(2). □

**Lemma 11.4.3** *Let $K$ be a henselian discrete valuation field and let $L$ be a finite separable extension. Let*

$$\begin{array}{ccc} \operatorname{Spec} O_L = T & \longrightarrow & Q \\ \downarrow & & \downarrow \\ \operatorname{Spec} O_K = S & \longrightarrow & P \end{array} \tag{11.4.6}$$

*be a cartesian diagram where the horizontal arrows are immersion to smooth schemes over S. Let $r > 0$ be a rational number. Let $K'$ be an extension of a discrete valuation field K of ramification index e such that er is an integer, and* $S' = \operatorname{Spec} O_{K'}$.

(1) *The scheme $P^{(r)}_{S'}$ is smooth over $S'$ and the closed fiber $P^{(r)}_{F'}$ is isomorphic to an affine space $\mathbf{A}^n_{F'}$.*
(2) *We consider the commutative diagram*

$$\begin{array}{ccc} T^{\natural}_{S'} & \longrightarrow & Q^{(r)}_{S'} \\ \downarrow & & \downarrow \\ S' & \longrightarrow & P^{(r)}_{S'} \end{array} \tag{11.4.7}$$

*defined as in* (11.4.3). *Then the underlying set of the inverse image by the right vertical arrow of the image of $S' \to P^{(r)}_{S'}$ equals the image of $T^{\natural}_{S'} \to Q^{(r)}_{S'}$.*

*Proof* (1) By Lemma 11.4.2(2) applied to $Q' = P \to Q = S$ and $T = S$, the morphism $P^{(r)}_{S'} \to S'$ is smooth. Further, the closed fiber $P^{(r)}_{F'}$ is isomorphic to $\mathbf{A}^n_{F'}$ if $P \to S$ is of relative dimension $n$.

(2) By Lemma 11.4.2(1) and Lemma 11.3.3, we may assume that $Q \to P$ is finite flat, the horizontal arrows in (11.4.6) are closed immersions, and $P$ and $Q$ are irreducible. Since the diagram (11.4.6) is cartesian, the finite morphism $Q \to P$ is generically étale and faithfully flat. The morphism $Q^{[r]}_{S'} \to P^{[r]}_{S'}$ is its base change by Lemma 11.1.2(2), and is finite flat and generically étale. Hence, the morphism $Q^{(r)}_{S'} \to P^{(r)}_{S'}$ of the normalizations is also finite. Since $\operatorname{Spec} K' \subset S'$ is dense and $T_{K'} = S_{K'} \times_P Q$ is dense in $T^{\natural}_{S'}$, the inverse image $S' \times_{P^{(r)}_{S'}} Q^{(r)}_{S'}$ equals the image of $T^{\natural}_{S'}$ by Corollary 11.2.4(1). □

**Proposition 11.4.4** *Let $T = \operatorname{Spec} O_L \to Q$ be an immersion to a smooth scheme over S. Let $r > 0$ be a rational number. Let $K'$ be an extension of discrete valuation field K of ramification index e such that er is an integer.*

(1) *The mapping $T^{\natural}_{S'} \to Q^{(r)}_{S'}$ defines a surjection $T^{\natural}_{F'} \to \pi_0(Q^{(r)}_{F'})$ to the set of connected components.*
(2) *Let $0 \leqslant s < r$ be a rational number such that es is also an integer. Then the image of the closed fiber $Q^{(r)}_{F'}$ by $Q^{(r)}_{S'} \to Q^{(s)}_{S'}$ equals the image of $T^{\natural}_{F'} \to Q^{(s)}_{S'}$.*

*Proof* By Lemma 11.4.2.(1), we may replace $Q$ by an étale neighborhood of $T$. We apply Lemma 11.3.3 to $T \to S$ and the immersions $T \to Q$, $1_S \colon S \to S$. Then, after replacing $Q$ by an étale neighborhood of $T$, we obtain a cartesian diagram (11.4.6). By Lemma 11.3.3(2), we may assume that $Q \to P$ is finite and faithfully flat. Since $T \to S$ is étale at the generic point of $S$ and (11.4.6) is cartesian, $Q \to P$ is étale on a dense open. Hence, $Q_{S'}^{(r)} \to P_{S'}^{(r)}$ is also finite.

(1) Let $C$ be a connected component of the closed fiber $Q_{F'}^{(r)}$. Since $Q_{F'}^{(r)}$ is the inverse image of $P_{F'}^{(r)}$, and since $P_{F'}^{(r)}$ is isomorphic to an affine space $\mathbf{A}^n_{F'}$ by Lemma 11.4.3(1) and is connected, the restriction $C \to P_{F'}^{(r)}$ of the morphism $Q_{F'}^{(r)} \to P_{F'}^{(r)}$ is a surjection by Proposition 11.2.3(2). Since the image of $T_{F'}^{\natural} \to Q_{F'}^{(r)}$ equals the inverse image of the origin $0 \in \mathbf{A}^n_{F'}$ by $Q_{F'}^{(r)} \to P_{F'}^{(r)} = \mathbf{A}^n_{F'}$, by Lemma 11.4.3(2), the assertion follows.

(2) We consider the commutative diagram

$$\begin{array}{ccccc} T_{S'}^{\natural} & \longrightarrow & Q_{S'}^{(r)} & \longrightarrow & Q_{S'}^{(s)} \\ \downarrow & & \downarrow & & \downarrow \\ S' & \longrightarrow & P_{S'}^{[r]} & \longrightarrow & P_{S'}^{[s]} \end{array}$$

Since the morphism $P_{S'}^{[r]} \to P_{S'}^{[s]}$ is identified with the dilatation of $P_{S'}^{[s]}$ with respect to the closed subscheme $S'$ and the Cartier divisor defined by $\mathfrak{m}_{K'}^{e(s-r)}$ by Lemma 11.1.3(1), the image of the closed fiber $P_{F'}^{[r]}$ in $P_{S'}^{[s]}$ equals the image of the closed point $s' \in S'$. Hence, the upper line induces $T_{F'}^{\natural} \to Q_{F'}^{(r)} \to Q_{S'}^{(s)} \times_{P_{S'}^{[s]}} s'$. Since $Q_{S'}^{(s)} \times_{P_{S'}^{[s]}} s'$ is finite and discrete, the second arrow induces $\pi_0(Q_{F'}^{(r)}) \to Q_{S'}^{(s)} \times_{P_{S'}^{[s]}} s'$. Hence, the assertion follows from (1). □

**Definition 11.4.5** Let $K$ be a henselian discrete valuation field, $L$ be a finite separable extension of $K$, and let $T = \operatorname{Spec} O_L \to Q$ be an immersion to a smooth scheme over $S = \operatorname{Spec} O_K$. Let $K'$ be an extension of a discrete valuation field of $K$ of ramification index $e = e_{K'/K}$, and $r > 0$ be a rational number such that $er$ is an integer. If the geometric closed fiber of $Q_{S'}^{(r)} \to Q' \to S'$ is reduced, we call $Q_{S'}^{(r)}$ the stable integral model and $Q_{F'}^{(r)}$ the stable closed fiber. For an algebraic closure $\overline{F}$ of $F'$, we call $Q_{\overline{F}}^{(r)} = Q_{F'}^{(r)} \times_{\operatorname{Spec} F'} \operatorname{Spec} \overline{F}$ the stable geometric closed fiber.

By the reduced fiber theorem, a stable integral model is defined over a finite separable extension.

**Proposition 11.4.6** *Let $K$ be a henselian discrete valuation field and $L$ be a finite separable extension. Let $r > 0$ be a rational number.*

(1) *There exists a finite separable extension $K'$ of ramification index $e = e_{K'/K}$*

*such that er is an integer and that the geometric closed fiber of* $Q^{(r)}_{S'} \to Q_{S'} \to S'$ *is reduced.*

(2) *Let* $K'$ *be an extension of* $K$ *satisfying the condition in (1) and let* $K''$ *be an extension of* $K'$. *Then the geometric closed fiber of* $Q^{(r)}_{S''} \to S''$ *is reduced and the canonical morphism* $Q^{(r)}_{S''} \to Q^{(r)}_{S'} \times_{S'} S''$ *is an isomorphism.*

*Proof* (1) Let $K_1$ be a finite separable extension of $K$ of ramification index $e_1 = e_{K_1/K}$ such that $e_1 r$ is an integer. It suffices to apply Proposition 9.3.5 to $Q^{[r]}_{S_1} \to S_1$.

(2) By Corollary 9.3.4, $Q^{(r)}_{S'} \times_{S'} S''$ is normal. Hence, the canonical morphism $Q^{(r)}_{S''} \to Q^{(r)}_{S'} \times_{S'} S''$ is an isomorphism and the geometric closed fiber of $Q^{(r)}_{S''}$ is reduced. □

We postpone the proof of Proposition 11.4.8 to the end of the next chapter as it requires some geometric machinery. To state and prove the proposition, we introduce some terminology on surjections of sets.

**Definition 11.4.7** Let $A \to B$ and $A \to C$ be surjections of sets. If there exists a commutative diagram

$$\begin{array}{ccc} A & \longrightarrow & B \\ & \searrow & \downarrow \\ & & C \end{array}$$

we say that $A \to B$ is *finer* than $A \to C$ and that $A \to C$ is *coarser* than $A \to B$. We say that $A \to B$ and $A \to C$ are equivalent if the vertical arrow is a bijection.

**Proposition 11.4.8** *Let* $L$ *be a finite separable extension of* $K$, *and let* $T = \operatorname{Spec} O_L \to Q$ *be an immersion over* $S = \operatorname{Spec} O_K$ *to a smooth scheme over* $S$. *For a rational number* $r > 0$, *let* $Q^{(r)}_{\bar{F}}$ *denote the stable geometric closed fiber and let* $T^{\natural}_{\bar{F}} \to Q^{(r)}_{\bar{F}}$ *be the canonical morphism. For* $r = 0$, *we set* $Q = Q^{(r)}$. *Let* $0 < m$ *be an integer. Then there exist rational numbers* $r_0 = 0 < r_1 < \cdots < r_n = m$ *such that, for each* $i = 1, \ldots, n$, *the surjections* $T^{\natural}_{\bar{F}} \to \pi_0(Q^{(r)}_{\bar{F}})$ *for* $r \in (r_{i-1}, r_i]$ *and* $T^{\natural}_{\bar{F}} \to \operatorname{Im}(T^{\natural}_{\bar{F}} \to Q^{(s)}_{\bar{F}})$ *for* $s \in [r_{i-1}, r_i)$ *are equivalent.*

**Exercise 11.4** Let $K$ be a discrete valuation field and let $L$ be a quadratic Galois extension of $K$. Let $O_L = O_K[x]/(x^2 + ax + b)$ and $T = \operatorname{Spec} O_L \to Q = \mathbf{A}^1_S = \operatorname{Spec} O_K[x]$ be the closed immersion to a smooth scheme over $S = \operatorname{Spec} O_K$. Let $d = a^2 - 4b$ be the discriminant and set $r = \operatorname{ord}_K d$. Show that the normalization $Q^{(r)}_T$ is smooth over $T$.

## 11.5 Monogenic Case

We compute explicitly the construction in the previous section for separable extensions $L$ of $K$ assuming that the ring $O_L$ of integers is generated by a single element.

**Lemma 11.5.1** *Let $L$ be a finite extension of $K$, and let $E$ be the residue field of $L$. If one of the following conditions is satisfied, then $O_L$ is generated by a single element over $O_K$.*

(1) *$E$ is a separable extension of $F$.*

(2) *The residue field $E$ is generated by a single element over $F$, and $e_{L/K} = 1$.*

*Proof* (1) This case is proved in Lemma 1.8.3(1).

(2) If $E = F(\overline{u})$ for $u \in O_L$ and $e_{L/K} = 1$, then we have $O_L = O_K[u]$ by Nakayama's lemma. □

The following example shows that the converse does not hold.

**Example 11.5.2** Let $K$ be a discrete valuation field, and assume that the residue field $F$ is a rational function field $k(u)$ over a perfect field $k$ of characteristic $p > 0$. Let $t$ be a uniformizer of $K$, and let $u \in O_K$ denote a lifting of $u \in F$ abusively. Let $M$ be the totally ramified extension of $K$ defined by $s^p - ts = t$, and let $L$ be the extension of $M$ defined by $x^p - tx = u + s$. Then $O_L = O_K[x]$, but the residue field $E$ is a purely inseparable extension $F(x)$ of $F$ of degree $p$ and $e_{L/K} = p$.

Assume $O_L = O_K[a]$ and let $f \in O_K[X]$ be the minimal polynomial of $a$. Let $O_K[X]/(f) \to O_L$ be the isomorphism defined by $a \in O_L$, and define a closed immersion

$$T = \text{Spec } O_K[X]/(f) \to Q = \mathbf{A}^1_{O_K} = \text{Spec } O_K[X].$$

We compute the stable integral model using the rescaling function defined in Definition 11.5.5.

**Definition 11.5.3** Let $a \in O_L = O_K[X]/(f)$ be the class of $X$ and define a polynomial $g = f(Y + a) = Y^n + c_1 Y^{n-1} + \cdots + c_{n-1} Y \in O_L[Y]$. Let $v\colon [0, n-1] \to [0, \text{ord}_L c_{n-1}]$ be the largest piecewise linear convex continuous function such that the graph is below the points $(0, 0)$ and $(k, \text{ord}_L c_k)$ for $k = 1, \ldots, n-1$. We call the graph of $v$ the Newton polygon of the polynomial $g$.

**Lemma 11.5.4** *Let $K'$ be an extension of a discrete valuation field $K$, and assume that $f$ is decomposed into the product $f = \prod_{i=1}^n (X - a_i)$ in $O_{K'}[X]$.*

*Extend the normalized valuation* $\mathrm{ord}_L$ *to* $K'$. *Changing the numbering if necessary, we assume that the valuations* $s_i = \mathrm{ord}_L(a_i - a_n)$ *are increasing in* $i = 1, \ldots, n-1$.

(1) *For* $i = 1, \ldots, n-1$, *the slope of the Newton polygon of* $g = f(Y + a_n)$ *on the interval* $[i-1, i]$ *equals* $s_i$.
(2) *The slopes* $s_1, \ldots, s_{n-1}$ *are independent of the choice of* $a_n$.

*Proof* (1) Set $b_i = a_i - a_n \in O_{K'}$ for $i = 1, \ldots, n-1$. Since $g(Y) = Y \cdot \prod_{i=1}^{n-1}(Y - b_i)$, its coefficients $c_i$ are $(-1)^i$ times the sum of products $b_{j_1} \cdots b_{j_i}$ for $1 \leqslant j_1 < \cdots < j_i \leqslant n-1$ for each $i = 1, \ldots, n-1$. Hence, we have $\mathrm{ord}_L c_i \geqslant \sum_{j=1}^{i} s_j$ for $i = 1, \ldots, n-1$ and the equality holds if $s_i < s_{i+1}$. This implies that $v(x) = \sum_{j=1}^{i-1} s_j + s_i(x - (i-1))$ on the interval $[i-1, i]$.

(2) Since $a_1, \ldots, a_n$ are conjugate to each other, the assertion follows. □

**Definition 11.5.5** For $i = 1, \ldots, n-1$, let $s_i = \mathrm{ord}_L(a_i - a_n)$ be the slope of the Newton polygon on the interval $[i-1, i]$. We define a rescaling function $\varphi\colon [0, \infty) \to [0, \infty)$ by

$$\varphi(s) = \frac{1}{e_{L/K}}\Bigl(\sum_{i=1}^{n-1} \min(s, s_i) + s\Bigr) \tag{11.5.1}$$

and set $r_{L/K} = \varphi(s_{n-1})$.

We extend $s_1 \leqslant \cdots \leqslant s_{n-1}$ by $s_0 = 0$ and $s_n = \infty$. Then, the rescaling function is given by

$$\varphi(s) = \frac{1}{e_{L/K}}\Bigl(\sum_{i=1}^{k} s_i + (n-k)s\Bigr) \tag{11.5.2}$$

on $[s_k, s_{k+1})$ for $k = 0, \ldots, n-1$, and hence is a piecewise linear concave function. In the case where the residue field extension is separable, the rescaling function defined here is slightly modified from that studied in Chapter 6, as follows.

**Lemma 11.5.6** *Let* $K$ *be a henselian discrete valuation field, and let* $L$ *be a finite Galois extension of* $K$ *such that the residue field* $E$ *is a separable extension of the residue field* $F$ *of* $K$. *Let* $G_i \subset G = \mathrm{Gal}(L/K)$ *be the lower numbering filtration* (1.6.1). *Let* $\varphi_{\log}\colon \mathbf{N} \to \mathbf{Q}$ *denote the rescaling function defined by*

$$\varphi_{\log}(i) = \frac{1}{\#G_0}\Bigl(\sum_{\sigma \in I - G_i} i_G(\sigma) + \sum_{\sigma \in G_i} i\Bigr). \tag{11.5.3}$$

*Then* $\varphi(i+1) = \varphi_{\log}(i) + 1$ *for* $i \in \mathbf{N}$.

The function $\varphi_{\log}$ defined in (11.5.3) is the same as $\varphi$ defined in (6.2.3).

*Proof* Since $i_G(\sigma) = \mathrm{ord}_L(\sigma(t) - t) - 1$, the assertion follows. □

In the case where the ramification index $e_{L/K}$ is 1, the rescaling function defined here is the same as that studied in Chapter 7 without modification.

The rescaling function $\varphi$ is a modified Legendre transform of the function $v$ defining the Newton polygon of the polynomial of $g = f(Y + a_n)$ in the following sense: Let $s$ be in the interval $[s_{i-1}, s_i]$ and $P$ be the vertex on the Newton polygon such that the slope is $s_{i-1}$ on the left of $P$ and $s_i$ on the right of $P$. Then, for the linear function $\lambda$ of slope $s$ such that $P$ is on the graph of $\lambda$, we have $\varphi(s) = \lambda(n)/e_{L/K}$. More concretely, if $P_0 = (0,0)$ and $P_i = (u_i, v_i)$, $i = 1, \ldots, m$, are the vertices of the Newton polygon, and if the slopes of the edges $P_{i-1}P_i$ are $s_i$ for $i = 1, \ldots, m$, then the graph of $\varphi$ on $[0, s_m]$ is the polygon with vertices $Q_0 = (0,0)$ and $Q_i = (s_i, (v_i + s_i(n - u_i))/e_{L/K})$ for $i = 1, \ldots, m$.

Since the different $D_{L/K} \subset O_L$ is generated by $f'(a_n) = \prod_{i=1}^{n-1}(a_n - a_i)$ by Proposition 1.9.5(1), its valuation $d_{L/K} = \mathrm{ord}_L(D_{L/K})$ equals $\sum_{i=1}^{n-1} s_i$ and we have

$$r_{L/K} = \frac{1}{e_{L/K}}(d_{L/K} + s_{n-1}). \tag{11.5.4}$$

**Lemma 11.5.7** *Let $L$ be a finite separable extension of a henselian discrete valuation field $K$. Assume that $O_L$ is generated by a single element over $O_K$.*

(1) *The following conditions are equivalent:*

   (a) *$L$ is an unramified extension of $K$.*
   (b) $r_{L/K} = 0$.
   (c) $r_{L/K} < 1$.

(2) *The following conditions are equivalent:*

   (a) *$L$ is a tamely ramified extension of $K$.*
   (b) $r_{L/K} = 0$ *or* $1$.
   (c) $r_{L/K} \leqslant 1$.

*Proof* By Proposition 1.9.1, we have $d_{L/K} \geqslant e_{L/K} - 1$ and the equality holds if and only if $L$ is tamely ramified over $K$. We have $s_{n-1} \geqslant 0$ and the equality holds if and only if $L$ is unramified. If $L$ is ramified, we have $s_{n-1} \geqslant 1$ and the equality holds if $L$ is tamely ramified. The assertions follow. □

**Proposition 11.5.8** *Let $r = \varphi(s) > 0$ be a rational number, and let $t, c \in O_{K'}$ be elements satisfying $\mathrm{ord}_K t = r$ and $\mathrm{ord}_L c = s$.*

(1) *The morphisms* $U_j = \mathbf{A}^1_{O_{K'}} = \operatorname{Spec} O_{K'}[Y_j] \to Q = \mathbf{A}^1_{O_K} = \operatorname{Spec} O_K[X]$ *defined by sending* $X$ *to* $cY_j + a_j$ *for* $j = 1, \ldots, n$ *induce open immersions* $U_j \to Q^{(r)}_{S'}$ *and give an open covering of the stable integral model* $Q^{(r)}_{S'}$.
(2) *We regard* $U_j$ *as open subschemes of* $Q^{(r)}_{S'}$ *by the open immersions in (1). For* $j, j' = 1, \ldots, n$, *we have* $U_j = U_{j'}$ *if* $\operatorname{ord}_L(a_j - a_{j'}) \geqslant s$. *Otherwise, the intersection* $U_j \cap U_{j'}$ *equals the generic fiber* $Q_{K'}$.

*Proof* (1) Let $A = O_{K'}[X, f/t] \subset K'[X]$ be the subring so that we have $Q^{[r]}_{S'} = \operatorname{Spec} A$. We have $A = O_{K'}[X, Z]/(f - tZ)$, and $A$ is finite flat over the subring $C = O_{K'}[Z]$. Let $w$ be the discrete valuation of $K'(Z)$ defined by the generic point of the closed fiber of $W = \operatorname{Spec} C = \mathbf{A}^1_{S'}$.

For $j = 1, \ldots n$, let $Y_j = (X - a_j)/c \in K'[X]$ and

$$B_j = O_{K'}[Y_j] \subset K'[X] = A \otimes_{O_{K'}} K' = B_j \otimes_{O_{K'}} K'$$

be the subrings so that we have $U_j = \operatorname{Spec} B_j$. Let $v_j$ be the discrete valuation of $K'(X)$ defined by the generic point of the closed fiber of $U_j = \mathbf{A}^1_{S'}$. We have $B_j = \{x \in K'[X] \mid v_j(x) \geqslant 0\}$.

To compute the normalization of $A$, we show the following claims.

**Claim 1** *$A$ is a subring of $B_j$ for $j = 1, \ldots, n$.*

**Claim 2** *For $j, j' = 1, \ldots, n$, the equality $v_j = v_{j'}$ is equivalent to* $\operatorname{ord}_L(a_j - a_{j'}) \geqslant s$.

**Claim 3** *The valuation $v_j$ is an extension of $w$ and the residual degree $f_{v_j/w}$ is $\#\{i \mid \operatorname{ord}_L(a_i - a_j) \geqslant s\}$ for $j = 1, \ldots, n$.*

*Proof of Claims 1–3* Fixing $j$ and changing the numbering, we may assume that $j = n$ and $s_i = \operatorname{ord}_K b_i$ for $b_i = a_i - a_n$ is increasing in $i = 1, \ldots, n$. We set $s_0 = 0$ and $s_n = \infty$, and let $0 \leqslant k < n$ be the integer satisfying $s_k < s \leqslant s_{k+1}$.

Claim 1: Since

$$f = \prod_{i=1}^{n} (cY_n - b_i) = \prod_{i \leqslant k} \left( \frac{c}{b_i} Y_n - 1 \right) \prod_{i > k} \left( Y_n - \frac{b_i}{c} \right) \cdot \prod_{i \leqslant k} b_i \cdot c^{n-k} \quad (11.5.5)$$

and $\operatorname{ord}_K \prod_{i \leqslant k} b_i \cdot c^{n-k} = \varphi(s) = r$, the fraction $f / \prod_{i \leqslant k} b_i \cdot c^{n-k}$ is an element of $B_n$. Hence, the generators $Z$ and $X = cY_n + a_n$ of $A$ are elements of $B_n$ and we have an inclusion $A \subset B_n$.

Claim 2: For $j = k + 1, \ldots, n$, we have $v_j = v_n$ since $a_j - a_n$ is divisible by $c$. For $j = 1, \ldots, k$, we have $v_j(X - a_n) = v_j(cY_j - (a_n - a_j)) = \operatorname{ord}_{K'} b_j < \operatorname{ord}_{K'} c = v_n(cY_n) = v_n(X - a_n)$.

Claim 3: The morphism $F'[Z] \to F'[Y_n]$ induced by the composition $C \to A \to B_n$ maps $Z$ to $\prod_{i > k} (Y_n - (b_i/c)) \equiv Y_n^{n-k} \bmod \mathfrak{m}_{K'}$. Hence, this defines a finite flat morphism $U_{n,F'} = \operatorname{Spec} F'[Y_n] \to W_{F'} = \operatorname{Spec} F'[Z]$ of degree $n - k$. □

By Claim 1, $C = O_{K'}[Z] \subset A = O_{K'}[X,Z]/(f - tZ)$ are subrings of the intersection $B = \bigcap_{n=1}^{n} B_j$. Since $A$ is finite flat of rank $n$ over $C$, the valuations $v_1, \dots, v_n$ exhaust the discrete valuations of $K'[X]$ dominating $w$ by Claims 2 and 3 and [14, Chapter VI, Section 8.5, Theorem 2]. Hence, $B = \bigcap_{n=1}^{n} B_j = \{x \in K'[X] \mid v_j(x) \geqslant 0 \text{ for } j = 1, \dots, n\}$ is the integral closure of $C$ in $K(X)$ by Claim 1 and [14, Chapter VII, Section 1.3, Corollary of Theorem 2]. Since $A$ is finite over $C$, the ring $B$ is the normalization of $A$. Thus, we have $Q_{S'}^{(r)} = \operatorname{Spec} B$.

We show that the inclusions $B \to B_j$ induce open immersions $U_j \to Q_{S'}^{(r)}$. We fix $j$ and changing the numbering, and assume $j = n$.

**Claim 4** *Assume that $s_i = \operatorname{ord}_K b_i$ for $b_i = a_i - a_n$ is increasing in $i = 1, \dots, n$, and let*

$$h = \prod_{i \leqslant k} \frac{X - a_i}{b_i} = \prod_{i \leqslant k} \left( \frac{c}{b_i} Y_n - 1 \right) \in B_n.$$

*Then, for every $j = 1, \dots, n$, we have $h \in B_j$ and $B_n[1/h] \subset B_j[1/h]$.*

*Proof of Claim 4* For $k + 1 \leqslant j \leqslant n$, we have $B_j = B_n$. For $1 \leqslant j \leqslant k$, by the proof of Claim 2, we have $h \in \mathfrak{m}_{K'} B_j$ and $B_j[1/h] = K'[X][1/h] \supset B_n[1/h]$. □

The inclusion $B \to B_n$ induces an isomorphism $B[1/t] = K'[X] \to B_n[1/t]$. By Claim 4, we have $h \in B$ and the inclusion $B \to B_n$ induces an isomorphism $B[1/h] \to B_n[1/h]$. Since

$$h = \prod_{i \leqslant k} \left( \frac{c}{b_i} Y_n - 1 \right) \equiv \pm 1 \bmod \mathfrak{m}_{K'},$$

we have $(h, t) = B_n$. Hence, the inclusion $B \to B_n$ induces an isomorphism $U_n = \operatorname{Spec} B_n \to \operatorname{Spec} B - V(h,t) \subset Q_{S'}^{(r)}$ to the complement of the closed subset defined by the ideal $(h, t)$.

We show that $U = \bigcup_{j=1}^{n} U_j$ equals $Q_{S'}^{(r)}$. It suffices to show the equality $U_{F'} = Q_{F'}^{(r)}$ of the closed fibers. Since $U_{j,F'} = \operatorname{Spec} F'[Y_j] \to W_{F'} = \operatorname{Spec} F'[Z]$ are finite, $U_{F'}$ is a closed subset of $Q_{F'}^{(r)}$. Since $Q_{S'}^{(r)} \to W = \operatorname{Spec} O_{K'}[Z]$ is a finite morphism of normal schemes and $U_{F'} \subset Q_{F'}^{(r)}$ contains the inverse image $\{v_1, \dots, v_n\}$ of the generic point $w \in W_{F'}$, we have $U_{F'} = Q_{F'}^{(r)}$ by Proposition 11.2.3(1).

(2) In the proof of (1), we have shown that $U_j = U_{j'}$ if $\operatorname{ord}_L(a_j - a_{j'}) \geqslant s$. The generic fiber $U_j \times_{S'} \operatorname{Spec} K'$ equals $Q_{K'}$. Since the closed fibers $U_{j,F}$ are

normal and irreducible, the intersection $U_{j,F} \cap U_{j',F}$ is empty if and only if $v_j \neq v_{j'}$. This is equivalent to $\mathrm{ord}_L(a_i - a_j) < s$ by Claim 2. □

**Corollary 11.5.9** (1) *For $j = 1, \ldots, n$, the open subset $U_{j,F'} \subset Q^{(r)}_{F'}$ is a connected component of $Q^{(r)}_{F'}$. For $j, j' = 1, \ldots, n$, the following conditions are equivalent:*

(a) $U_{j,F'} = U_{j',F'}$.

(b) $\mathrm{ord}_K(a_j - a'_j) \geqslant s$.

(2) *For $j, j' = 1, \ldots, n$, the following conditions are equivalent:*

(a) *The images of the sections $S' \to Q^{(r)}_{S'}$ defined by $a_j$ and $a_{j'}$ share the closed point.*

(b) $\mathrm{ord}_K(a_j - a'_j) > s$.

*Proof* (1) This is proved in Proposition 11.5.8(2).

(2) Since (a) implies (1) (a) and (b) implies (1) (b), it suffices to prove the equivalence assuming $U_j = U_{j'}$. Further, changing the numbering if necessary, we may assume that $U_j = U_{j'} = U_n$ and hence $k + 1 \leqslant j, j' \leqslant n$ in the notation of the proof of Claim 4. Then, the images of the sections of $U_n$ defined by sending $Y_n$ to $b_j/t, b'_j/t$ share the closed point if and only if the valuation of $b_j - b_{j'} = a_j - a_{j'}$ is strictly larger than that of $t$. □

**Example 11.5.10** Let $K$ be a henselian discrete valuation field of characteristic $p > 0$, $F$ be the residue field, and $t \in K$ be a uniformizer.

(1) Let $L$ be the Artin–Schreier extension defined by $x^p - x = 1/t^n$ for $n = pm - 1 > 0$, and $s = t^m x \in L$ be a uniformizer as in Example 1.4.4(1). Let $f = X^p - t^{(p-1)m}X - t \in O_K[X]$ be the minimal polynomial and set $S = \mathrm{Spec}\, O_K$ and $T = \mathrm{Spec}\, O_L$. Define a finite flat morphism $Q = \mathbf{A}^1_S = \mathrm{Spec}\, O_K[X] \to P = \mathbf{A}^1_S = \mathrm{Spec}\, O_K[Z]$ by sending $Z$ to $f$. Since $f = (X-s)(X-(s+t^m)) \cdots (X-(s+(p-1)t^m))$, by setting $Y = (X-s)/t^m$ we have $f/t^{pm} = Y(Y-1) \cdots (Y-(p-1)) = Y^p - Y$. Hence, for $r = pm = n+1$, the morphism $Q^{(r)}_T = \mathbf{A}^1_T = \mathrm{Spec}\, O_L[Y] \to P^{(r)}_T = \mathbf{A}^1_T = \mathrm{Spec}\, O_L[W]$ is an étale morphism sending $W$ to $Y^p - Y$.

(2) Let $L$ be the Artin–Schreier extension defined by $x^p - x = v/t^n$ for $n = pm > 0$ and for $v \in O_K$ such that $\bar{v} \in F - F^p$. Let $u = t^m x$ be the generator of the residue field $E = F(u)$ as in Example 1.4.4(2). Let $f = X^p - t^{(p-1)m}X - v \in O_K[X]$ be the minimal polynomial and set $S = \mathrm{Spec}\, O_K$ and $T = \mathrm{Spec}\, O_L$. Define a finite flat morphism $Q = \mathbf{A}^1_S = \mathrm{Spec}\, O_K[X] \to P = \mathbf{A}^1_S = \mathrm{Spec}\, O_K[Z]$ by sending $Z$ to $f$. Since $f = (X-u)(X-(u+t^m)) \cdots (X-(u+(p-1)t^m))$, by setting $Y = (X-u)/t^m$ we have $f/t^{pm} = Y(Y-1) \cdots (Y-(p-1)) = Y^p - Y$. Hence, for $r = pm = n$,

the morphism $Q_T^{(r)} = \mathbf{A}_T^1 = \operatorname{Spec} O_L[Y] \to P_T^{(r)} = \mathbf{A}_T^1 = \operatorname{Spec} O_L[W]$ is an étale morphism sending $W$ to $Y^p - Y$.

**Exercise 11.5** Let $p$ be a prime number and $n \geqslant 1$ be an integer.

(1) Let $f = \sum_{i=0}^{p-1} X^{p^{n-1}i} = (X^{p^n} - 1)/(X^{p^{n-1}} - 1) \in \mathbf{Q}_p[X]$ be the minimal polynomial of a primitive $p^n$th root $\zeta_{p^n}$ of 1. Find the vertices of the Newton polygon of the polynomial $g = f(Y + \zeta_{p^n})$ and those of the graph of the rescaling function $\varphi$.

(2) Let $K$ be the fraction field of the henselization of the localization $\mathbf{Z}_p[T]_{(p)}$ of the localization at the prime ideal generated by $p$, and set $f = X^{p^n} - T$. Find the vertices of the Newton polygon of the polynomial $g = f(Y + u)$ for a solution $u$ of $f = 0$, and those of the graph of the rescaling function $\varphi$.

## Historical Notes

Dilatations are studied systematically in [4] and [71]. The monogenic case in Proposition 11.5.8 is studied in [1] using rigid geometry.

# 12

# Upper Ramification Subgroups

## 12.1 Construction of Functors

Let $K$ be a henselian discrete valuation field, $F$ be the residue field, and $K_s$ be a separable closure. For a finite separable extension $L$ of $K$, let $F(L)$ denote the finite set $\mathrm{Mor}_K(L, K_s)$ and $T = \mathrm{Spec}\, O_L$. Further, for an extension $K' \subset K_s$ of $K$ containing a Galois closure of $L$ and $S' = \mathrm{Spec}\, O_{K'}$, we identify the normalization $T^{\natural}_{S'}$ of $T_{S'} = T \times_S S'$ with the disjoint union of copies of $S' = \mathrm{Spec}\, O_{K'}$ indexed by $F(L)$ and identify the closed fiber $T^{\natural}_{F'}$ of $T^{\natural}_{S'}$ with $F(L)$.

**Lemma 12.1.1** *Let $L$ be a finite separable extension of $K$ and let $E$ be the residue field. Let $K_s$ be a separable closure of $K$ and let $\overline{F}$ be the residue field of $K_s$. Let $F(L)$ and $F^{0+}(L)$ denote the finite set $\mathrm{Mor}_K(L, K_s)$ of morphisms over $K$ and the finite set $\mathrm{Mor}_F(E, \overline{F})$ of morphisms over $F$ respectively. Let $r > 0$ be a rational number.*

(1) *There exist unique quotient sets $F^r(L)$ and $F^{r+}(L)$ of $F(L)$ and a surjection $F^{r+}(L) \to F^r(L)$ of quotient sets satisfying the following properties:*

*Let $Q$ be a smooth scheme over $S = \mathrm{Spec}\, O_K$, and let $T = \mathrm{Spec}\, O_L \to Q$ be an immersion over $S$. Let $K' \subset K_s$ be an extension of ramification index $e = e_{K'/K}$ of discrete valuation field $K$ containing a Galois closure of $L$ over $K$ such that $er$ is an integer and that $Q^{(r)}_{S'}$ is a stable integral model. Let $T^{\natural}_{S'} \to Q^{(r)}_{S'}$ be the morphism over $S'$ induced by $T \to Q$. Then we have* *a commutative diagram*

$$\begin{array}{ccccc} F(L) = T^{\natural}_{F'} & \longrightarrow & F^{r+}(L) & \longrightarrow & \mathrm{Im}(T^{\natural}_{F'} \to Q^{(r)}_{F'}) \\ & \searrow & \downarrow & & \downarrow \\ & & F^{r}(L) & \longrightarrow & \pi_0(Q^{(r)}_{F'}) \end{array}$$

*of mappings where the horizontal arrows in the square are bijections and the right vertical arrow maps a point to the connected component containing it.*

(2) *Let $0 \leqslant s < r$ be a rational number. Then $F(L) \to F^{r}(L)$ is finer than $F(L) \to F^{s+}(L)$.*

*Proof* (1) By Proposition 11.4.4(1), the mapping $T^{\natural}_{S'} \to Q^{(r)}_{S'}$ defines a surjection $F(L) = T^{\natural}_{F'} \to \pi_0(Q^{(r)}_{F'})$. The mapping $T^{\natural}_{S'} \to Q^{(r)}_{S'}$ also defines a surjection $F(L) = T^{\natural}_{F'} \to \mathrm{Image}(T^{\natural}_{F'} \to Q^{(r)}_{F'})$.

Let $T \to Q'$ be another immersion to a smooth scheme over $S$. Then it suffices to show that the surjections $F(L) \to \pi_0(Q^{(r)}_{F'})$ and $F(L) \to \pi_0(Q'^{(r)}_{F'})$ (resp. $F(L) \to \mathrm{Im}(T^{\natural}_{F'} \to Q^{(r)}_{F'})$ and $F(L) \to \mathrm{Im}(T^{\natural}_{F'} \to Q'^{(r)}_{F'})$) are equivalent. By considering the product $T \to Q \times_S Q'$ and projections, we may assume that there exists a smooth morphism $Q \to Q'$ compatible with the immersions of $T$. Then the assertion follows from Lemma 11.4.2(2).

(2) This follows from Proposition 11.4.4(2). □

We study the functoriality of the construction of Lemma 12.1.1.

**Lemma 12.1.2** *Let $K_1$ be an extension of a henselian discrete valuation field $K$ of ramification index $e$, and fix a morphism $K_s \to K_{1s}$ of separable closures. Let $L_1$ be a finite separable extension of $K_1$, and $L \to L_1$ be a morphism of fields compatible with $K \to K_1$. Let $F_1(L_1) = \mathrm{Mor}_{K_1}(L_1, K_{1s}) \to F(L) = \mathrm{Mor}_K(L, K_s)$ be the canonical mapping. Let $r > 0$ be a rational number and $r_1 = er > 0$.*

(1) *There exist unique mappings $F_1^{r_1}(L_1) \to F^{r}(L)$ and $F_1^{r_1+}(L_1) \to F^{r+}(L)$ such that the diagram*

$$\begin{array}{ccccccc} F_1(L_1) & \longrightarrow & F_1^{r_1+}(L_1) & \longrightarrow & F_1^{r_1}(L_1) & \longrightarrow & F_1^{0+}(L_1) \\ \downarrow & & \downarrow & & \downarrow & & \downarrow \\ F(L) & \longrightarrow & F^{r+}(L) & \longrightarrow & F^{r}(L) & \longrightarrow & F^{0+}(L) \end{array} \qquad (12.1.1)$$

*is commutative.*

(2) *Let*

$$\begin{array}{ccccc} T_1 & \longrightarrow & Q_1 & \longrightarrow & S_1 \\ \downarrow & & \downarrow & & \downarrow \\ T & \longrightarrow & Q & \longrightarrow & S \end{array} \tag{12.1.2}$$

*be a commutative diagram such that the right horizontal arrows are smooth and the left horizontal arrows are immersions. Let* $K' \to K'_1$ *be a morphism of extensions of discrete valuation fields of* $K$ *and* $K_1$ *compatible with* $K \to K_1$ *such that* $Q^{(r)}_{S'}$ *and* $Q^{(r_1)}_{1S'_1}$ *are stable integral models. Then the diagrams*

$$\begin{array}{ccc} F_1^{r_1}(L_1) & \longrightarrow & \pi_0(Q^{(r_1)}_{1F'_1}) \\ \downarrow & & \downarrow \\ F^r(L) & \longrightarrow & \pi_0(Q^{(r)}_{F'}), \end{array} \qquad \begin{array}{ccc} F_1^{r_1+}(L_1) & \longrightarrow & \mathrm{Im}(T^{\natural}_{1F'_1} \to Q^{(r_1)}_{1F'_1}) \\ \downarrow & & \downarrow \\ F^{r+}(L) & \longrightarrow & \mathrm{Im}(T^{\natural}_{F'} \to Q^{(r)}_{F'}) \end{array}$$

*are commutative.*

*Proof* (1) Let $T_1 = \operatorname{Spec} O_{L_1} \to Q_1$ be an immersion to a smooth scheme over $S_1 = \operatorname{Spec} O_{K_1}$. Then, after replacing $Q_1$ by an étale neighborhood of $T_1$ if necessary, we may assume that there exists a morphism $Q_1 \to Q$ such that the diagram

$$\begin{array}{ccccc} T_1 & \longrightarrow & Q_1 & \longrightarrow & S_1 \\ \downarrow & & \downarrow & & \downarrow \\ T & \longrightarrow & Q & \longrightarrow & S \end{array} \tag{12.1.3}$$

is commutative by Lemma 11.3.2. Let $K'_1 \subset K_{1s}$ be a finite separable extension such that $Q^{(r_1)}_{S'_1}$ is a stable integral model. We may further assume that there exists a morphism $K' \to K'_1$ of fields extending $K \to K_1$. Then, by the functoriality of dilatations and normalizations, the commutative diagram (12.1.3) induces a commutative diagram

$$\begin{array}{ccc} T^{\natural}_{1S'_1} & \longrightarrow & Q^{(r_1)}_{S'_1} \\ \downarrow & & \downarrow \\ T^{\natural}_{S'} & \longrightarrow & Q^{(r)}_{S'} \end{array} \tag{12.1.4}$$

by the functoriality (11.4.3) of the construction. This defines the vertical arrows in (12.1.1).

(2) The assertion follows from the commutative diagram (12.1.4). □

By the functoriality on $K'$ in Lemma 12.1.2, the finite sets $F^r(L)$ and $F^{r+}(L)$

carry a continuous action of the absolute Galois group $G_K = \mathrm{Gal}(K_s/K)$, and the surjections $F(L) \to F^{r+}(L) \to F^r(L)$ are compatible with the $G_K$-actions. Further, by the functoriality on $L$, we may make the following definition.

**Definition 12.1.3** (1) Let $\mathbf{Q}_+$ be the disjoint union $\{r \in \mathbf{Q} \mid r > 0\} \amalg \{r+ \mid r \in \mathbf{Q}, r \geqslant 0\}$. We define an injection $\mathbf{Q}_+ \to P(\mathbf{Q})$ to the power set by sending $r$ to $(-\infty, r)$ and $r+$ to $(-\infty, r]$, and define a total order on $\mathbf{Q}_+$ to be the pull-back of the inclusion on $P(\mathbf{Q})$.

(2) Let $K_s$ be a separable closure of $K$, let $G_K = \mathrm{Gal}(K_s/K)$ be the absolute Galois group, and let

$$F\colon (\text{finite separable extensions}/K) \to (\text{finite } G_K\text{-sets with continuous action})$$

be the fiber functor sending $L$ to $\mathrm{Mor}_K(L, K_s)$. We define a projective system $(F^r)_{r\in\mathbf{Q}_+}$ of functors and a projective system $(F \to F^r)_{r\in\mathbf{Q}_+}$ of surjections of functors by the construction in Lemmas 12.1.1 and 12.1.2.

(3) Let $L$ be a finite separable extension of $K$ and $r \in \mathbf{Q}_+$. We say that the ramification of $L$ over $K$ is bounded by $r$ if the surjection $F(L) \to F^r(L)$ is a bijection.

We identify canonically the set $\mathbf{Q}_+$ with a set of ideals of the valuation ring $O_{K_s} \subset K_s$ of a separable closure. The following property is a key ingredient in the construction of the upper ramification subgroups in Section 12.3.

**Proposition 12.1.4** *Let $L \subset L'$ be finite separable extensions of $K$ and $r \in \mathbf{Q}_+$. Then the diagram*

$$\begin{array}{ccc} F(L') & \longrightarrow & F^r(L') \\ \downarrow & & \downarrow \\ F(L) & \longrightarrow & F^r(L) \end{array}$$

*induces a surjection $F(L') \to F(L) \times_{F^r(L)} F^r(L')$.*

Proposition 12.1.4 is a consequence of the going down theorem in commutative algebra.

*Proof* Applying Proposition 11.3.5 to $T' = \mathrm{Spec}\, O_{L'} \to T = \mathrm{Spec}\, O_L$, we obtain a cartesian diagram (11.3.4) such that the vertical arrows are finite and faithfully flat. Then the morphism $Q'^{(r)}_{S'} \to Q^{(r)}_{S'}$ of stable integral models is finite and is faithfully flat on the generic fiber. We apply Corollary 11.2.4 to the

commutative diagram

$$\begin{array}{ccccc} T'^{\natural}_{S'} & \longrightarrow & Q'^{(r)}_{S'} & \longleftarrow & Q'_{K'} \\ \downarrow & & \downarrow & & \downarrow \\ T^{\natural}_{S'} & \longrightarrow & Q^{(r)}_{S'} & \longleftarrow & Q_{K'} \end{array}$$

Then the commutative diagram

$$\begin{array}{ccccc} T'^{\natural}_{F'} & \longrightarrow & \mathrm{Im}(T'^{\natural}_{F'} \to Q'^{(r)}_{F'}) & \longrightarrow & \pi_0(Q'^{(r)}_{F'}) \\ \downarrow & & \downarrow & & \downarrow \\ T^{\natural}_{F'} & \longrightarrow & \mathrm{Im}(T^{\natural}_{F'} \to Q^{(r)}_{F'}) & \longrightarrow & \pi_0(Q^{(r)}_{F'}) \end{array} \qquad (12.1.5)$$

induces surjections $T'^{\natural}_{F'} \to T^{\natural}_{F'} \times_{\pi_0(Q^{(r)}_{F'})} \pi_0(Q'^{(r)}_{F'})$ and $T'^{\natural}_{F'} \to T^{\natural}_{F'} \times_{\mathrm{Im}(T^{\natural}_{F'} \to Q^{(r)}_{F'})} \mathrm{Im}(T'^{\natural}_{F'} \to Q'^{(r)}_{F'})$ as required. □

**Corollary 12.1.5** *Let $L$ and $L'$ be finite separable extensions of $K$ and $r \in \mathbf{Q}_+$.*

(1) *Assume that $L \subset L'$, and that the ramification of $L'$ over $K$ is bounded by $r$. Then the ramification of $L$ over $K$ is bounded by $r$.*

(2) *Assume that the ramification of $L$ and $L'$ over $K$ is bounded by $r$. Then the ramification of $LL'$ over $K$ is bounded by $r$.*

*Proof* (1) Since $F(L') \to F(L) \times_{F^r(L)} F^r(L')$ is a surjection by Proposition 12.1.4, if $F(L') \to F^r(L')$ is a bijection, the projection $F(L) \times_{F^r(L)} F^r(L') \to F^r(L')$ is a bijection. Since $F(L') \to F(L)$ is a surjection, the induced mapping $F^r(L') \to F^r(L)$ is also a surjection. Thus, $F(L) \to F^r(L)$ is a bijection since its base change by a surjection $F^r(L') \to F^r(L)$ is a bijection.

(2) We consider the commutative diagram

$$\begin{array}{ccc} F(LL') & \longrightarrow & F^r(LL') \\ \downarrow & & \downarrow \\ F(L) \times F(L') & \longrightarrow & F^r(L) \times F^r(L') \end{array}$$

Since the left vertical arrow is an injection, if the lower horizontal arrow is an injection then the upper horizontal arrow is also an injection. □

Let $K'$ be an extension of a discrete valuation field of ramification index $e$, and fix a morphism $K_s \to K'_s$ of separable closures. Let $r, r'$ be elements of $\mathbf{Q}_+$ satisfying $r' \geqslant er$. Let $L$ (resp. $L'$) be a finite separable extension of $K$ (resp. of $K'$) and let $L \to L'$ be a morphism compatible with $K \to K'$. Then we have a commutative diagram

$$\begin{array}{ccc} F'(L') & \longrightarrow & F'^{r'}(L') \\ \big\downarrow & & \big\downarrow \\ F(L) & \longrightarrow & F^r(L) \end{array} \tag{12.1.6}$$

by the functoriality (11.4.3) of the construction.

**Lemma 12.1.6** *Let $L$ be a finite separable extension of $K$, and $r > 0$ be a rational number such that the ramification of $L$ over $K$ is bounded by $r+$. Let $K'$ be an extension of discrete valuation field $K$ of ramification index $e$, and let $L' = LK'$ be a composition field and $r' = er$. Then the ramification of $L'$ over $K'$ is bounded by $r'+$.*

*Proof* In the commutative diagram (12.1.6), the left vertical arrow is an injection since $L' = LK'$. Hence, if the lower horizontal arrow is an injection, then the upper horizontal arrow is also an injection. □

In the case where the ring of integers is generated by a single element, the sets $F^r(L)$ for $r \in \mathbf{Q}_+$ are computed as follows.

**Proposition 12.1.7** *Assume that $O_L = O_K[a]$. Let $a_1, \ldots, a_n$ be the conjugates of $a$ in a separable closure $K_s$, and we identify $F(L) = \{a_1, \ldots, a_n\}$. Let $\varphi$ be the rescaling function (Definition 11.5.5) and $r = \varphi(s) > 0$ be a rational number.*

(1) *The images of $a_j, a_{j'} \in F(L)$ in $F^r(L)$ are the same if and only if $\operatorname{ord}_L(a_j - a_{j'}) \leqslant s$.*
(2) *The images of $a_j, a_{j'} \in F(L)$ in $F^{r+}(L)$ are the same if and only if $\operatorname{ord}_L(a_j - a_{j'}) < s$.*
(3) *Assume that $L$ is a ramified extension of $K$, and let $r = r_{L/K} > 0$ be as in (11.5.4). Then the ramification of $L$ over $K$ is bounded by $r_{L/K}+$ but not bounded by $r_{L/K}$.*

*Proof* (1) Let $Q_{F'}^{(r)}$ be the stable closed fiber. Since $\pi_0(Q_{F'}^{(r)}) = \{U_{j,F'} \mid j = 1, \ldots, n\}$, the assertion follows from Corollary 11.5.9(1).

(2) The normalization $T_{S'}^{\natural}$ of $T \times_S S'$ is the disjoint union of $n$ copies of $S'$, and the canonical morphism $T_{S'}^{\natural} \to Q_{S'}^{(r)}$ is defined by $a_j$ on each component. Hence, the assertion follows from Corollary 11.5.9(2).

(3) Let $s = s_{n-1} > 0$. Then the ramification of $L$ over $K$ is bounded by $r+$ by (2) but not by $r$ by (1). □

**Exercise 12.1** Let $p$ be a prime number and $n \geqslant 1$ be an integer.

(1) Let $L = \mathbf{Q}_p(\zeta_{p^n})$ and identify the Galois group $G = \mathrm{Gal}(\mathbf{Q}_p(\zeta_{p^n})/\mathbf{Q}_p)$ with $(\mathbf{Z}/p^n\mathbf{Z})^\times$. Identify $F(L) = G$ and express $F^r(L)$ for $r \in \mathbf{Q}_+$ as a quotient of $(\mathbf{Z}/p^n\mathbf{Z})^\times$.
(2) Let $K$ be the fraction field of the henselization of $\mathbf{Z}_p[T]_{(p)}$, and $L = K(u)$ be the extension generated by a solution of $X^{p^n} = T$. Identify $F(L) = \{v \in L(\zeta_{p^n}) \mid v^{p^n} = T\}$ and compute its quotient set $F^r(L)$ for $r \in \mathbf{Q}_+$.

## 12.2 Characterization of Tamely Ramified Extensions

We give characterizations of unramified extensions and tamely ramified extensions.

**Lemma 12.2.1** *Let $L$ be a finite separable extension of a henselian discrete valuation field $K$ such that the residue field $F$ is of characteristic $p > 0$. Then there exists a sequence $K = K_0 \subset K_1 \subset \cdots \subset K_n$ of separable extensions of degree $p$ satisfying the following properties:*

*The residue field $F_n$ of $K_n$ is a purely inseparable extension of $F$ and the ramification index $e_{K_n/K}$ is 1. The residue field $E_n$ of $L_n = LK_n$ is a separable extension of $F_n$.*

*Proof* Let $(u_i)_{i\in I}$ be a $p$-basis of the residue field $F$ so that $F(u_i^{1/p^n}; i \in I, n \geqslant 0)$ is a perfect closure $F^{1/p^\infty}$ of $F$. Let $(u_{i,0})_{i\in I}$ be a lifting to $O_K$ and let $t \in O_K$ be a uniformizer. For each $i \in I$, we define inductively a sequence $K_{i,n} = K_{i,n-1}(u_{i,n})$ of separable extensions of degree $p^n$ of $K = K_{i,0}$ of ramification index 1 by the equation $u_{i,n}^p - tu_{i,n} = u_{i,n-1}$. Then the residue field $F_{i,n}$ of $K_{i,n}$ is $F(u_i^{1/p^n})$.

For a finite subset $J \subset I$ and an integer $n \geqslant 0$, let $K_{J,n}$ be the composite field $\bigotimes_{i\in J} K_{i,n}$. Then the inductive limit $\widetilde{K} = \varinjlim_{J,n} K_{J,n}$ is a henselian discrete valuation field with perfect residue field $F^{1/p^\infty}$, and the ramification index $e_{\widetilde{K}/K}$ is 1. Since the residue field of $\widetilde{L} = L\widetilde{K}$ is a separable extension of $F^{1/p^\infty}$, there exists a finite subset $J \subset I$ and an integer $n \geqslant 0$ such that the residue field of $L_{J,n} = LK_{J,n}$ is a separable extension of the residue field $F_{J,n}$ of $K_{J,n}$. Hence, the assertion follows. □

**Proposition 12.2.2** ([1, Proposition 6.8]) *Let $L$ be a finite separable extension of a henselian discrete valuation field $K$.*

(1) *The following conditions are equivalent:*

 (a) *$L$ is an unramified extension of $K$.*
 (b) *The ramification of $L$ over $K$ is bounded by* 1.

(c) *The ramification of L over K is bounded by* 0+.

(2) *The following conditions are equivalent:*

(a) *L is a tamely ramified extension of K.*

(b) *The ramification of L over K is bounded by* 1+.

*Proof* (1) (a) $\Rightarrow$ (c): Since the identity $T = \operatorname{Spec} O_L \to T$ is an immersion to a smooth scheme over $S$, the mapping $F(L) \to F^{0+}(L)$ is a bijection.

(2) (a) $\Rightarrow$ (b): Since $O_L$ is generated by a single element over $O_K$ by Lemma 11.5.1, this follows from Lemma 11.5.7 and Proposition 12.1.7(3).

(1) (c) $\Rightarrow$ (b): Since the surjection $F(L) \to F^{0+}(L)$ is coarser than $F(L) \to F^1(L)$ by Lemma 12.1.1(2), the assertion follows.

(1, 2) (b) $\Rightarrow$ (a): If the ring $O_L$ of integers is generated by a single element over $O_K$, the assertion follows from Proposition 12.1.7(3) and Lemma 11.5.7. We prove the general case by reducing to this case.

By Lemma 12.2.1, there exists a sequence $K \subset K_1 \subset \cdots \subset K_n = K'$ such that $K_i$ is an extension of $K_{i-1}$ of degree $p$ of ramification index 1 with inseparable residue field extension for each $i = 1, \dots, n$, and that the residue field $E'$ of $L' = LK'$ is separable over the residue field $F'$ of $K'$. Assume that the ramification of $L$ over $K$ is bounded by 1 (resp. by 1+). We show that the extension $L_i = LK_i$ over $K_i$ is unramified (resp. tamely ramified) by descending induction on $i$. By Lemma 12.1.6, the ramification of $L_i$ over $K_i$ is bounded by 1 (resp. by 1+) since $e_{K_i/K} = 1$.

First, we show the case where $i = n$. Since the residue field $E'$ is separable over $F'$, the valuation ring $O_{L'}$ is generated by a single element over $O_{K'}$ by Lemma 11.5.1. Hence, $L'$ is unramified (resp. tamely ramified) over $K'$.

We may assume that $n = 1$. We show that $[L' : L] = p$. To show this, let $M_t \subset L'$ be the maximal tamely ramified subextension of $K$. Then, by the already proved (a) $\Rightarrow$ (b), the ramification of $M_t$ over $K$ is bounded by 1+. By Corollary 12.1.5(2), the ramification of $LM_t$ over $K$ is bounded by 1+. On the other hand, since $O_{K'}$ is generated by a single element over $O_K$ by Lemma 11.5.1 and $K'$ is wildly ramified over $K$, the ramification of $K'$ over $K$ is not bounded by 1+ by the already proved case of (b) $\Rightarrow$ (a). Hence, we have $K' \not\subset LM_t$ by Corollary 12.1.5(1) and $L \subset LM_t \subsetneqq L'$. Since $1 < [L' : LM_t] \leqslant [L' : L] \leqslant [K' : K] = p$ and $[L' : LM_t] \mid [L' : M_t]$ is a power of $p$, we have $[L' : L] = p$ and $M_t \subset L$.

Let $M_s \subset L'$ be the maximal unramified subextension over $K$. Since $L'$ is unramified (resp. tamely ramified) over $K'$, we have $L' = K'M_s$ by Corollary 1.5.4(2) (resp. $L' = K'M_t$ by Corollary 1.5.6(2)). Since $p = [L' : L] \leqslant$

$[L' : M_s] = [K'M_s : M_s] \leqslant [K' : K] = p$ (resp. $p = [L' : L] \leqslant [L' : M_t] = [K'M_t : M_t] \leqslant [K' : K] = p$), we obtain $L = M_s$ (resp. $L = M_t$). □

**Exercise 12.2** Let $K$ be a henselian discrete valuation field, and let $L$ and $K'$ be finite separable extensions. Assume that the ramification of $L$ over $K$ is bounded by $r$ for $r \in \mathbf{Q}_+$, and that $K'$ is a tamely ramified extension of $K$. Show that either $L$ is unramified over $K$ or the ramification of $L' = LK'$ over $K$ is bounded by $r$.

## 12.3 Upper Ramification Groups

We define upper ramification groups by applying the following proposition to the system of functors constructed in Section 12.1. Let $G$ be a finite group. We say that a $G$-set $X$ is connected if $X$ admits a surjection $G \to X$ of $G$-sets. A $G$-set $X$ is connected if and only if there exist a subgroup $H \subset G$ and an isomorphism $G/H \to X$ of $G$-sets. A morphism $X \to Y$ of connected $G$-sets is always a surjection.

**Proposition 12.3.1** *Let $G$ be a finite group and let $C$ be the category of connected $G$-sets. Let $F = 1_C$ be the identity functor, $F' : C \to C$ be another functor, and $F \to F'$ be a morphism of functors.*

(1) *The following conditions are equivalent:*

  (a) *For every morphism $X \to Y$ in $C$, the diagram*

$$\begin{array}{ccc} F(X) & \longrightarrow & F'(X) \\ \downarrow & & \downarrow \\ F(Y) & \longrightarrow & F'(Y) \end{array} \tag{12.3.1}$$

  *of $G$-sets is cocartesian.*

  (b) *For every morphism $X \to Y$ in $C$, the diagram* (12.3.1) *induces a surjection $F(X) \to F(Y) \times_{F'(Y)} F'(X)$ of $G$-sets.*

  (c) *There exists a normal subgroup $N \lhd G$ such that $F \to F'$ induces an isomorphism $F(X)/N \to F'(X)$ of $G$-sets for every object $X$ of $C$.*

(2) *If the equivalent conditions* (a)–(c) *in* (1) *are satisfied, the normal subgroup $N \lhd G$ in* (c) *is the stabilizer of each element of $F'(G)$.*

*Proof* (1) (a) $\Rightarrow$ (c): Since $F(G) = G \to F'(G)$ is a surjection of $G$-sets, this induces an isomorphism $G/N \to F'(G)$ of $G$-sets for a subgroup $N \subset G$. Since the right translations of $G$ induce automorphisms of the $G$-set $F'(G) = G/N$, the stabilizer of every point of $G/N$ equals $N$. Hence, the subgroup $N \subset G$ is a normal subgroup.

Let $X$ be a connected $G$-set. Let $G \to X$ be a morphism of $G$-sets. Then the cocartesian diagram (12.3.2) for the morphism $G \to X$ is

$$\begin{array}{ccc} G & \longrightarrow & G/N \\ \downarrow & & \downarrow \\ X & \longrightarrow & F'(X) \end{array} \tag{12.3.2}$$

If the left vertical arrow $G \to X$ induces an isomorphism $G/H \to X$ for a subgroup $H$, then the co-fiber product of (12.3.2) in the category of $G$-sets represents the functor sending a $G$-set $Y$ to the intersection $Y^H \cap Y^N = Y^{HN}$ of the fixed parts. Hence, the co-fiber product is isomorphic to $X/N$ and the bottom horizontal arrow of (12.3.2) induces an isomorphism $X/N \to F'(X)$.

(b) $\Rightarrow$ (a): Since $F(Y) \to F'(Y)$ and $F'(X) \to F'(Y)$ are surjections, the diagram (12.3.1) with $F(X)$ replaced by $F(Y) \times_{F'(Y)} F'(X)$ is cocartesian. Hence, if $F(X) \to F(Y) \times_{F'(Y)} F'(X)$ is a surjection then the diagram (12.3.1) of $G$-sets is cocartesian.

(c) $\Rightarrow$ (b): Let $X \to Y$ be a morphism in $C$. Take a morphism $G \to X$ of $G$-sets. Then $X \to Y$ is identified with $G/K \to G/H$ for subgroups $K \subset H \subset G$. Hence, it suffices to show that the morphism $G \to G/H \times_{G/NH} G/NK$ is surjective. Let $g, g' \in G$ and assume that there exist $n \in N$ and $h \in H$ such that $g = g'nh$. Then $gh^{-1} = g'n$ and the surjectivity follows.

(2) Since $G/N \to F'(G)$ is an isomorphism, the assertion follows. □

Let $K_s$ be a separable closure of $K$, and let $G_K = \mathrm{Gal}(K_s/K)$ be the absolute Galois group. For a finite separable extension $L$ and for $r \in \mathbf{Q}_+$, we have a natural action of the absolute Galois group $G_K = \mathrm{Gal}(K_s/K)$ on $F^r(L)$ by the functoriality, and the surjection $F(L) \to F^r(L)$ is compatible with the $G_K$-actions. Thus, we may regard $F$ and $F^r$ as functors

$$F^r : (\text{finite separable extensions of } K) \to \begin{pmatrix} \text{connected finite left } G_K\text{-sets} \\ \text{with continuous action} \end{pmatrix}$$

and a morphism $F \to F^r$ of functors is defined.

**Theorem 12.3.2** *Let $L$ be a finite Galois extension of $K$, and $G = \mathrm{Gal}(L/K)$. Let $K_s$ be a separable closure of $K$, and fix a morphism $L \to K_s$ over $K$.*

(1) *For every subextension $M \subset L$ and $r \in \mathbf{Q}_+$, the action of $G_K$ on $F^r(M)$ factors through $G$.*
(2) *There exists a unique family $(G^r)_{r \in \mathbf{Q}_+}$ of normal subgroups of $G$ such that,*

*for every subextension $M \subset L$, the morphism $F(M) \to F^r(M)$ of $G$-sets induces a bijection $F(M)/G^r \to F^r(M)$.*

*Proof* (1) Let $M \subset L$ be a subextension and $r \in \mathbf{Q}_+$. The morphisms $F(L) \to F(M) \to F^r(M)$ of $G_K$-sets are surjections. Since the action of $G_L \subset G_K$ on $F(L) = \mathrm{Mor}_K(L, K_s)$ is trivial, the assertion follows.

(2) If we identify the category of subextensions $M \subset L$ with $C$ in the notation of Proposition 12.3.1, by (1), the functors $F$ and $F^r$ define functors from $C$ to $C$. The morphism $F \to F^r$ of functors satisfies condition (b) in Proposition 12.3.1(1) by Proposition 12.1.4. Hence, the assertion follows by Proposition 12.3.1(1). □

The subgroups $G^r$ are called upper ramification groups. The filtrations by upper ramification groups are compatible with quotients.

**Corollary 12.3.3** *Let $L$ be a Galois extension of $K$ of Galois group $G$, and let $L' \subset L$ be the sub-Galois extension corresponding to a normal subgroup $N$. Then, for $r \in \mathbf{Q}_+$, the subgroup $G'^r$ of $G' = G/N$ equals the image of $G^r \subset G$.*

*Proof* The $G$-set $F^r(L')$ is identified with $G/NG^r$ and the $G'$-set $F^r(L')$ is identified with $G'/G'^r$. Hence, $G'^r \subset G'$ equals the image of $G^r$. □

**Corollary 12.3.4** *Let $L$ be a finite Galois extension of $K$ of Galois group $G$. Then $G^1$ is the inertia group $I = \mathrm{Ker}(G \to \mathrm{Aut}(E))$ and $G^{1+}$ is the $p$-Sylow subgroup $P$ of $I$.*

*Proof* For a subextension $M \subset L$ corresponding to a subgroup $H \subset G$, the inclusion $H \supset G^1$ (resp. $H \supset G^{1+}$) is equivalent to the condition that the ramification of $M$ over $K$ is bounded by 1 (resp. by 1+). The inclusion $H \supset I$ (resp. $H \supset P$) is equivalent to the condition that $M$ is unramified (resp. tamely ramified) over $K$. Hence, the assertion follows from Proposition 12.2.2. □

**Corollary 12.3.5** *There exist finitely many rational numbers $1 = r_0 < r_1 < \cdots < r_n$ such that $G^r$ are constant on the intervals $[r_i+, r_{i+1}]$ for $i = 0, \ldots, n-1$, and $G^r = 1$ on $[r_n+, \infty)$.*

*Proof* By Lemma 11.4.1, there exists an integer $m > 0$ such that $G^{m+} = 1$. Hence, the assertion follows from Proposition 11.4.8 (to be proved in Section 12.5). □

In the case where $O_L$ is generated by a single element, the lower ramification groups and the upper ramification groups are related by the rescaling function as follows. Define the rescaling function $\varphi\colon [0, \infty) \to [0, \infty)$ by (11.5.2). First, we consider the case where the extension of the residue field is separable.

**Corollary 12.3.6** *Let $L$ be a Galois extension of Galois group $G$. Assume that the residue field $E$ of $L$ is a separable extension of $F$. For integers $i \geqslant 1$, let $G_{i-1} = \mathrm{Ker}(G \to \mathrm{Aut}(O_L/\mathfrak{m}_L^i))$ be the lower ramification groups.*

(1) *For integers $i \geqslant 1$, $G_{i-1} = G^{\varphi(i)}$ and $G_i = G^{\varphi(i)+}$.*

(2) *Let $s \geqslant 1$ be the smallest integer such that $G_s = 1$, and let $d_{L/K} = \mathrm{ord}_L(D_{L/K})$ be the valuation of the different $D_{L/K}$. Then the rational number*

$$r = \varphi(s) = \frac{1}{e_{L/K}}(d_{L/K} + s)$$

*satisfies $G^r \neq 1$ and $G^{r+} = 1$.*

(3) *Assume further that $L$ is an abelian extension of $K$. Then, for each integer $n \geqslant 1$ and $r \in \mathbf{Q}_+$ satisfying $n-1 < r \leqslant n$, we have $G^n_{\mathrm{coh}} = G^{r+1}$.*

*Proof* (1) Let $i \geqslant 1$ be an integer and set $r = \varphi(i)$. Proposition 12.1.7(1) and 12.1.7(2) mean that $F^r(L)$ (resp. $F^{r+}(L)$) is the quotient of $F(L)$ by $G_{i-1}$ (resp. $G_i$). Hence, the assertion follows.

(2) By (1), we have $G^r = G_{s-1} \supsetneqq G^{r+} = G_s = 1$.

(3) Since both $G^n_{\mathrm{coh}}$ and $G^r$ are compatible with quotients by Corollary 12.3.3, we may assume that $L$ is a cyclic extension. Further, it suffices to show the equality $n = r$ for the largest integer $n$ and the largest rational number $r$ such that $G^n_{\mathrm{coh}} \neq 1$ and $G^{r+1} \neq 1$. This follows from the Hasse–Arf theorem, Theorem 5.3.4, and (2). □

Similarly, if the ramification index is 1, the following holds.

**Corollary 12.3.7** *Let $L$ be a Galois extension of Galois group $G$. Assume that $e_{L/K} = 1$ and that the residue field $E$ is generated by a single element over $F$. For integers $i \geqslant 1$, let $G_i = \mathrm{Ker}(G \to \mathrm{Aut}(O_L/\mathfrak{m}_L^i))$ be the lower ramification groups.*

(1) *For integers $i \geqslant 1$, $G_i = G^{\varphi(i)}$ and $G_{i+1} = G^{\varphi(i)+}$.*

(2) *Let $s \geqslant 1$ be the smallest integer such that $G_s = 1$, and let $d_{L/K} = \mathrm{ord}_L(D_{L/K})$ be the valuation of the different $D_{L/K}$. Then the rational number*

$$r = \varphi(s) = \frac{1}{e_{L/K}}(d_{L/K} + s)$$

*satisfies $G^r \neq 1$ and $G^{r+} = 1$.*

(3) *Assume further that $L$ is an abelian extension of $K$. Then, for each integer $n \geqslant 1$ and $r \in \mathbf{Q}_+$ satisfying $n-1 < r \leqslant n$, we have $G^n_{\mathrm{coh}} = G^r$.*

**Example 12.3.8** Let $K$ be a henselian discrete valuation field of characteristic $p > 0$, and $t \in K$ be a uniformizer.

(1) Let $L$ be the Artin–Schreier extension defined by $x^p - x = 1/t^n$ for $n = pm - 1 > 0$ as in Example 1.4.4(1). For $r = pm = n + 1$, in the notation of Example 11.5.10(1), the morphism $Q_T^{(r)} = \mathbf{A}_T^1 = \operatorname{Spec} O_L[Y] \to P_T^{(r)} = \mathbf{A}_T^1 = \operatorname{Spec} O_L[W]$ is the finite étale morphism sending $W$ to $Y^p - Y$. Hence, $F^r(L)$ is a singleton and $F(L) \to F^{r+}(L)$ is a bijection. Thus, we have $G^r = G$ and $G^{r+} = 1$ for $r = n + 1$.

(2) Let $L$ be the Artin–Schreier extension defined by $x^p - x = v/t^n$ for $n = pm > 0$ and for $v \in O_K$ such that $\bar{v} \in F - F^p$. Let $u = t^m x$ be the generator of the residue field $E = F(u)$ as in Example 1.4.4(2). For $r = pm = n$, in the notation of Example 11.5.10(2), the morphism $Q_T^{(r)} = \mathbf{A}_T^1 = \operatorname{Spec} O_L[Y] \to P_T^{(r)} = \mathbf{A}_T^1 = \operatorname{Spec} O_L[W]$ is the finite étale morphism sending $W$ to $Y^p - Y$. Hence, $F^r(L)$ is a singleton and $F(L) \to F^{r+}(L)$ is a bijection. Thus, we have $G^r = G$ and $G^{r+} = 1$ for $r = n$.

**Lemma 12.3.9** *Let $K'$ be an extension of a discrete valuation field $K$ of ramification index $e$. Let $L'$ be a Galois extension of $K'$ containing $LK'$. Let $r, r' \in \mathbf{Q}_+$ such that $r' \geqslant er$.*

(1) *The canonical morphism $G' = \operatorname{Gal}(L'/K') \to G$ induces $G'^{r'} \to G^r$.*

(2) *Assume that $O_L \otimes_{O_K} O_{K'} \to O_{L'}$ is an isomorphism and $r' = er$. Then the induced morphism $G'^{r'} \to G^r$ is an isomorphism.*

*Proof* (1) By the commutative diagram

$$\begin{array}{ccc} F'(L') & \longrightarrow & F'^{r'}(L') \\ \downarrow & & \downarrow \\ F(L) & \longrightarrow & F^r(L) \end{array} \tag{12.3.3}$$

the morphism $G' \to G$ induces $G'/G'^{r'} \to G/G^r$.

(2) Let $T \to Q$ be an immersion to a smooth scheme over $S$. Then, by the assumption, we have an induced immersion $T' \to Q'$ to a smooth scheme over $S'$. Since the formation of stable integral models commutes with base change by Proposition 11.4.6(2), the vertical arrows in the diagram (12.3.3) are bijections for $r \in \mathbf{Q}_+$. □

**Exercise 12.3** Let $p$ be a prime number and $n \geqslant 1$ be an integer.

(1) Determine the upper ramification subgroups $G^r$ of the Galois group $G = \operatorname{Gal}(\mathbf{Q}_p(\zeta_{p^n})/\mathbf{Q}_p)$ for $r \in \mathbf{Q}_+$.

(2) Let $K$ be the fraction field of the henselization of $\mathbf{Z}_p[T]_{(p)}$, and let $L = K(\zeta_{p^n}, T^{1/p^n})$ be the Galois closure of the extension $K(u)$ in Exercise 12.1(2). Identify the Galois group $G = \operatorname{Gal}(L/K)$ with the semi-direct

product $(\mathbf{Z}/p^n\mathbf{Z})^\times \ltimes \mathbf{Z}/p^n\mathbf{Z}$ with respect to multiplication, and determine the upper ramification subgroups $G^r \subset G = \mathrm{Gal}(L/K)$ for $r \in \mathbf{Q}_+$.

## 12.4 Flat Morphisms with Geometrically Reduced Fibers

We will prove Proposition 11.4.8 at the end of the next section. As a preliminary, we prepare some facts on the sets of connected components of geometric fibers of flat morphisms with geometrically reduced fibers, and a generalization of the reduced fiber theorem, Proposition 9.3.5.

**Definition 12.4.1** ([31, Définition (6.8.1)]) Let $f: X \to S$ be a flat morphism locally of finite presentation of schemes. We say that $f$ is *reduced* if, for every geometric point $s$ of $S$, the geometric fiber $X_s$ is *reduced*.

In [32, Exposé X, Definition 1.1], reduced morphism is called separable morphism. A separated morphism $f$ of finite presentation is étale if and only if $f$ is quasi-finite, flat, and reduced.

First, we prove the following proposition on connected components.

**Proposition 12.4.2** *Let $S$ be a noetherian scheme and $f: X \to Y$ be a morphism of schemes of finite type over $S$. Assume that $X \to S$ is étale and that $Y \to S$ is flat and reduced. Then there exists an open subset $U \subset X \times_S X$ satisfying the following condition:*

*For every geometric point $s$ of $S$, the geometric fiber $U_s \subset X_s \times X_s$ equals the fiber product $X_s \times_{\pi_0(Y_s)} X_s$ of the mapping $X_s \to \pi_0(Y_s)$ to the set of connected components of the geometric fiber.*

We recall some facts on the sets of connected components of schemes.

**Proposition 12.4.3** *Let $X \to S$ be a morphism of finite presentation of schemes.*

(1) [31, Théorème (9.7.7)(ii)] *Assume that $S$ is irreducible and let $\eta$ be the generic point. Assume that the generic fiber $X_\eta$ is geometrically connected. Then there exists a dense open subscheme $U \subset S$ such that every geometric fiber of $X \times_S U \to U$ is connected.*
(2) [31, Corollaire (18.9.11)] *Assume that $S$ is a normal noetherian scheme and that the morphism $X \to S$ is flat and reduced. Then, for every dense open subset $U \subset S$, the open immersion $X \times_S U \to X$ induces a bijection $\pi_0(X \times_S U) \to \pi_0(X)$.*

**Corollary 12.4.4** *Let $X \to S$ be a morphism of finite presentation of schemes.*

*Assume that $S$ is integral and let $\eta$ be the generic point. Then there exists a dense open subscheme $U \subset S$, a finite étale scheme $P$ over $U$, and a morphism $X_U = X \times_S U \to P$ over $U$ such that every geometric fiber of $X \times_S U \to P$ is connected.*

*Proof* First, we show the case where $S = \operatorname{Spec} K$ for a field $K$. Let $K_s$ be a separable closure and $\operatorname{Gal}(K_s/K)$ be the absolute Galois group. Then the finite set $\pi_0(X_{K_s})$ of connected components of the geometric fiber has a canonical continuous action of $\operatorname{Gal}(K_s/K)$ and defines a finite étale scheme $P$ over $S$. The canonical mapping $X_{K_s} \to \pi_0(X_{K_s})$ is compatible with the Galois actions and defines a morphism $X \to P$ of schemes over $S$. Every geometric fiber of $X \to P$ is connected.

Let $P_\eta$ be a finite étale scheme over $\eta$, and $X_\eta \to P_\eta$ be a morphism over $\eta$ such that every geometric fiber of $X_\eta \to P_\eta$ is connected. Then there exists a dense open subscheme $U \subset S$, a finite étale scheme $P$ over $U$, and a morphism $X_U = X \times_S U \to P$ over $U$ such that the base change by $\eta \to U$ equals $X_\eta \to P_\eta$. By Proposition 12.4.3(1), there exists a dense open subset $V \subset P$ such that every geometric fiber of $X_U \times_P V \to V$ is connected. Since $P_\eta \subset V$, it suffices to replace $U$ by the complement of the image of the complement $P - V$. □

Let $E$ be a subset of the underlying set of a noetherian scheme $X$. Recall that $E$ is constructible if and only if $E$ is a finite union of locally closed subsets [30, Corollaire (9.1.7)]. We say that a subset $E \subset X$ is closed under generalization (resp. specialization) if, for every $x, y \in X$ such that $x \in \overline{\{y\}}$, the condition $x \in E$ implies $y \in E$ (resp. $y \in E$ implies $x \in E$).

**Lemma 12.4.5** *Let $X$ be a noetherian scheme and $E$ be a subset.*

(1) [30, Proposition (9.2.3)] *$E$ is constructible if and only if, for every irreducible closed subset $Z \subset X$, either the intersection $E \cap Z$ or its complement $Z - E \cap Z$ contains a dense open subset of $Z$.*
(2) [30, Proposition (9.2.5)] *Assume that $E$ is constructible. Then $E$ is an open subset of $X$ if and only if $E$ is stable under generalization.*

*Proof of Proposition* 12.4.2 Let $t \in S$ be a point. Then, by Corollary 12.4.4, there exists a finite étale scheme $P_t$ over $t$ and a morphism $Y_t \to P_t$ over $t$ such that every geometric fiber of $Y_t \to P_t$ is connected. Let $U \subset X \times_S X$ be the union of the subsets $U_t = X_t \times_{P_t} X_t \subset X_t \times_t X_t$ for $t \in S$. For a geometric point $s$ above $t \in S$, the geometric fiber $U_s$ equals $X_s \times_{\pi_0(Y_s)} X_s$. Hence, it suffices to show that $U \subset X \times_S X$ is an open subset.

We show that $U$ is constructible. By Corollary 12.4.4, if $S$ is integral, there exist a dense open subscheme $V \subset S$, a finite étale scheme $P_V \to V$, and a morphism $Y \times_S V \to P_V$ over $V$ such that every geometric fiber of $Y \times_S V \to P_V$ is connected. Then, the inverse image of $U$ to $X_V \times_V X_V$ equals the open subset $X_V \times_{P_V} X_V$. Hence, by Lemma 12.4.5(1), $U$ is constructible. Thus, by Lemma 12.4.5(2), it suffices to show that $U$ is stable under generalization.

First we reduce the assertion to the case where $S = \operatorname{Spec} A$ for a discrete valuation ring $A$. Note that the formation of $U \subset X \times_S X$ commutes with every base change $S' \to S$ of noetherian schemes. Let $s, t \in S$ be distinct points, and assume that $s$ is in the closure of $\{t\}$. By replacing $S$ by the closure of $t$, we may assume that $S$ is integral and that $t$ is the generic point of $S$. By replacing $S$ by the blow-up $S'$ at the closure of $\{s\}$ with reduced closed subscheme structure, and replacing $s$ by a maximal point of the inverse image of $s$, we may assume that $\dim \mathcal{O}_{S,s} = 1$. Replacing $S$ by $\operatorname{Spec} \mathcal{O}_{S,s}$, we may assume that $S = \operatorname{Spec} A$ for a local noetherian domain $A$ of dimension 1. By the Krull–Akizuki theorem, the normalization of $A$ is a Dedekind domain. Hence, by replacing $A$ by its normalization, we may assume that $A$ is a discrete valuation ring.

Let $K$ be the fraction field of a discrete valuation ring $A$. By Corollary 12.4.4, there exists a finite separable extension $L$ of $K$ such that $X_L$ is isomorphic to a disjoint union of copies of $\operatorname{Spec} L$, and that every connected component of $Y_L$ is geometrically connected over $L$. By further replacing $S$ by the localization of the normalization of $A$ in $L$, we may assume that $X_K$ is isomorphic to a disjoint union of copies of $\operatorname{Spec} K$, and that every connected component of $Y_K$ is geometrically connected over $K$.

Let $\bar{s}$ and $\bar{t}$ be geometric points above $s$ and $t$, and consider the commutative diagram

$$\begin{array}{ccccccc} X_{\bar{s}} & \longrightarrow & \pi_0(X) & \longleftarrow & X_t & \longleftarrow & X_{\bar{t}} \\ \downarrow & & \downarrow & & \downarrow & & \downarrow \\ \pi_0(Y_{\bar{s}}) & \longrightarrow & \pi_0(Y) & \longleftarrow & \pi_0(Y_t) & \longleftarrow & \pi_0(Y_{\bar{t}}) \end{array} \tag{12.4.1}$$

where the vertical arrows are induced by the morphism $X \to Y$. By the assumption, the right horizontal arrows are bijections. By Proposition 12.4.3(2), the middle horizontal arrows are bijections. Thus, in the diagram

$$U \times_S \bar{s} = X_{\bar{s}} \times_{\pi_0(Y_{\bar{s}})} X_{\bar{s}} \to \pi_0(X) \times_{\pi_0(Y)} \pi_0(X) \leftarrow X_{\bar{t}} \times_{\pi_0(Y_{\bar{t}})} X_{\bar{t}} = U \times_S \bar{t}$$

the right arrow is a bijection. Hence, $U$ is stable under generalization. □

Let $s$ and $t$ be geometric points of $S$, and let $S_{(s)}$ denote the strict localization of $S$. We call a morphism $t \to S_{(s)}$ over $S$ a specialization, and write $t \to s$.

**Corollary 12.4.6** *Let $S$ be a noetherian scheme. Let $X$ be a finite set, and let*

*$X_S$ denote the disjoint union of copies of $S$ indexed by $X$. Let $f\colon X_S \to Y$ be a morphism of schemes over $S$. For each geometric point $s$ of $S$, let $f_s\colon X \to Y_s$ be the base change of $f$ and let $f_{s*}\colon X \to \pi_0(Y_s)$ be the induced mapping to the set of connected component.*

(1) *Assume that the morphism $Y \to S$ is flat and reduced. For each geometric point $s \in S$, let $A_s$ be the image of the mapping $f_{s*}\colon X \to \pi_0(Y_s)$. Then, for each specialization $t \to s$, the surjection $X \to A_s$ is finer than $X \to A_t$.*

(2) *Assume that $Y$ is a separated scheme over $S$. For each geometric point $s \in S$, let $B_s$ be the image of the mapping $f_s\colon X \to Y_s$. Then, for each specialization $t \to s$, the surjection $X \to B_s$ is coarser than $X \to B_t$.*

*Proof* (1) Let $U \subset X \times_S X$ be the open subset in Proposition 12.4.2. Then, for a geometric point $s$ of $S$, the surjection $X \to A_s$ is the quotient of $X = X_s$ by the equivalence relation whose graph is $U_s$. By Proposition 12.4.2 and Lemma 12.4.5(2), $U$ is stable under generalization and the assertion follows.

(2) By the assumption that $Y \to S$ is separated, the immersion $X_S \times_Y X_S \subset X_S \times_S X_S = (X \times X)_S$ is a closed immersion. Hence, by Lemma 12.4.5(2), $X_S \times_Y X_S$ is stable under specialization and the assertion follows. □

**Remark 12.4.7** Let $S = \operatorname{Spec} O_K$ for a discrete valuation ring $O_K$, and let $X = \operatorname{Spec} A$ be an affine scheme of finite type over $S$. Let $\bar{s} \to S$ be a geometric closed point. Let $\mathfrak{X} = \operatorname{Spf} \widehat{A}$ be the formal completion along the closed fiber, and let $\mathfrak{X}_{K_s} = \operatorname{Sp} \widehat{A} \otimes_{O_K} K_s$ be the associated affinoid variety over an algebraic closure $K_s$ of the fraction field $K$ of $O_K$. If $X$ is flat and reduced over $S$, the cospecialization mapping $\pi_0(X_{\bar{s}}) \to \pi_0(\mathfrak{X}_{K_s})$ is a bijection, by Hensel's lemma applied to the separable polynomial $X^2 - X = X(X-1)$.

The following statement is a combination of the reduced fiber theorem and the flattening theorem.

**Theorem 12.4.8** ([13, Theorem 2.1′], [68, Théorème (5.2.2)]) *Let $S$ be a noetherian scheme and let $U \subset S$ be a schematically dense open subscheme. Let $X$ be a scheme of finite type over $S$ such that $X_U = X \times_S U$ is schematically dense in $X$ and that $X_U \to U$ is flat and reduced. Then there exists a commutative diagram*

$$\begin{array}{ccc} X & \longleftarrow & X' \\ \downarrow & & \downarrow \\ S & \longleftarrow & S' \end{array} \tag{12.4.2}$$

*of schemes satisfying the following conditions:*

(1) *The morphism $S' \to S$ is the composition of a blow-up $S^* \to S$ with center*

*supported in $S - U$ and a faithfully flat morphism $S' \to S^*$ of finite type such that $U' = S' \times_S U \to U$ is étale.*

(2) *The morphism $X' \to S'$ is* flat *and* reduced. *The induced morphism $X' \to X \times_S S'$ is finite and its restriction $X' \times_{S'} U' \to X \times_S U'$ is an isomorphism.*

**Corollary 12.4.9** *Let $K$ be a henselian discrete valuation field. Let $X \to C$ be a morphism of flat schemes of finite type over $S = \operatorname{Spec} O_K$. Assume that the generic fiber $C_K = C \times_S \operatorname{Spec} K$ is a smooth curve over $K$, and that the morphism $X_K \to C_K$ is flat and reduced. Then there exist a finite separable extension $K'$ of discrete valuation field $K$ and a commutative diagram*

$$\begin{array}{ccl} X & \longleftarrow & X' \\ \downarrow & & \downarrow \\ C & \longleftarrow & C' \\ \downarrow & & \downarrow \\ S & \longleftarrow & S' = \operatorname{Spec} O_{K'} \supset U' = \operatorname{Spec} K' \end{array}$$

*of schemes satisfying the following conditions:*

(1) *The scheme $C'$ is a nodal curve over $S'$, and the morphism $C' \to C \times_S S'$ is the composition $C' \to C'_1 \to C'_0 \to C \times_S S'$ of a proper modification $C'_0 \to C \times_S S'$ such that $C'_0 \times_{S'} U' \to C \times_S U'$ is an isomorphism, a faithfully flat morphism $C'_1 \to C'_0$ such that $C'_1 \times_{S'} U' \to C'_0 \times_{S'} U'$ is étale, and a proper modification $C' \to C'_1$ such that $C' \times_{S'} U' \to C'_1 \times_{S'} U'$ is an isomorphism.*

(2) *The morphism $X' \to C'$ is flat and reduced, the morphism $X' \to X \times_C C'$ is finite, and $X' \times_{S'} U' \to X \times_C C' \times_{S'} U'$ is an isomorphism.*

*Proof* By the reduced fiber theorem, Theorem 12.4.8, applied to $X \to C$ and $C_K \subset C$, there exists a commutative diagram

$$\begin{array}{ccc} X & \longleftarrow & X_1 \\ \downarrow & & \downarrow \\ C & \longleftarrow & C_1 \end{array}$$

satisfying conditions (1) and (2). Since $C_{1,K} \to C_K$ is étale and $C_1$ is flat over $S$, by the stable reduction theorem, Theorem 10.1.8, there exist a finite separable extension $K'$ of $K$, a nodal curve $C' \to S'$, and a commutative diagram

$$\begin{array}{ccc} C_1 & \longleftarrow & C' \\ \downarrow & & \downarrow \\ S & \longleftarrow & S' \end{array}$$

such that $C' \to C_1$ is proper and $C_1 \times_S U' \to C'_{K'}$ is an isomorphism.

We show that $X' = X_1 \times_{C_1} C' \to C' \to S'$ satisfy the required conditions. Since $C_1 \to C$ is obtained by applying Theorem 12.4.8, the composition $C' \to C'_1 = C_1 \times_S S' \to C \times_S S'$ satisfies the condition in (1). The base change $X' \to C'$ of a flat and reduced morphism $X_1 \to C_1$ is flat and reduced. Since $X_1 \to X \times_C C_1$ is obtained by applying Theorem 12.4.8, the morphism $X' = X_1 \times_{C_1} C' \to X \times_C C'$ satisfies condition (2). □

**Exercise 12.4** Let $O_K$ be a discrete valuation ring and $t \in O_K$ be a uniformizer. Let $X = \operatorname{Spec} O_K[x]/(x^2 - x)$ be the disjoint union of two copies of $S = \operatorname{Spec} O_K$. Define $Y = \operatorname{Spec} O_K[u, v]/(uv - t) - \{y\}$ to be the complement of the singular point $y = (u, v)$, and $Z = \operatorname{Spec} O_K[z]/(z(z - t))$. Define an immersion $X \to Y$ by sending $u, v$ to $1 - x + tx, t(1 - x) + x$ and a finite morphism $X \to Z$ by sending $z$ to $tx$. Compute the open subset $U \subset X \times_S X$ in Proposition 12.4.2 and the closed subset $X \times_Z X \subset X \times_S X$.

## 12.5 Semi-continuity of Filtration

In this section we prove Proposition 11.4.8. The following lemma means that a singular point of a nodal curve defines an annulus in the language of rigid geometry.

**Lemma 12.5.1** *Let $K$ be a discrete valuation field, and let $C$ be a nodal curve over $S = \operatorname{Spec} O_K$. Let $C_1, C_2$ be two irreducible components of $C_F$, and let $x$ be a point of the intersection $C_1 \cap C_2$. Let $R \subset C$ be an effective Cartier divisor such that $R = r_1 C_1 + r_2 C_2$ on a neighborhood of $x$, and assume $r_1 < r_2$. Let $K'$ be an extension of discrete valuation field $K$, and extend $\operatorname{ord}_K$ to a valuation of $K'$.*

(1) *Let $S' = \operatorname{Spec} O_{K'} \to C$ be a morphism over $S$ such that the image of the closed point $s' \in S'$ equals $x$, and let $a \in O_{K'}$ be an element such that the pull-back of $R$ equals the divisor of $a$. Then $r_1 < \operatorname{ord}_K a < r_2$.*
(2) *Conversely, let $r$ be a rational number satisfying $r_1 < r < r_2$, and $a \in O_{K'}$ be an element satisfying $r = \operatorname{ord}_K a$. Then, after replacing $K'$ by a finite separable extension, there exists a morphism $S' = \operatorname{Spec} O_{K'} \to C$ such*

*that the image of the closed point $s' \in S'$ equals $x$, and the pull-back of $R$ by $S' \to C$ is the divisor of $a$.*

*Proof* (1) Let $t \in O_K$ be a uniformizer. Then, since $\operatorname{div} t^{r_1} < R < \operatorname{div} t^{r_2}$, we have $r_1 < \operatorname{ord}_K a < r_2$.

(2) By Proposition 10.1.6, we may assume that there exists an étale morphism $C \to C_0 = \operatorname{Spec} O_K[u,v]/(uv - t^m)$ for a uniformizer $t \in \mathfrak{m}_K$ and an integer $m \geqslant 1$. We may assume that $C_1$ is defined by $u = t = 0$, and $C_2$ is defined by $v = t = 0$.

After replacing $K'$ by a finite separable extension, we may assume that there exists an element $b \in O_{K'}$ satisfying $0 < \operatorname{ord}_K b = m \cdot (\operatorname{ord}_K a - r_1)/(r_2 - r_1) < m$. By sending $u, v$ to $t^m/b, b \in \mathfrak{m}_{K'}$, we define a morphism $S' \to C_0$. The image of the closed point $s' \in S'$ is the maximal ideal $(u, v, t)$. By replacing $K'$ by an unramified extension, we lift it to $S' \to C$.

Since $mR = \operatorname{div} u^{r_1} v^{r_2} = mr_1 \cdot \operatorname{div} t + (r_2 - r_1) \cdot \operatorname{div} v$, its pull-back to $S'$ equals $mr_1 \cdot \operatorname{div} t + (r_2 - r_1) \cdot \operatorname{div} b = m \cdot \operatorname{div} a$ as required. □

*Proof of Proposition* 11.4.8 Let $T = \operatorname{Spec} O_L \to Q$ be an immersion as in the statement of Proposition 11.4.8. We may assume that $Q$ is affine and hence is separated over $S$.

For $0 \leqslant s < r \leqslant m$, the surjection $T^{\natural}_{\bar{F}} \to \operatorname{Im}(T^{\natural}_{\bar{F}} \to Q^{(r)}_{\bar{F}})$ is finer than $T^{\natural}_{\bar{F}} \to \pi_0(Q^{(r)}_{\bar{F}})$ by Proposition 11.4.4(1), and the surjection $T^{\natural}_{\bar{F}} \to \pi_0(Q^{(r)}_{\bar{F}})$ is finer than $T^{\natural}_{\bar{F}} \to \operatorname{Im}(T^{\natural}_{\bar{F}} \to Q^{(s)}_{\bar{F}})$ by Proposition 11.4.4(2). Hence, it suffices to show that there exist rational numbers $0 \leqslant b_j \leqslant c_j \leqslant m$, $j = 1, \ldots, l$, satisfying the following conditions:

(1) $[0, m] = [b_1, c_1] \cup \cdots \cup [b_l, c_l]$.

(2) For each $i = 1, \ldots, l$, the surjections $T^{\natural}_{\bar{F}} \to \pi_0(Q^{(r)}_{\bar{F}})$ are equivalent to each other for $r \in (b_j, c_j]$.

(3) For each $i = 1, \ldots, l$, the surjections $T^{\natural}_{\bar{F}} \to \operatorname{Im}(T^{\natural}_{\bar{F}} \to Q^{(s)}_{\bar{F}})$ are equivalent to each other for $s \in [b_j, c_j)$.

First, we construct a universal family of dilatations parametrized by a nodal curve. Let $t \in O_K$ be a uniformizer and define a nodal curve

$$C = \operatorname{Spec} O_K[x, y]/(xy - t^m).$$

Let $R \subset C$ be the effective Cartier divisor defined by $x$. Let $T_C = T \times_S C \to Q_C = Q \times_S C$ be the base change, and $R_{Q_C}$ be the pull-back of $R \subset C$. We define a dilatation $Q_C^{[R]}$ to be $Q_C^{[R_{Q_C} \cdot T_C]}$.

Applying Corollary 12.4.9 to $Q_C^{[R]} \to C \to S$, we obtain a commutative

diagram

$$\begin{array}{ccc} Q_C^{[R]} & \longleftarrow & Q_{C'}^{(R')} \\ \downarrow & & \downarrow \\ C & \longleftarrow & C' \\ \downarrow & & \downarrow \\ S & \longleftarrow & S' \ = \operatorname{Spec} O_{K'} \end{array}$$

such that $C'$ is a nodal curve, $R' \subset C'$ is the pull-back of the Cartier divisor $R \subset C$, and the base change by $S \leftarrow \operatorname{Spec} K$ is cartesian. We may assume that $K'$ contains a Galois closure of $L$. Let $T_{C'}^{\natural}$ be the normalization of $T \times_S C'$. By the universality of dilatations, the morphism $T_C \to Q_C$ is lifted to $T_C \to Q_C^{[R]}$. Further, by the functoriality of normalizations, the base change $T \times_S C' \to Q_C^{[R]} \times_C C'$ is lifted to $T_{C'}^{\natural} \to Q_{C'}^{(R')}$.

Let $V = \{v_1, \dots, v_k\}$ denote the set of irreducible components of the closed fiber $C'_{F'}$, and let $E = \{u_1, \dots, u_l\}$ be the set of its singular points. For each irreducible component $v_i$, extend the normalized valuation $\operatorname{ord}_K$ to the local field at the generic point and let $a_i = \operatorname{ord}_K R' \in \mathbf{Q}$ denote the valuation. Let $u_j$ be a singular point. If $u_j$ is in the intersection of two distinct components $v_i$ and $v_{i'}$, set $b_j = \min(a_i, a_{i'}) \leqslant c_j = \max(a_i, a_{i'})$. If $u_j$ is a singular point of a component $v_i$, set $b_j = c_j = a_i$.

We show that conditions (1), (2), and (3) are satisfied. Let $0 \leqslant r \leqslant m$ be a rational number. Then there exist an extension $K_1$ of the discrete valuation field $K$ and an element $q \in O_{K_1}$ satisfying $\operatorname{ord}_K q = r$ for the extension of $\operatorname{ord}_K$ to $K_1$. Then, the morphism $O_K[x, y]/(xy - t^m) \to O_{K_1}$ sending $x$ to $q$ defines a morphism $S_1 = \operatorname{Spec} O_{K_1} \to C$ over $S$. Since $C' \to C$ satisfies condition (1) in Corollary 12.4.9, there exists an extension $K'_1$ of the discrete valuation field $K_1$ and a commutative diagram

$$\begin{array}{ccc} C' & \longleftarrow & S'_1 = \operatorname{Spec} O_{K'_1} \\ \downarrow & & \downarrow \\ C & \longleftarrow & S_1 \end{array} \tag{12.5.1}$$

by the valuative criterion of properness.

If the image of the closed point $s'_1$ of $S'_1$ is on a smooth point of $v_i$ not on the other components, then we have $r = a_i$. If the image of $s'_1$ is $u_j$, then we have $r \in [b_j, c_j]$ by Lemma 12.5.1(1). Thus, we have $[0, m] \cap \mathbf{Q} = (\{a_1, \dots, a_k\} \cup [b_1, c_1] \cup \dots \cup [b_l, c_l]) \cap \mathbf{Q}$. This implies the equality in (1), $[0, m] = [b_1, c_1] \cup \dots \cup [b_l, c_l]$.

We show (2) and (3). Let $j = 1, \ldots, l$, and $r \in (b_j, c_j)$ be a rational number. Then, by Lemma 12.5.1(2), there exists a finite extension $K''$ and a morphism $S'' = \operatorname{Spec} O_{K''} \to C'$ such that $\operatorname{ord}_K R = r$ and the image of the closed point of $S''$ is $u_j$. For the composition $S'' \to C' \to C$, we have a cartesian diagram

$$\begin{array}{ccc} Q_C^{[R]} & \longleftarrow & Q_{S''}^{[r]} \\ \downarrow & & \downarrow \\ C & \longleftarrow & S'' \end{array} \tag{12.5.2}$$

Since flat and reduced morphisms are preserved by base change, Proposition 9.3.3(2) shows that the cartesian diagram

$$\begin{array}{ccc} Q_{C'}^{(R')} & \longleftarrow & Q_{S''}^{(r)} \\ \downarrow & & \downarrow \\ C' & \longleftarrow & S'' \end{array} \tag{12.5.3}$$

lifting (12.5.2) defines a stable integral model $Q_{S''}^{(r)}$. Further, the morphism $T_{S''}^{\natural} \to Q_{S''}^{(r)}$ is the base change of $T_{C'}^{\natural} \to Q_{C'}^{(R')}$ by $S'' \to C'$. Hence, the stable closed fiber $Q_{F''}^{(r)}$ is the base change of the fiber $Q_{C'}^{(R')} \times_{C'} u_j$ at the singular point $u_j \in C'$ and is independent of $r \in (b_j, c_j)$. Thus, the following weaker conditions are satisfied:

(2′) For each $j = 1, \ldots, l$, the surjections $T_{\bar{F}}^{\natural} \to \pi_0(Q_{\bar{F}}^{(r)})$ are equivalent for $r \in (b_j, c_j)$.

(3′) For each $j = 1, \ldots, l$, the surjections $T_{\bar{F}}^{\natural} \to \operatorname{Im}(T_{\bar{F}}^{\natural} \to Q_{\bar{F}}^{(s)})$ are equivalent for $s \in (b_j, c_j)$.

By Corollary 12.4.6(1), for $r \in (b_j, c_j)$ the surjection $T_{\bar{F}}^{\natural} \to \pi_0(Q_{\bar{F}}^{(b_j)})$ is finer than $T_{\bar{F}}^{\natural} \to \pi_0(Q_{\bar{F}}^{(r)})$. Since $b_j < r$, this implies that they are equivalent and condition (2) is satisfied.

By Corollary 12.4.6(2), for $r \in (b_j, c_j)$ the surjection $T_{\bar{F}}^{\natural} \to \operatorname{Im}(T_{\bar{F}}^{\natural} \to Q_{\bar{F}}^{(s)})$ is finer than $T_{\bar{F}}^{\natural} \to \operatorname{Im}(T_{\bar{F}}^{\natural} \to Q_{\bar{F}}^{(c_j)})$. Since $s < c_j$, this implies that they are equivalent and condition (3) is satisfied. □

**Exercise 12.5** Let $K$ be a discrete valuation field and $a, b, c \in \mathfrak{m}_K$ be non-zero elements satisfying $c = a^2b^2$. Define a closed immersion $T = \operatorname{Spec} O_K[x]/(x^2 - ax) \to Q = \mathbf{A}_S^1 = \operatorname{Spec} O_K[x]$ to a smooth scheme over $S = \operatorname{Spec} O_K$ and a nodal curve $C = \operatorname{Spec} O_K[u, v]/(uv - c)$ over $S$.

Let $T_C = T \times_S C \to Q_C = Q \times_S C$ be the base change of the immersion. Let $R \subset C$ be the effective Cartier divisor defined by $v$, and $R_Q = R \times_C Q \subset Q_C$ be the pull-back. Let $Q_C^{[R]} = Q_C^{[R_Q \cdot T_C]} \to Q_C$ be the dilatation.

Let $C' = \operatorname{Spec} O_K[u', v']/(u'v' - c') \to C$ be the finite morphism of the nodal curves over $S$ defined by $c' = ab, u = u'^2$ and $v = v'^2$. Let $C'' \to C'$ be the blow-up by the ideal $(v', a)$, $Q^{[R]}_{C''} = Q^{[R]}_C \times_C C''$ be the base change, and let $Q^{(R)}_{C''}$ be the normalization. Let $T^\natural = S \amalg S \to T$ be the normalization, and $T^\natural_{C''} = C'' \amalg C'' \to Q^{(R)}_{C''}$ be the morphism lifting $T_C \to Q_C$.

(1) Show that $C''$ is a nodal curve over $S$, and that the morphism $Q^{(R)}_{C''} \to C''$ is smooth. Compute the morphism $T^\natural_{C''} = C'' \amalg C'' \to Q^{(R)}_{C''}$ and its image. For each geometric point $x$ of the closed fiber $C''_F$, compute the finite sets $\operatorname{Image}(T^\natural_x \to Q^{(R)}_{C'',x})$ and $\pi_0(Q^{(R)}_{C'',x})$.

(2) Let $K'$ be an extension of $K$, and let $S' = \operatorname{Spec} O'_K \to C'$ be the morphism defined by sending $v'$ to $w \in O'_K$ satisfying $r' = \operatorname{ord}_K w \leqslant \operatorname{ord}_K c'$. Show that, for the morphism $S' \to C''$ lifting $S' \to C'$ defined by $w$, the base change $Q^{(R)}_{C''} \times_{C''} S'$ equals $Q^{(2r')}_{S'}$.

## Historical Notes

The definition of upper ramification subgroups is given in [1], using the terminology of rigid geometry. It is then rephrased in the language of schemes in [71]. The condition for a quotient of a fiber functor to be defined by a normal subgroup is also given in [1] in a slightly different formulation. The semi-continuity is also proved in [1].

# 13

# Logarithmic Variant and Artin–Schreier–Witt Extensions

We introduce a logarithmic variant of upper ramification groups in Section 13.3. This explains the shift by 1 in the definition of ramification groups in [78]. The definition is a modification of that in the non-logarithmic case in the previous chapter, by replacing immersions to smooth schemes by exact immersions to log smooth schemes. We show in Proposition 13.3.4(2) that the definition of logarithmic filtration is reduced to that of the non-logarithmic case after eliminating the ramification index by taking a log smooth base extension. In Sections 13.1 and 13.2, we define and study basic properties of log smooth schemes and log smooth extensions of discrete valuation fields.

For abelian extensions in the positive characteristic case, we show that the logarithmic filtration is given by the filtration on Witt vectors defined by Brylinski, recalled in Section 13.4. The key ingredient of the proof is computation of the differences of the pull-backs of Witt vectors by the two projections from dilatations of the self product studied in Section 13.5. Relating the computation to a covering of dilatation appearing in the definition of ramification groups, we complete the proof in Section 13.6.

## 13.1 Log Smooth Schemes

We define and study elementary properties of log smooth schemes over $S = \operatorname{Spec} O_K$. We consider only the case where the scheme is regular and the reduced closed fiber is smooth. We will not introduce general terminology on log schemes.

**Definition 13.1.1** Let $K$ be a discrete valuation field with residue field $F$ of characteristic $p \geqslant 0$, and $S = \operatorname{Spec} O_K$. Let $Q$ be a regular and flat scheme locally of finite type over $S$ such that the generic fiber $Q_K$ is smooth over $K$ and the reduced closed fiber $D = Q_{F,\mathrm{red}}$ is smooth over $F$.

(1) Let $t$ be a uniformizer of $O_K$. Let $m \geqslant 1$ be an integer. We define a scheme $Q_m = Q_{m,S}$ by $Q_m = \operatorname{Spec} O_K[x, u^{\pm 1}]/(ux^m - t)$. If $m$ is invertible in $O_K$, we also define $Q'_m = \operatorname{Spec} O_K[x]/(x^m - t)$.
(2) We say that $Q$ is log smooth over $S$ if, for every point $x \in D$ where the multiplicity $m$ of $D$ in the closed fiber $Q_F$ is divisible by $p$, there exist an open neighborhood $U$ of $x$ and a smooth morphism $U \to Q_m$ over $S$.

The scheme $Q'_m$ over $S$ depends on $t$, but $Q_m$ is independent up to an isomorphism over $S$. Consequently, the definition in Definition 13.1.1.2 is independent of the choice of uniformizer $t$ in Definition 13.1.1(1). Log smooth schemes are locally characterized as follows.

**Proposition 13.1.2** *Let $m \geqslant 1$ be an integer.*

(1) *The scheme $Q_m$ is log smooth over $S$. If $m$ is invertible in $O_K$, then $Q'_m$ is log smooth over $S$.*
(2) *Let $Q$ be a regular and flat scheme locally of finite type over $S$, and let $D$ be the reduced closed fiber $Q_{F,\mathrm{red}}$. Assume that $Q_F = mD$ and that $m \geqslant 1$ is prime to $p$. Then the following conditions are equivalent:*

 (a) *$Q$ is log smooth over $S$.*

 (a′) *The generic fiber $Q_K$ is smooth over $K$ and the reduced closed fiber $D$ is smooth over $F$.*

 (b) *Étale locally on $Q$, there exists a smooth morphism $Q \to Q'_m$ over $S$.*
 (c) *Let $Q'$ be the normalization of $Q \times_S Q'_m$. Then $Q' \to Q$ is finite étale and $Q' \to Q'_m$ is smooth.*
 (d) *Let $Q'$ be the normalization of $Q \times_S Q_m$. Then $Q' \to Q$ and $Q' \to Q_m$ are smooth.*

(3) *Let $Q$ be a regular and flat scheme locally of finite type over $S$, and let $D$ be the reduced closed fiber $Q_{F,\mathrm{red}}$. Assume that $Q_F = mD$, and $m \geqslant 1$ is divisible by $p$. Let $t$ be a uniformizer of $K$ and $x$ be a uniformizer of $D$, and define $u = t/x^m$. Then the following conditions are equivalent:*

 (a) *$Q$ is log smooth over $S$.*
 (b) *The generic fiber $Q_K$ is smooth over $K$, the reduced closed fiber $D$ is smooth over $F$, and $du \in \Gamma(D, \Omega^1_{D/F})$ vanishes nowhere on $D$.*
 (c) *Locally on a neighborhood of each point of the closed fiber of $Q$, there exists a smooth morphism to $Q_m$ over $S$. The generic fiber $Q_K$ is smooth over $K$.*

*Proof* (1) Since the morphism

$$Q_m = \operatorname{Spec} O_K[x, u^{\pm 1}]/(ux^m - t) \to Q_1 = \operatorname{Spec} O_K[x_1, u^{\pm 1}]/(ux_1 - t)$$

defined by sending $x_1$ to $x^m$ is flat, and $Q_1 = \operatorname{Spec} O_K[u^{\pm 1}]$ is flat over $O_K$, $Q_m$ is flat over $S$. The generic fiber $Q_{m,K} = \operatorname{Spec} K[x, u^{\pm 1}]/(ux^m - t) = \operatorname{Spec} K[x^{\pm 1}]$ is smooth over $K$, and the reduced closed fiber $D = Q_{m,F,\mathrm{red}} = \operatorname{Spec} F[u^{\pm 1}]$ is smooth over $F$. Since $Q_m$ is flat over $S$ and $x$ divides $t$, the closed subscheme $D \subset Q_m$ is a Cartier divisor defined by a non-zero divisor $x$. Since $D$ is regular, the scheme $Q_m$ is regular.

If $p \nmid m$, then $O_K[x]/(x^m - t)$ is the valuation ring of a tamely ramified extension of $K$, and $Q'_m$ is log smooth over $S$.

(2) (a) $\Rightarrow$ (a′): Clear from the definition.

(a′) $\Rightarrow$ (c): Let $t$ be a uniformizer of $K$. We may assume that $Q = \operatorname{Spec} A$ is affine and $D \subset Q$ is defined by $x \in A$. Let $u = t/x^m \in A^\times$. Then $Q' = \operatorname{Spec} A'$ for $A' = A[v]/(v^m - u)$ and $Q' \to Q$ is finite étale. Since $D' = D \times_Q Q'$ is finite étale over $D$, it is smooth over $F$. Hence, the morphism $Q' \to Q'_m$ of flat schemes over $S$ is smooth on a neighborhood of $D'$ by [31, Proposition (17.8.2)]. Since the generic fiber of $Q'$ is smooth over $K$, the generic fiber of $Q' \to Q'_m$ is also smooth.

(c) $\Rightarrow$ (b): Since $Q' \to Q$ is an étale covering, (c) implies (b).

(b) $\Rightarrow$ (d): We may assume that $Q = Q'_m$. Since $Q_m$ is log smooth by (1), the assertion follows from (a) $\Rightarrow$ (c) applied to $Q = Q_m$.

(d) $\Rightarrow$ (a): Since $Q_m$ is log smooth by (1), condition (d) implies that $Q'$ is also log smooth. Since $Q' \to Q$ in (d) is faithfully flat, the log smoothness of $Q'$ implies that $Q$ is log smooth.

(3) (a) $\Rightarrow$ (b): If $Q$ is log smooth over $S$, then $Q_K$ is smooth over $K$ and $D$ is smooth over $F$. Since the last condition on $du$ is independent of the choice of $t$ and $x$, we may assume that $Q = Q_m = \operatorname{Spec} O_K[u^{\pm 1}, x]/(ux^m - t)$. Then $D = \operatorname{Spec} F[u^{\pm 1}]$, and $du \in \Gamma(D, \Omega^1_{D/F})$ is a basis of $\Omega^1_{D/F}$.

(b) $\Rightarrow$ (c): We may assume that $Q = \operatorname{Spec} A$ is affine and define $Q \to Q_m$ by $x \in A$ and $u \in A^\times$. Then, since $du \neq 0$, the morphism $D = \operatorname{Spec} A/xA \to Q_{m,F,\mathrm{red}} = \operatorname{Spec} F[u^{\pm 1}]$ on the reduced closed fibers is smooth. Hence, the morphism $Q \to Q_m$ of flat schemes over $S$ is smooth on a neighborhood of the closed fiber by [31, Proposition (17.8.2)].

(c) $\Rightarrow$ (a): Since $Q_m$ is log smooth, (c) implies that $Q$ is log smooth. □

The normalization of the fiber product of log smooth schemes is again log smooth.

**Corollary 13.1.3** (1) *Let $Q$ and $Q'$ be log smooth schemes over $S$. Then the normalization $Q''$ of $Q \times_S Q'$ is log smooth. If $Q_F = m \cdot Q_{F,\mathrm{red}}$ and*

*$Q'_F = e \cdot Q'_{F,\mathrm{red}}$, and if $n$ is the least common multiple of $m$ and $e$, then $Q''_F = n \cdot Q''_{F,\mathrm{red}}$. If $e$ is divisible by $m$, then $Q''$ is smooth over $Q'$.*

(2) *For a log smooth scheme $Q$ over $S$, the morphism $Q \to S$ is smooth if and only if the closed fiber $Q_F$ is reduced.*

*Proof* (1) By Proposition 13.1.2, we may assume that $Q = Q_m$ and $Q' = Q_e$ for integers $m, e \geqslant 1$. Let $A_m = O_K[x, u^{\pm 1}]/(ux^m - t)$ and $A_e = O_K[y, v^{\pm 1}]/(vy^e - t)$. Let $d = (m, e)$ be the greatest common divisor and set $m = m'd$, $e = e'd$. Let $a, b$ be integers satisfying $d = am + be$. Then the normalization of $A_m \otimes_{O_K} A_e = O_K[x, y, u^{\pm 1}, v^{\pm 1}]/(ux^m - t, vy^e - t)$ contains $z = x^b y^a$ and $w = y^{e'}/x^{m'}$ since they satisfy $z^{e'} = y^{ae'} x^{1-am'} = x \cdot w^a$ and $w^d = u/v$. Since $z^{m'} w^b = y^{am'+be'} x^{bm'-bm'} = y$, we set $B = O_K[z, v^{\pm 1}, w^{\pm 1}]/(v(w^b z^{m'})^e - t)$.

Since $Q'' = \operatorname{Spec} B$ is smooth over $Q_{m'e} = \operatorname{Spec} O_K[s^{\pm 1}, z]/(sz^{m'e} - t)$, it follows that $Q''$ is log smooth over $S$ and is regular. Hence, $B$ is the normalization of $A_m \otimes_{O_K} A_e$ by the morphism $A_m \otimes_{O_K} A_e \to B$ defined by sending $x, y, u$ to $z^{e'}/w^a, z^{m'} w^b, vw^d$. Thus, the normalization $Q''$ of $Q \times_S Q'$ is log smooth over $S$. The product $m'e$ is the least common multiple $n$ of $m$ and $e$.

If $e$ is divisible by $m$, we may take $a = 1, b = 0$ and we have $d = m, m' = 1$. Then we have $Q_{m'e} = Q_e$ by $z = y$, $s = v$ and $Q''$ is smooth over $Q'$.

(2) This follows from (a) $\Leftrightarrow$ (c) in Proposition 13.1.2(2). □

**Definition 13.1.4** Let $L$ be an extension of a discrete valuation field $K$ with residue field $E$, and let $Q$ be a log smooth scheme over $S = \operatorname{Spec} O_K$. We say that a morphism $T = \operatorname{Spec} O_L \to Q$ over $S$ is strict if the inverse image $T \times_Q Q_{F,\mathrm{red}}$ of the reduced closed fiber $Q_{F,\mathrm{red}}$ equals $\operatorname{Spec} E$. Further, if $T \to Q$ is an immersion, we say $T \to Q$ is exact.

A base change of log smooth schemes satisfies the following properties.

**Proposition 13.1.5** *Let $Q$ be a log smooth scheme over $S$, and let $K'$ be an extension of a discrete valuation field $K$. Let $Q_{S'}$ be the normalization of $Q \times_S S'$.*

(1) *Let $Q'$ be another log smooth scheme over $S$, and let $S' = \operatorname{Spec} O_{K'} \to Q'$ be a strict morphism over $S$. Let $Q''$ be the normalization of $Q \times_S Q'$. Then the normalization $Q_{S'}$ equals the fiber product $Q'' \times_{Q'} S'$ and is log smooth over $S'$. The reduced closed fiber $(Q_{S'} \times_{S'} \operatorname{Spec} F')_{\mathrm{red}} \subset Q_{S'}$ equals the inverse image of the reduced closed fiber $(Q'' \times_S \operatorname{Spec} F)_{\mathrm{red}} \subset Q''$.*

(2) *Assume $Q_F = m \cdot Q_{F,\mathrm{red}}$, let $e = e_{K'/K}$ be the ramification index, and set $d = \gcd(m, e)$ and $m' = m/d$. Then the closed fiber $Q_{F'} = Q_{S'} \times_{S'} \operatorname{Spec} F'$ equals $m' \cdot Q'_{F',\mathrm{red}}$. In particular, if $e$ is divisible by $m$, then $Q_{S'}$ is smooth over $S'$.*

(3) *The inseparable degree of the finite flat morphism* $Q_{F',\mathrm{red}} \to Q_{F,\mathrm{red}} \times_F F'$ *is the* $p$*-part of* $d$.

*Proof* (1) Since the assertion is étale local on $S'$, we may assume that $Q' = Q_e$ for the ramification index $e = e_{K'/K}$. We may further assume that $Q = Q_m$ for an integer $m \geqslant 1$. In the notation of the proof of Proposition 13.1.3(1), let $A_e \to O_{K'}$ be the morphism sending $y$ to a uniformizer $t' \in O_{K'}$. Then $Q'' \times_{Q'} S' = \operatorname{Spec} C$ for $C = B \otimes_{A_e} O_{K'} = O_{K'}[z, w^{\pm 1}]/(w^b z^{m'} - t')$.

We show that $Q'' \times_{Q'} S' = \operatorname{Spec} C$ is log smooth over $S'$. First, we assume $p \nmid m'$. Then $C' = C[u']/(u'^{m'} - w) = O_{K'}[u'^{\pm 1}, z]/(u'^b z)^{m'} - t')$ is finite étale over $C$ and is smooth over $Q'_{m',S'}$. Hence, $Q'' \times_{Q'} S'$ is log smooth over $S'$ in the case $p \nmid m'$. Next, we assume $p \mid m'$. Since $b$ and $m'$ are relatively prime to each other, we have $p \nmid b$. Then $Q'' \times_{Q'} S' = \operatorname{Spec} C$ is finite étale over $Q'_{m',S'} = \operatorname{Spec} O_{K'}[x', u'^{\pm 1}]/(x'^{m'} u - t')$, and is log smooth over $S'$ in this case also.

(2) Let $Q'$ be another log smooth scheme over $S$, and let $S' \to Q'$ be a strict morphism over $S$. Since $Q_{S'}$ is $Q'' \times_{Q'} S'$ in the notation of (1), the assertion follows from (1) and Corollary 13.1.3.

(3) We may assume that $Q = Q_m$. Let the notation be as in the proof of Corollary 13.1.3(1) and of (1). The purely inseparable degree of the finite flat morphism $(A_m \otimes_{O_K} F)_{\mathrm{red}} \otimes_F F' = F'[u^{\pm 1}] \to (C \otimes_{O_{K'}} F')_{\mathrm{red}} = F'[w^{\pm 1}]$ is the $p$-part of $d$ since the morphism is defined by sending $u$ to $vw^d$. □

The terminology "log smooth scheme" is explained by the following property.

**Corollary 13.1.6** *Let $Q$ be a log smooth scheme over $S = \operatorname{Spec} O_K$. Let $K'$ be an extension of a discrete valuation field, $L'$ be a finite extension of $K'$, and $T' = \operatorname{Spec} O_{L'} \to Q'$ be an exact immersion to a log smooth scheme over $S' = \operatorname{Spec} O_{K'}$. Let $T' \to Q$ be a morphism compatible with $S' \to S$. Then, on an étale neighborhood of $T'$ in $Q'$, there exists a morphism $Q' \to Q$ over $S$ lifting $T' \to Q$.*

*Proof* Since the normalization $Q_{S'}$ of $Q \times_S S'$ is log smooth over $S'$ by Proposition 13.1.5(2), by replacing $Q \to S$ by $Q_{S'} \to S'$, we may assume $S' = S$. We may assume that $Q_F = mD$ and $Q'_F = eD'$ for the reduced closed fibers $D \subset Q$ and $D' \subset Q'$ and integers $m$ and $e$. Since $T' \to Q'$ is an exact immersion, $m$ divides $e$. Hence, the normalization $Q''$ of $Q \times_S Q'$ is smooth over $Q'$ by Corollary 13.1.3(1). The section $T' \to Q''$ induced by $T \to Q$ is lifted to an étale neighborhood of $T$ in $Q'$. □

**Corollary 13.1.7** *Let $L$ be a finite separable extension of a henselian discrete valuation field $K$. Then the following conditions are equivalent:*

(1) *$L$ is a log monogenic extension of $K$.*

(2) *There exists an exact closed immersion $T = \operatorname{Spec} O_L \to Q$ to a log smooth scheme $Q$ over $O_K$ of relative dimension* 1.

*Proof* (1) $\Rightarrow$ (2) Let $M \subset L$ be the maximal tamely ramified extension of $K$, and set $e = e_{M/K}$. Let $s$ and $t$ be uniformizers of $L$ and of $M$ such that $u = s^e/t$ is a generator of the residue field of $L$ over the residue field of $M$. Then we have an exact closed immersion $T = \operatorname{Spec} O_L \to Q = \operatorname{Spec} O_M[u^{\pm 1}, s]/(ut - s^e)$ to a log smooth scheme $Q$ over $O_M$. It suffices to show that $Q$ is log smooth over $O_K$.

If $e = 1$, the assertion follows from Proposition 13.1.2(2). Assume $e > 1$. Let $t_1$ be a uniformizer of $K$ and $m = e_{L/K}$, $m_1 = e_{M/K}$. Then $u_1 = t_1/s^m$ is a constant multiple of $u^{m_1}$. Let $F$ be the residue field of $K$, and let $D$ be the reduced part of the closed fiber $Q_F$. Then, since $du$ is a basis of $\Omega^1_{D/F}$ and $m_1$ is prime to $p$, $du_1$ is also a basis of $\Omega^1_{D/F}$. Hence, by Proposition 13.1.2(3) (b) $\Rightarrow$ (a), $Q$ is log smooth over $S = \operatorname{Spec} O_K$.

(2) $\Rightarrow$ (1) Let $T = \operatorname{Spec} O_L \to Q$ be an exact closed immersion to a log smooth scheme $Q$ over $S = \operatorname{Spec} O_K$ of relative dimension 1, and let $M \subset L$ be the maximal tamely ramified extension of $K$. Let $S' = \operatorname{Spec} O_M$ and $T' \to Q'$ be the normalizations of the base change $T \times_S S' \to Q \times_S S'$. Then $T \to S'$ induces an open and closed immersion $T \to T'$, and $Q'$ is log smooth over $S'$ by Proposition 13.1.5(1). Further, by Proposition 13.1.5(2), the induced immersion $T \to Q'$ is an exact closed immersion. Hence, we may assume $M = K$.

Let $t$ be a uniformizer of $K$. If the ramification index $e_{L/K} = m$ is a non-trivial power of $p$, we may assume that there exists an étale morphism $Q \to Q_m = \operatorname{Spec} O_K[u^{\pm 1}, s]/(us^m - t)$. If $e_{L/K} = 1$, the scheme $Q$ is smooth over $S$ and hence we may assume that there exists an étale morphism $Q \to Q_1 = \operatorname{Spec} O_K[u^{\pm 1}, s]/(us - t)$. Since the residue field $E$ of $L$ is a purely inseparable extension of the residue field $F$ of $K$, and since the image of $s$ is a uniformizer of $L$, the composition $T \to Q_m$ defines a surjection $O_K[u^{\pm 1}, s]/(us^m - t) \to O_L$. Since the image of $u$ is a generator of $E$ over $F$, the extension $L$ is log monogenic. □

**Exercise 13.1** Let $K$ be a discrete valuation field and let $X$ be a regular flat curve over $S = \operatorname{Spec} O_K$. Assume that the closed fiber $X_F$ consists of two irreducible components $C_1$ and $C_2$ meeting transversally at a closed point $x$ with residue field $F$, and that $X_F = mC_1 + nC_2$ for integers $m$ prime to the characteristic $p$ of the residue field and $n$ divisible by $p$. Show that there exists an open neighborhood $U$ of $x$ such that $U - \{x\}$ is log smooth over $S$.

## 13.2 Log Smooth Extensions

**Definition 13.2.1** Let $K$ be a discrete valuation field. We say that a separable extension $K'$ of $K$ is log smooth over $K$ if there exist an étale morphism $O_{K'} \to O_{K''}$ of discrete valuation rings, a log smooth scheme $Q$ over $S = \operatorname{Spec} O_K$, a point $\xi$ of the closed fiber, and a morphism $O_{Q,\xi} \to O_{K''}$ over $O_K$ inducing an isomorphism on the completions.

**Lemma 13.2.2** *Let $K$ be a discrete valuation field and let $K'$ be a separable extension. Let $F$ and $F'$ be the residue fields of $K$ and $K'$. Then the following conditions are equivalent:*

(1) *The extension $K'$ is a log smooth extension of $K$.*
(2) *The extension $F'$ of $F$ is a finitely generated separable extension. Further, if the ramification index $m = e_{K'/K}$ is divisible by the characteristic $p$ of the residue field $F$, and if $t$ and $t'$ are uniformizers of $K$ and $K'$, then the class $u \in F'$ of $t/t'^m \in O_{K'}^\times$ is transcendental over $F$ and $F'$ is a separable extension of $F(u)$.*

*Proof* (1) $\Rightarrow$ (2): The fraction field of the local ring at the generic point of the closed fiber of $Q_m$ for $m \geqslant 1$ and $Q'_m$ for $p \nmid m$ satisfies the condition in (2). Since condition (2) is independent of the choice of uniformizers, the assertion follows.

(2) $\Rightarrow$ (1): Let $t$ and $t'$ be uniformizers of $K$ and $K'$, and set $u = t/t'^m \in O_{K'}^\times$.

First, consider the case where the ramification index $m = e_{K'/K}$ is prime to $p$. Let $u_1, \ldots, u_n, v \in O_{K'}$ be elements such that $\bar{u}_1, \ldots, \bar{u}_n \in F_0 = F(\bar{u}_1, \ldots, \bar{u}_n) \subset F'$ form a transcendental basis over $F$, and $F' = F_0(\bar{v})$ is a finite separable extension of $F_0$. Let $A_0$ be the localization of the polynomial ring $O_K[U_1, \ldots, U_n]$ at the prime ideal generated by $\mathfrak{m}_K$ with residue field $F_0$. Take a monic polynomial $f \in A_0[V]$ lifting the minimal polynomial of $\bar{v} \in F'$ over $F_0$, and define a finite étale morphism $A_0 \to A_1 = A_0[V]/f(V)$ of discrete valuation rings. The residue field of $A_1$ is $F_1$. Set $A_2 = A_1[T]/(T^m - t)$.

Define a morphism $A_0 \to O_{K'}$ by $u_1, \ldots, u_n \in O_{K'}$ and let $O_{K'_1}$ be the local ring of $O_{K'} \otimes_{A_0} A_1$ at the maximal ideal defined by $A_1 \to F'$ sending $V$ to $\bar{v}$. The normalization of $O_{K'_1} \otimes_{A_1} A_2$ is the ring $O_{K'_1}[W]/(W^m - u)$, étale over $O_{K'_1}$, where $A_2 \to O_{K'_1}[W]/(W^m - u)$ is defined by sending $T$ to $Wt'$. Define a discrete valuation ring $O_{K'_2}$ to be the local ring of the normalization $O_{K'_1}[W]/(W^m - u)$ at a maximal ideal. Then the morphisms of $O_{K'} \to O_{K'_1} \to O_{K'_2}$ of discrete valuation rings are étale.

Set $Q_0 = \operatorname{Spec} O_K[U_1, \ldots, U_n]$ and extend the morphism $A_0 \to A_1$ to an étale morphism $Q_1 \to Q_0$ of smooth schemes over $S$. Further, the morphism $A_1 \to A_2$ is extended to a quasi-finite flat morphism $Q_2 \to Q_1$ of log smooth schemes over $S$. Then the morphism $A_2 \to O_{K'_2}$ induces an isomorphism on the completions and induces $\operatorname{Spec} O_{K'_2} \to Q_2$. Hence, the assertion follows in this case.

Assume $p \mid m$. Let $u_1 = t/t'^m, u_2, \ldots, u_n, v \in O_{K'}$ be elements such that $\overline{u}_1, \ldots, \overline{u}_n \in F_0 = F(\overline{u}_1, \ldots, \overline{u}_n) \subset F'$ form a transcendental basis over $F$ and $F' = F_0(v)$ is a finite separable extension of $F_0$. Let $A_0$ be the localization of $O_K[U_1, \ldots, U_n, T']/(U_1 T'^m - t)$ at the prime ideal $(T')$. Take a monic polynomial $f \in A_0[V]$ lifting the minimal polynomial of $\overline{v} \in F'$ over $F_0$, and define a finite étale morphism $A_0 \to A_1 = A_0[V]/f(V)$ of discrete valuation rings.

Define a morphism $A_0 \to O_{K'}$ of discrete valuation rings by sending $U_1, \ldots, U_n, T'$ to $u_1, \ldots, u_n, t'$, and define an étale morphism $O_{K'} \to O_{K'_1}$ of discrete valuation rings as in the case $p \nmid m$.

The scheme $Q_0 = \operatorname{Spec} O_K[U_1, \ldots, U_n, T']/(U_1 T'^m - t)$ is smooth over $Q_m$ and hence is log smooth over $S$. The étale morphism $A_0 \to A_1$ is extended to an étale morphism $Q_1 \to Q_0$ of log smooth schemes. The rest of the proof is the same as the last part of the case $p \nmid m$. □

**Example 13.2.3** Let $t$ be a uniformizer of $K$. If $K'$ is an unramified extension of the fraction field of the localization of $O_K[u, t', x_1, \ldots, x_m]/(ut'^e - t)$ at the prime ideal $(t')$, or if $K'$ is a tamely ramified extension of $K$, then $K'$ is a log smooth extension of $K$.

**Proposition 13.2.4** *Let $K$ be a discrete valuation field and let $L$ be an extension of $K$ of ramification index $m = e_{L/K}$. Let $K'$ be a log smooth extension of $K$ of ramification index $e = e_{K'/K}$, and $S' = \operatorname{Spec} O_{K'}$.*

(1) *A composition field $L' = LK'$ is a log smooth extension of $L$ and we have $e_{L'/K'} = m/\gcd(m, e)$. In particular, if $m$ divides $e$, then the ramification index $e_{L'/K'}$ is 1.*

    *Let $F', E, E'$ be the residue fields of $K', L, L'$. Then the inseparable degree $[E' : EF']_{\mathrm{insep}}$ is the $p$-part of $d = \gcd(m, e)$.*

(2) *Let $T = \operatorname{Spec} O_L \to Q$ be an exact immersion to a log smooth scheme over $S$, and let $Q_{S'}$ be the normalization of $Q \times_S S'$. Then $T \times_Q Q_{S'}$ is the normalization $T'$ of $T \times_S S'$, and the normalization $T' \to Q_{S'}$ of the base change of $T \to Q$ is an exact immersion to a log smooth scheme over $S'$.*

*Proof* (1) We may assume that $K'$ is the fraction field of the local ring of a log smooth scheme $Q'$ over $S$ at the generic point of an irreducible component of the closed fiber. Then, by Proposition 13.1.5(2), the normalization $Q'_T$ of $T \times_S Q'$ is log smooth over $T$ and is regular. Further, the normalization $T'$ of $T \times_S S'$ equals $Q'_T \times_{Q'} S'$. Hence, $L'$ is a log smooth extension of $L$.

We have $e_{L'/K'} = m/\gcd(m, e)$ by Proposition 13.1.5(2). The assertion on the inseparable degree $[E' : EF']_{\text{insep}}$ follows from Proposition 13.1.5(3).

(2) Let $S' \to Q'$ be as in the proof of (1), and let $Q''$ be the normalization of $Q \times_S Q'$. Then, by Proposition 13.1.5(1), the normalization of $T \times_S Q'$ equals $T \times_Q Q''$ and the normalization $Q_{S'}$ of $Q \times_S S'$ equals $Q'' \times_{Q'} S'$. Hence, the normalization $T'$ of $T \times_S S'$ equals $T \times_Q Q'' \times_{Q'} S' = T \times_Q Q_{S'}$.

Further, the immersion $T' \to Q_{S'}$ is exact by Proposition 13.1.5(1). □

**Corollary 13.2.5** *Let $L$ be a finite separable extension. Let $K'$ be a log smooth extension of $K$ of ramification index $e = e_{K'/K}$, and let $L' = LK'$ be a composition field.*

*Let $L_1 \subset L$ be the maximal tamely ramified extension over $K$. Then $L'_1 = L_1 K' \subset L'$ is the maximal tamely ramified extension over $K'$, and $L \otimes_{L_1} L'_1 \to L'$ is an isomorphism.*

*Proof* By replacing $K$ by $L_1$, we may assume $L_1 = K$. Namely, we assume that the ramification index $m = e_{L/K}$ is a power of $p$ and that the residue field extension $E$ over $F$ is purely inseparable. Let $d = (m, e)$ be the greatest common divisor. Then, by Proposition 13.2.4(1), the ramification index $e_{L'/K'} = e_{L/K}/d$ is a power of $p$ and we have $[E' : EF']_{\text{insep}} = d$. Further, since $F'$ is a separable extension of $F$, the degree $[E' : F'] = [E' : EF'] \cdot [EF' : F']$ is divisible by $[E' : F']_{\text{insep}} = d \cdot [E : F]$. Hence, $[L' : K'] = [L : K]$ and $[E' : F'] = [E' : F']_{\text{insep}}$. □

**Exercise 13.2** Let $K$ be a discrete valuation field and $L$ be a log smooth extension of $K$. Let $K' \subset L$ be a subextension. Show that $K'$ is a log smooth extension of $K$.

## 13.3 Logarithmic Upper Ramification Groups

For finite separable extensions $L$ of a henselian discrete valuation field $K$, we study exact immersions (Definition 13.1.4) $T = \operatorname{Spec} O_L \to Q$ to log smooth schemes over $S = \operatorname{Spec} O_K$.

**Example 13.3.1** (1) Let $K$ be a henselian discrete valuation field, and $L$ be a finite separable extension of ramification index $m$. Let $t$ be a uniformizer of $O_K$, and let $a_1, \ldots, a_n \in O_L$ be a system of generators over $O_K$ such that $a_1$ is a uniformizer. Define a unit $u \in O_L^\times$ by $t = ua_1^m$ and a surjection $O_K[X_1, \ldots, X_n, U^{\pm 1}]/(UX_1^m - t) \to O_L$ by sending $X_1, \ldots, X_n, U$ to $a_1, \ldots, a_n, u$. Then this surjection defines an exact closed immersion

$$T = \operatorname{Spec} O_L \to Q = \operatorname{Spec} O_K[X_1, \ldots, X_n, U^{\pm 1}]/(UX_1^m - t)$$

to a log smooth scheme $Q$ over $S$.

One can slightly generalize the construction. Let $P$ be a smooth scheme over $S$ and $D \subset P$ be a divisor smooth over $S$. Let $m \geqslant 1$ be an integer and $P_F$ be the closed fiber. Let $P^{[P_F \cdot mD]}$ be the dilatation and let $Q \subset P^{[P_F \cdot mD]}$ be the complement of the proper transform of $D$. Then $Q$ is log smooth over $S$ and $Q_E = m \cdot Q_{E,\mathrm{red}}$. Let $T \to P$ be a closed immersion such that $T \times_P D = \operatorname{Spec} E$. Then $T \to P$ is lifted to an exact closed immersion $T \to Q$.

(2) Let $k$ be a field and $X$ be a smooth scheme over $k$. Let $D \subset X$ be a smooth irreducible divisor, and set $O_K = O_{X,\xi}$ for the generic point $\xi$ of $D$. Let $Y \to X$ be a quasi-finite flat morphism of smooth schemes over $k$ such that $V = Y \times_X U \to U = X - D$ is étale and $V = Y - E$ is the complement of an irreducible smooth divisor over $k$. Then the fraction field $L$ of the local ring $O_L = O_{Y,\eta}$ at the generic point $\eta \in E$ is a finite extension of $K$. The ramification index $m = e_{L/K}$ is the integer satisfying $D \times_X Y = mE$.

Let $S = \operatorname{Spec} O_K$. We apply the construction to a smooth scheme $P = Y \times_k S$ over $S$ and the divisor $mE \times_k S \subset P$ to define a log smooth scheme $Q$ over $S$. The canonical morphism $T = \operatorname{Spec} O_L \to Y$ induces an exact immersion $T \to Q$ over $S$.

Let $K$ be a henselian discrete valuation field, and let $L$ be a finite separable extension. We consider the following construction for exact immersions $T = \operatorname{Spec} O_L \to Q$ to log smooth schemes over $S$, similarly to the non-logarithmic case. Let $K'$ be an extension of $K$ of ramification index $e$ such that $er$ is an integer. Let $T' = T \times_S S' \subset Q' = Q \times_S S'$ be the base change, and let $D' \subset Q'$ be the Cartier divisor defined by $\mathfrak{m}_{K'}^{er} O_{Q'}$. Then we define the dilatation $Q_{S'}^{[r]} = Q^{[D' \cdot T']}$ and its normalization

$$Q_{S'}^{(r)} \longrightarrow Q_{S'}^{[r]}.$$

By the functoriality of normalizations, the immersion $T' \to Q'$ is lifted to a morphism $T_{S'}^{\natural} \to Q_{S'}^{(r)}$ from the normalization.

This logarithmic construction has a functoriality similar to the non-logarithmic case. Let $K'$ be an extension of henselian discrete valuation field $K$ of ramification index $e$. Let $L$ and $L'$ be finite separable extensions of $K$ and of $K'$ respectively, and let $L \to L'$ be a morphism of fields compatible with $K \to K'$. Set $S = \operatorname{Spec} O_K$, $S' = \operatorname{Spec} O_{K'}$, $T = \operatorname{Spec} O_L$, $T' = \operatorname{Spec} O_{L'}$, and let $T \to Q$ and $T' \to Q'$ be exact immersions to log smooth schemes over $S$ and over $S'$ respectively. Then, by Corollary 13.1.6, after replacing $Q'$ by an étale neighborhood of $T'$ if necessary, we have a commutative diagram

$$\begin{array}{ccc} T' & \longrightarrow & Q' \\ \downarrow & & \downarrow \\ T & \longrightarrow & Q \end{array} \tag{13.3.1}$$

compatible with $S' \to S$.

Let (13.3.1) be a commutative diagram of schemes such that the horizontal arrows are exact immersions to log smooth schemes over $S$ and over $S'$ respectively, and the vertical arrows are compatible with $S' \to S$. Let $r > 0$ and $r' > er$ be rational numbers. Let $K_1$ and $K'_1$ be extensions of discrete valuation fields $K$ and $K'$ of ramification indices $e_1$ and $e'_1$ such that $e_1 r$ and $e'_1 r'$ are integers. Let $K_1 \to K'_1$ be morphisms of fields compatible with $K \to K'$, and set $S_1 = \operatorname{Spec} O_{K_1}$, $S'_1 = \operatorname{Spec} O_{K'_1}$. Then, similarly to (11.4.3), the diagram (13.3.1) defines a commutative diagram

$$\begin{array}{ccccccc} T'^{\natural}_{S'_1} & \longrightarrow & Q'^{(r')}_{S'_1} & \longrightarrow & Q'^{[r']}_{S'_1} & \longrightarrow & Q' \times_{S'} S'_1 \\ \downarrow & & \downarrow & & \downarrow & & \downarrow \\ T^{\natural}_{S_1} & \longrightarrow & Q^{(r)}_{S_1} & \longrightarrow & Q^{[r]}_{S_1} & \longrightarrow & Q \times_S S_1 \end{array} \tag{13.3.2}$$

compatible with $S'_1 \to S_1$ by the functoriality of dilatations Lemma 11.1.2 and normalizations Lemma 11.2.2.

We study the relation with the non-logarithmic construction.

**Proposition 13.3.2** *Let $T = \operatorname{Spec} O_L \to P$ be an immersion to a smooth scheme $P$ over $S = \operatorname{Spec} O_K$, and $T \to Q$ be an exact immersion to a log smooth regular flat scheme over $S$. Let $P^{[1]} = P^{[P_F \cdot T]} \to P$ be the dilatation for the closed fiber $P_F = P \times_S F$.*

(1) *The scheme $P^{[1]}$ is flat over $S$ and the immersion $T \to P^{[1]}$ is a regular immersion. The morphism $N_{T/P} \to N_{T/P^{[1]}}$ induces an isomorphism $N_{T/P} \otimes_{O_K} \mathfrak{m}_K^{-1} \to N_{T/P^{[1]}}$.*
(2) *On an étale neighborhood of $T \to Q$, there exists a morphism $Q \to P$ over $S$ compatible with the immersions of $T$. On an étale neighborhood of $T \to P^{[1]}$, there exists a morphism $P^{[1]} \to Q$ over $S$ compatible with the immersions of $T$.*
(3) *Assume that the residue field $E$ of $L$ is a separable extension of $F$, and that the smooth morphism $P \to S$ is of relative dimension 1. Then the dilatation $P^{[1]}$ is log smooth over $S$ and the immersion $T \to P^{[1]}$ is exact.*

*Proof* Since the assertion is local on a neighborhood of $T$, we may assume that $P = \operatorname{Spec} A$ is affine and $O_L = A/I$. Let $t \in O_K$ be a uniformizer. Then $P^{[1]} = \operatorname{Spec} A_1$ for the subring $A_1 = A[I/t] \subset A[1/t]$.

(1) A subring $A_1 = A[I/t]$ of $A[1/t]$ is flat over $O_K$. Since the immersion $T \to P$ is a regular immersion, we may further assume that $I$ is generated by a regular sequence $f_1, \dots, f_n$. Then, since the kernel $(I/t)A_1$ of the surjection $A_1 \to O_L$ is generated by a regular sequence $f_1/t, \dots, f_n/t$, the immersion $T \to P^{[1]}$ is a regular immersion and the morphism $N_{T/P} \to N_{T/P^{[1]}}$ induces an isomorphism $N_{T/P} \otimes_{O_K} \mathfrak{m}_K^{-1} \to N_{T/P^{[1]}}$.

(2) Since $P$ is smooth over $S$, the immersion $T \to P$ is extended to a morphism $Q \to P$ after replacing $Q$ by an étale neighborhood of $T$.

We show that there exists an open neighborhood $P^{[1]\prime}$ of $T \subset P^{[1]}$ such that the normalization $Q_{P^{[1]\prime}}$ of $Q \times_S P^{[1]\prime}$ is smooth over $P^{[1]\prime}$. Let $s \in A$ be a lifting of a uniformizer of $L$, and let $m = e_{L/K}$ be the ramification index. Since $T \to Q$ is an exact immersion, we may assume that $Q$ is smooth over $Q_m = \operatorname{Spec} O_K[u^{\pm 1}, x]/(ux^m - t)$ by replacing $Q$ by the normalization of $Q \times_S Q_m$ by Proposition 13.1.2(2) in the case $p \nmid m$. After shrinking $P$, we may assume that there exists a unit $v \in A^\times$ satisfying $t = vs^m$ in $O_L$. Then, since $t - vs^m \in I = \operatorname{Ker}(A \to O_L)$, the element $a = (t - vs^m)/t \in A_1$ satisfies $vs^m = t(1-a)$. Since $at = t - vs^m \in \operatorname{Ker}(A_1 \to O_L)$, we have $a \in \operatorname{Ker}(A_1 \to O_L)$. Hence, $P^{[1]\prime} = \operatorname{Spec} A_1' \subset P^{[1]}$ for $A_1' = A_1[(1-a)^{-1}]$ is an open neighborhood of $T \subset P^{[1]}$. Let $w = v/(1-a) \in A_1'^\times$ so that we have $ws^m = t$ in $A_1'$. Then the normalization of $P^{[1]\prime} \times_S Q_m$ is $\operatorname{Spec} A_1'[z^{\pm 1}]$ by sending $x, u$ to $zs, z^{-m}w$, and is smooth over $P^{[1]\prime}$.

Since the normalization $Q_{P^{[1]\prime}}$ is smooth over $P^{[1]\prime}$, the section $T \to Q_{P^{[1]\prime}}$ induced by $T \to Q$ by the functoriality of normalizations is lifted to an étale neighborhood of $T$ in $P^{[1]\prime}$.

(3) If $L$ is unramified over $K$, then $P^{[1]}$ is smooth over $S$ and the assertion follows. Assume that $m = e_{L/K} > 1$. We may assume that $I = (f)$ is generated by a single element, and let $s \in A$ be a lifting of a uniformizer of $L$. Then the maximal ideal $\mathfrak{m}_x$ at the image $x \in P$ of the closed point of $T$ equals $(f, s)$. Since $(t, f) = (s^m, f)$ and $t \in \mathfrak{m}_x - \mathfrak{m}_x^2$, we have $\mathfrak{m}_x = (t, s)$. Since the residue field $E = O_{P,x}/\mathfrak{m}_x$ is a separable extension of $F = O_K/\mathfrak{m}_K$, and since the assertion is local at $x$, after replacing $P$ by an open neighborhood of $x$, we may assume that a morphism $P \to \mathbf{A}^1_S = \operatorname{Spec} O_K[s]$ over $S$ is defined by $s$ and is étale.

Since the quotients by the ideals $(f, t) = (f, t, s^m) \supset (t, s^m)$ of the local ring $O_{P,x}$ are of length $m$, we may further assume that an equality $(f, t) = (t, s^m)$ of ideals of $O_P$ holds. The dilatation of $\mathbf{A}^1_S$ with respect to the Cartier divisor defined by the ideal $(t)$ and the closed subscheme defined by the ideal $(s^m)$ is a log smooth scheme $Q_m = \operatorname{Spec} O_K[u^{\pm 1}, s]/(t - us^m)$ over $S$. Since $P^{[1]} =$

$P \times_{\mathbf{A}^1_S} Q_m$ is étale over $Q_m$, it is log smooth over $S$. Since the reduced closed fiber of $P^{[1]} \subset P$ is defined by $s$, the immersion $T \to P^{[1]}$ is exact. □

Let the notation be as in Proposition 13.3.2. By the functoriality of dilatations and normalizations, we have morphisms

$$P^{(r+1)}_{S'} \to Q^{(r)}_{S'} \to P^{(r)}_{S'} \tag{13.3.3}$$

compatible with the morphisms from $T^{\natural}_{S'}$. If the residue field $E$ is a separable extension of $F$, there exists an immersion $T \to P = \mathbf{A}^1_S$ over $S$ by Lemma 1.8.3(1). Then, for $Q = P^{[1]}$ as in Proposition 13.3.2(3), the first morphism $P^{(r+1)}_{S'} \to Q^{(r)}_{S'}$ is an isomorphism. If $e = 1$, for $Q = P$ the second morphism $Q^{(r)}_{S'} \to P^{(r)}_{S'}$ is an isomorphism.

We go back to the general case and let $T = \operatorname{Spec} \mathcal{O}_L \to Q$ be an exact immersion to a log smooth scheme $Q$ over $S = \operatorname{Spec} \mathcal{O}_K$. By the reduced fiber theorem, for sufficiently large $K'$ the geometric closed fiber $Q^{[r]}_{\overline{F}}$ of $Q^{[r]}_{S'} \to Q^{[r]}_{S'}$ is reduced. For two exact immersions $T \to Q$ and $T \to Q'$ to log smooth schemes over $S$ and the normalization $Q''$ of $Q \times_S Q'$, the projections induce smooth morphisms $Q'' \to Q$ and $Q'' \to Q'$ compatible with the immersions of $T$ by Corollary 13.1.3(1). Hence, similarly to Lemma 11.4.2(2), the smooth morphisms induce bijections $\pi_0(Q''^{(r)}_{S'}) \to \pi_0(Q^{(r)}_{S'}), \pi_0(Q''^{(r)}_{S'}) \to \pi_0(Q'^{(r)}_{S'})$ on the sets of connected components, and also bijections $\operatorname{Im}(T^{\natural}_{S'} \to Q''^{(r)}_{S'}) \to \operatorname{Im}(T^{\natural}_{S'} \to Q^{(r)}_{S'}), \operatorname{Im}(T^{\natural}_{S'} \to Q'^{(r)}_{S'}) \to \operatorname{Im}(T^{\natural}_{S'} \to Q^{(r)}_{S'})$.

Thus, similarly to Lemma 12.1.1, we obtain a projective system $(F^r_{\log})$ of functors indexed by $r \in \mathbf{Q}_+$ and a projective system $(F \to F^r_{\log})$ of morphisms of functors. The morphisms (13.3.3) induce morphisms of functors

$$F^{r+1} \to F^r_{\log} \to F^r \tag{13.3.4}$$

for $r \in \mathbf{Q}_+$. We say that the logarithmic ramification of a finite separable extension $L$ is bounded by $r \in \mathbf{Q}_+$ if $F(L) \to F^r_{\log}(L)$ is a bijection.

**Corollary 13.3.3** *Let $L$ be a finite separable extension of $K$ and $r \in \mathbf{Q}_+$.*

(1) *We consider the conditions:*

 (a) *The ramification of $L$ is bounded by $r$.*
 (b) *The logarithmic ramification of $L$ is bounded by $r$.*
 (c) *The ramification of $L$ is bounded by $r + 1$.*

 *We have* (1) ⇒ (2) ⇒ (3).

(2) *The following conditions are equivalent:*

 (a) *$L$ is tamely ramified over $K$.*

(b) *The logarithmic ramification of* $L$ *is bounded by* $0+$.

(3) $F^{r+1} = F^r_{\log}$ *if the extension* $E$ *of the residue field* $F$ *is separable.*

(4) $F^r_{\log} = F^r$ *if* $e_{L/K} = 1$.

*Proof* (1) This follows from (13.3.4).

(2) (a) $\Rightarrow$ (b): Since the identity $T \to T$ is an exact immersion to a log smooth scheme over $S$, the mapping $F(L) \to F^{0+}_{\log}(L)$ is a bijection.

(b) $\Rightarrow$ (a): By (1) (b) $\Rightarrow$ (c), the ramification of $L$ is bounded by $1+$. Hence, the assertion follows from Proposition 12.2.2(2).

(3) The ring $O_L$ is generated by a single element over $O_K$ in this case by Lemma 11.5.1. Hence, there exists a closed immersion $T = \operatorname{Spec} O_L \to P = \mathbf{A}^1_S = \operatorname{Spec} O_K$ over $O_K$ and the assertion follows from Proposition 13.3.2(3).

(4) Let $T = \operatorname{Spec} O_L \to Q$ be an immersion to a smooth scheme $Q$ over $O_K$. Then $Q$ is log smooth and the immersion $T \to Q$ is exact. Hence, the assertion follows. □

Let $K'$ be an extension of a henselian discrete valuation field $K$ of ramification index $e = e_{K'/K}$, and let $L \to L'$ and $K_s \to K'_s$ be respectively a morphism of finite separable extensions and a morphism of separable closures compatible with $K \to K'$. For $r, r' \in \mathbf{Q}_+$ such that $r' \geqslant er$, the canonical morphism $F'(L') = \operatorname{Mor}_{K'}(L', K'_s) \to F(L) = \operatorname{Mor}_K(L, K_s)$ induces $F'^{r'}(L') \to F^r(L)$ by the functoriality (13.3.2).

**Proposition 13.3.4** *Let* $L$ *be a finite separable extension. Let* $K'$ *be a log smooth extension of* $K$ *of ramification index* $e = e_{K'/K}$, *and let* $L' = LK'$ *be a composition field.*

(1) *The injection* $F'(L') \to F(L)$ *induces an injection* $F'^{er}_{\log}(L') \to F^r_{\log}(L)$.

(2) *In particular, if* $e = e_{K'/K}$ *is divisible by the ramification index* $m = e_{L/K}$, *the injection* $F'(L') \to F(L)$ *induces an injection* $F'^{er}(L') \to F^r_{\log}(L)$.

*Proof* (1) Let $T \to Q$ be an exact immersion to a log smooth scheme $Q$ over $S$. Then, taking the normalization of $T \times_S S' \to Q \times_S S'$, we obtain an exact closed immersion $T' \to Q_{S'}$ to a log smooth scheme over $S'$ by Proposition 13.2.4(2). Hence, $F'^{er}_{\log}(L')$ is computed by using $T' \to Q_{S'}$, and the assertion follows from Lemma 11.1.2(2).

(2) The assumption $e = e_{L/K} \mid m = e_{K'/K}$ implies $e_{L'/K'} = 1$ by Proposition 13.1.5(2). Hence, $F'^{er}_{\log}(L') = F'^{er}(L')$ by Corollary 13.3.3(2). □

Similarly to the non-logarithmic case, we obtain a filtration $G^r_{\log}$ of $G$ by normal subgroups indexed by $r \in \mathbf{Q}_+$. We call them the logarithmic upper ramification groups.

**Lemma 13.3.5** *Let $L$ be a finite Galois extension of $K$ of Galois group $G = \mathrm{Gal}(L/K)$, and $r = \mathbf{Q}_+$.*

(1) *We have*

$$G^{r+1} \subset G^r_{\log} \subset G^r. \tag{13.3.5}$$

(2) $G^{0+}_{\log} = P$.

(3) *If the extension $E$ of the residue field $F$ is separable, we have $G^{r+1} = G^r_{\log}$.*

(4) *If $e = 1$, $G^r_{\log} = G^r$.*

*Proof* (1) This follows from (13.3.4).

(2) This follows from Corollary 13.3.3(2).

(3) This follows from Corollary 13.3.3(3).

(4) This follows from Corollary 13.3.3(4). □

Let $K'$ be an extension of a henselian discrete valuation field $K$ of ramification index $e = e_{K'/K}$, and $L \to L'$ be a morphism of Galois extensions compatible with $K \to K'$. For $r, r' \in \mathbf{Q}_+$ such that $r' \geqslant er$, the canonical morphism $G' = \mathrm{Gal}(L'/K') \to G = \mathrm{Gal}(L/K)$ induces $G'^{r'}_{\log} \to G^r_{\log}$ by the functoriality (13.3.2). If $K \to K'$ induces an isomorphism on completions, and if $L' = LK'$, the isomorphism $G' \to G$ induces an isomorphism $G'^r_{\log} \to G^r_{\log}$.

After taking a log smooth base extension killing the ramification index, the definition of logarithmic ramification groups is reduced to that of non-logarithmic ramification groups.

**Proposition 13.3.6** *Let $L$ be a finite Galois extension of Galois group $G$, and let $K'$ be a log smooth extension of $K$. Let $L' = LK'$, and $G' = \mathrm{Gal}(L'/K') \to G$ be the canonical injection. Let $P \subset G$ and $P' \subset G'$ be the wild inertia subgroups.*

(1) *The injection $G' \to G$ induces an isomorphism $P' \to P$.*

(2) *Let $e = e_{K'/K}$ be the ramification index, and $r \in \mathbf{Q}_+$. Then the injection $G' \to G$ induces an isomorphism $G'^{er}_{\log} \to G^r_{\log}$. In particular, if $e$ is divisible by the ramification index $e_{L/K}$, the injection $G' \to G$ induces an isomorphism $G'^{er} \to G^r_{\log}$.*

*Proof* (1) Since the wild inertia subgroups $P \subset G$ and $P' \subset G'$ correspond to the maximal tamely ramified extensions $L_1 \subset L$ and $L'_1 \subset L'$, the assertion follows from Corollary 13.2.5.

(2) Since the injection $G' = F'(L') \to G = F(L)$ induces an injection $G'/G'^{er}_{\log} = F'^{er}_{\log}(L') \to G/G^r_{\log} = F^r_{\log}(L)$ by Proposition 13.3.4(1), $G'^{er}_{\log}$ is the inverse image of $G^r_{\log}$. Since $G^r_{\log} \subset G^{0+}_{\log} = P$ and $G'^{er}_{\log} \subset G'^{0+}_{\log} = P'$ by

Lemma 13.3.5(2), the first assertion follows from (1). If $e$ is divisible by $e_{L/K}$, similarly to Proposition 13.3.4(2), we have $G'^{er}_{\log} = G'^{er}$. □

**Exercise 13.3** Let $k$ be a perfect field of characteristic $p > 0$, and $K = k(X)((t))$. Find a Galois extension $L$ of $K$ such that the Galois group $G = \mathrm{Gal}(L/K)$ is isomorphic to $\mathbf{F}_p^2$, satisfying the following property: For $r \in \mathbf{Q}_+$, either $G^r = G$ or $G^r = 1$. There exists $s \in \mathbf{Q}_+$ such that $1 \subsetneqq G^s_{\log} \subsetneqq G$.

## 13.4 Filtration on Witt Vectors

We give an explicit description of the logarithmic upper numbering filtration for abelian extensions of characteristic $p > 0$ in terms of the Artin–Schreier–Witt theory.

Let $K$ be a field of characteristic $p > 0$ and $m \geqslant -1$ be an integer. The exact sequence $0 \to \mathbf{Z}/p^{m+1}\mathbf{Z} = W_{m+1}(\mathbf{F}_p) \to W_{m+1} \xrightarrow{\mathrm{F}-1} W_{m+1} \to 0$ defines a canonical isomorphism

$$W_{m+1}(K)/(\mathrm{F}-1)W_{m+1}(K) \to X_K[p^{m+1}]. \tag{13.4.1}$$

We also define a morphism

$$\mathbf{d}\colon W_{m+1}(K) \to \Omega^1_K \tag{13.4.2}$$

by $\mathbf{d}(a_0, \ldots, a_m) = \sum_{i=0}^m a_i^{p^{m-i}-1} da_i$.

Assume that $K$ is a discrete valuation field. Following Brylinski [16], define an increasing filtration $F_n W_{m+1}(K) \subset W_{m+1}(K)$ indexed by $n \in \mathbf{Z}$ by

$$\begin{aligned} &F_n W_{m+1}(K) \\ &\quad = \{(a_0, \ldots, a_m) \in W_{m+1}(K) \mid p^{m-i}\mathrm{ord}_K a_i \geqslant -n \text{ for } i = 0, \ldots, m\}. \end{aligned} \tag{13.4.3}$$

The injection $V\colon W_{m+1}(K) \to W_{m+2}(K)$ defined by

$$V(a_0, \ldots, a_m) = (0, a_0, \ldots, a_m)$$

induces $F_n W_{m+1}(K) \to F_n W_{m+2}(K)$ and $\mathrm{Gr}^F_n W_{m+1}(K) \to \mathrm{Gr}^F_n W_{m+2}(K)$.

We define an increasing filtration on $\Omega^1_K$ by sub-$O_K$-modules by $F_n\Omega^1_K = \mathfrak{m}_K^{-n} \cdot d\log K^\times \subset \Omega^1_K$ and the graded quotients $\mathrm{Gr}^F_n \Omega^1_K = F_n\Omega^1_K / F_{n-1}\Omega^1_K$. Then the morphism $\mathbf{d}\colon W_{m+1}(K) \to \Omega^1_K$ (13.4.2) induces $F_n W_{m+1}(K) \to F_n\Omega^1_K$ and $\mathrm{Gr}^F_n W_{m+1}(K) \to \mathrm{Gr}^F_n \Omega^1_K$, also denoted by $\mathbf{d}$. By Lemma 3.1.3(3), we have a canonical isomorphism $\mathrm{Gr}^F_0 \Omega^1_K \to \Omega^1_F(\log)$. By this isomorphism, we identify

$$\mathrm{Gr}^F_n \Omega^1_K = \mathrm{Hom}_F(\mathfrak{m}^n_K/\mathfrak{m}^{n+1}_K, \Omega^1_F(\log)). \tag{13.4.4}$$

**Lemma 13.4.1** *Let $K$ be a discrete valuation field, and $n \geqslant 1$ and $m \geqslant 0$ be integers.*

(1) *The diagram*

$$\begin{array}{ccc} \mathrm{Gr}_n^F W_{m+1}(K) & & \\ {\scriptstyle V}\downarrow & \searrow{\scriptstyle \mathbf{d}} & \\ \mathrm{Gr}_n^F W_{m+2}(K) & \xrightarrow{\ \mathbf{d}\ } & \mathrm{Gr}_n^F \Omega_K^1 \end{array}$$

*is commutative.*

(2) [50, 3.2, 3.3], [4, Proposition 10.14] *If $n$ is an integer prime to $p$, then $\mathbf{d}\colon \mathrm{Gr}_n^F W_{m+1}(K) \to \mathrm{Gr}_n^F \Omega_K^1$ is an injection. If $n$ is divisible by $p$, then the sequence*

$$0 \longrightarrow \mathrm{Gr}_{n/p}^F W_{m+1}(K) \xrightarrow{\ \ \mathrm{F}\ \ } \mathrm{Gr}_n^F W_{m+1}(K) \xrightarrow{\ \ \mathbf{d}\ \ } \mathrm{Gr}_n^F \Omega_K^1 \tag{13.4.5}$$

*is exact.*

We omit the proof of Lemma 13.4.1(2). Lemma 13.4.1(1) is clear from the definition of $\mathbf{d}$.

**Definition 13.4.2** Let $K$ be a henselian discrete valuation field of characteristic $p > 0$, and let $K_t \subset K_s$ denote the maximal tamely ramified extension of $K$ in a separable closure. Let $X_K = \mathrm{Hom}_{\mathrm{cont}}(G_K^{\mathrm{ab}}, \mathbf{Q}/\mathbf{Z})$ be the character group of the absolute Galois group $G_K = \mathrm{Gal}(K_s/K)$, and $n \geqslant 0$ be an integer.

For each integer $m \geqslant 0$, define a subgroup $F_n^B X_K[p^{m+1}] \subset X_K[p^{m+1}] = \mathrm{Hom}_{\mathrm{cont}}(G_K^{\mathrm{ab}}, (1/p^{m+1})\mathbf{Z}/\mathbf{Z})$ by

$$F_n^B X_K[p^{m+1}] = \mathrm{Im}(F_n W_{m+1}(K) \to X_K[p^{m+1}]) \tag{13.4.6}$$

with respect to the surjection $W_{m+1}(K) \to X_K[p^{m+1}] \subset X_K$. Let $F_n^B X_K \subset X_K$ be the sum

$$F_n^B X_K = \bigcup_{m \geqslant 0} F_n^B X_K[p^{m+1}] + F_0 X_K \tag{13.4.7}$$

of the union with $F_0 X_K = \mathrm{Ker}(X_K \to X_{K_t})$. We set $F_{-1}^B X_K = 0$ and

$$\mathrm{Gr}_n^B X_K = F_n^B X_K / F_{n-1}^B X_K$$

for $n \geqslant 0$.

We will sketch the proof of the equality $F_n X_K = F_n^B X_K$ in Corollary 16.3.5.

**Proposition 13.4.3** *Let $K$ be a henselian discrete valuation field and $n \geqslant 1$ be an integer.*

(1) *There exists a unique injection* $\mathrm{Gr}_n^B X_K \to \mathrm{Gr}_n^F \Omega_K^1$ *such that the diagram*

$$\begin{array}{ccc} \mathrm{Gr}_n^F W_{m+1}(K) & & \\ \downarrow & \searrow^{\mathbf{d}} & \\ \mathrm{Gr}_n^B X_K & \longrightarrow & \mathrm{Gr}_n^F \Omega_K^1 \end{array}$$

*is commutative for every* $m \geqslant 0$. *In particular,* $\mathrm{Gr}_n^B X_K$ *is annihilated by* $p$.

(2) *For* $m \geqslant 0$, $F_{n-1}^B X_K \cap X_K[p^{m+1}] = F_{n-1}^B X_K[p^{m+1}]$.

We deduce Proposition 13.4.3 from Lemma 13.4.1.

*Proof* (1) The surjection $F_n W_{m+1}(K) \to F_n^B X_K[p^{m+1}]$ induces an isomorphism

$$\mathrm{Gr}_n^F W_{m+1}(K) \to \mathrm{Gr}_n^B X_K[p^{m+1}] \qquad \text{if } n \text{ is prime to } p, \tag{13.4.8}$$
$$\mathrm{Coker}(\mathrm{F}\colon \mathrm{Gr}_{n/p}^F W_{m+1}(K) \to \mathrm{Gr}_n^F W_{m+1}(K)) \to \mathrm{Gr}_n^B X_K[p^{m+1}] \qquad \text{if } n \text{ is divisible by } p.$$

Hence, the assertion follows from Lemma 13.4.1.

(2) Further, by the isomorphism (13.4.8) and Lemma 13.4.1, the inclusion $F_n^B X_K[p^{m+1}] \to F_n^B X_K$ induces an injection $\mathrm{Gr}_n^B X_K[p^{m+1}] \to \mathrm{Gr}_n^B X_K$ for every $n \geqslant 1$. Hence, the morphism $X_K[p^{m+1}]/F_{n-1}^B X_K[p^{m+1}] \to X_K/F_{n-1}^B X_K$ is an injection and we have $F_{n-1}^B X_K \cap X_K[p^{m+1}] = F_{n-1}^B X_K[p^{m+1}]$. □

In the rest of this chapter we prove the following description of the logarithmic upper numbering filtration for abelian extensions.

**Theorem 13.4.4** *Let K be a henselian discrete valuation field of characteristic* $p > 0$, *and let L be a finite abelian extension of Galois group G. Let* $n \geqslant 1$ *be an integer.*

(1) *For* $r \in \mathbf{Q}_+$ *satisfying* $(n-1)+ \leqslant r \leqslant n$, *we have*

$$G_{\log}^r = \{\sigma \in G \mid \chi_a(\sigma) = 0 \textit{ for every } m \geqslant 0,\ a \in F_{n-1}W_{m+1}(K)\}. \tag{13.4.9}$$

(2) *The graded quotient* $\mathrm{Gr}_{\log}^n G = G_{\log}^n / G_{\log}^{n+1}$ *is an* $\mathbf{F}_p$*-vector space and there exists a unique injection*

$$\mathrm{Hom}(\mathrm{Gr}_{\log}^n G, \mathbf{F}_p) \to \mathrm{Hom}_F(\mathfrak{m}_K^n/\mathfrak{m}_K^{n+1}, \Omega_F^1(\log)) \tag{13.4.10}$$

*satisfying the following property: For every* $a \in F_n W_{m+1}(K)$ *and the character* $G/G_{\log}^{n+1} \to \mathbf{Z}/p^{m+1}\mathbf{Z}$ *induced by* $\chi_a \in X_K$, *the restriction* $\chi_a|_{\mathrm{Gr}_{\log}^n G} \in \mathrm{Hom}(\mathrm{Gr}_{\log}^n G, \mathbf{F}_p)$ *is sent to*

$$\mathbf{d}a \in \mathrm{Gr}_n^F \Omega_K^1 = \mathrm{Hom}_F(\mathfrak{m}_K^n/\mathfrak{m}_K^{n+1}, \Omega_F^1(\log)).$$

By Theorem 16.3.4, for which we omit the detail of the proof, the logarithmic upper numbering filtration $G^n_{\log}$ equals the cohomological filtration $G^n_{\mathrm{coh}}$ and the refined Swan conductor of a character $\chi_a$ is given by $-\mathbf{d}a$.

We prepare reduction steps for the proof of Theorem 13.4.4.

**Lemma 13.4.5** *Let $K$ be a henselian discrete valuation field.*

(1) *Let $\widehat{K}$ be the completion. Then Theorem* 13.4.4 *holds for $K$ if and only it holds for $\widehat{K}$.*

(2) *Assume that $K$ is the inductive limit $\varinjlim_\lambda K_\lambda$ of henselian discrete valuation fields such that $e_{K/K_\lambda} = 1$. If Theorem* 13.4.4 *holds for every $K_\lambda$, then it also holds for $K$.*

*Proof* (1) The inclusion $K \to \widehat{K}$ induces an isomorphism $G_{\widehat{K}} \to G_K$ on the absolute Galois group by Proposition 1.4.8. Let $L$ be a finite abelian extension of $K$, and $\widehat{L} = L\widehat{K}$ be the completion. Then, since $O_L \otimes_{O_K} O_{\widehat{K}} \to O_{\widehat{L}}$ is an isomorphism, the isomorphism $G' = \mathrm{Gal}(\widehat{L}/\widehat{K}) \to G = \mathrm{Gal}(L/K)$ induces isomorphisms $G'^r_{\log} \to G^r_{\log}$ for every $r \in \mathbf{Q}_+$, similarly to Lemma 12.3.9(2).

The morphism $F_n W_{m+1} K / F_{-1} W_{m+1} K \to F_n W_{m+1} \widehat{K} / F_{-1} W_{m+1} \widehat{K}$ induced by the inclusion $W_{m+1} K \to W_{m+1} \widehat{K}$ is an isomorphism. Since the Witt vectors in $F_{-1} W_{m+1} \widehat{K}$ define the trivial character by Hensel's lemma, the canonical isomorphism $X_K \to X_{\widehat{K}}$ induces an isomorphism $F_n^B X_K \to F_n^B X_{\widehat{K}}$ for every integer $n \geqslant 0$. Hence, the assertion follows.

(2) Let $L$ be a finite abelian extension. Since $K = \varinjlim_\lambda K_\lambda$, there exist $\lambda$, a finite abelian extension $L_\lambda$ of $K_\lambda$, and an isomorphism $O_{L_\lambda} \otimes_{O_{K_\lambda}} O_K \to O_L$. Then, as in the proof of (1), the isomorphism $G' = \mathrm{Gal}(L/K) \to G_\lambda = \mathrm{Gal}(L_\lambda/K_\lambda)$ induces isomorphisms $G^r_{\log} \to G^r_{\lambda,\log}$ for every $r \in \mathbf{Q}_+$.

Since $F_n W_{m+1} K = \varinjlim_\lambda F_n W_{m+1} K_\lambda$ for every $n$ and $m$, we have $F_n^B X_K = \varinjlim_\lambda F_n^B X_{K_\lambda}$. Hence, the assertion follows. □

**Exercise 13.4** Show that the intersection $F_n X_K[p] = F_n X_K \cap X_K[p]$ equals $F_n^B X_K \cap X_K[p]$ for $n \geqslant 0$.

## 13.5 Dilatation and Witt Vectors

To prove Theorem 13.4.4, we study the filtration on Witt vectors in a geometric situation using dilatation. Let $X = \mathrm{Spec}\, A$ be a smooth affine integral scheme over a perfect field $k$, and let $D \subset X$ be an irreducible divisor, smooth over $k$, defined by a non-zero divisor $t \in A$. Let $A_U = A[1/t]$ so that $U = \mathrm{Spec}\, A_U = X - D$. We consider the discrete valuation of the fraction field $K$ defined by

$D$. We consider $W_{m+1}A_U$ as a subring of $W_{m+1}K$ and define a filtration by $F_nW_{m+1}A_U = W_{m+1}A_U \cap F_nW_{m+1}K$.

Let $\Omega^1_{A/k}(\log) \subset \Omega^1_{A_U/k}$ denote the $A$-module generated by $\Omega^1_{A/k}$ and $d\log t$ defining $\Omega^1_X(\log D)$. Define also an increasing filtration $F_n\Omega^1_{A_U/k} \subset \Omega^1_{A_U/k}$ by

$$F_n\Omega^1_{A_U/k} = t^{-n}\Omega^1_{A/k}(\log D). \tag{13.5.1}$$

We have $F_n\Omega^1_{A_U/k} = \Omega^1_{A_U/k} \cap F_n\Omega^1_K$. Set $\overline{A} = A/tA$ and $\Omega^1_{\overline{A}/k}(\log) = \Omega^1_{A/k}(\log D) \otimes_A \overline{A}$. We identify

$$\mathrm{Gr}^F_n\Omega^1_{A_U/k} = \mathrm{Hom}_{\overline{A}}(t^nA/t^{n+1}A, \Omega^1_{\overline{A}/k}(\log)). \tag{13.5.2}$$

The morphism $\mathbf{d}\colon W_{m+1}(K) \to \Omega^1_K$ (13.4.2) induces $\mathbf{d}\colon W_{m+1}A_U \to \Omega^1_{A_U/k}$. This further induces $F_nW_{m+1}A_U \to F_n\Omega^1_{A_U/k}$ and

$$\mathrm{Gr}^F_n W_{m+1}A_U \to \mathrm{Gr}^F_n\Omega^1_{A_U/k} = \mathrm{Hom}_{\overline{A}}(t^nA/t^{n+1}A, \Omega^1_{\overline{A}/k}(\log))$$

on graded quotients, also denoted by $\mathbf{d}$.

**Lemma 13.5.1** *Let $X$ be a scheme, smooth over a field $k$, and $D \subset X$ be an irreducible divisor smooth over $k$. Let $(X \times_k X)' \to X \times_k X$ be the blow-up at the closed subscheme $D \times_k D$ of codimension 2, and let $P \subset (X \times_k X)'$ be the complement of the proper transform of the union $(X \times_k D) \cup (D \times_k X)$.*

1. *The scheme $P$ is smooth over $k$, and the compositions $p_1, p_2\colon P \to X$ with the projections $X \times_k X \to X$ are smooth.*
2. *The dense open subset $U \times U \subset P$ is the complement of a divisor $D_P \subset P$, smooth over $k$. We have $D_P = p_1^*D = p_1^*D$.*
3. *The diagonal morphism $\delta\colon X \to X \times_k X$ is uniquely lifted to an immersion $e\colon X \to P$. The conormal sheaf $N_{X/P}$ is canonically identified with $\Omega^1_X(\log D)$.*

*Proof* (2) The exceptional divisor $E \subset (X \times_k X)'$ is a $\mathbf{P}^1$-bundle over $D \times_k D$ and is smooth over $k$. Hence, $U \times_k U = P - (E \cap P)$ is the complement of a divisor $D_P = E \cap P \subset P$, smooth over $k$. We have $D_P = p_1^*D = p_1^*D$.

(1) Since $D_P$ is smooth over $D$ and the projections $U \times_k U \to U$ are smooth, the morphisms $p_1, p_2\colon P \to X$ are smooth and $P$ is smooth over $k$.

(3) By the universality of blow-up, the immersion $\delta\colon X \to X \times_k X$ is uniquely lifted to an immersion $X \to P$. Let $j\colon U \to X$ denote the open immersion. Since the projections $P \to X$ are smooth, the immersion $X \to P$ is a regular immersion extending the diagonal morphism $U \to U \times U$. Hence, the conormal sheaf $N_{X/P}$ is a locally free submodule of $j_*\Omega^1_U$. Since the assertion is local, we may assume that $D \subset X$ is defined by a section $t$ of $O_X$. Then the subsheaf

$N_{X/P}$ is generated by $\Omega^1_X$, and $(p_1^*t/p_2^*t) - 1$. Since the image of the latter is $d\log t$, the assertion follows. □

We give a concrete description of $P$ in the case where $X = \operatorname{Spec} A$ is affine in Example 13.5.2.

Let $n \geqslant 1$ be an integer, and let $P^{[n]}$ be the dilatation $P^{[nD_P \cdot X]}$. Since the projections $p_1, p_2 \colon P \to X$ are smooth, the compositions $p_1, p_2 \colon P^{[n]} \to P \to X$ with the projections are also smooth by Lemma 11.1.4(1). By Lemma 13.5.1(3) and Lemma 11.1.5, the base change $P^{[n]}_D = P^{[n]} \times_X D$ by either of the projections is canonically identified with the vector bundle over $D$ defined by the dual of the locally free $\mathcal{O}_D$-module $\Omega^1_X(\log D)(nD) \otimes_{\mathcal{O}_X} \mathcal{O}_D$.

**Example 13.5.2** Assume that $X = \operatorname{Spec} A$ is affine and that $D \subset X$ is defined by a non-zero divisor $t \in A$. Let $A_U = A[1/t]$, and let $u, v \colon A_U \to A_U \otimes_k A_U$ be the morphisms sending $a$ to $a \otimes 1$, $1 \otimes a$ respectively. Then $U = \operatorname{Spec} A_U$ and $P = \operatorname{Spec} A_P$ for the subring

$$A_P = (A \otimes_k A)\left[\frac{u(t)}{v(t)}, \frac{v(t)}{u(t)}\right] \subset A_U \otimes_k A_U.$$

Let $I \subset A_P$ be the kernel of the surjection $A_P \to A$ induced by $A \otimes_k A \to A$ sending $u(t)/v(t)$ and $v(t)/u(t)$ to 1. Then $P^{[n]} = \operatorname{Spec} B$ and $P^{[n]}_D = \operatorname{Spec} \overline{B}$ for the subring $B = A_P[I/v(t)^n] \subset A_U \otimes_k A_U$, and $\overline{B} = B/v(t)B$.

The surjection $B \to A$ defining the closed immersion $X \to P^{[n]}$ induces an isomorphism $B/(1/v(t)^n)IB \to A$. By Lemma 13.5.1(3), the $\mathcal{O}_X$-module $\Omega^1_X(\log D)(nD)$ is defined by the $A$-module $(1/t^n) \cdot I/I^2$. Let $\overline{A} = A/tA$. The $\overline{A}$-module $\mathrm{Gr}_n \Omega^1_{A_U/k} = \operatorname{Hom}_{\overline{A}}(t^n A/t^{n+1}A, \Omega^1_{\overline{A}}(\log))$ is identified with $(1/t^n) \cdot I/I^2 \otimes_A \overline{A}$. The extension of the derivation $d \colon \mathcal{O}_X \to \Omega^1_X$ defines $d \colon (1/t^n)A \to (1/t^n) \cdot I/I^2$. The identification of $P^{[n]}_D$ as a vector bundle over $D$ above defines an isomorphism $S_{\overline{A}}((1/t^n) \cdot I/I^2 \otimes_A \overline{A}) \to \overline{B}$ from the symmetric algebra.

**Lemma 13.5.3** *Let $X = \operatorname{Spec} A$ and $D = \operatorname{Spec} A/tA \subset X$ be as in Example 13.5.2. Let $r \geqslant n \geqslant 1$ be integers. For $a \in (1/t^r)A$, the difference $u(a) - v(a) \in A_U \otimes_k A_U$ is an element of $(1/v(t)^{r-n})B$, and $t^r \cdot da \in \Omega^1_{A_U}$ is an element of $\Omega^1_A(\log) = I/I^2$. Further, we have a congruence*

$$v(t)^{r-n} \cdot (u(a) - v(a)) \equiv \frac{1}{t^n} \cdot t^r\, da \bmod v(t)B \tag{13.5.3}$$

*under the identification $B/v(t)B = \overline{B} = S_A((1/t^n) \cdot I/I^2) \otimes_A \overline{A}$.*

*Proof* First, we show the case $a = (1/t^r)$. Set

$$\frac{1}{v(t)^n} \cdot \left(\frac{u(t)}{v(t)} - 1\right) = s \in B.$$

Then $u(t)/v(t) = 1 + s \cdot v(t)^n$ and the difference

$$\frac{1}{u(t)^r} - \frac{1}{v(t)^r} = -\frac{1}{u(t)^r}\left(\frac{u(t)^r}{v(t)^r} - 1\right) = -\left(\frac{v(t)}{u(t)}\right)^r \frac{1}{v(t)^r}((1 + s \cdot v(t)^n)^r - 1)$$

is an element of $(1/v(t)^{r-n})B$ by binomial expansion. Further, its product with $v(t)^{r-n}$ is congruent to $-r \cdot s \bmod v(t)B$. Since $s \equiv t^{-n} \cdot d\log t$ and $t^r \cdot d(1/t^r) = -r \cdot d\log t$, the assertion follows in this case.

We show the general case. Set $t^r a = b \in A$. Then

$$u(a) - v(a) = v(b)\left(\frac{1}{u(t)^r} - \frac{1}{v(t)^r}\right) + \frac{1}{u(t)^r}(u(b) - v(b)).$$

Since $u(b) - v(b) \in B$ is divisible by $v(t)^n$, and $(1/v(t)^n) \cdot (u(b) - v(b))$ is congruent to $(1/t^n) \cdot db \bmod v(t)B$, the right-hand side is an element of $(1/v(t)^{r-n})B$ and its product with $v(t)^{r-n}$ is congruent to

$$b \cdot \frac{1}{t^n} \cdot t^r d\frac{1}{t^r} + \frac{1}{t^n} \cdot db = \frac{1}{t^n} \cdot t^r da. \qquad \square$$

We define a sequence of polynomials $D_m \in \mathbf{Z}[1/p][X_0, \ldots, X_m, Y_0, \ldots, Y_m]$ indexed by integers $m \in \mathbf{N}$ inductively by

$$\sum_{i=0}^{m} p^i D_i^{p^{m-i}} = \sum_{i=0}^{m} p^i (X_i + Y_i)^{p^{m-i}} - \sum_{i=0}^{m} p^i X_i^{p^{m-i}}. \tag{13.5.4}$$

We have $D_0 = Y_0$ and

$$D_1 = Y_1 + \sum_{i=1}^{p-1} \frac{1}{p}\binom{p}{i} X_0^{p-i} Y_0^i,$$

for example. The polynomials $D_m$ satisfy the following property.

**Lemma 13.5.4** (1) *The polynomial $D_m \in \mathbf{Z}[1/p][X_0, \ldots, X_m, Y_0, \ldots, Y_m]$ is contained in the ideal $I = (Y_0, \ldots, Y_m)$ of $\mathbf{Z}[X_0, \ldots, X_m, Y_0, \ldots, Y_m]$ and satisfies the congruence*

$$D_m \equiv \sum_{i=0}^{m} X_i^{p^{m-i}-1} Y_i \bmod I^2. \tag{13.5.5}$$

(2) *If we give the weight $p^i$ on $X_i$ and $Y_i$ for $0 \leqslant i \leqslant m$, then the polynomial $D_m$ is homogeneous of weight $p^m$.*

(3) *Let $A$ be a ring over $\mathbf{F}_p$ and $x = (x_0, \ldots, x_m), y = (y_0, \ldots, y_m) \in A^{m+1}$. Set $z_i = x_i + y_i$ and $z = (z_0, \ldots, z_m)$. Then we have an equality*

$$z - x = (D_0(x, y), \ldots, D_m(x, y))$$

*in the ring $W_{m+1}(A)$ of Witt vectors.*

*Proof* (1) Let $S_m \in \mathbf{Z}[X_0, \ldots, X_m, Y_0, \ldots, Y_m]$ be the sequence of polynomials defining the addition of Witt vectors. Then (13.5.4) means $S_m(X_0, \ldots, X_m, D_0, \ldots, D_m) = X_m + Y_m$. Since the difference $S_m(X_0, \ldots, X_m, D_0, \ldots, D_m) - (X_m + D_m)$ is a polynomial of $X_0, \ldots, X_{m-1}, D_0, \ldots, D_{m-1}$, we have $D_m \in \mathbf{Z}[X_0, \ldots, X_m, Y_0, \ldots, Y_m]$ by induction on $m$. It suffices to show the congruence (13.5.5) in $\mathbf{Z}[1/p][X_0, \ldots, X_m, Y_0, \ldots, Y_m]$, and then it follows from (13.5.4).

(2) This also follows from (13.5.4).

(3) This follows from (13.5.4) and the definition of the addition of Witt vectors. □

In the notation of Example 13.5.2, by the identification $S_{\overline{A}}((1/t^n) \cdot I/I^2 \otimes_A \overline{A}) \to \overline{B}$, we regard $\mathrm{Gr}_n \Omega^1_{A_U/k} = \mathrm{Hom}_{\overline{A}}(t^n A/t^{n+1} A, \Omega^1_{\overline{A}}(\log)) = (1/t^n) \cdot I/I^2 \otimes_A \overline{A}$ as a sub-$\overline{A}$-module of $\overline{B}$.

**Proposition 13.5.5** *Let $X = \mathrm{Spec}\, A$ and the notation be as in Example 13.5.2. Let $a = (a_0, \ldots, a_m) \in F_n W_{m+1}(A_U)$.*

(1) *Define $b = (b_0, \ldots, b_m) \in B[1/v(t)]^{m+1}$ by $b_i = u(a_i) - v(a_i)$ for $i = 0, \ldots, m$. Then $D_i(v(a), b)$ are elements of $B$ for $i = 0, \ldots, m$ and we have the congruences*

$$D_i(v(a), b) \equiv \begin{cases} 0 & \text{for } i = 0, \ldots, m-1, \\ \sum_{j=0}^m a_j^{p^{m-j}-1} b_j & \text{for } i = m \end{cases} \mod v(t)B. \tag{13.5.6}$$

(2) *Let $\mathbf{d}\colon \mathrm{Gr}_n^F W_{m+1} A_U \to \mathrm{Gr}_n \Omega^1_{A_U/k} = \mathrm{Hom}_{\overline{A}}(t^n A/t^{n+1} A, \Omega^1_{\overline{A}}(\log)) \subset \overline{B}$ be the morphism induced on the graded quotients by $\mathbf{d}\colon W_{m+1} A_U \to \Omega^1_{A_U/k}$, and let $V\colon \overline{B} \to W_{m+1} \overline{B}$ be the Verschibung. Then $u(a) - v(a) \in W_{m+1}(B)$ and*

$$u(a) - v(a) \equiv V(\mathbf{d}(a)) \mod v(t)B. \tag{13.5.7}$$

*Proof* (1) The assumption $a \in F_n W_{m+1}(A_U)$ implies $D_i(v(a), b) \in B$ by Lemma 13.5.3 and Lemma 13.5.4(2). Further, by the congruences (13.5.3) and (13.5.5), we obtain (13.5.6).

(2) Define $b = (b_0, \ldots, b_m) \in B[1/v(t)]^{m+1}$ as in (1). Then

$$u(a) - v(a) = (D_0(v(a), b), \ldots, D_m(v(a), b))$$

by (13.5.4), and the right-hand side is contained in $W_{m+1}(B)$. Further, by (13.5.6) and Lemma 13.5.3(2), we have the congruence (13.5.7). □

**Corollary 13.5.6** *Let $G = \mathbf{Z}/p^{m+1}\mathbf{Z}$ and let $V \to U$ be the $G$-torsor defined by $a \in F_n W_{m+1}(A_U)$. Let $(V \times V)/G$ be the quotient of $V \times V$ by the diagonal action of $G$, and regard $(V \times V)/G$ as a $G$-torsor by the action on the first factor.*

(1) *The $G$-torsor $(V \times V)/G \to U \times U$ is extended to a $G$-torsor $W^{(n)} \to P^{[n]}$.*
(2) *The restriction $W^{(n)} \times_{P^{[n]}} P^{[n]}_D$ is the disjoint union of $p^m$ copies of the Artin–Schreier covering of $P^{[n]}_D$ defined by $\mathbf{d}(a) \in \overline{B}$.*

*Proof* (1) The cyclic covering $(V \times V)/G \to U \times U$ is defined by the difference $p_1^* a - p_2^* a \in W_{m+1}(A_U \otimes_k A_U)$. Hence, by Proposition 13.5.5(1), the normalization $W^{(n)} \to P^{[n]}$ is the cyclic étale covering defined by $p_1^* a - p_2^* a \in W_{m+1}(B)$.

(2) By Proposition 13.5.5(2), the restriction of $W^{(n)} \to P^{[n]}$ on the closed fiber $P^{[n]}_D$ is a disjoint union of copies of the Artin–Schreier covering defined by the non-zero linear form $\mathbf{d}(a) \in \overline{B}$. □

**Exercise 13.5** Let $k$ be a field and $n \geqslant 1$ be an integer. Let $A = k[x, y]$, and identify $A \otimes_k A$ and its elements $x \otimes 1,\ y \otimes 1,\ 1 \otimes x,\ 1 \otimes y$ with $k[x, y, s, t]$ and $x, y, s, t$ respectively. Further, identify the subring $B$ of the fraction field $k(x, y, s, t)$ generated by $u = ((x \otimes 1)/(1 \otimes x))$, its inverse,

$$v = \frac{1}{1 \otimes x^n}\left(\frac{x \otimes 1}{1 \otimes x} - 1\right),$$

and $w = 1/(1 \otimes x^n) \cdot (y \otimes 1 - 1 \otimes y)$ over $A \otimes_k A$ with $k[x, y, s, t][u^{\pm 1}, v, w](x - us, u - (1 + s^n v), y - (t + s^n w)) = k[s, t, v, w][(1 + s^n v)^{-1}]$. Show that the difference $y/x^n - t/s^n \in k(x, y, s, t)$ lies in $B$ and compute its image in $B/(s) = k[t, v, w]$.

## 13.6 Dilatation and Finite Abelian Coverings

We consider a commutative diagram

$$\begin{array}{ccccc} E & \longrightarrow & Y & \longleftarrow & V \\ \downarrow & & \downarrow & & \downarrow \\ D & \longrightarrow & X & \longleftarrow & U \end{array}$$

of smooth schemes over a perfect field $k$. Assume that $U = X - D$ and $V = Y - E$ are the complements of smooth irreducible divisors, and the right vertical arrow $V \to U$ is finite étale. Assume further that $Y$ is the normalization of $X$

in $V$, and $Y \to X$ is finite flat. Let $P$ be as in Lemma 13.5.1. The fiber product $Q = Y \times_X P$ with respect to the first projection $P \to X$ is the complement in the blow-up $(Y \times X)'$ of $Y \times X$ at $E \times D$ of the proper transforms of the union of $Y \times D$ and $E \times X$ since $Y \times X \to X \times X$ is flat.

**Lemma 13.6.1** (1) *The scheme $Q$ is smooth over $k$ and is the normalization of $P$ in $V \times U \to U \times U$.*

(2) *Let $\xi$ be the generic point of an irreducible component of $D$, and let $O_K$ be the henselization $O^{\mathrm{h}}_{X,\xi}$. Let $P_S = P \times_X S \to S = \operatorname{Spec} O_K$ and $Q_S = Q \times_X S \to S$ be the base changes with respect to the second projections $P \to X$ and $Q \to X$. Then $P_S \to S$ is smooth and $Q_S \to S$ is log smooth.*

*Proof* (1) The first projection $Q \to Y$ is the base change of the projection $P \to X$ and hence is smooth. Since $Y$ is smooth over $k$, the scheme $Q$ is smooth over $k$ and hence is normal. Since $Q \to P$ is finite, it is the normalization of $P$ in $V \times U \to U \times U$.

(2) Since $P \to X$ is smooth, its base change $P_S \to S$ is also smooth. We show that $Q_S$ is log smooth. Since $Q$ is regular, its base change $Q_S$ by a localization $S \to X$ is also regular. The generic fiber of $Q_S \to S$ is $V \times_k \operatorname{Spec} K$ and is smooth over $K$. The residue field $F$ of $K$ is the function field of the divisor $D \subset X$. Since the exceptional divisor $Z$ of $(Y \times_k X)'$ is a $\mathbf{P}^1$-bundle over $E \times_k D$, the reduced closed fiber $D_{Q_S} = Q_{F,\mathrm{red}}$ is an open subscheme of a $\mathbf{P}^1$-bundle over $E \times_k \operatorname{Spec} F$ and is smooth over $F$. Define an integer $e \geqslant 1$ by $D \times_X Y = eE$. Then $Q_F = eD_{Q_S}$.

We show the condition on $du$ in Proposition 13.1.2(3) (b) in the case $p \mid e$. Let $\operatorname{Spec} A \subset X$ and $\operatorname{Spec} B \subset Y$ be affine open subsets, and assume that $D$ and $E$ are defined by $t \in A$ and $t' \in B$ respectively. Then $t$ defines a uniformizer of $K$, and the reduced closed fiber $D_{Q_S} \subset Q_S$ is defined by $t'$. Since $u = t/t'^e$ defines a coordinate of the $\mathbf{P}^1$-bundle $Z \subset (Y \times_k X)'$, the condition on $du$ is satisfied. □

We deduce Theorem 13.4.4 from the following proposition.

**Proposition 13.6.2** *Let $K$ be the fraction field of the henselization of a local ring of a smooth scheme over a perfect field. Let $n \geqslant 1$ be an integer and $a \in F_n W_{m+1}(K)$. Let $L$ be the cyclic extension defined by the character $\chi_a$.*

(1) *The log ramification of $L$ is bounded by $n+$.*
(2) *The following conditions are equivalent:*
  (a) *The log ramification of $L$ is bounded by $n$.*
  (b) $\mathbf{d}a = 0$ *in* $\mathrm{Gr}^F_n \Omega^1_K$.

*Proof* Let $X$ be a smooth scheme over a perfect field, and let $D \subset X$ be an irreducible divisor. Let $\xi$ be the generic point of $D$, and assume that $O_K$ is the henselization $O^{\rm h}_{X,\xi}$. By replacing $X$ by an open neighborhood of $\xi$, we may assume that $X = \operatorname{Spec} A$ is affine, and $D$ is defined by $t \in A$ and is smooth over $k$. By replacing $X$ by an étale neighborhood of $\xi$, we may assume that $a \in F_n W_{m+1} A_U$.

After replacing $X$ by an étale neighborhood of $\xi$, let $Y \to X$ be the normalization in a finite étale scheme $V \to U$ such that $Y \times_X S = T = \operatorname{Spec} O_L$. Further replacing $X$ by an open neighborhood of $\xi$, we may assume that $Y$ is smooth over $k$, that $V = Y - E$ is the complement of a smooth irreducible divisor $E$, and that $Y \to X$ is finite flat. As in Lemma 13.6.1, we define a smooth scheme $P_S$ and a log smooth scheme $Q_S$ over $S$, and finite flat morphism $Q_S \to P_S$ over $S$. The immersion $Y \to Y \times_k X$ induces an exact immersion $T \to Q_S$.

Let $D_n \subset Q_S$ be the divisor defined by the ideal $\mathfrak{m}_K^n O_{Q_S}$, and let $Q_S^{[n]}$ be the dilatation $Q_S^{[D_n \cdot T]}$. Since $Y \to X$ is flat, we have the cartesian diagram

$$\begin{array}{ccccccc} T & \longrightarrow & Q_S^{[n]} & \longrightarrow & Q_S & \longrightarrow & Y \\ \downarrow & & \downarrow & & \downarrow & & \downarrow \\ S & \longrightarrow & P_S^{[n]} & \longrightarrow & P_S & \longrightarrow & X \end{array}$$

(1) Let $W^{(n)} \to P^{[n]}$ be the $G = \mathbf{Z}/p^{m+1}\mathbf{Z}$-torsor as in Corollary 13.5.6(1). Since the pull-back $W^{(n)} \times_{P^{[n]}} P_T^{[n]} \to P_T^{[n]}$ is isomorphic to $Q_T^{(n)} \to P_T^{[n]}$, the morphism $Q_T^{(n)} \to P_T^{[n]}$ is finite étale by Corollary 13.5.6(1). Hence, the ramification is bounded by $n+$.

(2) By Corollary 13.5.6(2), the restriction $Q_E^{(n)} \to P_E^{[n]}$ is a disjoint union of copies of the Artin–Schreier coverings defined by the linear form $\mathbf{d}(a)$ on $P_E^{[n]}$. Hence, the étale covering $Q_E^{(n)} \to Q_{E,\rm red}^{[n]}$ splits if and only if $\mathbf{d}(a) = 0$. □

*Proof of Theorem* 13.4.4 By Lemma 13.4.5, we may assume that $K$ is complete. By Corollary 9.1.6(2), there exist an inductive system $(O_{K_\lambda})$ of henselian discrete valuation rings essentially smooth over the henselization $O_{K_0} = k[t]^{\rm h}_{(t)}$ of the local ring of the affine line $\mathbf{A}^1_k$ at the origin, and an isomorphism from the completion of $\varinjlim_{\lambda\in\Lambda} O_{K_\lambda}$ to $O_K$. Hence, by Lemma 13.4.5 again, it is reduced to the case where $K$ is the fraction field of the henselization of a local ring of a smooth scheme over a perfect field $k$, and we may apply Proposition 13.6.2.

(1) It suffices to show that the following four conditions on $\chi \in X_K$ inducing a character on $G$ are equivalent. We say that the log ramification of $\chi$ is bounded by $r$ for $r \in \mathbf{Q}_+$ if $\chi(G^r) = 0$.

(a) There exists $a \in F_{n-1} W_{m+1}(K)$ such that $\chi = \chi_a$.
(b) There exists $a \in F_n W_{m+1}(K)$ such that $\chi = \chi_a$ and $\mathbf{d}a = 0$ in $\operatorname{Gr}_n^F \Omega_K^1$.

(c) The log ramification of $\chi$ is bounded by $(n-1)+$.
(d) The log ramification of $\chi$ is bounded by $n$.

We have (a) $\Rightarrow$ (c) by Proposition 13.6.2(1). The implication (c) $\Rightarrow$ (d) is obvious. The implication (b) $\Rightarrow$ (a) follows from the injectivity of $\mathrm{Gr}^B_n X_K \to \mathrm{Gr}^F_n \Omega^1_K$ in Proposition 13.4.3.

We show the implication (d) $\Rightarrow$ (b). Since the upper ramification groups are compatible with quotients, we may assume that $L$ is the cyclotomic extension defined by $\chi$. Let $m$ be an integer and $a \in W_{m+1}(K)$ be an element such that $\chi = \chi_a$. Assume that the ramification of $\chi$ is bounded by $n$. There exists an integer $n' \geqslant n$ such that $a \in F_{n'} W_{m+1}(K)$. We show that (b) holds by induction from above.

Since the implication (b) $\Rightarrow$ (a) is already proved, we may assume $n' = n$ by the induction hypothesis. By Proposition 13.6.2(2), $\mathbf{d}a = 0$ and (b) holds.

(2) By (1) and Proposition 13.4.3, the injection $\mathrm{Hom}(G, \mathbf{Q}/\mathbf{Z}) \to X_K$ induces an injection $\mathrm{Hom}(\mathrm{Gr}^n_{\log} G, \mathbf{Q}/\mathbf{Z}) \to \mathrm{Gr}^F_n X_K \to \mathrm{Gr}^F_n \Omega^1_K$. Hence, $\mathrm{Gr}^n_{\log} G$ is an $\mathbf{F}_p$-vector space and the assertion follows. □

**Exercise 13.6** Let $k$ be a field of characteristic $p > 0$ and $n \geqslant 1$ be an integer. Let $X = \mathbf{A}^2 = \operatorname{Spec} k[x, y] \supset U = X - D = \operatorname{Spec} k[x^{\pm 1}, y]$ and $\chi \in \mathrm{H}^1(U, \mathbf{F}_p)$ defined by the Artin–Schreier covering $T^p - T = y/x^n$. Let $X \times_k X = \operatorname{Spec} k[x, y, s, t] \subset U \times_k U$ and $\chi_1, \chi_2 \in \mathrm{H}^1(U \times_k U, \mathbf{F}_p)$ be the pull-backs by the projections. Let $(X \times_k X)' \to X \times_k X$ be the blow-up at $D \times_k D \subset X \times_k X$, and $P = \operatorname{Spec} k[x, y, s, t, u^{\pm 1}]/(x - us) \subset (X \times_k X)'$ be the complement of the proper transforms of $D \times_k X$ and $X \times_k D$. Let $X \to P$ be the lifting of the diagonal $X \to X \times_k X$, and let $D_P \subset P$ be the pull-back of $D \subset X$ by either projection. Let $P_n = P^{[nD_P \cdot X]}$ be the dilatation, and $D_n \subset P_n$ be the pull-back of $D \subset X$ by either projection.

Show that the difference $\chi_1 - \chi_2 \in \mathrm{H}^1(U \times_k U, \mathbf{F}_p)$ is extended to $\widetilde{\chi} \in \mathrm{H}^1(P_n, \mathbf{F}_p)$ and compute its restriction as an element of $\mathrm{H}^1(D_n, \mathbf{F}_p)$.

## Historial notes

The logarithmic filtration was also defined in [1]. A reformulation using log geometry is given in [2] and [71]. The dilatation at the logarithmic diagonal was introduced in [3]. The equality in Theorem 13.4.4 between the filtration of Brylinski and the logarithmic upper numbering filtration for abelian extensions is proved in [4]. The non-logarithmic variant of Brylinski's filtration was defined by Matsuda [62]. A large part of the non-logarithmic version of Theorem 13.4.4 was also proved in [4]; the remaining part was proved by Yatagawa [84]. The property that the definition of logarithmic filtration is reduced to that of non-logarithmic filtration by taking log smooth extensions is also observed in [53].

# PART VI

# GRADED QUOTIENTS AND CHARACTERISTIC FORMS

We study the graded quotients of the upper numbering filtration. We prove that the graded quotients $\mathrm{Gr}^r G$ for $r > 1$ are abelian groups annihilated by $p$ in Chapter 14, and construct an injection from the character groups to certain vector spaces in Chapter 15. We study a logarithmic variant in Chapter 16, along with the relation with the cohomological filtration and the refined Swan conductor for abelian extensions. Since the upper numbering filtration is compatible with quotients, it suffices to study the last piece of the filtration.

# 14

# Graded Quotients

We prove in Section 14.6 that the last piece $G^r$ of the upper ramification groups is an $\mathbf{F}_p$-vector space if $r > 1$, by reduction to the case where the residue field is perfect. Since the upper ramification groups are compatible with quotients, this implies that the graded quotients $\mathrm{Gr}^r G$ for $r > 1$ are $\mathbf{F}_p$-vector spaces.

The first step toward the proof is to show in Section 14.4 that the closed fiber of the stable integral model at the critical multiplicity defines a finite étale covering. In Section 15.3 we study this covering explicitly in the case where the ring of integers is generated by a single element.

As a second step, we show the functoriality of the étale morphisms in Section 14.5. By taking minimal immersions, we identify the targets of the morphisms with the Tor groups for the relative $\Omega^1$.

On the étale covering, we have two mutually commutative actions of Galois groups. One is defined by the functoriality. The other comes from the Galois group of the base field extension to define the stable integral model. By comparing these two actions, we deduce that the last piece of the ramification group is abelian in Proposition 14.6.1.

To show that the Galois action on the Tor group is unramified, in Section 14.2 we define an injection to the first homology group of the cotangent complex studied in Section 14.1. This group is an extension of $\Omega^1$ of the residue field by the maximal ideal modulo squared, and may be considered as a Frobenius twist of "the" space of absolute differentials, which is not known to be defined in general.

The last step of the proof is the reduction to the case where the residue field is perfect. We construct an extension with perfect residue field such that the morphism on the first homology groups of cotangent complexes is injective. This injectivity, together with the computation of the first étale cohomology of a vector space in Section 14.3, enables us to complete the reduction.

## 14.1 First Homology of Cotangent Complexes

For a smooth scheme $X$ over a perfect field $k$, the cotangent space $T_x^*X$ at a closed point $x$ is defined as the fiber $\Omega_X^1 \otimes_{\mathcal{O}_X} k(x)$ and is canonically identified with $\mathfrak{m}_x/\mathfrak{m}_x^2$. We consider a flat regular scheme $X$ of finite type over a discrete valuation ring $O_K$ with perfect residue field $k$. Then, similarly, the cotangent space at a closed point $x$ of the closed fiber may be defined as the $k(x)$-vector space $\mathfrak{m}_x/\mathfrak{m}_x^2$. We may ask if there is a vector bundle $T^*X$ on $X$ such that the fiber at each $x$ is canonically isomorphic to $\mathfrak{m}_x/\mathfrak{m}_x^2$. Although such a vector bundle is not known, at least on the closed fiber $X \times_{\operatorname{Spec} O_K} \operatorname{Spec} k$, one can construct "the Frobenius pull-back" of such a bundle using the cohomology sheaf of the cotangent complex. In this section, we consider this construction at the closed point of the spectrum of a discrete valuation ring. Another construction is given in [75].

We recall some basic properties on cotangent complexes from [41, Chapitres II and III]. For a morphism of schemes $X \to S$, the cotangent complex $L_{X/S}$ is defined [41, Chapitre II, 1.2.3] as a chain complex of flat $\mathcal{O}_X$-modules whose cohomology sheaves are quasi-coherent. There is a canonical isomorphism $\mathcal{H}_0(L_{X/S}) \to \Omega^1_{X/S}$ [41, Chapitre II, Proposition 1.2.4.2]. This induces a canonical morphism $L_{X/S} \to \Omega^1_{X/S}[0]$.

If $X \to S$ is a closed immersion defined by the ideal sheaf $\mathcal{I}_X \subset \mathcal{O}_S$ and if $N_{X/S} = \mathcal{I}_X/\mathcal{I}_X^2$ denotes the conormal sheaf, there exists a canonical isomorphism

$$\mathcal{H}_1(L_{X/S}) \to N_{X/S} \tag{14.1.1}$$

[41, Chapitre III, Corollaire 1.2.8.1]. This induces a canonical morphism $L_{X/S} \to N_{X/S}[1]$.

For a commutative diagram

$$\begin{array}{ccc} X' & \longrightarrow & S' \\ {\scriptstyle f}\big\downarrow & & \big\downarrow \\ X & \longrightarrow & S \end{array} \tag{14.1.2}$$

we have a canonical morphism $Lf^*L_{X/S} \to L_{X'/S'}$ [41, Chapitre II, (1.2.3.2)′]. For a morphism $f: X \to Y$ of schemes over a scheme $S$, a distinguished triangle

$$Lf^*L_{Y/S} \to L_{X/S} \to L_{X/Y} \to \tag{14.1.3}$$

is defined [41, Chapitre II, Proposition 2.1.2].

**Lemma 14.1.1** (1) [41, Chapitre III, Proposition 1.2.9] *Let $f: X \to Y$ be an immersion of schemes over a scheme $S$. Then the boundary morphism*

$\partial\colon N_{X/Y} \to f^*\Omega^1_{Y/S}$ *of the distinguished triangle* $Lf^*L_{Y/S} \to L_{X/S} \to L_{X/Y} \to$ *sends* $g$ *to* $-dg$.

(2) [41, Chapitre III, Proposition 3.1.2 (i) ⇒ (ii)] *If* $X \to S$ *is a smooth morphism, the canonical morphism* $L_{X/S} \to \Omega^1_{X/S}[0]$ *is a quasi-isomorphism.*

(3) [41, Chapitre III, Proposition 3.2.4 (iii)] *If* $X \to S$ *is a regular immersion, the canonical morphism* $L_{X/S} \to N_{X/S}[1]$ *is a quasi-isomorphism.*

(4) [41, Chapitre II, Proposition 2.2.1] *If the diagram* (14.1.2) *is cartesian, and if* $O_X \otimes^L_{O_S} O_{S'} \to O_{X'}$ *is an isomorphism, the morphism* $Lf^*L_{X/S} \to L_{X'/S'}$ *is a quasi-isomorphism.*

**Lemma 14.1.2** *Let* $E$ *be an extension of a field* $F$ *of characteristic* $p > 0$.

(1) *If* $E$ *is a separable extension of* $F$*, the canonical morphism* $L_{E/F} \to \Omega^1_{E/F}[0]$ *is a quasi-isomorphism.*

(2) *In general, the distinguished triangle* $L_{F/\mathbf{F}_p} \otimes_F E \to L_{E/\mathbf{F}_p} \to L_{E/F} \to$ *defines an exact sequence*

$$0 \to \mathrm{H}_1(L_{E/F}) \longrightarrow \Omega^1_{F/\mathbf{F}_p} \otimes_F E \longrightarrow \Omega^1_{E/\mathbf{F}_p} \longrightarrow \Omega^1_{E/F} \to 0 \tag{14.1.4}$$

*and* $\mathrm{H}_q(L_{E/F}) = 0$ *for* $q \geqslant 2$.

(3) *Assume that* $E$ *is generated over* $F$ *by a single element. Let* $Z_{F\backslash E} = \mathrm{Ker}(\Omega^1_F \to \Omega^1_E)$*. Then the exact sequence* (14.1.4) *induces an isomorphism*

$$Z_{F\backslash E} \otimes_F E \to \mathrm{H}_1(L_{E/F}). \tag{14.1.5}$$

*Proof* (1) Assume that $E$ is a separable extension. Since the construction of cotangent complexes commutes with filtered inductive limits, we may assume that $E$ is of finite type over $F$. Then there exists a smooth irreducible scheme $X$ over $F$ such that $E$ is the function field. Since the canonical morphism $L_{X/F} \to \Omega^1_{X/F}[0]$ is a quasi-isomorphism, by taking inductive limits, the morphism $L_{E/F} \to \Omega^1_{E/F}[0]$ is a quasi-isomorphism.

(2) Since $E$ and $F$ are separable extensions of a perfect field $\mathbf{F}_p$, the assertion follows from (1).

(3) Since $Z_{F\backslash E} \otimes_F E \to \mathrm{Ker}(\Omega^1_F \otimes_F E \to \Omega^1_E)$ is an isomorphism by Proposition 4.4.5(1), the exact sequence (14.1.4) defines an isomorphism (14.1.5). □

**Proposition 14.1.3** *Let* $O_K$ *be a discrete valuation ring. Let* $S = \mathrm{Spec}\, O_K$ *and* $E$ *be an extension of the residue field* $F = O_K/\mathfrak{m}_K$.

(1) *The morphisms* $\mathrm{Spec}\, E \to \mathrm{Spec}\, F \to S$ *define an exact sequence*

$$0 \longrightarrow \mathfrak{m}_K/\mathfrak{m}_K^2 \otimes_F E \longrightarrow \mathrm{H}_1(L_{E/S}) \longrightarrow \Omega^1_F \otimes_F E \longrightarrow \Omega^1_E \tag{14.1.6}$$

*of $E$-vector spaces.*

(2) *If $E$ is a separable extension of $F$, the first morphism*

$$\mathfrak{m}_K/\mathfrak{m}_K^2 \otimes_F E \to \mathrm{H}_1(L_{E/S}) \tag{14.1.7}$$

*in* (14.1.6) *is an isomorphism.*

(3) *If $E$ is an extension of $F^{1/p}$, the sequence* (14.1.6) *defines an exact sequence*

$$0 \longrightarrow \mathfrak{m}_K/\mathfrak{m}_K^2 \otimes_F E \longrightarrow \mathrm{H}_1(L_{E/S}) \longrightarrow \Omega_F^1 \otimes_F E \longrightarrow 0. \tag{14.1.8}$$

(4) *The closed immersion $S_2 = \operatorname{Spec} O_K/\mathfrak{m}_K^2 \to S$ induces an isomorphism $\mathrm{H}_1(L_{E/S}) \to \mathrm{H}_1(L_{E/S_2})$. If $S_2$ is a scheme over $\mathbf{F}_p$, the morphism $\operatorname{Spec} E \to S_2$ over $\mathbf{F}_p$ defines an injection*

$$\mathrm{H}_1(L_{E/S_2}) \to \Omega^1_{S_2/\mathbf{F}_p} \otimes_{O_{S_2}} E. \tag{14.1.9}$$

If the fiber of the cotangent bundle of $S = \operatorname{Spec} O_K$ at the closed point existed, it should be an $F$-vector space $T^*$ fitting in an exact sequence

$$0 \longrightarrow \mathfrak{m}_K/\mathfrak{m}_K^2 \longrightarrow T^* \longrightarrow \Omega_F^1 \longrightarrow 0 \tag{14.1.10}$$

of $F$-vector spaces. The exact sequence (14.1.8) shows that $T^* \otimes_F E$ is defined as $\mathrm{H}_1(L_{E/S})$ for $E \supset F^{1/p}$, although $T^*$ itself may not exist. If the residue field $F$ is of characteristic $p > 0$ and if $p$ is not a uniformizer, Proposition 14.1.3(4) shows that $T^* = \Omega^1_{S_2/\mathbf{F}_p} \otimes_{O_{S_2}} F$ gives such a vector space.

*Proof* (1) The morphisms $\operatorname{Spec} E \to \operatorname{Spec} F \to S$ define a distinguished triangle $L_{F/S} \otimes_F E \to L_{E/S} \to L_{E/F} \to$. Since $\operatorname{Spec} F \to S$ is a regular immersion, the canonical morphism $L_{F/S} \to \mathfrak{m}_K/\mathfrak{m}_K^2[1]$ is an isomorphism and we obtain an exact sequence $\mathrm{H}_2(L_{E/F}) \to \mathfrak{m}_K/\mathfrak{m}_K^2 \to \mathrm{H}_1(L_{E/S}) \to \mathrm{H}_1(L_{E/F}) \to 0$. Hence, the assertion follows from Lemma 14.1.2(2).

(2) If $E$ is a separable extension of $F$, the last morphism $\Omega_F^1 \otimes_F E \to \Omega_E^1$ in (14.1.6) is an injection.

(3) If $E$ is an extension of $F^{1/p}$, the last morphism $\Omega_F^1 \otimes_F E \to \Omega_E^1$ in (14.1.6) is the 0-morphism.

(4) The distinguished triangle $L_{S_2/S} \otimes^L_{O_{S_2}} E \to L_{E/S} \to L_{E/S_2} \to$ defines an exact sequence $N_{S_2/S} \otimes_{O_{S_2}} E \to \mathrm{H}_1(L_{E/S}) \to \mathrm{H}_1(L_{E/S_2}) \to 0$. The first morphism is 0 since $\mathfrak{m}_K^2/\mathfrak{m}_K^4 \otimes_{O_{S_2}} E \to \mathfrak{m}_K/\mathfrak{m}_K^2 \otimes_F E \subset \mathrm{H}_1(L_{E/S})$ is 0.

The morphisms $\operatorname{Spec} E \to \operatorname{Spec} F \to S_2$ over $\mathbf{F}_p$ define a commutative

diagram

$$\begin{array}{ccccccc} 0 \longrightarrow \mathfrak{m}_K/\mathfrak{m}_K^2 \otimes_F E & \longrightarrow & \mathrm{H}_1(L_{E/S_2}) & \longrightarrow & \mathrm{H}_1(L_{E/F}) \longrightarrow 0 \\ \| & & \downarrow & & \downarrow \\ 0 \longrightarrow \mathfrak{m}_K/\mathfrak{m}_K^2 \otimes_F E & \longrightarrow & \Omega^1_{S_2/\mathbf{F}_p} \otimes_{\mathcal{O}_{S_2}} E & \longrightarrow & \Omega^1_F \otimes_F E \longrightarrow 0 \end{array}$$

of exact sequences. Since the right vertical arrow is an injection, the middle vertical arrow is an injection. □

**Corollary 14.1.4** *Let $E'$ be an extension of $E$. Then the morphism*

$$\mathrm{H}_1(L_{E/S}) \otimes_E E' \to \mathrm{H}_1(L_{E'/S}) \tag{14.1.11}$$

*is an injection. If $E \supset F^{1/p}$, the morphism (14.1.11) is an isomorphism and, in particular, the morphism $\mathrm{H}_1(L_{F^{1/p}/S}) \otimes_{F^{1/p}} E \to \mathrm{H}_1(L_{E/S})$ is an isomorphism.*

*Proof* By Proposition 14.1.3(1), we obtain a commutative diagram

$$\begin{array}{ccccccc} 0 & \longrightarrow & \mathfrak{m}_K/\mathfrak{m}_K^2 \otimes_F E' & \longrightarrow & \mathrm{H}_1(L_{E/S}) \otimes_E E' & \longrightarrow & \Omega^1_F \otimes_F E' \\ & & \| & & \downarrow & & \| \\ 0 & \longrightarrow & \mathfrak{m}_K/\mathfrak{m}_K^2 \otimes_F E' & \longrightarrow & \mathrm{H}_1(L_{E'/S}) & \longrightarrow & \Omega^1_F \otimes_F E' \end{array} \tag{14.1.12}$$

of exact sequences. Hence, (14.1.11) is an injection. For $E = F^{1/p}$, the morphism $\mathrm{H}_1(L_{E/S}) \to \Omega^1_F \otimes_F E$ is a surjection. Hence, the vertical arrow in (14.1.12) is an isomorphism. □

For an element $a \in \mathcal{O}_K$ such that $\bar{a} \in F$ has a $p$th root in $E$, we define an element $w(a) \in \mathrm{H}_1(L_{E/S})$ as follows.

**Lemma 14.1.5** *Let $a \in \mathcal{O}_K$ and let $F(\sqrt[p]{a})$ be the extension generated by a $p$th root of $\bar{a} \in F$. Define a closed immersion* $\mathrm{Spec}\, F(\sqrt[p]{a}) \to \mathbf{A}^1_{\mathcal{O}_K} = \mathrm{Spec}\, \mathcal{O}_K[X]$ *by sending $X$ to $\sqrt[p]{a}$, and let $\mathrm{H}_1(L_{F(\sqrt[p]{a})/\mathcal{O}_K}) \to N_{F(\sqrt[p]{a})/\mathbf{A}^1_{\mathcal{O}_K}}$ be the morphism defined by $L_{F(\sqrt[p]{a})/\mathcal{O}_K} \to L_{F(\sqrt[p]{a})/\mathbf{A}^1_{\mathcal{O}_K}}$. Then there is a unique element*

$$w(a) \in \mathrm{H}_1(L_{F(\sqrt[p]{a})/\mathcal{O}_K}) \tag{14.1.13}$$

*such that the image is $X^p - a \in N_{F(\sqrt[p]{a})/\mathbf{A}^1_{\mathcal{O}_K}}$.*

For an extension $E$ of $F(\sqrt[p]{a})$, let $w(a) \in \mathrm{H}_1(L_{E/\mathcal{O}_K})$ also denote the image of $w(a) \in \mathrm{H}_1(L_{F(\sqrt[p]{a})/\mathcal{O}_K})$.

*Proof* The morphism $\mathrm{Spec}\, F(\sqrt[p]{a}) \to \mathbf{A}^1_{\mathcal{O}_K}$ over $\mathrm{Spec}\, \mathcal{O}_K$ defines a distinguished triangle $L_{\mathbf{A}^1_{\mathcal{O}_K}/\mathcal{O}_K} \otimes^L_{\mathcal{O}_{\mathbf{A}^1_{\mathcal{O}_K}}} F(\sqrt[p]{a}) \to L_{F(\sqrt[p]{a})/\mathcal{O}_K} \to L_{F(\sqrt[p]{a})/\mathbf{A}^1_{\mathcal{O}_K}} \to$.

This defines an exact sequence

$$0 \to \mathrm{H}_1(L_{F(\sqrt[p]{a})/O_K}) \to N_{F(\sqrt[p]{a})/\mathbf{A}^1_{O_K}} \xrightarrow{-d} \Omega^1_{\mathbf{A}^1_{O_K}/O_K} \otimes_{O_{\mathbf{A}^1_{O_K}}} F(\sqrt[p]{a}).$$

Since $X^p - a \in N_{F(\sqrt[p]{a})/\mathbf{A}^1_{O_K}}$ goes to $-d(X^p - a) = 0$ in $\Omega^1_{\mathbf{A}^1_{O_K}/O_K} \otimes_{O_{\mathbf{A}^1_{O_K}}} F(\sqrt[p]{a})$, the unique element $w(a)$ is defined. □

**Proposition 14.1.6** *Let $E$ be an extension of $F$, and let $a, b \in O_K$ be elements such that $E$ contains $p$th roots of $\bar{a}, \bar{b} \in F$.*

(1) *If $a \in \mathfrak{m}_K$, the element $w(a)$ equals the image of the class of $-a$ by the morphism $\mathfrak{m}_K/\mathfrak{m}_K^2 \to \mathrm{H}_1(L_{E/S})$ in (14.1.6).*
(2) *Let $P(X,Y) \in \mathbf{Z}[X,Y]$ be the unique polynomial satisfying $(X+Y)^p = X^p + Y^p + p \cdot P(X,Y)$. Then*

$$w(a+b) = w(a) + w(b) - P(\sqrt[p]{\bar{a}}, \sqrt[p]{\bar{b}}) \cdot w(p), \tag{14.1.14}$$
$$w(ab) = aw(b) + bw(a). \tag{14.1.15}$$

(3) *The image of $w(a)$ by $\mathrm{H}_1(L_{E/S}) \to \Omega^1_F \otimes_F E$ (14.1.6) equals $da$.*
(4) *If $S_2 = \operatorname{Spec} O_K/\mathfrak{m}_K^2$ is a scheme over $\mathbf{F}_p$, the image of $w(a)$ by the injection $\mathrm{H}_1(L_{E/S}) \to \Omega^1_{S_2/\mathbf{F}_p} \otimes_{O_K/\mathfrak{m}_K^2} E$ (14.1.9) equals $da$.*

*Proof* (1) We may assume that $E = F = F(\sqrt[p]{0})$. Let $\operatorname{Spec} F \to \mathbf{A}^1_S = \operatorname{Spec} O_K[X]$ be the closed immersion defined by sending $X$ to 0. Then, since $X^p \equiv 0$ in $N_{F/\mathbf{A}^1_{O_K}}$ and since the composition $\mathfrak{m}_K/\mathfrak{m}_K^2 \to \mathrm{H}_1(L_{E/S}) \to N_{F/\mathbf{A}^1_{O_K}}$ is the canonical morphism, we have $w(a) \equiv -a$.

(2) It suffices to prove the case where $E = F(\sqrt[p]{\bar{a}}, \sqrt[p]{\bar{b}})$. Define a closed immersion $\operatorname{Spec} E \to \mathbf{A}^2_{O_K} = \operatorname{Spec} O_K[X,Y]$ by sending $X, Y$ to $\sqrt[p]{\bar{a}}, \sqrt[p]{\bar{b}}$. Then, by the functoriality for the morphism $\mathbf{A}^2_{O_K} \to \mathbf{A}^1_{O_K}$ defined by sending $X$ to $X+Y$, we see that $w(a+b) \in \mathrm{H}_1(L_{E/S})$ is the unique element such that the image in $N_{E/\mathbf{A}^2_{O_K}}$ equals $(X+Y)^p - (a+b)$. Hence, the equality (14.1.14) follows from

$$(X+Y)^p - (a+b) = (X^p - a) + (Y^p - b) + P(X,Y) \cdot p$$

and 1 for $a = p$. Similarly, the equality (14.1.15) follows from

$$(XY)^p - ab = a(Y^p - b) + b(X^p - a) + (X^p - a)(Y^p - b).$$

(3) We may assume that $E = F(\sqrt[p]{a})$. The commutative diagram

$$\begin{array}{ccccc} \operatorname{Spec} E & \longrightarrow & \operatorname{Spec} F & \longrightarrow & S \\ \| & & \uparrow & & \uparrow \\ \operatorname{Spec} E & \longrightarrow & \mathbf{A}^1_F & \longrightarrow & \mathbf{A}^1_S \end{array}$$

defines a commutative diagram

$$\begin{array}{ccccc} \mathrm{H}_1(L_{E/S}) & \longrightarrow & \mathrm{H}_1(L_{E/F}) & \longrightarrow & \Omega^1_F \otimes_F E \\ \downarrow & & \downarrow & & \downarrow \\ N_{E/\mathbf{A}^1_S} & \longrightarrow & N_{E/\mathbf{A}^1_F} & \xrightarrow{-d} & \Omega^1_{\mathbf{A}^1_F} \otimes_{O_{\mathbf{A}^1_F}} E \end{array}$$

by Proposition 14.1.1(1). The composition of the upper arrows is (14.1.6). Since $-d(X^p - a) = da$ and the right vertical arrow is an injection, the assertion follows.

(4) We may assume that $E = F(\sqrt[p]{a})$. The immersion $\operatorname{Spec} E \to S_2$ defines a commutative diagram

$$\begin{array}{ccc} \mathrm{H}_1(L_{E/S}) & \xrightarrow{(14.1.9)} & \Omega^1_{S_2/\mathbf{F}_p} \\ \downarrow & & \downarrow \\ N_{E/\mathbf{A}^1_{S_2}} & \xrightarrow{-d} & \Omega^1_{\mathbf{A}^1_{S_2}/\mathbf{F}_p} \end{array}$$

by Proposition 14.1.1(1). The canonical morphism $N_{E/\mathbf{A}^1_{S_2}} \to N_{E/\mathbf{A}^1_S}$ is an isomorphism. Since $-d(X^p - a) = da$ and the right vertical arrow is an injection, the assertion follows. □

The following construction will be used in the reduction to the perfect residue field case of the proof of the property that the graded quotients are abelian groups annihilated by $p$.

**Lemma 14.1.7** *Let $K$ be a discrete valuation field with residue field $F$. Let $(u_i)_{i \in I}$ be a family of elements of $O_K$ such that $(du_i)_{i \in I}$ is a basis of $\Omega^1_F$, and let $t \in O_K$ be a uniformizer.*

(1) *Let $K'$ be an extension of the discrete valuation field $K$ with residue field $F'$. Let $E$ be an extension of $F$ containing $F^{1/p}$ as a subfield, and let $E \to E'$ be a morphism of extensions of $F$ and $F'$ compatible with $F \to F'$. We identify $\mathfrak{m}_{K'}/\mathfrak{m}^2_{K'} \otimes_{F'} E'$ with a subspace of $\mathrm{H}_1(L_{E'/O_K'})$ by the injection in (14.1.6). Then the following conditions are equivalent:*

(a) *The morphism* $\mathrm{H}_1(L_{E/O_K}) \to \mathrm{H}_1(L_{E'/O'_K})$ *induces a morphism*

$$\mathrm{H}_1(L_{E/O_K}) \to \mathfrak{m}_{K'}/\mathfrak{m}_{K'}^2 \otimes_{F'} E' \subset \mathrm{H}_1(L_{E'/O'_K});$$

*further, the morphism* $\mathrm{S}_E\mathrm{H}_1(L_{E/O_K}) \to \mathrm{S}_{F'}(\mathfrak{m}_{K'}/\mathfrak{m}_{K'}^2) \otimes_{F'} E'$ *is an injection.*

(b) *The ramification index* $e_{K'/K}$ *is* 1. *There exist families* $(v_i)_{i\in I}$ *and* $(w_i)_{i\in I}$ *of elements of* $O_{K'}$ *satisfying* $u_i = w_i^p + v_i t$ *for* $i \in I$ *such that the family* $(v_i)_{i\in I}$ *of elements of* $F' \subset E'$ *is algebraically independent over* $E$.

(2) *There exists an extension* $K'$ *of the discrete valuation field* $K$ *with perfect residue field* $F'$ *such that, for every morphism* $\overline{F} = E \to \overline{F}' = E'$ *of algebraic closures, the equivalent conditions* (a) *and* (b) *in* (1) *are satisfied.*

*Proof* (1) By the assumption that $E$ contains $F^{1/p}$ as a subfield, and by Proposition 14.1.3(3) and (1), $\mathrm{H}_1(L_{E/O_K}) \to \mathrm{H}_1(L_{E'/O'_K})$ induces $\mathrm{H}_1(L_{E/O_K}) \to \mathfrak{m}_{K'}/\mathfrak{m}_{K'}^2 \otimes_{F'} E' \subset \mathrm{H}_1(L_{E'/O'_K})$ if and only if $\Omega^1_F \to \Omega^1_{F'}$ is the 0-mapping. This condition is also equivalent to the existence of families $(v_i)_{i\in I}$ and $(w_i)_{i\in I}$ of elements of $O_{K'}$ satisfying $u_i = w_i^p + v_i t$. The ramification index $e_{K'/K}$ is 1 if and only if the image of $t$ in $\mathfrak{m}_{K'}/\mathfrak{m}_{K'}^2$ is non-zero. Hence, it suffices to show that the injectivity in (a) is equivalent to the algebraic independence in (b).

Further, by Proposition 14.1.3(3), $w(t), (w(u_i))_{i\in I}$ form a basis of the $E$-vector space $\mathrm{H}_1(L_{E/O_K})$. By Proposition 14.1.6(2), we have $w(u_i) = w(w_i^p + v_i t) = w(w_i^p) + w(v_i t) = p w_i^{p-1} w(w_i) + v_i w(t) = v_i w(t)$ in $\mathrm{H}_1(L_{E'/O_{K'}})$. Hence, the injectivity is equivalent to the algebraic independence.

(2) Let $O_{K_0}$ be the localization of the polynomial ring $O_K[v_i; i \in I]$ at the prime ideal $\mathrm{Ker}(O_K[v_i; i \in I] \to F[v_i; i \in I])$. The residue field $F_0$ of $K_0$ is the rational function field $F(v_i; i \in I)$, and $(du_i, dv_i)_{i\in I}$ is a basis of $\Omega^1_{F_0}$.

We write $v_i = v_{i,0}$ and set $w_{i,0} = u_i - v_i t \in O_{K_0}$ for $i \in I$. For integers $n \geqslant 1$, define $O_{K_n}$ inductively to be $O_{K_{n-1}}[v_{i,n}, w_{i,n}; i \in I]/(v_{i,n}^p - v_{i,n-1}, w_{i,n}^p - w_{i,n-1}; i \in I)$ and set $K' = \varinjlim_n K_n$. Then the residue field $F_n$ of $K_n$ is $F_0^{1/p^n}$, and the residue field $F'$ of $K'$ is a perfect closure $F_0^{1/p^\infty}$.

Since $F'$ is perfect, the morphism $\Omega^1_F \to \Omega^1_{F'} = 0$ is the 0-mapping. The ramification index $e_{K'/K}$ is 1. Since $(v_{i,0}; i \in I)$ is algebraically independent over $F$, the assertion follows. □

**Exercise 14.1** Let $K$ be a discrete valuation field with residue field $F$ of characteristic $p > 0$, and set $S = \mathrm{Spec}\, O_K$. Let $E$ be an extension of $F$.

(1) For integers $a$, show that $w(a) = ((a - a^p)/p) \cdot w(p)$ in $\mathrm{H}^1(L_{E/S})$.

(2) Assume that $E$ is a purely inseparable extension of degree $p$ of $F$. Let

$v \in O_K$ be an element such that $E = F(u)$ is generated by a $p$th root $u$ of $\bar{v} \in F$, and $t$ be a uniformizer of $K$. Show that the $E$-vector space $\mathrm{H}^1(L_{E/S})$ is of dimension 2 and that $w(v), w(t)$ is a basis.

(3) Assume that $p$ is not a uniformizer, and let $S_2 = \operatorname{Spec} O_K/\mathfrak{m}_K^2$ as in Proposition 14.1.3(4). Show that the injection $\mathrm{H}_1(L_{E/S}) = \mathrm{H}_1(L_{E/S_2}) \to \Omega^1_{S_2/\mathbf{F}_p} \otimes_{O_{S_2}} E$ maps $w(a)$ to $da$ for $a \in O_K$ such that $E$ contains $\sqrt[p]{\bar{a}}$.

## 14.2 Tor Groups of Modules of Differentials

A link connecting the "cotangent space" $\mathrm{H}_1(L_{E/S})$ to the conormal space $N_{T/Q} \otimes_{O_T} E$ for an immersion $T = \operatorname{Spec} O_L \to Q$ to a smooth scheme over $S = \operatorname{Spec} O_K$ is given by $\mathrm{Tor}_1^{O_L}(\Omega^1_{T/S}, E)$.

**Lemma 14.2.1** *Let $L$ be a finite separable extension of a henselian discrete valuation field $K$. Then the canonical morphism*

$$L_{O_L/O_K} \to \Omega^1_{O_L/O_K}[0] \tag{14.2.1}$$

*is a quasi-isomorphism.*

*Proof* Let $T = \operatorname{Spec} O_L \to Q$ be a closed immersion to a smooth scheme $Q$ over $S$. Then, since the immersion $T \to Q$ is a regular immersion, the distinguished triangle $L_{Q/S} \otimes^L_{O_Q} O_T \to L_{T/S} \to L_{T/Q} \to$ defines a distinguished triangle $\Omega^1_{Q/S} \otimes_{O_Q} O_T \to L_{T/S} \to N_{T/Q}[1] \to$. Since $L$ is a separable extension, the morphism $N_{T/Q} \to \Omega^1_{Q/S} \otimes_{O_Q} O_T$ of free $O_T$-modules of the same rank is an injection. Hence, we have $\mathrm{H}_q(L_{T/S}) = 0$ for $q \geqslant 1$ and the assertion follows. □

**Proposition 14.2.2** *Let $L$ be a finite separable extension of a henselian discrete valuation field $K$. Let $T = \operatorname{Spec} O_L$ and $E$ be the residue field.*

(1) *The immersion $\operatorname{Spec} E \to T$ over $S$ defines an exact sequence*

$$\begin{aligned} 0 \to \mathrm{Tor}_1^{O_T}(\Omega^1_{O_T/O_S}, E) &\to \mathrm{H}_1(L_{E/S}) \to N_{E/T} \\ &\to \Omega^1_{O_T/O_S} \otimes_{O_L} E \to \Omega^1_{E/F} \to 0 \end{aligned} \tag{14.2.2}$$

*of $E$-vector spaces.*

(2) *Assume that $E$ is a separable extension of $F$, and that $L$ is a ramified extension of $K$. Then the first arrow $\mathrm{Tor}_1^{O_L}(\Omega^1_{T/S}, E) \to \mathrm{H}_1(L_{E/S})$ in (14.2.2) is an isomorphism of $E$-vector spaces of dimension 1.*

(3) *Assume that the ramification index $e_{L/K}$ is 1. Then (14.2.2) defines an exact*

*sequence* $0 \to \mathrm{Tor}_1^{O_T}(\Omega^1_{O_T/O_S}, E) \to \mathrm{H}_1(L_{E/S}) \to N_{E/T} \to 0$*, and the composition*

$$\mathrm{Tor}_1^{O_T}(\Omega^1_{O_T/O_S}, E) \to \mathrm{H}_1(L_{E/S}) \to \mathrm{H}_1(L_{E/F}) \tag{14.2.3}$$

*is an isomorphism.*

*Proof* (1) Since $\mathrm{Spec}\, E \to T$ is a regular immersion, the canonical morphism $L_{E/T} \to N_{E/T}[1]$ is an isomorphism. Hence, by Lemma 14.2.1, the distinguished triangle $L_{T/S} \otimes^L_{O_T} E \to L_{E/S} \to L_{E/T} \to$ defined by $\mathrm{Spec}\, E \to T \to S$ gives (14.2.2).

(2) If $E$ is separable over $F$, the third arrow $\Omega^1_F \otimes_F E \to \Omega^1_E$ in (14.1.6) is an isomorphism and the first arrow $\mathfrak{m}_K/\mathfrak{m}_K^2 \to \mathrm{H}_1(L_{E/S})$ is an isomorphism of $E$-vector spaces of dimension 1. Further, we have $\Omega^1_{E/F} = 0$ and hence the third arrow $N_{E/T} \to \Omega^1_{O_L/O_K} \otimes_{O_L} E$ in (14.2.2) is a surjection. If $L$ is ramified over $K$, then we have $\Omega^1_{O_L/O_K} \otimes_{O_L} E \neq 0$ and hence $N_{E/T} \to \Omega^1_{O_L/O_K} \otimes_{O_L} E$ is an isomorphism of $E$-vector spaces of dimension 1.

(3) Since $T \to S$ is flat and $E = O_L \otimes_{O_K} F$ by $e_{L/K} = 1$, the last arrow $\Omega^1_{O_T/O_S} \otimes_{O_L} E \to \Omega^1_{E/F}$ in (14.2.2) and the composition (14.2.3) are isomorphisms by Lemma 14.1.1(4). □

**Corollary 14.2.3** *Assume that $L$ is a Galois extension of the Galois group $G = \mathrm{Gal}(L/K)$. Then the action of the inertia group $I = \mathrm{Ker}(G \to \mathrm{Aut}_F E)$ on $\mathrm{Tor}_1^{O_T}(\Omega^1_{O_T/O_S}, E)$ is trivial.*

*Proof* Since the action of $I$ on $E$ is trivial, the action on $\mathrm{H}_1(L_{E/S})$ is trivial. Hence, the assertion follows from the injectivity of $\mathrm{Tor}_1^{O_T}(\Omega^1_{O_T/O_S}, E) \to \mathrm{H}_1(L_{E/S})$ (14.2.2). □

**Lemma 14.2.4** *Let $T \to Q$ be an immersion to a smooth scheme over $S$. Then the morphisms $\mathrm{Spec}\, E \to T \to Q$ over $S$ define a commutative diagram*

$$\begin{array}{ccccccc}
0 \longrightarrow & \mathrm{Tor}_1^{O_L}(\Omega^1_{O_L/O_K}, E) & \longrightarrow & N_{T/Q} \otimes_{O_L} E & \longrightarrow & \Omega^1_{Q/S} \otimes_{O_Q} E \\
& \downarrow & & \downarrow & & \| \\
0 \longrightarrow & \mathrm{H}_1(L_{E/S}) & \longrightarrow & N_{E/Q} & \longrightarrow & \Omega^1_{Q/S} \otimes_{O_Q} E
\end{array} \tag{14.2.4}$$

*of exact sequences of $E$-vector spaces. The composition $\mathfrak{m}_K/\mathfrak{m}_K^2 \to N_{E/Q}$ of the first arrow in* (14.1.6) *with the first arrow in the lower line is the canonical morphism.*

*Proof* The immersions $\mathrm{Spec}\, E \to T \to Q$ over $S$ define a commutative

diagram

$$\begin{array}{ccccccc} L_{Q/S}\otimes^{L}_{O_L} E & \longrightarrow & L_{T/S}\otimes^{L}_{O_L} E & \longrightarrow & L_{T/Q}\otimes^{L}_{O_L} E & \longrightarrow \\ \| & & \downarrow & & \downarrow & \\ L_{Q/S}\otimes^{L}_{O_Q} E & \longrightarrow & L_{E/S} & \longrightarrow & L_{E/Q} & \longrightarrow \end{array} \tag{14.2.5}$$

of distinguished triangles. By Lemma 14.1.1(2) and (3), we have isomorphisms $L_{Q/S} \to \Omega^1_{Q/S}[0]$, $L_{T/Q} \to N_{T/Q}[1]$, and $L_{E/Q} \to N_{E/Q}[1]$. By the isomorphism $L_{T/S} \to \Omega^1_{T/S}[0]$ (14.2.1), we obtain (14.2.4).

The composition $\mathfrak{m}_K/\mathfrak{m}_K^2 \to N_{E/Q}$ is the canonical morphism defined by the commutative diagram

$$\begin{array}{ccc} \operatorname{Spec} E & \longrightarrow & Q \\ \downarrow & & \downarrow \\ \operatorname{Spec} F & \longrightarrow & S \end{array}$$

by the functoriality of cotangent complexes. □

**Example 14.2.5** Let $K$ be a henselian discrete field with residue field $F$ of characteristic $p > 0$.

(1) Let $L$ be a totally ramified extension of $K$ of degree $n > 1$, and let $f \in O_K[X]$ be the minimal polynomial of a uniformizer $s$ of $L$. Define a closed immersion by $T = \operatorname{Spec} O_K[X]/(f) \to Q = \operatorname{Spec} O_K[X]$. Then the first top arrow $\operatorname{Tor}_1^{O_L}(\Omega^1_{O_L/O_K}, E) \to N_{T/Q}\otimes_{O_L} E$ in (14.2.4) is an isomorphism and the image of the basis $f$ of $N_{T/Q}\otimes_{O_L} E$ in $N_{E/Q}$ equals the image of the constant term $f(0) \in \mathfrak{m}_K/\mathfrak{m}_K^2$ since $f$ is an Eisenstein polynomial. Let $\widetilde{f} \in \operatorname{Tor}_1^{O_L}(\Omega^1_{O_L/O_K}, E)$ denote the inverse image of $f \in N_{T/Q}\otimes_{O_L} E$. Then the left vertical arrow in (14.2.4) maps the basis $\widetilde{f}$ to $-w(f(0)) \in \mathrm{H}_1(L_{E/S})$ by Proposition 14.1.6(1).

(2) Let $L$ be an extension of $K$ of ramification index $e_{L/K} = 1$, and assume that the residue field $E = F(u)$ is a purely inseparable extension generated by the image of a single element $u \in O_L$. Let $f \in O_K[X]$ be the minimal polynomial of $u$, and define a closed immersion by $T = \operatorname{Spec} O_K[X]/(f) \to Q = \operatorname{Spec} O_K[X]$. Then we have a commutative diagram

$$\begin{array}{ccccc} \operatorname{Tor}_1^{O_L}(\Omega^1_{O_L/O_K}, E) & \longrightarrow & N_{T/Q}\otimes_{O_L} E & \longrightarrow & N_{E/Q_F} \\ \downarrow & & \downarrow & & \downarrow{\scriptstyle -d} \\ \mathrm{H}_1(L_{E/S}) & \longrightarrow & \mathrm{H}_1(L_{E/F}) & \xrightarrow{\subset} & \Omega^1_F\otimes_F E \end{array}$$

by Proposition 14.1.1(1) and the upper horizontal arrows are isomorphisms.

**Exercise 14.2** Let $K$ be a henselian discrete valuation field with residue field $F$ of characteristic $p > 0$, and $L$ be a ramified separable extension of degree $p$. Let $x \in O_L = O_K[x]$ be a generator of the ring of integers, and $f \in O_K[X]$ be the minimal polynomial of $x$. Define a closed immersion $T = \operatorname{Spec} O_L \to Q = \mathbf{A}^1_S = \operatorname{Spec} O_K[X]$ to a smooth scheme over $S = \operatorname{Spec} O_K$ by sending $X$ to $x$.

(1) Assume that $L$ is a totally ramified extension and $x$ is a uniformizer of $L$. Show that the morphism $\operatorname{Tor}_1^{O_T}(\Omega^1_{T/S}, F) \to N_{T/P} \otimes_{O_T} F$ is an isomorphism and compute the image by $\operatorname{Tor}_1^{O_T}(\Omega^1_{T/S}, F) \to \mathrm{H}_1(L_{F/S}) = N_{F/S} = \mathfrak{m}_K/\mathfrak{m}_K^2$ (14.2.2) of the inverse image of the basis $f$ of $N_{T/P} \otimes_{O_T} F$ by $\operatorname{Tor}_1^{O_T}(\Omega^1_{T/S}, F) \to N_{T/P} \otimes_{O_T} F$.

(2) Assume that the residue field $E$ of $L$ is an inseparable extension. Show that the morphism $\operatorname{Tor}_1^{O_T}(\Omega^1_{T/S}, F) \to N_{T/P} \otimes_{O_T} F$ is an isomorphism and compute the image by $\operatorname{Tor}_1^{O_T}(\Omega^1_{T/S}, F) \to \mathrm{H}_1(L_{F/S})$ (14.2.2) of the inverse image of the basis $f$ of $N_{T/P} \otimes_{O_T} F$ by $\operatorname{Tor}_1^{O_T}(\Omega^1_{T/S}, F) \to N_{T/P} \otimes_{O_T} F$ as a linear combination of the basis $w(-f(0)), w(t)$ for the constant term $f(0)$ and a uniformizer $t$ of $K$.

## 14.3 Artin–Schreier Coverings of Vector Spaces

For a scheme $X$ over $\mathbf{F}_p$ and $A = \Gamma(X, O_X)$, the Artin–Schreier sequence

$$0 \longrightarrow \mathbf{F}_p \longrightarrow \mathbf{G}_a \xrightarrow{\mathrm{F}-1} \mathbf{G}_a \longrightarrow 0 \tag{14.3.1}$$

defines an injection

$$A/(\mathrm{F}-1)A \to \mathrm{H}^1(X, \mathbf{F}_p) \tag{14.3.2}$$

to the étale cohomology. If $X$ is affine, (14.3.2) is an isomorphism since $\mathrm{H}^1(X, O_X) = 0$. If we regard $f \in A$ as a morphism $X \to \mathbf{G}_a$, the image of $f$ in $\mathrm{H}^1(X, \mathbf{F}_p)$ is the pull-back by $f$ of the class $[\mathrm{AS}] \in \mathrm{H}^1(\mathbf{G}_a, \mathbf{F}_p)$ of the $\mathbf{F}_p$-torsor defined by the Artin–Schreier sequence (14.3.1).

Let $k$ be a field of characteristic $p > 0$, and let $V$ be a $k$-vector space of finite dimension. We regard the dual space $V^\vee = \operatorname{Hom}_k(V, k)$ as a smooth algebraic group $V^\vee = \operatorname{Spec} \mathrm{S}_k V$ over $k$. Then the isomorphism (14.3.2) for $X = V^\vee$ is induced by the morphism

$$[\mathrm{AS}] \colon \mathrm{S}_k V = \Gamma(V^\vee, O_{V^\vee}) = \operatorname{Mor}_k(V^\vee, \mathbf{G}_a) \to \mathrm{H}^1(V^\vee, \mathbf{F}_p) \tag{14.3.3}$$

sending a morphism $f \colon V^\vee \to \mathbf{G}_a$ to the class $f^*[\mathrm{AS}]$.

**Proposition 14.3.1** *Let $k$ be a field of characteristic $p > 0$, and let $V$ be a $k$-vector space of finite dimension. Let $V^\vee = \operatorname{Spec} \mathrm{S}_k V$ be the dual space regarded as a scheme over $k$. The isomorphism* (14.3.3) *induces an isomorphism*

$$\bigoplus_{p\nmid m} \varinjlim_{\mathrm{F}} \mathrm{S}_k^{p^e m} V \to \mathrm{H}^1(V^\vee, \mathbf{F}_p)/\mathrm{H}^1(k, \mathbf{F}_p). \tag{14.3.4}$$

*Proof* The decomposition $\mathrm{S}_k V = k \oplus \bigoplus_{p\nmid m}(\bigoplus_{e=0}^\infty \mathrm{S}_k^{p^e m} V)$ is stable under the action of the Frobenius F. We have an isomorphism

$$\operatorname{Coker}(\mathrm{F} - 1 \colon k \to k) \to \mathrm{H}^1(k, \mathbf{F}_p).$$

For each integer $m \geqslant 1$ prime to $p$, the cokernel $\operatorname{Coker}(\mathrm{F}-1 \colon \bigoplus_{e=0}^\infty \mathrm{S}_k^{p^e m} V \to \bigoplus_{e=0}^\infty \mathrm{S}_k^{p^e m} V)$ is identified with the inductive limit $\varinjlim_{\mathrm{F}} \mathrm{S}_k^{p^e m} V$. □

**Corollary 14.3.2** *Let $k \to k'$ be a morphism of separably closed fields of characteristic $p > 0$, and let $V$ and $V'$ be a $k$-vector space and a $k'$-vector space respectively of finite dimension. Let $V \to V'$ be a linear mapping compatible with $k \to k'$, and let $V'^\vee \to V^\vee$ be the morphism of schemes defined by $\mathrm{S}_k V \to \mathrm{S}_{k'} V'$. Then the following conditions are equivalent:*

(1) *The morphism $\mathrm{H}^1(V^\vee, \mathbf{Q}_p/\mathbf{Z}_p) \to \mathrm{H}^1(V'^\vee, \mathbf{Q}_p/\mathbf{Z}_p)$ is an injection.*
(2) *The mapping $\mathrm{S}_k V \to \mathrm{S}_{k'} V'$ is an injection.*

*Proof* Since $\mathrm{H}^1(V^\vee, \mathbf{F}_p)$ is the $p$-torsion part of $\mathrm{H}^1(V^\vee, \mathbf{Q}_p/\mathbf{Z}_p)$, condition (1) is equivalent to the injectivity of $\mathrm{H}^1(V^\vee, \mathbf{F}_p) \to \mathrm{H}^1(V'^\vee, \mathbf{F}_p)$. Since $k$ and $k'$ are separably closed, this is equivalent to the injectivity of

$$\bigoplus_{p\nmid m} \varinjlim_{\mathrm{F}} \mathrm{S}_k^{p^e m} V \to \bigoplus_{p\nmid m} \varinjlim_{\mathrm{F}} \mathrm{S}_{k'}^{p^e m} V' \tag{14.3.5}$$

by Proposition 14.3.1. Since $\mathrm{S}_k^{p^e m} V \to \varinjlim_{\mathrm{F}} \mathrm{S}_k^{p^e m} V$ is an injection, this implies the injectivity of $\mathrm{S}_k V \to \mathrm{S}_{k'} V'$. Conversely, if $\mathrm{S}_k V \to \mathrm{S}_{k'} V'$ is an injection, then (14.3.5) is an injection. □

**Lemma 14.3.3** *Let $G$ be a finite abelian group of order $n \geqslant 1$, and let $G^\vee = \operatorname{Hom}(G, \mathbf{Z}/n\mathbf{Z})$ be the dual group. For a $G$-torsor $P$ over a connected scheme $X$, the following conditions are equivalent:*

(1) *$P$ is connected.*
(2) *The morphism $[P] \colon G^\vee \to \mathrm{H}^1(X, \mathbf{Z}/n\mathbf{Z})$ sending a character $\chi \colon G \to \mathbf{Z}/n\mathbf{Z}$ to the class of $\chi_* P$ is an injection.*

*Proof* (1) ⇒ (2): Assume that $P$ is connected. Let $\chi$ be a non-trivial character and $G' \subset \mathbf{Z}/n\mathbf{Z}$ be the image. Then, since $\mathrm{H}^1(X, G') \to \mathrm{H}^1(X, \mathbf{Z}/n\mathbf{Z})$ is an injection and $X$ is connected, the class $\chi_*[P]$ of a connected $G'$-torsor defines a non-trivial element of $\mathrm{H}^1(X, \mathbf{Z}/n\mathbf{Z})$.

(2) $\Rightarrow$ (1): Let $P^\circ \subset P$ be a connected component of $P$ and let $G^\circ$ be its stabilizer. Then, since $X$ is connected, we have a bijection $G/G^\circ \to \pi_0(P)$ to the set of connected components. Hence, $P/G^\circ$ is a trivial $G/G^\circ$-torsor over $X$ and, for every character $\chi\colon G/G^\circ \to \mathbf{Z}/n\mathbf{Z}$, we have $\chi_*[P] = 0$. Since $[P]$ is an injection, we have $G/G^\circ = 0$. Namely, we have $G = G^\circ$ and $P = P^\circ$ as required. □

**Exercise 14.3** Let $k \to k'$ be a morphism of algebraically closed fields of characteristic $p > 0$, and let $t \in k'$ be a transcendental element over $k$. Define a linear mapping $V = k^2 \to V' = k'$ compatible with $k \to k'$ by sending the canonical basis $e_1, e_2 \in V$ to $1, t \in V' = k'$. Show that the inclusion $\mathrm{Im}(V \to \mathrm{H}^1(V^\vee, \mathbf{F}_p)) \subset$ (inverse image of $\mathrm{Im}(V' \to \mathrm{H}^1(V'^\vee, \mathbf{F}_p)))$ with respect to the commutative diagram

$$\begin{array}{ccc} V & \longrightarrow & V' \\ \downarrow & & \downarrow \\ \mathrm{H}^1(V^\vee, \mathbf{F}_p) & \longrightarrow & \mathrm{H}^1(V'^\vee, \mathbf{F}_p) \end{array}$$

is not an equality.

## 14.4 Etale Morphisms on the Closed Fibers

**Lemma 14.4.1** *Let $K$ be a discrete valuation field, and let $K'$ be an extension of the discrete valuation field $K$ of ramification index $e = e_{K'/K}$. Let $X$ be a flat scheme of finite type over $S' = \mathrm{Spec}\, O_{K'}$, and $D \subset X$ be the Cartier divisor defined by the ideal $\mathfrak{m}_{K'}^n O_X$ for an integer $n \geqslant 1$. Let $\mathcal{M} = \mathrm{Coker}(\mathcal{E} \to \mathcal{F})$ be the cokernel of a morphism $\mathcal{E} \to \mathcal{F}$ of locally free $O_X$-modules of the same finite rank.*

*Assume that there exists an open subscheme $U \subset X$, smooth over $S'$, such that the closed fiber $U_{F'}$ is non-empty and the restriction $\mathcal{E}|_U \to \mathcal{F}|_U$ is an injection that induces a morphism $\mathcal{F}|_U \to \mathcal{E}(D)|_U$ such that $\mathcal{F}|_U \otimes_{O_U} O_{U_{F'}} \to \mathcal{E}(D)|_U \otimes_{O_U} O_{U_{F'}}$ is the 0-morphism. Assume further that there exist another finite extension $L$ of the discrete valuation field $K$, a morphism $D \to T = \mathrm{Spec}\, O_L$ over $S = \mathrm{Spec}\, O_K$, an $O_L$-module $N$ of finite type, and an isomorphism $\mathcal{M} \otimes_{O_X} O_D \to N \otimes_{O_L} O_D$ of $O_D$-modules.*

*Assume $n > e = e_{K'/K}$.*

(1) *The ramification index $e_{L/K}$ divides $e_{K'/K}$. Let $m$ be the largest integer*

*satisfying* $m \cdot e_{K'/K}/e_{L/K} < n$. *Then the* $O_L$*-module* $N$ *is annihilated by* $\mathfrak{m}_L^m$.

(2) *There exists an open neighborhood* $W \subset X$ *of the closed fiber* $X_{F'}$ *such that the restriction* $\mathcal{M}|_W$ *is an* $O_W$*-module annihilated by* $\mathfrak{m}_{K'}^{n-1}$.

*Proof* (1) By replacing $K'$ by an unramified extension, we may assume that there exists a section $S' \to U$. By the assumption $n > e_{K'/K}$, the morphism $O_L \to O_{D\cap U} \to O_{K'}/\mathfrak{m}_{K'}^n$ over $O_K$ maps a uniformizer of $K$ to the class of an element $x$ satisfying $\mathrm{ord}_{K'}x = e_{K'/K}$. Hence, $e_{L/K}$ divides $e_{K'/K}$. Further, if $y \in O_L$ satisfies $a = \mathrm{ord}_L y \cdot e_{K'/K}/e_{L/K} < n$, its image is a generator of $\mathfrak{m}_{K'}^a/\mathfrak{m}_{K'}^n$.

The exact sequence $0 \to \mathcal{M}|_U \to \mathcal{E}(D)|_U/\mathcal{E}|_U \to \mathcal{E}(D)|_U/\mathcal{F}|_U \to 0$ of $O_U$-modules remains exact after $\otimes_{O_U} O_{K'}$ since the injection $\mathcal{F}|_U \to \mathcal{E}(D)|_U$ remains an injection after $\otimes_{O_U} O_{K'}$. By the assumption that $\mathcal{F}|_U \otimes_{O_U} O_{U_{F'}} \to \mathcal{E}(D)|_U \otimes_{O_U} O_{U_{F'}}$ is the 0-morphism, the surjection $\mathcal{E}(D)|_U \otimes_{O_U} F' \to \mathcal{E}(D)|_U/\mathcal{F}|_U \otimes_{O_U} F'$ is an isomorphism. Hence, the $O_{K'}$-module $\mathcal{M} \otimes_{O_X} O_{K'} \subset \mathcal{E}(D) \otimes_{O_X} O_{K'}/\mathfrak{m}_{K'}^n$ is annihilated by $\mathfrak{m}_{K'}^{n-1}$. The pull-back of the isomorphism $\mathcal{M} \otimes_{O_X} O_D \to N \otimes_{O_L} O_D$ defines an isomorphism $\mathcal{M} \otimes_{O_X} O_{K'}/\mathfrak{m}_{K'}^n \to N \otimes_{O_L} O_{K'}/\mathfrak{m}_{K'}^n$ of $O_{K'}$-modules. Hence, $N \otimes_{O_L} O_{K'}/\mathfrak{m}_{K'}^n$ is also annihilated by $\mathfrak{m}_{K'}^{n-1}$. Since $m$ is the largest integer satisfying $m \cdot e_{K'/K}/e_{L/K} < n$, the $O_L$-module $N$ is annihilated by $\mathfrak{m}_L^m$.

(2) By the isomorphism $\mathcal{M} \otimes_{O_X} O_D \to N \otimes_{O_L} O_D$, the $O_D$-module $\mathcal{M} \otimes_{O_X} O_D$ is annihilated by the ideal $\mathfrak{m}_{K'}^{n-1} O_X$. Since $\mathcal{M}/\mathfrak{m}_{K'}^n\mathcal{M}$ is annihilated by $\mathfrak{m}_{K'}^{n-1}$, we have $\mathfrak{m}_{K'}^{n-1}\mathcal{M} = \mathfrak{m}_{K'} \cdot \mathfrak{m}_{K'}^{n-1}\mathcal{M}$. Hence, by Nakayama's lemma applied to $\mathfrak{m}_{K'}^{n-1}\mathcal{M}$, there exists an open neighborhood $W$ of the closed fiber $X_{F'}$ where $\mathfrak{m}_{K'}^{n-1}\mathcal{M}|_W = 0$. □

We recall the Zariski–Nagata purity theorem.

**Theorem 14.4.2** ([32, Exposé X, Théorème 3.1]) *Let*

$$\begin{array}{ccc} X & \longleftarrow & U \\ \downarrow & & \downarrow \\ Y & \longleftarrow & V \end{array}$$

*be a cartesian diagram of normal noetherian schemes such that* $X \to Y$ *is quasi-finite and the horizontal arrows are dense open immersions. Assume that* $Y$ *is regular and* $V \subset Y$ *contains every point of codimension* 1. *Then, if the right vertical arrow* $U \to V$ *is finite étale, the morphism* $X \to Y$ *is étale.*

**Lemma 14.4.3** *Let* $K$ *be a henselian discrete valuation field, and* $L$ *be a finite separable extension of* $K$. *Let* $T = \mathrm{Spec}\, O_L \to Q$ *be an immersion to a smooth* *scheme over $S = \operatorname{Spec} O_K$ and $r > 0$ be a rational number. Let $K'$ be a finite extension of $K$ containing a Galois closure of $L$ over $K$ such that $n = e_{K'/K} r$ is an integer.*

(1) *Let $U \subset Q_{S'}^{(r)}$ be an open subscheme, smooth over $S'$. Then the composition $\Omega^1_{Q/S} \otimes_{O_Q} O_U \to \Omega^1_{U/S'} \to \Omega^1_{U_{F'}/F'}$ is the 0-morphism.*

(2) *Assume that $T = S$. Then $\Omega^1_{Q_{S'}^{(r)}/S'} = \mathfrak{m}_{K'}^{-n}(\Omega^1_{Q/S} \otimes_{O_Q} O_{Q_{S'}^{(r)}})$.*

*Proof* (1) By the assumption that $K'$ contains a Galois closure of $L$ over $K$, the residue fields of $E \otimes_F F'$ are isomorphic to $F'$. By Proposition 11.4.4(2) for $s = 0$, we have the commutative diagram

$$\begin{array}{ccc} Q_{S'}^{(r)} & \longleftarrow & U_{F'} \\ \downarrow & & \downarrow \\ Q_{S'} & \longleftarrow & \operatorname{Spec}(E \otimes_F F')_{\mathrm{red}} \end{array}$$

Hence, the composition factors through $\Omega^1_{(E \otimes_F F')_{\mathrm{red}}/F'} = 0$.

(2) Since the assertion is étale local on $Q$, by Lemma 11.4.2(1) we may assume that $Q = \mathbf{A}^N_S = \operatorname{Spec} O_K[X_1, \ldots, X_N]$ and that $S \to Q$ is the 0-section. Then $Q_{S'}^{(r)} = \mathbf{A}^N_{S'} = \operatorname{Spec} O_{K'}[X'_1, \ldots, X'_N]$ and the morphism $Q_{S'}^{(r)} \to Q$ is defined by sending $X_i$ to $t \cdot X'_i$ for an element $t \in O_{K'}$ of valuation $\operatorname{ord}_K t = n$. Since $\Omega^1_{Q/S}$ is a free $O_Q$-module of basis $dX_1, \ldots dX_N$, and $\Omega^1_{Q_{S'}^{(r)}/S'}$ is a free $O_{Q_{S'}^{(r)}}$-module of basis $dX'_1, \ldots dX'_N$, the assertion follows. □

**Theorem 14.4.4** *Let $L$ be a finite separable extension of a henselian discrete valuation field $K$, and $r > 0$ be a rational number. We consider the following conditions:*

(1) *The ramification of $L$ over $K$ is bounded by $r+$.*

(2) *Let*

$$\begin{array}{ccc} T & \longrightarrow & Q \\ \downarrow & & \downarrow \\ S & \longrightarrow & P \end{array} \qquad (14.4.1)$$

*be a cartesian diagram such that the horizontal arrows are immersions to smooth schemes over $S$ and the vertical arrows are quasi-finite and flat. Let $K'$ be an extension of $K$ containing a Galois closure of $L$ such that the stable integral model $Q_{S'}^{(r)}$ is defined. Then the morphism $Q_{S'}^{(r)} \to P_{S'}^{(r)}$ is étale on a neighborhood of the image of $T^{\natural}_{S'}$.*

(3) *Let $T = \operatorname{Spec} O_L \to Q$ be an immersion to a smooth scheme over $S =$*

*Spec $O_K$, and let $K'$ be as in (2). Then the finite morphism $Q^{(r)}_{F'} \to Q^{[r]}_{F',\mathrm{red}}$ is étale.*

*We have (1) ⇔ (2) ⇐ (3). If $r > 1$, the three conditions are equivalent. If the equivalent conditions (1) and (2) are satisfied, we have the cartesian diagram*

$$\begin{array}{ccc} T^{\natural}_{F'} & \longrightarrow & Q^{(r)}_{F'} \\ \downarrow & & \downarrow \\ \mathrm{Spec}\,(E \otimes_F F')_{\mathrm{red}} & \longrightarrow & Q^{[r]}_{F',\mathrm{red}} \end{array} \tag{14.4.2}$$

*Further, if $r > 1$, the $O_L$-module $\Omega^1_{O_L/O_K}$ is annihilated by $\mathfrak{m}_L^m$ for the largest integer $m$ satisfying $m < e_{L/K} \cdot r$.*

*Proof* By Lemma 11.3.3, there exists a cartesian diagram (14.4.1) satisfying the condition in (2). By Lemma 11.4.2, the assertion is étale local on $Q$. Hence, by Lemma 11.3.3, we may assume that the vertical arrows in (14.4.1) are finite and flat. Since the diagram (14.4.1) is cartesian and since $K'$ contains a Galois closure of $L$, the reduced part of $Q \times_P \mathrm{Spec}\, F' = T \times_S \mathrm{Spec}\, F'$ and $T^{\natural}_{F'}$ are étale over $F'$. Since $Q \to P$ is flat, we have $Q^{[r]}_{S'} = P^{[r]}_{S'} \times_P Q$ and the morphism $Q^{[r]}_{F',\mathrm{red}} \to P^{(r)}_{F'}$ is finite étale.

(1) ⇔ (2): Since the image of $T^{\natural}_{S'} \to Q^{(r)}_{S'}$ equals the inverse image of $S' \to P^{(r)}_{S'}$ by Lemma 11.4.3(2), condition (2) is equivalent to the equality $\#\mathrm{Im}(T^{\natural}_{F'} \to Q^{(r)}_{F'}) = [Q : P]$ by [31, Théorème (18.10.16)]. Since the diagram (14.4.1) is cartesian, the equality is equivalent to $\#F^{r+}(L) = \#F(L)$ and hence to (1).

The underlying diagram (14.4.2) of sets is cartesian by Lemma 11.4.3(2). Since the left vertical arrow is a morphism of schemes étale over $F'$, if the right vertical arrow is étale on a neighborhood of the image of $T^{\natural}_{S'}$, the diagram (14.4.2) of schemes is cartesian.

(3) ⇒ (2): Since $Q^{[r]}_{F',\mathrm{red}} \to P^{(r)}_{F'}$ is finite étale, condition (3) is equivalent to the condition that the finite morphism $Q^{(r)}_{S'} \to P^{(r)}_{S'}$ of flat schemes over $S'$ is étale on a neighborhood of $Q^{(r)}_{F'}$. Thus, we have the implication (3) ⇒ (2).

(2) ⇒ (3): We assume $r > 1$ and set $X = Q^{(r)}_{S'}$. We show that $\Omega^1_{O_L/O_K}$ is annihilated by $\mathfrak{m}_L^m$, and that the finite morphism $X = Q^{(r)}_{S'} \to P^{(r)}_{S'}$ is étale on a neighborhood of the closed fiber $Q^{(r)}_{F'}$. We assume that an open neighborhood $U$ of the image of $T^{\natural}_{S'} \to Q^{(r)}_{S'}$ is étale over $P^{(r)}_{S'}$.

We consider the $O_X$-module $\mathcal{M} = \Omega^1_{Q/P} \otimes_{O_Q} O_X = \mathrm{Coker}(\mathcal{E} \to \mathcal{F})$ for $\mathcal{E} = \Omega^1_{P/S} \otimes_{O_P} O_X$ and $\mathcal{F} = \Omega^1_{Q/S} \otimes_{O_Q} O_X$. To apply Lemma 14.4.1 to $\mathcal{M}$, we consider the commutative diagram

$$\begin{array}{ccc} \Omega^1_{Q/S} \otimes_{\mathcal{O}_Q} \mathcal{O}_X & \longrightarrow & \Omega^1_{X/S'} \\ \uparrow & & \uparrow \\ \Omega^1_{P/S} \otimes_{\mathcal{O}_P} \mathcal{O}_X & \longrightarrow & \Omega^1_{P^{(r)}_{S'}/S'} \otimes_{\mathcal{O}_{P^{(r)}_{S'}}} \mathcal{O}_X \end{array}$$

Since we assume $r > 1$, we have $n = e_{K'/K} r > e_{K'/K}$. Let $D \subset X$ be the Cartier divisor defined by the ideal sheaf $\mathfrak{m}^n_{K'} \mathcal{O}_X$. The lower horizontal arrow induces an isomorphism $\Omega^1_{P/S}(D) \otimes_{\mathcal{O}_P} \mathcal{O}_X \to \Omega^1_{P^{(r)}_{S'}/S'} \otimes_{\mathcal{O}_{P^{(r)}_{S'}}} \mathcal{O}_X$ by Lemma 14.4.3(2). The right vertical arrow is an isomorphism on $U \subset X = Q^{(r)}_{S'}$ since $U \to P^{(r)}_{S'}$ is étale. The upper horizontal arrow induces the 0-morphism on the pull-back to the closed fiber $U_{F'}$ by Lemma 14.4.3(1).

We consider the $\mathcal{O}_L$-module $N = \Omega^1_{\mathcal{O}_L/\mathcal{O}_K}$. Since the restriction of $X \to Q$ to $D$ factors through $T \subset Q$, we have an isomorphism $\Omega^1_{Q/P} \otimes_{\mathcal{O}_Q} \mathcal{O}_D \to \Omega^1_{\mathcal{O}_L/\mathcal{O}_K} \otimes_{\mathcal{O}_L} \mathcal{O}_D$ of $\mathcal{O}_D$-modules by the cartesian diagram (14.4.1). Thus, the assumptions in Lemma 14.4.1 are satisfied. Hence, the $\mathcal{O}_L$-module $\Omega^1_{\mathcal{O}_L/\mathcal{O}_K}$ is annihilated by $\mathfrak{m}^m_L$ by Lemma 14.4.1(1). Further, by Lemma 14.4.1(2), there exists an open neighborhood $W \subset X$ of the closed fiber $Q^{(r)}_{F'}$ such that $\Omega^1_{Q/P} \otimes_{\mathcal{O}_Q} \mathcal{O}_W$ is annihilated by $\mathfrak{m}^{n-1}_{K'}$.

Since $\Omega^1_{Q/P} \otimes_{\mathcal{O}_Q} \mathcal{O}_W$ is annihilated by $\mathfrak{m}^{n-1}_{K'}$, and since the morphism $X = Q^{(r)}_{S'} \to P^{(r)}_{S'}$ is flat on the generic fiber, the morphism $X \to P^{(r)}_{S'}$ is étale on $W - Q^{(r)}_{F'}$. Since $Q^{(r)}_{S'} \to P^{(r)}_{S'}$ is finite, the complement $V$ of the image of $Q^{(r)}_{S'} - W$ is a neighborhood of $P^{(r)}_{F'}$. Since $P^{(r)}_{S'}$ is smooth over $S'$ by Lemma 11.4.3(1), the scheme $P^{(r)}_{S'}$ is regular. Since $X = Q^{(r)}_{S'} \to P^{(r)}_{S'}$ is also étale on $U$, it is étale on $W$ by Theorem 14.4.2 applied to $V \cap P^{(r)}_{K'}$. □

**Corollary 14.4.5** *Let $r > 1$ be a rational number and assume that the ramification of $L$ over $K$ is bounded by $r+$. Assume that $K'$ is a Galois extension of $K$ containing $L$, and that the stable integral model $Q^{(r)}_{S'}$ is defined over $K'$. Then the action of the Galois group $G = \mathrm{Gal}(K'/K)$ on the stable closed fiber $Q^{(r)}_{F'}$ is trivial on the subgroup $G^{r+}$.*

*Proof* The action of $G_K$ on $N_{T/Q} \otimes_F F'$ is trivial on the inertia group $I \subset G$. Hence, the action of $I$ on $Q^{[r]}_{F',\mathrm{red}}$ is trivial. The morphism $Q^{(r)}_{F'} \to Q^{[r]}_{F',\mathrm{red}}$ is finite étale by Theorem 14.4.4, and $Q^{[r]}_{F',\mathrm{red}}$ is a vector bundle $(N_{T/Q} \otimes_F F') \otimes_{E \otimes_F F'} (E \otimes_F F')_{\mathrm{red}}$ over $T_{F',\mathrm{red}} = \mathrm{Spec}\,(E \otimes_F F')_{\mathrm{red}}$. Hence, the kernel of the action of $I$ on $Q^{(r)}_{F'}$ is the same as that on the fiber product $Q^{(r)}_{F'} \times_{Q^{[r]}_{F',\mathrm{red}}} T_{F',\mathrm{red}}$ with respect to the 0-section $T_{F',\mathrm{red}} \to Q^{[r]}_{F',\mathrm{red}}$. Since the fiber product is identified canonically with the finite set $T^{\natural}_{F'} = \mathrm{Mor}_K(L, K') = F(L)$, and since $F(L) \to F^{r+}(L)$ is a bijection by the assumption that the ramification is bounded by $r+ \geqslant 1$, the action of $G^{r+} \subset I = G^1$ is trivial on $Q^{(r)}_{F'}$. □

**Corollary 14.4.6** *Let $r > 1$ be a rational number. Then the following conditions are equivalent:*

(1) *The ramification of $L$ over $K$ is bounded by $r+$ but not by $r$.*
(2) *The finite morphism $Q^{(r)}_{F'} \to Q^{[r]}_{F',\mathrm{red}}$ is étale but does not split completely.*

*Proof* Since $r > 1$, the ramification of $L$ over $K$ is bounded by $r+$ if and only if the finite morphism $Q^{(r)}_{F'} \to Q^{[r]}_{F',\mathrm{red}}$ is étale by Theorem 14.4.4. Further, the ramification of $L$ over $K$ is bounded by $r$ if and only if the finite étale morphism $Q^{(r)}_{F'} \to Q^{[r]}_{F',\mathrm{red}}$ splits completely. □

By Lemma 11.1.5, we have a canonical isomorphism

$$Q^{[r]}_{F',\mathrm{red}} \to \mathrm{Hom}_F(N_{T/Q} \otimes_{O_T} E, \mathfrak{m}^{er}_{K'}/\mathfrak{m}^{er+1}_{K'}) \otimes_{E \otimes_F F'} (E \otimes_F F')_{\mathrm{red}}. \tag{14.4.3}$$

We will study the finite étale morphism $Q^{(r)}_{F'} \to Q^{[r]}_{F',\mathrm{red}}$ in Section 14.6. We give an explicit description in Section 15.3 in the case where $O_L$ is generated by a single element over $O_K$.

**Exercise 14.4** Let $k$ be a perfect field of characteristic $p > 0$, and $X = \mathrm{Spec}\, A$ be a smooth affine scheme over $k$. Let $D \subset X$ be an irreducible divisor, smooth over $k$, defined by $t \in A$, and $U = X - D = \mathrm{Spec}\, A_U$ for $A_U = A[1/t]$ be the complement. Let $n = pm \geqslant 1$ be an integer divisible by $p$, $a \in A$, and $V \to U$ be the Artin–Schreier covering defined by $x^p - x = f = a/t^n$. Assume that the normalization $Y$ of $X$ in $V$ is smooth over $k$. Let $K$ be the fraction field of the henselization of the local ring at the generic point of $D$, and assume that $V \times_U \mathrm{Spec}\, K = \mathrm{Spec}\, L$ for a ramified extension $L$ of $K$. Let $S = \mathrm{Spec}\, O_K$, $P = X \times_{\mathrm{Spec}\, k} S$, and let $T = \mathrm{Spec}\, O_L \to Q = Y \times_{\mathrm{Spec}\, k} S$ be the immersion to a smooth scheme over $S$ induced by the canonical morphism $T = \mathrm{Spec}\, O_L \to Y$.

Show that the ramification of $L$ over $K$ is bounded by $n+$ and compute the finite étale morphism $Q^{(n)}_E \to Q^{[n]}_{E,\mathrm{red}}$.

## 14.5 Functoriality of the Étale Morphisms

To obtain an étale covering of closed fibers independent of $Q$, we study the $E$-vector space $N_{T/Q} \otimes_{O_T} E$. Let $T = \mathrm{Spec}\, O_L \to Q$ be an immersion to a smooth scheme over $S = \mathrm{Spec}\, O_K$. Then we have an exact sequence

$$0 \longrightarrow N_{T/Q} \longrightarrow \Omega^1_{Q/S} \otimes_{O_Q} O_T \longrightarrow \Omega^1_{T/S} \longrightarrow 0 \tag{14.5.1}$$

of $O_T$-modules. Since this defines a free resolution of an $O_T$-module $\Omega^1_{T/S}$, we obtain an injection

$$\mathrm{Tor}_1^{O_T}(\Omega^1_{T/S}, E) \to N_{T/Q} \otimes_{O_T} E. \tag{14.5.2}$$

Let $K'$ be an extension of a henselian discrete valuation field $K$, let $L'$ be a finite separable extension of $K'$, and let $L \to L'$ be a morphism of fields compatible with $K \to K'$. Let

$$\begin{array}{ccc} T' = \mathrm{Spec}\, O_{L'} & \longrightarrow & Q' \\ \downarrow & & \downarrow \\ T = \mathrm{Spec}\, O_L & \longrightarrow & Q \end{array}$$

be a commutative diagram of schemes where the horizontal arrows are immersions to smooth schemes over $S$ and $S' = \mathrm{Spec}\, O_{K'}$ respectively and the vertical arrows are compatible with $S' \to S$. Then we have a commutative diagram

$$\begin{array}{ccc} \mathrm{Tor}_1^{O_T}(\Omega^1_{T/S}, E) & \longrightarrow & N_{T/Q} \otimes_{O_T} E \\ \downarrow & & \downarrow \\ \mathrm{Tor}_1^{O_{T'}}(\Omega^1_{T'/S'}, E') & \longrightarrow & N_{T'/Q'} \otimes_{O_{T'}} E' \end{array} \tag{14.5.3}$$

**Lemma 14.5.1** *Let $T = \mathrm{Spec}\, O_L \to Q$ be an immersion to a smooth scheme over $S = \mathrm{Spec}\, O_K$. Then the following conditions are equivalent:*

(1) *The surjection $\Omega^1_{Q/S} \otimes_{O_Q} O_T \to \Omega^1_{T/S}$ induces an isomorphism $\Omega^1_{Q/S} \otimes_{O_Q} E \to \Omega^1_{T/S} \otimes_{O_T} E$.*
(2) *The injection $\mathrm{Tor}_1^{O_T}(\Omega^1_{T/S}, E) \to N_{T/Q} \otimes_{O_T} E$ (14.5.2) is an isomorphism.*
(3) *The morphism $N_{T/Q} \otimes_{O_T} E \to \Omega^1_{Q/S} \otimes_{O_Q} E$ is 0.*

*Proof* The assertion follows from the exact sequence $0 \to \mathrm{Tor}_1^{O_T}(\Omega^1_{T/S}, E) \to N_{T/Q} \otimes_{O_T} E \to \Omega^1_{Q/S} \otimes_{O_Q} E \to \Omega^1_{T/S} \otimes_{O_T} E \to 0$ defined by the distinguished triangle $L_{Q/S} \otimes_{O_Q} O_T \to L_{T/S} \to L_{T/Q} \to$. □

**Definition 14.5.2** Let $L$ be a finite separable extension of $K$.

(1) Let $r > 1$ be a rational number. For an extension $K'$ of a discrete valuation field $K$ with residue field $F'$ and of ramification index $e = e_{K'/K}$ such that $er$ is an integer, we define $\Theta^{(r)}_{L/K,F'}$ to be the vector bundle over $\mathrm{Spec}\,(E \otimes_F F')_{\mathrm{red}}$ defined by the $E \otimes_F F'$-module $\mathrm{Tor}_1^{O_L}(\Omega^1_{O_L/O_K}, E) \otimes_F \mathfrak{m}_{K'}^{-er}/\mathfrak{m}_{K'}^{-er+1}$.
(2) We say that an immersion $T \to Q$ to a smooth scheme over $S = \mathrm{Spec}\, O_K$ is minimal if the equivalent conditions in Lemma 14.5.1 are satisfied.

**Lemma 14.5.3** *Let L be a finite separable extension of K.*

(1) *There exists a minimal immersion $T \to Q$ to a smooth scheme over $S = \operatorname{Spec} O_K$.*

(2) *Let $T \to Q$ and $T \to Q'$ be immersions to smooth schemes over $S = \operatorname{Spec} O_K$, and assume that $T \to Q$ is minimal. Let $Q' \to Q$ be a morphism over $S$ compatible with the immersions of $T$. Then $Q' \to Q$ is smooth on a neighborhood of $T$.*

*Proof* (1) We take a system of generators of $O_L$ over $O_K$ and define an immersion $T \to Q' = \mathbf{A}^n_S = \operatorname{Spec} O_K[X_1, \ldots, X_n]$. Let $I \subset O_K[X_1, \ldots, X_n]$ be the kernel of the surjection $O_K[X_1, \ldots, X_n] \to O_L$ and take a lifting $f_1, \ldots, f_r \in I$ of a basis of the image of $I \otimes_{O_K[X_1,\ldots,X_n]} E = N_{T/Q'} \otimes_{O_T} E \to \Omega^1_{Q'/S} \otimes_{O_{Q'}} E$. Then the closed subscheme $Q$ of $Q'$ defined by the ideal $(f_1, \ldots, f_r)$ is smooth over $S$ on a neighborhood of $T$. Since $T \to Q$ satisfies condition (3) in Lemma 14.5.1, it suffices to replace $Q$ by a neighborhood of $T$.

(2) By condition (1) in Lemma 14.5.1, the morphism $\Omega^1_{Q/S} \otimes_{O_Q} E \to \Omega^1_{Q'/S} \otimes_{O_{Q'}} E$ is an injection. Hence, $Q' \to Q$ is smooth on a neighborhood of $T$. □

**Lemma 14.5.4** *Let $L$ be a finite separable extension of $K$, and let $T = \operatorname{Spec} O_L \to Q$ be a minimal immersion to a smooth scheme over $S = \operatorname{Spec} O_K$. Let $E$ and $F$ be the residue fields of $L$ and $K$.*

(1) *Assume that $E$ is a separable extension of $F$, and $e_{L/K} \neq 1$. Then there exists an isomorphism*

$$N_{T/Q} \otimes_{O_T} E \to \mathfrak{m}_K/\mathfrak{m}_K^2 \otimes_F E \tag{14.5.4}$$

*of $E$-vector spaces of dimension 1 fitting in the commutative diagram*

$$\begin{array}{ccc} N_{T/Q} \otimes_{O_T} E & \longrightarrow & \mathfrak{m}_K/\mathfrak{m}_K^2 \otimes_F E \\ {\scriptstyle (14.5.2)}\uparrow & & \downarrow \\ \operatorname{Tor}_1^{O_T}(\Omega^1_{T/S}, E) & \xrightarrow{(14.2.2)} & \mathrm{H}_1(L_{E/S}) \end{array}$$

(2) *Assume $e_{L/K} = 1$, and set $Z_{F\backslash E} = \operatorname{Ker}(\Omega^1_F \to \Omega^1_E)$. Then there exists an isomorphism*

$$N_{T/Q} \otimes_{O_T} E \to Z_{F\backslash E} \otimes_F E \tag{14.5.5}$$

*fitting in the commutative diagram*

$$\begin{array}{ccc} N_{T/Q} \otimes_{O_T} E & \longrightarrow & Z_{F\backslash E} \otimes_F E \\ {\scriptstyle (14.5.2)}\uparrow & & \downarrow{\scriptstyle (14.1.5)} \\ \operatorname{Tor}_1^{O_T}(\Omega^1_{T/S}, E) & \xrightarrow{(14.2.3)} & \mathrm{H}_1(L_{E/F}) \end{array}$$

*Proof* By condition (2) in Lemma 14.5.1, the morphism $\mathrm{Tor}_1^{\mathcal{O}_T}(\Omega^1_{T/S}, E) \to N_{T/Q} \otimes_{\mathcal{O}_T} E$ (14.5.2) is an isomorphism.

(1) By Proposition 14.2.2(2) and its proof, the morphism $\mathrm{Tor}_1^{\mathcal{O}_L}(\Omega^1_{T/S}, E) \to \mathrm{H}_1(L_{E/S})$ (14.2.2) and the first arrow $\mathfrak{m}_K/\mathfrak{m}_K^2 \otimes_F E \to \mathrm{H}_1(L_{E/S})$ in (14.1.6) are isomorphisms of $E$-vector spaces of dimension 1. Hence, we obtain an isomorphism (14.5.4).

(2) By Proposition 14.2.2(3), the morphism $\mathrm{Tor}_1^{\mathcal{O}_L}(\Omega^1_{T/S}, E) \to \mathrm{H}_1(L_{E/F})$ (14.2.3) is an isomorphism. Hence, we obtain (14.5.5). □

**Proposition 14.5.5** *Let $L$ be a finite separable extension of $K$. Let $r > 1$ be a rational number such that the ramification of $L$ over $K$ is bounded by $r+$. Let $K'$ be an extension of $K$ such that the stable integral model $Q^{(r)}_{S'}$ is defined. Then there exists a unique cartesian diagram*

$$\begin{array}{ccc} T^{\natural}_{F'} & \longrightarrow & \Phi^{(r)}_{L/K,F'} \\ \downarrow & & \downarrow \\ \mathrm{Spec}\,(E \otimes_F F')_{\mathrm{red}} & \xrightarrow{0} & \Theta^{(r)}_{L/K,F'} \end{array} \qquad (14.5.6)$$

*of schemes over $(E\otimes_F F')_{\mathrm{red}}$ up to a unique isomorphism satisfying the following properties: The vertical arrows are finite étale and the bottom horizontal arrow is the 0-section of a vector bundle. For every immersion $T \to Q$ to a smooth scheme over $S$ and the stable closed fiber $Q^{(r)}_{F'}$, the diagram (14.5.6) is divided into a cartesian diagram*

$$\begin{array}{ccccc} T^{\natural}_{F'} & \longrightarrow & Q^{(r)}_{F'} & \longrightarrow & \Phi^{(r)}_{L/K,F'} \\ \downarrow & & \downarrow & & \downarrow \\ \mathrm{Spec}\,(E \otimes_F F')_{\mathrm{red}} & \longrightarrow & Q^{[r]}_{F',\mathrm{red}} & \longrightarrow & \Theta^{(r)}_{L/K,F'} \end{array} \qquad (14.5.7)$$

*where the lower right horizontal arrow is the smooth morphism induced by the injection $\mathrm{Tor}_1^{\mathcal{O}_L}(\Omega^1_{T/S}, E) \to N_{T/Q} \otimes_{\mathcal{O}_T} E$ (14.5.2).*

*Proof* We may take a minimal immersion $T \to Q_0$ to a smooth scheme over $S$. Then, by condition (1) in Lemma 14.5.1, the morphism $Q^{[r]}_{0,F',\mathrm{red}} \to \Theta^{(r)}_{L/K,F'}$ is an isomorphism. We define a finite étale scheme $\Phi^{(r)}_{L/K,F'} \to \Theta^{(r)}_{L/K,F'}$ by the right cartesian square in (14.5.7) with $Q$ replaced by $Q_0$. Then we obtain a cartesian diagram (14.5.6) by Theorem 14.4.4. This is a unique cartesian diagram up to unique isomorphism satisfying the condition for $T \to Q_0$.

We show that this satisfies the condition for any immersion $T \to Q$ to a smooth scheme over $S$. The left square of (14.5.7) is cartesian by Theorem 14.4.4. By Lemma 11.4.2(1), we may replace $Q$ by an étale neighborhood of $Q$. By Lemma 11.3.2, we may take a morphism $Q \to Q_0$ compatible with the immersion of $T$. Then, by the functoriality of the construction, we obtain the right square of (14.5.7). □

Let $L$ be a finite separable extension of a henselian discrete valuation field $K$, and let $r > 1$ be a rational number such that the ramification of $L$ over $K$ is bounded by $r+$. In the cartesian diagram (14.5.6), we identify $\operatorname{Spec}(E \otimes_F F')_{\mathrm{red}} = \operatorname{Mor}_F(E, F')$ and $T^{\natural}_{\overline{F}} = \operatorname{Mor}_K(L, K')$ with the sets of morphisms over $F$ and over $K$ respectively. Then we obtain a cartesian diagram

$$\begin{array}{ccc} \operatorname{Mor}_K(L, K') & \longrightarrow & \Phi^{(r)}_{L/K,F'} \\ \downarrow & & \downarrow \\ \operatorname{Mor}_F(E, F') & \longrightarrow & \Theta^{(r)}_{L/K,F'} \end{array} \tag{14.5.8}$$

The lower line induces a bijection $\operatorname{Mor}_F(E, F') \to \pi_0(\Theta^{(r)}_{L/K,F'})$ to the set of connected components.

If $K'$ is a Galois extension of $K$, the commutative diagram (14.5.8) is compatible with the action of the Galois group $\operatorname{Gal}(K'/K)$. In particular, we have canonical actions of $\operatorname{Gal}(K'/K)$ on $\Theta^{(r)}_{L/K,F'}$ and $\Phi^{(r)}_{L/K,F'}$. We call these actions the arithmetic actions to distinguish them from another action defined below. The inertia group $I' \subset \operatorname{Gal}(K'/K)$ acts trivially on the set $\operatorname{Mor}_F(E, F') = \pi_0(\Theta^{(r)}_{L/K,F'})$ of connected components. Since the wild inertia group $P' \subset \operatorname{Gal}(K'/K)$ acts trivially on $\mathfrak{m}^{er}_{K'}/\mathfrak{m}^{er+1}_{K'}$, the action of $P'$ on $\Theta^{(r)}_{L/K,F'} = \operatorname{Hom}_F(\operatorname{Tor}_1^{O_L}(\Omega^1_{O_L/O_K}, E), \mathfrak{m}^{er}_{K'}/\mathfrak{m}^{er+1}_{K'})$ is trivial. By Corollary 14.4.5, the actions are trivial on the subgroup $\operatorname{Gal}(K'/K)^{r+} \subset \operatorname{Gal}(K'/K)$.

For a rational number $r > 1$, let $C_K^{\leqslant r+}$ be the category of finite separable extensions $L$ such that the ramification is bounded by $r+$. The category $C_K^{\leqslant r+}$ is equivalent to the category of finite sets with continuous action of the quotient $G_K/G_K^{r+}$ of $G_K = \operatorname{Gal}(K_s/K)$ that are connected with respect to the action. By attaching $\Phi^{(r)}_{L/K,\overline{F}} \to \Theta^{(r)}_{L/K,\overline{F}}$ to $L$, we obtain a functor from $C_K^{\leqslant r+}$ to the category of finite étale morphisms with continuous action of $G_K/G_K^{r+}$.

Assume that $L$ is a Galois extension of $K$. Then, by the functoriality of the construction of $\Phi^{(r)}_{L/K,F'} \to \Theta^{(r)}_{L/K,F'}$, we have another action of the Galois group $G = \operatorname{Gal}(L/K)$ on $\Phi^{(r)}_{L/K,F'}$ and $\Theta^{(r)}_{L/K,F'}$. We call this action the geometric action. The Galois group $G$ acts on the cartesian diagram (14.5.8) where $\operatorname{Mor}_K(L, K')$ is a $G$-torsor. Since the inertia group $I \subset G$ acts trivially on $\operatorname{Tor}_1^{O_L}(\Omega^1_{O_L/O_K}, E)$ by Corollary 14.2.3, the action of $I \subset G$ on

$$\Theta^{(r)}_{L/K,F'} = \operatorname{Hom}_F(\operatorname{Tor}_1^{O_L}(\Omega^1_{O_L/O_K}, E), \mathfrak{m}^{er}_{K'}/\mathfrak{m}^{er+1}_{K'})$$

is trivial.

Note that the arithmetic action and the geometric action are both right actions since the Galois groups act on the extensions on the left and the functor Spec is contravariant.

**Lemma 14.5.6** *Let $L$ be a Galois extension of a henselian discrete valuation field $K$ of Galois group $G = \mathrm{Gal}(L/K)$. Let $K'$ be another Galois extension of $K$ of Galois group $G' = \mathrm{Gal}(K'/K)$ of ramification index $e$ containing $L$ as a subfield. Let $r > 1$ be a rational number such that the ramification of $L$ over $K$ is bounded by $r+$. Assume that $er$ is an integer, and that the stable integral model $Q^{(r)}_{S'}$ is defined over $S' = \mathrm{Spec}\, O_{K'}$.*

(1) *The arithmetic action and the geometric action are commutative to each other.*
(2) *Under the identification $T^{\natural}_{F'} = \mathrm{Mor}_K(L, K')$ and the canonical morphism $T^{\natural}_{F'} \to \Theta^{(r)}_{L/K,F'}$, the geometric action of $\mathrm{Gal}(L/K)$ is compatible with the action on $\mathrm{Mor}_K(L, K')$ defined as the composition $\mathrm{Gal}(L/K) \times \mathrm{Mor}_K(L, K') \to \mathrm{Mor}_K(L, K')$. The arithmetic action of $\mathrm{Gal}(K'/K)$ is compatible with the action on $\mathrm{Mor}_K(L, K')$ defined as the composition $\mathrm{Gal}(K'/K) \times \mathrm{Mor}_K(L, K') \to \mathrm{Mor}_K(L, K')$ of the inverse.*

*Proof* (1) The assertion follows from the commutative diagram (11.4.4).

(2) Since the geometric action of $G = \mathrm{Gal}(L/K)$ is compatible with its action on $T = \mathrm{Spec}\, O_L$, we obtain the first compatibility. To show the compatibility for the arithmetic action, we may assume that $Q = \mathrm{Spec}\, A$ is affine. Then the arithmetic action is induced by the action of $\mathrm{Gal}(K'/K)$ on $A \otimes_K K'$. Let $f : A \to K'$ be a morphism over $O_K$, and $\sigma \in \mathrm{Gal}(K'/K)$. Then, since the diagram

$$\begin{array}{ccc} A \otimes_K K' & \xrightarrow{(f, 1_{K'})} & K' \\ {\scriptstyle 1\otimes\sigma^{-1}}\uparrow & & \downarrow{\scriptstyle \sigma} \\ A \otimes_K K' & \xrightarrow{(f, 1_{K'})} & K' \end{array}$$

is commutative, the right arithmetic action is compatible with the composition with the inverse. □

**Exercise 14.5** Let $K$ be a henselian discrete valuation with residue field $F$ of characteristic $p > 0$. Let $L$ be a cyclic extension of degree $p$ of $K$, and assume that the residue field $E$ is a purely inseparable extension of degree $p$ of $F$. Let $n = \mathrm{Sw}\,\chi \geqslant 1$ be the Swan conductor of a faithful character $\chi$ of $G = \mathrm{Gal}(L/K)$. Show that the finite étale covering $\Phi^{(n)}_{L/K,E} \to \Theta^{(n)}_{L/K,E}$ is defined by an Artin–Schreier covering of an affine line.

## 14.6 Graded Quotients

Let $L$ be a finite separable extension of a henselian discrete valuation field $K$, and let $r > 1$ be a rational number such that the ramification of $L$ over $K$ is bounded by $r+$. Let $T = \operatorname{Spec} O_L \to Q$ be an immersion to a smooth scheme over $S = \operatorname{Spec} O_K$, and let $K'$ be a finite separable extension of $K$ of ramification index $e$ such that the stable integral model $Q^{(r)}_{S'}$ is defined. We fix a morphism $L \to K'$ over $K$ and consider the corresponding point of $T^{\natural}_{\bar{F}}$. We regard its images by $T^{\natural}_{\bar{F}} \to \Phi^{(r)}_{L/K,F'} \to \Theta^{(r)}_{L/K,F'}$ as a geometric point $0 \in \Theta^{(r)}_{L/K,F'}$ and its lifting to $\Phi^{(r)}_{L/K,F'}$. Let $\Theta^{(r)\circ}_{L/K,F'}$ and $\Phi^{(r)\circ}_{L/K,F'}$ be the connected components containing 0 and the lifting. The scheme $\Theta^{(r)\circ}_{L/K,F'}$ is the dual of the $F'$-vector space $\operatorname{Tor}_1^{O_L}(\Omega^1_{O_L/O_K}, \mathfrak{m}^{er}_{K'}/\mathfrak{m}^{er+1}_{K'})$ with respect to the composition $O_L \to E \to F'$ induced by $L \to K'$.

**Proposition 14.6.1** *Assume that $L$ is a Galois extension of $K$ of Galois group $G$, and let $r > 1$ be a rational number such that the ramification of $L$ over $K$ is bounded by $r+$.*

(1) *By the geometric action of $G$ on $\Phi^{(r)}_{L/K,F'}$, the connected component $\Phi^{(r)\circ}_{L/K,F'}$ is a $G^r$-torsor over $\Theta^{(r)\circ}_{L/K,F'}$.*
(2) *The group $G^r$ is an abelian $p$-group.*

*Proof* (1) The fiber $\Phi^{(r)\circ}_{L/K,F'} \times_{\Theta^{(r)\circ}_{L/K,F'}} 0$ is canonically identified with the intersection $\Phi^{(r)\circ}_{L/K,F'} \cap \operatorname{Mor}_K(L, K')$. Since the latter is a $G^r$-torsor, the connected scheme $\Phi^{(r)\circ}_{L/K,F'}$ itself is a $G^r$-torsor over $\Theta^{(r)\circ}_{L/K,F'}$.

(2) By (1), we have an isomorphism

$$G^r \to \operatorname{Aut}(\Phi^{(r)\circ}_{L/K,F'}/\Theta^{(r)\circ}_{L/K,F'}). \tag{14.6.1}$$

We may assume that $K'$ is a Galois extension of $K$ containing $L$ as a subfield. Then, by the arithmetic action, the ramification group $\operatorname{Gal}(K'/K)^r$ stabilizes the connected components $\Theta^{(r)\circ}_{L/K,F'}$ and $\Phi^{(r)\circ}_{L/K,F'}$ by Lemma 14.5.6(2). Hence, we obtain a morphism

$$\operatorname{Gal}(K'/K)^r \to \operatorname{Aut}(\Phi^{(r)\circ}_{L/K,F'}/\Theta^{(r)\circ}_{L/K,F'}). \tag{14.6.2}$$

Since the surjection $\operatorname{Gal}(K'/K) \to \operatorname{Gal}(L/K) = G$ induces a surjection $\operatorname{Gal}(K'/K)^r \to G^r$ by Corollary 12.3.3, the group $\operatorname{Gal}(K'/K)^r$ acts transitively on the fiber $\Phi^{(r)\circ}_{L/K,F'} \times_{\Theta^{(r)\circ}_{L/K,F'}} 0$ by Lemma 14.5.6(2) again. Hence, the morphism (14.6.2) is a surjection. Since the arithmetic action and the geometric action commute with each other by Lemma 14.5.6(1), and since (14.6.2) is a surjection and (14.6.1) is an isomorphism, $G^r$ is commutative.

The group $G^r$ is a $p$-group by Proposition 12.2.2. □

**Corollary 14.6.2** *Let $L$ be a finite Galois extension of $K$, and $r > 1$ be a rational number such that the ramification of $L$ over $K$ is bounded by $r+$. Let $G^{r\vee} = \mathrm{Hom}(G^r, \mathbf{Q}_p/\mathbf{Z}_p)$ denote the dual group of the finite abelian $p$-group $G^r$.*

(1) *The $G^r$-torsor $\Phi^{(r)\circ}_{L/K,\overline{F}}$ over $\Theta^{(r)\circ}_{L/K,\overline{F}}$ defines an injection*

$$G^{r\vee} \longrightarrow \mathrm{H}^1(\Theta^{(r)\circ}_{L/K,\overline{F}}, \mathbf{Q}_p/\mathbf{Z}_p). \tag{14.6.3}$$

(2) *Let $K'$ be a finite Galois extension of $K$ containing $L$ as a subfield such that the stable integral model $Q^{(r)}_{S'}$ is defined. Let $\mathrm{Gal}(K'/K) \to G = \mathrm{Gal}(L/K)$ act on $G^{r\vee}$ by the conjugate action on $G^r$. Let $\mathrm{Gal}(K'/K) \to G = \mathrm{Gal}(L/K)$ act diagonally on $\Theta^{(r)\circ}_{L/K,\overline{F}}$ by the composite of the mutually commutative arithmetic action and the geometric action. Then the morphism (14.6.3) is compatible with the actions of $\mathrm{Gal}(K'/K)$.*

*Proof* (1) Since $\Phi^{(r)\circ}_{L/K,\overline{F}}$ is a connected $G^r$-torsor by Proposition 14.6.1(1), the assertion follows from Lemma 14.3.3.

(2) Let $\sigma \in \mathrm{Gal}(K'/K)$ and $\tau \in G^r$, and let $\tau_*, \tau^*$ denote the arithmetic and geometric actions on $\Phi^{(r)\circ}_{L/K,\overline{F}}$ respectively. Then, since $\sigma_*$ and $\sigma^*$ commute with each other by Lemma 14.5.6, the diagram

$$\begin{array}{ccc} \Phi^{(r)\circ}_{L/K,\overline{F}} & \xrightarrow{\sigma_*\sigma^{*-1}} & \Phi^{(r)\circ}_{L/K,\overline{F}} \\ {\scriptstyle \tau^*}\big\downarrow & & \big\downarrow{\scriptstyle (\sigma\tau\sigma^{-1})^*} \\ \Phi^{(r)\circ}_{L/K,\overline{F}} & \xrightarrow{\sigma_*\sigma^{*-1}} & \Phi^{(r)\circ}_{L/K,\overline{F}} \end{array}$$

is commutative. Hence, the assertion follows. □

**Corollary 14.6.3** *Let $L$ be a finite Galois extension of $K$ of Galois group $G$, and $r > 1$ be a rational number. Then $[P, G^r] \subset G^{r+}$.*

*Proof* Since the upper ramification group is compatible with quotient by Corollary 12.3.3, by replacing $L$ by the subextension corresponding to $G^{r+}$, we may assume $G^{r+} = 1$. Let the notation be as in Corollary 14.6.2(2). Then the action of the wild ramification subgroup $P' \subset \mathrm{Gal}(K'/K)$ on the target $\mathrm{H}^1(\Theta^{(r)\circ}_{L/K,\overline{F}}, \mathbf{Q}_p/\mathbf{Z}_p)$ of (14.6.3) is trivial. Since (14.6.3) is an injection, the action on $G^{r\vee}$ is also trivial. Since $G^r$ is a finite abelian group, the assertion follows. □

**Lemma 14.6.4** *Let $L$ be a finite Galois extension of $K$, and $r > 0$ be a rational* *number such that the ramification of L over K is bounded by r+. Let K′ be an extension of discrete valuation field K of ramification index e, and let $L' = LK'$ be a composition field and $r' = er$. Let $G = \mathrm{Gal}(L/K)$ and $G' = \mathrm{Gal}(L'/K')$ be the Galois groups and let $\overline{F} \to \overline{F}'$ be a morphism of algebraic closures compatible with $E \to E'$. If the morphism*

$$\mathrm{Tor}_1^{O_L}(\Omega^1_{O_L/O_K}, E) \to \mathrm{Tor}_1^{O_{L'}}(\Omega^1_{O_{L'}/O_{K'}}, E') \tag{14.6.4}$$

*induces an injection $\mathrm{S}_{\overline{F}}\mathrm{Tor}_1^{O_L}(\Omega^1_{O_L/O_K}, \overline{F}) \to \mathrm{S}_{\overline{F}'}\mathrm{Tor}_1^{O_{L'}}(\Omega^1_{O_{L'}/O_{K'}}, \overline{F}')$ on the symmetric algebras, the canonical injection $G' \to G$ induces an isomorphism $G'^{r'} \to G^r$.*

*Proof* Let $G^{r\vee} = \mathrm{Hom}(G^r, \mathbf{Q}_p/\mathbf{Z}_p)$ and $G'^{r'\vee} = \mathrm{Hom}(G'^{r'}, \mathbf{Q}_p/\mathbf{Z}_p)$ denote the dual groups of the finite abelian $p$-groups $G^r$ and $G'^{r'}$. Then the morphism

$$\begin{array}{ccc} \Phi^{(r')\circ}_{L'/K',\overline{F}'} & \longrightarrow & \Theta^{(r')\circ}_{L'/K',\overline{F}'} \\ \downarrow & & \downarrow \\ \Phi^{(r)\circ}_{L/K,\overline{F}} & \longrightarrow & \Theta^{(r)\circ}_{L/K,\overline{F}} \end{array}$$

from a $G'^{r'}$-torsor to a $G^r$-torsor compatible with $G'^{r'} \to G^r$ defines a commutative diagram

$$\begin{array}{ccc} G'^{r'\vee} & \longrightarrow & \mathrm{H}^1(\Theta^{(r')\circ}_{L'/K',\overline{F}'}, \mathbf{Q}_p/\mathbf{Z}_p) \\ \uparrow & & \uparrow \\ G^{r\vee} & \longrightarrow & \mathrm{H}^1(\Theta^{(r)\circ}_{L/K,\overline{F}}, \mathbf{Q}_p/\mathbf{Z}_p) \end{array}$$

of injections (14.6.3). By Corollary 14.3.2, the injectivity of the symmetric algebras means that the right vertical arrow is an injection. Hence, the left vertical arrow is also an injection. Thus, the injection $G'^{r'} \to G^r$ is a surjection. □

Lemma 14.6.4 asserts that the ramification groups of $G'$ determine the ramification groups of $G$. Lemma 14.1.7 asserts that we may take a discrete valuation field with perfect residue field as $K'$ in Lemma 14.6.4. They give a characterization of ramification groups by reduction to the classical case where the residue field is perfect [77].

**Theorem 14.6.5** *Let L be a finite Galois extension of a henselian discrete valuation field K of Galois group $G = \mathrm{Gal}(L/K)$, and let $r > 1$ be a rational number. Then the graded quotient $\mathrm{Gr}^r G = G^r/G^{r+}$ is abelian and is an $\mathbf{F}_p$-vector space.*

*Proof* Since the upper ramification group is compatible with quotients by Corollary 12.3.3, by replacing $L$ by the subextension corresponding to $G^{r+}$, we may assume that $\mathrm{Gr}^r G = G^r$. By Proposition 14.6.1(2), the last piece $G^r$ of the filtration is an abelian $p$-group.

Let $K'$ be an extension as in Lemma 14.1.7, and let $L' = LK'$ be the composition field. Then, since the morphism $\mathrm{S}_{\overline{F}}\mathrm{H}_1(L_{\overline{F}/O_K}) \to \mathrm{S}_{\overline{F}'}\mathrm{H}_1(L_{\overline{F}'/O'_K})$ of symmetric algebras is an injection, the morphism $\mathrm{S}_{\overline{F}}\mathrm{Tor}_1^{O_L}(\Omega^1_{O_L/O_K}, E) \to \mathrm{S}_{\overline{F}'}\mathrm{Tor}_1^{O_{L'}}(\Omega^1_{O_{L'}/O_{K'}}, E')$ is also an injection by Proposition 14.2.2(1). Hence, by Lemma 14.6.4, the morphism $G'^r \to G^r$ is an isomorphism.

By Corollary 11.5.9, the upper ramification group $G'^r$ equals the lower ramification group $\mathrm{Ker}(G' \to \mathrm{Aut}(O_{L'}/\mathfrak{m}^s_{L'}))$ and is an $\mathbf{F}_p$-vector space for the integer satisfying $r = \varphi(s)$. Hence, the last graded piece $G^r$ is also an $\mathbf{F}_p$-vector space. □

**Corollary 14.6.6** *The graded quotient* $\mathrm{Gr}^r_{\log} G$ *is abelian and is an* $\mathbf{F}_p$*-vector space.*

*Proof* Let $m = e_{L/K}$ be the ramification index, and let $K'$ be a log smooth extension of $K$ of ramification index $e_{K'/K}$ divisible by $m$. Using Proposition 13.3.6(2), by replacing $L, K$ by $L' = LK', K'$, we may assume that $m = 1$. Then $\mathrm{Gr}^r_{\log} G = \mathrm{Gr}^r G$ and the assertion follows from Theorem 14.6.5. □

We define a generalization of a conductor and prove the integrality. Let $L$ be a finite Galois extension of a henselian discrete valuation field $K$ of Galois group $G = \mathrm{Gal}(L/K)$. Let $\Lambda$ be a field of characteristic different from the characteristic $p$ of the residue field of $K$, and let $V$ be a representation of $G$ on a $\Lambda$-vector space of finite dimension. Then, since $P = G^{1+}$ is a finite $p$-group, we have a unique decomposition

$$V = \bigoplus_{r \geqslant 1} V^{(r)} \tag{14.6.5}$$

such that $V^{G^{r+}} = \bigoplus_{1 \leqslant s \leqslant r} V^{(s)}$ for every rational number $r \geqslant 1$.

**Definition 14.6.7** Let $L$ be a finite Galois extension of a henselian discrete valuation field $K$ of Galois group $G = \mathrm{Gal}(L/K)$. Let $\Lambda$ be a field of characteristic different from the characteristic of the residue field of $K$. Let $V$ be a representation of $G$ on a $\Lambda$-vector space of finite dimension; we consider the decomposition (14.6.5). Then we define the total dimension $\mathrm{dimtot}\, V \in \mathbf{Q}$ by $\mathrm{dimtot}\, V = \sum_{r \geqslant 1} r \cdot \dim V^{(r)}$.

**Lemma 14.6.8** (1) *We have* $\mathrm{dimtot}\, V \geqslant \dim V$. *The equality is equivalent to the condition that the action of* $P$ *on* $V$ *is trivial.*

(2) *If the residue field extension is separable, we have* $\operatorname{dimtot} V = \dim V + \operatorname{Sw} V \in \mathbf{N}$.

*Proof* (1) We have $\operatorname{dimtot} V = \sum_{r\geqslant 1} r \cdot \dim V^{(r)} \geqslant \sum_{r\geqslant 1} \dim V^{(r)} = \dim V$. The equality is equivalent to $V = V^{(1)} = V^P$.

(2) If the residue field extension is separable, we have $G^r_{\log} = G^{r+1}$ by Lemma 13.3.5. Hence, we have $\operatorname{dimtot} V = \dim V + \operatorname{Sw} V \in \mathbf{Z}$. Since $\operatorname{Sw} V$ is an integer by Theorem 6.4.2(1), the sum $\operatorname{dimtot} V = \dim V + \operatorname{Sw} V$ is also an integer. □

**Theorem 14.6.9** *The total dimension* $\operatorname{dimtot} V$ *is an integer.*

This is a generalization of the Hasse–Arf theorem (Corollary 6.3.9(2)).

*Proof* Let $K'$ be an extension of henselian discrete valuation of $K$ with perfect residue field $F'$ as in the proof of Theorem 14.6.5. Then we have $G^r = G'^r$. Hence, we have $\operatorname{dimtot}_G V = \operatorname{dimtot}_{G'} V$ and the assertion follows from Lemma 14.6.8(2). □

**Exercise 14.6** Let $p$ be a prime number and $n \geqslant 1$ be an integer. Let $K$ be the fraction field of the henselization of $\mathbf{Z}_p[T]_{(p)}$, and let $L = K(\zeta_{p^n}, T^{1/p^n})$ as in Exercise 12.3. Identify the Galois group $\operatorname{Gal}(L/K)$ with the semi-direct product $G = (\mathbf{Z}/p^n\mathbf{Z})^\times \ltimes \mathbf{Z}/p^n\mathbf{Z}$.

(1) Verify that $\operatorname{Gr}^r G$ are $\mathbf{F}_p$-vector spaces for $r > 1$.
(2) Let $\chi\colon H = \mathbf{Z}/p^n\mathbf{Z} \to \Lambda^\times$ be a faithful character and $V = \operatorname{Ind}_H^G \chi$ be the induced representation. Compute the total dimension $\operatorname{dim\,tot} V$.

## Historical Notes

The cotangent complex was introduced by Illusie in [41]. The cohomology sheaf $\mathcal{H}_1(L_{X/S})$ of the cotangent complex is closely related to the module of imperfection studied earlier in [31, Chapitre 0, Section 20.6]. Later, Grothendieck introduced the truncation $\tau_{[-1,0]} L_{X/S}$ denoted $L^{X/S}$ in his prenotes [35] and called $\mathcal{H}_1(L^{X/S})$ the conormal module of $X$ relatively to $S$.

The application in ramification theory of the homology group $\mathrm{H}_1(L_{E/S})$ of the cotangent complex was found during the author's stay at the Chinese Academy of Science in September 2019 in search of the proof of Corollary 14.2.3, and is developed in [76]. Gabber pointed out to the author that the same group constructed in a different way was used in [28] in their study of the regularity of local rings. Another construction was found earlier by Dupuy, Katz, Robinov, and Zureick-Brown [24] and is developed in [75].

The proof of the étaleness of the morphism is the same as that in [1]. The proof of the property that the graded quotients are abelian is a modification of that for the logarithmic variant in [2], where $\mathrm{H}_1(L_{E/S})$ was missing. The proof of the property that the graded quotients are $\mathbf{F}_p$-vector spaces by reduction to the perfect residue case using the injectivity of étale cohomology $\mathrm{H}^1(-, \mathbf{F}_p)$ is a modification of the first proof using the pro-$p$ quotient of the fundamental group found during the same stay in Beijing.

The Hasse–Arf theorem (Theorem 14.6.9) and its logarithmic variant were proved with some restriction in [81, 82] by a different method using $p$-adic differential equations.

# 15

# Characteristic Forms

Section 15.6 introduces canonical injections of the character groups of the graded quotients to the Tor groups or to the homology group of the cotangent complex. Using the description of the groups of extensions of vector spaces by finite groups in Section 15.2, the definition is derived from the property that the finite étale covering of the closed fiber studied in the previous chapter defines an extension of a vector space by the last graded piece.

We give a proof of this property only in the equal characteristic case and for integral indices. In this case, the group structure arises from a geometric groupoid structure defined on a dilatation of the product constructed in Section 15.5. We relate this groupoid to the geometric construction given in Section 11.4.

As preliminaries, we study endomorphisms of the additive group in Section 15.1 and extensions by finite groups in Section 15.2. We prepare general terminology on groupoids in Section 15.4.

In the case where the ring of integers of the extension is generated by a single element, we compute the extension class in Section 15.3 and observe that the class is the same as the refinement of the rescaling function defined in Part III.

## 15.1 Additive Polynomials

Let $k$ be a field of characteristic $p > 0$. Let $\mathbf{G}_a = \operatorname{Spec} k[X]$ be the additive group scheme. For a $k$-vector space of finite dimension $V$, we regard the dual space $V^\vee = \operatorname{Spec} \mathrm{S}_k V$ as a smooth group scheme over $k$ defined by the symmetric algebra $\mathrm{S}_k V = \bigoplus_{n=0}^{\infty} \mathrm{S}_k^n V$. The group scheme $V^\vee$ is isomorphic to a direct sum of finitely many copies of $\mathbf{G}_a$. In the case $V = k$, we have $V^\vee = \mathbf{G}_a$.

We compute the group $\operatorname{Hom}(V^\vee, \mathbf{G}_a)$ of morphisms of group schemes over $k$. For an integer $e \geqslant 0$, the iterated absolute Frobenius $\mathrm{F}^e \colon \mathrm{S}_k V \to \mathrm{S}_k V$

induces a morphism $V \to S_k^{p^e} V$ compatible with the iterated absolute Frobenius $F^e: k \to k$. Identifying $F^e: k \to k$ with the inclusion $k \to k^{1/p^e}$, we obtain a morphism

$$V \otimes_k k^{1/p^e} \to S_k^{p^e} V. \tag{15.1.1}$$

**Proposition 15.1.1** *Let $k$ be a field of characteristic $p > 0$, and $V$ be a $k$-vector space of finite dimension.*

(1) *The subgroup* $\mathrm{End}(\mathbf{G}_a)$ *of* $\mathrm{Mor}_k(\mathbf{G}_a, \mathbf{G}_a) = k[X]$ *is identified with*

$$\bigoplus_{e=0}^{\infty} kX^{p^e}.$$

(2) *The morphism* (15.1.1) *defines an isomorphism*

$$\bigoplus\nolimits_{e=0}^{\infty}(V \otimes_k k^{1/p^e}) \to \mathrm{Hom}(V^\vee, \mathbf{G}_a) \subset \mathrm{Mor}_k(V^\vee, \mathbf{G}_a) = S_k V \tag{15.1.2}$$

*to the subgroup consisting of morphisms of group schemes.*

*Proof* (1) Let $f: \mathbf{G}_a \to \mathbf{G}_a$ be a morphism of schemes over $k$ defined by a polynomial $f \in k[X]$. Then the morphism $f$ is an endomorphism of a group scheme $\mathbf{G}_a$ if and only if $f(X+Y) = f(X) + f(Y)$. Comparing the terms total-degree-wise, we see that the equality $f(X+Y) = f(X)+f(Y)$ means that $f$ is a linear combination of $X^{p^e}$ for integers $e \geqslant 0$.

(2) Since $V$ is isomorphic to the direct sum of finitely many copies of $\mathbf{G}_a$, it suffices to prove the case where $V = \mathbf{G}_a = \mathrm{Spec}\, k[X]$. In this case, the source $\bigoplus_{e=0}^{\infty}(V \otimes_k k^{1/p^e})$ is identified with $\bigoplus_{e=0}^{\infty} kX^{p^e}$ by taking the $p^e$th power on each component. Hence, the assertion follows from (1). □

We call a polynomial $f \in \bigoplus_{e=0}^{\infty} kX^{p^e}$ an additive polynomial.

**Corollary 15.1.2** *Let $k$ be a field of characteristic $p > 0$ and let $k^{1/p^\infty} = \varinjlim k^{1/p^e}$ be the perfection of $k$.*

(1) *The inductive limit*

$$V \otimes_k k^{1/p^\infty} \to \varinjlim_{\mathrm{F}} S_k^{p^e} V \tag{15.1.3}$$

*of* (15.1.1) *is an injection.*

(2) *The morphism* $\bigoplus_{e=0}^{\infty}(V \otimes_k k^{1/p^e}) \to \varinjlim_{\mathrm{F}} S_k^{p^e} V \subset \bigoplus_{p \nmid m} \varinjlim_{\mathrm{F}} S_k^{p^e m} V$ *induced by the sum of* (15.1.1) *makes the diagram*

$$\begin{array}{ccc} \bigoplus_{e=0}^{\infty}(V \otimes_k k^{1/p^e}) & \longrightarrow & \bigoplus_{p \nmid m} \varinjlim_{\mathrm{F}} S_k^{p^e m} V \\ {\scriptstyle (15.1.2)}\downarrow & & \downarrow{\scriptstyle (14.3.4)} \\ \mathrm{Hom}(V^\vee, \mathbf{G}_a) & \xrightarrow{[\mathrm{AS}]} & \mathrm{H}^1(V^\vee, \mathbf{F}_p)/\mathrm{H}^1(k, \mathbf{F}_p) \end{array} \tag{15.1.4}$$

*commutative, where the vertical arrows are isomorphisms. Its image is identified with* $V \otimes_k k^{1/p^\infty}$.

(3) *For the additive group* $\mathbf{G}_a$, *the image of* $\mathrm{End}(\mathbf{G}_a) \to \mathrm{H}^1(\mathbf{G}_a, \mathbf{F}_p)$ *is isomorphic to* $\varinjlim_F kX^{p^e} = k^{1/p^\infty}X$.

*Proof* (1) Since (15.1.1) are injections for all $e \geqslant 0$, their inductive limit is also an injection.

(2) Since an element of $V \otimes_k k^{1/p^e}$ is regarded as an element of $\Gamma(V^\vee, \mathcal{O}_{V^\vee}) = \mathrm{Mor}_k(V^\vee, \mathbf{G}_a)$ by taking its $p^e$th power in the definition of (15.1.2), the diagram is commutative. Since $k^{1/p^\infty} = \varinjlim k^{1/p^e}$, the canonical morphism

$$\bigoplus_{e=0}^{\infty}(V \otimes_k k^{1/p^e}) \to V \otimes_k k^{1/p^\infty} \tag{15.1.5}$$

is a surjection. Hence, the assertion follows from (1).

(3) By Proposition 15.1.1(1), the image of $\mathrm{End}(\mathbf{G}_a)$ equals the image of $\bigoplus_{e=0}^{\infty} kX^{p^e}$. Since $\mathrm{H}^1(\mathbf{G}_a, \mathbf{F}_p) = k[X]/(\mathrm{F} - 1)$, its image is $\varinjlim_F kX^{p^e}$. Since $F^e : kX \to kX^{p^e}$ induces an isomorphism $k^{1/p^e}X \to kX^{p^e}$, we have an isomorphism $k^{1/p^\infty}X \to \varinjlim_F kX^{p^e}$. □

**Proposition 15.1.3** *Let* $k$ *be a field of characteristic* $p > 0$, *and* $\mathbf{G}_a = \mathrm{Spec}\, k[X]$ *be the additive group over* $k$. *Let* $p_1, p_2, +\colon \mathbf{G}_a \times \mathbf{G}_a \to \mathbf{G}_a$ *be the projections and the addition. Let* $f \in k[X]$ *be a polynomial such that the constant term is* 0, *and let* $\chi \in \mathrm{H}^1(\mathbf{G}_a, \mathbf{F}_p)$ *be the image of* $f$ *by the boundary mapping* $k[X] \to \mathrm{H}^1(\mathbf{G}_a, \mathbf{F}_p)$ *defined by the Artin–Schreier sequence* (14.3.1). *Then the following conditions are equivalent:*

(1) $f$ *is an additive polynomial.*
(2) *There exists an extension* $0 \to \mathbf{F}_p \to E \to \mathbf{G}_a \to 0$ *of smooth group schemes over* $k$ *such that* $\chi$ *is the class of an* $\mathbf{F}_p$*-torsor* $E$ *over* $\mathbf{G}_a$.
(3) $+^*\chi = p_1^*\chi + p_2^*\chi$ *in* $\mathrm{H}^1(\mathbf{G}_a \times \mathbf{G}_a, \mathbf{F}_p)$.

*Proof* (1) ⇒ (2): An additive polynomial $f \in k[X]$ defines an endomorphism $f\colon \mathbf{G}_a \to \mathbf{G}_a$ and $\chi = f^*[\mathrm{AS}]$ is the class of the extension of $E$ by $\mathbf{F}_p$ defined by pulling back the Artin–Schreier sequence (14.3.1).

(2) ⇒ (3): The sum $p_1^*\chi + p_2^*\chi$ is the class of the $\mathbf{F}_p$-torsor on $\mathbf{G}_a \times \mathbf{G}_a$ defined by the quotient of $E \times E$ by the anti-diagonal action of $\mathbf{F}_p$. Hence, the addition $E \times E \to E$ defines an isomorphism of $\mathbf{F}_p$-torsors on $\mathbf{G}_a \times \mathbf{G}_a$ that implies the equality $+^*\chi = p_1^*\chi + p_2^*\chi$.

(3) ⇒ (1): By replacing $k$ by an algebraic closure, we may assume that $k$ is algebraically closed. Then $k[X] \to \mathrm{H}^1(\mathbf{G}_a, \mathbf{F}_p)$ induces an isomorphism $\bigoplus_{p \nmid m} kX^m \to \mathrm{H}^1(\mathbf{G}_a, \mathbf{F}_p)$. Hence, we may further assume that the non-zero terms of $f$ are of degree prime to $p$.

Since $+^*\chi = \mathrm{pr}_1^*\chi + \mathrm{pr}_2^*\chi$, we have $f(X+Y) \equiv f(X) + f(Y) \bmod \mathrm{F} - 1$. Since $\bigoplus_{p\nmid m} \mathrm{S}_k^m(kX \oplus kY) \to \mathrm{H}^1(\mathbf{G}_a \times \mathbf{G}_a, \mathbf{F}_p)$ is an injection, it is reduced to the case where $f = aX^m$ is a monomial for $p \nmid m$ satisfying the equality $f(X+Y) = f(X) + f(Y)$. Then, $m = 1$ and the assertion follows. □

**Lemma 15.1.4** *Let $k$ be a field of characteristic $p > 0$, and let $G \subset k$ be a finite subgroup of the additive group. The polynomial*

$$a_G = \prod_{\sigma \in G} (X - \sigma) \tag{15.1.6}$$

*$\in k[X]$ is a unique additive separable monic polynomial such that the sequence*

$$0 \longrightarrow G \longrightarrow \mathbf{G}_a \xrightarrow{a_G} \mathbf{G}_a \longrightarrow 0 \tag{15.1.7}$$

*is exact.*

*Proof* Since $G = a_G^{-1}(0)$, it suffices to show that $a_G$ is an additive polynomial. We show this by induction on $\dim_{\mathbf{F}_p} G$. If $G = 0$, then $a_0 = X$ is an additive polynomial. Let $H \subsetneqq G$ be a subgroup of index $p$ and $\tau \in G - H$. Then $a_H = \prod_{\sigma \in H}(X - \sigma) \in k[X]$ is an additive polynomial by the induction hypothesis, and

$$a_G = \prod_{i \in \mathbf{F}_p} a_H(X - i\tau) = \prod_{i \in \mathbf{F}_p} (a_H(X) - ia_H(\tau)) = a_H(X)^p - a_H(\tau)^{p-1} a_H(X) \tag{15.1.8}$$

is also an additive polynomial. □

**Exercise 15.1** Let $a, b \in k$ be elements linearly independent over $\mathbf{F}_p$. Find the monic additive separable polynomial $f \in k[X]$ such that $\operatorname{Ker} f$ is generated by $a, b \in k$.

## 15.2 Extensions of Vector Spaces by Finite Groups

We recall some elementary facts on extensions of vector spaces by finite groups. Let $k$ be a field of characteristic $p > 0$, $V$ be a $k$-vector space of finite dimension, and consider the dual $V^\vee = \mathrm{Hom}_k(V, k)$ as a smooth group scheme over $k$. Let $G$ be an $\mathbf{F}_p$-vector space of finite dimension. Let $\mathrm{Ext}(V^\vee, G)$ denote the group of isomorphism classes of extensions $0 \to G \to E \to V^\vee \to 0$. Since $V^\vee$ is isomorphic to a direct sum of copies of $\mathbf{G}_a$, and $G$ is isomorphic to a direct sum of copies of $\mathbf{F}_p$, the pull-back by linear forms $V^\vee \to \mathbf{G}_a$ and the push-forward by $\mathbf{F}_p \to G$ define a canonical isomorphism $V \otimes_k \mathrm{Ext}(\mathbf{G}_a, \mathbf{F}_p) \otimes_{\mathbf{F}_p} G \to \mathrm{Ext}(V^\vee, G)$.

Let F denote the Frobenius endomorphism of the additive group scheme $\mathbf{G}_a = \operatorname{Spec} k[X]$ defined by the additive polynomial $X^p$, and let $[\mathrm{AS}] \in \operatorname{Ext}(\mathbf{G}_a, \mathbf{F}_p)$ be the class of the Artin–Schreier extension

$$0 \longrightarrow \mathbf{F}_p \longrightarrow \mathbf{G}_a \xrightarrow{\mathrm{F}-1} \mathbf{G}_a \longrightarrow 0 \tag{15.2.1}$$

of smooth group schemes.

**Proposition 15.2.1** *Let $k$ be a field of characteristic $p > 0$, and let $V^\vee = \operatorname{Hom}_k(V, k)$ denote the dual space of $V$.*

(1) *The morphism*

$$\operatorname{Ext}(V^\vee, \mathbf{F}_p) \to \mathrm{H}^1(V^\vee, \mathbf{F}_p) \tag{15.2.2}$$

*sending the class of extension $0 \to \mathbf{F}_p \to E \to V^\vee \to 0$ to the class of $\mathbf{F}_p$-torsor $E$ over $V^\vee$ induces an injection*

$$\operatorname{Ext}(V^\vee, \mathbf{F}_p) \to \mathrm{H}^1(V^\vee, \mathbf{F}_p)/\mathrm{H}^1(k, \mathbf{F}_p).$$

(2) *The morphism*

$$\operatorname{Hom}(V^\vee, \mathbf{G}_a) \to \operatorname{Ext}(V^\vee, \mathbf{F}_p) \tag{15.2.3}$$

*sending a morphism $f : V^\vee \to \mathbf{G}_a$ to the pull-back $f^*[\mathrm{AS}]$ is a surjection.*

*Proof* (1) Let $0 \to \mathbf{F}_p \to E \to V^\vee \to 0$ be an extension. If the underlying $\mathbf{F}_p$-torsor is trivial, then the connected component of $E$ containing 0 defines a section and the extension is trivial. Hence, the morphism (15.2.2) is an injection. Since the composition of (15.2.2) with the pull-back $\mathrm{H}^1(V^\vee, \mathbf{F}_p) \to \mathrm{H}^1(k, \mathbf{F}_p)$ by the 0-section is the 0-mapping, the intersection of the image of (15.2.2) and $\mathrm{H}^1(k, \mathbf{F}_p) \subset \mathrm{H}^1(V^\vee, \mathbf{F}_p)$ is 0. Hence, $\operatorname{Ext}(V^\vee, \mathbf{F}_p) \to \mathrm{H}^1(V^\vee, \mathbf{F}_p)/\mathrm{H}^1(k, \mathbf{F}_p)$ is an injection.

(2) Since $V$ is isomorphic to the direct sum of finitely many copies of $k$, we may assume that $V^\vee = \mathbf{G}_a = \operatorname{Spec} k[X]$. Since

$$k[X]/k \to \mathrm{H}^1(\mathbf{G}_a, \mathbf{F}_p)/\mathrm{H}^1(k, \mathbf{F}_p)$$

is a surjection, the assertion follows from Proposition 15.1.3 (2) $\Rightarrow$ (1). □

**Corollary 15.2.2** *There exists a unique isomorphism*

$$V \otimes_k k^{1/p^\infty} \longrightarrow \operatorname{Ext}(V^\vee, \mathbf{F}_p) \tag{15.2.4}$$

*that keeps the diagram*

$$\begin{array}{ccccc} \bigoplus_{e=0}^{\infty}(V \otimes_k k^{1/p^e}) & \xrightarrow{(15.1.5)} & V \otimes_k k^{1/p^\infty} & \xrightarrow{(15.1.3)} & \bigoplus_{p \nmid m} \varinjlim_{\mathrm{F}} \mathrm{S}_k^{p^e m} V \\ {\scriptstyle (15.1.2)}\downarrow & & & & \downarrow{\scriptstyle (14.3.4)} \\ \mathrm{Hom}(V^\vee, \mathbf{G}_a) & \xrightarrow{(15.2.3)} & \mathrm{Ext}(V^\vee, \mathbf{F}_p) & \xrightarrow{(15.2.2)} & \mathrm{H}^1(V^\vee, \mathbf{F}_p)/\mathrm{H}^1(k, \mathbf{F}_p) \end{array} \tag{15.2.5}$$

*commutative, where the vertical arrows are isomorphisms.*

*Proof* The vertical arrows (15.1.2) and (14.3.4) are isomorphisms by Propositions 15.1.1(2) and 14.3.1. The compositions of the horizontal arrows are the same as those in (15.1.4). The arrows (15.1.5) and (15.2.3) are surjections and the arrows (15.1.3) and (15.2.2) are injections by Proposition 15.2.1 and Corollary 15.1.2. Hence, the assertion follows. □

**Corollary 15.2.3** *Let $k \to k'$ be a morphism of fields of characteristic $p > 0$. Let $V$ and $V'$ be a $k$-vector space and a $k'$-vector space respectively, and let $V \otimes_k k' \to V'$ be an injection. Let $V'^\vee \to V^\vee$ be the morphism defined by $\mathrm{S}_k V \to \mathrm{S}_k V'$. Let $G$ be an $\mathbf{F}_p$-vector space of finite dimension. Then the commutative diagram*

$$\begin{array}{ccc} \mathrm{Ext}(V^\vee, G) & \xrightarrow{(15.2.2)} & \mathrm{H}^1(V^\vee, G)/\mathrm{H}^1(k, G) \\ \downarrow & & \downarrow \\ \mathrm{Ext}(V'^\vee, G) & \xrightarrow{(15.2.2)} & \mathrm{H}^1(V'^\vee, G)/\mathrm{H}^1(k', G) \end{array}$$

*is cartesian.*

*Proof* It suffices to consider the cases where $V \otimes_k k' \to V'$ is an isomorphism and where $k \to k'$ is an isomorphism respectively. First, assume $k = k'$. Then there is a splitting $V' \to V$ of the injection $V \to V'$ and the assertion follows in this case.

Next, we consider the case when $V \otimes_k k' \to V'$ is an isomorphism. Since $G$ is isomorphic to a direct sum of finitely many copies of $\mathbf{F}_p$, we may assume that $G = \mathbf{F}_p$. By the isomorphisms $\bigoplus_{p \nmid m} \varinjlim_{\mathrm{F}} \mathrm{S}_k^{p^e m} V \to \mathrm{H}^1(V^\vee, \mathbf{F}_p)/\mathrm{H}^1(k, \mathbf{F}_p)$ (14.3.4) and $V \otimes_k k^{1/p^\infty} \to \mathrm{Ext}(V^\vee, \mathbf{F}_p)$ (15.2.4), it suffices to show that the diagram

$$\begin{array}{ccc} V \otimes_k k^{1/p^\infty} & \longrightarrow & \varinjlim_{\mathrm{F}} \mathrm{S}_k^{p^e} V \\ \downarrow & & \downarrow \\ V' \otimes_{k'} k'^{1/p^\infty} & \longrightarrow & \varinjlim_{\mathrm{F}} \mathrm{S}'^{p^e}_k V' \end{array}$$

is cartesian. Since the lower line is identified with the base change $V \otimes_k k'^{1/p^\infty} \to (\varinjlim_{\mathrm{F}} \mathrm{S}_k^{p^e} V) \otimes_{k^{1/p^\infty}} k'^{1/p^\infty}$ of the upper line, the assertion follows. □

**Example 15.2.4** Let $X = \operatorname{Spec} A$ be a smooth affine scheme over a perfect field $k$ of characteristic $p > 0$, and $U = \operatorname{Spec} A_U = X - D$ be the complement of an irreducible smooth divisor over $k$. Let $K$ be the henselization of the fraction field of the local ring $\mathcal{O}_{X,\xi}$ at the generic point $\xi$ of $D$, and let $F = k(\xi)$ be the function field of $D$. Let $V \to U$ be the cyclic covering defined by a Witt vector $a \in F_n W_{m+1}(A_U)$ for $m \geqslant 0$ and $n \geqslant 1$. Assume that the injection $\chi\colon G = \operatorname{Gal}(V/U) \to \mathbf{Z}/p^{m+1}\mathbf{Z}$ is an isomorphism and let $\chi$ also denote its image in $X_K$. Assume further that $\operatorname{Sw}\chi = n$.

Define the dilatation $P^{[n]} = \operatorname{Spec} B$ as in Example 13.5.2 and let $W \to P^{[n]}$ be the normalization of the cyclic covering of $(V \times V)/G \to U \times U$ defined by the Witt vector $u(a) - v(a) \in W_{m+1}(A_U \otimes_k A_U)$. Let $X \to W$ be the canonical lifting of the extension $X \to P^{[n]}$ of the diagonal $U \to U \times U$. Identify the complement $P^{[n]} - U \times U$ with the twisted tangent bundle $TX(-nD) \times_X D$ restricted to $D$. Let $T_\xi$ denote its generic fiber $TX(-nD) \times_X \xi$ and $W_\xi = (W \times_{P^{[n]}} T_\xi)^\circ$ be the connected component meeting the section $X \to W$. Then, by Proposition 13.5.5, $W_\xi$ is an Artin–Schreier covering of the vector space $T_\xi$ over $F$. Further, it is an extension of $T_\xi$ by $\mathbf{F}_p$ and its class in $\operatorname{Ext}(T_\xi, \mathbf{F}_p)$ is the image of the linear form $\mathbf{d}a \in \operatorname{Gr}_n^F \Omega_K^1 = \operatorname{Hom}(\mathfrak{m}_K^n/\mathfrak{m}_K^{n+1}, \Omega_F^1(\log))$ of $T_\xi$.

We will see in Corollary 16.3.5(2) that the refined Swan conductor $\operatorname{rsw}\chi$ equals $-\mathbf{d}a$.

**Proposition 15.2.5** *Let $G$ be an $\mathbf{F}_p$-vector space of finite dimension, and let $G^\vee = \operatorname{Hom}_{\mathbf{F}_p}(G, \mathbf{F}_p)$ denote the dual space. There exists a unique isomorphism*

$$\operatorname{Ext}(V^\vee, G) \to \operatorname{Hom}_{\mathbf{F}_p}(G^\vee, V \otimes_k k^{1/p^\infty}) \tag{15.2.6}$$

*satisfying the following property: Let $0 \to G \to E \to V^\vee \to 0$ be an extension of group schemes, let $[E]\colon G^\vee \to V \otimes_k k^{1/p^\infty}$ denote the morphism defined as the image of the class of $E$ by* (15.2.6)*, and let $\chi\colon G \to \mathbf{F}_p$ be a character and $\varphi\colon V^\vee \to \mathbf{G}_a$ be a morphism of smooth group schemes. Then $[E](\chi) \in V \otimes_k k^{1/p^\infty}$ equals the image of $\varphi$ by the composition* $\operatorname{Hom}(V^\vee, \mathbf{G}_a) \to V \otimes_k k^{1/p^\infty}$ *of* (15.2.3) *and the inverse of the isomorphism* (15.2.4) *if and only if there exists a commutative diagram*

$$\begin{array}{ccccccccc} 0 & \longrightarrow & G & \longrightarrow & E & \longrightarrow & V^\vee & \longrightarrow & 0 \\ & & \chi\downarrow & & \downarrow & & \downarrow\varphi & & \\ 0 & \longrightarrow & \mathbf{F}_p & \xrightarrow{\subset} & \mathbf{G}_a & \xrightarrow{\mathrm{F}-1} & \mathbf{G}_a & \longrightarrow & 0 \end{array} \tag{15.2.7}$$

*of extensions of group schemes.*

*Proof* The mapping sending the pair $(\chi, [E])$ of character $\chi: G \to \mathbf{F}_p$ and the class of extension $0 \to G \to E \to V^\vee \to 0$ to $\chi_*[E] \in \mathrm{Ext}(V^\vee, \mathbf{F}_p)$ is biadditive; define a morphism

$$G^\vee \otimes_{\mathbf{F}_p} \mathrm{Ext}(V^\vee, G) \to \mathrm{Ext}(V^\vee, \mathbf{F}_p). \tag{15.2.8}$$

We show that the morphism (15.2.6) defined as the adjoint of the composition of (15.2.8) with the inverse $\mathrm{Ext}(V^\vee, \mathbf{F}_p) \to V \otimes_k k^{1/p^\infty}$ of the isomorphism (15.2.4) satisfies the required property.

We claim that the morphism (15.2.6) defined as the adjoint is an isomorphism. The isomorphism (15.2.4) defines the lower horizontal isomorphism in

$$\begin{array}{ccc} \mathrm{Ext}(V^\vee, G) & \longrightarrow & \mathrm{Hom}_{\mathbf{F}_p}(G^\vee, V \otimes_k k^{1/p^\infty}) \\ \uparrow & & \uparrow \\ \mathrm{Ext}(V^\vee, \mathbf{F}_p) \otimes G & \longleftarrow & (V \otimes_k k^{1/p^\infty}) \otimes_{\mathbf{F}_p} G \end{array} \tag{15.2.9}$$

Since $G$ is isomorphic to the product of finitely many copies of $\mathbf{F}_p$, we obtain the left vertical isomorphism sending the tensor product $[E] \otimes g$ of the class of extension $E$ and $g \in G$ to the class $[g_*E]$ by the morphism $\mathbf{F}_p \to G$ sending 1 to $g$. The right vertical arrow sending the tensor product $v \otimes g \in (V \otimes_k k^{1/p^\infty}) \otimes_{\mathbf{F}_p} G$ to the morphism $G^\vee \to V \otimes_k k^{1/p^\infty}$ sending character $\chi$ to $\chi(g) \cdot v$ is an isomorphism. Since the upper horizontal isomorphism (15.2.6) making the diagram (15.2.9) commutative is the adjoint, the adjoint is an isomorphism as claimed.

We show that the isomorphism (15.2.6) defined as the adjoint satisfies the required property. Let $\chi: G \to \mathbf{F}_p$ be a character and $0 \to G \to E \to V^\vee \to 0$ be an extension. Then (15.2.8) maps $\chi \otimes [E]$ to the class $\chi_* E$. Hence, for a morphism $\varphi: V^\vee \to \mathbf{G}_a$, its image by $\mathrm{Hom}(V^\vee, \mathbf{G}_a) \to V \otimes_k k^{1/p^\infty}$ equals $[E](\chi)$ if and only if $\chi_*[E] = \varphi^*[\mathrm{AS}] \in \mathrm{Ext}(V^\vee, \mathbf{F}_p)$. This equality is equivalent to the existence of the commutative diagram (15.2.7) of extensions. Hence, the morphism (15.2.6) defined as the adjoint satisfies the required property.

Since $\mathrm{Hom}(V^\vee, \mathbf{G}_a) \to \mathrm{Ext}(V^\vee, \mathbf{F}_p)$ is a surjection by Proposition 15.2.1(2), the morphism satisfying the condition is unique. □

**Corollary 15.2.6** *Let $0 \to G \to E \to V^\vee \to 0$ be an extension of group schemes. Then the following conditions are equivalent:*

(1) *$E$ is connected.*

(2) *The morphism $[E]: G^\vee \to V \otimes_k k^{1/p^\infty}$ defined as the image of the class of $E$ by* (15.2.6) *is an injection.*

*Proof* The morphism $\mathrm{Ext}(V^\vee, \mathbf{F}_p) \to \mathrm{H}^1(V^\vee, \mathbf{F}_p)$ is an injection by Proposition 15.2.1(1). Hence, the assertion follows from Lemma 14.3.3 and the isomorphism (15.2.4). □

In the case where $V^\vee$ is the additive group $\mathbf{G}_a$, the isomorphism (15.2.6) is computed as follows.

**Lemma 15.2.7** *Let $k$ be a field of characteristic $p > 0$, and let $G \subset k$ be a finite subgroup of the additive group. Let $[E] \in \mathrm{Ext}(G, \mathbf{G}_a)$ be the class of the extension (15.1.7). Then its image $\mathrm{Hom}(G^\vee, k^{1/p^\infty})$ by the isomorphism (15.2.6) sends a non-trivial character $\chi\colon G \to \mathbf{F}_p$ to the inverse of $\prod_{\sigma \in G, \chi(\sigma)=1} \sigma^p \in k^\times$.*

*Proof* Let $H = \mathrm{Ker}\, \chi \subsetneq G$ be the kernel and let $\tau \in G$ be an element such that $\chi(\tau) = 1$. Let $a_H$ and $a_{\mathbf{F}_p} = X^p - X$ be the additive polynomials for the subgroups $H, \mathbf{F}_p \subset k$. Since $a_H(\tau) = \prod_{\sigma \in H}(\tau - \sigma) = \prod_{\sigma \in G, \chi(\sigma)=1} \sigma$, it suffices to show that the diagram

$$\begin{array}{ccccccccc} 0 & \longrightarrow & G & \longrightarrow & \mathbf{G}_a & \xrightarrow{a_G} & \mathbf{G}_a & \longrightarrow & 0 \\ & & \downarrow{\scriptstyle \chi} & & \downarrow{\scriptstyle a_H/a_H(\tau)} & & \downarrow{\scriptstyle 1/a_H(\tau)^p} & & \\ 0 & \longrightarrow & \mathbf{F}_p & \xrightarrow{\subset} & \mathbf{G}_a & \xrightarrow{a_{\mathbf{F}_p}} & \mathbf{G}_a & \longrightarrow & 0 \end{array} \tag{15.2.10}$$

of extensions is commutative by Corollary 15.2.6. The left square is commutative, since both compositions are 0 on $H$ and send $\tau \in G$ to 1. The commutativity of the right square follows from (15.1.8). □

**Lemma 15.2.8** *Let $k$ be a field and $V$ be a smooth commutative connected group scheme over $k$. Let $G$ be a finite group.*

(1) *Let $E$ be another smooth connected group scheme and let $E \to V$ be a morphism of group schemes over $k$. Assume that $E$ is a $G$-torsor over $V$. Then $E$ and $G$ are also commutative and $E$ is an extension of $V$ by $G$.*
(2) *Let $E$ be a connected $G$-torsor over $V$, and $e \in E \times_V 0$ be a $k$-valued point above the origin $0 \in V$. Then there exists at most one group scheme structure on $E$ such that $e$ is the unit and the morphism $E \to V$ is a morphism of group schemes.*

*Proof* (1) Since $V$ is commutative, the pair of the operation $ab$ on $E$ and its opposite $ba$ defines a morphism $E \times E \to E \times_V E$. Since $E$ is a $G$-torsor, the target is identified with $E \times G$. Since $E \times E$ is connected, the composition with the second projection is a constant morphism to $G$. Since its value at the origin is 0, the group scheme $E$ is commutative.

For $g \in G$, define a morphism $E \to E$ by sending $x$ to $g(x) - x$. Since

$g$ acts on $E$ as an automorphism over $V$, the same argument shows that we have $g(x) = x + g(0)$. Hence, the mapping $G \to \mathrm{Ker}(E \to V)$ sending $g$ to $g(0)$ is an isomorphism of group schemes and the action of $G$ on $E$ is defined by the translation. Hence, $G$ is commutative and we obtain an exact sequence $0 \to G \to E \to V \to 0$.

(2) Two group structures on $E$ such that $E \to V$ is a morphism of group schemes define a morphism $E \times E \to E \times_V E$. Since $E$ is a $G$-torsor, the target is identified with $E \times G$. Since $E \times E$ is connected, the composition with the second projection is a constant morphism to $G$. If $e$ is the unit element, the image is the origin $0 \in V$. □

**Exercise 15.2** Let $a, b \in k$ be elements linearly independent over $\mathbf{F}_p$. Let $\chi_a \colon G = \langle a, b\rangle \to \mathbf{F}_p$ be the character defined by $\chi_a(a) = 1$, $\chi_a(b) = 0$. Let $f \in k[X]$ be the solution of Exercise 15.1, and $[E] \colon G^\vee \to \mathrm{Ext}(\mathbf{G}_a, \mathbf{F}_p) = k^{1/p^\infty}$ denote the morphism defined by the extension $0 \to G \to \mathbf{G}_a \xrightarrow{f} \mathbf{G}_a \to 0$. Find the image $[E](\chi_a) \in k^{1/p^\infty}$ of $\chi_a \in G^\vee$.

## 15.3 Monogenic Case

We study the monogenic case. Let $K$ be a henselian discrete valuation field with residue field $F$. Let $L$ be a Galois extension of $K$ of degree $n$. Let $G$ be the Galois group and $E$ be the residue field of $L$. Assume that $L$ is wildly ramified over $K$ and let $e = e_{L/K}$ be the ramification index. Assume that $O_L$ is generated over $O_K$ by a single element $a \in O_L$.

Let $a_1, \ldots, a_n \in O_L$ be the conjugates of $a$ and set $b_j = a_j - a_n$ for $j = 1, \ldots, n$. By changing the numbering if necessary, we assume that $s_j = \mathrm{ord}_L b_j \in \mathbf{N} \cup \{\infty\}$ is increasing. The mapping $G \to \{a_1, \ldots, a_n\}$ sending $\sigma \in G$ to $\sigma(a_n)$ is a bijection. Let $\sigma_j \in G$ denote the element satisfying $\sigma_j(a_n) = a_j$ for $j = 1, \ldots, n$.

Set $s = s_{n-1}$ and define a rational number $r$ by

$$r = \varphi(s) = \frac{1}{e_{L/K}} \left( \sum_{i=1}^{n-1} s_i + s_{n-1} \right). \tag{15.3.1}$$

By the assumption that $L$ is wildly ramified over $K$, we have $s > 1$. Since $\sum_{i=1}^{n-1} s_i = \mathrm{ord}_L D_{L/K} > e_{L/K} - 1$ by Proposition 1.2.2 and Lemma 1.9.4, we have $r > 1$. Let $0 \leqslant k < n - 1$ be the smallest integer such that $s_{k+1} = s$. Then the subgroup $G^r \subset G$ equals $\{\sigma_{k+1}, \ldots, \sigma_n\}$ by Proposition 12.1.7(1).

The minimal polynomial $f \in O_K[X]$ of $a \in O_L$ decomposes as $f =$

$\prod_{i=1}^{n}(X - a_i)$ in $O_L[X]$. Let $T = \operatorname{Spec} O_K[X]/(f) \to Q = \mathbf{A}^1_S = \operatorname{Spec} O_K[X]$ be the closed immersion. By Lemma 11.1.5, we have a canonical isomorphism

$$Q^{[r]}_{E,\mathrm{red}} \to \operatorname{Hom}_F(N_{T/Q} \otimes_{O_T} E, \mathfrak{m}_L^{er}/\mathfrak{m}_L^{er+1}) \otimes_{E \otimes_F E} (E \otimes_F E)_{\mathrm{red}} \qquad (15.3.2)$$

of vector bundles over $(E \otimes_F E)_{\mathrm{red}}$ for $S' = T$. For the connected component $Q^{[r]\circ}_{E,\mathrm{red}} \subset Q^{[r]}_{E,\mathrm{red}}$ corresponding to the identity $E \to E$, we have an isomorphism

$$Q^{[r]\circ}_{E,\mathrm{red}} \longrightarrow \operatorname{Hom}_E(N_{T/Q} \otimes_{O_T} E, \mathfrak{m}_L^{er}/\mathfrak{m}_L^{er+1}). \qquad (15.3.3)$$

Let $Q_E^{(r)\circ} \subset Q_E^{(r)}$ be the connected component containing the point corresponding to $a_n$.

We regard $\mathfrak{m}_L^u/\mathfrak{m}_L^{u+1}$ for integers $u$ as group schemes over $E$ isomorphic to $\mathbf{G}_a$.

**Proposition 15.3.1** (1) *The mapping*

$$b\colon G^r \longrightarrow \mathfrak{m}_L^s/\mathfrak{m}_L^{s+1} \qquad (15.3.4)$$

*sending $\sigma$ to $\sigma(a_n) - a_n$ is an injection of abelian groups.*

(2) *Let $0 \leqslant k < n-1$ be the smallest integer such that $s_{k+1} = s_{n-1} = s$. The mapping $g\colon \mathfrak{m}_L^s/\mathfrak{m}_L^{s+1} \to \mathfrak{m}_L^{er}/\mathfrak{m}_L^{er+1}$ sending $x$ to $(-1)^k \prod_{j=1}^{k} b_j \cdot \prod_{j=k+1}^{n}(x - b_j)$ is a morphism of group schemes over E. The sequence*

$$0 \longrightarrow G^r \xrightarrow{(15.3.4)} \mathfrak{m}_L^s/\mathfrak{m}_L^{s+1} \xrightarrow{g} \mathfrak{m}_L^{er}/\mathfrak{m}_L^{er+1} \longrightarrow 0 \qquad (15.3.5)$$

*is an exact sequence of smooth group schemes. The polynomial $h = \prod_{j=k+1}^{n}(X - b_j/b_{n-1}) \in E[X]$ is an additive polynomial.*

(3) *Let $\mathrm{ev}_f\colon \operatorname{Hom}_E(N_{T/Q} \otimes_{O_L} E, \mathfrak{m}_L^{er}/\mathfrak{m}_L^{er+1}) \to \mathfrak{m}_L^{er}/\mathfrak{m}_L^{er+1}$ denote the isomorphism defined by the evaluation at the basis $f \in N_{T/Q} \otimes_{O_L} E$. Let $U_n = \mathbf{A}^1_T \subset Q_T^{(r)}$ be the open subscheme defined in Proposition 11.5.8. Define an element $t = (-1)^k \prod_{j=1}^{k} b_j \cdot b_{n-1}^{n-k}$ of $O_L$ satisfying $\operatorname{ord}_L t = e_{L/K} \cdot r$. Then there exists an isomorphism $\mathfrak{m}_L^s/\mathfrak{m}_L^{s+1} \longrightarrow U_{n,E} = Q_E^{(r)\circ}$ of schemes over E fitting in the commutative diagram*

$$\begin{array}{ccccccc}
\mathbf{G}_a & \xrightarrow{b_{n-1}} & \mathfrak{m}_L^s/\mathfrak{m}_L^{s+1} & \longrightarrow & U_{n,E} & = & Q_E^{(r)\circ} \\
{\scriptstyle h}\downarrow & & \downarrow{\scriptstyle g} & & & & \downarrow \\
\mathbf{G}_a & \xrightarrow{\cdot t} & \mathfrak{m}_L^{er}/\mathfrak{m}_L^{er+1} & \xleftarrow{\mathrm{ev}_f} & \operatorname{Hom}_E(N_{T/Q} \otimes_{O_L} E, \mathfrak{m}_L^{er}/\mathfrak{m}_L^{er+1}) & \xleftarrow{(15.3.3)} & Q^{[r]\circ}_{E,\mathrm{red}}
\end{array}$$

*Proof* (1) By Corollary 11.5.9, $G^r = \operatorname{Ker}(G \to \operatorname{Aut}(O_L/\mathfrak{m}_L^s))$. Hence, similarly to Proposition 1.6.5, for $\sigma, \tau \in G^r$ we have $\sigma\tau(a_n) - a_n - ((\sigma(a_n) - a_n) + \tau(a_n) - a_n) = \sigma(\tau(a_n) - a_n) - (\tau(a_n) - a_n)$. Since the action of $G^r$

on $\mathfrak{m}_L^s/\mathfrak{m}_L^{s+1}$ is trivial, the mapping $G^r \to \mathfrak{m}_L^s/\mathfrak{m}_L^{s+1}$ is a morphism of groups. Since the kernel consists of the unit, the assertion follows.

(2) For $x = b_{n-1}y \in \mathfrak{m}_L^s$, we have

$$\prod_{j=1}^{k}(-b_j) \cdot \prod_{j=k+1}^{n}(x - b_j) \equiv h(y) \cdot t \tag{15.3.6}$$

mod $\mathfrak{m}_L^{er+1}$. This means that the right square in the diagram

$$\begin{array}{ccccccccc} 0 & \longrightarrow & G^r & \xrightarrow{(15.3.4)} & \mathfrak{m}_L^s/\mathfrak{m}_L^{s+1} & \xrightarrow{g} & \mathfrak{m}_L^{er}/\mathfrak{m}_L^{er+1} & \longrightarrow & 0 \\ & & \| & & \uparrow b_{n-1} & & \uparrow t & & \\ 0 & \longrightarrow & G^r & \longrightarrow & \mathbf{G}_a & \xrightarrow{h} & \mathbf{G}_a & \longrightarrow & 0 \end{array} \tag{15.3.7}$$

is commutative. By (1), $G^r \to \mathbf{G}_a$ defined by the commutativity of the left square is an injection of groups. Hence, the image $\{b_j/b_{n-1} \mid j = k+1, \dots, n\}$ is a subgroup of $E$ and the polynomial $h \in E[X]$ is a separable additive polynomial by Lemma 15.1.4. Since the lower sequence is exact and the vertical arrows are isomorphisms, the upper sequence is also exact.

(3) By (15.3.7), the left square is commutative. Let $T = \operatorname{Spec} O_L \to Q = \operatorname{Spec} O_K[X]$ be the immersion defined by $a = a_n$. The morphism

$$\mathbf{G}_a = \operatorname{Spec} E[Z] \to Q_T^{[r]} = \operatorname{Spec} O_L[X, f/t] = O_L[X, Z]/(tZ - f) \tag{15.3.8}$$

defined by sending $X$ to 0 and $Z$ to $Z$ defines the composition of the bottom line. Let $\mathbf{G}_a \to U_{n,E}$ be the restriction to the closed fiber of the isomorphism $\mathbf{A}_T^1 \to U_n$. Since $f(x + a_n) = \prod_{j=1}^n (x - b_j) \equiv \prod_{j=1}^k (-b_j) \cdot \prod_{j=k+1}^n (x - b_j)$ by (11.5.5), the diagram is commutative. □

By Proposition 15.3.1, we have an extension

$$0 \longrightarrow G^r \longrightarrow Q_E^{(r)\circ} \longrightarrow Q_{E,\mathrm{red}}^{[r]\circ} \longrightarrow 0 \tag{15.3.9}$$

of smooth group schemes over $E$. We identify $Q_{E,\mathrm{red}}^{[r]\circ}$ with $\operatorname{Hom}_E(N_{T/Q} \otimes_{O_T} E, \mathfrak{m}_L^{er}/\mathfrak{m}_L^{er+1})$ by the isomorphism (15.3.2). By the isomorphism

$$\begin{aligned} &\operatorname{Ext}(\operatorname{Hom}_E(N_{T/Q} \otimes_{O_T} E, \mathfrak{m}_L^{er}/\mathfrak{m}_L^{er+1}), G^r) \\ &\quad \to \operatorname{Hom}(G^{r\vee}, \operatorname{Hom}_E(\mathfrak{m}_L^{er}/\mathfrak{m}_L^{er+1}, N_{T/Q} \otimes_{O_T} E^{1/p^\infty})) \end{aligned} \tag{15.3.10}$$

(15.2.6), the class of the extension (15.3.9) defines a morphism

$$[\Phi] \colon G^{r\vee} \to \operatorname{Hom}_E(\mathfrak{m}_L^{er}/\mathfrak{m}_L^{er+1}, N_{T/Q} \otimes_{O_T} E^{1/p^\infty}). \tag{15.3.11}$$

Since $Q_E^{(r)\circ}$ is connected, the morphism $[\Phi]$ (15.3.11) is an injection by Corollary 15.2.6. Thus, for a non-trivial character $\chi\colon G^r \to \mathbf{F}_p$, its image $[\Phi](\chi)$ defines an injection $\mathfrak{m}_L^{er}/\mathfrak{m}_L^{er+1} \to N_{T/Q} \otimes_{O_T} E^{1/p^\infty}$.

The morphism $[\Phi]$ in (15.3.11) is related to the refinement of the rescaling function introduced in Part III, as we will see in Proposition 15.3.3.

**Lemma 15.3.2** *Let $L$ be a wildly ramified finite Galois extension of $K$ of degree $n$ of Galois group $G$. Assume that $O_L$ is generated by a single element $a$ over $O_K$, and let $f \in O_K[X]$ be the minimal polynomial. Let $r > 1$ be the rational number satisfying $G^r \neq 1$ and $G^{r+} = 1$, and $i > 1$ be the integer satisfying $G_i = \{\sigma \in G \mid \sigma(a)/a \equiv 1 \bmod \mathfrak{m}_L^i\} \neq 1$ and $G_{i+1} = 1$. Let $\chi\colon G^r \to \mathbf{F}_p$ be a non-trivial character and define $\varphi(\chi, a) \in L^\times$ by*

$$\varphi(\chi, a) = \prod_{j=1}^{k} b_j \cdot \prod_{k+1 \leqslant j < n, \chi(\sigma_j) \neq 0} b_j \cdot \prod_{k+1 \leqslant j < n, \chi(\sigma_j) = 1} b_j. \tag{15.3.12}$$

(1) *The valuation is* $\mathrm{ord}_L \varphi(\chi, a) = e_{L/K} \cdot r$.

*Let*

$$N_{T/Q} \otimes_{O_T} E \to \mathfrak{m}_L^{er}/\mathfrak{m}_L^{er+1} \tag{15.3.13}$$

*be the morphism sending the basis $f \in N_{T/Q} \otimes_{O_T} E$ to the basis $\varphi(\chi, a) \in \mathfrak{m}_L^{er}/\mathfrak{m}_L^{er+1}$.*

(2) *Assume that $L$ is a totally ramified extension of $K$ of degree $n$, and that $a \in L$ is a uniformizer. We identify $N_{T/Q} \otimes_{O_T} E$ with $\mathfrak{m}_K/\mathfrak{m}_K^2$ by the isomorphism* (14.5.4), *and let $\varphi_{\log}\colon \mathbf{N} \to \mathbf{Q}$ be the function defined in* (11.5.3). *Then $r = \varphi_{\log}(i) + 1$, $G^r = G_i$, $G^{r+} = G_{i+1} = 1$, and the morphism* (15.3.13) *is the multiplication by $(-1)^n \varphi(i, \chi) \in R_L$* (6.2.17).

(3) *Assume that the ramification index $e_{L/K}$ is 1, and that the residue field $E$ is a purely inseparable extension of $F$. We identify $N_{T/Q} \otimes_{O_T} E$ with $Z_{F\backslash E} \otimes_F E = \mathrm{Ker}(\Omega_F^1 \to \Omega_E^1) \otimes_F E$ by the isomorphism* (14.5.5). *Then $r = \varphi(i)$, $G^r = G_i$, $G^{r+} = G_{i+1} = 1$, and the morphism* (15.3.13) *equals $\varphi(i, \chi) \in \mathrm{Isom}_E(Z_{F\backslash E} \otimes_F E, \mathfrak{m}_L^r/\mathfrak{m}_L^{r+1}) \subset T_{L/K}$* (7.2.5).

*Proof* (1) We have $\mathrm{ord}_L \varphi(\chi, a) = e_{L/K} \cdot r$ by (15.3.1).

(2) By Corollary 12.3.7 and Lemma 11.5.6, for $r = \varphi_{\log}(i) + 1 = \varphi(i+1)$, we have $G^r = G_i \neq 1$ and $G^{r+} = G_{i+1} = 1$.

Since $f$ is an Eisenstein polynomial, the basis $f \in N_{T/Q} \otimes_{O_T} E$ is identified with the constant term $f(0) = -a_n^n \in \mathfrak{m}_K/\mathfrak{m}_K^2$ by the isomorphism (14.5.4), by Example 14.2.5(1). By (15.3.12) and (6.2.18), we have

$$\begin{aligned}\varphi(\chi, a) &= \prod_{j=1}^{k}(a_i - a_n) \cdot \prod_{k+1\leqslant j<n-1,\, \chi(\sigma_j)\neq 0} (\sigma_j(a_n) - a_n) \\ &\quad \cdot \prod_{k+1\leqslant j<n-1,\, \chi(\sigma_j)=1} (\sigma_j(a_n) - a_n) \\ &= (-1)^n \cdot (-a_n^n) \cdot \varphi(i, \chi).\end{aligned}$$

The assertion follows.

(3) By Corollary 12.3.7, for $r = \varphi(i)$ we have $G^r = G_i \neq 1$ and $G^{r+} = G_{i+1} = 1$. Let $b \in F$ be the element satisfying $a^n \equiv b \bmod \mathfrak{m}_K$ for $n = [L : K] = [E : F]$. Since $f \equiv (X - a)^n = X^n - b \bmod \mathfrak{m}_K$, the basis $f \in N_{T/Q} \otimes_{O_T} E$ is identified with $db \in Z_{F\backslash E} \otimes_F E$ by the isomorphism (14.5.5) and Proposition 14.1.1(2). We have

$$\begin{aligned}\varphi(\chi, a) &= \prod_{j=1}^{k}(a_i - a_n) \cdot \prod_{k+1\leqslant j<n-1,\, \chi(\sigma)\neq 0} (\sigma_j(a_n) - a_n) \\ &\quad \cdot \prod_{k+1\leqslant j<n-1,\, \chi(\sigma)=1} (\sigma_j(a_n) - a_n) \\ &= \varphi(i, \chi)((da)^{\otimes n})\end{aligned}$$

by (15.3.12) and (7.2.5). Since the canonical isomorphism $(\Omega^1_{E/F})^{\otimes n} \to Z_{F\backslash E} \otimes_F E$ maps $(da)^{\otimes n}$ to $db$, the assertion follows. □

In Lemma 15.3.2(3) the sign is cancelled by that in Proposition 14.1.1(1), but in Lemma 15.3.2(2) it is not.

**Proposition 15.3.3** *Assume that $O_L$ is generated by a single element $a \in O_L$ over $O_K$. Let $f \in O_K[X]$ be the minimal polynomial of $a$. Let $\chi\colon G^r \to \mathbf{F}_p$ be a non-trivial character, and let*

$$[\Phi](\chi)\colon \mathfrak{m}_L^{er}/\mathfrak{m}_L^{er+1} \to N_{T/Q} \otimes_{O_T} E^{1/p^\infty} \tag{15.3.14}$$

*be the injection defined by the image of $\chi$ by the injection $[\Phi]$ (15.3.11). Then the injection $[\Phi](\chi)$ (15.3.14) sends the basis $(-1)^n\varphi(\chi, a) \in \mathfrak{m}_L^{er}/\mathfrak{m}_L^{er+1}$ (15.3.12) to the class of $f$ in $N_{T/Q} \otimes_{O_T} E \subset N_{T/Q} \otimes_{O_T} E^{1/p^\infty}$.*

*Proof* Let $T = \operatorname{Spec} O_L \to Q = \mathbf{A}^1_S = \operatorname{Spec} O_K[X]$ be the closed immersion defined by $a$ as in Proposition 15.3.1. Let $c\colon G^r \to \mathbf{G}_a$ be the injection of groups defined by $c(\sigma) = (\sigma(a_n) - a_n)/b_{n-1}$, and $h = \prod_{\sigma\in G^r}(X - c(\sigma)) \in E[X]$ be the separable additive polynomial. By Proposition 15.3.1, we have an isomorphism

$$\begin{array}{ccccccccc} 0 & \longrightarrow & G^r & \longrightarrow & Q_E^{(r)\circ} & \longrightarrow & Q_{E,\mathrm{red}}^{[r]\circ} & \longrightarrow & 0 \\ & & \| & & \uparrow & & \uparrow & & \\ 0 & \longrightarrow & G^r & \xrightarrow{c} & \mathbf{G}_a & \xrightarrow{h} & \mathbf{G}_a & \longrightarrow & 0 \end{array} \tag{15.3.15}$$

of extensions where the right vertical arrow is an isomorphism defined by the morphism sending the basis $f \in N_{T/Q} \otimes_{O_T} E$ to $t = (-1)^k \prod_{j=1}^k b_j \cdot b_{n-1}^{n-k} \in \mathfrak{m}_L^{er}/\mathfrak{m}_L^{er+1}$.

Let $\chi\colon G^r \to \mathbf{F}_p$ be a non-trivial character. Then, by Lemma 15.2.7, we have a commutative diagram

$$\begin{array}{ccccccccc} 0 & \longrightarrow & G^r & \xrightarrow{c} & \mathbf{G}_a & \xrightarrow{h} & \mathbf{G}_a & \longrightarrow & 0 \\ & & \chi\downarrow & & \downarrow & & \downarrow u^{-p} & & \\ 0 & \longrightarrow & \mathbf{F}_p & \longrightarrow & \mathbf{G}_a & \xrightarrow{\mathrm{F}-1} & \mathbf{G}_a & \longrightarrow & 0 \end{array} \tag{15.3.16}$$

where $u = \prod_{k+1 \leqslant j < n, \chi(\sigma_j)=1} c(\sigma_j)$. Combining (15.3.15) and (15.3.16), we obtain a commutative diagram

$$\begin{array}{ccccccccc} 0 & \longrightarrow & G^r & \longrightarrow & Q_E^{(r)\circ} & \longrightarrow & Q_{E,\mathrm{red}}^{[r]\circ} & \longrightarrow & 0 \\ & & \chi\downarrow & & \downarrow & & \downarrow & & \\ 0 & \longrightarrow & \mathbf{F}_p & \longrightarrow & \mathbf{G}_a & \xrightarrow{\mathrm{F}-1} & \mathbf{G}_a & \longrightarrow & 0 \end{array} \tag{15.3.17}$$

of exact sequences where the inverse of the right vertical arrow is defined by the morphism sending $f$ to $tu^p$.

For $k+1 \leqslant j < n$ such that $\chi(\sigma_j) = 1$, we have $(b_{n-1}c(\sigma_j))^p = (\sigma_j(a_n) - a_n)^p \equiv (-1)^p \prod_{i=1}^{p-1} (\sigma_j^i(a_n) - a_n) \cdot (\sigma_j(a_n) - a_n)$ since $(\sigma_j^i(a_n) - a_n)/(\sigma_j(a_n) - a_n) \equiv i$ and $(-1)^p(p-1)! \equiv 1 \bmod p$. Since the mapping $\{\sigma_j \mid k+1 \leqslant j < n,\ \chi(\sigma_j) = 1\} \times \{i \mid i = 1, \ldots, p-1\} \to \{\sigma_j \mid k+1 \leqslant j < n,\ \chi(\sigma_j) \neq 0\}$ sending $(\sigma, i)$ to $\sigma^i$ is a bijection and $n - k = p \cdot \#\{\sigma_j \mid k+1 \leqslant j < n,\ \chi(\sigma_j) = 1\}$, we have

$$tu^p = (-1)^k \prod_{j=1}^k b_j \cdot \prod_{k+1 \leqslant j < n, \chi(\sigma_j)=1} (b_{n-1}c(\sigma_j))^p = (-1)^n \varphi(\chi, a)$$

and the assertion follows from Proposition 15.2.5. □

For abelian extensions, we show that the cohomological filtration and the logarithmic filtration are the same in the two cases (i) and (ii) specified below. We will give an outline proof of how to remove the assumption in Section 15.6.

Let $L$ be a finite abelian extension of $K$ of Galois group $G = \mathrm{Gal}(L/K)$. For an integer $n \geqslant 0$, let $G^n_{\mathrm{coh}} \subset G$ be the subgroup defined by

$$G^n_{\mathrm{coh}} = \{g \in G \mid \chi(g) = 0 \text{ for every } \chi \in F_{n-1}X_K\}. \tag{15.3.18}$$

Assume one of the following conditions is satisfied:

(i) $E$ is a separable extension of $F$.
(ii) The ramification index $e_{L/K}$ is 1 and $E$ is generated by a single element over $F$.

For $r \in \mathbf{Q}_+$, in case (i) (resp. (ii)), we have $G^r_{\log} = G^{r+1}$ (resp. $G^r_{\log} = G^r$) by Lemma 13.3.5.

**Corollary 15.3.4** *Let $L$ be a finite abelian extension of $K$ of of Galois group $G = \mathrm{Gal}(L/K)$. Let $n \geqslant 1$ be an integer.*

*Assume one of the following conditions is satisfied:*

*(i) $E$ is a separable extension of $F$.*
*(ii) The ramification index $e_{L/K}$ is 1 and $\dim \Omega^1_{E/F} = 1$.*

(1) *Let $r \in \mathbf{Q}_+$ and $n \in \mathbf{N}$. For $(n-1)+ \leqslant r \leqslant n$, $G^r_{\log} = G^n_{\mathrm{coh}}$.*
(2) *Let $r > 1$ be the integer satisfying $G^r \neq 1$ and $G^{r+} = 1$, and let*

$$\Phi\colon \mathrm{Hom}(G^r, \mathbf{F}_p) \to \mathrm{Hom}_F(\mathfrak{m}^r_K/\mathfrak{m}^{r+1}_K, N_{T/Q} \otimes_E \overline{F}) \tag{15.3.19}$$

*be the injection* (15.3.11). *Let $\chi\colon G \to \mathbf{Q}/\mathbf{Z}$ be a character such that the restriction $\chi|_{G^r}$ is non-trivial and let*

$$\Phi(\chi|_{G^r})\colon \mathfrak{m}^r_K/\mathfrak{m}^{r+1}_K \to N_{T/Q} \otimes_E \overline{F}$$

*be the image by $\Phi$ of $\chi|_{G^r}$ regarded as a morphism $G^r \to \mathbf{F}_p$. Let $n = r-1$ (resp. $n = r$) in case* (i) *(resp.* (ii)*) and let*

$$\mathrm{rsw}\,\chi\colon \mathfrak{m}^n_K/\mathfrak{m}^{n+1}_K \to \Omega^1_F(\log)$$

*be the refined Swan conductor.*

*In case* (i), *we identify $N_{T/Q} \otimes_{O_T} E$ with $\mathfrak{m}_K/\mathfrak{m}^2_K$ by the isomorphism* (14.5.4), *and regard $\Phi(\chi|_{G^r})$ as a morphism $\mathfrak{m}^r_K/\mathfrak{m}^{r+1}_K \to \mathfrak{m}_K/\mathfrak{m}^2_K \otimes_F \overline{F}$. Then $\Phi(\chi|_{G^r})$ is induced by the composition of* $\mathrm{rsw}\,\chi$ *with the residue morphism $\Omega^1_F(\log) \to F$.*

*In case* (ii), *we identify $N_{T/Q} \otimes_{O_T} E$ with $Z_{F\backslash E} \otimes_F E$ by the isomorphism* (14.5.5), *and regard $\Phi(\chi|_{G^r})$ as a morphism $\mathfrak{m}^r_K/\mathfrak{m}^{r+1}_K \to Z_{F\backslash E} \otimes_F \overline{F} \subset \Omega^1_F \otimes_F \overline{F} \subset \Omega^1_F(\log) \otimes_F \overline{F}$. Then $\Phi(\chi|_{G^r})$ is defined by* $-\mathrm{rsw}\,\chi$.

*Proof* By replacing $K$ by the largest unramified extension in $L$, we may assume that $E$ is a purely inseparable extension of $F$. Thus, in case (i), $L$ is a totally ramified extension.

(1) Since the upper numbering filtration and the cohomological filtrations are compatible with quotient, it suffices to show that the rational number $r$ satisfying $G^r_{\log} \neq 1$ and $G^{r+}_{\log} = 1$ is an integer and satisfies $G^r_{\mathrm{coh}} \neq 1$ and $G^{r+1}_{\mathrm{coh}} = 1$. Thus, the assertion follows from Proposition 11.5.8 and Theorem 5.3.4.

(2) Since the upper numbering filtrations and constructions of morphisms are compatible with quotients, we may further assume that $L$ is a cyclic extension such that $\chi\colon G \to \mathbf{Q}/\mathbf{Z}$ is an injection. Then, in case (i), the composition is given by the inverse of $\varphi(i, \chi|_{G^r}) \in R_K$ by Proposition 6.2.9(2). In case (ii) $\mathrm{rsw}\,\chi$ is given by the inverse of $\varphi(i, \chi|_{G^r}) \in S_K$ by Proposition 7.2.5(2). Hence, the assertion follows from Proposition 15.3.3 and Lemma 15.3.2. □

**Example 15.3.5** Let $K$ be a henselian discrete valuation field of characteristic $p > 0$, $F$ be the residue field, and $t \in K$ be a uniformizer.

(1) Let $L$ be the Artin–Schreier extension defined by $x^p - x = 1/t^n$ for $n = pm - 1 > 0$ as in Example 1.4.4(1). Let $S = \operatorname{Spec} O_K$ and $T = \operatorname{Spec} O_L$. Define a finite flat morphism $Q = \mathbf{A}^1_S = \operatorname{Spec} O_K[X] \to P = \mathbf{A}^1_S = \operatorname{Spec} O_K[Z]$ by sending $Z$ to the minimal polynomial $f = X^p - t^{(p-1)m}X - t \in O_K[X]$ as in Example 11.5.10(1). For $r = pm = n+1$, in the notation of Example 11.5.10(1), the morphism $Q^{(r)}_T = \mathbf{A}^1_T = \operatorname{Spec} O_L[Y] \to P^{(r)}_T = \mathbf{A}^1_T = \operatorname{Spec} O_L[W]$ is the finite étale morphism sending $W$ to $Y^p - Y$. Thus, we have a commutative diagram

$$\begin{array}{ccccccccc} 0 & \longrightarrow & G & \longrightarrow & Q^{(r)}_E & \longrightarrow & Q^{[r]}_{E,\mathrm{red}}(= P^{(r)}_E) & \longrightarrow & 0 \\ & & \downarrow & & \downarrow & & \downarrow & & \\ 0 & \longrightarrow & \mathbf{F}_p & \longrightarrow & \mathbf{A}^1_E & \xrightarrow{\mathrm{F}-1} & \mathbf{A}^1_E & \longrightarrow & 0 \end{array} \tag{15.3.20}$$

The vertical arrows are defined by the generator $\sigma \in G$ sending $x$ to $x+1$, $Y = (X - s)/t^m$, and $W = Z/t^{n+1}$ respectively. By the isomorphism $Q^{[r]}_{E,\mathrm{red}} = \operatorname{Hom}_E(\mathfrak{m}^{n+1}_K/\mathfrak{m}^{n+2}_K, N_{T/Q} \otimes_{O_L} E) \to \operatorname{Hom}_E(\mathfrak{m}^{n+1}_K/\mathfrak{m}^{n+2}_K, \mathfrak{m}_K/\mathfrak{m}^2_K)$, the linear form $W = Z/t^{n+1}$ sending $t^{n+1}$ to $f$ is sent to the basis sending $t^{n+1}$ to the constant term $-t$ of $f$. Since the refined Swan conductor $\mathrm{rsw}\,\chi$ of the character $\chi$ sending $\sigma$ to $1/p$ sends $t^n$ to $-d\log t$, the right vertical arrow is the map defined by $\mathrm{rsw}\,\chi$.

Since $s^p \equiv t$, $\varphi(\chi, s) = (p-1)!t^{pm}$ (15.3.12) equals $(-1)^{p-1}s^p$ times $\varphi(n, \chi) = -t^n$, as in Example 6.2.10(1).

(2) Let $L$ be the Artin–Schreier extension defined by $x^p - x = v/t^n$ for $n = pm > 0$ and for $v \in O_K$ such that $\bar{v} \in F - F^p$. Let $u = t^m x$ be the generator of the residue field $E = F(u)$ as in Example 1.4.4(2). Let $S = \operatorname{Spec} O_K$ and $T = \operatorname{Spec} O_L$. Define a finite flat morphism $Q = \mathbf{A}^1_S = \operatorname{Spec} O_K[X] \to P = \mathbf{A}^1_S = \operatorname{Spec} O_K[Z]$ by sending $Z$ to the minimal

polynomial $f = X^p - t^{(p-1)m}X - v \in O_K[X]$ as in Example 11.5.10(2). For $r = pm = n$, in the notation of Example 11.5.10(2), the morphism $Q_T^{(r)} = \mathbf{A}^1_T = \operatorname{Spec} O_L[Y] \to P_T^{(r)} = \mathbf{A}^1_T = \operatorname{Spec} O_L[W]$ is the finite étale morphism sending $W$ to $Y^p - Y$. Thus, we have the same commutative diagram (15.3.20) where the vertical arrows are defined by the generator $\sigma \in G$ sending $x$ to $x + 1$, $Y = (X - u)/t^m$, and $W = Z/t^n$ respectively. By the isomorphism $Q^{[r]}_{E,\mathrm{red}} = \operatorname{Hom}_E(\mathfrak{m}_L^{n+1}/\mathfrak{m}_L^{n+2}, N_{T/Q} \otimes_{O_L} E) \to \operatorname{Hom}_E(\mathfrak{m}_L^{n+1}/\mathfrak{m}_L^{n+2}, Z_{F\backslash E})$, the basis $W = Z/t^n$ sending $t^n$ to $f$ is sent to the basis sending $t^n$ to $-d(f) = dv$ since the diagram

$$\begin{array}{ccc} N_{E/Q_F} & \xrightarrow{-d} & \Omega^1_{Q_F} \otimes_{O_{Q_F}} E \\ \uparrow & & \uparrow \\ Z_{F\backslash E} \otimes_F E & \longrightarrow & \Omega^1_F \otimes_F E \end{array}$$

is commutative by Proposition 14.1.1(1). Since the refined Swan conductor $\operatorname{rsw}\chi$ of the character $\chi$ sending $\sigma$ to $1/p$ sends $t^n$ to $-dv$, the right vertical arrow is the map defined by $-\operatorname{rsw}\chi$.

Since $u^p \equiv v$, $\varphi(\chi, u) = (p-1)!t^n = -t^n$ (15.3.12) equals $\varphi(m, \chi)(dv) = (p-1)!t^n = -t^n$ as in Example 7.2.6.

**Exercise 15.3** Let $p$ be a prime number and $n \geqslant 2$ be an integer. Let $G^n = \operatorname{Gal}(\mathbf{Q}_p(\zeta_{p^n})/\mathbf{Q}_p(\zeta_{p^{n-1}})) \subset G = \operatorname{Gal}(\mathbf{Q}_p(\zeta_{p^n})/\mathbf{Q}_p)$ be the smallest nontrivial upper ramification subgroup and identify it with $1+(p^{n-1}) \subset (\mathbf{Z}/p^n\mathbf{Z})^\times$. Let $\chi \in G^{n\vee}$ be the dual basis of $1 + p^{n-1} \in 1 + (p^{n-1})$. Compute the image of $\chi$ by the injection $\Phi\colon G^{n\vee} \to \operatorname{Hom}_{\mathbf{F}_p}((p^n)/(p^{n+1}), (p)/(p^2)) = \operatorname{Hom}_{\mathbf{F}_p}((p^{n-1})/(p^n), \mathbf{F}_p)$ (15.3.11) and verify Corollary 15.3.4.

## 15.4 Groupoids and Étale Morphisms

Let $S$ be a scheme and $s, t\colon M \to X$ be morphisms of schemes over $S$. To ease the notation, let $M \times_X M$ denote the fiber product $M \times_{s,X,t} M$ with respect to $s\colon M \to X$ on the first factor and $t\colon M \to X$ on the second factor. Although the morphisms $s, t$ do not appear in the notation, we consider $s$ for the first factor and $t$ for the second factor. Similarly, we write $M \times_X M \times_X M$ for the fiber product $M \times_{s,X,t} M \times_{s,X,t} M$.

**Definition 15.4.1** Let $S$ be a scheme and $s, t\colon M \to X$ be morphisms of schemes over $S$. Let $\mu\colon M \times_X M \to M$, $e\colon X \to M$, and $i\colon M \to M$ be morphisms of schemes over $S$. Let $\delta\colon X \to X \times_S X$ be the diagonal morphism and let $i\colon X \times_S X \to X \times_S X$ be the isomorphism switching the two factors. We say abusively that $(M, X)$ is a groupoid over $S$ if the following six diagrams are commutative:

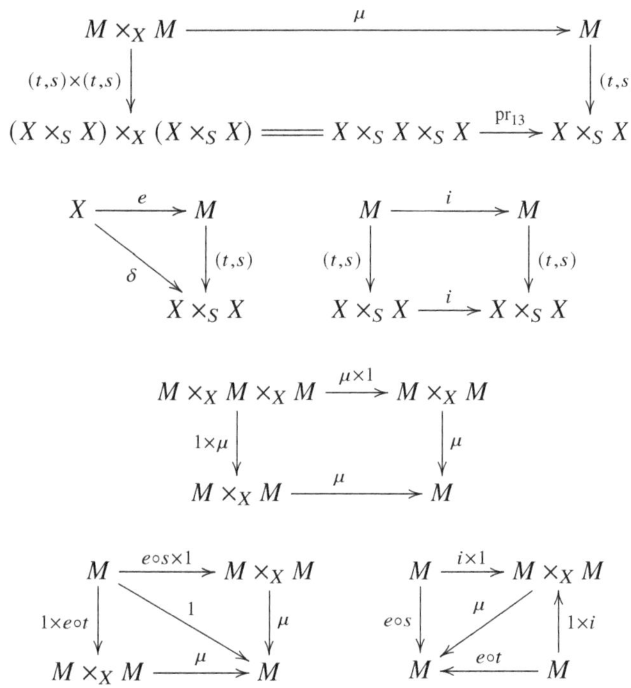

A groupoid $(M, X)$ over $S$ represents a functor from the category of schemes over $S$ to the category of groupoids. Conversely, a functor from the category of schemes over $S$ to the category of groupoids defines a groupoid over $S$ if the underlying functors are representable. If $s = t$, a groupoid $(M, X)$ over $S$ is a group scheme $M$ over a scheme $X$ over $S$.

**Lemma 15.4.2** *Let $(M, X)$ be a groupoid over S.*

(1) *The morphism $i\colon M \to M$ is an involution.*
(2) *If $s\colon M \to X$ is smooth, then $t$ and $\mu$ are also smooth.*

*Proof* (1) The composition of $(1, i, i \circ i)\colon M \to M \times_X M \times_X M$ with the 4th commutative diagram shows that $i \circ i\colon M \to M$ is 1.

(2) Since $i\colon M \to M$ is an isomorphism by (1), the composition $t = s \circ i$ is also smooth. Since $\nu = (\mathrm{pr}_1, \mu)\colon M \times_X M \to M \times_X M$ is an isomorphism and $s$ is smooth, the morphism $\mu = \mathrm{pr}_2 \circ \nu\colon M \times_X M \to M$ is smooth. □

Let $(M', X')$ be another groupoid over $S$. We say that a pair of morphisms

$X' \to X$ and $M' \to M$ of schemes over $S$ is a morphism $(M', X') \to (M, X)$ of groupoids over $S$ if they are compatible with $s, t, \mu, e, i$.

**Example 15.4.3** Let $X \to S$ be a morphism of schemes.

(1) Let $s = \mathrm{pr}_1$ and $t = \mathrm{pr}_2$ be the projections $X \times_S X \to X$. Let

$$\mu = \mathrm{pr}_{13}\colon (X \times_S X) \times_X (X \times_S X) = X \times_S X \times_S X \to X \times_S X$$

be the projection, $e\colon X \to X \times_S X$ be the diagonal, and $i\colon X \times_S X \to X \times_S X$ be the morphism switching the two factors. Then $(X \times_S X, X)$ is a groupoid over $S$.

If $(M, X)$ is a groupoid over $S$, the first three commutative diagrams in Definition 15.4.1 mean that the morphism $(t, s)\colon M \to X \times_S X$ defines a morphism of groupoids from $(M, X)$ to $(X \times_S X, X)$ over $S$. A morphism $X \to Y$ of schemes over $S$ defines a morphism $(X \times_S X, X) \to (Y \times_S Y, Y)$ of groupoids.

(2) Let $G$ be a finite group and let $Y \to X = Y/G$ be a $G$-torsor. Let $W = (Y \times_S Y)/G$ be the quotient of $Y \times_S Y$ by the diagonal action of $G$, and let $s', t'\colon W \to X$, $e'\colon X \to W$, and $i'\colon W \to W$ be the morphisms on the quotients induced by $s, t, e, i$ for $Y$. Let

$$\mu'\colon W \times_X W = (Y \times_S Y \times_X Y \times_S Y)/(G \times G) \to (Y \times_S Y)/G = W$$

be the morphism on the quotients induced by $\mathrm{pr}_{14}$ and the diagonal $G \to G \times G$. Then $(W, X)$ is a groupoid over $S$. The morphism $W \to X \times_S X$ induced by $Y \times_S Y \to X \times_S X$ defines a morphism $(W, X) \to (X \times_S X, X)$ of groupoids.

**Lemma 15.4.4** *Let*

$$\begin{array}{ccccc} X' & \stackrel{\supset}{\longleftarrow} & U' & \longleftarrow & V' \\ \downarrow & & \downarrow & & \downarrow \\ X & \stackrel{\supset}{\longleftarrow} & U & \longleftarrow & V \end{array}$$

*be a commutative diagram of normal schemes. The left horizontal arrows are dense open immersions and the right horizontal arrows are finite étale. Let $Y$ and $Y'$ be the normalizations of $X$ and $X'$ in $V$ and $V'$, and let $W \subset Y$ and $W' \subset Y'$ be the largest open subschemes étale over $X$ and over $X'$ respectively. If $X' \to X$ is* smooth, *then the morphism $Y' \to Y$ sends $W'$ to $W$.*

*Proof* By the assumption that $X' \to X$ is smooth, by replacing $X \leftarrow U \leftarrow V$ by $X' \leftarrow U' \leftarrow V \times_U U'$, we may assume that $X' = X$ and $U' = U$. It suffices to show that $Y' \times_X W' \to Y \times_X W'$ maps the image of the section $W' \to Y \times_X W'$ to the largest open subset $W \times_X W' \subset Y \times_X W'$ étale over $W'$.

Hence, by replacing the schemes by the base changes by $W' \to X$, we may assume that $W' \to X$ admits a section $X \to W'$ and it suffices to show that the morphism $Y' \to Y$ maps the image of the section $X \to W'$ to $W \subset Y$. The section $X \to W' \subset Y' \to Y$ defines a decomposition $V = U \amalg V_1$ and hence $Y = X \amalg Y_1$. Since $W = X \amalg (W \cap Y_1)$ and the image of the section $X \to W'$ is included in $X \subset W$, the assertion follows. □

**Lemma 15.4.5** *Let $(M, X)$ be a groupoid over a scheme $S$. Let $U \subset X$ be an open subscheme and assume $s^{-1}(U) = t^{-1}(U)$ so that they are equal to the inverse image $M_U \subset M$ of $U \times_S U$ by $(t, s)\colon M \to X \times_S X$.*

(1) *$(M_U, U)$ is a groupoid over $S$.*
(2) *Assume that $X$ is normal and that the morphism $s\colon M \to X$ is smooth. Let $(V, U)$ be a groupoid over $S$, and let $(V, U) \to (M_U, U)$ be a morphism of groupoids over $S$ such that the underlying morphism $V \to M_U$ is finite étale. Let $N \to M$ be the normalization of $M$ in $V$, and let $W \subset N$ be the largest open subscheme étale over $M$. Then the following conditions are equivalent:*

    (a) *$(W, X)$ has a groupoid structure over $S$ such that the morphism $(V, U) \to (M_U, U)$ of groupoids is extended to $(W, X) \to (M, X)$.*
    (b) *$e'\colon U \to V$ is extended to $e_W\colon X \to W$.*

    *If these equivalent conditions are satisfied, the groupoid structure on $(W, X)$ in* (a) *is unique.*

*Proof* (1) Since $(t, s)\colon M \to X \times_S X$ defines a morphism of groupoids, the morphisms $\mu, i$ for $(M, X)$ induce $\mu, i$ for $(M_U, U)$. By the assumption $M_U = s^{-1}(U) = t^{-1}(U)$, the morphisms $s, t, e$ for $(M, X)$ induce $s, t, e$ for $(M_U, U)$. The assertion follows.

(2) It is obvious that condition (b) is a necessary condition for (a). To show the converse, we first show that $s, t, \mu, i$ are defined for $(W, X)$. This is clear for $s$ and $t$. Since $s\colon M \to X$ is smooth, $M$, $M \times_X M$, and $W$ are also normal. By Lemma 15.4.2, the morphisms $t, \mu$ for $(M, X)$ are smooth and $i$ is an isomorphism. By Lemma 15.4.4, $W \times_X W \subset N \times_X N$ is the largest open subscheme étale over $M \times_X M$. Hence, by Lemma 15.4.4, the morphisms $\mu, i$ for $(V, U)$ are extended to morphisms $\mu\colon W \times_X W \to W$ and $i\colon W \times_X W \to W \times_X W$.

Assume that $e_W\colon X \to W$ is defined. Since the morphism $W \to M$ is separated, the morphisms $s, t, \mu, i, e$ for $W$ are uniquely determined by their restrictions on $V$. The morphisms $s, t, \mu, i, e$ for $(W, X)$ satisfy the groupoid axioms since they are satisfied on the dense open subschemes $(V, U)$. Thus, (b) implies (a). Since $W \to M$ is separated, the groupoid structure on $W$ compatible with $W \to M$ is unique. □

**Exercise 15.4** Let $k$ be a field and $X = \operatorname{Spec} k[T]$ be the affine line.

(1) Let $M = \operatorname{Spec} k[T,S][U^{\pm 1}]/(S - UT) = \operatorname{Spec} k[S,U^{\pm 1}]$, and define $(t,s)\colon M \to X \times_k X = \operatorname{Spec} k[T,S]$ to be the morphism defined by the inclusion $k[T,S] \to k[T,S][U^{\pm 1}]$. Show that $(M,X)$ is a groupoid over $k$ such that $(t,s)$ defines a morphism $(M,X) \to (X \times_k X, X)$ of groupoids over $k$.
(2) Let $n \geqslant 2$ be an integer and let $M = \operatorname{Spec} k[T,S][V,(1+VS^{n-1})^{-1}]/(T - (1+VS^{n-1})S) = \operatorname{Spec} k[S,V,(1+VS^{n-1})^{-1}]$; define $(t,s)\colon M \to X \times_k X = \operatorname{Spec} k[T,S]$ to be the morphism defined by $k[T,S] \to k[T,S][V,(1+VS^{n-1})^{-1}]$. Show the same conclusion as in (1).

## 15.5 Dilatations and Groupoids

Let $k$ be a field and $X$ be a smooth scheme over $k$. Let $D$ be an irreducible smooth divisor and $U = X - D$ be the complement. We consider the product $U \times U$ together with projections $\mathrm{pr}_1, \mathrm{pr}_2\colon U \times U \to U$ as a groupoid by the projection $\mathrm{pr}_{13}\colon (U \times U) \times_U (U \times U) = U \times U \times U \to U \times U$ as in Example 15.4.3(2).

Let $n \geqslant 1$ be an integer and let $\overline{P}^{[n]} \to X \times X$ be the blow-up at the divisor $nD \subset X \subset X \times X$ in the diagonal; define an open subscheme $P^{[n]} \subset \overline{P}^{[n]}$ to be the complement of the proper transforms of $X \times D$ and $D \times X$. Then the inverse image of the open subset $U \times U \subset X \times X$ is canonically identified with $U \times U \subset P^{[n]}$.

**Proposition 15.5.1** (1) *There exist unique morphisms*

$$\mu\colon P^{[n]} \times_X P^{[n]} \to P^{[n]}, \quad e\colon X \to P^{[n]}, \quad i\colon P^{[n]} \to P^{[n]}$$

*of schemes over $k$ extending the projection $\mathrm{pr}_{13}\colon (U \times U) \times_U (U \times U) = U \times U \times U \to U \times U$, the diagonal $\delta\colon U \to U \times U$, and the morphism $i\colon U \times U \to U \times U$ switching the two factors. These morphisms together with the canonical morphism $P^{[n]} \to X \times X$ define a groupoid $(P^{[n]}, X)$ over $k$ and a morphism $(P^{[n]}, X) \to (X \times X, X)$ of groupoids over $k$.*

(2) *The open subscheme $U \times U \subset P^{[n]}$ equals the complement of the inverse image of $D \subset X \subset X \times X$ in the diagonal by the canonical morphism $P^{[n]} \to X \times X$. The group scheme $P^{[n]} \times_{X \times X} D$ over $D$ is canonically identified with the restriction $TX(-nD)|_D$ of the twisted tangent bundle on $D$ defined by the dual of $\Omega^1_X(nD) \otimes_{\mathcal{O}_X} \mathcal{O}_D$.*

*Proof* (1) Let $\overline{P}_2^{[n]} \to X \times X \times X$ be the blow-up at the diagonal $D \subset X \subset X \times X \times X$ in the diagonal, and define $P_2^{[n]} \subset \overline{P}_2^{[n]}$ to be the complement of the union of the proper transforms of $D \times X \times X$, $X \times D \times X$, and $X \times X \times D$. Then the canonical isomorphism $X \times X \times X \to (X \times X) \times_X (X \times X)$ induces an isomorphism $P_2^{[n]} \to P^{[n]} \times_X P^{[n]}$ to the fiber product. Hence, by the universality of dilatations, the morphism $\mathrm{pr}_{13} \colon X \times X \times X \to X \times X$ induces $\mu \colon P^{[n]} \times_X P^{[n]} = P_2^{[n]} \to P^{[n]}$. Again by the universality of dilatation, $X \to X \times X$ is uniquely lifted to $X \to P^{[n]}$. The isomorphism $i \colon X \times X \to X \times X$ induces $i \colon P^{[n]} \to P^{[n]}$. They satisfy the axiom of groupoids and the assertion follows.

(2) The canonical morphism $P^{[n]} \to X \times X$ is an isomorphism on the complement $U \times U$ of the inverse image $P^{[n]} \times_{X \times X} D$ of $D$ in the diagonal $X \subset X \times X$. Since the conormal sheaf $N_{X/X \times X}$ is $\Omega^1_X$, the isomorphism $P^{[n]} \times_{X \times X} D \to TX(-nD)|_D$ follows from Lemma 11.1.5. □

**Proposition 15.5.2** ([6, Lemma 2.7], [73, Lemma 1.20]) *Let $X$ be a smooth scheme over an irreducible normal noetherian scheme $S$, let $U \subset X$ be a dense open subscheme, and let $V \to U$ be a finite étale morphism. Let $p \colon S' \to S$ be a quasi-finite dominant morphism of irreducible normal noetherian schemes, and let*

$$\begin{array}{ccccccc} S' & \longleftarrow & X' & \longleftarrow & U' & \longleftarrow & V' \\ {\scriptstyle p}\downarrow & & \downarrow{\scriptstyle q} & & \downarrow & & \downarrow \\ S & \longleftarrow & X & \longleftarrow & U & \longleftarrow & V \end{array}$$

*be a cartesian diagram of normal noetherian schemes. Let $Y$ and $Y'$ be the normalizations of $X$ and of $X'$ in $V$ and $V'$. Let $g \colon S \to Y$ be a section such that $V \cap g(S) \neq \varnothing$, and let $g' \colon S' \to Y'$ be the section induced by $g$. Consider the commutative diagram*

$$\begin{array}{ccccc} X' & \xleftarrow{f'} & Y' & \xleftarrow{g'} & S' \\ {\scriptstyle q}\downarrow & & {\scriptstyle h}\downarrow & & \downarrow{\scriptstyle p} \\ X & \xleftarrow{f} & Y & \xleftarrow{g} & S \end{array} \tag{15.5.1}$$

*Let $s' \in S'$ and assume that $f' \colon Y' \to X'$ is étale at $y' = g'(s') \in Y'$. Then $f \colon Y \to X$ is étale at $y = h(y') \in Y$.*

Proposition 15.5.2 is a special case of [6, Lemma 2.7] and [73, Lemma 1.20]. For the sake of convenience, we give a proof here.

*Proof* By replacing $X$, $Y$, $S$, and $S'$ by étale neighborhoods of $x = f(y)$, $y$, $s = p(s')$, and of $s'$, we may assume that $f^{-1}(x) = \{y\}$, $p^{-1}(s) = \{s'\}$, and

$p: S' \to S$ is finite surjective by Theorem 11.3.4. Since $X' = X \times_S S'$, we have $q^{-1}(x) = \{x'\}$ for $x' = f'(y')$.

We show that $h^{-1}(y) = \{y'\}$. Let $\eta \in V \subset Y$ be the image of the generic point of $S$ by $g$. Since $V' = V \times_S S' \subset Y'$, the inverse image $h^{-1}(V \cap g(S)) \subset Y'$ equals $V' \cap g'(S') \neq \varnothing$. Since $Y' \to Y$ is a surjective finite morphism of normal schemes, the inverse image $h^{-1}(g(S))$ equals the closure $g'(S')$ of $V' \cap g'(S')$ by the going down theorem, Proposition 11.2.3. Hence, we have $h^{-1}(y) = \{y'\}$ and consequently $f'^{-1}(x') = \{y'\}$.

Since the morphism $k(x') \to k(y')$ of the residue fields is an isomorphism, the finite étale morphism $\mathcal{O}_{X',x'} \to \mathcal{O}_{Y',y'}$ is an isomorphism. Since $\mathcal{O}_{Y',y'}$ is the normalization of $\mathcal{O}_{Y,y} \otimes_{\mathcal{O}_{X,x}} \mathcal{O}_{X',x'}$, the degree $[\operatorname{Frac} \mathcal{O}_{Y,y} : \operatorname{Frac} \mathcal{O}_{X,x}] = [\operatorname{Frac} \mathcal{O}_{Y',y'} : \operatorname{Frac} \mathcal{O}_{X',x'}]$ is 1 and $\mathcal{O}_{X,x} \to \mathcal{O}_{Y,y}$ is an isomorphism. □

**Proposition 15.5.3** *Let $X$ be a smooth scheme over a perfect field $k$, and $U = X - D$ be the complement of an irreducible divisor $D$ smooth over $k$. Let $G$ be a finite group and let $V \to U$ be a $G$-torsor. Let $K$ be the fraction field of the henselization $\mathcal{O}_K$ of the local ring $\mathcal{O}_{X,\xi}$ at the generic point $\xi$ of $D$, and assume that $V \times_U \operatorname{Spec} K$ is $\operatorname{Spec} L$ for a finite Galois extension $L$ of $K$.*

*Let $n \geqslant 1$ be an integer and let $(P^{[n]}, X)$ be the groupoid as in Proposition 15.5.1; let $((V \times V)/G, U)$ be the groupoid as in Example 15.4.3(2). Let $R^{(n)}$ be the normalization of $P^{[n]}$ in $(V \times V)/G \to U \times U$, and let $W \subset R^{(n)}$ be the largest open subset étale over $X$. Then the following conditions are equivalent:*

(1) *The ramification of $L$ over $K$ is bounded by $n+$.*
(2) *After replacing $X$ by a neighborhood of $\xi$, the pair $(W, X)$ is a groupoid over $k$ and the morphism $((V \times V)/G, U) \to (U \times U, U)$ of groupoids is extended to a morphism $(W, X) \to (P^{[n]}, X)$.*

*Proof* Let $Y$ be the normalization of $X$ in $V$. By replacing $X$ by a neighborhood of $\xi$, we may assume that $Y$ is smooth over $k$. Let $T = \operatorname{Spec} \mathcal{O}_L \to Q = Y \times S$ be the immersion to a smooth scheme over $S = \operatorname{Spec} \mathcal{O}_K$, and let $K'$ be a finite extension of $K$ such that $Q^{(n)}_{S'}$ is a stable integral model. We may assume that $K'$ contains $L$ as a subfield and fix a morphism $L \to K'$. We consider the cartesian diagram

$$\begin{array}{ccccc} P^{[n]} & \longleftarrow & P^{[n]}_S & \longleftarrow & P^{[n]}_{S'} \\ {\scriptstyle \mathrm{pr}_2}\downarrow & & \downarrow & & \downarrow \\ X & \longleftarrow & S & \longleftarrow & S' \end{array}$$

where the vertical arrows are smooth. By Theorem 14.4.4, condition (1) is equivalent to the following condition:

(1′) The finite morphism $Q^{(n)}_{S'} \to P^{[n]}_{S'}$ is étale on a neighborhood of the section $S' \to P^{[n]}_{S'}$ induced by the diagonal $X \to X \times X$.

Since the Galois group $G$ acts transitively on the inverse image of the section $S' \to P^{[n]}_{S'}$ by $Q^{(n)}_{S'} \to P^{[n]}_{S'}$, condition (1′) is equivalent to the following condition:

(1″) The finite morphism $Q^{(n)}_{S'} \to P^{[n]}_{S'}$ is étale on a neighborhood of the section $S' \to Q^{(n)}_{S'}$ induced by the morphism $L \to K'$.

Let $X \to R^{(n)}$ be the morphism extending the morphism $e\colon U = V/G \to (V \times V)/G$. Then, by Lemma 15.4.5(2), condition (2) is equivalent to the condition that the morphism $R^{(n)} \to P^{[n]}$ is étale on a neighborhood of the image of the generic point of $D \subset X$ by $X \to R^{(n)}$. Let $S \to R^{(n)}_S = R^{(n)} \times_X S \to P^{[n]}_S$ be the base change of $X \to R^{(n)} \to P^{[n]}$. Then, since $S$ is the henselization, condition (2) is further equivalent to the following condition:

(2′) The finite morphism $R^{(n)}_S \to P^{[n]}_S$ is étale on a neighborhood of the image of the section $S \to R^{(n)}_S$.

Since $S'$ is finite over $S$, condition (2′) is equivalent to the following condition by Proposition 15.5.2:

(2″) The finite morphism $R^{(n)}_{S'} \to P^{[n]}_{S'}$ is étale on a neighborhood of the image of the section $S' \to R^{(n)}_{S'}$.

We show that $R^{(n)}_{S'}$ is isomorphic to an open subscheme of $Q^{(n)}_{S'}$ to prove that conditions (1″) and (2″) are equivalent to each other. Since $((V \times_k V)/G) \times_U V = V \times_k V$, the normalization $R^{(n)}_T$ of the base change $R^{(n)}_S \times_S T$ by $T = \operatorname{Spec} O_L \to S$ is isomorphic to an open subscheme of the normalization $Q^{(n)}_T$. Since the morphism $\operatorname{Spec} K' \to U$ factors through $\operatorname{Spec} K' \to \operatorname{Spec} L \to V$, the normalization $R^{(n)}_{S'}$ of $R^{(n)}_S \times_S S'$ is isomorphic to an open subscheme of $Q^{(n)}_{S'}$. □

In the case of Proposition 15.5.3, the stable integral model is in fact defined over $T$.

**Exercise 15.5** Show that the groupoids $M$ in Exercise 15.4 are $P^{[n]}$ for $X = \mathbf{A}^1_k$, $D = (0)$, $n = 1$, and $n \geqslant 2$.

## 15.6 Characteristic Forms

**Theorem 15.6.1** *Let $K$ be a henselian discrete valuation field and $L$ be a finite Galois extension of $K$ of Galois group $G$. Let $r > 1$ be a rational number* *and assume that* $G^{r+} = 1$. *Fix a morphism* $L \to K_s$ *to a separable closure over* $K$, *and let the notation be as in Proposition* 14.6.1. *Then there exists a unique group scheme structure on* $\Phi^{(r)\circ}_{L/K,\overline{F}}$ *such that*

$$0 \longrightarrow G^r \longrightarrow \Phi^{(r)\circ}_{L/K,\overline{F}} \longrightarrow \Theta^{(r)\circ}_{L/K,\overline{F}} \longrightarrow 0 \qquad (15.6.1)$$

*is an extension of smooth group schemes over the residue field* $\overline{F}$ *of* $K_s$.

We give a proof under the assumption that $K$ is of characteristic $p > 0$ and that $r > 1$ is an integer $n \geqslant 2$, using the geometric construction in the previous section and an elementary lemma below. We omit the reduction to the case where $r = n \geqslant 2$ is an integer. In the mixed characteristic case, the proof in [76] is a modification of that in the equal characteristic case.

*Proof of Theorem* 15.6.1 *for* char $K = p > 0$ *and* $r = n \geqslant 2$ *an integer* The uniqueness follows from Lemma 15.2.8(2). Similarly to Lemma 13.4.5, by Corollary 9.1.6(2) we may assume that $K$ is the local field of a smooth scheme $X$ over a perfect field $k$ at the generic point of an irreducible smooth divisor $D$, and that the Galois extension $L$ is defined by a $G$-torsor $V \to U = X - D$ on the complement.

Let the notation be as in Proposition 15.5.3 and its proof. Then we obtain a morphism $(W, X) \to (P^{[n]}, X)$ of groupoids over $k$. The base change by $\operatorname{Spec} F \to D \to X \to X \times X$ defines a morphism $W_F \to P^{[n]}_F$ of group schemes over the residue field $F$. Let $W^\circ_F \subset W_F$ be the connected component containing the origin. Since $W^\circ_F \to P^{[n]}_F$ and $Q^{(n)\circ}_{F'} \to P^{[n]}_{F'}$ are finite étale, the open immersion $W^\circ_{F'} \to Q^{\circ(n)}_{F'}$ is an isomorphism. Applying Lemma 15.2.8(1) to the connected component $W^\circ_{F'} = Q^{\circ(n)}_{F'}$, we obtain an extension $0 \to G^n \to Q^{\circ(n)}_{F'} \to P^{[n]}_{F'} \to 0$ of smooth group schemes over $F'$. Since the diagram

$$\begin{array}{ccc} Q^{\circ(n)}_{\overline{F}} & \longrightarrow & P^{[n]}_{\overline{F}} \\ \downarrow & & \downarrow \\ \Phi^{(n)\circ}_{L/K,\overline{F}} & \longrightarrow & \Theta^{(n)\circ}_{L/K,\overline{F}} \end{array}$$

is cartesian and the right vertical arrow $P^{[n]}_{\overline{F}} \to \Theta^{(n)\circ}_{L/K,\overline{F}}$ admits a section, the assertion follows. □

For a rational number $r > 0$, let $\mathfrak{m}^r_{K_s} \supset \mathfrak{m}^{r+}_{K_s}$ be the ideals $\{x \in K_s \mid \operatorname{ord}_K x \geqslant r\} \supset \{x \in K_s \mid \operatorname{ord}_K x > r\}$ for the extension of the valuation of $K$ to a separable closure $K_s$. The quotient $\mathfrak{m}^r_{K_s}/\mathfrak{m}^{r+}_{K_s}$ is an $\overline{F}$-vector space of dimension 1. Since the ramification group $G^r$ is an $\mathbf{F}_p$-vector space by

Theorem 14.6.5, Theorem 15.6.1 implies that the injection (14.6.3) factors through an injection

$$\begin{aligned}\Phi\colon G^{r\vee} = \mathrm{Hom}_{\mathbf{F}_p}(G^r, \mathbf{F}_p) &\to \mathrm{Ext}(\Theta^{(r)\circ}_{L/K,\overline{F}}, \mathbf{F}_p) \\ &= \mathrm{Hom}_{\overline{F}}(\mathfrak{m}^r_{K_s}/\mathfrak{m}^{r+}_{K_s}, \mathrm{Tor}_1^{O_L}(\Omega^1_{O_L/O_K}, \overline{F}))\end{aligned} \tag{15.6.2}$$

by Corollary 15.2.6.

If the extension $E$ of the residue field is separable over $F$, (15.6.2) defines an injection

$$\begin{aligned}\Phi\colon \mathrm{Hom}_{\mathbf{F}_p}(G^r, \mathbf{F}_p) &\to \mathrm{Hom}_{\overline{F}}(\mathfrak{m}^r_{K_s}/\mathfrak{m}^{r+}_{K_s}, \mathfrak{m}_K/\mathfrak{m}_K^2 \otimes_F \overline{F}) \\ &= \mathrm{Hom}_{\overline{F}}(\mathfrak{m}^{r-1}_{K_s}/\mathfrak{m}^{(r-1)+}_{K_s}, \overline{F})\end{aligned} \tag{15.6.3}$$

by the isomorphism $\mathrm{Tor}_1^{O_L}(\Omega^1_{O_L/O_K}, \overline{F}) \to \mathfrak{m}_K/\mathfrak{m}_K^2 \otimes_F \overline{F}$ (14.1.7). By taking the composition of (15.6.2) with the morphism induced by the composition $\mathrm{Tor}_1^{O_L}(\Omega^1_{O_L/O_K}, E) \to \mathrm{H}_1(L_{E/S}) \to \Omega^1_F \otimes_F E$, we obtain

$$\overline{\Phi}\colon \mathrm{Hom}_{\mathbf{F}_p}(G^r, \mathbf{F}_p) \longrightarrow \mathrm{Hom}_{\overline{F}}(\mathfrak{m}^r_{K_s}/\mathfrak{m}^{r+}_{K_s}, \Omega^1_F \otimes_F \overline{F}). \tag{15.6.4}$$

If the ramification index $e_{L/K}$ is 1, the morphism $\overline{\Phi}$ is an injection since $\mathrm{Tor}_1^{O_L}(\Omega^1_{O_L/O_K}, E) \to \Omega^1_F \otimes_F E$ is an injection by Proposition 14.2.2(3).

The injection (15.6.2) satisfies the following functoriality. Let $K'$ be an extension of henselian discrete valuation field $K$, and let $e = e_{K'/K}$ be the ramification index. Let $L'$ be a finite separable extension of $K'$, and let $L \to L'$ be a morphism compatible with $K \to K'$. Let $K_s \to K'_s$ be a morphism of separable closures compatible with $K \to K'$. We also fix $L' \to K'_s$ compatible with $L \to L'$ and with $L \to K_s$. Then, for $r' = er$, the diagram

$$\begin{array}{ccc}\mathrm{Hom}_{\mathbf{F}_p}(G^r, \mathbf{F}_p) & \xrightarrow{\Phi} & \mathrm{Hom}_{\overline{F}}(\mathfrak{m}^r_{K_s}/\mathfrak{m}^{r+}_{K_s}, \mathrm{Tor}_1^{O_L}(\Omega^1_{O_L/O_K}, \overline{F})) \\ \downarrow & & \downarrow \\ \mathrm{Hom}_{\mathbf{F}_p}(G'^{r'}, \mathbf{F}_p) & \xrightarrow{\Phi} & \mathrm{Hom}_{\overline{F}}(\mathfrak{m}^{r'}_{K'_s}/\mathfrak{m}^{r'+}_{K'_s}, \mathrm{Tor}_1^{O_{L'}}(\Omega^1_{O_{L'}/O_{K'}}, \overline{F}'))\end{array} \tag{15.6.5}$$

is commutative.

If the residue field of $K'$ is perfect and if $K'$ satisfies the condition in Lemma 14.1.7, the right vertical arrow with $\mathrm{Tor}_1^{O_{L'}}(\Omega^1_{O_{L'}/O_{K'}}, \overline{F}')$ replaced by $\mathfrak{m}_{K'}/\mathfrak{m}_{K'}^2 \otimes_{F'} \overline{F}'$ is an injection. Hence, the morphism (15.6.2) in general is characterized by the functoriality (15.6.5) and the morphism (15.6.3) in the case where the residue field is perfect.

The composition with the injection $\mathrm{Tor}_1^{O_L}(\Omega^1_{O_L/O_K}, \overline{F}) \to \mathrm{H}_1(L_{\overline{F}/O_K})$ (14.2.2) defines an injection

$$\mathrm{ch}\colon \mathrm{Hom}_{\mathbf{F}_p}(G^r, \mathbf{F}_p) \longrightarrow \mathrm{Hom}_{\overline{F}}(\mathfrak{m}^r_{K_s}/\mathfrak{m}^{r+}_{K_s}, \mathrm{H}_1(L_{\overline{F}/O_K})). \tag{15.6.6}$$

Removing the condition $G^{r+} = 1$ and applying the construction to the subextension $M$ over $K$ corresponding to the subgroup $G^{r+}$, we obtain an injection

$$\mathrm{ch}\colon \mathrm{Hom}_{\mathbf{F}_p}(\mathrm{Gr}^r G, \mathbf{F}_p) \longrightarrow \mathrm{Hom}_{\overline{F}}(\mathfrak{m}^r_{K_s}/\mathfrak{m}^{r+}_{K_s}, \mathrm{H}_1(L_{\overline{F}/O_K})). \quad (15.6.7)$$

The composition with the surjection $\mathrm{H}_1(L_{\overline{F}/O_K}) \to \Omega^1_F \otimes_F \overline{F}$ defines

$$\overline{\mathrm{ch}}\colon \mathrm{Hom}_{\mathbf{F}_p}(\mathrm{Gr}^r G, \mathbf{F}_p) \longrightarrow \mathrm{Hom}_{\overline{F}}(\mathfrak{m}^r_{K_s}/\mathfrak{m}^{r+}_{K_s}, \Omega^1_F \otimes_F \overline{F}). \quad (15.6.8)$$

If the ramification index $e_{L/K}$ is 1, this morphism $\overline{\mathrm{ch}}$ is also an injection. The functoriality (15.6.5) implies that these morphisms satisfy a similar functoriality.

For a non-trivial character $\chi\colon \mathrm{Gr}^r G \to \mathbf{F}_p$, its image by the injection (15.6.7) defines an injection

$$\mathrm{ch}\,\chi\colon \mathfrak{m}^r_{K_s}/\mathfrak{m}^{r+}_{K_s} \to \mathrm{H}_1(L_{\overline{F}/O_K}). \quad (15.6.9)$$

We call this morphism the characteristic form of $\chi$. If the ramification index $e_{L/K}$ is 1, the induced morphism

$$\overline{\mathrm{ch}}\,\chi\colon \mathfrak{m}^r_{K_s}/\mathfrak{m}^{r+}_{K_s} \to \Omega^1_F \otimes_F \overline{F} \quad (15.6.10)$$

is also an injection since $\mathrm{Tor}_1^{O_L}(\Omega^1_{O_L/O_K}, \overline{F}) \to \Omega^1_F \otimes_F \overline{F}$ is an injection by Proposition 14.2.2(3).

The logarithmic variant of $\mathrm{ch}\,\chi$ introduced in the next chapter is a non-abelian generalization of the refined Swan conductor.

**Example 15.6.2** Let $K$ be a henselian discrete valuation field of characteristic $p > 0$, $F$ be the residue field, and $t \in K$ be a uniformizer.

(1) Let $L$ be the Artin–Schreier extension defined by $x^p - x = 1/t^n$ for $n = pm-1 > 0$ as in Example 1.4.4(1). Let $\sigma \in G = \mathrm{Gal}(L/K)$ be the generator defined by $\sigma(x) = x + 1$, and let $\chi \in \mathrm{Hom}(G, \mathbf{F}_p)$ be the character defined by $\chi(\sigma) = 1$. Since the isomorphism $\mathbf{A}^1_E \to Q^{[r]}_{E,\mathrm{red}}$ for $r = n + 1$ in Example 15.3.5(1) is defined by the basis of $\mathrm{Hom}_E(\mathfrak{m}^{n+1}_K/\mathfrak{m}^{n+2}_K, \mathfrak{m}_K/\mathfrak{m}^2_K)$ sending $t^{n+1}$ to $-t$, the morphism $\mathrm{ch}\colon \mathrm{Hom}_{\mathbf{F}_p}(G^r, \mathbf{F}_p) \to \mathrm{Hom}_{\overline{F}}(\mathfrak{m}^r_{K_s}/\mathfrak{m}^{r+}_{K_s}, \mathrm{H}_1(L_{\overline{F}/O_K}))$ (15.6.6) sends $\chi$ to the morphism sending $t^{n+1}$ to the image $w(t)$ of $-t$ by $\mathfrak{m}_K/\mathfrak{m}^2_K \to \mathrm{H}_1(L_{\overline{F}/O_K})$, by Proposition 14.1.6(1).

(2) Let $L$ be the Artin–Schreier extension defined by $x^p - x = v/t^n$ for $n = pm > 0$ and for $v \in O_K$ such that $\bar{v} \in F - F^p$. Let $u = t^m x$ be the generator of the residue field $E = F(u)$ as in Example 1.4.4(2). Let $\sigma \in G = \mathrm{Gal}(L/K)$ be the generator defined by $\sigma(x) = x + 1$, and let $\chi \in \mathrm{Hom}(G, \mathbf{F}_p)$ be the character defined by $\chi(\sigma) = 1$. The isomorphism $\mathbf{A}^1_E \to Q^{[r]}_{E,\mathrm{red}}$ for $r = n$ in Example 15.3.5(2) is defined by the basis of

$\mathrm{Hom}_E(\mathfrak{m}_K^{n+1}/\mathfrak{m}_K^{n+2}, N_{T/Q} \otimes_{O_L} E)$ sending $t^n$ to $f$. In the commutative diagram

$$\begin{array}{ccc} N_{T/Q} \otimes_{O_L} E & \longrightarrow & N_{E/Q} \\ \uparrow & & \uparrow \\ \mathrm{Tor}_1^{O_L}(\Omega^1_{T/S}, E) & \longrightarrow & \mathrm{H}_1(L_{E/S}) \end{array}$$

the vertical arrows are injections and the basis $f = X^p - t^{(p-1)m}X - v$ of $N_{T/Q}$ is sent to $X^p - v$ unless $n = p = 2$, and to the linear combination $(X^p - v) - u \cdot t$ of the basis $X^p - v, t$ of $N_{E/Q}$ in the exceptional case $n = p = 2$. Hence, the morphism $\mathrm{ch}\colon \mathrm{Hom}_{\mathbf{F}_p}(G^r, \mathbf{F}_p) \to \mathrm{Hom}_{\overline{F}}(\mathfrak{m}^r_{K_s}/\mathfrak{m}^{r+}_{K_s}, \mathrm{H}_1(L_{\overline{F}/O_K}))$ (15.6.6) sends $\chi$ to the morphism sending $t^n$ to $w(v)$ unless $n = p = 2$ (resp. to $w(v) + vw(t)$ if $n = p = 2$).

**Exercise 15.6** Let $K$ be a henselian discrete valuation with residue field $F$ of characteristic $p > 0$. Let $L$ be a cyclic extension of degree $p$ of $K$, and assume that the residue field $E$ is a purely inseparable extension of degree $p$ of $F$ as in Exercise 14.5. Let $\chi\colon G = \mathrm{Gal}(L/K) \to \mathbf{F}_p$ be a faithful character and $n = \mathrm{Sw}\,\chi \geqslant 1$ be the Swan conductor.

(1) Show that $G^n = G$ and $G^{n+} = 1$.
(2) Compute the characteristic form $\mathrm{ch}\,\chi\colon \mathfrak{m}_K^n/\mathfrak{m}_K^{n+1} \to \mathrm{H}_1(L_{E/S})$.
(3) Show that the composition $\mathfrak{m}_K^n/\mathfrak{m}_K^{n+1} \to \mathrm{H}_1(L_{E/S}) \to \Omega^1_F \otimes_F E$ is induced by *minus* the refined Swan conductor $\mathrm{rsw}\,\chi$.

## Historical Notes

The description of the extension group of $\mathbf{G}_a$ by $\mathbf{F}_p$ is given in [29]. The construction of the characteristic form given in [76] is a modification of that in the equal characteristic case using the groupoid structure in [73]. The monogenic case is also studied in [1]. The $F$-rationality of the characteristic form for abelian extensions is proved in [64] by establishing and using a relation with the refined Swan conductor.

Another construction of the characteristic form and its logarithmic variant using a $p$-adic differential equation was given in [83].

# 16

# Logarithmic Characteristic Forms and the Refined Swan Conductor

We construct a logarithmic variant of characteristic form in Section 16.2. As a preliminary, in Section 16.1 we study a logarithmic variant of the constructions in Section 14.2. For Artin–Schreier–Witt extensions, we give a description of the logarithmic variant of the injection in terms of Witt vectors in Section 16.3.

Graded quotients of the cohomological filtration are captured by the refined Swan conductor, whose existence is stated as Theorem 16.3.1 in Section 16.3.

In the mixed characteristic case, we sketch an outline of part of Kato's original proof of Theorem 16.3.1 in Section 16.4, based on a detailed study of $p$-adic cohomology. We show the existence and the characterizing property, Theorem 16.3.1, of refined Swan conductors by introducing the integral parts of the Brauer group and the space of differential forms of the residue field with log poles.

In the positive characteristic case, the refined Swan conductor is constructed using Witt vectors and Brylinski's filtration. In this case, the cohomological filtration and the logarithmic filtration are the same by this construction and Theorem 13.4.4. In the mixed characteristic case, the equality of the two filtrations is proved by reduction to the case where the ramification index is 1 and the ring of integers of the corresponding cyclic extension is generated by a single element. To show the reduction, we use the fact that the refined Swan conductor and the logarithmic characteristic form share the same characterization up to sign.

## 16.1 Differential Forms with Log Poles

We recall the definition and some basic properties of the sheaves of differential forms with logarithmic poles.

Let $X$ be a regular noetherian scheme and let $D \subset X$ be a regular divisor. Let

$j: U = X - D \to X$ be the open immersion of the complement and let $\mathcal{M}_X$ denote the sheaf $\mathcal{M}_X = \mathcal{O}_X \cap j_*\mathcal{O}_U^\times \subset j_*\mathcal{O}_U$ of multiplicative monoids. Let $i_1: \mathcal{M}_X \to \mathcal{O}_X$ and $i_2: \mathcal{M}_X \to j_*\mathcal{O}_U^\times$ be the injections, and let $d: \mathcal{M}_X \to \Omega^1_X$ denote the composition of $d: \mathcal{O}_X \to \Omega^1_X$ and $i_1: \mathcal{M}_X \to \mathcal{O}_X$. We define the sheaf $\Omega^1_X(\log D)$ of differentials with logarithmic poles to be the quotient

$$\Omega^1_X(\log D) = (\Omega^1_X \oplus (\mathcal{O}_X \otimes_{\mathbf{Z}_X} j_*\mathcal{O}_U^\times))/\mathcal{O}_X \cdot \mathrm{Im}(d, -i_1 \otimes i_2).$$

For a section $a$ of $j_*\mathcal{O}_U^\times$, let $d\log a$ denote the image of $(0, 1 \otimes i_2(a))$. The relation defining $\Omega^1_X(\log D)$ implies the equality $da = a \cdot d\log a$ for a section $a$ of $\mathcal{M}_X$.

**Lemma 16.1.1** *Let $X$ be a regular noetherian scheme and let $D \subset X$ be a regular divisor.*

(1) *Assume that $D$ is defined by a global section $t \in \Gamma(X, \mathcal{O}_X)$. Then the $\mathcal{O}_X$-module $\Omega^1_X(\log D)$ is the quotient $(\Omega^1_X \oplus \mathcal{O}_X \cdot d\log t)/(dt - t \cdot d\log t)$.*
(2) *Let $i: D \to X$ be the closed immersion. Then the morphism $j_*\mathcal{O}_U^\times \to i_*\mathbf{Z}_D$ of sheaves defined by the valuation at $D$ induces an $\mathcal{O}_X$-linear morphism $\mathrm{res}: \Omega^1_X(\log D) \to i_*\mathcal{O}_D$ called the residue morphism, and the sequence*

$$\Omega^1_X \longrightarrow \Omega^1_X(\log D) \xrightarrow{\mathrm{res}} i_*\mathcal{O}_D \longrightarrow 0 \tag{16.1.1}$$

*is exact. The sequence* (16.1.1) *induces an exact sequence*

$$0 \longrightarrow \Omega^1_D \longrightarrow \Omega^1_X(\log D) \otimes_{\mathcal{O}_X} \mathcal{O}_D \xrightarrow{\mathrm{res}} \mathcal{O}_D \longrightarrow 0 \tag{16.1.2}$$

*of $\mathcal{O}_D$-modules.*
(3) *Let $f: X \to Y$ be a smooth morphism of regular noetherian schemes. Let $E \subset Y$ be a regular divisor and assume that $D \subset X$ is the inverse image. Then we have an exact sequence $0 \to f^*\Omega^1_Y(\log E) \to \Omega^1_X(\log D) \to \Omega^1_{X/Y} \to 0$.*

*Proof* (1) Since the quotients $\mathcal{M}_X/\mathcal{O}_X^\times$ and $j_*\mathcal{O}_U^\times/\mathcal{O}_X^\times$ by the subsheaf $\mathcal{O}_X^\times$ are isomorphic to $i_*\mathbf{N}_D$ and $i_*\mathbf{Z}_D$ generated by $t$, the assertion follows.

(2) The morphism $\Omega^1_X(\log D)/\Omega^1_X \to i_*\mathcal{O}_D$ induced by the residue morphism is an isomorphism by (1). Again by (1), locally on $D$, $\Omega^1_X(\log D) \otimes_{\mathcal{O}_X} \mathcal{O}_D$ is isomorphic to the quotient of $(\Omega^1_X \otimes_{\mathcal{O}_X} \mathcal{O}_D) \oplus \mathcal{O}_D \cdot d\log t$ divided by $dt$. Hence, the exact sequence $N_{D/X} \to \Omega^1_X \otimes_{\mathcal{O}_X} \mathcal{O}_D \to \Omega^1_D \to 0$ implies (16.1.2).

(3) We have a locally split exact sequence $0 \to f^*\Omega^1_Y \to \Omega^1_X \to \Omega^1_{X/Y} \to 0$. We also have a description of $\Omega^1_Y(\log E)$ as in (1) with common local uniformizer $t$. Hence, the assertion follows. □

Let $f: X \to Y$ be a morphism of regular noetherian schemes. Let $D \subset X$ and $E \subset Y$ be regular divisors such that we have an equality $D = f^{-1}(E)$

of the underlying sets. Then the canonical morphism $f^*\Omega^1_Y \to \Omega^1_X$ induces $f^*\Omega^1_Y(\log E) \to \Omega^1_X(\log D)$. Let $\Omega^1_{X/Y}(\log D/\log E)$ denote the cokernel $\mathrm{Coker}(f^*\Omega^1_Y(\log E) \to \Omega^1_X(\log D))$.

For a regular flat noetherian scheme $X$ over $S = \mathrm{Spec}\, O_K$ such that the reduced closed fiber $D = X_{F,\mathrm{red}} \subset X$ is a regular divisor, we drop $D$ in the notation and write $\Omega^1_X(\log) = \Omega^1_X(\log D)$. We also write $\Omega^1_{X/S}(\log/\log) = \mathrm{Coker}(\Omega^1_S(\log) \otimes_{O_S} O_X \to \Omega^1_X(\log))$.

**Proposition 16.1.2** *Let $Q$ be a log smooth scheme over $S = \mathrm{Spec}\, O_K$.*

(1) *The morphism $\Omega^1_S(\log) \otimes_{O_S} O_Q \to \Omega^1_Q(\log)$ is a locally split injection and its cokernel*

$$\Omega^1_{Q/S}(\log/\log) = \mathrm{Coker}(\Omega^1_S(\log) \otimes_{O_S} O_Q \to \Omega^1_Q(\log)) \tag{16.1.3}$$

*is a locally free $O_Q$-module of finite rank.*

(2) *Let $S' \to S$ be a morphism of the spectrums of discrete valuation rings, and let $Q_{S'}$ be the normalization of $Q \times_S S'$. Then $\Omega^1_{Q/S}(\log/\log) \otimes_{O_Q} O_{Q_{S'}} \to \Omega^1_{Q_{S'}/S'}(\log/\log)$ is an isomorphism.*

*Proof* By Lemma 16.1.1(2), we may assume $Q = Q_m$. We consider the case where $m$ is divisible by $p$. The case where $m$ is prime to $p$ is similar and easier, and we omit the proof. Set $A = O_K[X, U^{\pm 1}]/(UX^m - t)$.

(1) $\Omega^1_A = (\Omega^1_{O_K} \otimes_{O_K} A \oplus A \cdot dX \oplus A \cdot dU)/(dt - mX^{m-1} \cdot dX - X^m \cdot dU)$. Hence, by Lemma 16.1.1(1), we have

$$\begin{aligned}\Omega^1_A(\log) &= \frac{\Omega^1_{O_K}(\log) \otimes_{O_K} A \oplus A \cdot d\log X \oplus A \cdot dU}{d\log t - m \cdot d\log X - d\log U}\\ &= \Omega^1_{O_K}(\log) \otimes_{O_K} A \oplus A \cdot d\log X,\end{aligned}$$

and the assertion follows.

(2) Let $Q'$ be another log smooth scheme over $S$, and let $S' \to Q$ be a strict morphism over $S$. Let $Q''$ be the normalization of $Q \times_S Q'$. Then, by Proposition 13.1.5(2), $Q_{S'} = Q'' \times_{Q'} S'$ and $\Omega^1_{Q''/Q'}(\log/\log) \otimes_{O_{Q''}} O_{Q_{S'}} \to \Omega^1_{Q_{S'}/S'}(\log/\log)$ is an isomorphism. Hence, it suffices to show that

$$\Omega^1_{Q/S}(\log/\log) \otimes_{O_Q} O_{Q''} \to \Omega^1_{Q''/Q'}(\log/\log)$$

is an isomorphism.

We may assume that $Q = Q_m = \mathrm{Spec}\, A$ and $Q' = Q_e = \mathrm{Spec}\, B$ for $A = O_K[x, u^{\pm 1}]/(ux^m - t)$ and $B = O_K[y, v^{\pm 1}]/(vy^e - t)$, as in the proof of Corollary 13.1.3(1). In the notation of its proof, we have $Q'' = \mathrm{Spec}\, C$ for $C = B[z, w^{\pm 1}]/(w^b z^{m'} - y)$. Then, $\Omega^1_{Q/S}(\log/\log)$ is a free $O_Q$-module of rank 1 generated by $d\log x$, and $\Omega^1_{Q''/Q'}(\log/\log)$ is a free $O_{Q''}$-module of rank 1 generated by $d\log w, d\log z$ satisfying the relation $b\cdot d\log w + m'\cdot d\log z = 0$. Since the morphism $\Omega^1_{Q/S}(\log/\log)\otimes_{O_Q} O_{Q''} \to \Omega^1_{Q''/Q'}(\log/\log)$ sends the basis $d\log x$ to $e'\cdot d\log z - a'\cdot d\log w$ and $am' + be' = 1$, the assertion follows. □

**Corollary 16.1.3** *Let $Q$ be a log smooth scheme over $S$, and let $X \subset Q$ be a regular closed subscheme flat over $S$. Then the following conditions are equivalent:*

(1) *$X$ is log smooth over $S$.*

(2) *The morphism $N_{X/Q} \to \Omega^1_{Q/S}(\log/\log)\otimes_{O_Q} O_X$ is a locally split injection.*

*Proof* (1) ⇒ (2): Since the sequence $N_{X/Q} \to \Omega^1_{Q/S}(\log/\log)\otimes_{O_Q} O_X \to \Omega^1_{X/S}(\log/\log) \to 0$ of locally free $O_X$-modules is exact, and since $N_{X/Q} \to \Omega^1_{Q/S}(\log/\log)\otimes_{O_Q} O_X$ is an injection on the generic fiber $X_K$, the assertion follows.

(2) ⇒ (1): We may assume that $Q$ is smooth over $Q_m$ for $m \geqslant 1$ divisible by $p$ or over $Q'_m$ for $m \geqslant 1$ prime to $p$. First, assume that $Q$ is smooth over $Q_m$ for $m \geqslant 1$ divisible by $p$. Then the reduced closed fibers $D = Q_{F,\mathrm{red}}$ and $D_0 = Q_{m,F,\mathrm{red}}$ are smooth over $F$ and we have an exact sequence $0 \to \Omega^1_{D/D_0} \to \Omega^1_{Q/S}(\log/\log)\otimes_{O_Q} O_D \overset{\mathrm{res}}{\to} O_D \to 0$. Since the composition of $N_{X/Q}\otimes_{O_X} O_{X\cap D} \to \Omega^1_{D/D_0}\otimes_{O_D} O_{X\cap D}$ with the residue morphism is the 0-morphism, this is a locally split injection and hence $X$ is smooth over $Q_m$ on a neighborhood of $X\cap D$. Thus $X$ is log smooth over $S$ in this case.

In the case where $Q$ is smooth over $Q'_m$ for $m \geqslant 1$ prime to $p$, since the canonical morphism $\Omega^1_{Q/S}(\log/\log) \to \Omega^1_{Q/Q'_m}$ is an isomorphism, the scheme $X$ is smooth over $Q'_m$. Hence, $X$ is also log smooth over $S$ in this case. □

**Lemma 16.1.4** *Let $L$ be a finite separable extension of $K$, and let $T = \operatorname{Spec} O_L \to Q$ be an exact immersion to a log smooth scheme over $S = \operatorname{Spec} O_K$.*

(1) *The sequence*

$$0 \to N_{T/Q}\otimes_{O_T} E \to \Omega^1_Q(\log)\otimes_{O_Q} E \to \Omega^1_E(\log) \to 0 \qquad (16.1.4)$$

*is exact.*

(2) *The sequence*

$$0 \to N_{T/Q} \to \Omega^1_{Q/S}(\log/\log)\otimes_{O_Q} O_T \to \Omega^1_{T/S}(\log/\log) \to 0 \quad (16.1.5)$$

*is exact and defines an exact sequence*

$$0 \to \mathrm{Tor}_1^{\mathcal{O}_L}(\Omega^1_{T/S}(\log/\log), E) \to N_{T/Q} \otimes_{\mathcal{O}_T} E$$
$$\to \Omega^1_{Q/S}(\log/\log) \otimes_{\mathcal{O}_Q} E \to \Omega^1_{T/S}(\log/\log) \otimes_{\mathcal{O}_Q} E \to 0. \quad (16.1.6)$$

To be compatible with the sign in Proposition 14.1.1(1), we normalize the canonical morphism $N_{T/Q} \otimes_{\mathcal{O}_T} E \to \Omega^1_Q(\log) \otimes_{\mathcal{O}_Q} E$ (16.1.4) to be $-d$.

*Proof* (1) Let $D$ be the reduced part of the closed fiber $Q \times_S \operatorname{Spec} F$. By Lemma 14.1.2(1), the distinguished triangle $L_{D/\mathbf{F}_p} \otimes^L_{\mathcal{O}_D} E \to L_{E/\mathbf{F}_p} \to L_{E/D} \to$ defines an exact sequence $0 \to N_{E/D} \to \Omega^1_D \otimes_{\mathcal{O}_D} E \to \Omega^1_E \to 0$. By the exact sequence $0 \to \Omega^1_D \to \Omega^1_Q(\log) \otimes_{\mathcal{O}_Q} \mathcal{O}_D \to \mathcal{O}_D \to 0$ (16.1.2), the assertion follows from the isomorphism $N_{T/Q} \otimes_{\mathcal{O}_T} E \to N_{E/D}$.

(2) By the description in Lemma 16.1.1(1), we obtain an exact sequence (16.1.5) except for the injectivity of the morphism

$$N_{T/Q} \to \Omega^1_{Q/S}(\log/\log) \otimes_{\mathcal{O}_Q} \mathcal{O}_T$$

of free $\mathcal{O}_L$-modules. Since this is injective after $\otimes_{\mathcal{O}_L} L$, it is itself injective.

Since (16.1.5) is a free resolution, we obtain (16.1.6). □

**Lemma 16.1.5** *Let $L$ be a finite separable extension of $K$, and let $T = \operatorname{Spec} \mathcal{O}_L \to Q$ be an exact immersion to a log smooth scheme over $S = \operatorname{Spec} \mathcal{O}_K$.*

(1) *On a neighborhood of $T \subset Q$, there exists a cartesian diagram*

$$\begin{array}{ccc} T & \longrightarrow & Q \\ \downarrow & & \downarrow \\ S & \longrightarrow & P \end{array} \quad (16.1.7)$$

*of schemes over $S$ where the lower horizontal arrow $S \to P$ is an immersion to a smooth scheme over $S$ and the vertical arrows are quasi-finite flat.*

(2) *Let* (16.1.7) *be a cartesian diagram satisfying the conditions in* (1). *Then we have an exact sequence*

$$0 \to \Omega^1_{P/S} \otimes_{\mathcal{O}_P} \mathcal{O}_T \to \Omega^1_{Q/S}(\log/\log) \otimes_{\mathcal{O}_Q} \mathcal{O}_T \to \Omega^1_{T/S}(\log/\log) \to 0 \quad (16.1.8)$$

*of $\mathcal{O}_T$-modules.*

*Proof* (1) The immersion $T \to Q$ is a regular immersion. Let $n$ be the codimension. On a neighborhood of $T \subset Q$, define a morphism $Q \to P = \mathbf{A}^n_S$ by taking a system of generators of the ideal. Then, for the 0-section $S \to P$, we obtain a cartesian diagram (16.1.7). Since $T \to S$ is finite flat, on a neighborhood of $T$ the morphism $Q \to P$ is quasi-finite and flat, similarly to Lemma 11.3.3(2).

(2) The exact immersion $T \to Q$ defines an exact sequence $0 \to N_{T/Q} \to \Omega^1_{Q/S}(\log/\log) \otimes_{\mathcal{O}_Q} \mathcal{O}_T \to \Omega^1_{T/S}(\log/\log) \to 0$ (16.1.5). Since the vertical arrows are flat in the cartesian diagram (16.1.7), the canonical morphism $N_{S/P} \otimes_{\mathcal{O}_S} \mathcal{O}_T \to N_{T/Q}$ is an isomorphism. By combining the isomorphism $N_{S/P} \to \Omega^1_{P/S} \otimes_{\mathcal{O}_P} \mathcal{O}_S$, we obtain (16.1.8). □

We construct a logarithmic variant of the first arrow $\mathrm{Tor}_1^{\mathcal{O}_L}(\Omega^1_{T/S}, E) \to \mathrm{H}_1(L_{E/S})$ in (14.2.2).

**Proposition 16.1.6** *Let $L$ be a finite separable extension of $K$ with residue field $E$, and set $T = \mathrm{Spec}\, \mathcal{O}_L$.*

(1) *There exists a unique morphism*

$$\mathrm{Tor}_1^{\mathcal{O}_L}(\Omega^1_{T/S}(\log/\log), E) \to \Omega^1_F(\log) \otimes_F E \tag{16.1.9}$$

*satisfying the following condition: Let $T \to Q$ be an exact immersion to a log smooth scheme over $S$. Then we have a commutative diagram*

$$\begin{array}{ccccccc} 0 \longrightarrow \mathrm{Tor}_1^{\mathcal{O}_L}(\Omega^1_{T/S}(\log/\log), E) & \longrightarrow & N_{T/Q} \otimes_{\mathcal{O}_L} E & \longrightarrow & \Omega^1_{Q/S}(\log/\log) \otimes_{\mathcal{O}_Q} E \\ \downarrow{\scriptstyle (16.1.9)} & & \downarrow{\scriptstyle (16.1.4)} & & \| \\ 0 \longrightarrow \quad \Omega^1_F(\log) \otimes_F E & \longrightarrow & \Omega^1_Q(\log) \otimes_{\mathcal{O}_Q} E & \longrightarrow & \Omega^1_{Q/S}(\log/\log) \otimes_{\mathcal{O}_Q} E \end{array} \tag{16.1.10}$$

*of exact sequences. In* (16.1.10)*, the upper line is part of* (16.1.6)*, the lower line is defined by* (16.1.3)*, and the right vertical arrow is the first arrow in* (16.1.4).

(2) *The sequence*

$$0 \longrightarrow \mathrm{Tor}_1^{\mathcal{O}_L}(\Omega^1_{T/S}(\log/\log), E) \xrightarrow{(16.1.9)} \Omega^1_F(\log) \otimes_F E \longrightarrow \Omega^1_E(\log) \tag{16.1.11}$$

*is exact.*

*Proof* Let $T \to Q$ be an exact immersion to a log smooth scheme over $S$.

(1) Since $0 \to \Omega^1_S(\log) \otimes_{\mathcal{O}_S} \mathcal{O}_Q \to \Omega^1_Q(\log) \to \Omega^1_{Q/S}(\log/\log) \to 0$ is a locally split exact sequence, the lower line of (16.1.10) is an exact sequence. Hence, we obtain a morphism (16.1.9) fitting in the diagram (16.1.10). The morphism (16.1.9) defined above is independent of the choice of exact immersion $T \to Q$ by the functoriality of the construction with respect to $Q$. Hence, the assertion follows.

(2) By the exact sequence (16.1.4), we obtain (16.1.11) by applying the snake lemma to (16.1.10). □

We study the relation of the morphism (16.1.9) with its non-logarithmic variant $\mathrm{Tor}_1^{O_L}(\Omega^1_{T/S}, E) \to \mathrm{H}_1(L_{E/S})$ in (14.2.2). We define a morphism

$$\mathrm{H}_1(L_{E/S}) \to \Omega^1_F(\log) \otimes_F E \tag{16.1.12}$$

to be the composition of the canonical morphism $\mathrm{H}_1(L_{E/S}) \to \Omega^1_F \otimes_F E$ and the morphism induced by the inclusion $\Omega^1_F \to \Omega^1_F(\log)$.

**Proposition 16.1.7** *Let $L$ be a finite separable extension of $K$ with residue field $E$, and set $T = \operatorname{Spec} O_L$.*

(1) *Let $T = \operatorname{Spec} O_L \to P$ be an immersion to a smooth scheme $P$ over $S = \operatorname{Spec} O_K$, and $T \to Q$ be an exact immersion to a log smooth regular flat scheme over $S$. Let $Q \to P$ be a morphism over $S$ compatible with the immersions of $T$. Then the diagram*

$$\begin{array}{ccc} \mathrm{Tor}_1^{O_L}(\Omega^1_{T/S}, E) & \longrightarrow & N_{T/P} \otimes_{O_T} E \\ \downarrow & & \downarrow \\ \mathrm{Tor}_1^{O_L}(\Omega^1_{T/S}(\log/\log), E) & \longrightarrow & N_{T/Q} \otimes_{O_T} E \end{array} \tag{16.1.13}$$

*is commutative.*

(2) *The diagram*

$$\begin{array}{ccc} \mathrm{Tor}_1^{O_L}(\Omega^1_{T/S}, E) & \xrightarrow{(14.2.2)} & \mathrm{H}_1(L_{E/S}) \\ \downarrow & & \downarrow {\scriptstyle (16.1.12)} \\ \mathrm{Tor}_1^{O_L}(\Omega^1_{T/S}(\log/\log), E) & \xrightarrow{(16.1.9)} & \Omega^1_F(\log) \otimes_F E \end{array} \tag{16.1.14}$$

*is commutative.*

*Proof* (1) The morphism $Q \to P$ induces a morphism

$$\begin{array}{ccccccccc} 0 \longrightarrow & N_{T/P} & \longrightarrow & \Omega^1_{P/S} \otimes_{O_P} O_T & \longrightarrow & \Omega^1_{T/S} & \longrightarrow 0 \\ & \downarrow & & \downarrow & & \downarrow & \\ 0 \longrightarrow & N_{T/Q} & \longrightarrow & \Omega^1_{Q/S}(\log/\log) \otimes_{O_Q} O_T & \longrightarrow & \Omega^1_{T/S}(\log/\log) & \longrightarrow 0 \end{array}$$

of free resolutions. The assertion follows.

(2) Let $T = \operatorname{Spec} O_L \to P$ be an immersion to a smooth scheme $P$ over $S = \operatorname{Spec} O_K$, and $T \to Q$ be an exact immersion to a log smooth regular flat scheme over $S$. Since $P$ is smooth over $S$, after replacing $Q$ by an étale neighborhood of $T$, we may assume that there exists a morphism $Q \to P$ over $S$ compatible with the immersions of $T$.

By the commutative diagram

$$\begin{array}{ccc} \mathrm{Tor}_1^{\mathcal{O}_L}(\Omega^1_{T/S}, E) & \longrightarrow & \mathrm{H}_1(L_{E/S}) \\ \downarrow & & \downarrow \\ N_{T/P} \otimes_{\mathcal{O}_L} E & \longrightarrow & N_{E/P} \end{array} \tag{16.1.15}$$

(14.2.4) and by the left square in (16.1.10), it suffices to show that the diagram

$$\begin{array}{ccccccc} \mathrm{Tor}_1^{\mathcal{O}_L}(\Omega^1_{T/S}, E) & \longrightarrow & N_{T/P} \otimes_{\mathcal{O}_T} E & \longrightarrow & N_{E/P} & \leftarrow & \mathrm{H}_1(L_{E/S}) \\ \downarrow & & \downarrow & & \downarrow & & \downarrow \\ \mathrm{Tor}_1^{\mathcal{O}_L}(\Omega^1_{T/S}(\log/\log), E) & \xrightarrow{(16.1.6)} & N_{T/Q} \otimes_{\mathcal{O}_T} E & \xrightarrow{(16.1.6)} & \Omega^1_Q(\log) \otimes_{\mathcal{O}_Q} E & \leftarrow & \Omega^1_F(\log) \otimes_F E \end{array} \tag{16.1.16}$$

is commutative. The left square is commutative by (1). The middle square with the lower-right term $\Omega^1_Q(\log) \otimes_{\mathcal{O}_Q} E$ replaced by $N_{E/Q}$ is commutative. Hence, the middle square itself is commutative. We show the commutativity of the right square. Let $P_F = P \times_S \operatorname{Spec} F$ denote the closed fiber. Then the diagram

$$\begin{array}{ccccccc} \mathrm{H}_1(L_{E/S}) & \longrightarrow & \mathrm{H}_1(L_{E/F}) & \longrightarrow & \Omega^1_F \otimes_F E & \longrightarrow & \Omega^1_F(\log) \otimes_F E \\ \downarrow & & \downarrow & & \downarrow & & \downarrow \\ N_{E/P} & \longrightarrow & N_{E/P_F} & \longrightarrow & \Omega^1_{P_F} \otimes_{\mathcal{O}_{P_F}} E & \longrightarrow & \Omega^1_Q(\log) \otimes_{\mathcal{O}_Q} E \end{array} \tag{16.1.17}$$

is commutative by the functoriality of cotangent complexes. Hence, the right square of (16.1.16) is commutative by the definition of (16.1.12). □

In the case where the extension $E$ of the residue field $F$ is separable, we have the following extra compatibility. Define a surjection

$$\Omega^1_F(\log) \otimes_F E \to E = \mathrm{H}_1(L_{E/S}) \otimes_{\mathcal{O}_K} \mathfrak{m}_K^{-1} \tag{16.1.18}$$

to be the composition of $\mathrm{res} \otimes 1 \colon \Omega^1_F(\log) \otimes_F E \to E$ with the isomorphism induced by $\mathfrak{m}_K/\mathfrak{m}_K^2 \otimes_F E \to \mathrm{H}_1(L_{E/S})$.

**Lemma 16.1.8** *Let $L$ be a finite separable extension of $K$ such that the residue field $E$ is a separable extension of $F$. Assume that the ramification index $n = e_{L/K}$ is divisible by $p$. Let $T = \operatorname{Spec} \mathcal{O}_L \to P$ be a closed immersion to a smooth scheme $P$ of relative dimension* 1 *over $S = \operatorname{Spec} \mathcal{O}_K$. Let $D = P_F \subset P$ be the closed fiber and $Q = P^{[D \cdot T]}$ be the dilatation that is log smooth over $S$ by Proposition* 13.3.2(3). *Identify the conormal module $N_{T/Q}$ of the immersion $T \to Q$ lifting $T \to P$ with $N_{T/P} \otimes_{\mathcal{O}_K} \mathfrak{m}_K^{-1}$ by the isomorphism induced* *by the morphism $Q \to P$. Then the first arrow $\mathrm{Tor}_1^{O_L}(\Omega^1_{T/S}(\log/\log), E) \to N_{T/Q} \otimes_{O_T} E$ in (16.1.6) is an isomorphism and the morphism (16.1.9) defines*

$$N_{T/Q} \otimes_{O_T} E \to \Omega^1_F(\log) \otimes_F E. \tag{16.1.19}$$

*Further, the diagram*

$$\begin{array}{ccc} N_{T/Q} & \xrightarrow{(16.1.19)} & \Omega^1_F(\log) \otimes_F E \\ \downarrow & & \downarrow {\scriptstyle (16.1.18)} \\ N_{T/P} \otimes_{O_K} \mathfrak{m}_K^{-1} & \xrightarrow{(14.5.4)\otimes 1} & E = \mathrm{H}_1(L_{E/S}) \otimes_{O_K} \mathfrak{m}_K^{-1} \end{array} \tag{16.1.20}$$

*is commutative.*

*Proof* We show that the compositions send a basis of $N_{T/Q}$ to the same element. Let $t$ be a uniformizer of $K$. Since the assertion is local, we may assume that the closed immersion $T \to P$ is a divisor defined by a section $f$ of $O_P$. As in the proof of Proposition 13.3.2, we may assume that there exists an étale morphism $P \to \mathbf{A}^1_S = \mathrm{Spec}\, O_K[s]$ such that the image of $s$ in $O_T$ is a uniformizer and that we have an equality $(f, s^n) = (t, s^n)$ of ideals, and we may write $f = at + bs^n$ as a linear combination. Then $P \to \mathbf{A}^1_S$ induces an étale morphism $Q \to \mathrm{Spec}\, O_K[s,u]/(s^n - ut)$ on dilatations. Since $(f, s) = (t, s)$, we may assume that $a$ is a unit.

The immersion $T \to Q$ is defined by the section $g = f/t = a + bu$. Hence, $g$ defines a basis of $N_{T/Q}$. Since $n$ is divisible by $p$, we have $du = u \cdot d\log u = -u \cdot d\log t$ in $\Omega^1_Q(\log) \otimes_F E$. Hence, the image of the basis $g = a + bu \in N_{T/Q}$ in $\Omega^1_F(\log) \otimes_F E$ equals $da + udb - bu \cdot d\log t$. Since $da + udb \in \Omega^1_E = \Omega^1_F \otimes_F E$ and $a + bu = 0$ on $T$, the image of $da + udb - bu \cdot d\log t$ is the same as the image of $a \cdot d\log t$. Since $\Omega^1_F(\log) \to \mathrm{H}_1(L_{E/S}) \otimes_{O_K} \mathfrak{m}_K^{-1}$ maps $d\log t$ to $-w(t) \otimes t^{-1}$ by Proposition 14.1.6(1), the image via the upper-right is $-a \cdot w(t) \otimes t^{-1}$. Since $N_{E/P}$ has a basis $t, s$ and $n \geqslant 1$, the morphism $N_{T/P} \otimes_{O_T} E \to \mathrm{H}_1(L_{E/S})$ is an isomorphism sending a basis $f = at + bs^n$ to $-a \cdot w(t)$. Hence, the diagram is commutative. □

**Exercise 16.1** Let $K$ be a henselian discrete valuation field with residue field $F$ of characteristic $p > 0$, and $L$ be a ramified separable extension of degree $p$ as in Exercise 14.2. Let $x \in O_L = O_K[x]$ be a generator of the ring of integers, and $f \in O_K[X]$ be the minimal polynomial of $x$.

(1) Assume that $L$ is a totally ramified extension and $x$ is a uniformizer of $L$. Let $t \in O_K$ be a uniformizer of $K$ and write the Eisenstein polynomial $f$ as $f = X^p + t \cdot h$ for $h \in O_K[X]$ such that the constant term $h(0)$ is invertible. Define an exact closed immersion $T = \mathrm{Spec}\, O_L \to Q =$

$\operatorname{Spec} O_K[X, U^{\pm 1}]/(UX^p - t)$ to a log smooth scheme over $S = \operatorname{Spec} O_K$ by sending $X, U$ to $x, u = -h(x)^{-1}$, and define a closed immersion $T \to P = \mathbf{A}^1_S = \operatorname{Spec} O_K[X]$ to a smooth scheme over $S$ by sending $X$ to $x$. Let $D = P_F$ be the closed fiber. Show that $Q$ is an open subscheme of the dilatation $P^{[D \cdot T]}$, and that the morphism $\operatorname{Tor}_1^{O_T}(\Omega^1_{T/S}(\log/\log), E) \to N_{T/Q} \otimes_{O_T} E$ is an isomorphism. Verify the commutativity of the diagram (16.1.20) by computing the images of a basis of $N_{T/Q}$ in two ways.

(2) Assume that the residue field $E$ of $L$ is an inseparable extension, and let $T = \operatorname{Spec} O_L \to Q = \mathbf{A}^1_S = \operatorname{Spec} O_K[X]$ be an exact closed immersion defined by sending $X$ to $x$. Verify the commutativity of the diagram (16.1.14) by computing the images of the basis of $\operatorname{Tor}_1^{O_L}(\Omega^1_{T/S}, E)$ obtained in Exercise 14.2 in two ways.

## 16.2 Logarithmic Characteristic Forms

We give a logarithmic variant of the construction of the extension (15.6.1). A logarithmic variant of Theorem 14.4.4 holds for $r > 0$.

**Theorem 16.2.1** *Let $L$ be a finite separable extension of a henselian discrete valuation field $K$, and $r > 0$ be a rational number. Then the following conditions are equivalent:*

(1) *The log ramification of $L$ over $K$ is bounded by $r+$.*

(2) *Let*

$$\begin{array}{ccc} T & \longrightarrow & Q \\ \downarrow & & \downarrow \\ S & \longrightarrow & P \end{array} \tag{16.2.1}$$

*be a cartesian diagram such that the upper (resp. lower) horizontal arrow is an exact immersion to a log smooth scheme (resp. an immersion to a smooth scheme) over $S$, and the vertical arrows are quasi-finite and flat. Let $K'$ be an extension of $K$ containing a Galois closure of $L$ such that the stable integral model $Q^{(r)}_{S'}$ is defined. Then the morphism $Q^{(r)}_{S'} \to P^{(r)}_{S'}$ is étale on a neighborhood of the image of $T^{\natural}_{S'}$.*

(3) *Let $T = \operatorname{Spec} O_L \to Q$ be an exact immersion to a log smooth scheme over $S = \operatorname{Spec} O_K$, and let $K'$ be as in (2). Then the finite morphism $Q^{(r)}_{F'} \to Q^{[r]}_{F',\mathrm{red}}$ is étale.*

*Further, we have a cartesian diagram*

$$\begin{array}{ccc} T^{\natural}_{F'} & \longrightarrow & Q^{(r)}_{F'} \\ \downarrow & & \downarrow \\ \mathrm{Spec}\,(E\otimes_F F')_{\mathrm{red}} & \longrightarrow & Q^{[r]}_{F',\mathrm{red}} \end{array} \tag{16.2.2}$$

*The $O_L$-module $\Omega^1_{O_L/O_K}(\log/\log)$ is annihilated by $\mathfrak{m}_L^m$ for the largest integer $m$ satisfying $m < e_{L/K}\cdot r$.*

*Proof* The proof is similar to that of Theorem 14.4.4. We replace Lemma 11.3.3 by Lemma 16.1.5 and $\Omega^1_{Q/S}$ by $\Omega^1_{Q/S}(\log/\log)$. On $D \subset X = Q^{(r)}_{S'}$, the exact sequence (16.1.8) induces an isomorphism $\mathrm{Coker}(\Omega^1_{P/S}\otimes_{O_P} O_Q \to \Omega^1_{Q/S}(\log/\log))\otimes_{O_Q} O_D \to \Omega^1_{T/S}(\log/\log)\otimes_{O_T} O_D$. Then, if we assume $r > 1$, the proof goes in the same way as the proof of Theorem 14.4.4.

We reduce the general case to the case $r > 1$. Let $d$ be an integer prime to $p$ such that $r_1 = dr > 1$, and $K_1$ be a totally ramified extension of degree $d$ of $K$. Let $S_1 = \mathrm{Spec}\, O_{K_1}$, $L_1 = L\otimes_K K_1$, and $T_1 = \mathrm{Spec}\, O_{L_1}$. Since $K_1$ is a log smooth extension of $K$, the normalization $T_1 \to Q_1$ of the base change of $T \to Q$ by $S_1 \to S$ is an exact immersion to a log smooth scheme over $S_1$ by Proposition 13.2.4(2). Since $r_1 > 0$, the assertion is reduced to the case $r > 1$. □

Let $L$ be a finite separable extension of $K$. Similarly to Definition 14.5.2, we say that an exact immersion $T = \mathrm{Spec}\, O_L \to Q$ to a log smooth scheme $Q$ over $S$ is minimal if the injection $\mathrm{Tor}_1^{O_T}(\Omega^1_{T/S}(\log/\log), E) \to N_{T/Q}\otimes_{O_T} E$ induced by the free resolution $0 \to N_{T/Q} \to \Omega^1_{Q/S}(\log/\log) \to \Omega^1_{T/S}(\log/\log) \to 0$ is an isomorphism. By the same argument as in Lemma 14.5.3(1), it follows from Corollary 16.1.3 that there exists a minimal exact immersion $T \to Q$ to a log smooth scheme over $S$.

For each rational number $r > 0$, we define a vector bundle $\Theta^{(r)}_{L/K,\log,F'}$ over $\mathrm{Spec}\,(E\otimes_F F')_{\mathrm{red}}$ to be that defined by the dual of the free $E\otimes_F F'$-module

$$\mathrm{Hom}_F(\mathfrak{m}_{K'}^{er}/\mathfrak{m}_{K'}^{er+1}, \mathrm{Tor}_1^{O_T}(\Omega^1_{T/S}(\log/\log), E)).$$

Assume that the log ramification of $L$ over $K$ is bounded by $r+$. Then, by Theorem 16.2.1, similarly to Proposition 14.5.5, we have a functorial construction of finite étale morphisms $\Phi^{(r)}_{L/K,\log,F'} \to \Theta^{(r)}_{L/K,\log,F'}$. By fixing a morphism $L \to K'$ over $K$, we obtain a finite étale morphism

$$\Phi^{(r)\circ}_{L/K,\log,F'} \to \Theta^{(r)\circ}_{L/K,\log,F'} \tag{16.2.3}$$

of connected smooth schemes. If $L$ is a Galois extension of $K$, then $\Phi^{(r)\circ}_{L/K,\log,F'}$

is a $G^r_{\log}$-torsor over $\Theta^{(r)\circ}_{L/K,\log,F'}$. Similarly to Theorem 15.6.1, we have the following.

**Proposition 16.2.2** *Let $r > 0$ be a rational number and assume that $G^{r+}_{\log} = 1$. Fix a morphism $L \to K_s$ over $K$. Then there exists a unique group structure on $\Phi^{(r)\circ}_{L/K,\log,\overline{F}}$ such that*

$$0 \longrightarrow G^r_{\log} \longrightarrow \Phi^{(r)\circ}_{L/K,\log,\overline{F}} \longrightarrow \Theta^{(r)\circ}_{L/K,\log,\overline{F}} \longrightarrow 0 \qquad (16.2.4)$$

*is an extension of smooth group schemes over $\overline{F}$.*

*Proof* Let $K'$ be a log smooth extension of $K$ such that the ramification index $e = e_{K'/K}$ is divisible by $m = e_{L/K}$. Since $\Omega^1_F(\log) \otimes_F F' \to \Omega^1_{F'}(\log)$ is an injection, we may assume that $e_{L/K} = 1$ by Corollary 15.2.3. In the case $e_{L/K} = 1$, the étale morphism $\Phi^{(r)\circ}_{L/K,\log,\overline{F}} \to \Theta^{(r)\circ}_{L/K,\log,\overline{F}}$ equals $\Phi^{(r)\circ}_{L/K,\overline{F}} \to \Theta^{(r)\circ}_{L/K,\overline{F}}$, and the assertion follows from Theorem 15.6.1. □

In the same way as in the non-logarithmic case, the image of the class of the extension (16.2.4) defines an injection

$$\Phi_{\log} : \mathrm{Hom}_{\mathbf{F}_p}(G^r_{\log}, \mathbf{F}_p) \to \mathrm{Hom}_{\overline{F}}(\mathfrak{m}^r_{K_s}/\mathfrak{m}^{r+}_{K_s}, \mathrm{Tor}_1^{O_L}(\Omega^1_{O_L/O_K}(\log/\log), \overline{F})). \qquad (16.2.5)$$

By taking the composition with the injection $\mathrm{Tor}_1^{O_L}(\Omega^1_{O_L/O_K}(\log/\log), E) \to \Omega^1_F(\log) \otimes_F E$ (16.1.9) and applying the construction to the subextension corresponding to $G^{r+}_{\log}$, we obtain an injection

$$\mathrm{ch}_{\log} : \mathrm{Hom}_{\mathbf{F}_p}(\mathrm{Gr}^r_{\log} G, \mathbf{F}_p) \longrightarrow \mathrm{Hom}_{\overline{F}}(\mathfrak{m}^r_{K_s}/\mathfrak{m}^{r+}_{K_s}, \Omega^1_F(\log) \otimes_F \overline{F}) \qquad (16.2.6)$$

for $r > 0$ without the assumption that the logarithmic ramification of $L$ is bounded by $r+$. Let $K'$ be an extension of $K$ of ramification index $e$. Then, for $L' = LK'$, we have an injection $G' = \mathrm{Gal}(L'/K') \to G$. For $r' = er$, similarly to (15.6.5), we have a commutative diagram

$$\begin{array}{ccc} \mathrm{Hom}(G^r_{\log}, \mathbf{F}_p) & \xrightarrow{\mathrm{ch}_{\log}} & \mathrm{Hom}_F(\mathfrak{m}^r_{K_s}/\mathfrak{m}^{r+}_{K_s}, \Omega^1_F(\log) \otimes_F \overline{F}) \\ \downarrow & & \downarrow \\ \mathrm{Hom}(G'^{r'}_{\log}, \mathbf{F}_p) & \xrightarrow{\mathrm{ch}_{\log}} & \mathrm{Hom}_F(\mathfrak{m}^{r'}_{K'_s}/\mathfrak{m}^{r'+}_{K'_s}, \Omega^1_{F'}(\log) \otimes_{F'} \overline{F}') \end{array} \qquad (16.2.7)$$

We study the relation between the logarithmic and non-logarithmic constructions. Assume that the ramification of $L$ over $K$ is bounded by $r+$. Then the surjection $F^{r+}_{\log}(L) \to F^{r+}(L)$ is a bijection and the log ramification of $L$ over $K$ is also bounded by $r+$. Applying Lemma 16.1.4(2) to minimal immersions

$T \to P$ to a smooth scheme $P$ over $S$ and $T \to Q$ to a log smooth scheme $Q$ over $S$, we obtain a commutative diagram

$$\begin{array}{ccc} \Phi^{(r)\circ}_{L/K,\log,F'} & \longrightarrow & \Theta^{(r)\circ}_{L/K,\log,F'} \\ \downarrow & & \downarrow \\ \Phi^{(r)\circ}_{L/K,F'} & \longrightarrow & \Theta^{(r)\circ}_{L/K,F'} \end{array} \tag{16.2.8}$$

If $L$ is a Galois extension, the diagram (16.2.8) is compatible with the actions of $G^r_{\log} \subset G^r$ and defines a commutative diagram

$$\begin{array}{ccc} \mathrm{Hom}_{\mathbf{F}_p}(G^r, \mathbf{F}_p) & \xrightarrow{\Phi} & \mathrm{Hom}_{\overline{F}}(\mathfrak{m}^r_{K_s}/\mathfrak{m}^{r+}_{K_s}, \mathrm{Tor}_1^{O_L}(\Omega^1_{O_L/O_K}, \overline{F})) \\ \downarrow & & \downarrow \\ \mathrm{Hom}_{\mathbf{F}_p}(G^r_{\log}, \mathbf{F}_p) & \xrightarrow{\Phi_{\log}} & \mathrm{Hom}_{\overline{F}}(\mathfrak{m}^r_{K_s}/\mathfrak{m}^{r+}_{K_s}, \mathrm{Tor}_1^{O_L}(\Omega^1_{O_L/O_K}(\log/\log), \overline{F})) \end{array} \tag{16.2.9}$$

where the left vertical arrow is induced by the inclusion $G^r_{\log} \to G^r$ and the right vertical arrow is induced by the canonical morphism $\Omega^1_{O_L/O_K} \to \Omega^1_{O_L/O_K}(\log/\log)$. If $e_{L/K} = 1$, the vertical arrows are isomorphisms. Further, by Proposition 16.1.7(2) we obtain a commutative diagram

$$\begin{array}{ccc} \mathrm{Hom}_{\mathbf{F}_p}(\mathrm{Gr}^r G, \mathbf{F}_p) & \xrightarrow{\mathrm{ch}} & \mathrm{Hom}_{\overline{F}}(\mathfrak{m}^r_{K_s}/\mathfrak{m}^{r+}_{K_s}, \mathrm{H}_1(L_{\overline{F}/S})) \\ \downarrow & & \downarrow \\ \mathrm{Hom}_{\mathbf{F}_p}(\mathrm{Gr}^r_{\log} G, \mathbf{F}_p) & \xrightarrow{\mathrm{ch}_{\log}} & \mathrm{Hom}_{\overline{F}}(\mathfrak{m}^r_{K_s}/\mathfrak{m}^{r+}_{K_s}, \Omega^1_F(\log) \otimes_F \overline{F}) \end{array} \tag{16.2.10}$$

for $r > 1$ without the assumption that the ramification of $L$ is bounded by $r+$. The right vertical arrow is induced by (16.1.12). In the case that $e_{L/K}$ is 1, the left vertical arrow is an isomorphism. If the residue field $E$ of $L$ is a separable extension of $F$, we have $G^r_{\log} = G^{r+1}$ by Lemma 13.3.5(3) and, further, by Lemma 16.1.8 we have a commutative diagram

$$\begin{array}{ccc} \mathrm{Hom}_{\mathbf{F}_p}(\mathrm{Gr}^r_{\log} G, \mathbf{F}_p) & \xrightarrow{\mathrm{ch}_{\log}} & \mathrm{Hom}_{\overline{F}}(\mathfrak{m}^r_{K_s}/\mathfrak{m}^{r+}_{K_s}, \Omega^1_F(\log) \otimes_F \overline{F}) \\ \| & & \downarrow \scriptstyle{(16.1.18)} \\ \mathrm{Hom}_{\mathbf{F}_p}(\mathrm{Gr}^{r+1} G, \mathbf{F}_p) & \xrightarrow{\mathrm{ch}} & \mathrm{Hom}_{\overline{F}}(\mathfrak{m}^r_{K_s}/\mathfrak{m}^{r+}_{K_s}, \overline{F}) \end{array} \tag{16.2.11}$$

for $r > 0$.

For Artin–Schreier–Witt extensions, the logarithmic characteristic form is computed by Witt vectors.

**Proposition 16.2.3** *Let $K$ be a henselian discrete valuation field of characteristic $p > 0$. Let $\chi \in X_K$ be a character of order $p^{m+1}$, and $n > 0$ be the smallest integer such that $\chi$ is contained in the image of $F_n W_{m+1}(K)$. Let $\mathbf{d}\colon \mathrm{Gr}_n^F W_{m+1}(K) \to \mathrm{Gr}^F \Omega^1_K = \mathrm{Hom}_F(\mathfrak{m}_K^n/\mathfrak{m}_K^{n+1}, \Omega^1_F(\log))$ denote the morphism induced by $\mathbf{d}\colon W_{m+1}(K) \to \Omega^1_K$. Let $a \in F_n W_{m+1}(K)$ be an element whose image equals $\chi$.*

*Let $L$ be the cyclic extension trivializing $\chi$, and let $G = \mathrm{Gal}(L/K)$ be the Galois group. Then the injection*

$$\mathrm{ch}_{\log}\colon \mathrm{Hom}(G^n_{\log}, \mathbf{F}_p) \to \mathrm{Hom}_F(\mathfrak{m}_K^n/\mathfrak{m}_K^{n+1}, \Omega^1_F(\log) \otimes_F \overline{F})$$

*sends the restriction $\chi|_{G^n_{\log}}$ to $\mathbf{d}(a) \in \mathrm{Hom}_F(\mathfrak{m}_K^n/\mathfrak{m}_K^{n+1}, \Omega^1_F(\log))$.*

*Proof* By the same argument as in the proof of Theorem 13.4.4, it is reduced to the following case: $O_K$ is the henselization of the local ring at the generic point of a smooth divisor $D$ of a smooth scheme $X$ over a perfect field $k$. For a finite cyclic covering $Y \to X$ of smooth schemes over $k$ of Galois group $G$ étale over the complement $U = X - D$, we have a cartesian diagram

$$\begin{array}{ccc} T = \mathrm{Spec}\, O_L & \longrightarrow & Y \\ \downarrow & & \downarrow \\ S = \mathrm{Spec}\, O_K & \longrightarrow & X \end{array}$$

Let the notation be as in Corollary 13.5.6. Then the $G$-torsor $(V \times V)/G \to U \times U$ is extended to a $G$-torsor $W^{(n)} \to P^{[n]}$ of schemes over $X \times X$ by Corollary 13.5.6(1). We consider its base change $\mathrm{Spec}\, F \to D \to X \to X \times X$ at the generic point of $D$ embedded in the diagonal. Then the fiber of $P^{[n]}$ is the dual of the vector space $\Omega^1_X(\log D)(nD) \otimes_{O_X} F$, and the connected component of the fiber of $W^{(n)}$ is the Artin–Schreier covering defined by the linear form $\mathbf{d}(a)$ by Corollary 13.5.6(2).

Let $T \to Q = Y \times_k S$ be the immersion to a smooth scheme over $S$. Since the finite étale morphism $Q_E^{(n)} \to Q_{E,\mathrm{red}}^{[n]}$ is its base change as in the proof of case (1) in Proposition 13.6.2, the assertion follows. □

**Exercise 16.2** Let $K$ be a henselian discrete valuation field with residue field $F$ of characteristic $p > 0$. Let $L$ be a totally ramified cyclic extension of $K$ of degree $p$. Let $G$ be the Galois group and let $\chi\colon G \to \mathbf{F}_p$ be a faithful character. Let $n = \mathrm{Sw}\,\chi \geqslant 1$ be the Swan conductor. Show that the logarithmic characteristic form $\mathrm{ch}_{\log}\chi\colon \mathfrak{m}_K^n/\mathfrak{m}_K^{n+1} \to \Omega^1_F(\log) \otimes_F \overline{F}$ equals minus the refined Swan character.

## 16.3 Refined Swan Conductor

In Theorem 5.3.4, we constructed the refined Swan conductor for characters corresponding to log monogenic extensions. It is known to exist in general. More strongly, the following holds.

**Theorem 16.3.1** *Let $K$ be a henselian discrete valuation field with residue field $F$ of characteristic $p > 0$, and let $K_1$ be the extension as in Definition* 2.3.1. *Let $\chi \in X_K$ be a character and assume that the Swan conductor $n = \operatorname{Sw}\chi$ satisfies $n \geqslant 1$. Then there exists a unique morphism $f\colon \mathfrak{m}_K^n/\mathfrak{m}_K^{n+1} \to \Omega_F^1(\log)$ of $F$-vector spaces satisfying the following property:*

*Let $K'$ be any extension of the henselian discrete valuation field of $K$ and let $F'$ be the residue field. Let $f_{K'}$ denote the induced $F'$-linear morphism $\mathfrak{m}_K^n O_{K'}/\mathfrak{m}_K^n\mathfrak{m}_{K'} = \mathfrak{m}_K^n/\mathfrak{m}_K^{n+1} \otimes_F F' \to \Omega_F^1(\log) \otimes_F F' \to \Omega_{F'}^1(\log)$. Then, for every $a \in \mathfrak{m}_K^n O_{K'}$, we have an equality*

$$\{\chi, 1 + a\} = \beta_{K'}(f_{K'}(\overline{a})) \tag{16.3.1}$$

*in* $\operatorname{Br} K'$.

*In particular, the refined Swan conductor*

$$f = \operatorname{rsw}\chi \in \operatorname{Hom}_F(\mathfrak{m}_K^n/\mathfrak{m}_K^{n+1}, \Omega_F^1(\log))$$

*exists.*

Recall that the equality (16.3.1) for $K' = K$ and $K_1$ in Definition 2.3.1 is the defining property of the refined Swan conductor in Definition 2.5.5. Theorem 16.3.1 asserts that the equality holds for every extension of henselian discrete valuation fields. We sketch an outline of the proof in the case where $K$ is of characteristic 0 in Section 16.4. In the case where $K$ is of characteristic $p > 0$, we have a concrete description using Witt vectors as in Theorem 16.3.4.

For $n \geqslant 1$, the morphism

$$\operatorname{rsw}\colon F_n X_K \to \operatorname{Hom}_F(\mathfrak{m}_K^n/\mathfrak{m}_K^{n+1}, \Omega_F^1(\log)) \tag{16.3.2}$$

sending $\chi$ to its refined Swan conductor is also called the refined Swan conductor.

**Corollary 16.3.2** *Let $K$ be a henselian discrete valuation field, and $\chi \in X_K$. For an extension $K'$ of the henselian discrete valuation field of $K$, let $\chi_{K'} \in X_{K'}$ denote the image of $\chi$.*

(1) *For every integer $n \geqslant 0$, the following conditions are equivalent:*

   (a) *For every extension $K'$ of the henselian discrete valuation field $K$, we have $\{\chi_{K'}, 1 + \mathfrak{m}_K^n\mathfrak{m}_{K'}\}_{K'} = 0 \subset \operatorname{Br} K'$.*

(b) *For every extension $K'$ of the henselian discrete valuation field $K$ of ramification index* 1, *we have* $\{\chi_{K'}, 1 + \mathfrak{m}_{K'}^{n+1}\}_{K'} = 0 \subset \mathrm{Br}\, K'$.

(c) $\chi$ *is an element of* $F_n X_K$.

(2) *For* $n \geqslant 1$, *the morphism* $\mathrm{rsw}\colon F_n X_K \to \mathrm{Hom}_F(\mathfrak{m}_K^n/\mathfrak{m}_K^{n+1}, \Omega_F^1(\log))$ *induces an injection*

$$\mathrm{rsw}\colon \mathrm{Gr}_n^F X_K = F_n X_K / F_{n-1} X_K \to \mathrm{Hom}_F(\mathfrak{m}_K^n/\mathfrak{m}_K^{n+1}, \Omega_F^1(\log)). \tag{16.3.3}$$

*Proof* (1) The implications (a) $\Rightarrow$ (b) $\Rightarrow$ (c) are clear. The implication (c) $\Rightarrow$ (a) follows from Theorem 16.3.1.

(2) It suffices to show that the kernel of the morphism $\mathrm{rsw}\colon F_n X_K \to \mathrm{Hom}_F(\mathfrak{m}_K^n/\mathfrak{m}_K^{n+1}, \Omega_F^1(\log))$ (16.3.2) is $F_{n-1}X_K$. Let $\chi \in F_n X_K$. If $\{\chi, 1 + aT\} = 0$ in $\mathrm{Br}\, K_1$ for every $a \in \mathfrak{m}_K^n$, we have $\mathrm{rsw}\,\chi = 0$ by Lemma 2.5.6(1). Conversely, if $\mathrm{rsw}\,\chi = 0$, then, for every $a \in \mathfrak{m}_{K_1}^n$, we have $\{\chi, 1 + a\} = \beta_{K_1}(\mathrm{rsw}\,\chi(\overline{a})) = 0$ by (16.3.1). □

**Corollary 16.3.3** *Let $K \to K'$ be an extension of a henselian discrete valuation field of ramification index $e$.*

(1) *For integers $n \geqslant 1$, the canonical morphism $X_K \to X_{K'}$ induces $F_n X_K \to F_{en} X_{K'}$, and the diagram*

$$\begin{array}{ccc} \mathrm{Gr}_n^F X_K & \xrightarrow{\mathrm{rsw}} & \mathrm{Hom}_F(\mathfrak{m}_K^n/\mathfrak{m}_K^{n+1}, \Omega_F^1(\log)) \\ \downarrow & & \downarrow \\ \mathrm{Gr}_{ne}^F X_{K'} & \xrightarrow{\mathrm{rsw}} & \mathrm{Hom}_{F'}(\mathfrak{m}_{K'}^{ne}/\mathfrak{m}_{K'}^{ne+1}, \Omega_{F'}^1(\log)) \end{array} \tag{16.3.4}$$

*is commutative.*

(2) *If $\Omega_F^1(\log) \to \Omega_{F'}^1(\log)$ is injective, the arrows in* (16.3.4) *are injections and the diagram*

$$\begin{array}{ccc} F_n X_K & \xrightarrow{\subset} & X_K \\ \downarrow & & \downarrow \\ F_{en} X_{K'} & \xrightarrow{\subset} & X_{K'} \end{array} \tag{16.3.5}$$

*is cartesian.*

For a log smooth extension $K'$ over $K$, the assumption that $\Omega_F^1(\log) \to \Omega_{F'}^1(\log)$ is injective is satisfied.

*Proof* (1) By Corollary 16.3.2(1), the canonical morphism $X_K \to X_{K'}$ induces $F_n X_K \to F_{en} X_{K'}$. The commutative diagram (16.3.4) follows from Theorem 16.3.1 and the functoriality of cup-product.

(2) If $\Omega^1_F(\log) \to \Omega^1_{F'}(\log)$ is injective, the right vertical arrow in (16.3.4) is an injection. Since the horizontal arrows are injections by Corollary 16.3.2(2), the left vertical arrow is also an injection.

Since $X_K = \bigcup_{n \geqslant 0} F_n X_K$ by Proposition 2.3.3(1), the injectivity of the left vertical arrow in (16.3.4) implies that the diagram (16.3.5) is cartesian. □

If $\operatorname{char} K = p > 0$, the cohomological filtration $F_n X_K$ equals Brylinski's filtration $F^B_n X_K$ (13.4.7).

**Theorem 16.3.4** *Let $K$ be a henselian discrete valuation ring of characteristic $p > 0$. Let $n \geqslant 1$ and $m \geqslant 0$ be integers and $a \in F_n W_{m+1}(K)$, and let $\chi_a \in X_K$ be the image of $a \in F_n W_{m+1}(K)$ by the canonical morphism $W_{m+1}(K) \to X_K$. Then, for $y \in \mathfrak{m}^n_K$,*

$$\{\chi_a, 1 + y\} = \beta_K(-y\mathbf{d}a) \tag{16.3.6}$$

*in* $\operatorname{Br} K$.

**Corollary 16.3.5** (1) [50, Theorem (3.2) (1)] $F_n X_K = F^B_n X_K$ *for every* $n \geqslant 0$.

(2) [50, Theorem (3.2) (3)] *Let $n \geqslant 1$ and $m \geqslant 0$ be integers and $a \in F_n W_{m+1}(K)$ such that its image $\chi_a \in F^B_n X_K$ is not contained in $F^B_{n-1} X_K$. Then the morphism $f = -\mathbf{d}a \in \operatorname{Hom}_F(\mathfrak{m}^n_K/\mathfrak{m}^{n+1}_K, \Omega^1_F(\log)) = \operatorname{Gr}^F_n \Omega^1_K$* (13.4.4) *satisfies Theorem* 16.3.1. *Consequently,* $\operatorname{rsw} \chi_a = -\mathbf{d}a$.

*Proof that Theorem* 16.3.4 ⇒ *Corollary* 16.3.5 It suffices to show the first assertion in (2) for $a \in F_n W_{m+1}(K)$ and $\chi_a \in X_K$. Let $K'$ be an extension of henselian discrete valuation field $K$ of ramification index $e$, and set $n' = en$. Then the assertion follows from the commutative diagram

$$\begin{array}{ccc} F_n W_{m+1}(K) & \xrightarrow{-\mathbf{d}} & F_n \Omega^1_K \\ \downarrow & & \downarrow \\ F_{n'} W_{m+1}(K') & \xrightarrow{-\mathbf{d}} & F_{n'} \Omega^1_{K'} \end{array}$$

and Theorem 16.3.4. □

We sketch an outline of the proof of Theorem 16.3.4 without details. Let $d\colon W_{m+1}K \to W_{m+1}\Omega^1_K$ be a part of the de Rham–Witt complex and $V_m\colon K \to W_{m+1}K$ be the morphism sending $x$ to $(0, \ldots, 0, x)$. Let $\mathrm{F}\colon W_{m+1}\Omega^1_K \to W_m\Omega^1_K$ and $\mathrm{F}\colon W_{m+1}\Omega^1_K \to W_{m+1}\Omega^1_K/dV_m K$ be the Frobenius morphisms [42, Théorème 2.17, Proposition 3.3]. We also have a morphism $d\log\colon K^\times \to W_{m+1}\Omega^1_K$ defined by sending $x$ to $d\log x = (x^{-1}, 0, \ldots, 0) \cdot d(x, 0, \ldots, 0)$ [42, (3.23.1)] lifting $d\log\colon K^\times \to \Omega^1_K$. Similarly to Theorem 2.4.1(2), we have the following.

**Proposition 16.3.6** ([17, 1.4, Lemme 2]) *Let $K$ be a field of characteristic $p > 0$ and $m \geqslant 0$ be an integer. Then the sequence*

$$0 \longrightarrow K^\times/K^{\times p^{m+1}} \xrightarrow{d\log} W_{m+1}\Omega^1_K \xrightarrow{\mathrm{F}-1} W_{m+1}\Omega^1_K/dV_mK$$

*is exact. If $K$ is separably closed, the last arrow is a surjection.*

By Proposition 16.3.6, the morphism $b\colon \Omega^1_K \to \mathrm{Br}\,K[p]$ (2.4.4) is extended to a morphism

$$b\colon W_{m+1}\Omega^1_K \to \mathrm{Br}\,K[p^{m+1}] \tag{16.3.7}$$

satisfying $b(x \cdot d\log y) = a(x) \cup \{y\}$ for $x \in W_{m+1}K$ and $y \in K^\times$. Lemma 3.1.3 is extended as follows.

**Lemma 16.3.7** *Let $K$ be a henselian discrete valuation field of characteristic $p > 0$. Let $n, i \geqslant 0$ be integers.*

(1) [50, Proof of (3.5)] *For $y \in \mathfrak{m}^i_K$,*

$$(1 + y, 0, \ldots, 0) \equiv (1, 0, \ldots, 0) + (0, \ldots, 0, y) \bmod F_{-2i}W_{m+1}K.$$

(2) [50, (3.5), $s = 1$] *For $x \in F_nW_{m+1}K$ and $y \in \mathfrak{m}^i_K$, we have $x \cdot d\log(1+y) \in \mathrm{Im}(F_{n-i}W_{m+1}K \otimes K^\times \to W_{m+1}\Omega^1_K) + dW_{m+1}K$ and*

$$x \cdot d\log(1 + y) \equiv -y \cdot \mathbf{d}x \bmod \mathrm{Im}(F_{n-2i}W_{m+1}K \otimes K^\times \to W_{m+1}\Omega^1_K) + dW_{m+1}K. \tag{16.3.8}$$

Lemma 16.3.7(1) follows from the definition of addition of Witt vectors.

*Proof of Theorem* 16.3.4 Since the morphism $b\colon W_{m+1}\Omega^1_K \to \mathrm{Br}\,K$ (16.3.7) annihilates $\mathrm{Im}(F_{-1}W_{m+1}K \otimes K^\times \to W_{m+1}\Omega^1_K) + dW_{m+1}K$, for $y \in \mathfrak{m}^n_K$ the cup-product $\{\chi_a, 1+y\} = b(a \cdot d\log(1+y))$ equals $-b(y\mathbf{d}a) = -\beta_K(y\mathbf{d}a)$ in $\mathrm{Br}\,K$ by Lemma 16.3.7(2). □

In the positive characteristic case, Corollary 16.3.5(2) implies that the diagram

$$\begin{array}{ccc}
\mathrm{Gr}^F_n W_{m+1}(K) & \xrightarrow{-\mathbf{d}} & \mathrm{Gr}^F_n \Omega^1_K \\
\downarrow & & \big\| \,(13.4.4) \\
\mathrm{Gr}^F_n X_K & \xrightarrow{\mathrm{rsw}} & \mathrm{Hom}_F(\mathfrak{m}^n_K/\mathfrak{m}^{n+1}_K, \Omega^1_F(\log))
\end{array} \tag{16.3.9}$$

is commutative for every $n \geqslant 1$ and $m \geqslant 0$.

For an abelian extension $L$ over $K$ and an integer $n \geqslant 0$, the cohomological filtration $G^n_{\mathrm{coh}} \subset G = \mathrm{Gal}(L/K)$ is defined in (15.3.18). For $n \geqslant 1$, the morphism

$$\mathrm{rsw}\colon \mathrm{Hom}(\mathrm{Gr}^n_{\mathrm{coh}}G, \mathbf{F}_p) \to \mathrm{Hom}_F(\mathfrak{m}^n_K/\mathfrak{m}^{n+1}_K, \Omega^1_F(\log)) \tag{16.3.10}$$

defined by the refined Swan conductor is an injection by Corollary 16.3.3(2).

In the positive characteristic case, Theorem 13.4.4 and Corollary 16.3.5 imply the equality of the cohomological filtration and the logarithmic upper numbering filtration. They also imply the equality of minus the refined Swan conductor and the logarithmic characteristic form. We prove that the same thing also holds in the mixed characteristic case in Theorem 16.3.9, using the following functorial property of refined Swan conductors similar to (16.2.7). In the rest of this section we admit Theorem 16.3.1 even in the mixed characteristic case.

**Proposition 16.3.8** *Let $L$ be a finite abelian extension of $K$ of Galois group $G = \mathrm{Gal}(L/K)$. Let $K'$ be an extension of $K$ of ramification index $e = e_{K'/K}$, and let $L' = LK'$.*

(1) *Assume that $\Omega^1_F(\log) \to \Omega^1_{F'}(\log)$ is an injection. Then, for each integer $n \geqslant 1$, the canonical injection $G' \to G$ induces isomorphisms $G'^{en}_{\mathrm{coh}} \to G^n_{\mathrm{coh}}$ and $\mathrm{Gr}^{en}_{\mathrm{coh}} G' \to \mathrm{Gr}^n_{\mathrm{coh}} G$.*
(2) *Assume that $\Omega^1_F \to \Omega^1_{F'}$ is an injection and that, for every integer $n \geqslant 1$, the image of the morphism rsw (16.3.10) is a subgroup of $\mathrm{Hom}_F(\mathfrak{m}^n_K/\mathfrak{m}^{n+1}_K, \Omega^1_F)$. Then, for each integer $n \geqslant 1$, the canonical injection $G' \to G$ induces isomorphisms $G'^{en}_{\mathrm{coh}} \to G^n_{\mathrm{coh}}$ and $\mathrm{Gr}^{en}_{\mathrm{coh}} G' \to \mathrm{Gr}^n_{\mathrm{coh}} G$.*

*Proof* (1) By descending induction, we may assume that $n \geqslant 1$ is the largest integer such that $G^n_{\mathrm{coh}} \neq 1$. We have a commutative diagram

$$\begin{array}{ccc} \mathrm{Hom}(\mathrm{Gr}^n_{\mathrm{coh}} G, \mathbf{F}_p) & \xrightarrow{\mathrm{rsw}} & \mathrm{Hom}_F(\mathfrak{m}^n_K/\mathfrak{m}^{n+1}_K, \Omega^1_F(\log)) \\ \downarrow & & \downarrow \\ \mathrm{Hom}(\mathrm{Gr}^{en}_{\mathrm{coh}} G', \mathbf{F}_p) & \xrightarrow{\mathrm{rsw}} & \mathrm{Hom}_F(\mathfrak{m}^{en}_{K'}/\mathfrak{m}^{en+1}_{K'}, \Omega^1_{F'}(\log)) \end{array} \tag{16.3.11}$$

Since the horizontal arrows are injection and the right vertical arrow is an injection by assumption, the left vertical arrow is also an injection. Consequently, the induced morphism $\mathrm{Gr}^{en}_{\mathrm{coh}} G' \to \mathrm{Gr}^n_{\mathrm{coh}} G = G^n_{\mathrm{coh}}$ is a surjection for every $n \geqslant 1$. Hence, the injection $G' \to G$ induces an isomorphism $G'^{en}_{\mathrm{coh}} \to G^n_{\mathrm{coh}}$ as required.

(2) By assumption, the composition of the arrows in (16.3.11) via the upper-right corner is an injection. Hence, the same proof as in (1) works. □

**Theorem 16.3.9** *Let $L$ be a finite abelian extension of $K$ of Galois group $G = \mathrm{Gal}(L/K)$. Let $n \geqslant 1$ be an integer.*

(1) *Let $r \in \mathbf{Q}_+$ and assume $(n-1)+ \leqslant r \leqslant n$. Then $G^r_{\log} = G^n_{\mathrm{coh}}$.*
(2) *Let $r = n$ be the integer satisfying $G^r_{\log} = G^n_{\mathrm{coh}} \neq 1$ and $G^{r+}_{\log} = G^{n+1}_{\mathrm{coh}} = 1$.*

*Let* $\chi \in \mathrm{Hom}(G, \mathbf{Q}/\mathbf{Z})$ *be a character. Then the image*

$$\mathrm{ch}_{\log}(\chi|_{G^r_{\log}}) : \mathfrak{m}^r_K/\mathfrak{m}^{r+1}_K \to \Omega^1_F(\log) \otimes_F \overline{F}$$

*of the restriction* $\chi|_{G^r_{\log}}$ *by the injection* (16.2.6) *equals the composition of minus the refined Swan conductor* $-\mathrm{rsw}\,\chi : \mathfrak{m}^r_K/\mathfrak{m}^{r+1}_K \to \Omega^1_F(\log)$ *with the injection* $\Omega^1_F(\log) \to \Omega^1_F(\log) \otimes_F \overline{F}$.

*Proof* Let $m = e_{L/K}$ be the ramification index. Let $K'$ be a log smooth extension of $K$ of ramification index $e$ divisible by $pm$, and set $L' = LK'$ and $G' = \mathrm{Gal}(L'/K')$. Then, by Proposition 13.3.6, we have $G^r_{\log} = G'^{er}_{\log}$. Further, we have the commutative diagram (16.2.7). Similarly, we have $G^n_{\mathrm{coh}} = G'^{en}_{\mathrm{coh}}$ and the commutative diagram (16.3.11). Thus, by replacing $K, L$ by $K', L'$, we may assume that $e_{L/K} = 1$ by Proposition 13.1.5(2). Further, since the image of $\Omega^1_F(\log) \to \Omega^1_{F'}(\log)$ is contained in $\Omega^1_{F'}$, we may assume that rsw defines a morphism $\mathrm{Hom}(G^r, \mathbf{F}_p) \to \mathrm{Hom}_F(\mathfrak{m}^r_K/\mathfrak{m}^{r+1}_K, \Omega^1_F)$. Then, the logarithmic filtration $G^r_{\log}$ equals the non-logarithmic filtration $G^r$ and the injection (16.2.6) is induced by

$$\overline{\mathrm{ch}} : \mathrm{Hom}(G^r, \mathbf{F}_p) \to \mathrm{Hom}_F(\mathfrak{m}^r_{K_s}/\mathfrak{m}^{r+}_{K_s}, \Omega^1_F \otimes_F \overline{F}). \tag{16.3.12}$$

Thus, if we further assume $[F : F^p] = p$, both assertions (1) and (2) are proved in case (ii) of Corollary 15.3.4.

We reduce the general $e_{L/K} = 1$ case to this $[F : F^p] = p$ case. Let $K'$ be an extension as in Corollary 9.2.8 and let $e = e_{K'/K}$ be the ramification index. Let $r' = er$ and consider the commutative diagram

$$\begin{array}{ccc} \mathrm{Hom}(G^r, \mathbf{F}_p) & \xrightarrow{\overline{\mathrm{ch}}} & \mathrm{Hom}_F(\mathfrak{m}^r_{K_s}/\mathfrak{m}^{r+}_{K_s}, \Omega^1_F \otimes_F \overline{F}) \\ \downarrow & & \downarrow \\ \mathrm{Hom}(G'^{r'}, \mathbf{F}_p) & \xrightarrow{\overline{\mathrm{ch}}} & \mathrm{Hom}_F(\mathfrak{m}^{r'}_{K'_s}/\mathfrak{m}^{r'+}_{K'_s}, \Omega^1_{F'} \otimes_{F'} \overline{F}') \end{array} \tag{16.3.13}$$

in Proposition 16.3.8(2). By the injectivity condition $\Omega^1_F \otimes_F \overline{F} \to \Omega^1_{F'} \otimes_{F'} \overline{F}'$ in Corollary 9.2.8, the right vertical arrow is an injection. Since the horizontal arrows are injections, the left vertical arrow is also an injection. Hence, the injection $G'^{r'} \to G^r$ is a surjection. Similarly, by the assumption that rsw defines a morphism $\mathrm{Hom}(G^r, \mathbf{F}_p) \to \mathrm{Hom}_F(\mathfrak{m}^r_K/\mathfrak{m}^{r+1}_K, \Omega^1_F)$, the injection $G'^{n'}_{\mathrm{coh}} \to G^n_{\mathrm{coh}}$ is an isomorphism and we have a commutative diagram (16.3.11) with (log) removed. Thus, the general $e_{L/K} = 1$ case is reduced to the case $[F : F^p] = p$. □

**Exercise 16.3** Let $K$ be a henselian discrete valuation field with residue field $F$ of characteristic $p > 0$. Let $L$ be a ramified cyclic extension of degree

$p$ of $K$, and $G = \mathrm{Gal}(L/K)$ be the Galois group. Let $\chi\colon G \to \Lambda^\times$ be a faithful character and $n = \mathrm{Sw}\,\chi$ be the Swan conductor. Show directly, without using Theorem 16.3.9, that $G^r_{\mathrm{coh}} = G^r_{\log}$ for every $r \in \mathbf{Q}_+$, and the equality $-\mathrm{rsw}\,\chi = \mathrm{ch}_{\log}\chi$ of morphisms $\mathfrak{m}_K^n/\mathfrak{m}_K^{n+1} \to \Omega^1_F(\log) \subset \Omega^1_F(\log) \otimes_F \overline{F}$.

## 16.4 Integral Part

In this section we sketch the proof of Theorem 16.3.1 in the mixed characteristic case in [50], admitting Proposition 16.4.1, Lemma 16.4.3, and Proposition 16.4.4.

To prove Theorem 16.3.1, we introduce the integral part $\mathrm{Br}^\circ K_1$ (16.4.9) of $\mathrm{Br}\,K_1 = \mathrm{H}^2(K_1, \mu_\infty)$ using the following universal object. Let $A$ be the henselization [67, Chapitre XI] of the polynomial ring $O_K[T]$ with respect to the ideal $\mathfrak{m}_K O_K[T]$. We have a canonical morphism $O_K[T] \to A$ inducing an isomorphism $F[T] \to A/\mathfrak{m}_K A$. For an extension $K'$ of henselian discrete valuation field $K$, the mapping

$$\mathrm{Mor}_{O_K}(A, O_{K'}) \to O_{K'} \tag{16.4.1}$$

sending a morphism $f\colon A \to O_{K'}$ over $O_K$ to the image $f(T) \in O_{K'}$ is a bijection by the functoriality of henselizations.

We define an algebraic analogue $D_K = \mathrm{Spec}\,A \otimes_{O_K} K$ of the rigid analytic closed disc of dimension 1 and radius 1. Let $\mathrm{H}^2(D_K, \mu_\infty)$ denote the inductive limit $\varinjlim_n \mathrm{H}^2(D_K, \mu_n)$ of étale cohomology. The canonical morphism $\mathrm{H}^2(D_K, \mu_p) \to \mathrm{H}^2(D_K, \mu_\infty)[p]$ is an isomorphism by [50, Proposition (1.8)]. A morphism $A \to O_{K'}$ over $O_K$ induces

$$\mathrm{H}^2(D_K, \mu_\infty) \to \mathrm{H}^2(K', \mu_\infty) = \mathrm{Br}\,K' \tag{16.4.2}$$

by the functoriality of étale cohomology.

We set $\mathrm{H}^1(F) = \mathrm{Coker}(\mathrm{F} - 1\colon \Omega^1_F(\log) \to \Omega^1_F(\log)/dF)$. We define a sub-$F[T]$-module $\Omega^1_{F[T]}(\log) \subset \Omega^1_{F(T)}(\log)$ to be that generated by $\Omega^1_{F[T]}$ and $d\log t$ for $t \in K^\times$. This fits in an exact sequence

$$0 \to \Omega^1_{F[T]} \to \Omega^1_{F[T]}(\log) \to F[T] \to 0$$

similarly to $\Omega^1_{F(T)}(\log)$. The derivation $d\colon F[T] \to \Omega^1_{F[T]}(\log)$ and the Frobenius $\mathrm{F}\colon \Omega^1_{F[T]}(\log) \to \Omega^1_{F[T]}(\log)/dF[T]$ are defined. We set

$$\mathrm{H}^1(F[T]) = \mathrm{Coker}(\mathrm{F} - 1\colon \Omega^1_{F[T]}(\log) \to \Omega^1_{F[T]}(\log)/dF[T]).$$

Note that $\mathrm{H}^1(F)$ and $\mathrm{H}^1(F[T])$ depend not only on $F$ or $F[T]$ but also on $K$ or $K_1$, although they do not appear in the notation.

For an extension $K'$ of the discrete valuation field $K$, a morphism $O_K[T] \to O_{K'}$ over $O_K$ induces a morphism

$$\mathrm{H}^1(F[T]) \to \mathrm{H}^1(F') \tag{16.4.3}$$

compatible with the induced morphism $F[T] \to F'$. In particular, the morphism $O_K[T] \to O_{K_1}$ sending $T$ to $T$ induces an injection

$$\mathrm{H}^1(F[T]) \to \mathrm{H}^1(F(T)). \tag{16.4.4}$$

**Proposition 16.4.1** (1) *There exists a unique morphism*

$$\beta^{\circ}_{K_1} : \mathrm{H}^1(F[T]) \to \mathrm{H}^2(D_K, \mu_p) \tag{16.4.5}$$

*satisfying the following property: Let $K'$ be an extension of the henselian discrete valuation field $K$, and $A \to O_{K'}$ be a morphism over $O_K$. Then the diagram*

$$\begin{array}{ccc} \mathrm{H}^1(F[T]) & \xrightarrow{\beta^{\circ}_{K_1}} & \mathrm{H}^2(D_K, \mu_p) \\ {\scriptstyle (16.4.3)}\downarrow & & \downarrow{\scriptstyle (16.4.2)} \\ \mathrm{H}^1(F') & \xrightarrow{\beta_{K'}} & \mathrm{H}^2(K', \mu_p) = \mathrm{Br}\, K'[p] \end{array} \tag{16.4.6}$$

*is commutative.*

(2) [50, (1.8.1)] *As $A \to O_{K'}$ in* (1), *take the canonical morphism $A \to O_{K_1}$ sending $T$ to $T$. Then the diagram*

$$\begin{array}{ccc} \mathrm{H}^1(F[T]) & \xrightarrow{\beta^{\circ}_{K_1}} & \mathrm{H}^2(D_K, \mu_p) \\ {\scriptstyle (16.4.4)}\downarrow & & \downarrow{\scriptstyle (16.4.2)} \\ \mathrm{H}^1(F(T)) & \xrightarrow{\beta_{K_1}} & \mathrm{H}^2(K_1, \mu_p) = \mathrm{Br}\, K_1[p] \end{array} \tag{16.4.7}$$

*is cartesian and the morphisms are injections. The morphism*

$$\mathrm{H}^2(D_K, \mu_\infty) \to \mathrm{H}^2(K_1, \mu_\infty) = \mathrm{Br}\, K_1 \tag{16.4.8}$$

*is an injection.*

We define the integral part

$$\mathrm{Br}^{\circ} K_1 \subset \mathrm{Br}\, K_1 \tag{16.4.9}$$

to be the image of the injection (16.4.8). We omit the proof of Proposition 16.4.1 and give only a sketch here.

The morphism $\beta^{\circ}_{K_1}$ in (16.4.5) is defined as follows. Let

$$\mathbf{A}^1_F = \operatorname{Spec} F[T] \xrightarrow{\ i\ } \operatorname{Spec} A \xleftarrow{\ j\ } D_K$$

be the closed immersion and the open immersion. Then, we consider morphisms

$$\mathrm{H}^1(F[T]) \to \mathrm{H}^1(\mathbf{A}^1_F, i^*R^1j_*\mu_p) \leftarrow \mathrm{H}^1(\operatorname{Spec} A, R^1j_*\mu_p) \to \mathrm{H}^2(D_K, \mu_p).$$

The first arrow is defined similarly to (2.5.5) using a computation of nearby cycles in [12, Corollary (1.4.1)] generalizing Lemma 3.2.1 and is an isomorphism. The second arrow is an isomorphism by a theorem of Gabber [27, Theorem 1]. The last arrow is the edge morphism of the spectral sequence $\mathrm{E}_2^{s,t} = \mathrm{H}^s(\operatorname{Spec} A, R^tj_*\mu_p) \to \mathrm{H}^{s+t}(D_K, \mu_p)$ and is an injection. The morphism $\beta^{\circ}_{K_1}$ in (16.4.5) is defined as the composition and is an injection. The commutativity of the diagram (16.4.6) follows from the functoriality of the construction.

The injectivity of the right vertical arrow $\mathrm{H}^2(D_K, \mu_p) \to \mathrm{H}^2(K_1, \mu_p) = \operatorname{Br} K_1[p]$ in the diagram (16.4.7) and the $p$-part of (16.4.8) are proved in [50, (1.8.1)]. Then, since the horizontal arrows are injective, the commutativity of the diagram (16.4.7) implies that the left vertical arrow is also an injection. The property that the diagram (16.4.7) is cartesian means the injectivity of the morphism $\operatorname{Coker} \beta^{\circ}_{K_1} \to \operatorname{Coker} \beta_{K_1}$. This is also a consequence of the computation of nearby cycles in [12, Corollary (1.4.1)]. The injectivity of the prime-to-$p$ part of (16.4.8) follows from the absolute purity.

**Corollary 16.4.2** (1) *Let $K'$ be an extension of $K$ such that $e_{K'/K} = 1$. Then, for $\chi \in F^nX_K$, $\{\chi, 1 + \mathfrak{m}_{K'}^{n+1}\} = 0$.*

(2) *The morphism* $\operatorname{Br} K \to \operatorname{Br} K_1$ *is an injection.*

*Proof* (1) Let $t \in K$ be a uniformizer and let $a = t^{n+1}b$ be an element of $\mathfrak{m}_{K'}^{n+1}$. Then the morphism $A \to O_{K'}$ defined by sending $T$ to $b$ maps $\{\chi, 1+t^{n+1}T\} = 0$ to $\{\chi, 1 + t^{n+1}b\} = \{\chi, 1 + a\}$. Hence, the assertion follows.

(2) It suffices to show that $\operatorname{Br} K \to \operatorname{Br}^{\circ} K_1$ is an injection. Since the morphism $A \to O_K$ defined by sending $T$ to 0 induces a section $\operatorname{Br}^{\circ} K_1 \to \operatorname{Br} K$ of $\operatorname{Br} K \to \operatorname{Br}^{\circ} K_1$, the assertion follows. □

We define a generalization of the morphism $\beta\colon \Omega^1_F(\log) \to \operatorname{Br} K$. Let $K$ be a henselian discrete valuation field with residue field $F$ of characteristic $p > 0$. Let $\Omega^2_F(\log)$ be the second exterior power of the $F$-vector space $\Omega^1_F(\log)$. The differential $d\colon \Omega^1_F(\log) \to \Omega^2_F(\log)$ and the Frobenius operator $\mathrm{F}\colon \Omega^2_F(\log) \to \Omega^2_F(\log)/d\Omega^1_F(\log)$ are defined similarly to $\Omega^1_F(\log)$. Let $\mathrm{H}^2(F)$ denote the cokernel of $\mathrm{F} - 1\colon \Omega^2_F(\log) \to \Omega^2_F(\log)/d\Omega^1_F(\log)$, and similarly for $\mathrm{H}^1(F)$.

Then, as an analogue of the injection $\mathrm{H}^1(F) \to \mathrm{Br}\, K[p]$ (2.5.7), a canonical injection

$$\gamma\colon \mathrm{H}^2(F) \to \mathrm{H}^3(K, \mu_p^{\otimes 2}) \tag{16.4.10}$$

is defined [50, (1.4)], [47, Theorem 3].

We also introduce its integral parts. Let $\Omega^2_{F[T]}(\log)$ be the second exterior power of the $F[T]$-module $\Omega^1_{F[T]}(\log)$ fitting in an exact sequence $0 \to \Omega^2_{F[T]} \to \Omega^2_{F[T]}(\log) \to \Omega^1_{F[T]} \to 0$. Similarly to $\mathrm{H}^1(F[T])$, we define $\mathrm{H}^2(F[T]) = \mathrm{Coker}(\mathrm{F} - 1\colon\ \Omega^2_{F[T]}(\log) \to \Omega^2_{F[T]}(\log)/d\Omega^1_{F[T]}(\log))$. The canonical morphism $\mathrm{H}^2(F[T]) \to \mathrm{H}^2(F(T))$ is also an injection [50, (1.8.1)]. Thus, we obtain an injection

$$\gamma^\circ_{K_1} : \mathrm{H}^2(F[T]) \to \mathrm{H}^3(K_1, \mu_p^{\otimes 2}) \tag{16.4.11}$$

similar to $\beta^\circ_{K_1}$. By these injections and that in Lemma 2.5.6(1), we identify $T \cdot \Omega^1_F(\log)$, $\mathrm{H}^1(F[T])$ and $\mathrm{H}^2(F[T])$ with subgroups of $\mathrm{Br}\, K_1[p]$ and $\mathrm{H}^3(K_1, \mu_\infty^{\otimes 2})$.

We have two variable variants

$$\mathrm{H}^1(F[T_1,T_2]) = \mathrm{Coker}(\mathrm{F} - 1\colon \Omega^1_{F[T_1,T_2]}(\log) \to \Omega^1_{F[T_1,T_2]}(\log)/dF[T_1,T_2]),$$
$$\mathrm{H}^2(F[T_1,T_2]) = \mathrm{Coker}(\mathrm{F} - 1\colon \Omega^2_{F[T_1,T_2]}(\log) \to \Omega^2_{F[T_1,T_2]}(\log)/dF[T_1,T_2])$$

and injections

$$\mathrm{H}^1(F[T_1,T_2]) \to \mathrm{H}^1(F(T_1,T_2)), \qquad \mathrm{H}^2(F[T_1,T_2]) \to \mathrm{H}^2(F(T_1,T_2))$$

[50, (1.8.1)]. Let $p_1, p_2, s\colon F[T] \to F[T_1,T_2]$ be morphisms over $F$ defined by $p_1(T) = T_1$, $p_2(T) = T_2$, and $s(T) = T_1 + T_2$. By the functoriality, the morphisms $p_1^*, p_2^*, s^* : \mathrm{H}^q(F[T]) \to \mathrm{H}^q(F[T_1,T_2])$ for $q = 1, 2$ are defined.

**Lemma 16.4.3** ([50, Lemma 5.7]) *Let* $c \in \mathrm{Ker}(p_1^* + p_2^* - s^* : \mathrm{H}^1(F[T]) \to \mathrm{H}^1(F[T_1,T_2]))$. *If there exists an element* $c' \in \mathrm{Ker}(p_1^* + p_2^* - s^* : \mathrm{H}^2(F[T]) \to \mathrm{H}^2(F[T_1,T_2]))$ *such that* $c \wedge (dT/T) = c'$ *in* $\mathrm{H}^2(F(T))$, *then* $c$ *is in the image of the injection* $T\cdot : \Omega^1_F(\log) \to \mathrm{H}^1(F[T])$ (2.4.14).

We also omit the proof of Lemma 16.4.3.

We will use the following analogue of Corollary 5.2.3(1) in the proof of Theorem 16.3.1. For $c \in \mathrm{Br}\, K[p] = \mathrm{H}^2(K, \mu_p)$ and $a \in K_1^\times$, let $\{c, a\}$ denote the cup-product $c \cup \{a\} \in \mathrm{H}^3(K_1, \mu_p^{\otimes 2})$.

**Proposition 16.4.4** ([50, Proposition (4.1)(1)]) *The image of* $\beta\colon \Omega^1_F(\log) \to \mathrm{Br}\, K$ *equals* $\{c \in \mathrm{Br}\, K[p] \mid \{c, 1 + \mathfrak{m}_{K_1}\} = 0\}$.

We omit the proof of Proposition 16.4.4; the idea is similar to that of Corollary 5.2.3(1) but is more involved, using the computation in [12] of the graded quotients of $\mathrm{H}^2(K, \mu_p^{\otimes 2})$.

*Proof of Theorem* 16.3.1 *in the mixed characteristic case* Let $\chi$ be an element of $F_n X_K$ and $a \in \mathfrak{m}_K^n$. By Proposition 2.3.3(2), the cup-product $c = \{\chi, 1+aT\} \in \mathrm{Br}\, K_1$ is killed by $p$. Since $1+aT \in A$ is invertible, it follows that $c$ is the image of $\chi \cup \{1+aT\} \in \mathrm{H}^2(D_K, \mu_\infty)[p] = \mathrm{H}^2(D_K, \mu_p)$. We show that $c \in \mathrm{Br}^\circ K_1[p]$ is in the image of $T \cdot \Omega_F^1(\log) \subset \mathrm{H}^1(F[T])$ by $\beta_{K_1}^\circ$.

Let $K_2$ be the fraction field of the henselization $O_{K_2}$ of the local ring of $O_K[T_1, T_2]$ at the prime ideal $\mathfrak{m}_K O_K[T_1, T_2]$, and define a morphism $K_1 \to K_2$ by sending $T$ to $T_1$. Then $O_{K_2}$ is the henselization of the local ring of $O_{K_1}[T_2]$ at the prime ideal $\mathfrak{m}_{K_1} O_{K_1}[T_2]$, and we have $K_2 = (K_1)_1$.

For an integer $N \geqslant 1$ annihilating $\chi$, we have

$$\begin{aligned}\{c, 1+\mathfrak{m}_{K_2}\} = \{\chi, 1+aT, U^1 K_2^\times\} &\subset \{\chi, U^{n+1}\mathrm{H}^2(K_2, \mu_N^{\otimes 2})\} \\ &= \{\chi, 1+\mathfrak{m}_{K_2}^{n+1}, K_2^\times\}\end{aligned}$$

in $\mathrm{H}^3(K_2, \mu_N^{\otimes 2})$ by Lemma 3.2.5. Since $\{\chi, 1+\mathfrak{m}_{K_2}^{n+1}, K_2^\times\} = 0$ by Corollary 16.4.2(1), we have $\{c, 1+\mathfrak{m}_{K_2}\} = 0$. Hence, by Proposition 16.4.4, $c \in \mathrm{Br}\, K_1[p]$ is in the image of $\beta_{K_1} : \mathrm{H}^1(F(T)) \to \mathrm{Br}\, K_1[p]$. Since $c$ is an element of the integral part $\mathrm{Br}^\circ K_1[p]$, $c$ is further in the image of the injection $\beta_{K_1}^\circ : \mathrm{H}^1(F[T]) \to \mathrm{Br} K_1[p]$ by Proposition 16.4.1(2).

In the following, we regard $c$ as an element of $\mathrm{H}^1(F[T])$ by abuse of notation. Since $(1+aT_1)(1+aT_2) \equiv 1 + a(T_1+T_2) \bmod \mathfrak{m}_{K_2}^{n+1}$, we have an equality $(p_1^* + p_2^* - s^*)(c) = \{\chi, (1+aT_1)(1+aT_2)/(1+a(T_1+T_2))\} = 0$ in $\mathrm{Br}\, K_2$ by Corollary 16.4.2(1). Since $\beta_{K_2}^\circ : \mathrm{H}^1(F[T_1, T_2]) \to \mathrm{Br}\, K_2$ is an injection, $c$ is an element of $\mathrm{Ker}(p_1^* + p_2^* - s^* : \mathrm{H}^1(F[T]) \to \mathrm{H}^1(F[T_1, T_2]))$.

Let $c' = -c \wedge d\log(-a) \in \mathrm{H}^2(F[T])$. Since $(p_1^* + p_2^* - s^*)(c) = 0$ in $\mathrm{H}^1(F[T_1, T_2]))$, we also have $(p_1^* + p_2^* - s^*)(c') = 0$ in $\mathrm{H}^2(F[T_1, T_2]))$. Since $\{c, -aT\} = \{\chi, 1+aT, -aT\} = 0$, we have $c' = -\{c, -a\} = \{c, T\} \in \mathrm{H}^3(K_1, \mu_p^{\otimes 2})$. Since $\gamma_{K_1} : \mathrm{H}^2(F(T)) \to \mathrm{H}^3(K_1, \mu_p^{\otimes 2})$ is an injection, the assumption $c' = c \wedge (dT/T)$ in $\mathrm{H}^2(F(T))$ in Lemma 16.4.3 is satisfied. Hence, $c$ is in the image of the injection $T \cdot : \Omega_F^1(\log) \to \mathrm{H}^1(F[T])$. Let $c$ also denote the corresponding element of $T \cdot \Omega_F^1(\log)$, by abuse of notation.

Since $(1+aT)(1+bT) \equiv 1+(a+b)T \bmod 1+\mathfrak{m}_{K_1}^{n+1}$, the mapping $\widetilde{f} : \mathfrak{m}_K^n \to \Omega_F^1(\log) \subset \mathrm{Br}\, K_1[p]$ defined by $T \cdot \widetilde{f}(a) = \{\chi, 1+aT\}$ is additive by Corollary 16.4.2 and induces a morphism $f : \mathfrak{m}_K^n/\mathfrak{m}_K^{n+1} \to \Omega_F^1(\log)$ of groups.

Let $u \in O_K^\times$ be a unit and $u_* : K_1 \to K_1$ denote the automorphism over $K$ sending $T$ to $uT$. Then $u_*\{\chi, 1+aT\} = \{\chi, 1+uaT\}$ and we have a commutative

diagram

$$\begin{array}{ccc} T\cdot\Omega^1_F(\log) & \xrightarrow{\beta_{K_1}} & \operatorname{Br} K_1 \\ {\scriptstyle \bar{u}\cdot}\downarrow & & \downarrow{\scriptstyle u_*} \\ T\cdot\Omega^1_F(\log) & \xrightarrow{\beta_{K_1}} & \operatorname{Br} K_1 \end{array}$$

by the functoriality of the morphism $\beta$. Hence, the morphism $f\colon \mathfrak{m}^n_K/\mathfrak{m}^{n+1}_K \to \Omega^1_F(\log)$ is $F$-linear.

We show that $f\colon \mathfrak{m}^n_K/\mathfrak{m}^{n+1}_K \to \Omega^1_F(\log)$ satisfies the property in Theorem 16.3.1. Let $K'$ be an extension of the discrete valuation field $K$, and $a = t^n b \in \mathfrak{m}^n_K \mathcal{O}_{K'}$. Then the morphism $A \to \mathcal{O}_{K'}$ sending $T$ to $b$ induces a morphism $\mathrm{H}^2(D_K, \mu_p) \to \operatorname{Br} K'[p]$. This maps $\{\chi, 1+t^nT\} = T\cdot \operatorname{rsw}\chi(t^n) \in T\cdot\Omega^1_F(\log) \subset \mathrm{H}^2(D_K,\mu_p)$ to $\{\chi, 1+a\} = b\cdot \operatorname{rsw}\chi(t^n) \in \mathrm{H}^2(K',\mu_p) = \operatorname{Br} K'[p]$ by the commutative diagram (16.4.6) in Proposition 16.4.1(1). Hence, the assertion follows. □

**Exercise 16.4** Let $\chi \in X_{K_1}$ be an element in the image of $\mathrm{H}^1(D_K, \mathbf{Q}/\mathbf{Z}) \to X_{K_1}$, and $a \in A[1/p]^\times \subset K_1^\times$. Show that $\{\chi, a\} \in \operatorname{Br} K_1$ is an element of the integral part $\operatorname{Br}^\circ K_1$.

## Historical Notes

The logarithmic variant of the characteristic form was constructed earlier than the characteristic form in the equal characteristic case [71] and in the mixed characteristic case [72]. The description of the logarithmic variant of characteristic from for Artin–Schreier–Witt extensions in terms of Witt vectors is taken from [4].

The last two sections sketch the contents in [50] together with the proof of the equality between the cohomological filtration and the logarithmic upper numbering filtration for abelian extensions taken from [52]. The computation of cohomology of complete discrete valuation fields and its generalization to $p$-adic vanishding cycles are given in [47] and in [12] respectively.

# Solutions to Exercises

## Chapter 1

**Exercise 1.1** Since $\mathbf{Z}[1/p][X]/(X^p - 1) = \mathbf{Z}[\zeta_p][1/p] \times \mathbf{Z}[1/p]$ and $(X^p - 1)' = pX^{p-1}$ is invertible in $\mathbf{Z}[1/p][X]/(X^p - 1)$, we see that $\mathbf{Z}[\zeta_p][1/p]$ is étale over $\mathbf{Z}$ by Lemma 1.1.6. Since $\mathbf{Z}[\zeta_p] \otimes_{\mathbf{Z}} \mathbf{F}_p$ is isomorphic to $\mathbf{F}_p[X]/(X^{p-1})$ and is not étale over $\mathbf{F}_p$ for $p - 1 > 1$, we see that $\mathbf{Z}[\zeta_p]$ is not étale over $\mathbf{Z}$ at the prime ideal $(\zeta_p - 1)$.

**Exercise 1.2** We have $\mathrm{Tr}_{B/A} = P'(b) \cdot f_n = n \cdot f_1 + (n-1)a_1 \cdot f_2 + \cdots + a_{n-1} \cdot f_n$ by Proposition 1.2.2. Hence, $\mathrm{Tr}_{B/A} c_i = (n - i + 1)a_{i-1}$ for $i = 1, \ldots, n$ and $a_0 = 1$. The equality

$$\begin{pmatrix} c_1 \\ c_2 \\ \vdots \\ c_n \end{pmatrix} = \boldsymbol{A} \begin{pmatrix} 1 \\ b \\ \vdots \\ b^{n-1} \end{pmatrix}$$

in the proof of Proposition 1.2.2(1) implies that

$$\begin{pmatrix} 1 \\ b \\ \vdots \\ b^{n-1} \end{pmatrix} = \boldsymbol{A}^{-1} \begin{pmatrix} c_1 \\ c_2 \\ \vdots \\ c_n \end{pmatrix}.$$

By applying $\mathrm{Tr}_{B/A}$ on both sides, we obtain the equality.

**Exercise 1.3** Since $\mathbf{Q}_p$ is complete, $\mathbf{Z}_p$ is a henselian local ring by Corollary 1.3.4. The polynomial $X^{p-1} - 1 \in \mathbf{Z}_p[X]$ is decomposed as $X^{p-1} - 1 = \prod_{a \in \mathbf{F}_p^\times} (X - a)$ in $\mathbf{F}_p[X]$. Hence, we may apply Proposition 1.3.3(2′).

**Exercise 1.4** The cyclotomic polynomial

$$P = (X^{p^n} - 1)/(X^{p^{n-1}} - 1) = X^{p^{n-1}(p-1)} + \cdots + X^{p^{n-1}} + 1$$

satisfies $P(X+1) \equiv X^{(p-1)p^{n-1}} \bmod p$ and $P(1) = p$. Hence, $P(X+1)$ is an Eisenstein polynomial. By Lemma 1.4.3(1), $P(X+1)$ is the minimal polynomial of a uniformizer $\zeta_{p^n} - 1$ of $\mathbf{Q}_p(\zeta_{p^n})$, and $\mathbf{Q}_p(\zeta_{p^n})$ is an extension of $\mathbf{Q}_p$ of degree $(p-1)p^{n-1}$.

**Exercise 1.5** (1) Since $X^m - 1 \in \mathbf{F}_p[X]$ is a separable polynomial, the extension $\mathbf{Q}_p(\zeta_m)$ of $\mathbf{Q}_p$ is an unramified extension and $\mathbf{Z}_p[\zeta_m]$ is the ring of integers. Since $\mathbf{Q}_p(\zeta_{p^e})$ is a totally ramified extension of $\mathbf{Q}_p$ and $\mathbf{Z}_p[\zeta_{p^e}]$ is the ring of integers by Exercise 1.4, the morphisms $\mathbf{Q}_p(\zeta_m) \otimes_{\mathbf{Q}_p} \mathbf{Q}_p(\zeta_{p^e}) \to \mathbf{Q}_p(\zeta_n)$ and $\mathbf{Z}_p[\zeta_m] \otimes_{\mathbf{Z}_p} \mathbf{Z}_p[\zeta_{p^e}] \to \mathbf{Z}_p[\zeta_n]$ are isomorphisms. Hence, $\mathbf{Q}_p(\zeta_m) \subset \mathbf{Q}_p(\zeta_n)$ is the maximal unramified extension. Since $e_{\mathbf{Q}_p(\zeta_p)/\mathbf{Q}_p} = p-1$ is prime to $p$ and $e_{\mathbf{Q}_p(\zeta_{p^e})/\mathbf{Q}_p(\zeta_p)} = p^{e-1}$, the composite extension $\mathbf{Q}_p(\zeta_m, \zeta_p) = \mathbf{Q}_p(\zeta_{mp})$ is the maximal tamely ramified extension.

(2) By (1), we have $G = \langle p \rangle \times (\mathbf{Z}/p^e\mathbf{Z})^\times \subset (\mathbf{Z}/m\mathbf{Z})^\times \times (\mathbf{Z}/p^e\mathbf{Z})^\times$ and $I \supset P$ are identified with $(\mathbf{Z}/p^e\mathbf{Z})^\times \supset (1 + p\mathbf{Z})/(1 + p^e\mathbf{Z})$.

**Exercise 1.6** For $1 < a < p^n$ prime to $p$, let $\sigma_a \in G$ be the element corresponding to $a \in (\mathbf{Z}/p^n\mathbf{Z})^\times$. Then, for $0 \leqslant m = \mathrm{ord}_p(a-1) < n$, we have $i_G(\sigma_a) = \mathrm{ord}((\zeta_{p^n}^a - \zeta_{p^n})/(\zeta_{p^n} - 1)) = \mathrm{ord}(\zeta_{p^n}^{a-1} - 1) - 1 = \mathrm{ord}(\zeta_{p^n}^{p^m} - 1) - 1 = p^m - 1$. Hence, $G_i = 1 + (p^m) \subset (\mathbf{Z}/p^n\mathbf{Z})^\times = \mathrm{Gal}(\mathbf{Q}_p(\zeta_{p^n})/\mathbf{Q}_p)$ for $p^{m-1} \leqslant i < p^m$, $1 \leqslant m < n$. We have $G_i = 1$ for $i \geqslant p^{n-1}$ and $G = G_0$.

**Exercise 1.7** The extension $K'$ is generated over $K$ by a solution of $x^p - x = a = v/t^p$, and the ring $O_{K'} = O_K[u]$ of integers is generated by $u = tx$ satisfying $u^p - t^{p-1}u = v$. The extension $L$ is generated over $K'$ by a solution of $z^p - z = b - c$ for $b = 1/t^{p^3-1}$ and $c = ((x+a)^p - x^p - a^p)/p \equiv a^{p-1}x = v^{p-1}u/t^{(p(p-1)+1)}$, and $L$ is a totally ramified extension of $K'$ generated by a uniformizer $s = t^{p^2}z$. For $\tau \in N - \{1\}$, we have $i_G(\tau) = \mathrm{ord}_L(\tau(s)/s - 1) = \mathrm{ord}_L(\tau(z)/z - 1) = -\mathrm{ord}_L z = p^3 - 1$. Thus, $\varphi(j) = j$ for $j \leqslant p^3 - 1$ and $\varphi(j) = (j + (p-1)(p^3-1))/p$ for $j \geqslant p^3 - 1$.

For $\sigma \in \overline{G} - \{1\}$, we have $i_{\overline{G}}(\sigma) = \mathrm{ord}_{K'}(\sigma(u) - u) = \mathrm{ord}_K t = 1$. For $\tau \in \sigma \subset G$, we have $i_G(\tau) = \min(\mathrm{ord}_L(\tau(u) - u), \mathrm{ord}_L(\tau(s)/s - 1))$. We have $\mathrm{ord}_L(\tau(u) - u) = \mathrm{ord}_L t = p$. Since $w = \tau(z) - z$ is a solution of $w^p - w = c - \tau(c)$ and $\mathrm{ord}_L(c - \tau(c)) = \mathrm{ord}_L a^{p-1} = -p^2(p-1)$, we have $\mathrm{ord}_L(\tau(s)/s - 1) = \mathrm{ord}_L(\tau(z) - z)/z = p^3 - 1 - p(p-1) > p = \mathrm{ord}_L t = \mathrm{ord}_L(\tau(u) - u)$. Thus, $i_G(\tau) = p$ and $j(\sigma) = p$. We have $p \cdot i_{\overline{G}}(\sigma) = p < p^2$ in (1.7.9) and $i_{\overline{G}}(\sigma) = 1 < \varphi(p) = p$ in (1.7.10).

**Exercise 1.8** (1) Define cyclic extensions $L_1, L_2$ of degree $p^2$ by the respective

Artin–Schreier–Witt equations $(x^p, y^p)-(x, y) = (a, 0)$ and $(x^p, y^p)-(x, y) = (a, b)$ for $b = 1/t^{p^3-1}$. The unit $u = tx \in M$ is a solution of $u^p - t^{p-1}u = v$ and the residue field $E$ is generated over $F = \mathbf{F}_p(v)$ by a $p$th root of $v$. Let $c = ((x+a)^p - x^p - a^p)/p \equiv a^{p-1}x = v^{p-1}u/t^{(p(p-1)+1)} \in K'$. Then $L_1$ and $L_2$ are defined over $K'$ by the Artin–Schreier equations $y^p - y = -c$ and $y^p - y = b - c$ respectively. Since $\mathrm{N}_{L_1/K'}y = y^p - y \equiv -u/t$ and $\mathrm{N}_{L_2/K'}y \equiv t \bmod K'^{\times p}$, $L_1$ is log monogenic and $L_2$ is not log monogenic.

(2) Let $\rho$ be the generator of $\mathrm{Gal}(L/L')$ satisfying $\rho(y) = y+1$. Since $u = tx$ satisfies $u^p - t^{p-1}u = v$, we have $O_{K'} = O_K[u]$. Since $s = ty$ satisfies $s^p - t^{p-1}s = t$, $s$ is a uniformizer of $O_{L'}$. Since $\sigma(u)/u-1 = \sigma(x)/x-1 = 1/x = t/u$ and $\rho(s)/s - 1 = \rho(y)/y - 1 = 1/y = t/s$, we have $i_G(\sigma) = p > i_G(\rho) = p-1$ and $G = G_{p-1} \supsetneqq G_p = \langle\sigma\rangle \supsetneqq G_{p+1} = 1$. Hence, $\varphi(j) = j$ for $j \leqslant p-1$ and $\varphi(j) = p - 2 + (j+1)/p$ for $j \geqslant p-1$. We also have $j(\overline{\sigma}) = i_G(\sigma) = p$.

Since $\overline{G} = \overline{G}_1 \supsetneqq \overline{G}_2 = 1$, we have $i_{\overline{G}}(\overline{\sigma}) = 1 < \varphi(j(\sigma)) = p - 1 + 1/p$. Hence, by Corollary 1.8.8(1), $L$ is not log monogenic over $K$. For $i = \varphi(j) = j = 1$, $\overline{G}_1 = \overline{G}$ is the image of $G_1 = G$. If $i = \varphi(j) = p - 2 + (j+1)/p \geqslant p$, then $j \geqslant p$, $j \equiv -1 \bmod p$. Hence, we have $j \geqslant 2p - 1 \geqslant p + 1$ and $\overline{G}_i = 1$ is the image of $G_j = 1$.

**Exercise 1.9** (1) The cyclotomic polynomial $\Phi_{p^n}$ is the minimal polynomial of $\zeta_{p^n}$. Since $\Phi'_p = ((X^p-1)/(X-1))' = (pX^{p-1}(X-1) - (X^p-1))/(X-1)^2$, we have $\Phi'_p(\zeta_p) = p/(\zeta_p(\zeta_p - 1))$. Since $\Phi_{p^n} = \Phi_p(X^{p^{n-1}})$, we have $\Phi'_{p^n} = \Phi'_p(X^{p^{n-1}}) \cdot p^{n-1}X^{p^{n-1}-1}$ and $\Phi'_{p^n}(\zeta_{p^n}) = \Phi'_p(\zeta_p) \cdot p^{n-1}\zeta_p/\zeta_{p^n} = p^n/(\zeta_{p^n}(\zeta_p - 1))$. Hence, by Proposition 1.9.5(1), the different $D_{\mathbf{Q}_p(\zeta_{p^n})/\mathbf{Q}_p}$ is generated by $p^n/(\zeta_p - 1)$.

We have $\mathrm{ord}_{\mathbf{Q}_p(\zeta_{p^n})} D^{\log}_{\mathbf{Q}_p(\zeta_{p^n})/\mathbf{Q}_p} = \mathrm{ord}_{\mathbf{Q}_p(\zeta_{p^n})} p^n/(\zeta_p - 1) - ([\mathbf{Q}_p(\zeta_{p^n}) : \mathbf{Q}_p] - 1) = n(p-1)p^{n-1} - p^{n-1} - ((p-1)p^{n-1} - 1) = (n-1)(p-1)p^{n-1} - (p^{n-1} - 1)$. By Exercise 1.6, for $G = \mathrm{Gal}(\mathbf{Q}_p(\zeta_{p^n})/\mathbf{Q}_p(\zeta_p))$, we have $\sum_{\sigma\in G-\{1\}} i_G(\sigma) = \sum_{m=1}^{n-1}(p^{n-m} - p^{n-m-1})(p^m - 1) = (p^n - p^{n-1})(n-1) - (p^{n-1} - 1)$.

(2) The solution of (1) implies that $b = -1/p^{n-1} \cdot \zeta_{p^n}(\zeta_p - 1)/(\zeta_{p^n} - 1) = -\sum_{i=1}^{p^{n-1}} \zeta_{p^n}^i/p^{n-1}$ is a generator of $(D^{\log}_{\mathbf{Q}_p(\zeta_{p^n})/\mathbf{Q}_p})^{-1}$. Since the trace $\mathrm{Tr}_{\mathbf{Q}_p(\zeta_{p^n})/\mathbf{Q}_p(\zeta_p)}\zeta_{p^n}^i$ is 0 for $i = 1, \ldots, p^{n-1} - 1$, we have $\mathrm{Tr}_{\mathbf{Q}_p(\zeta_{p^n})/\mathbf{Q}_p} b = -p^{n-1}\mathrm{Tr}_{\mathbf{Q}_p(\zeta_p)/\mathbf{Q}_p}\zeta_p/p^{n-1} = 1$.

**Exercise 1.10** We have $[\mathbf{Q}_p(\zeta_{p^n}^\times) : \mathbf{Q}_p] = \#(\mathbf{Z}_p/p^n\mathbf{Z}_p)^\times = (p-1)p^{n-1}$. If $p \neq 2$, we have inclusions $1 + p^n\mathbf{Z}_p = \mathrm{Ker}(\mathbf{Z}_p^\times \to (\mathbf{Z}_p/p^n\mathbf{Z}_p)^\times) \subset \mathbf{Q}_p^{\times(p-1)p^{n-1}} \subset \mathrm{N}_{\mathbf{Q}_p(\zeta_{p^n})/\mathbf{Q}_p}\mathbf{Q}_p(\zeta_{p^n})^\times$. If $p = 2$ and $n \geqslant 2$, since we have $\mathrm{N}_{\mathbf{Q}_2(\zeta_4)/\mathbf{Q}_2}(1 + 2\zeta_4) = 5$, we therefore have $1 + 2^n\mathbf{Z}_2 = \overline{\langle(1+4)^{2^{n-2}}\rangle} \subset$

$N_{\mathbf{Q}_2(\zeta_4)/\mathbf{Q}_2}\mathbf{Q}_2(\zeta_4)^{\times[\mathbf{Q}_2(\zeta_{2^n}):\mathbf{Q}_2(\zeta_4)]} \subset N_{\mathbf{Q}_2(\zeta_{2^n})/\mathbf{Q}_2}\mathbf{Q}_2(\zeta_{2^n})^\times$. Since we have $N_{\mathbf{Q}_p(\zeta_{p^n})/\mathbf{Q}_p}(\zeta_{p^n}-1) = (-1)^{(p-1)p^{n-1}}p = p$ unless $p^n = 2$, the inclusion follows.

# Chapter 2

**Exercise 2.1** By Proposition 2.1.1, we have an isomorphism $\mathbf{R}^\times/N_{\mathbf{C}/\mathbf{R}}\mathbf{C}^\times \to \operatorname{Br}\mathbf{R}$. Hence, the Brauer group $\operatorname{Br}\mathbf{R}$ is a cyclic group of order 2.

**Exercise 2.2** (1) By Exercise 1.9(3), we have an inclusion $(1+p^n\mathbf{Z}_p)\cdot\langle p\rangle \subset N_{\mathbf{Q}_p(\zeta_{p^n})/\mathbf{Q}_p}\mathbf{Q}_p(\zeta_{p^n})^\times$. Hence, we have a surjection $\mathbf{Q}_p^\times/((1+p^n\mathbf{Z}_p)\cdot\langle p\rangle) = (\mathbf{Z}_p/p^n\mathbf{Z}_p)^\times \to \mathbf{Q}_p^\times/N_{\mathbf{Q}_p(\zeta_{p^n})/\mathbf{Q}_p}\mathbf{Q}_p(\zeta_{p^n})^\times$. Since the reciprocity mapping induces an isomorphism

$$\mathbf{Q}_p^\times/N_{\mathbf{Q}_p(\zeta_{p^n})/\mathbf{Q}_p}\mathbf{Q}_p(\zeta_{p^n})^\times \to \operatorname{Gal}(\mathbf{Q}_p(\zeta_{p^n})/\mathbf{Q}_p) = (\mathbf{Z}_p/p^n\mathbf{Z}_p)^\times$$

by Theorem 2.2.5(1), the composition is an isomorphism $(\mathbf{Z}_p/p^n\mathbf{Z}_p)^\times \to \operatorname{Gal}(\mathbf{Q}_p(\zeta_{p^n})/\mathbf{Q}_p)$ and the equality $N_{\mathbf{Q}_p(\zeta_{p^n})/\mathbf{Q}_p}\mathbf{Q}_p(\zeta_{p^n})^\times = (1+p^n\mathbf{Z}_p)\cdot\langle p\rangle$ follows.

(2) Since $\bigcap_{n\geqslant 1}(1+p^n\mathbf{Z}_p) = 1$, the maximal abelian extension $\mathbf{Q}_p^{\mathrm{ab}}$ is the composition of the maximal unramified extension $\mathbf{Q}_p^{\mathrm{ur}}$ and $\mathbf{Q}_p(\zeta_{p^n}; n \geqslant 1)$ by Corollary 2.2.6(1).

**Exercise 2.3** (1) We prove it by induction on $n$. Assume either $p \neq 2$ and $n \geqslant 2$ or $p = 2$ and $n \geqslant 3$. Since $(1+p^{n-1}x)^p \equiv 1 + p^n x \bmod p^{n+1}$, the inclusion holds by induction on $n$. Since $((1+p^{n-1}X)^p - 1)/p^n \equiv X \bmod p$, the equality follows from Hensel's lemma by induction on $n$.

(2) If $n = 1$, then $\chi$ is tamely ramified and we have $\operatorname{Sw}\chi = 0$. Assume $n \geqslant 2$. Let $K = \mathbf{Q}_p$ and let $K_1$ be as in Definition 2.3.1. Assume $p \neq 2$. Since $(p-1)p^{n-1}\cdot\chi = 0$ and $1+\mathfrak{m}_{K_1}^n = (1+\mathfrak{m}_{K_1})^{p^{n-1}}$ by (1), we obtain $\{\chi, 1+\mathfrak{m}_{K_1}^n\} = 0$ and $\operatorname{Sw}\chi \leqslant n-1$.

Assume $p = 2$. First assume $n \geqslant 3$. Since $2^{n-2}\cdot\chi = 0$ and $1+\mathfrak{m}_{K_1}^n = (1+\mathfrak{m}_{K_1}^2)^{2^{n-2}}$ by (1), we also obtain $\{\chi, 1+\mathfrak{m}_{K_1}^n\} = 0$ and $\operatorname{Sw}\chi \leqslant n-1$. Let $n = 2$ and $\chi\colon \operatorname{Gal}(\mathbf{Q}_2(\zeta_4)/\mathbf{Q}_2) \to \mathbf{Q}/\mathbf{Z}$ be the unique non-trivial character. Let $L = \mathbf{Q}_2(\zeta_4)$ and define $L_1$ as in Definition 2.3.1. Since $N_{L_1/K_1}(1+2(1+\sqrt{-1})x) = (1+2x)^2+(2x)^2 = 1+4x+8x^2$ for $x \in O_{K_1}$, we have $N_{L_1/K_1}(1+\mathfrak{m}_{L_1}^3) \supset 1+\mathfrak{m}_{K_1}^2$ by Hensel's lemma. Hence, $\{\chi, 1+\mathfrak{m}_{K_1}^2\} = 0$ and $\operatorname{Sw}\chi \leqslant 1$.

(3) We may assume $n \geqslant 2$. By Exercise 2.1, the inverse image of the kernel $\operatorname{Ker}(\operatorname{Gal}(\mathbf{Q}_p(\zeta_{p^n})/\mathbf{Q}_p) \to \operatorname{Gal}(\mathbf{Q}_p(\zeta_{p^{n-1}})/\mathbf{Q}_p))$ by the restriction of the reciprocity mapping $\mathbf{Z}_p^\times \to \operatorname{Gal}(\mathbf{Q}_p(\zeta_{p^n})/\mathbf{Q}_p)$ equals $1+p^{n-1}\mathbf{Z}_p$. Hence, if $\chi$ does

not factor through the quotient $\mathrm{Gal}(\mathbf{Q}_p(\zeta_{p^n})/\mathbf{Q}_p) \to \mathrm{Gal}(\mathbf{Q}_p(\zeta_{p^{n-1}})/\mathbf{Q}_p)$, we have $\{\chi, 1+p^{n-1}\mathbf{Z}_p\} \neq 1$. This means that we have an equality $\mathrm{Sw}\,\chi = n-1$.

(4) The solution of (3) implies $G^m_{\mathrm{coh}} = G^m_{\mathrm{rec}}$ for $m = n-1$. Since this holds for every $n \geqslant 1$, the assertion follows from the compatibility of the cohomological filtration with quotients.

**Exercise 2.4** The residue mapping $\mathrm{Res}\colon \Omega^1_K \to F$ induces an isomorphism $\mathrm{Coker}(\mathrm{F}-1\colon \Omega^1_K \to \Omega^1_K/dK) \to \mathrm{Coker}(\mathrm{F}-1\colon F \to F)$. Hence, it suffices to show the equality for $x \in F$ and $y = t$. In this case $a(x) \in X_K$ is unramified and $\{a(x), y\} \in \mathrm{Br}\,K$ is identified with $a(x) \in X_F$ by Lemma 2.2.3(2). Hence, $\mathrm{inv}\{a(x), y\} = (1/p)\mathrm{Tr}_{F/\mathbf{F}_p}x$ by Lemma 2.2.1(2).

**Exercise 2.5** (1) We may assume that $a \notin L^{\times 2}$ and $L(\sqrt{a})$ is a quadratic extension of $L$. Then, since $\mathrm{N}_{L(\sqrt{a})/L}(1-\sqrt{a}) = (1-\sqrt{a})(1+\sqrt{a}) = 1-a$, we have $(a, 1-a) = 0$.

(2) Define an extension $K_1$ of $K = \mathbf{Q}_2$ as in Definition 2.3.1. For $x \in O_{K_1}$, we have $(-1, 1+2x) = (1-2, 1+2x) = (-1, 2(1+2x)) = (-(1-2-4x), 2(1+2x)) = (1+4x, 2(1+2x))$ by (1). Since $\kappa_{1+4x} = \alpha_x$ is unramified, the right-hand side equals $\{\alpha_x, 2(1+2x)\} = \beta(x \cdot d\log 2)$. Hence, the refined Swan conductor $\mathrm{rsw}\,\chi$ exists and is the unique non-trivial morphism $2\mathbf{Z}_2/4\mathbf{Z}_2 \to \mathbf{F}_2$ sending $2x$ to $x \cdot d\log 2 = x$.

# Chapter 3

**Exercise 3.1** (1) The description of $G^k_{\mathrm{coh}}$ follows from Proposition 3.1.5. Further, we have $\mathrm{Sw}\,\chi = pm < \mathrm{Sw}\,\chi' = pl$.

(2) Since $Z = t^n u$ satisfies $Z^p - t^{(p-1)n}Z = X$ and $e_{M/K} = 1$, we have $M = k(Z,Y)((t))$. The element $w' = w - Z/t^l$ satisfies $w'^p - w' = (X - Z^p)/t^{pl} + Z/t^l = -Z/t^{pl-(p-1)n} + Z/t^l = -Z(1-t^{(p-1)(l-n)})/t^{pl-(p-1)n}$. Hence, by Proposition 3.1.5, we have $\mathrm{Sw}\,\chi'_M = pl-(p-1)n < pm = \mathrm{Sw}\,\chi_M$.

(3) For $r = pl-(p-1)n < pm$ as in (2), we have $H = H^r_{\mathrm{coh}} \supsetneqq H^{r+1}_{\mathrm{coh}} = \mathrm{Gal}(L/M(w')) = H^{pm}_{\mathrm{coh}} \supsetneqq H^{pm+1}_{\mathrm{coh}} = 1$.

**Exercise 3.2** Since $\zeta_{p^{n-1}} - 1$ is a uniformizer of $K$ and $e = e_{K/\mathbf{Q}_p} = p^{n-2}(p-1)$, we have $\mathrm{Sw}\,\chi = ep/(p-1) - 1 = p^{n-1} - 1$ by Proposition 3.2.7(1). In the notation of Proposition 3.2.8(1), we have $x = \zeta_{p^{n-1}} - 1$, $u = \zeta_{p^n} - 1$, $v = (\zeta_p \cdot \zeta_{p^n} - 1)/(\zeta_{p^n} - 1) - 1 = (\zeta_p - 1)\zeta_{p^n}/(\zeta_{p^n} - 1)$, $w = (\zeta_p - 1)^p \zeta_{p^{n-1}}(-1)^{p-1}/(\zeta_{p^{n-1}} - 1) \equiv -p(\zeta_p - 1)/(\zeta_{p^{n-1}} - 1)$, and $z = \zeta_{p^{n-1}} - 1$. Hence, $\mathrm{rsw}\,\chi$ maps $-w \equiv p(\zeta_p - 1)/(\zeta_{p^{n-1}} - 1)$ to $d\log z = 1 \in \Omega^1_{\mathbf{F}_p}(\log) = \mathbf{F}_p$.

# Chapter 4

**Exercise 4.1** Let $O_K[X_1, \dots, X_n] \to O_L$ be a surjection over $O_K$. Since $O_L$ is regular, the kernel is locally generated by $n$ elements. For the residue fields $F$ and $E$, the kernel of the induced surjection $F[X_1, \dots, X_n] \to E$ is also generated by $n$ elements. Since $O_L$ is finite over $O_K$, condition (3) in Proposition 4.1.11 is satisfied.

**Exercise 4.2** The commutative diagram

$$\begin{array}{ccc} \widetilde{A}_0 = A[S] & \longrightarrow & \widetilde{B}_0 = \widetilde{A}_0 \otimes_A B \\ \downarrow & & \downarrow \\ \widetilde{A} & \longrightarrow & \widetilde{B} \end{array}$$

defines a commutative diagram

$$\begin{array}{ccc} \mathrm{Hom}_A(B, A) \otimes_B \mathrm{Hom}_B(N_{B/\widetilde{A}_0}, B) & \longrightarrow & \mathrm{Hom}_B(\Omega^1_{\widetilde{B}_0/B} \otimes_{\widetilde{B}_0} B, B) \\ \uparrow & & \uparrow \\ \mathrm{Hom}_A(J, I) \otimes_B \mathrm{Hom}_B(N_{B/\widetilde{A}}, B) & \longrightarrow & \mathrm{Hom}_B(\Omega^1_{\widetilde{B}/B} \otimes_{\widetilde{B}} B, B) \end{array}$$

of $B$-modules. By Lemma 4.2.6, the upper line maps $f_e \otimes dS^\vee$ to $P^\vee$. Since $P = t(U + Q(S)) \bmod S^e - Ut$ and $dS = s dX$ in $\Omega^1_{\widetilde{B}/B} \otimes_{\widetilde{B}} B$, the lower line maps $t/s \cdot f_e \otimes dX^\vee$ to $(U + Q(S))^\vee$.

**Exercise 4.3** The morphism (4.3.14) sends $d \log s \otimes (U + Q(S))$ to the projection to $\Omega^1_A(\log) \otimes_A (\Omega^1_{\widetilde{A}/A}(\log/\log) \otimes_{\widetilde{A}} B)$ of $-d(U + Q(S)) \wedge d \log S \in \Omega^2_{\widetilde{A}}(\log) \otimes_{\widetilde{A}} B$. Since $d \log t + d \log U = e \cdot d \log S$ in $\Omega^1_{\widetilde{A}}(\log)$, we have $-dU \wedge d \log S = d \log t \wedge U \cdot d \log S$ and $d(Q(S)) \wedge d \log S = \sum_i dq_i \wedge S^{e-i} d \log S$. Hence, the projection is $d \log t \otimes Q(s) \cdot d \log S - \sum_i dq_i \otimes s^{e-i} \cdot d \log S$.

Let $f_1, \dots, f_e$ be the dual basis of $\mathrm{Hom}_A(B, A)$ as in Lemma 1.2.1. Since the basis $d \log S$ of $\Omega^1_{\widetilde{A}/A}(\log/\log)$ is sent to $-dX$ in $\Omega^1_{\widetilde{B}/B} \otimes_{\widetilde{B}} B$, the image of $d \log s$ by the morphism (4.3.15) is $t/s \cdot (-Q(s) \cdot f_e \otimes d \log t + \sum_i s^{e-i} \cdot f_e \otimes dq_i)$ by Exercise 4.2. Hence, $\mathrm{Tr}_{B/A}(cs \otimes d \log s) = t \otimes \sum_i f_i(c) \cdot (-q_i d \log t + dq_i)$ and

$$\mathrm{Tr}_{B/A}(c \cdot d \log s) = \sum_i f_i(\frac{t}{s} \cdot c) \cdot (-q_i d \log t + dq_i).$$

**Exercise 4.4** It suffices to show that the $F$-vector space $\Omega^1_{E/F}$ of dimension $p$ is generated by the image of $d \log \colon E^\times \to \Omega^1_E$. It suffices to show that for distinct elements $a_1, \dots, a_p \in F^\times$, $1/(u - a_1), \dots, 1/(u - a_p) \in E$ are linearly independent over $F$. After multiplying by $u^p - a_1^p, \dots, u^p - a_p^p \in F$, it further suffices to show that $u^{p-1} + a_1 u^{p-2} + \cdots + a_1^{p-1}, \dots, u^{p-1} + a_p u^{p-2} + \cdots + a_p^{p-1}$ are linearly independent over $F$. Since $1, u, \dots, u^{p-1}$ is a basis of $E$ over $F$, this follows from $\det(a_i^j) = \prod_{1 \leqslant i < j \leqslant p}(a_j - a_i) \neq 0$.

## Chapter 5

**Exercise 5.1** Let $t = a_n$ be the constant term. Since $t/s \equiv -s^{n-1} \bmod \mathfrak{m}_K O_L$, by the solution of Exercise 4.3 we have $\mathrm{Tr}_{L/K} d \log s \equiv f_n(s^{n-1}) \cdot d \log t = d \log t$.

**Exercise 5.2** Let $\chi' \colon H = \mathrm{Gal}(\mathbf{Q}_p(\zeta_{p^2})/\mathbf{Q}_p(\zeta_p)) \to \mathbf{Q}/\mathbf{Z}$ be the restriction of $\chi$. Then, by Exercise 3.2, we have $\mathrm{Sw}\,\chi' = p - 1 = \mathrm{ord}_{\mathbf{Q}_p(\zeta_p)} p$ and $\mathrm{rsw}\,\chi'$ maps $p$ to $1 \in \Omega^1_{\mathbf{F}_p}(\log) = \mathbf{F}_p$. Hence, by Proposition 5.2.2, we have $\mathrm{Sw}\,\chi = 1$ and $\mathrm{rsw}\,\chi$ maps $p$ to $1/(p-1) = -1 \in \Omega^1_{\mathbf{F}_p}(\log) = \mathbf{F}_p$.

**Exercise 5.3** By Exercise 1.9, the valuation of the logarithmic different $D^{\log}_{\mathbf{Q}_p(\zeta_{p^n})/\mathbf{Q}_p}$ is $n \cdot (p-1)p^{n-1} - p^{n-1} - ((p-1)p^{n-1} - 1) = (n-1) \cdot (p-1)p^{n-1} - (p^{n-1} - 1)$. By Exercise 1.6, the largest integer $i$ such that $G_i \neq 0$ is $i = p^{n-1} - 1$. Hence, by Theorem 5.3.4, we have $\mathrm{Sw}\,\chi = n - 1$.

The morphism $d_\sigma \colon \mathbf{F}_p \to \mathfrak{m}^i_{\mathbf{Z}_p[\zeta_{p^n}]}/\mathfrak{m}^{i+1}_{\mathbf{Z}_p[\zeta_{p^n}]}$ maps 1 to $\sigma(\zeta_{p^n} - 1)/(\zeta_{p^n} - 1) - 1 = (\zeta_p - 1)\zeta_{p^n}/(\zeta_{p^n} - 1) = \sum_{i=1}^{p^{n-1}} \zeta^i_{p^n}$. The right vertical arrow $\mathrm{Tr}_{L/K} \colon \mathbf{F}_p \to \mathbf{F}_p$ in (5.3.13) is the identity by Exercise 5.1. Since

$$\mathrm{Tr}_{\mathbf{Q}_p(\zeta_{p^n})/\mathbf{Q}_p} \zeta^i_{p^n} = \mathrm{Tr}_{\mathbf{Q}_p(\zeta)/\mathbf{Q}_p} \mathrm{Tr}_{\mathbf{Q}_p(\zeta_{p^n})/\mathbf{Q}_p(\zeta_p)} \zeta^i_{p^n} = 0$$

for $i = 1, \dots, p^{n-1} - 1$ and $\mathrm{Tr}_{\mathbf{Q}_p(\zeta_{p^n})/\mathbf{Q}_p} \zeta^{p^{n-1}}_{p^n} = p^{n-1} \cdot \mathrm{Tr}_{\mathbf{Q}_p(\zeta_p)/\mathbf{Q}_p} \zeta_p = -p^{n-1}$, the left vertical arrow $\mathrm{Tr}_{L/K} \colon \mathfrak{m}^i_{\mathbf{Z}_p[\zeta_{p^n}]}/\mathfrak{m}^{i+1}_{\mathbf{Z}_p[\zeta_{p^n}]} \to \mathfrak{m}^{n-1}_{\mathbf{Z}_p}/\mathfrak{m}^n_{\mathbf{Z}_p}$ in (5.3.13) sends $\sigma(\zeta_{p^n} - 1)/(\zeta_{p^n} - 1) - 1 = (\zeta_p - 1) \cdot \zeta_{p^n}/(\zeta_{p^n} - 1)$ to $-p^{n-1}$. Hence, $\mathrm{rsw}\,\chi \colon \mathfrak{m}^{n-1}_{\mathbf{Z}_p}/\mathfrak{m}^n_{\mathbf{Z}_p} \to \mathbf{F}_p$ maps $p^{n-1}$ to $-1$ by Theorem 5.3.4(2).

(The minus sign is consistent with the inverse in the description of the reciprocity mapping at the end of Exercise 2.3.)

## Chapter 6

**Exercise 6.1** (1) We identify $\mathrm{Gal}(\mathbf{Q}_p(\zeta_{p^e})/\mathbf{Q}_p) = (\mathbf{Z}/p^e\mathbf{Z})^\times$. For $i \in (\mathbf{Z}/p^e\mathbf{Z})^\times = \mathbf{Z}/p^e\mathbf{Z} - p\mathbf{Z}/p^e\mathbf{Z}$, let $\tau_i \in \mathrm{Gal}(\mathbf{Q}_p(\zeta_{p^e})/\mathbf{Q}_p)$ denote the corresponding element. Then $\mathrm{Tr}^{\mathrm{Br}}(\sigma^i : M) = \tau_i(\mathrm{Tr}^{\mathrm{Br}}(\sigma : M))$ and the right-hand side of (6.1.5) equals the middle term of (6.1.6). Since $\sum_{i\in\mathbf{Z}/p^e\mathbf{Z}}\mathrm{Tr}^{\mathrm{Br}}(\sigma^i : M) = p^e\cdot\dim M^{\sigma}$ and $\sum_{i\in p\mathbf{Z}/p^e\mathbf{Z}}\mathrm{Tr}^{\mathrm{Br}}(\sigma^i : M) = p^{e-1}\cdot\dim M^{\sigma^p}$, the second equality follows.

(2) Since the left-hand side $\dim_\Lambda(P\otimes_{O_E} M)^G$ is a rational integer, the operator $(1/([\mathbf{Q}(\zeta_n):\mathbf{Q}]))\mathrm{Tr}_{\mathbf{Q}(\zeta_n)/\mathbf{Q}}$ acts as the identity.

**Exercise 6.2** Let $N = (p-1)p^{n-1}$ be the degree $[\mathbf{Q}_p(\zeta_{p^n}):\mathbf{Q}_p]$ and $\Phi_{p^n}$ be the cyclotomic polynomial. By Exercise 1.4, $t = \zeta_{p^n} - 1$ is a uniformizer of $\mathbf{Q}_p(\zeta_{p^n})$ and its minimal polynomial $P = \Phi_{p^n}(X+1) = \Phi_p((X+1)^{p^{n-1}})$ over $\mathbf{Q}_p$ is an Eisenstein polynomial. By Lemma 6.2.5(3), we have $\Delta_{\mathbf{Q}_p(\zeta_{p^n})/\mathbf{Q}_p} = P'(t)/t^{N-1}$. As in the solution of Exercise 1.9, we have $P'(t) = p^n/((\zeta_p - 1)\cdot\zeta_{p^n})$.

Let $\sigma\in G$ be the image of $1+p^{n-1}$ by the isomorphism $(\mathbf{Z}/p^n\mathbf{Z})^\times\to G$. Then we have $\delta(\sigma) = \sigma(\zeta_{p^n}-1)/(\zeta_{p^n}-1)-1 = (\zeta_p-1)\zeta_{p^n}/t$ and $i = \mathrm{ord}\,(\zeta_p-1)\zeta_{p^n}/t = p^{n-1}-1$. Since $\Phi_{p^n}(1) = \Phi_p(1) = p$, we have $t^N \equiv -p \bmod pt$. Since $\Delta_{\mathbf{Q}_p(\zeta_{p^n})/\mathbf{Q}_p}\cdot\delta(\sigma) = p^n/((\zeta_p-1)\cdot\zeta_{p^n}t^{N-1})\cdot(\zeta_p-1)\zeta_{p^n}/t = p^n/t^N \equiv -p^{n-1}$, we have $\varphi(i) = \mathrm{ord}_{\mathbf{Q}_p}p^{n-1} = n-1$ and $\varphi(i,\chi) = -p^{n-1}$.

By Exercise 5.3, we have $\mathrm{Sw}\,\chi = n-1$ and $\mathrm{rsw}\,\chi = -1/p^{n-1}\in R_{\mathbf{Q}_p}$.

**Exercise 6.3** By Exercise 6.2, we have $\mathrm{Sw}\,\rho = n-1$ and $\mathrm{rsw}_\zeta\,\rho = -p^{n-1}\in R_{\mathbf{Q}_p}$. On the other hand, for $\sigma_i\in P$ corresponding to $i\in(1+p\mathbf{Z}/1+p^n\mathbf{Z})-\{1\}$, we have $\delta(\sigma_i) = \sigma_i(\zeta_{p^n}-1)/(\zeta_{p^n}-1)-1 = (\sigma_i(\zeta_{p^n})-\zeta_{p^n})/(\zeta_{p^n}-1) = (\zeta_{p^n}^{i-1}-1)\zeta_{p^n}/(\zeta_{p^n}-1)\equiv(\zeta_{p^n}^{i-1}-1)/(\zeta_{p^n}-1)$. We also have $\widetilde{s}(1) = \Delta_{\mathbf{Q}_p(\zeta_{p^n})/\mathbf{Q}_p} = (p-1)p^{n-1}\cdot\zeta_p/(\zeta_{p^n}(\zeta_{p^n}-1)^{p^{n-1}-1})\equiv -p^{n-1}(\zeta_{p^n}-1)/(\zeta_p-1)$ by Exercise 6.2. For $i\not\equiv 1 \bmod p^{n-1}$, since $\widetilde{s}(\sigma_i)$ depends only on $i \bmod p^{n-1}$ and since $\chi$ is primitive, we have $\sum_{j=1}^{p}\widetilde{s}(\sigma_{i+jp^{n-1}})\otimes\widetilde{\chi}(\sigma_{i+jp^{n-1}}) = \widetilde{s}(\sigma_i)\otimes\sum_{j=1}^{p}\zeta^j = 0$. Thus, the right-hand side equals $\sum_{j=0}^{p-1}\widetilde{s}(\sigma_{1+jp^{n-1}})\otimes\widetilde{\chi}(\sigma_{1+jp^{n-1}}) + \varepsilon(\zeta) = (-p^{n-1})(\zeta_{p^n}-1)/(\zeta_p-1)\otimes 1 - \sum_{j=1}^{p-1}((\zeta_p^j-1)/j(\zeta_{p^n}-1)j)\otimes\zeta^j = (-p^{n-1})\otimes 1 - \sum_{j=0}^{p-1}(\zeta_p-1)/(\zeta_{p^n}-1)\otimes\zeta^j = (-p^{n-1})\otimes 1$.

**Exercise 6.4** By Exercise 1.9, we have $d_{\mathbf{Q}_p(\zeta_{p^n})/\mathbf{Q}_p} = (n-1)(p-1)p^{n-1} - p^{n-1}+1$. Similarly to the solution of Exercise 6.2, we have $\Delta_{\mathbf{Q}_p(\zeta_{p^n})/\mathbf{Q}_p} = -p^{n-1}(\zeta_{p^n}-1)/(\zeta_p-1)$. By (6.4.3) and (6.4.6), we have $\mathrm{Sw}\,\rho = d_{\mathbf{Q}_p(\zeta_{p^n})/\mathbf{Q}_p} = (n-1)(p-1)p^{n-1}-p^{n-1}+1$. Since $\mathrm{N}_{\mathbf{Q}_p(\zeta_{p^n})/\mathbf{Q}_p(\zeta_{p^{n-1}})}(1-\zeta_{p^n}) = 1-\zeta_{p^{n-1}}$ for $n\geqslant 2$, and $\mathrm{N}_{\mathbf{Q}_p(\zeta_p)/\mathbf{Q}_p}(1-\zeta_p) = p$ and $\Delta_{\mathbf{Q}_p(\zeta_{p^n})/\mathbf{Q}_p} - [m_{\mathbf{Q}_p(\zeta_{p^n})/\mathbf{Q}_p}] = (-p^{n-1})(\zeta_{p^n}-1)/(\zeta_p-1) - [p-1] = p^{n-1}(\zeta_{p^n}-1)/(\zeta_p-1)$, we have $\mathrm{rsw}_G\,\rho = \mathrm{N}_{\mathbf{Q}_p(\zeta_{p^n})/\mathbf{Q}_p}(\Delta_{\mathbf{Q}_p(\zeta_{p^n})/\mathbf{Q}_p} - [m_{\mathbf{Q}_p(\zeta_{p^n})/\mathbf{Q}_p}]) = p^N$ for $N = \mathrm{Sw}\,\rho$. We also have $\mathrm{Sw}\,\rho = \sum_{\chi:G\to\Lambda^\times}\mathrm{Sw}\,\chi = \sum_{i=2}^{n}((p-1)p^{i-1}-(p-1)p^{i-2})(i-1) = (p-1)(p^{n-1}(n-1)-\sum_{i=0}^{n-2}p^i) = (p-1)p^{n-1}(n-1)-(p^{n-1}-1)$ and $\mathrm{rsw}\,\rho = \sum_{\chi:G\to\Lambda^\times}\mathrm{rsw}\,\chi = \sum_{i=2}^{n}(p^{i-1}-p^{i-2})\sum_{a\in\mathbf{F}_p^\times}([a]+(-p^{i-1})) = p^N$.

## Chapter 7

**Exercise 7.1** Since $v = yt$ satisfies $v^p - y^{p-1}v = u$, the residue field $E = F(v)$ of $L$ is the purely inseparable extension of the residue field $F$ of $K$ generated by a $p$th root $v$ of $u$. Since the minimal polynomial of $v$ is $P = X^p - y^{p-1}X - u$, the different $\Delta_{L/K} \in T_{L/K}$ is the isomorphism $\Omega^{1\otimes(p-1)}_{E/F} \to (\mathfrak{m}_L/\mathfrak{m}_L^2)^{\otimes(p-1)}$ sending $(dv)^{\otimes p-1}$ to $P'(u) = -y^{p-1}$.

For a non-trivial element $\sigma$ of $G = \mathrm{Gal}(L/K)$ such that $\sigma(t) = t + a$, $a \in \mathbf{F}_p$, we have $\sigma(v) = \sigma(yt) = v + ay$ and $d(\sigma)\colon \Omega^1_{E/F} \to \mathfrak{m}_L/\mathfrak{m}_L^2$ is the morphism sending $dv$ to $ay$. Hence, the right-hand side of (7.1.13) is also the morphism sending $dv^{\otimes(p-1)}$ to $(p-1)! \cdot y^{p-1} = -y^{p-1}$, and the equality (7.1.13) holds.

**Exercise 7.2** For the generator $v = yt$ of $O_L$, we have $\sigma(v) = v + y$. Hence, we have $G_1 = G$, $G_2 = G_{1,\chi} = 0$, and $M_{1,\chi} = L$. Thus, $\varphi(1,\chi) = \Delta_{L/K} + d(\sigma) = [-y^{p-1}] + [y] - p[dv] = [-y^p] - p[dv]$ in $T_{L/K}$. This is the image of $[-y^p] - [du]$ in $S_{L/K}$.

By Proposition 3.1.5(1), $\mathrm{rsw}\,\chi\colon \mathfrak{m}_K^p/\mathfrak{m}_K^{p+1} \to \Omega^1_F(\log)$ is the morphism sending $y^p$ to $-du$. Hence, $\varphi(1,\chi) \in T_{L/K}$ is the image of its inverse by $S_{L/K} \to T_{L/K}$.

**Exercise 7.3** Since $\mathrm{rsw}_\zeta\chi\colon \mathfrak{m}_K^p/\mathfrak{m}_K^{p+1} \to \Omega^1_{F/k}$ is the isomorphism sending $y^p$ to $-du$ by Exercise 7.2, we have $\mathrm{ord}_x \mathrm{rsw}\, V = \mathrm{ord}_x du$. Since $V^P = 0$, we have $\dim V/V^P = 1$.

Let $C'$ be the smooth curve with function field $E$ and let $w \in C'$ be the unique point above $x \in C$. Since $\widetilde{s}(\sigma) \in T_{L/K}$ is the inverse of the morphism $d(\sigma)\colon \Omega^1_{E/F} \to \mathfrak{m}_L/\mathfrak{m}_L^2$ sending $dv$ to $y$, we have $\mathrm{ord}'_w \widetilde{s}(\sigma) = \mathrm{ord}'_w dv$. Hence, the right-hand side of (7.3.9) is $-(1/p)\sum_{a\in\mathbf{F}_p^\times} \mathrm{ord}'_w dv \cdot (\zeta_p^a - 1) = \mathrm{ord}'_w dv$. Since $\mathrm{ord}_x du + 1 = \mathrm{ord}'_x du = \mathrm{ord}'_w dv$, the equality (7.3.9) holds.

## Chapter 8

**Exercise 8.1** The finite morphism $C^\circ \to \mathbf{A}^1_k = \operatorname{Spec} k[t]$ is a cyclic étale covering of degree $p$ totally ramified at $\infty$ and the Galois group is generated by $\sigma$. Hence, the fixed point of $\sigma$ on $C$ is reduced to the unique point $\infty$ of the complement $C - C^\circ$, and the intersection multiplicity of $\Gamma_\sigma \cap \Delta_C$ at $\infty$ is $-\mathrm{ord}_\infty t^n + 1 = n + 1$ by Proposition 3.1.2(1). Since $\sigma^*$ acts trivially on $\mathrm{H}^0$ and $\mathrm{H}^2$, we have $\mathrm{Tr}(\sigma : \mathrm{H}^1(C_{\bar{k}}, \Lambda)) = 1 + 1 - (n+1) = 1 - n$ by the Lefschetz trace formula.

**Exercise 8.2** Let $\mathcal{F}$ be the locally constant sheaf of $\Lambda$-modules of rank 1 on $\mathbf{A}^1_k = \operatorname{Spec} k[t]$ defined by the Artin–Schreier equation $u^p - u = t^n$. Then the

Swan conductor of $\mathcal{F}$ at $\infty$ is $n$, and hence we have $\chi_c(\mathbf{A}^1, \mathcal{F}) = \chi_c(\mathbf{A}^1) - \mathrm{Sw}_\infty \mathcal{F} = 1 - n$. Since $2 - 2g = \chi(X) = \chi(\mathbf{P}^1_k) + (p-1)\chi_c(\mathbf{A}^1, \mathcal{F}) = 2 + (p-1)(1-n)$, we obtain $g = (p-1)(n-1)/2$.

Alternatively, let $\mathcal{G}$ be the locally constant sheaf of $\Lambda$-modules of rank 1 on $U = \operatorname{Spec} k[u, (u^p - u)^{-1}] \subset \mathbf{A}^1_k$ defined by the Kummer equation $t^n = u^p - u$. Then the Swan conductor of $\mathcal{G}$ at $D = \mathbf{P}^1_{\mathbf{F}_p} \subset \mathbf{P}^1_k$ is 0 and hence we have $\chi_c(U, \mathcal{G}) = \chi_c(U) - \sum_{x \in D} \mathrm{Sw}_x \mathcal{G} = 1 - p$. Since $2 - 2g = \chi(X) = \chi(\mathbf{P}^1_k) + (n-1)\chi_c(U, \mathcal{G}) = 2 + (n-1)(1-p)$, we also obtain $g = (p-1)(n-1)/2$.

# Chapter 9

**Exercise 9.1** Let $K$ be the fraction field of the localization $B_{\mathfrak{q}}$ of $B = O_{K_0}[u]$ at the prime ideal $\mathfrak{q} = tB$, and define a morphism $B = O_{K_0}[u] \to A$ over $O_{K_0}$ by sending $u$ to $y^n$. Then $K$ is formally smooth over $K_0$ and $O_L = O_K[x]/(x^m - t/u)$ is finite over $O_K$.

**Exercise 9.2** (1) The composite extension $L'$ is a cyclic extension of $K' = K'_0(x)$ defined by the Artin–Schreier equation $t^p - t = x/y'^p$. By Proposition 3.1.2(2), the ramification index $e_{L'/K'}$ is 1 and the residue field is the purely inseparable extension $k(x')$ of $k(x)$ of degree $p$ generated by a $p$th root $x' \equiv y't$ of $x$. Hence, $L'$ is formally smooth over $K'_0$.

(2) Let $K'_1$ be the cyclic extension of $K_0$ defined by the Artin–Schreier equation $s^p - s = 1/y$. Then $K'_1$ is a totally ramified extension of $K_0$ and $y' = 1/s$ is a uniformizer and is a root of the Eisenstein polynomial $Y^p + yY^{p-1} - y$. The composite extension $L'_1$ is a cyclic extension of $K'_1(x)$ defined by the Artin–Schreier equation $u^p - u = x(1 - y'^{p-1})/y'^p$. By Proposition 3.1.2(2), the ramification index $e_{L'_1/K'_1(x)}$ is 1 and the residue field is the purely inseparable extension $k(x')$ of $k(x)$ of degree $p$ generated by a $p$th root $x' \equiv y'u$ of $x$. Hence, $L'_1$ is also formally smooth over $K'_1$.

**Exercise 9.3** (1) The scheme $X'$ is the normalization of $\mathbf{A}^1_{S'} = \operatorname{Spec} O_{K'}[x]$ in the Artin–Schreier covering $X_{K'}$ of $\mathbf{A}^1_{K'}$ defined by $t^p - t = x/y'^p$. Setting $u = y't$, we get the equation $u^p - y'^{p-1}u = x$. Hence, we obtain $X' = \operatorname{Spec} O_{K'}[u]$ and $X'$ is smooth over $S'$.

(2) Let $K'_1$ be as in Exercise 9.2. Then the scheme $X'_1$ is the normalization of $\mathbf{A}^1_{S'_1} = \operatorname{Spec} O_{K'_1}[x]$ in the Artin–Schreier covering $X_{K'_1}$ of $\mathbf{A}^1_{K'_1}$ defined by $u^p - u = x(1 - y'^{p-1})/y'^p$. Setting $w = y'u$, we get the equation $w^p - y'^{p-1}w = x(1 - y'^{p-1})$. Since $1 - y'^{p-1}$ is invertible, we obtain $X'_1 = \operatorname{Spec} O_{K'_1}[w]$ and $X'_1$ is smooth over $S'_1$.

## Chapter 10

**Exercise 10.1** (1) The product $v = uy$ is a solution of $v^p - y^{p-1}v = t^n$. Since $A = O_K[v,t]/(v^p - y^{p-1}v - t^n)$ satisfies the conditions $(R_1)$ and $(S_2)$, we have $X = \operatorname{Spec} A$. Since the closed fiber $X_k = \operatorname{Spec} k[v,t]/(v^p - t^n)$ has a unique singular point at $(v,t)$, $X$ has a unique isolated singular point.

(2) Consider the finite flat morphism $X \to Y = \mathbf{A}^1_S = \operatorname{Spec} O_K[v]$. The blow-up $Y' \to Y$ at the closed point $(v, y)$ is the union of $V = \operatorname{Spec} O_K[u]$ such that $v = uy$ and $W_1 = \operatorname{Spec} O_K[v,w]/(vw - y)$. The union of $V$ and $W = \operatorname{Spec} C \subset W_1$ for $C = O_K[v, w, (1 - w^{p-1})^{-1}, (1 - 2w^{p-1})^{-1}]/(vw - y)$ is an open neighborhood of the exceptional divisor. We show that the normalizations $X'_V$ and $X'_W$ of $X \times_Y V \times_S S'$ and $X \times_Y W \times_S S'$ are nodal curves over $S'$.

Substituting $v = uy$, $y = y'^n$ into $v^p - y^{p-1}v - t^n$, dividing the result by $y'^{pn}$, and setting $t' = t/y'^p$, we obtain $u^p - u - t'^n$. Hence, $X'_V$ is $\operatorname{Spec} O_{K'}[u,x]/(u^p - u - t'^n)$ and is a smooth curve over $S'$.

Substituting $y = vw$ into $v^p - y^{p-1}v - t^n$, dividing the result by $v^{p-1}$, and setting $t'' = t/v^d$ for $p - 1 = nd$, we obtain $v - t''^n/(1 - w^{p-1})$. Substituting this and $y = y'^n$ into $y = vw$, dividing the result by $t''^n$ and setting $z = y'/t''$, we obtain $z^n - w/(1-w^{p-1})$. Since $d(w - (1-w^{p-1})z^n) = (1 - z^n w^{p-2})dw = (1 - 2w^{p-1})/(1 - w^{p-1}) \cdot dw$, the scheme $C' = O_{K'}[w, (1 - w^{p-1})^{-1}, (1 - 2w^{p-1})^{-1}, x'', z]/(x''z - y', z^n - w/(1-w^{p-1}))$ is étale over $O_{K'}[x'', z]/(x''z - y')$. Hence, $X'_W = \operatorname{Spec} C'$ is a nodal curve over $S'$.

**Exercise 10.2** By Exercise 10.1, $X'$ is semi-stable over $S'$ with a unique singular point $x'$ of $X'$ over $S'$. Hence, by Theorem 10.1.8, we have $R^0\psi'\Lambda = \Lambda$ and $R^1\psi'\Lambda = \Lambda(-1)_{x'}$ for the sheaves of nearby cycles on $X'$. By the spectral sequence $E_2^{i,j} = H^i(C_{\bar{k}}, R\psi'\Lambda|_C) \Rightarrow R^{i+j}\psi\Lambda_x$ and by the vanishing $R^2\psi\Lambda_x = 0$, we obtain isomorphisms $H^0(C_{\bar{k}}, \Lambda) \to R^0\psi\Lambda_x$, $R^1\psi'\Lambda_{x'} \to H^2(C_{\bar{k}}, \Lambda)$ and $H^1(C_{\bar{k}}, \Lambda) \to R^1\psi\Lambda_x$.

**Exercise 10.3** Let $S' = \operatorname{Spec} k((y'))$ for $y'^n = y$ and $f\colon X' \to X_{S'}$ be the nodal resolution of singularity as in the solution of Exercise 10.1. By Exercise 10.2, for the inverse image $C = f^{-1}(x)$, we have a canonical isomorphism $H^1(C_{\bar{k}}, \Lambda) \to R^1\psi\Lambda_x$. Since the projective smooth curve $C$ defined by $t^n = v^p - v$ is of genus $(p-1)(n-1)/2$ by Exercise 8.2, we have $\psi_x = \dim R^1\psi\Lambda_x = (p-1)(n-1)$.

The closed fiber $X_s$ is $\operatorname{Spec} A$ for $A = k[v,t]/(v^p - t^n)$, and the normalization $A^\natural = k[z]$ of $A$ is generated by $z = t/v^d$ for $p - 1 = nd$. Hence, we have $\#P_x = 1$ and $\gamma_x = 0$. Since $\delta_x = \dim A^\natural/A = \sum_{i=1}^{n-1} \dim k[v]z^i/k[v]t^i = \sum_{i=1}^{n-1} di = dn(n-1)/2 = (p-1)(n-1)/2$, the equality $\psi_x + \gamma_x = 2\delta_x$ is satisfied.

**Exercise 10.4** As in the solution of Exercise 10.3, we identify $R^1\psi\Lambda$ with the cohomology $\mathrm{H}^1(C_{\bar{k}}, \Lambda)$ of the projective smooth curve defined by $v^p - v = t^n$. Hence, by Exercise 8.1 we have $\psi_x(\sigma) = \mathrm{Tr}(\sigma : \mathrm{H}^1(C_{\bar{k}}, \Lambda)) = 1 - n$.

Since $P_x$ consists of a unique point $x^\natural$ of the normalization $X_s^\natural$ by the solution of Exercise 10.3, we have $\gamma_x(\sigma) = 0$.

Since $X = \operatorname{Spec} A$ for $A = O_K[v,t]/(v^p - y^{p-1}v - t^n)$ and $\sigma(v) = \sigma(uy) = v + y$, the $\sigma$-fixed part $X^\sigma$ is defined by the ideal $(y)$. Let $\xi$ be the generic point of the closed fiber $X_s$, and $k(\xi) = k(v,t) = k(t)[v]/(v^p - t^n)$ be the residue field. Then length $O_{X^\sigma,\xi} = 1$ and the morphism $d_\sigma \colon \Omega^1_F \to \mathfrak{m}_K/\mathfrak{m}_K^2 \otimes_k F$ maps the basis $dv$ to $y$. Since the valuation of $v \in k(\xi)$ at the closed point $x^\natural$ is $n$, we have $\operatorname{ord}_{x^\natural} dv = n - 1$. Hence, $\alpha_x(\sigma) = 1 - n$. Since $\sigma$ has no fixed point on the generic fiber $X_\eta$, we have $\beta_x(\sigma) = 0$. Thus, $\psi_x(\sigma) + \gamma_x(\sigma) = (1-n) + 0$ equals $\alpha_x(\sigma) + \beta_x(\sigma) = (1-n) + 0$ and the equality (10.4.9) holds.

**Exercise 10.5** The pull-back $\mathcal{L}'$ is defined by the Artin–Schreier covering $t^p - t = x/y'^p$. Let $p_1', p_2' \colon \mathbf{A}^2_k = \operatorname{Spec} k[x, y'] \to \mathbf{A}^1_k$ be the projections. Then the refined Swan conductor $\operatorname{rsw}(p_1'^* j_! \mathcal{F} \otimes j_{2!}' \mathcal{L}')$ equals $\operatorname{rank} \mathcal{F} \cdot \operatorname{rsw}(j_{2!}'\mathcal{L}')$. Since $\operatorname{rsw}(j_{2!}'\mathcal{L}')$ is the isomorphism sending $y'^p$ to $dx$, its valuation is 0. Hence, by the Deligne–Kato formula we have

$$\dim R^1\psi'(p_1'^* j_! \mathcal{F} \otimes j_{2!}'\mathcal{L}')_{0,0} = \operatorname{rank} \mathcal{F} + \operatorname{Sw}_0 \mathcal{F}.$$

# Part V

**Exercise V.1** We may assume that the valuations $s_i = \operatorname{ord}_L(a_i - a_n)$ are increasing in $i = 1, \ldots, n-1$. Since $a_i$ are conjugates to each other, the sequence $s_1, \ldots, s_{n-1}$ and hence the function $\varphi$ are independent of the choice of $a_n$. Let $x \in D$ be an element different from $a_1, \ldots, a_n$ and set $s = \max(\operatorname{ord}_L(x - a_i) : i = 1, \ldots, n)$. We may assume that $s = \operatorname{ord}_L(x - a_n)$. Then $\operatorname{ord}_L(x - a_i) = s_i$ if $s_i < s$ and $\operatorname{ord}_L(x - a_i) = s$ otherwise. Hence, the valuation $\operatorname{ord}_K f(x)$ of $f(x) = \prod_{i=1}^n (x - a_i)$ equals $\varphi(s)$. Thus, for $r = \varphi(s)$ we have $f^{-1}(D^r) = \{x \in D \mid \max(\operatorname{ord}_L(x - a_i) : i = 1, \ldots, n) \geqslant s\}$ as required.

# Chapter 11

**Exercise 11.1** We have $P^{[n]} = \operatorname{Spec} A$ for $A = O_K[x, (x^p - ax)/t^n] = O_K[x,y]/(x^p - ax - t^n y)$. If $(p-1)m \leqslant \operatorname{ord}_K a$, setting $z = x/t^m$ and $b = a/t^{(p-1)m}$, we see that the normalization of $A$ is $O_K[z,y]/(z^p - bz - y) = O_K[z]$ and $P^{(n)} = \operatorname{Spec} O_K[z] = \mathbf{A}^1_S$ is smooth over $S$. The closed fiber $P^{(n)}_F = \mathbf{A}^1_F$ is connected.

Assume $\operatorname{ord}_K a = (p-1)l < (p-1)m$, and set $z = x/t^l$ and $u = a/t^{(p-1)l} \in O_K^\times$. Then $A^\natural = O_K[z,y]/(z^p - uz - t^{(p-1)(m-l)}y)$ is étale over $O_K[y]$ and is the normalization of $A$. Hence, $P^{(n)} = \operatorname{Spec} A^\natural$ is smooth over $S$ and the closed fiber $\operatorname{Spec} F[z,y]/(z^p - uz)$ is an affine line over a finite étale ring $F[z]/(z^p - uz) = F \times F[z]/(z^{p-1} - u)$ over $F$ and is not connected.

**Exercise 11.2** Since $y = x/t^l$ satisfies $y^p - y = 0$ in $O_K[x]/(x^p - ax) \otimes_{O_K} K = K[x]/(x^p - ax)$, and since $O_K[y]/(y^p - y)$ is finite étale over $O_K$, the ring $O_K[y]/(y^p - y)$ is the normalization of $O_K[x]/(x^p - ax)$. Further, by Hensel's lemma, $O_K[y]/(y^p - y)$ is isomorphic to the product ring $\prod_{i=0}^{p-1} O_K$. Hence, $T = \operatorname{Spec} O_K[x]/(x^p - ax)$ is the union of $p$ sections $S \to Q$. Thus, each irreducible component of $T_n$ meets $E_n$ at a single point and $\pi_0(T_n \cap V)$ consists of $p$ elements.

By the solution of Exercise 11.1, if $m < l$, $T_n \cap E_n$ and $\pi_0(E_n)$ consists of single elements. If $m = l$, $\pi_0(T_n \cap V) \to T_n \cap E_n$ is a bijection and $\pi_0(E_n)$ consists of a single element. If $m > l$, $\pi_0(T_n \cap V) \to T_n \cap E_n \to \pi_0(E_n)$ are bijections.

**Exercise 11.3** Let $f \in O_K[Y]$ be the minimal polynomial of $a$ and define the vertical arrow $Q \to P$ by sending $X$ to $f$. Then $O_K[Y] = O_K[X][Y]/(f - X)$ is finite flat over $O_K[X]$ and the diagram is cartesian.

**Exercise 11.4** Let $A = O_L[x, (x^2 + ax + b)/d] = O_L[x,y]/(x^2 + ax + b - dy)$. Then the dilatation $Q_T^{[r]}$ is $\operatorname{Spec} A$. Let $v \in O_L$ be the image of $x$. Then, since $b = -v(a+v)$ and $d = a^2 + 4v(a+v) = (a+2v)^2$, we have $A = O_L[x,y]/((x-v)(x+a+v) - (a+2v)^2 y)$. Hence, the normalization of $A$ is $O_L[y,x']/(x'^2 - x' - y) = O_L[x']$ for $x' = -(x-v)/(a+2v)$ and is smooth over $O_L$.

**Exercise 11.5** (1) For $i = 1, \dots, n$, set $z_i = \zeta_{p^n}^{p^{n-i}}$. Then

$$f = \prod_{i=2}^{p-1} (X^{p^{n-1}} - z_1^i) \cdot \frac{X^{p^{n-1}} - z_1}{X^{p^{n-2}} - z_2} \cdots \frac{X^{p^2} - z_{n-2}}{X^p - z_{n-1}} \cdot \frac{X^p - z_{n-1}}{X - z_n} \cdot (X - z_n).$$

The polynomial $((Y+a)^{p^{i+1}} - a^{p^{i+1}})/((Y+a)^{p^i} - a^{p^i})$ is congruent to $Y^{(p-1)p^i}$ modulo $p$ and its constant term equals $p^{i+1}a^{p^{i+1}-1}/p^i a^{p^i-1} = p \cdot a^{(p-1)p^i}$. Hence, the polynomial $g_i$ of $Y$ obtained by substituting $X = Y + z_n$ in $(X^{p^{n-i+1}} - z_{i-1})/(X^{p^{n-i}} - z_i)$ is congruent to $Y^{p^{n-i}(p-1)} + p z_i^{p-1}$ modulo $pY$. Hence, if $P_0 = (0,0)$ and $P_i = (u_i, v_i)$, $i = 1, \dots, n$ denote the vertices of the Newton polygon, we have $P_1 = ((p-2)p^{n-1}, (p-2)\operatorname{ord}(z_1 - 1)) = ((p-2)p^{n-1}, (p-2)p^{n-1})$ and $(u_i, v_i) - (u_{i-1}, v_{i-1}) = ((p-1)p^{n-i}, \operatorname{ord} p) = ((p-1)p^{n-i}, (p-1)p^{n-1})$ for $i = 2, \dots, n$. Thus, we obtain $P_i = ((p-1)p^{n-1} - p^{n-i}, i \cdot (p-1)p^{n-1} - p^{n-1})$ for $i = 1, \dots, n$.

The slope $s_1$ of $P_0P_1$ is $(\operatorname{ord}(z_1 - 1))/p^{n-1} = 1$, and $s_i$ of $P_{i-1}P_i$ for $i = 2, \dots, n$ are $(\operatorname{ord} p)/((p-1)p^{n-i}) = p^{i-1}$. We have

$$\varphi(s_i) = \varphi(p^{i-1}) = \frac{1}{e_{\mathbf{Q}_p(\zeta_{p^n})/\mathbf{Q}_p}}(i \cdot (p-1)p^{n-1} - p^{n-1} + p^{i-1} \cdot p^{n-i}) = i$$

for $i = 1, \dots, n$

(2) Define a polynomial $\Phi(X,Y) = (X^p - Y^p)/(X - Y)$. Then we have $f = \prod_{i=1}^n \Phi(X^{p^{i-1}}, u^{p^{i-1}}) \cdot (X - u)$ and $g = \prod_{i=1}^n \Phi((Y+u)^{p^{i-1}}, u^{p^{i-1}}) \cdot Y$. Since $\Phi(X,Y) \equiv (X-Y)^{p-1} \bmod p$ and $\Phi(X,X) = pX^{p-1}$, we have $\Phi((Y+u)^{p^{i-1}}, u^{p^{i-1}}) \equiv Y^{p^{i-1}(p-1)} + p \cdot u^{p^{i-1}(p-1)} \bmod pY$. Hence, if $P_0 = (0,0)$ and $P_i = (u_i, v_i), i = 1, \dots, n$ are the vertices of the Newton polygon, we have $(u_i, v_i) - (u_{i-1}, v_{i-1}) = ((p-1)p^{n-i}, \operatorname{ord} p) = ((p-1)p^{n-i}, (p-1)p^{n-1})$ for $i = 1, \dots, n$ and we obtain $P_i = (p^n - p^{n-i},\ i \cdot (p-1)p^{n-1})$ for $i = 1, \dots, n$.

The slopes $s_i$ of $P_{i-1}P_i$ for $i = 1, \dots, n$ are $(\operatorname{ord} p)/(p^{n-i}(p-1)) = p^{i-1}$. We have

$$\varphi(s_i) = \varphi(p^{i-1}) = \frac{1}{e_{\mathbf{Q}_p(\zeta_{p^n})/\mathbf{Q}_p}}(i \cdot (p-1)p^{n-1} + p^{i-1} \cdot p^{n-i}) = i + \frac{1}{(p-1)}$$

for $i = 1, \dots, n$.

# Chapter 12

**Exercise 12.1** (1) By Exercise 11.5, the rescaling function $\varphi(s)$ is the piecewise linear function satisfying $\varphi(0) = 0$ and $\varphi(s) = \varphi(p^{n-1}) + (s - p^{n-1})$ for $s \geqslant p^{n-1}$ such that the edges of the graph are determined by $\varphi(p^{i-1}) = i$ for $i = 1, \dots, n$. For $a \in (\mathbf{Z}/p^n\mathbf{Z})^\times$ and $i = \operatorname{ord}_p(a-1)$, we have $\operatorname{ord}_L(\zeta_{p^n}^a - \zeta_{p^n}) = p^i$. Hence, by Proposition 12.1.7, we have $F^r(L) = 1$ for $r \leqslant 1$, $F^r(L) = (\mathbf{Z}/p^i\mathbf{Z})^\times$ for $i = 1, \dots, n-1$, and $i+ \leqslant r \leqslant i+1$ and $F^r(L) = (\mathbf{Z}/p^n\mathbf{Z})^\times$ for $r \geqslant n+$.

(2) For $i = 0, 1, \dots, n$, regard $F_i = \{w \in L(\zeta_{p^n}) \mid w^{p^i} = T\}$ as a quotient set of $F(L)$ by the surjection sending $v \in F(L)$ to $v^{p^{n-i}}$. Then, by Exercise 11.5, we have $F^r(L) = F_0$ for $0+ \leqslant r \leqslant 1+1/(p-1)$, $F^r(L) = F_i$ for $i = 1, \dots, n-1$, and $(i+1/(p-1))+ \leqslant r \leqslant i+p/(p-1)$ and $F^r(L) = F_n$ for $r \geqslant (n+1/(p-1))+$.

**Exercise 12.2** If $r \leqslant 1$, then $L$ is an unramified extension of $K$ by Proposition 12.2.2(1) (b) $\Rightarrow$ (a). Assume $r \geqslant 1+$. Then, by Proposition 12.2.2(2) (a) $\Rightarrow$ (b), the ramification of $K'$ over $K$ is bounded by $r$. Hence, by Corollary 12.1.5(1), the ramification of $L'$ over $K$ is bounded by $r$.

**Exercise 12.3** (1) Set $G_k = \operatorname{Ker}(G \to \operatorname{Aut}(\mathbf{Z}_p[\zeta_{p^n}]/(\zeta_{p^n} - 1)^k))$. For $i = 0, \dots, n-1$, we have $\operatorname{ord}(\zeta_{p^{n-i}} - 1) = p^i$. Hence, for $i = 1, \dots, n-1$ and

$p^{i-1} < k \leqslant p^i$, we have $G_k = \mathrm{Gal}(\mathbf{Q}_p(\zeta_{p^n})/\mathbf{Q}_p(\zeta_{p^i}))$. Then $G_1 = G$ and $G_k = 1$ for $k > p^{n-1}$.

For $i = 1, \ldots, n-1$ and $i+ \leqslant r \leqslant i+1$, we have $G^r = \mathrm{Gal}(\mathbf{Q}_p(\zeta_{p^n})/\mathbf{Q}_p(\zeta_{p^i}))$ by Exercise 11.5 and Corollary 12.3.6. For $r \leqslant 1$ we have $G^r = G$, and for $r \geqslant n+$ we have $G^r = 1$.

We may also reach the same conclusion by Exercise 12.1.

(2) The image of $G^r$ by the projection $G = (\mathbf{Z}/p^n\mathbf{Z})^\times \ltimes \mathbf{Z}/p^n\mathbf{Z} \to (\mathbf{Z}/p^n\mathbf{Z})^\times$ is determined by Exercise 12.1(1), and the image of $G^r \cdot (\mathbf{Z}/p^n\mathbf{Z})^\times$ by the projection $G = (\mathbf{Z}/p^n\mathbf{Z})^\times \ltimes \mathbf{Z}/p^n\mathbf{Z} \to \mathbf{Z}/p^n\mathbf{Z}$ is determined by Exercise 12.1(2). A normal subgroup $N \lhd G$ such that the image in $(\mathbf{Z}/p^n\mathbf{Z})^\times$ equals the kernel $P_i = \mathrm{Ker}((\mathbf{Z}/p^n\mathbf{Z})^\times \to (\mathbf{Z}/p^i\mathbf{Z})^\times)$ of the canonical surjection is $P_i \ltimes (p^j\mathbf{Z}/p^n\mathbf{Z})$ for $j = 0, \ldots, i$. Hence, we have $G^r = G$ for $0+ \leqslant r \leqslant 1$, $G^r = P_i \ltimes (p^{i-1}\mathbf{Z}/p^n\mathbf{Z})$ for $i = 1, \ldots, n$ and $i+ \leqslant r \leqslant i + 1/(p-1)$, $G^r = P_i \ltimes (p^i\mathbf{Z}/p^n\mathbf{Z})$ for $i = 1, \ldots, n-1$ and $i + 1/(p-1)+ \leqslant r \leqslant i+1$, and $G^r = 1$ for $r \geqslant n + 1/(p-1)+$.

**Exercise 12.4** For a geometric point $s$ above the closed point of $S$, the mapping $X_s \to \pi_0(Y_s)$ is a bijection. For a geometric point $\eta$ above the generic point of $S$, the mapping $X_\eta \to \pi_0(Y_\eta)$ is the two-to-one mapping. Hence, $U_s$ is the diagonal of $X_s \times X_s$ and $U_\eta = X_\eta \times X_\eta$.

For a geometric point $s$ above the closed point of $S$, the mapping $X_s \to Z_s$ is the two-to-one mapping. For a geometric point $\eta$ above the generic point of $S$, the mapping $X_\eta \to Z_\eta$ is a bijection. Hence, $X_s \times_{Z_s} X_s = X_s \times X_s$ and $X_\eta \times_{Z_\eta} X_\eta$ is the diagonal of $X_\eta \times X_\eta$.

**Exercise 12.5** (1) The blow-up $C'' = C''_1 \cup C''_2$ is the union of $C''_1 = \mathrm{Spec}\, B_1$ for $B_1 = O_K[u'', v']/(u''v' - a)$ and $C'_2 = \mathrm{Spec}\, B_2$ for $B_2 = O_K[u', v'']/(u'v'' - b)$, where $u'' = a/v'$ and $v'' = v'/a$. Hence, $C''$ is a nodal curve.

We have $Q_C^{[R]} = \mathrm{Spec}\, A$, where

$$A = O_K[u, v, x]/(uv - c)\left[\frac{x^2 - ax}{v}\right] = O_K[u, v, x, y]/(uv - c, x^2 - ax - vy).$$

Hence, the base change $Q_{C''}^{[R]} = Q_{C''_1}^{[R]} \cup Q_{C''_2}^{[R]}$ is the union of $Q_{C''_i}^{[R]} = \mathrm{Spec}\, A_i$, where $A_1 = B_1[x, y]/(x^2 - u''v'x - v'^2y)$ and $A_2 = B_2[x, y]/(x^2 - ax - a^2v''^2y)$. The normalization $Q_{C''}^{(R)} = Q_{C''_1}^{(R)} \cup Q_{C''_2}^{(R)}$ is the union of $Q_{C''_i}^{(R)} = \mathrm{Spec}\, A'_i$, where $A'_1 = B_1[x_1, y]/(x_1^2 - u''x_1 - y) = B_1[x_1]$ and $A'_2 = B_2[x_2, y]/(x_2^2 - x_2 - v''^2y)$ with $x_1 = x/v'$ and $x_2 = x/a$, and hence is smooth over $C''$.

On $C''_1$, the morphism $T^\natural_{C''_1} = C''_1 \amalg C''_1 \to Q_{C''_1}^{(R)} = \mathrm{Spec}\, B_1[x_1]$ is the disjoint union of the sections defined by $x_1 = 0$ and $x_1 = a/v' = u''$. The image of the mapping $T_x^\natural \to Q^{(R)}_{C_1'',x}$ consists of two points or one point according to whether $u'' \neq 0$ or not. The geometric fiber $Q^{(R)}_{C_1'',x}$ is irreducible.

On $C_2''$, the morphism $T^\natural_{C_2''} = C_2'' \amalg C_2'' \to Q^{(R)}_{C_2''} = \operatorname{Spec} B_2[x_2, y]/(x_2^2 - x_2 - v''^2 y)$ is the disjoint union of the sections defined by $x_2 = 0$ and $x_2 = 1$. The image of the mapping $T_x^\natural \to Q^{(R)}_{C_2'',x}$ consists of two points. The geometric fiber $Q^{(R)}_{C_2'',x}$ has two connected components or is irreducible according to whether $v'' = 0$ or not.

(2) If $r' \leqslant \operatorname{ord}_K a$, then the lifting $S' \to C''$ is defined by the morphism $B_1 \to O_{K'}$ sending $v'$ to $w$. Hence, the base change $Q^{(R)}_{C''} \times_{C''} S' = \operatorname{Spec} A_1' \otimes_{B_1} O_{K'} = \operatorname{Spec} O_{K'}[x_1]$ is the normalization $Q^{(2r')}_{S'}$ of

$$Q^{[2r']}_{S'} = \operatorname{Spec} O_{K'}\left[x, \frac{x^2 - ax}{w^2}\right].$$

If $\operatorname{ord}_K a \leqslant r' \leqslant \operatorname{ord}_K c'$, then the lifting $S' \to C''$ is defined by the morphism $B_2 \to O_{K'}$ sending $v''$ to $w' = w/a$. Hence, the base change $Q^{(R)}_{C''} \times_{C''} S' = \operatorname{Spec} A_2' \otimes_{B_2} O_{K'} = \operatorname{Spec} O_{K'}[x_2, y]/(x_2^2 - x_2 - w'^2 y)$ is the normalization $Q^{(2r')}_{S'}$ of $Q^{[2r']}_{S'}$.

# Chapter 13

**Exercise 13.1** We may assume that $X = \operatorname{Spec} A$ is affine, and $C_1$ and $C_2$ are defined by $u, v \in A$. Let $t \in O_K$ be a uniformizer. Then $t = wu^m v^n$ for $w \in A^\times$. Since the question is étale local and $m$ is invertible, we may assume that $w = z^n$ for $z \in A^\times$. By replacing $u$ by $zu$, we may assume that $w = 1$. Then the morphism $X \to \operatorname{Spec} O_K[u, v]/(u^m v^n - t)$ is étale on a neighborhood of $x$. Since $\operatorname{Spec} O_K[u^{\pm 1}, v]/(u^m v^n - t)$ and $\operatorname{Spec} O_K[u, v^{\pm 1}]/(u^m v^n - t)$ are log smooth over $S$, the assertion follows.

**Exercise 13.2** Let $F$, $F'$, and $E$ be the residue fields of $K$, $K'$, and $L$. Since $E$ is a finitely generated separable extension of $F$, the subextension $F'$ is also a finitely generated separable extension of $F$. Suppose that the ramification index $e = e_{K'/K}$ is divisible by the residue characteristic $p$. Then $m = e_{L/K}$ is also divisible by $p$. Let $t$, $t'$, and $x$ be uniformizers of $K$, $K'$, and $L$, and set $u = t/t'^e$, $v = t/x^m$, and $w = t'/x^{m/e}$. Then $v = u \cdot w^e$. Since $p \mid e$, if $dv = w^e \cdot du \in \Omega^1_{E/F}$ is non-zero, then $du \in \Omega^1_{F'/F}$ is non-zero. Hence, $K'$ is log smooth over $K$ by Lemma 13.2.2.

**Exercise 13.3** Let $n = pm \geqslant p$ be an integer, and let $L$ be the composition of the Artin–Schreier extensions defined by $x^p - x = 1/t^{n-1}$ and $x^p - x = X/t^n$. Cyclic subextensions $M \subset L$ of degree $p$ are defined by $x^p - x = (at + bX)/t^n$, $(a, b) \in \mathbf{F}_p^2 - \{0\}$. For any $M$ and $H = \mathrm{Gal}(M/K)$, we have $H^n = H$ and $H^{n+} = 0$ by Corollary 12.3.6(2) and Corollary 12.3.7(2). Since the upper ramification groups are compatible with quotients, we have $G^n = G$ and $G^{n+} = 1$.

On the other hand, by Lemma 13.3.5(3) and (4), $H^{n-1}_{\log} = H$ and $H^{(n-1)+}_{\log} = 0$ if $a \neq 0$, and $H^n_{\log} = H$ and $H^{n+}_{\log} = 0$ if $a = 0$. Since the logarithmic upper ramification groups are also compatible with quotients, we have $G^{n-1}_{\log} = G$, $G^{(n-1)+}_{\log} = G^n_{\log} = \mathrm{Gal}(L/M)$ for $a = 0$, and $G^{n+}_{\log} = 0$.

**Exercise 13.4** By Corollary 3.1.7, we have $F_n X_K[p] = F_n^B X_K[p]$. By Proposition 13.4.3(2), we have $F_n^B X_K[p] = X_K[p] \cap F_n^B X_K$.

**Exercise 13.5** Since $y/x^n = (t + s^n w)/u^n s^n = t/u^n s^n + w/u^n$, we have $y/x^n - t/s^n = t/s^n \cdot (1/u^n - 1) + w/u^n = t/u^n \cdot (1 - (1 + s^n v)^n)/s^n + w/u^n \in B$. Since $u \equiv 1 \bmod s$, the right-hand side is congruent to $-nt \cdot v + w$.

**Exercise 13.6** By the solution of Exercise 13.5, $\chi_1 - \chi_2$ is the restriction of $\widetilde{\chi} \in \mathrm{H}^1(P_n, \mathbf{F}_p)$ defined by $t/u^n \cdot (1 - (1 + s^n v)^n)/s^n + w/u^n \in B = k[s, t, v, w, (1 + s^n v)^{-1}] = \Gamma(P_n, O)$. Further, its restriction on $D_n$ is defined by $-nt \cdot v + w \in B/(s) = k[t, v, w] = \Gamma(D_n, O)$.

# Chapter 14

**Exercise 14.1** (1) By (14.1.15), $w(0) = 0$ and $w(1) = w(1) + w(1) = 0$. By (14.1.14) and the congruence $a^p \equiv a \bmod p$, we have $w(a + 1) = w(a) - P(a, 1) \cdot w(p)$. Hence, the equality follows by induction.

Alternative solution: Since $a^p \equiv a \bmod p$, we have $w(a) = w(a - a^p) + w(a^p)$ by (14.1.14) and $w(a - a^p) = ((a - a^p)/p) \cdot w(p)$ by (14.1.15). Since $X^p - a^p \equiv (X - a)^p \bmod p$, we have $w(a^p) = 0$ and the equality follows.

(2) The closed immersion $\mathrm{Spec}\, E \to \mathbf{A}^1_S = \mathrm{Spec}\, O_K[x]$ defined by sending $x$ to $u$ is a regular immersion of codimension 2 and the conormal space $N_{E/\mathbf{A}^1_S}$ is of dimension 2 with basis $x^p - v, t$. In the exact sequence $0 \to \mathrm{H}^1(L_{E/S}) \to N_{E/\mathbf{A}^1_S} \to \Omega^1_{\mathbf{A}^1_S/S} \otimes_{O_{\mathbf{A}^1_S}} E$, the last arrow is the 0-mapping. Hence, $\mathrm{H}^1(L_{E/S}) \to N_{E/\mathbf{A}^1_S}$ is an isomorphism and maps $w(v), w(t)$ to the basis $x^p - v, -t$ by Proposition 14.1.6(1).

Alternative solution: In the exact sequence $0 \to \mathfrak{m}_K/\mathfrak{m}_K^2 \to \mathrm{H}^1(L_{E/S}) \to \Omega^1_F \otimes_F E \to \Omega^1_E$ (14.1.6), the image of the basis $-t$ of $\mathfrak{m}_K/\mathfrak{m}_K^2$ in $\mathrm{H}^1(L_{E/S})$ is $w(t)$ and the image of $w(v) \in \mathrm{H}^1(L_{E/S})$ is the basis $dv$ of $\mathrm{Ker}(\Omega^1_F \otimes_F E \to \Omega^1_E)$.

(3) Let $a \in O_K$ and $E = F(\sqrt[p]{a})$. Define closed immersions $E \to Q_{S_2} = Q \times_S S_2 \to Q = \mathbf{A}^1_S = \operatorname{Spec} O_K[X]$ by sending $X$ to $\sqrt[p]{a}$. We have a commutative diagram

$$\begin{array}{ccccc} \mathrm{H}_1(L_{E/S}) & \longrightarrow & \mathrm{H}_1(L_{E/S_2}) & \longrightarrow & \Omega^1_{S_2/\mathbf{F}_p} \otimes_{O_{S_2}} E \\ \downarrow & & \downarrow & & \downarrow \\ N_{E/Q} & \longrightarrow & N_{E/Q_{S_2}} & \longrightarrow & \Omega^1_{Q_{S_2}/\mathbf{F}_p} \otimes_{O_{Q_{S_2}}} E \end{array}$$

Since $\Omega^1_{S_2/\mathbf{F}_p} \otimes_{O_{S_2}} O_{Q_{S_2}} \to \Omega^1_{Q_{S_2}/\mathbf{F}_p}$ is a splitting injection, the right vertical arrow is an injection. Since $X^p - a \in N_{E/Q}$ goes to $-d(X^p - a) = da \in \Omega^1_{Q_{S_2}/\mathbf{F}_p} \otimes_{O_{Q_{S_2}}} E$ by Lemma 14.1.1(1), the assertion follows.

**Exercise 14.2** (1) Since $f \equiv X^p \bmod \mathfrak{m}_K$ is an Eisenstein polynomial, we have $df = 0$ in $\Omega^1_{Q/S} \otimes_{O_Q} F$. Hence, the morphism $N_{T/Q} \otimes_{O_T} F \to \Omega^1_{Q/S} \otimes_{O_Q} F$ is 0 and the free resolution $N_{T/Q} \to \Omega^1_{Q/S} \otimes_{O_Q} O_T \to \Omega^1_{T/S} \to 0$ defines an isomorphism $\operatorname{Tor}_1^{O_T}(\Omega^1_{T/S}, F) \to N_{T/Q} \otimes_{O_T} F$. Since $f = f(0) \in N_{F/Q}$, the image is $f(0) \in \mathfrak{m}_K/\mathfrak{m}_K^2$.

(2) Since $f \equiv X^p + f(0) \bmod \mathfrak{m}_K$, we have $df = 0$ in $\Omega^1_{Q/S} \otimes_{O_Q} F$. Hence, similarly to (1), we obtain an isomorphism $\operatorname{Tor}_1^{O_T}(\Omega^1_{T/S}, F) \to N_{T/Q} \otimes_{O_T} F$. Define a polynomial $h \in O_K[X]$ by $f = X^p + t \cdot h + f(0)$. By the left square of injections in (14.2.4), the image of $f = (X^p - (-f(0))) + h \cdot t$ equals $w(-f(0)) + \bar{h}(x) \cdot w(t)$.

**Exercise 14.3** Set $V^\vee = \operatorname{Spec} k[X,Y]$ and $V'^\vee = \operatorname{Spec} k'[Z]$. Then the morphism $V'^\vee \to V^\vee$ of schemes is defined by $k[X,Y] \to k'[Z]$ sending $X$ to $Z$ and $Y$ to $tZ$. Let $i > 0, j > 0$ be integers satisfying $i + j = p$. Then the image of $X^iY^j \in k[X,Y]$ in $\mathrm{H}^1(V^\vee, \mathbf{F}_p)$ is not contained in the image of $V = kX + kY$, but the image of its image $t^jZ^p \in k'[Z]$ equals the image of $t^{j/p}Z \in V'$.

**Exercise 14.4** Let $I = \operatorname{Ker}(A \otimes_k A \to A)$ be the kernel of the morphism sending $y \otimes z$ to $yz$. Define $u, v \colon A_U \to B_U = A_U \otimes_k A_U$ by $u(s) = s \otimes 1$ and $v(s) = 1 \otimes s$ for $s \in A_U$, and set $B = A \otimes_k A[I/v(t^n)]$ and $\delta(s) = (u(s) - v(s))/v(t^n) \in B$ for $s \in A$. Let $W_K \to P_K$ be the Artin–Scherier covering defined by $x^p - x = u(a/t^n) - v(a/t^n)$. Then $Q_T^{(n)}$ is the normalization of $P_T^{[n]}$ in the base change $W_K \times_{P_K} P_L$. We have

$$u\left(\frac{a}{t^n}\right) - v\left(\frac{a}{t^n}\right) = \delta(a) + u(a) \cdot \left(\frac{1}{u(t^n)} - \frac{1}{v(t^n)}\right).$$

Since

$$\frac{u(t)}{v(t)} = 1 + \frac{u(t) - v(t)}{v(t^n)} \cdot v(t)^{n-1} \equiv 1 \bmod v(t)$$

and $p \mid n$, the difference

$$\frac{1}{u(t^n)} - \frac{1}{v(t^n)} = \frac{1}{u(t^n)}\frac{1}{v(t^n)}((v(t) + (u(t) - v(t)))^n - v(t)^n)$$

is integral on a neighborhood of the closed fiber. Hence, the normalization $W^{(n)}$ of $P_S^{[n]}$ in $W_K$ is étale over $P_S^{[n]}$ on a neighborhood of the closed fiber. Therefore, $Q_T^{(n)}$ is also étale over $P_T^{[n]}$ on a neighborhood of the closed fiber, and the ramification of $L$ over $K$ is bounded by $n+$.

Further, we have

$$u\left(\frac{a}{t^n}\right) - v\left(\frac{a}{t^n}\right) \equiv \delta(a) + \binom{n}{2} u(a)v(t)^{n-2}\delta(t)^2.$$

Hence, the finite étale morphism $Q_E^{(n)} \to Q_{E,\mathrm{red}}^{[n]}$ is the Artin–Schreier covering defined by $x^p - x = \delta(a)$ unless $p = n = 2$, and by $x^p - x = \delta(a) + u(a)\delta(t)^2$ if $p = n = 2$.

**Exercise 14.5** Let $u \in O_L = O_K[u]$ be a generator over $O_K$, and $t$ be a uniformizer of $K$. Let $f \in O_K[X]$ be the minimal polynomial of $u$. Let $\sigma$ be a generator of $G$. By (5.3.11), we have $n = p \cdot \mathrm{ord}_L(\sigma(u) - u)$ since $e_{L/K} = 1$. By the bases $(\sigma(u) - u)^p \in \mathfrak{m}_K^n/\mathfrak{m}_K^{n+1} \otimes_F E$ and $f \in \mathrm{Tor}_1^{O_L}(\Omega^1_{O_L/O_K}, E)$ of the $E$-vector spaces as in Exercise 14.2, we identify $\Theta^{(n)}_{L/K,E}$ with $\mathbf{A}^1$.

Define $Q = \mathbf{A}^1_S = \mathrm{Spec}\, O_K[X] \to P = \mathbf{A}^1_S = \mathrm{Spec}\, O_K[Z]$ by sending $Z$ to $f$. Then we have a cartesian diagram

$$\begin{array}{ccc} T & \longrightarrow & Q \\ \downarrow & & \downarrow \\ S & \longrightarrow & P \end{array}$$

Set $a_i = \sigma^i(u) \in O_L$ for $i = 1, \ldots, p$ so that $f = (X - a_1)\cdots(X - a_p)$. Further, set $b_i = a_i - a_p$, $c_i = b_i/b_1 \in O_L$, and $g = (Y - c_1)\cdots(Y - c_p) = f(b_1Y + a_p)/b_1^p \in O_L[Y]$. Then the morphism $Q_T^{(n)} \to P_T^{(n)}$ is given by $\mathbf{A}^1_T \to \mathbf{A}^1_T$ defined by $g$. Since $c_i \equiv i \bmod \mathfrak{m}_L$, we have $g \equiv Y^p - Y$ and the assertion follows.

**Exercise 14.6** (1) By the solution of Exercise 12.3, we have $G^1 = G \supset G^{1+} = (1+p\mathbf{Z})/(1+p^n\mathbf{Z}) \ltimes (\mathbf{Z}/p^n\mathbf{Z})$, for $i = 2, \ldots, n+1$ we have $G^i = (1+p^{i-1}\mathbf{Z})/(1+p^n\mathbf{Z}) \ltimes (p^{i-1}\mathbf{Z}/p^n\mathbf{Z}) \supset G^{i+} = (1+p^i\mathbf{Z})/(1+p^n\mathbf{Z}) \ltimes (p^{i-1}\mathbf{Z}/p^n\mathbf{Z})$, and for $i = 1, \ldots, n+1$ we have $G^{i+1/(p-1)} = (1+p^i\mathbf{Z})/(1+p^n\mathbf{Z}) \ltimes (p^{i-1}\mathbf{Z}/p^n\mathbf{Z}) \supset G^{i+1/(p-1)+} = (1+p^i\mathbf{Z})/(1+p^n\mathbf{Z}) \ltimes (p^i\mathbf{Z}/p^n\mathbf{Z})$. Hence, the graded quotients are isomorphic to $\mathbf{F}_p$.

(2) By the solution of Exercise 12.3, we have $\dim \mathrm{tot}\, V = [G : H] \cdot (n + 1/(p-1)) = (n(p-1)+1) \cdot p^{n-1}$.

# Chapter 15

**Exercise 15.1** $f = X^{p^2} - (a^{(p-1)p} + a^{(p-1)^2}b^{p-1} + \cdots + b^{(p-1)p})X^p + (a^{(p-1)p}b^{p-1} + a^{(p-1)^2}b^{(p-1)2} + \cdots + a^{p-1}b^{(p-1)p})X$ is an additive polynomial and has $a, b$ as roots.

**Exercise 15.2** By the commutative diagram

$$\begin{array}{ccccccccc}
0 & \longrightarrow & G & \xrightarrow{\ \subset\ } & \mathbf{G}_a & \xrightarrow{\ f\ } & \mathbf{G}_a & \longrightarrow & 0 \\
 & & {\scriptstyle \chi_a}\downarrow & & {\scriptstyle \frac{X^p - b^{p-1}X}{a^p - b^{p-1}a}}\downarrow & & \downarrow{\scriptstyle \frac{1}{(a^p - b^{(p-1)}a)^p}} & & \\
0 & \longrightarrow & \mathbf{F}_p & \xrightarrow{\ \subset\ } & \mathbf{G}_a & \xrightarrow{\mathrm{F}-1} & \mathbf{G}_a & \longrightarrow & 0
\end{array}$$

we have $[E](\chi_a) = 1/(a^p - b^{(p-1)}a)^p$.

**15.3.** As we saw in the solution of Exercise 11.5, the minimal polynomial $f = \sum_{i=0}^{p-1} X^{p^{n-1}i} = (X^{p^n} - 1)/(X^{p^{n-1}} - 1) \in \mathbf{Q}_p[X]$ of a primitive $p^n$th root $\zeta_{p^n}$ of 1 is decomposed as

$$f = \prod_{i=2}^{p-1}(X^{p^{n-1}} - z_1^i) \cdot \frac{X^{p^{n-1}} - z_1}{X^{p^{n-2}} - z_2} \cdots \frac{X^{p^2} - z_{n-2}}{X^p - z_{n-1}} \cdot (X^p - z_{n-1}),$$

where $z_i = \zeta_{p^n}^{p^{n-i}}$ for $i = 1, \ldots, n$. Then $g = f(Y + \zeta_{p^n})$ is decomposed as

$$g \equiv \prod_{i=1}^{p-2}(Y^{p^{n-1}} - i(z_1 - 1)) \cdot (Y^{p^{n-2}(p-1)} + pz_2^{p-1})$$
$$\cdots (Y^{p(p-1)} + pz_{n-1}^{p-1})(Y^p + pz_n^{p-1}Y).$$

Since $((\zeta_p - 1) + 1)^p - 1 = 0$, we have $(\zeta_p - 1)^{p-1} + p \equiv 0 \bmod p(\zeta_p - 1)$. Hence, for $Y = (\zeta_p - 1)Z$ we obtain

$$\begin{aligned} g((\zeta_p - 1)Z) &\equiv (-1)^{p-2}(p-2)!(\zeta_p - 1)^{p-2} \cdot p^{n-2} \cdot (\zeta_p - 1)^p(Z^p - Z) \\ &\equiv p^n(Z^p - Z). \end{aligned}$$

Thus, we obtain a commutative diagram

$$\begin{array}{ccccccccc}
0 & \longrightarrow & G^n & \xrightarrow{\sigma \mapsto \sigma(\zeta_{p^n})/\zeta_{p^n} - 1} & (\zeta_p - 1) & \xrightarrow{\ g\ } & p^n & \longrightarrow & 0 \\
 & & {\scriptstyle 1+p^{n-1}}\uparrow & & {\scriptstyle \zeta_p - 1}\uparrow & & \uparrow{\scriptstyle p^n} & & \\
0 & \longrightarrow & \mathbf{F}_p & \longrightarrow & \mathbf{G}_a & \xrightarrow{F-1} & \mathbf{G}_a & \longrightarrow & 0
\end{array}$$

This means that the image of $\chi \in G^{n\vee}$ by $\Phi$ maps the generator $p^n$ of $(p^n)/(p^{n+1})$ to $f(1) = p \in (p)/(p^2)$.

This is equal to $-\mathrm{rsw}\,\chi$ by Exercise 5.3.

**Exercise 15.4** (1) The schemes $X$ and $M$ represent functors sending rings $A$ over $k$ to $X(A) = A$ and $M(A) = \{(t, s, u) \in A \times A \times A^\times \mid t = us\}$ respectively. Let $\mu\colon M \times_X M \to M$ be the morphism sending $((t, s, u), (s, r, v)) \in M \times_X M(A)$ to $(t, r, uv) \in M(A)$. Let $e\colon X \to M$ be the morphism sending $t \in X(A)$ to $(t, t, 1) \in M(A)$, and $i\colon M \to M$ sending $(t, s, u) \in M(A)$ to $(s, t, u^{-1}) \in M(A)$. They define a functor from the category of rings over $k$ to the category of groupoids, and $(M, X)$ is a groupoid over $k$. We thus have a morphism $(M, X) \to (X \times_k X, X)$ of groupoids over $k$.

(2) The schemes $X$ and $M$ represent functors sending rings $A$ over $k$ to $X(A) = A$ and $M(A) = \{(t, s, v) \in A \times A \times A \mid 1 + vs^{n-1} \in A^\times, t = (1 + vs^{n-1})s\}$ respectively. Let $\mu\colon M \times_X M \to M$ be the morphism sending $((t, s, v), (s, r, w)) \in M \times_X M(A)$ to $(t, r, w + v(1 + wr^{n-1})^n) \in M(A)$. Let $e\colon X \to M$ be the morphism sending $t \in X(A)$ to $(t, t, 0) \in M(A)$, and $i\colon M \to M$ sending $(t, s, v) \in M(A)$ to $(s, t, -v/(1 + vs^{n-1})^n) \in M(A)$. They define a functor from the category of rings over $k$ to the category of groupoids, and the rest is the same as in the solution of (1).

**Exercise 15.5** The closed subscheme $D \subset \mathbf{A}^1$ in the diagonal of of $\mathbf{A}^1 \times \mathbf{A}^1 = \operatorname{Spec} k[T, S]$ is defined by the maximal ideal $(T, S)$. Hence, the dilatation $P^{[1]} = \operatorname{Spec} k[T, S][U^{\pm 1}]$ is an open subscheme of the blow-up $\overline{P}^{[1]}$. For $n \geqslant 2$, the closed subscheme $nD \subset \mathbf{A}^1$ in the diagonal of $\mathbf{A}^1 \times \mathbf{A}^1 = \operatorname{Spec} k[T, S]$ is defined by the ideal $(T - S, S^n)$. Hence, for $V = (T - S)/S^n$ we obtain an open subscheme $\operatorname{Spec} k[T, S, V]/(T - S - VS^n)$ of the blow-up $\overline{P}^{[n]}$.

**Exercise 15.6** (1) By Corollary 12.3.7, we have $G^r = G^m_{\mathrm{coh}}$ for $(m - 1)+ \leqslant r \leqslant m$. Hence, $G^n = G$ and $G^{n+} = 1$.

(2) Let $u \in O_L = O_K[u]$ be a generator over $O_K$, and $t$ be a uniformizer of $K$. Let $f \in O_K[X]$ be the minimal polynomial of $u$. Let $f = X^p + t \cdot h + v \in O_K[X]$ as in the solution of Exercise 14.2. For a generator $\sigma \in G$ satisfying $\chi(\sigma) = 1$, the characteristic form $\operatorname{ch}\chi\colon \mathfrak{m}_L^n/\mathfrak{m}_L^{n+1} \to \mathrm{H}_1(L_{E/S})$ is the morphism sending the basis $(\sigma(u) - u)^p$ to the image of $f \in \operatorname{Tor}_1^{O_L}(\Omega^1_{O_L/O_K}, E)$, by Exercise 14.5. The image of $f \in \operatorname{Tor}_1^{O_L}(\Omega^1_{O_L/O_K}, E)$ by (14.5.2) is $w(-v) - \overline{h(u)} \cdot w(t) \in \mathrm{H}_1(L_{E/S})$.

(3) The image of $w(-v) - \overline{h(u)} \cdot w(t) \in \mathrm{H}_1(L_{E/S})$ in $\Omega^1_F \otimes_F E$ equals $-dv$.

Since $\mathrm{N}_{L/K}(\sigma(u)/u) = 1$, we have $(\sigma(u) - u)^p = u^p \cdot \mathrm{N}_{L/K}(\sigma(u)/u - 1) \equiv -v \cdot -\mathrm{Tr}_{L/K}(\sigma(u)/u - 1)$ by Proposition 1.10.5(3). Hence, the composition is induced by minus the refined Swan conductor $\operatorname{rsw}\chi\colon \mathfrak{m}_K^n/\mathfrak{m}_K^{n+1} \to \Omega^1_F(\log)$ sending $\mathrm{Tr}_{L/K}(\sigma(u)/u - 1)$ to $d\log N_{L/K}u = d\log v$ by Theorem 5.3.4.

# Chapter 16

**Exercise 16.1** (1) Since $f$ is an Eisenstein polynomial, the ring $O_K[X][f/t]$ equals $O_K[X][X^p/t] \xleftarrow{\sim} O_K[X,V]/(X^p - Vt)$. Hence, the dilatation $P^{[D\cdot T]} =$ Spec $O_K[X][f/t]$ is isomorphic to Spec $O_K[X,V]/(X^p - Vt)$ and contains $Q$ as an open subscheme by sending $V$ to $U^{-1}$.

The closed subscheme $T \subset Q$ is defined by the ideal generated by $1+Uh$. The image of the basis $1+Uh$ of $N_{T/Q} \otimes_{O_L} E$ by $N_{T/Q} \otimes_{O_L} E \to \Omega^1_Q(\log) \otimes_{O_Q} E$ (16.1.4) is $-(h \cdot dU + U \cdot dh) = -U \cdot (h \cdot d\log t + dh)$. Hence, its image $-U \cdot h' X \cdot d\log X$ in $\Omega^1_{Q/S}(\log/\log) \otimes_{O_Q} E$ is 0. By the free resolution $N_{T/Q} \to \Omega^1_{Q/S}(\log/\log) \otimes_{O_Q} O_T \to \Omega^1_{T/S}(\log/\log) \to 0$, we obtain an isomorphism $\mathrm{Tor}_1^{O_T}(\Omega^1_{T/S}(\log/\log), E) \to N_{T/Q} \otimes_{O_L} E$.

Further, the image of the basis $(1+Uh) \otimes U^{-1}$ by $N_{T/Q} \otimes_{O_L} E \to \Omega^1_F(\log) \otimes_F E$ (16.1.19) is $-(h \cdot d\log t + dh) \otimes 1 = -(h(0) \cdot d\log t + dh(0)) \otimes 1$. For the sign, see the convention after Lemma 16.1.4. Its image by $\Omega^1_F(\log) \otimes_F E \to \mathrm{H}_1(L_{E/S}) \otimes_{O_K} \mathfrak{m}_K^{-1}$ (16.1.18) is $-h(0) \cdot w(t) \otimes t^{-1}$. On the other hand, the image of the basis $U^{-1}(1+Uh) = U^{-1} + h$ of $N_{T/Q}$ by $N_{T/Q} \to N_{T/P} \otimes_{O_K} \mathfrak{m}_K^{-1}$ is $(X^p + th) \otimes t^{-1} = f \otimes t^{-1}$. The morphism $N_{T/P} \to \mathrm{H}_1(L_{E/S})$ maps the basis $f$ to $-w(f(0)) = -h(0) \cdot w(t)$. Hence, the diagram is commutative.

(2) The morphism $\mathrm{Tor}_1^{O_T}(\Omega^1_{T/S}, E) \to N_{T/Q} \otimes_{O_L} E$ is an isomorphism by Exercise 14.2. The canonical morphism $\Omega^1_{T/S} \to \Omega^1_{T/S}(\log/\log)$ is an isomorphism. Let $f \in N_{T/Q}$ be the basis. Then, since its image by $N_{T/Q} \otimes_{O_L} E \to \Omega^1_Q(\log) \otimes_{O_Q} E$ (16.1.4) is $-df$, the image by $\mathrm{Tor}_1^{O_T}(\Omega^1_{T/S}(\log/\log), E) \to \Omega^1_F(\log) \otimes_F E$ (16.1.9) is $-df = -df(0) \in \Omega^1_F \otimes_F E \subset \Omega^1_F(\log) \otimes_F E$. Since the image by $\mathrm{Tor}_1^{O_T}(\Omega^1_{T/S}(\log/\log), E) \to \mathrm{H}_1(L_{E/S})$ (14.2.2) is $w(-f(0)) + \bar{h}(x) \cdot w(t)$ by Exercise 14.2, the diagram is commutative.

**Exercise 16.2** By Corollary 12.3.6, we have $G^{r+1} = G^m_{\mathrm{coh}}$ for $(m-1)+ \leqslant r \leqslant m$. Hence, by Lemma 13.3.5(3), $G^{n+1} = G^n_{\log} = G$, $G^{(n+1)+} = G^{n+}_{\log} = 1$.

Let $s \in O_L$ be a uniformizer. Then its norm $t = \mathrm{N}_{L/K} s$ is a uniformizer of $K$. The minimal polynomial $f \in O_K[x]$ of $s$ is an Eisenstein polynomial and its constant term equals $(-1)^p t \equiv -t$. Define a polynomial $h \in O_K[x]$ by $f = x^p + tx \cdot h + (-1)^p t$. As in Lemma 16.1.8, the immersion $T = \mathrm{Spec}\, O_L \to P = \mathbf{A}^1_S = \mathrm{Spec}\, O_K[x]$ defined by sending $x$ to $s$ is lifted to an exact immersion $T \to Q = \mathrm{Spec}\, O_K[u^{\pm 1}, x]/(x^p - ut)$ defined by $g = f/t = u + xh + (-1)^p$. We have $u = 1$ in $E$. As in the proof of Lemma 16.1.8, the isomorphism $-d\colon N_{T/Q} \to \Omega^1_F(\log) \otimes_F E$ (16.1.19) maps the basis $g = u + xh + (-1)^p$ to $-d(u + xh + (-1)^p) = u \cdot d\log t = d\log t$.

The $\mathbf{F}_p$-torsor $P^{(n+1)}_{\overline{F}} = Q^{(n)}_{\overline{F}} \to P^{[n+1]}_{\overline{F}} = Q^{[n]}_{\overline{F}}$ is the Artin–Schreier covering defined by the linear form $f/(\sigma(s) - s)^p = g/(\sigma(s)/s - 1)^p$.

Hence, the logarithmic characteristic form $\mathrm{ch}_{\log}\chi\colon \mathfrak{m}_K^n/\mathfrak{m}_K^{n+1} \to \Omega_F^1(\log)\otimes_F E$ maps the basis $\mathrm{N}_{L/K}(\sigma(s)/s - 1)$ to $-d\log t$. Since $\mathrm{N}_{L/K}(\sigma(s)/s) = 1$, we have $\mathrm{N}_{L/K}(\sigma(s)/s - 1) = -\mathrm{Tr}_{L/K}(\sigma(s)/s - 1)$ by Proposition 1.10.5(3). Hence, $\mathrm{ch}_{\log}\chi$ equals minus the refined Swan conductor $\mathrm{rsw}\,\chi\colon \mathfrak{m}_K^n/\mathfrak{m}_K^{n+1} \to \Omega_F^1(\log)\otimes_F E$ sending the basis $\mathrm{Tr}_{L/K}(\sigma(s)/s-1)$ to $-d\log t$ by Theorem 5.3.4.

**Exercise 16.3** In the case where the ramification index $e_{L/K}$ is 1, the assertion is proved in Exercise 15.6. In the case where $L$ is totally ramified over $K$, the assertion is proved in Exercise 16.2.

**Exercise 16.4** The assertion follows from the commutative diagram

$$\begin{array}{ccc}
\mathrm{H}^1(D_K, \mathbf{Q}/\mathbf{Z}) \times A[1/p]^\times & \longrightarrow & \mathrm{H}^2(D_K, \mu_\infty) \\
\downarrow & & \downarrow \\
X_{K_1} \times K_1^\times & \longrightarrow & \mathrm{Br}\, K_1
\end{array}$$

# References

[1] A. Abbes and T. Saito (2002). "Ramification of local fields with imperfect residue fields," *American Journal of Mathematics*, 124(5), 879–920.

[2] A. Abbes and T. Saito (2003). "Ramification of local fields with imperfect residue fields II," in S. Bloch, I. Fesenko, L. Illusie, M. Kurihara, S. Saito, T. Saito, and P. Schneider (eds.), *A Collection of Manuscripts Written in Honour of Kazuya Kato on the Occasion of His Fiftieth Birthday*. EMS Press, pp. 5–72.

[3] A. Abbes and T. Saito (2007). "The characteristic class and ramification of an $\ell$-adic étale sheaf," *Inventiones Mathematicae*, 168(3), 567–612.

[4] A. Abbes and T. Saito (2009). "Analyse micro-locale $\ell$-adique en caractéristique $p > 0$: Le cas d'un trait," *Publications of the Research Institute for Mathematical Sciences*, 45(1), 25–74.

[5] A. Abbes and T. Saito (2010). "Local Fourier transform and epsilon factors," *Compositio Mathematica*, 146(6), 1507–51.

[6] A. Abbes and T. Saito (2011). "Ramification and cleanliness," *Tohoku Mathematical Journal (2)*, 63(4) 775–853.

[7] C. Arf, "Untersuchungen über reinverzweigte Erweiterungen diskret bewerteter perfekter Körper," *Journal für die reine und angewandte Mathematik*, 181, 1–44.

[8] E. Artin and J. Tate (1968). *Class Field Theory*. Benjamin. Second printing, Addison-Wesley, 1974.

[9] M. Artin (1973). "Dimension cohomologique: Premiers résultats," in P. Deligne and M. Artin (eds.), *Théorie des Topos et Cohomologie Etale des Schémas*, Tome 3 (Lecture Notes in Mathematics 305). Springer, pp. 43–63.

[10] M. Atiyah and I. G. Macdonald (1969). *Introduction To Commutative Algebra*. Addison-Wesley.

[11] A. Bah (2022). "Variation of the Swan conductor of an $\mathbf{F}_\ell$-sheaf on a rigid annulus," *Tohoku Mathematical Journal*, 77, 499–537.

[12] S. Bloch and K. Kato (1986). "$p$-adic etale cohomology," *Publications mathématiques de l'IHÉS*, 63, 107–152.

[13] S. Bosch, W. Lütkebohmert, and M. Raynaud (1995). "Formal and rigid geometry IV: The reduced fibre theorem," *Inventiones Mathematicae*, 119(2), 361–98.

[14] N. Bourbaki (1985). *Algèbre commutative*. Masson.

[15] N. Bourbaki (2003). *Algebra II, Chapters 4–7*. Springer.

[16] J.-L. Brylinski (1983). "Théorie du corps de classes de Kato et revêtement abéliens de surfaces," *Annales de l'Institute Fourier*, 33, 23–38.

[17] J.-L. Colliot-Thélène, J.-J. Sansuc, and C. Soulé (1983). "Torsion dans le groupe de Chow de codimension deux," *Duke Mathematical Journal*, 50(3), 763–801.

[18] P. Colmez and A. Grothendieck (eds.) (2001). *Correspondance Grothendieck–Serre*. Société Mathématique de France.

[19] P. Deligne (1971). "Théorie de Hodge: II," *Publications Mathématiques de l'IHÉS*, 40, 5–57.

[20] P. Deligne (1973). "La formule de Picard–Lefschetz," in P. Deligne, N. Katz, and N. M. Katz (eds.), *Groupes de Monodromie en Géométrie Algébrique* (Lecture Notes in Mathematics 340). Springer, pp. 165–96.

[21] P. Deligne (1977). "Théorèmes de finitude en cohomologie $\ell$-adique," in P. Deligne, *Cohomologie étale* (Lecture Notes in Mathematics 569). Springer, pp. 233–51.

[22] P. Deligne (1984). "Les corps locaux de caractéristique $p$, limites de corps locaux de caractéristique 0," in *Representations of Reductive Groups over a Local Field*, travaux en cours. Hermann, pp. 119–57.

[23] P. Deligne and D. Mumford (1969). "The irreductibility of the space of curves of a given genus," *Publications mathématiques de l'IHÉS*, 36, 75–110.

[24] T. Dupuy, E. Katz, J. Rabinoff, and D. Zureick-Brown (2019). "Total $p$-differential on schemes over $\mathbf{Z}/p^2$," *Journal of Algebra*, 524, 110–23.

[25] H. P. Epp (1973). "Eliminating wild ramification," *Inventiones Mathematicae*, 19, 235–49.

[26] L. Fu (2015). *Etale Cohomology Theory*, revised edition (Nankai Tracts in Mathematics 14). World Scientific.

[27] O. Gabber (1994). "Affine analog of the proper base change theorem," *Israel Journal of Mathematics*, 87, 325–35.

[28] O. Gabber and L. Ramero (2004). "Foundations for almost ring theory – Release 7.5," arXiv:math/0409584.

[29] A. Grothendieck (1957). "Sous-groupes de Cartan, éléments réguliers. Groupes algébriques affines de dimension 1," *Séminaire Claude Chevalley*, 1(1956–1958), 1–9.

[30] A. Grothendieck (1961). "Éléments de géométrie algébrique III-1. Étude cohomologique des faisceaux cohérents," *Publications Mathématiques de l'IHÉS*, 11, 5–167.

[31] A. Grothendieck (1964–67). "Éléments de géométrie algébrique IV. Étude locale des schémas et des morphismes de schémas," *Publications Mathématiques de l'IHÉS* 20, 5–259; 24, 5–231; 28, 5–255; 32, 5–361.

[32] A. Grothendieck (1971). "Théorie de la spécialization du groupe fondamental," in A. Grothendieck and M. Raynaud (eds.), "Revêtements Étales et Groupe Fondamental" (Lecture Notes in Mathematics 224). Springer, pp. 261–84.

[33] A. Grothendieck (1971). *Éléments de géométrie algébrique I* (Die Grundlehren der mathematischen Wissenschaften in Einzeldarstellungen 165). Springer.

[34] A. Grothendieck (1977). "Formule d'Euler–Poincaré en cohomologie étale," in L. Illusie (ed.), *Cohomologie $\ell$-adique et Fonctions L* (Lecture Notes in Mathematics 589). Springer, pp. 372–406.

[35] A. Grothendieck (édité par R. Hartshorne) (2024). *Résidus et dualité, Prénotes pour un "Séminaire Hartshorne"*. Société Mathématique de France.

[36] R. Hartshorne (1966). *Residues and Duality* (Lecture Notes in Mathematics 20). Springer.

[37] H. Hasse (1930). "Führer, Diskriminante und Verzweigungskörper relativ-Abelscher Zahlkörper," *Journal für die reine und angewandte Mathematik*, 162, 169–84.

[38] J. Herbrand (1933). "Théorie arithmétique des corps de nombres de degré infini II, Extensions algébriques de degré infini," *Mathematische Annalen*, 108, 699–717.

[39] H. Hu (2015). "Ramification and nearby cycles for $\ell$-adic sheaves on relative curves," *Tohoku Mathematical Journal (2)*, 67(2), 153–94.

[40] O. Hyodo (1988). "A note on $p$-adic etale cohomology in the semi-stable reduction case," *Inventiones Mathematicae*, 91, 543–57.

[41] L. Illusie (1971). *Complexe cotangent et déformations I* (Lecture Notes in Mathematics 239). Springer.

[42] L. Illusie (1979). "Complexe de de Rham–Witt et cohomologie cristalline," *Annales scientifiques de l'ÉNS*, 12(4), 501–661.

[43] L. Illusie (1981). "Théorie de Brauer et caractéristique d'Euler–Poincaré, d'après Deligne," *Astérisque*, 82–83, 161–72.

[44] L. Illusie (2004). "On semistable reduction and the calculation of nearby cycles," in A. Adolphson, F. Baldassarri, P. Berthelot, N. Katz, and F. Loeser (eds.), *Geometric Aspects of Dwork's Theory*, vol. II. De Gruyter, pp. 785–803.

[45] L. Illusie (2015). "From Pierre Deligne's secret garden: Looking back at some of his letters," *Japanese Journal of Mathematics*, 10(2), 237–48.

[46] K. Iwasawa (1986). *Local Class Field Theory*. Oxford University Press.

[47] K. Kato (1982). "Galois cohomology of complete discrete valuation fields," in R. Keith Dennis (ed.), *Algebraic K-Theory* (Lecture Notes in Mathematics 967). Springer, 215–38.

[48] K. Kato (1987). "Vanishing cycles, ramification of valuations, and class field theory," *Duke Mathematical Journal*, 55(3), 629–59.

[49] K. Kato (1987). "Swan conductors with differential values," *Advanced Studies in Pure Mathematics*, 12, 315–42.

[50] K. Kato (1989). "Swan conductors for characters of degree one in the imperfect residue field case," *Contemporary Mathematics*, 83, 101–31.

[51] K. Kato (1989), "Logarithmic structures of Fontaine–Illusie," in J.-I. Igusa (ed.), *Algebraic Analysis, Geometry, and Number Theory*. Johns Hopkins University Press, pp. 191–224.

[52] K. Kato and T. Saito (2019). "Coincidence of two Swan conductors of abelian characters," *Épijournal de Géométrie Algébrique*, 3, 15.

[53] K. Kato and V. Thatte (2024). "Upper ramification groups for arbitrary valuation rings," *Tunisian Jourbal of Mathematics*, 6(4), 589–646.

[54] N. Katz (1970). "Nilpotent connections and the monodromy theorem," *Publications Mathématiques de l'IHÉS*, 39, 175–232.

[55] N. Katz (1988). *Gauss Sums, Kloosterman Sums, and Monodromy Groups* (Annals of Mathematics Studies 116). Princeton University Press.

[56] N. Katz and G. Laumon (1985). "Transformation de Fourier et majoration de sommes exponentielles," *Publications Mathématiques de l'IHÉS*, 62, 145–202.

[57] Y. Kawada (1953). "On the ramification theory of infinite algebraic extensions," *Annals of Mathematics*, 58(1), 24–47.

[58] F.-V. Kuhlmann (2003). "A correction to Epp's paper 'Elimination of wild ramification'," *Inventiones Mathematicae*, 153(3), 679–81.

[59] E. Kunz (1986). *Kähler Differentials*. Springer.

[60] G. Laumon (1981). "Semi-continuité du conducteur de Swan (d'après P. Deligne)," *Astérisque*, 82–83, 173–219.

[61] G. Laumon (1987). "Transformation de Fourier, constantes d'équations fonctionnelles et conjecture de Weil," *Publications Mathématiques de l'IHÉS*, 65, 131–210.

[62] S. Matsuda (1997). "On the Swan conductor in positive characteristic," *American Journal of Mathematics*, 119(4), 705–39.

[63] J. S. Milne (1980). *Etale Cohomology*. Princeton University Press.

[64] R. Ooe (2025). "$F$-characteristic cycle of a rank one sheaf on an arithmetic surface," *Épijournal de Géométrie Algébrique*, 9, 20.

[65] M. Rapoport and T. Zink (1980). "Über die lokale Zetafunktion von Shimura varietäten, Monodromiefiltrations und verschwindende Zyklen in ungeicher Characteristik," *Inventiones Mathematicae*, 68, 21–101.

[66] M. Raynaud (1965). "Caractéristique d'Euler–Poincaré d'un faisceau et cohomologie des variétés abéliennes," *Séminaire Bourbaki*, 9, 128–47.

[67] M. Raynaud (1970). *Anneaux Locaux Henséliens* (Lecture Notes in Mathematics 169). Springer.

[68] M. Raynaud and L. Gruson (1971). "Critères de platitude et de projectivité, techniques de 'platification' d'un module," *Inventiones Mathematicae*, 13, 1–89.

[69] T. Saito (1987). "Vanishing cycles and geometry of curves over a discrete valuation rings," *American Journal of Mathematics*, 109, 1043–85.

[70] T. Saito (1987). "Trace formula for vanishing cycles of curves," *Mathematische Annalen*, 276, 311–15.

[71] T. Saito (2009). "Wild ramification and the characteristic cycle of an $\ell$-adic sheaf," *Journal de l'Institut de Mathématiques de Jussieu*, 8(4), 769–829.

[72] T. Saito (2012). "Ramification of local fields with imperfect residue fields III," *Mathematische Annalen*, 352(3), 567–80.

[73] T. Saito (2017). "Wild ramification and the cotangent bundle," *Journal of Algebraic Geometry*, 26, 399–473.

[74] T. Saito (2017). "The characteristic cycle and the singular support of a constructible sheaf," *Inventiones Mathematicae*, 207(2), 597–695.

[75] T. Saito (2022). "Frobenius–Witt differentials and regularity," *Algebra & Number Theory*, 16(2), 369–91.

[76] T. Saito (2023). "Graded quotients of ramification groups of local fields with imperfect residue fields," *American Journal of Mathematics*, 145(5), 1389–464.

[77] T. Saito (2024). "A characterization of ramification groups of local fields with imperfect residue fields," in N. Fakhruddin, E. Ghate, A. Nair, C. S. Rajan, and S. Varma (eds.), *Arithmetic Geometry*. American Mathematical Society, pp. 421–33.

[78] J.-P. Serre (1968). *Corps locaux*, 3e éd. Hermann.

[79] J.-P. Serre (1971). *Représentations linéaires des groupes finis*, 2e éd. Hermann.

[80] M. Temkin, "Stable modification of relative curves," *Journal of Algebraic Geometry*, 19, 603–77.

[81] L. Xiao (2010). "On ramification filtrations and $p$-adic differential equations, I: Equal characterisitc case," *Algebra & Number Theory*, 4(8), 969–1027.
[82] L. Xiao (2012). "On ramification filtrations and $p$-adic differential equations, II: Mixed characteristic case," *Compositio Mathematica*, 148(2), 415–63.
[83] L. Xiao (2012). "On refined ramification filtrations in the equal characteristic case," *Algebra & Number Theory*, 6(8), 1579–667.
[84] Y. Yatagawa (2017). "Equality of two non-logarithmic ramification filtrations of abelianized Galois group in positive characteristic," *Documenta Mathematica*, 22, 917–52.

# Index

For EU product safety concerns, contact us at Calle de José Abascal, 56–1°, 28003 Madrid, Spain or eugpsr@cambridge.org.

www.ingramcontent.com/pod-product-compliance
Ingram Content Group UK Ltd.
Pitfield, Milton Keynes, MK11 3LW, UK
UKHW020222290926
15470UKWH00001B/330

* 9 7 8 1 0 0 9 6 1 7 5 3 6 *